Nanoparticles with Morphological and Functional Anisotropy

Glasgow, United Kingdom

4–6 July 2016

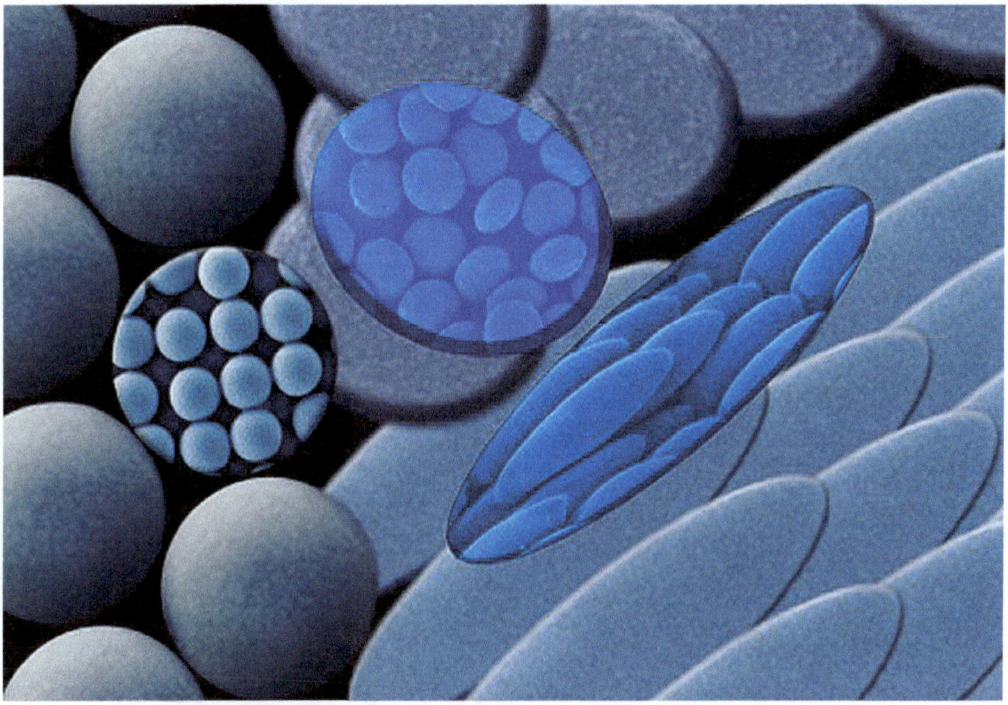

FARADAY DISCUSSIONS

Volume 191, 2016

ROYAL SOCIETY OF CHEMISTRY

Faraday Discussions (Print ISSN 1359-6640, Electronic ISSN 1364-5498) is published 10 times a year by the Royal Society of Chemistry, Thomas Graham House, Science Park, Milton Road, Cambridge, UK CB4 0WF.

Volume 191 ISBN 13: 978-1-78262-477-6

2016 annual subscription price: print+electronic £929 US $1635; electronic only £883, US $1553. Customers in Canada will be subject to a surcharge to cover GST. Customers in the EU subscribing to the electronic version only will be charged VAT. All orders, with cheques made payable to the Royal Society of Chemistry, should be sent to RSC Order Department, Royal Society of Chemistry, Thomas Graham House, Science Park, Milton Road, Cambridge, CB4 0WF, UK.

Tel +44 (0) 1223 432398; E-mail **orders@rsc.org**

If you take an institutional subscription to any RSC journal you are entitled to free, site-wide web access to that journal. You can arrange access via Internet Protocol (IP) address at **www.rsc.org/ip**. Customers should make payments by cheque in sterling payable on a UK clearing bank or in US dollars payable on a US clearing bank.

Printed in the UK

MIX
Paper from responsible sources
FSC www.fsc.org FSC® C013604

Faraday Discussions are unique international discussion meetings that focus on rapidly developing areas of chemistry and its interfaces with other scientific disciplines.

Nanoparticles with Morphological and Functional Anisotropy

Faraday Discussions

www.rsc.org/faraday_d

A General Discussion on Nanoparticles with Morphological and Functional Anisotropy was held in Glasgow, United Kingdom on the 4th, 5th and 6th of July 2016.

CONTENTS

ISSN 1359-6640; ISBN: 978-1-78262-477-6

Cover
See Vermant *et al.*, *Faraday Discuss.*, 2016, **191**, 325–349.

Mechanical deformations of PMMA particles can be used to generate a variety of morphological anisotropies, from spheroids to colloidal surfboards.

Front cover image courtesy of Professor Jan Vermant.

INTRODUCTORY LECTURE

PAPERS AND DISCUSSIONS

CONCLUDING REMARKS

ADDITIONAL INFORMATION

Faraday Discussions

PAPER

Seed mediated growth of gold nanorods: towards nanorod matryoshkas

Joshua G. Hinman, Andrew J. Stork, Jason A. Varnell, Andrew A. Gewirth and Catherine J. Murphy*

Received 6th June 2016, Accepted 8th June 2016
DOI: 10.1039/c6fd00145a

After a brief review of anisotropy on the nanoscale, experiments in which nanorod core–shell–shell particles are grown are presented. These "nanomatryoshkas" consist of a gold nanorod core, a silica shell, and a final gold shell. Calculation of the near-field properties of these structures using the discrete dipole approximation uncovers the change in location of local electric fields upon gold outer shell growth. Electrochemical experiments of the weak reducing agents used to grow the gold nanorod cores suggest a correlation between the strength of the reducing agent and its ability to promote longer nanorod growth. The final nanostructures do not exhibit a smooth outer shell, unlike their spherical counterparts.

1. Introduction: a brief review of anisotropy on the nanoscale

Anisotropy on the nanoscale—including anisotropically shaped nanoparticles, anisotropic surface chemistry and composition of nanoparticles, and the anisotropic assembly of nanoparticles—often has profound effects on the physical and chemical properties of nanoparticles and nanomaterials. The study of anisotropic nanostructures has generated a great deal of scientific interest because anisotropic nanoparticles are well suited for the study of fundamental science as well as in many applications. Parallels can be drawn between anisotropy in molecular systems, such as the behavior of surfactants[1] or, in biological systems, the assembly of lipids and proteins,[2] to the effects that anisotropy has on the assembly of colloidal nanoparticles. As nanoparticles become increasingly incorporated into new materials and technologies, control over their anisotropy will be important for directing their assembly and tuning their properties for different applications. The field is quite rich with examples of anisotropy on the nanoscale, but in general nanoparticles can exhibit anisotropy in surface chemistry, composition, shape, or a combination thereof. In the first part of this

Department of Chemistry, University of Illinois at Urbana-Champaign, 600 S. Mathews Ave., Urbana, Il, 61801, USA. E-mail: murphycj@illinois.edu

discussion, we examine methods that have been developed to introduce anisotropy on the nanoscale. Later we discuss our efforts to create anisotropic, multilayered, plasmonic nanoparticles.

1.1 Morphological anisotropy

There has been considerable effort towards the development of methods for shape control in nanoparticle syntheses because of their shape-dependent properties on the nanoscale. As an example, consider gold nanorods. Colloidal suspensions of gold nanoparticles display intense extinction in the visible region of the electromagnetic spectrum due to the coherent oscillation of free electrons with the electric field of the incident light, called a localized surface plasmon resonance (LSPR).[3] Gold nanospheres exhibit only a single LSPR, but the anisotropy of gold nanorods causes them to exhibit two LSPRs, one corresponding to the transverse axes of the nanorod and the other to the longitudinal axis. As demonstrated in Fig. 1, the longitudinal LSPR can be tuned throughout the visible region of the spectrum and into the near-infrared by changing the aspect ratio—the ratio of length to diameter—of the nanorod,[4] with greater aspect ratios causing the longitudinal LSPR to shift to longer wavelengths.

Such shape-dependent optical properties make anisotropic nanoparticles attractive for applications ranging from sensing[5] to imaging and photothermal therapy.[6] We recently published a perspective article that focuses on the mechanisms leading to the growth of anisotropic nanoparticles.[7] Here we will highlight some of the widely used techniques that are used to direct nanoparticle morphology.

Seed-mediated growth is one of the most important methods of shape-directed nanoparticle synthesis. In the seed-mediated growth strategy, the nucleation of particles and the growth of particles are separated in time and space.[7,8] Preformed seeds are added to a growth solution containing ligands and chemical precursors that are precipitated onto the seeds during nanoparticle growth. The seeds may be the same material as the growth material, but they do not necessarily have to be.[9] Because anisotropic particles have greater surface area than their isotropic counterparts, they are typically less thermodynamically stable and therefore the

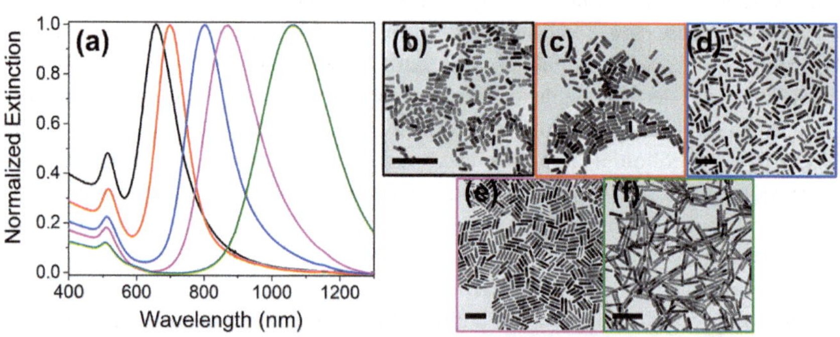

Fig. 1 (a) Normalized extinction spectra for gold nanorods with different aspect ratios. The black, red, blue, magenta, and green spectra were taken for gold nanorods with aspect ratios 2.4, 2.7, 3.6, 4.4, and 6.1, respectively, with corresponding TEM images in (b) through (f). Scale bars are 150 nm.

formation of anisotropic nanoparticles is usually governed by growth kinetics.[10,11] The advantage of using seed-mediated growth for anisotropic nanoparticle synthesis is that the separation of the nucleation and growth stages allows greater control over the kinetics of nanoparticle growth and particle morphology.[7] One possible mechanism behind the kinetic control of anisotropic nanocrystal growth is the selective adsorption of ligands and capping agents to specific crystal faces of the growing nanoparticle. If ligands bind more strongly to certain crystal faces, then those faces will be stabilized relative to the others and will consequently experience slower growth.[10] It has also been suggested that the mechanism of anisotropic growth of nanocrystals could occur as a result of the high chemical potential of precursors, which can lead to different rates of growth at crystal faces with different energies.[12]

We first reported a three-step seed-mediated synthesis of gold nanorods with aspect ratios >10.[13–15] In our synthesis, shown schematically in Fig. 2a, 3.5 nm citrate-capped gold seeds are added to a growth solution containing chloroauric acid ($HAuCl_4$), cetyltrimethylammonium bromide (CTAB), and the weak reducing agent ascorbic acid. After allowing some time for growth, an aliquot of nanoparticles from the first solution is removed and used as seeds in a second growth solution; after more time this is repeated as nanoparticles from the second growth solution are used to seed a third growth solution. Continued work on the synthesis of gold nanorods led to the silver-assisted seed-mediated synthesis of single crystalline gold nanorods, depicted in Fig. 2b, with aspect ratios of 2–5 by using smaller, 1.5 nm CTAB-capped gold seeds along with the addition of small amounts of silver nitrate to the growth solution, which allows precise control over the aspect ratio and results in high shape yield.[16,17] TEM images comparing the gold nanorods resulting from the silver-assisted and three-step syntheses are

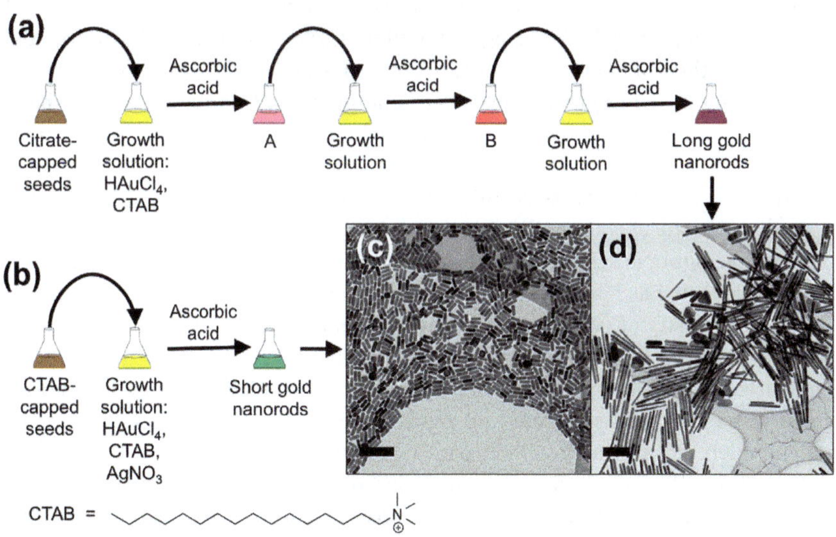

Fig. 2 Schematic representations of the (a) three-step seed-mediated synthesis of long gold nanorods and the (b) silver-assisted seed-mediated synthesis of short gold nanorods. TEM images of short and long gold nanorods are shown in (c) and (d) respectively. Scale bars are 200 nm.

shown in Fig. 2c and d. Most studies involving gold nanorods have since used the silver-assisted seed-mediated synthesis or its variations to prepare gold nanorods, highlighting the importance of the method.[8] The synthesis has been studied extensively and it has been shown that the anisotropic growth of the of the nanorods can be influenced by a number of factors including organic additives,[18] the concentration of surfactants and cosurfactants,[19] the halide counterion,[20] and reducing agents.[21] Besides gold, the seed-mediated method has also proven useful for the synthesis of anisotropic nanoparticles of many other materials. Xia and coworkers used the seed-mediated method to develop their polyol synthesis of silver nanowires using platinum seeds.[22,23] The polyol synthesis, which involves growth *via* the heating a polyol with a metal salt and a polymeric capping agent, has been used to produce anisotropic nanoparticles with a variety of different metals, including platinum,[24] palladium,[25] gold,[26] and rhodium.[27] Seed-mediated syntheses have also been used for the anisotropic growth of semiconductor nanoparticles.[9,28] Skrabalak *et al.* have shown that the use of seed-mediated growth can be coupled with co-reduction to lead to the growth of multimetallic nanoparticles.[29] They showed that the growth kinetics, and consequently particle morphology, can be controlled by changing the ratio of the different metal precursors to one another.[30]

While seed-mediated growth is the primary method used to prepare anisotropic gold nanoparticles, other methods have been demonstrated to lead to anisotropic gold nanoparticles, including oriented attachment.[31] In oriented attachment, individual nanocrystals, which may not be anisotropic themselves, assemble together with other nanocrystals to yield anisotropic nanoparticles. Kotov and coworkers have shown that oriented attachment can be used to produce nanostructures with a variety of shapes.[32] The assembly of nanoparticles is driven by the high surface-to-volume ratio of nanoparticles as well as their high surface energy. As with seed-mediated growth, the anisotropy of oriented attachment can result from the differences in energy between crystal faces of a nanoparticle; instead of leading to differential rates of deposition, in oriented attachment the difference in the energy of the crystal faces leads to a driving force for attachment being greater for higher energy faces. The technique has been applied to make nanostructures of a various materials, including TiO_2,[33] gold,[31] and CdTe.[34]

Given that the synthesis of anisotropic nanoparticles often depends on the anisotropy of the chemistry that is occurring on the surface during the growth of anisotropic nanoparticles, it is no surprise that the anisotropic surface chemistry of these objects has been explored. Among nanoparticle systems, the anisotropic surface chemistry of gold nanorods has been particularly well studied. El-Sayed and coworkers used thermogravimetric analysis and FTIR to show that CTAB on the surface of gold nanorods forms a bilayer.[35] The CTAB on the ends is postulated to be less tightly packed and more loosely bound than the CTAB on the sides, which may be due to tighter packing on of CTAB to the crystal faces on the sides of the rods compared to the ends[36] and to the greater curvature at the ends of the nanorods than at the sides.[37,38] The sparsity and/or greater lability of CTAB at the ends of gold nanorods has been corroborated by experiments demonstrating that gold nanorods are more reactive to cyanide dissolution at the ends than at the sides.[39] We demonstrated that gold nanorods can be functionalized somewhat preferentially at that ends with biotin disulfide by showing that the biotin-functionalized nanorods preferentially assembled end-to-end upon the addition of the multiple-biotin-binding protein streptavidin, as judged by transmission electron microscopy (TEM) of end-to-end nanorod assemblies.[40] Kumacheva and coworkers have made many important contributions, showing how gold nanorods end-functionalized with thiol-terminated polystyrene can be used to form analogues of block polymers that can be made to self assemble into complex structures.[41] Much of the evidence for the anisotropic functionalization of gold nanorods has been through TEM analysis showing nanorods assembled in specific arrangements.

Recently, the anisotropic surface chemistry of gold nanorods has been used for the regioselective coating of silica and titania.[38,42] By adding just enough silica precursor, Wang *et al.* showed that instead of making a complete silica shell, silica can be deposited just on the ends of the nanorod. They were also able to functionalize the ends of the nanorod to selectively functionalize the ends with poly(ethylene glycol) methyl ether thiol so that when they added silica precursors silica shells were formed only on the sides of the nanorod.[38] Wu and coworkers have shown that Wang *et al.*'s strategy for end-coating applies for titania growth on gold nanorods. They were able to improve the morphological yield of end-coated nanorods by replacing CTAB with alkytrimethylammonium compounds with shorter alkyl chain lengths, enhancing the lability of the ligands on the ends compared to the sides of the nanorod.[42] Of course, regioselective modification and functionalization of anisotropic nanoparticles is not limited to gold nanoparticles. Anisotropic surface chemistry has also been demonstrated on semiconductor nanocrystals on which metals and other semiconductors can be regioselectively deposited onto the tips of nanorods and tetrapods.[43–45]

1.2 Coating gold nanorods

While the anisotropic coating and functionalization of gold nanorods has been the focus of a number of studies, the majority of methods to coat gold nanorods lead to isotropic shell formation and functionalization. Controlling the surface chemistry of nanoparticles is critical in order to use them in different systems and for various applications. The use of gold nanoparticles in biological systems provides an especially apposite example, where their surface chemistry has been shown to effect cytotoxicity,[46] cell behavior,[47] and gene expression.[48] Many strategies have been reported for modifying the surface of gold nanorods. Ligand exchange can be conducted to replace the CTAB on the surface of gold nanorods with organic ligands, often thiols, which can possess different functional groups.[49] For nanoparticles that have a surface charge, electrostatic interactions can be used to wrap the nanoparticles with polyelectrolytes.[50] By alternating between the additions of positively and negatively charged polyelectrolytes, many layers can be wrapped around the nanoparticles, which is then useful to control the distance between molecules and the nanoparticle surface.[51] The surface chemistry can also be modified by forming shells on nanoparticles. On gold nanorods alone, the synthesis of a wide range of different kinds of shells has been reported, including metals,[52] nanocrystals for upconversion,[53] and even metal–organic frameworks.[54] Silica shells have been particularly important in large part because of the variety of modifications that can be made to their morphology and surface chemistry.

Two decades ago, Liz-Marzán and coworkers reported the synthesis of silica shells on gold nanoparticles.[55] In their synthesis, they made the gold nanoparticle surface vitreophilic using (3-aminopropyl)trimethoxysilane, and formed the silica shell by the addition of sodium silicate. We showed that silica shells could similarly be formed on CTAB-capped gold nanorods by modifying the nanorod surface with (3-mercaptopropyl)trimethoxysilane after excess CTAB is removed.[56] It was later demonstrated that it was not necessary to make the gold nanorod surface vitreophilic with functionalized silanes as the CTAB itself can act to template the growth of a mesoporous silica shell.[57–59] The silica coating of CTAB-capped nanoparticles is generally considered to occur in three stages: the hydrolysis and condensation of the silica precursor—tetraethyl orthosilicate (TEOS) in the present work—to form silica oligomers, the formation of silica/CTAB primary particles, and the CTAB-templated aggregation of those particles onto the nanoparticle surface.[57–59] Our group showed that the by carrying out silica coating at a CTAB concentration near 0.8 mM—approximately the critical micelle concentration for CTAB in water[60]—one could control the silica shell thickness from 11 nm to 26 nm by making small adjustments to the CTAB concentration from 0.4 mM to 1.2 mM. When the CTAB concentration is too high, then silica particles can form at free CTAB micelles apart from the nanorods, making control of the CTAB concentration critical for silica coating and the reduction of free silica particles.[59]

1.3 Nanomatryoshkas

Because of the emergent properties that often result from anisotropy on the nanoscale, we became interested in the possible effects that shape anisotropy would have on multilayered plasmonic nanoparticles, such as the nanorod matryoshka depicted in Fig. 3. Recently, it has been demonstrated that plasmonic nanoparticles consisting of alternating metallic and dielectric layers, called nanomatryoshkas by Halas *et al.*,[61] can be used to enhance molecular optical processes such as fluorescence[62] and Raman scattering,[63,64] and potentially for photothermal therapy.[63,65] The ability of nanomatryoshkas to enhance optical phenomena is due to electromagnetic enhancement that occurs in between the metallic layers. Theoretical calculations show that for nanomatryoshkas with multiple metallic layers, the nanomatryoshka acts like an optical condenser and the local electromagnetic field in the centermost dielectric layer can be enhanced multiplicatively with the number of layers in a nanomatryoshka.[66,67] The magnitude of near-field enhancement in a nanomatryoshka also depends on the geometry of the nanomatryoshka—*i.e.* the size of the core and the thickness of the metallic and dielectric layers[66]—and Kodali, *et al.* have developed computational

Fig. 3 The structure of a nanorod matryoshka, with a gold nanorod core, coated with a silica shell, coated with a gold shell.

methods to optimize the geometry of multilayered plasmonic particles to control the enhancement.[68] Additionally, by changing the geometry of nanomatryoshkas the wavelength of the plasmon can be tuned throughout the visible spectrum and into the near IR so that nanomatryoshkas can be tailored specifically for a variety of applications.[66,67]

Halas *et al.*, who were the first to publish the preparation of gold nanoshells, were also the first to report the synthesis of multilayered plasmonic nanstructures.[69] Later work by Xia *et al.*[70] and Halas *et al.*[67] resulted in gold@-silica@gold nanomatryoshkas in the nanoscale size regime. In most syntheses of nanomatryoshkas, a seed-mediated growth method is used to prepare the gold shell on silica that has been surface functionalized with (3-aminopropyl)trimethoxysilane (APTMS).[71] It is a multistep synthesis that requires several days for the aging of the gold seeds[71,72] and up to four more days for the attachment of the gold seeds to the APTMS-functionalized silica surface[62] in addition to difficulties in control over morphology and reproducibility, and colloidal stability,[63,73,74] especially for particles with smaller diameters.[75]

To date, the study of nanomatryoshkas has focused almost exclusively on spherical particles. Most of the experimental and computational work so far has been for nanomatryoshkas consisting of gold nanospheres with isotropic silica and gold shells. A notable exception is the work of Cong, *et al.* in which nanomatryoshkas were prepared from gold nanorod cores. Although the silica shells were quite thick in their nanorod matryoshkas and therefore rather spherically shaped, Cong *et al.* observed that for thinner silica shells the plasmon became less like that of the nanorod core but instead showed increasing plasmonic interaction between the core and the shell.[76] Since then, it has been shown that silica coating on CTAB-capped gold nanorods can be controlled for shells as thin as 11 nm,[59] making it possible to prepare nanorod matryoshkas, from silica coated-gold nanorods.

In this paper, we analyze the near-field properties of anisotropic nanomatryoshkas using the discrete dipole approximation (DDA). We show that, like nanorods, there are plasmon modes along the longitudinal and transverse directions of the nanorod matryoshkas but that, unlike nanorods, there is significant near-field enhancement along the sides of the nanorod core for the transverse mode, although the LSPRs are dependent on the configuration of the nanorod matryoshka. We also present our progress towards the synthesis of nanorod matryoshkas using methods adapted from the synthesis of nanomatryoshkas.[62] We also address some of the problems that we have encountered that merit consideration to move the field forward.

2. Experimental

2.1 Materials

Cetyltrimethylammonium bromide (CTAB, 99%) was purchased from Sigma. Hydrogen tetrachloroaurate ($HAuCl_4 \cdot 3H_2O$, 99.9%), tetraethyl orthosilicate (TEOS, 99.0%), (3-aminopropyl)trimethoxysilane (APTMS, 97%), and tetrakis-(hydroxymethyl)phosphonium chloride (THPC, 80% in water) were purchased from Aldrich. Sodium hydroxide (NaOH), silver nitrate ($AgNO_3$, 99.0%), hydroquinone (HQ, 99%) and formaldehyde (37 wt% in water) were purchased from Sigma Aldrich. Sodium borohydride ($NaBH_4$, 99%) was purchased from Fluka.

Anhydrous potassium carbonate (K_2CO_3, 99.0%) and methanol (MeOH, 99.9%) were purchased from Fisher Scientific. Ethanol (EtOH, 200 proof) was purchased from Decon Labs. Ammonium hydroxide (NH_4OH, 28–30% in water) was purchased from BDH Aristar. All chemicals were used as received without further purification.

2.1.1 Gold nanorod synthesis. The procedure used to synthesize gold nanorod matryoshkas was adapted from previously reported syntheses for nanomatryoshkas.[62] The process to make nanorod matryoshka can be broken into four steps as shown in Fig. 4: (i) synthesizing the gold nanorods for the cores of the nanorod matryoshkas, (ii) coating the nanorods with a silica shell functionalized with APTMS, (iii) attaching small gold seeds to the silica shell to make "seeded precursors," and finally (iv) the seed-mediated growth of a gold shell onto the silica-coated nanorod to complete the nanorod matryoshka.

Gold nanorods were prepared following the seed-mediated method described by Vigderman and Zubarev.[21] The procedure is similar to the seed-mediated synthesis of gold nanorods developed by Murphy *et al.*[17] and El-Sayed *et al.*,[16] but Vigderman and Zubarev have demonstrated that nanorods with a greater aspect ratio can be achieved when hydroquinone is used as the reducing agent instead of ascorbic acid. To prepare the seeds, 0.500 mL of 0.010 M $HAuCl_4$ was added to 9.5 mL of 0.10 M CTAB and the solution was stirred using a Teflon coated magnetic stir bar for 10 min. A 0.1 M $NaBH_4$ solution in 0.010 M NaOH was prepared: 45 mg of $NaBH_4$ were dissolved in 10 mL of ice cold 0.010 M NaOH. Quickly, 1 mL of the 0.1 M $NaBH_4$ in 0.010 M NaOH was diluted to 10 mL with the addition of ice cold 0.010 NaOH. Immediately after the dilution, 0.460 mL of 0.01 M $NaBH_4$ in 0.010 NaOH was added to the solution of $HAuCl_4$ and CTAB, causing the solution to turn dark brown. These gold seeds were stirred for 10 min and then aged for 1 h before use.

The CTAB-capped gold nanorods were prepared in 500 mL batches. For a typical synthesis, the growth solution was prepared by adding 25 mL of 0.010 M $HAuCl_4$ and 1.1 mL of freshly prepared 0.10 M $AgNO_3$ to 475 mL of 0.10 M CTAB.

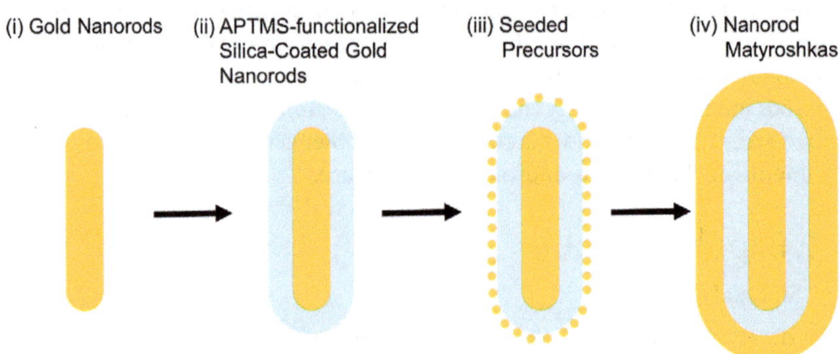

Fig. 4 A schematic representation of the steps in the synthesis in nanorod matyroshkas starting with (i) the synthesis of gold nanorods, followed by (ii) coating the gold nanorods with a silica shell and functionalizing the silica shell with APTMS, (iii) attaching small gold seeds to the silica shell to make seeded precursors, and finally (iv) the seed-mediated growth of a gold shell onto the silica-coated gold nanorod. Yellow layers are gold and gray layers are silica.

The solution was stirred using a Teflon coated magnetic stir bar as 25 mL of freshly prepared 0.1 M hydroquinone was added. The solution was stirred until it turned colorless—or mostly colorless as a slight yellow tint typically remained—after which 8 mL of gold seeds was added. The nanorods were stirred gently overnight. The following day the nanorods were purified by centrifugation at 7000 rcf for 30 min. The supernatant was discarded and the resulting pellet was dispersed in nanopure water.

2.1.2 Silica coating and APTMS functionalization. The synthesis of a mesoporous silica shell on the gold nanorods was conducted using methods described by Abadeer *et al.*[59] Typically, silica coating was done on a 10 mL scale. Since control of the CTAB concentration is important for silica coating, the gold nanorods were centrifuged again at 7000 rcf for 30 min and the supernatant was discarded. The gold nanorod pellet was dispersed in 10 mL of 0.8 mM CTAB such that the nanorod concentration was 1 nM, corresponding to an extinction near 10 for the longitudinal LSPR. The nanorods were gently mixed overnight to allow the CTAB to equilibrate on the surface of the nanorods. The following day 40 μL of 0.10 M NaOH was added to adjust the pH near 10. After 30 minutes, 90 μL of freshly prepared 20 vol% TEOS in MeOH was added and the solution was placed on a shaker at room temperature overnight. The silica-coated gold nanorods were purified by centrifugation at 7000 rcf for 30 min. The supernatant was discarded and the pellet was dispersed in 10 mL of EtOH. Centrifugation was repeated for three additional rounds. To functionalize the silica with APTMS, 50 μL of neat APTMS was added to the cleaned silica-coated nanorods and it was allowed to sit for 3 days. The APTMS-functionalized silica-coated nanorods were purified by four rounds of centrifugation as described above.

Two alternative methods of APTMS functionalization of the silica shell were also used. For the first, 50 μL of APTMS was added during the growth of the silica shell 1 h after the addition of TEOS. The following day any excess reactants were removed by four rounds of centrifugation at 7000 rcf for 30 min with the supernatant discarded after each round of centrifugation and the pellet dispersed in EtOH. The second alternative method of APTMS functionalization involved adding APTMS functionalized silica onto the mesoporous silica shell.[58] To 8 mL of purified silica-coated gold nanorods in EtOH was added 1 mL of water and 1 mL of 28% ammonium hydroxide. After 30 min 7.5 μL of TEOS and 7.5 μL of APTMS were added and these were allowed to react overnight. The APTMS-functionalized silica-coated nanorods were purified the following day by four rounds of centrifugation at 7000 rcf for 30 min. After each round the supernatant was discarded and the pellet was dispersed in 8 mL EtOH.

2.1.3 Seeded precursors synthesis. Following the procedure of Duff *et al.*,[77] 2–3 nm gold seeds, called the "Duff colloid" were synthesized. A Teflon coated stir bar was used to rapidly stir 45 mL of nanopure water. In quick succession, 300 μL of 1 M NaOH was added to the water followed by 1.00 mL 1.0% THPC. After two minutes 2 mL of 25 mM $HAuCl_4$ was added. As the gold seeds formed, the mixture instantly turned dark brown. The seeds were stored in the refrigerator at 4 °C.

To attach the gold seeds to the APTMS-functionalized silica-coated gold nanorods, 7 mL of Duff colloid and 150 μL of 1 M NaCl were simultaneously added to 3 mL of APTMS-functionalized 1 nM silica-coated gold nanorods in ethanol. The mixture was shaken vigorously and then sonicated for 30 min after which the mixture was placed on a shaker for 4 days. Each day the mixture was sonicated to break up any flocs that may have formed. To remove any excess seeds, the seeded precursors were centrifuged at 7000 rcf for 30 minutes. The supernatant, which contained excess seeds, was discarded and the pellet was dispersed in 3 mL of water. Centrifugation was repeated for two additional rounds, and samples were resuspended in 3 mL water each time to maintain a 1 nM concentration of the particles.

2.1.4 Gold shell synthesis. To prepare the gold shell growth solution, called "K-gold", 14 mg of K_2CO_3 were dissolved in 50 mL of nanopure water. Then, 750 mL of 25 mM $HAuCl_4$ were added and after mixing the solution it was aged in the dark overnight. Gold shell growth was carried out by adding 100 μL of seeded precursors to 1.5 mL of K-gold in a 1.7 mL microcentrifuge tube. A 20 μL drop of 37 wt% formaldehyde in water was placed on the cap of the microcentrifuge tube and then the tube was closed and rapidly shaken. Within a minute a color change indicated the growth of gold shells.

2.2 Characterization

Extinction spectra were taken using an Agilent Technologies Cary 5000 UV-Vis-NIR spectrophotometer. Transmission electron microscopy was performed using a JEOL 2100 cryo-TEM or a JEOL 2010 LaB_6 TEM operated at 200 kV. For the size analysis of the nanoparticles, at least 300 particles were measured; each measurement reported here is provided with its standard deviation. Zeta potential analysis was carried out using a Brookhaven Instruments Corporation ZetaPALs Zeta Potential Analyzer.

2.3 Electrochemical measurements

Cyclic voltammetry was carried out using a CH Instruments 760 D Electrochemical Workstation (Austin, TX) at room temperature using a Au disk working electrode, a "no-leak" Ag/AgCl (3 M KCl) reference electrode (EDAQ), and a carbon rod counter electrode. The Au disk was successively polished with diamond polish to a 3 μm finish, followed by rinsing and sonicating with water. All potentials were converted to SHE by measuring the open circuit potential of a Pt wire working electrode in H_2 saturated 1 M $HClO_4$. Solutions were prepared from stock solutions of ascorbic acid, hydroquinone, and CTAB in nanopure water and the pH was adjusted to 2.9 with HCl. Additional experiments were conducted with 0.5 M NaCl as supporting electrolyte to confirm that the solution conductivity and transport of electroactive species did not affect results.

3. Results and discussion

3.1 Calculated optical properties of nanorod matryoshkas

To study the near-field properties of nanorod matryoshkas, we modeled them using the Discrete Dipole Approximation (DDA). The calculations were done using the DDSCAT 7.0 source code developed by Draine and Flatau,[78] which has been made freely available. Nanorod matryoshkas consist of concentric cylinders with hemispherical endcaps and with the core and the outermost shell composed of gold and the inner shell of silica. The nanorod matryoshka was modeled as an array of point dipoles excited by an external electric field, E_0, on a cubic lattice in a medium with a refractive index of $n = 1.33$ to represent water. The dielectric constants for gold reported by Johnson and Christy[79] and the refractive index of $n = 1.456$ for silica on gold@silica nanoparticles reported by Liz-Marzán *et al.*[55] were used in this study. To model an unpolarized light source, each calculation was carried out twice, once with E_0 linearly polarized along the long axis of the particle and then again with E_0 linearly polarized along the short axis of the particle, and the results from the two calculations were averaged. The results were used to find the scattering, absorbance, and extinction efficiencies of the particles as well as to generate maps of the near-field intensity. Gaussian smoothing was used to mitigate anomalously high intensity near individual dipoles in the near-field maps.

In Fig. 5 the optical properties of gold nanorods are compared to those calculated for a nanorod matryoshka using DDA. The extinction spectra of gold nanorods show two LSPRs similar to the spectrum shown in Fig. 5a. The transverse LSPR usually has its maximum extinction between 500 and 530 nm while the location of the maximum extinction of the longitudinal LSPR, which is dependent on the aspect ratio of the nanorod (*i.e.* the ratio of its length to its diameter) can be tuned throughout the visible region of the electromagnetic spectrum and into the NIR.[4,80] The spectrum in Fig. 5a is typical for nanorods with an aspect ratio of 3.7 with the maximum extinction of the longitudinal LSPR occurring near 800 nm. The spectrum shown in Fig. 5b was calculated

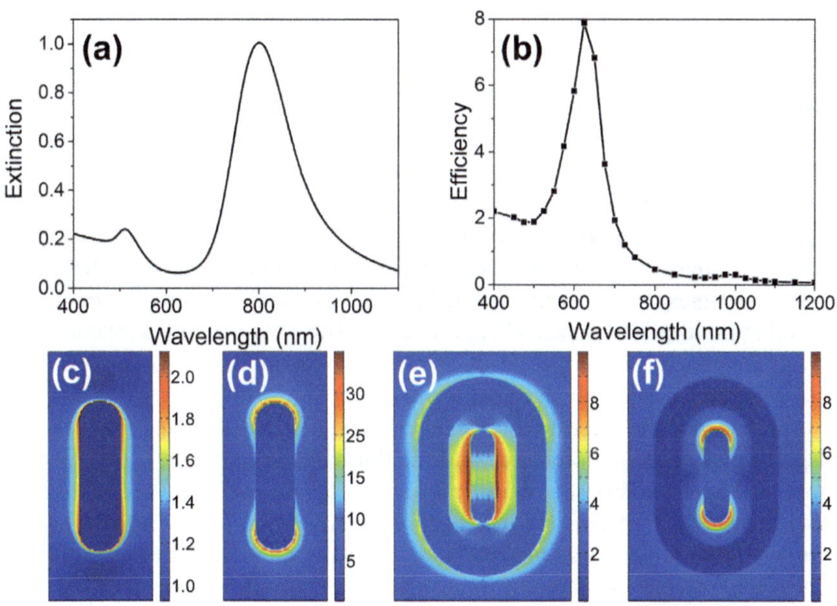

Fig. 5 (a) A normalized experimentally measured extinction spectrum of 74 × 20 nm gold nanorods with an aspect ratio of 3.7. (b) Calculated extinction spectrum using DDSCAT for a gold nanorod matryoshka with a 41 × 11 nm gold nanorod core, 10 nm silica shell, and 13 nm gold shell. The points in the spectrum show the extinction calculated by individual calculations with a line to guide the eye. In (c–f) electric field maps are shown with the color corresponding to the electric field enhancement, $|E|/|E_0|$. The (c) transverse and (d) longitudinal plasmons of a 41 × 11 nm gold nanorod at 507 nm and 800 nm, respectively, are shown as well as the (e) transverse and (f) longitudinal plasmon modes of the nanorod matryoshka, at 600 nm and 1000 nm, respectively.

using DDSCAT for a nanorod matryoshka with a 41 × 11 nm gold core (aspect ratio 3.7) with a 10 nm thick silica shell and a 13 nm thick gold shell. Like the gold nanorod spectrum, the nanorod matryoshka spectrum shows two plasmon resonances which can be attributed to a transverse plasmon for the peak near 600 nm and a longitudinal plasmon for peak near 1000 nm. In contrast to the gold nanorod spectra, the calculated nanorod matryoshka spectrum shows that the extinction of the transverse peak is greater than that of the longitudinal peak. Additionally, both of the nanorod matryoshka plasmons are red-shifted relative to the LSPRs exhibited by gold nanorods.

The most striking differences between the optical properties of gold nanorods and gold nanorod matryoshkas become apparent when comparing their near-field enhancement. In Fig. 5c and d electric field maps of the transverse and longitudinal LSPRs of gold nanorod are shown. Electric field maps for the transverse and longitudinal plasmon modes for a nanorod matryoshka are shown in Fig. 5e and f. Most of the electric field enhancement, $|E|/|E_0|$, for nanorod matryoshkas is located in the silica layer between the gold nanorod core and the gold shell. For the transverse plasmon mode the location with the greatest enhancement is near the sides of the nanorod core and for the longitudinal plasmon mode most of the near-field enhancement is located close to the ends of the nanorod core. While the location of the near-field enhancement is similar for gold nanorods, the magnitude is quite different. The near-field enhancement of the transverse LSPR is small for a gold nanorod, as shown in Fig. 5c, making it impractical to use the transverse LSPR for optical enhancement applications. Instead, the longitudinal LSPR is used because of the huge near-field enhancements at the ends—nearly 30 times the incident field intensity in some places—as shown in Fig. 5d. A caveat to that enhancement, though, is that only a fraction of the molecules attached to ends of the nanorods experience the bulk of the enhancement, while most of the dyes and reporters on its sides would experience a much lesser enhancement. Nanorod matryoshkas are unique because, as shown in Fig. 5e and f, there is significant near-field enhancement for both the transverse and longitudinal plasmons. Both plasmon modes have near-field intensities nearly ten times that of the incident field, which may make nanorod matryoshkas useful as multimodal probes. Along with control over the magnitude of near-field enhancement, the introduction of anisotropy to nanomatryoshkas allows the location of near-field enhancement as well.

Our calculations indicate that geometric parameters, including the size and aspect ratios of the nanorod core and the thickness and shape of the silica and gold shells, may affect the optical properties of nanorod matryoshkas. Fig. 6a shows calculated spectra for two nanorod matryoshkas with different aspect ratios. While both nanorod matryoshkas had 16 nm thick silica shells and 10 nm thick gold shells, the nanorod matryoshka that gave the red spectrum had a 66 × 26 nm gold nanorod core and the core for the nanorod matryoshka that resulted in the black spectrum had an 86 × 14 nm gold nanorod core. As the aspect ratio of the nanorod matryoshka is increased, it appears that the longitudinal plasmon peak diminishes with respect to the transverse plasmon. Both of the plasmon resonances appear to red shift as the aspect ratio of the nanorod matryoshka is increased. When comparing the electric field maps for the transverse plasmon modes of the nanorod matryoshka, the effect of the aspect ratio is even more pronounced. The electric field maps shown in Fig. 6 were evaluated at

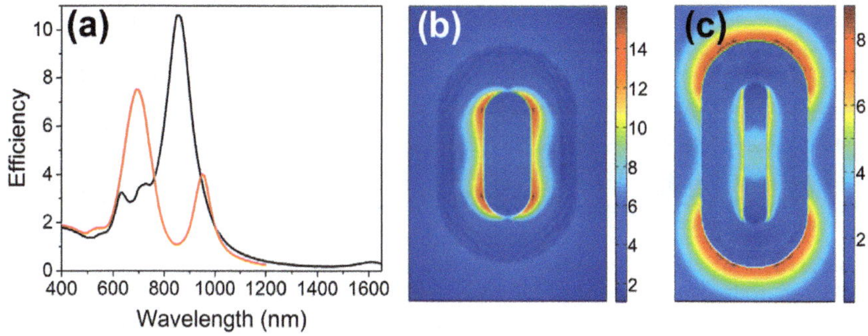

Fig. 6 (a) Calculated extinction spectra for nanorod matryoshkas of different geometries. The red spectrum corresponds to a nanorod matryoshka with a 66 × 26 nm gold nanorod core with 16 nm thick silica shell and a 10 nm thick gold shell and the electric field of the transverse plasmon at 690 nm is mapped in (b). The black spectrum corresponds to a nanorod matryoshka with an 86 × 14 nm gold nanorod core, a 16 nm thick silica shell and a 10 nm thick gold shell; its electric field map of the transverse plasmon at 855 nm is shown in (c).

the wavelength corresponding to the maximum extinction. The region with the greatest near-field enhancement intensity for the shorter nanorod matryoshka, shown in Fig. 6b, is concentrated near the sides of the nanorod core but for the longer nanorod matryoshka, shown in Fig. 6c, the greatest enhancement is located outside the nanorod matryoshka near the ends. Peaks on the shoulder of the transverse plasmon are more prominent for the longer nanorod matryoshka. From electric field maps (not shown) for each of the shoulder peaks, the near-field intensity distribution at each of the peaks is distinct, with varying enhancement along the sides of the nanorod core and along the sides on the outside of the nanorod matryoshka. This seems to indicate that the plasmonic interactions are not as simple as the transverse modes of the shell coupling solely to the transverse plasmon of the nanorod core. Furthermore, the degree of those interactions seems to depend on the geometry of the nanorod matryoshka. To maximize the plasmonic enhancement of nanorod matryoshkas for the possible future applications as fluorescent or Raman tags, methods like the genetic algorithms described by Kodali *et al.* will be useful to guide the synthesis of nanorod matryoshkas with optimal geometry.

3.2 Nanorod matryoshka synthesis

TEM images corresponding to the stages of nanorod matryoshka synthesis are shown in Fig. 7a–d. Several batches of gold nanorods were used in the present work. We synthesized gold nanorod cores using hydroquinone for the reducing agent, using a procedure adapted from the modifications to the seed-mediated synthesis of gold nanorods described by Vigderman and Zubarev.[21] Typical gold nanorods used in the preparation of gold nanorod matryoshkas are shown in the TEM image in Fig. 7a. Several batches of nanorods were prepared and their dimensions ranged between 75 and 85 nm long and 12 to 14 nm wide, but with aspect ratios near 6.1.

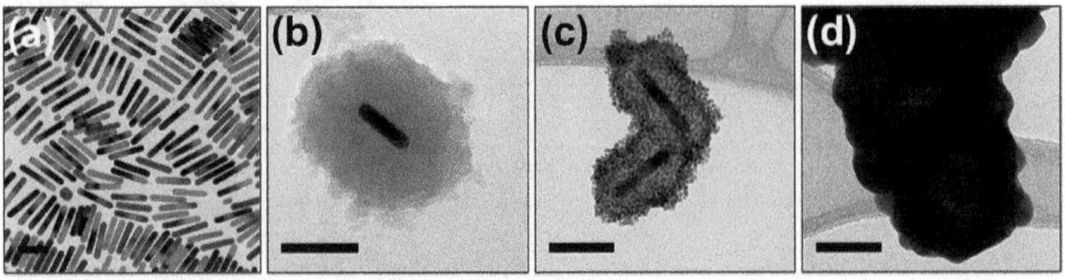

Fig. 7 The TEM images corresponding to the stages of nanorod matryoshka synthesis: (a) gold nanorods cores were synthesized in the first step and they were (b) coated with APTMS-functionalized silica, then (c) gold seeds were attached to the silica shell, which (d) seeded the growth of a gold shell. Scale bars are 50 nm.

We synthesized mesoporous silica shells on the CTAB-capped gold nanorod using methods previously published by our group.[59] Different methods were used to achieve the surface functionalization of the mesoporous silica shell. For the first method, the silica-coated nanorods were functionalized by adding neat APTMS to the nanorods in EtOH. The resulting silica shells were uniform and usually around 20 nm thick, as shown in Fig. 8a. APTMS functionalization was confirmed by the change of the zeta potential from slightly negative for the silica-coated nanorods to positive for the APTMS-functionalized silica-coated nanorods. The synthesis and functionalization of silica shells took a few days for each batch due to multiple rounds of centrifugation and the time allowed for silica growth and also silica functionalization. Alternatively, we also functionalized the silica shells by adding APTMS during the growth of the silica shells. If APTMS was added 30 minutes or less into silica shell formation, then aggregates of silica and nanorods result. When APTMS was added 45 min to an hour into silica shell formation, shells such as that shown in Fig. 8b result. A comparison of the histograms in Fig. 8b and c indicates that while the shells are of uniform thickness when APTMS is added after shell formation has completed, when APTMS is added during silica shell formation the shells are only half as thick at the ends, measured at 13 ± 4 nm for one batch *versus* 26 ± 4 nm on the sides. Despite the less uniform shells, silica shells formed and functionalized by adding APTMS during silica shell growth can be synthesized in a day and require fewer purification steps and are effective for use to make seeded precursors. We also have adapted the procedure of Gorelikov and Matsuura to grow silica on mesoporous silica.[58] Both TEOS and APTMS were added at the same time in order to form APTMS-functionalized silica on the silica shell and to block the pores. The resulting silica shells, like the one shown in Fig. 7b, had somewhat rough surfaces; however, they were quite effective for the attachment of gold seeds.

To make the seeded precursors, the APTMS-functionalized silica-coated nanorods were added to a dispersion of 2–3 nm gold seeds. The gold seeds are shown on the silica surface of a seeded precursor in Fig. 9a. Electrostatic attraction between the negative surface charge of the gold seeds,[81,82] positively charged APTMS-functionalized silica surface and electrostatic repulsion between the gold seeds affect the surface coverage of the gold seeds on the silica. Upon attachment of the seeds to the silica-coated gold nanorods, a visible color change was noticeable; the extinction spectrum in Fig. 9b shows that for the seeded precursor

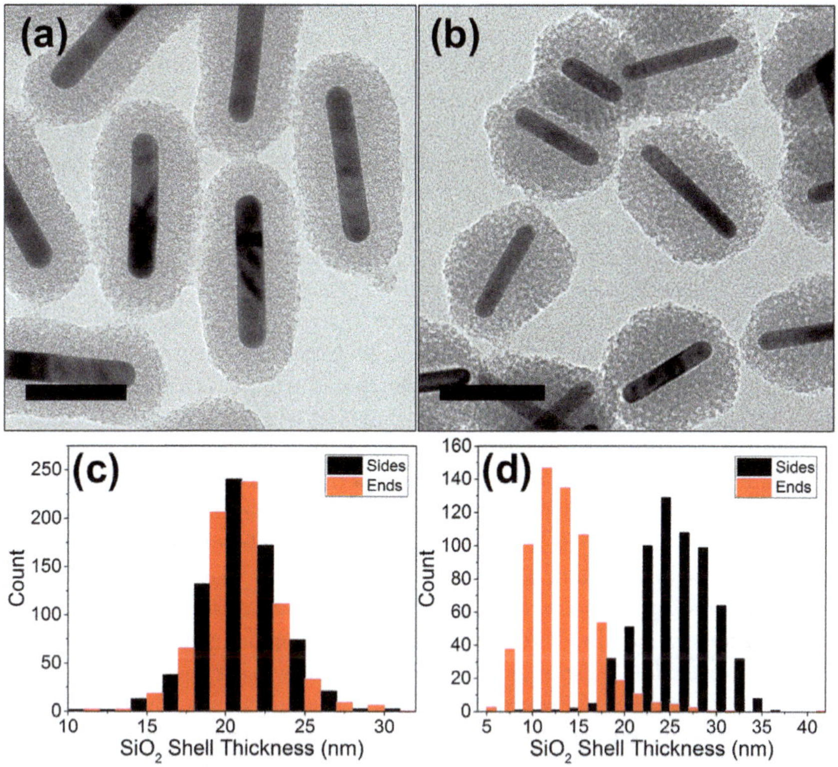

Fig. 8 TEM images of silica-coated and APTMS functionalized gold nanorods. In (a) the silica-coated nanorods were APTMS functionalized after the silica shells had been formed and the silica-coated nanorods had been purified. APTMS functionalization for the silica-coated nanorods in (b) was done by adding APTMS during silica shell formation. The histogram in (c) shows that the silica shells that were functionalized after silica shell formation have an isotropic thickness while the histogram in (d) demonstrates that adding APTMS before the completion of silica shell formation results in silica shells that are thinner at the ends than at the sides. Scale bars are 50 nm.

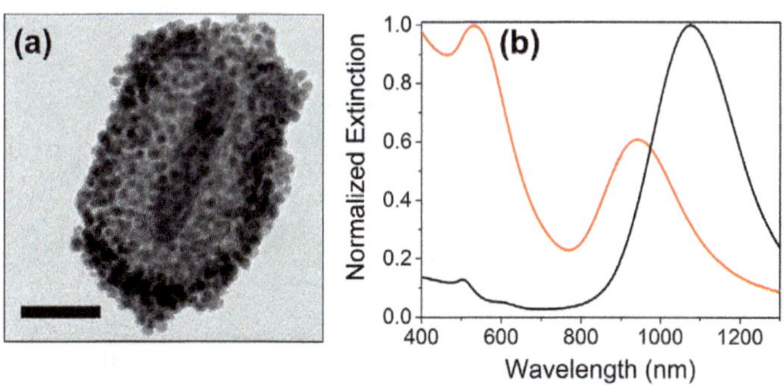

Fig. 9 (a) A TEM image showing a seeded precursor. Scale bar is 20 nm. (b) A comparison of a typical extinction spectrum for gold nanorod seeded precursors (red) to the extinction spectrum of gold nanorods (black).

the transverse LSPR becomes more intense relative to the longitudinal plasmon and the longitudinal LSPR blue shifts.

The formation of a gold shell onto the seeded precursors has posed the greatest synthetic challenge so far. In the synthesis of gold shells, formaldehyde is used to reduce the gold precursor $Au(OH)_4^-$, which results from the exchange of the chloride ligands of $AuCl_4^-$ since the K_2CO_3 is used to adjust the pH. In Fig. 10a, the measured extinction spectrum for the nanorod matryoshkas shown in Fig. 10b is compared to the theoretical extinction spectrum for a nanorod matryoshka calculated by DDSCAT. The measured extinction spectra for nanorod matryoshkas usually have very broad features. The roughness of the gold shells and the lack of uniformity between nanorod matryoshkas in any given sample likely contribute to broadening the plasmon resonance. Aggregation of the particles may also be one of the causes for broad resonances. TEM images of nanorod matryoshkas like that shown in Fig. 10b suggest that the gold shells appear to be a mosaic of many smaller gold particles, and it is not clear that the shells are contiguous. By increasing the ratio of the gold concentration to seeded precursor concentration, nanorod matryoshkas with more contiguous gold shells can be prepared, such as the nanorod matryoshka shown in Fig. 10c. However, the gold shells still appear to be rough.

To make anisotropic nanorod matryoshkas, it was important to start with high aspect ratio gold nanorod cores because silica coating and gold shell formation are both expected to be isotropic. Therefore, with increasing shell thickness, the overall aspect ratio of a given nanorod matryoshka decreases. Using our well-known seed-mediated synthesis of CTAB-capped gold nanorods,[17] the aspect ratio of the nanorods can be tuned by the addition of silver nitrate during the synthesis up to ~5. Vigderman and Zubarev have shown that by replacing ascorbic acid with hydroquinone as the reducing agent, the range of accessible aspect ratios that can be achieved by changing the silver nitrate concentration is extended up to 7.[21] Vigderman and Zubarev suggested that greater aspect ratios could be achieved when using hydroquinone as a reducing agent as opposed to ascorbic because hydroquinone is a weaker reducing agent, causing nanorod growth to take more time; these notions mix thermodynamic aspects with kinetic aspects.

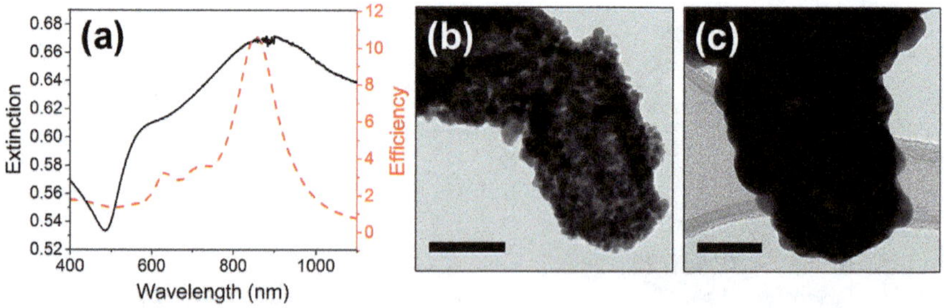

Fig. 10 Comparison of a measured extinction spectrum for nanorod matryoshkas to the extinction spectrum calculated using DDSCAT. (b) TEM image of a nanorod matryoshka from the sample that was measured to obtain the extinction spectrum in (a). (c) TEM image of nanorod matryoshka with thicker, more contiguous gold shell that was formed by increasing the gold-concentration to seeded-precursor-concentration ratio. Scale bars are 50 nm.

In order to study quantitatively the differences between the two reducing agents, we used cyclic voltammetry (CV) to measure the oxidation potential of both ascorbic acid and hydroquinone in solution. Fig. 11a shows cyclic voltammograms (CVs) for ascorbic acid and hydroquinone in pH 2.9 aqueous solutions without CTAB using an Au disk electrode. It is evident that the oxidation of ascorbic acid occurs at lower potential than the oxidation of hydroquinone, which suggests that on Au, ascorbic acid will function as a stronger reducing agent. Fig. 11b shows CVs of both ascorbic acid and hydroquinone in the presence of CTAB maintaining the same conditions as before. In this case, the onset for the oxidation of ascorbic acid again occurs at lower potential than for the oxidation of hydroquinone. The differences in the onset of oxidation observed for ascorbic acid and hydroquinone, approximately 250 mV without CTAB and 150 mV with CTAB, support Vigderman and Zubarev's statements that hydroquinone is a weaker reducing agent under synthetically relevant conditions. Interestingly, the oxidation of hydroquinone in the presence of CTAB looks similar to that of CTAB alone, suggesting that CTAB could be oxidized along with hydroquinone. How CTAB oxidation might affect nanorod synthesis is unknown at this point. Additionally, the lower onset potential observed for the oxidation of ascorbic acid in the presence of CTAB and the overlap of the onsets for the oxidation of hydroquinone in the presence of CTAB compared to CTAB only, suggest that ascorbic acid and hydroquinone might interact differently with the CTAB bilayer which affects their ability to be oxidized near the Au surface. The use of an additional supporting electrolyte resulted in little change in the redox behavior observed.

Once gold nanorods are synthesized, silica coating is undertaken to create nanorod matryoshkas. Using our methods for silica coating, we have achieved much greater control over silica shell thickness than Cong *et al.* previously reported in their synthesis of nanorod matryoshkas. Cong *et al.* identified the silica coating of CTAB-capped nanorods as one of the challenges in using nanorod cores for nanomatryoshkas.[76] Their approach was to remove excess CTAB from the

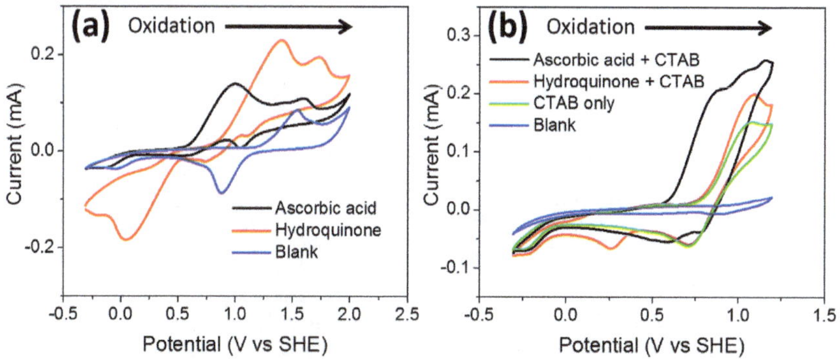

Fig. 11 (a) Cyclic voltammograms obtained from a solution containing 4.8 mM ascorbic acid (black) or 4.8 mM hydroquinone (red) in H_2O pH adjusted to 2.9 with HCl. The CV obtained from pH 2.9 H_2O (blue) is also shown. (b) Cyclic voltammograms obtained from solutions containing 4.8 mM ascorbic acid (black) or 4.8 mM hydroquinone (red), both in 0.1 M CTAB + pH 2.9 H_2O. CVs obtained from pH 2.9 H_2O with 0.1 M CTAB (green) and without CTAB (blue) are also shown.

nanorods to a concentration well below the critical micelle concentration (CMC) of CTAB by centrifuging the nanorods followed by silica coating in a mixture of water and isopropanol,[76] using a modified Stöber method.[83,84] In a recent publication it was shown that by adding poly(ethylene glycol)silane (PEG-silane) at different times during silica shell formation the silica surface could be functionalized and that the shell thickness could be controlled to form shells as thin as 2 nm by stopping silica growth.[85] As demonstrated by the silica-coated nanorods in Fig. 8b, the addition of APTMS caused the silica deposition to result in thinner shells on the ends of the gold nanorods than on their sides, or to disrupt silica shell deposition if it was added too soon into silica shell growth, rather than resulting in thinner shells as was reported with PEG-silane. Because APTMS is much smaller than PEG-silane, it may not be as effective at blocking the silica surface and stopping silica growth, therefore causing it to be poor for terminating silica shell growth to form thin shells.

We have struggled in the areas of control over the gold shell morphology, sample uniformity, and reproducibility while synthesizing our gold shells, with many syntheses ending in aggregation of the particles. These problems are illustrated well by the TEM image of nanorod matryoshkas in Fig. 12a in which the roughness of the gold shells is quite evident. Notice also that the nanomatryoshkas are clumped together. Though it is likely a result of the drying that occurred during TEM grid preparation, most of the nanomatryoshkas observed under TEM seem to occur on the grid in aggregates, although we cannot say for certain whether the nanorod matryoshkas form the aggregates in solution or if the aggregates on the TEM grid are the result of a drying effect. Some of our difficulties seem to be due to the small radii of our nanorods. While nanomatryoshkas prepared are prepared from gold nanosphere cores with diameters near 40 nm,[62] the typical diameters of our gold nanorod cores range from 11–14 nm. All gold nanoshells demonstrate some degree of roughness, but as Rasch *et al.* have shown, since the roughness is dependent on the size of the seeds, smaller shells made with the same size seeds will appear rougher. Furthermore, they showed that smaller shells are more prone to aggregation as

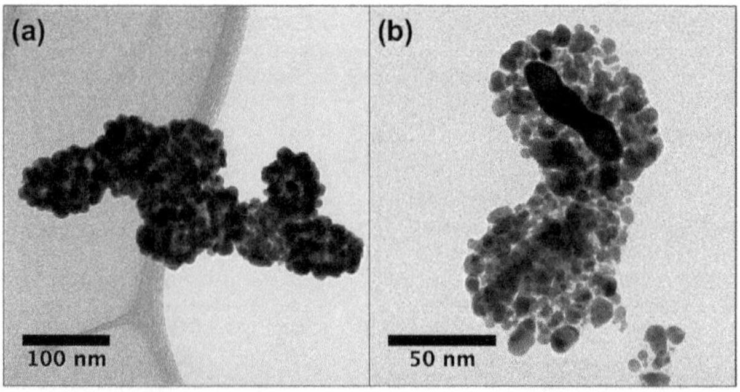

Fig. 12 TEM images depicting common problems encountered during gold shell synthesis for nanorod matryoshkas. (a) Several nanorod matryoshkas bunched together with rough gold shells. In the center of the nanorod matryoshkas the gold nanorods can be seen as dark rods surrounded by lighter gray of the silica shell with the gold shell appearing black and bumpy. (b) Nanorod matryoshkas with non-contiguous shells showing pronounced gold overgrowth on the nanorod core.

the tighter radius of curvature causes them to have a more diffuse charge double-layer.[75] Another problem that has been commonly encountered while trying to form gold shells is that metal overgrowth has been observed at the nanorod core, shown in the TEM image in Fig. 12b. This is likely due to the fact that the silica shell is mesoporous and that the pores were cleared of any remaining CTAB after several rounds of centrifugation in EtOH, allowing the gold ions to permeate through the silica shell and to be reduced at the nanorod surface. In a recent study by Liz-Marzán *et al.* mesoporous silica shells were recently shown to be an effective template for the growth of branched gold nanoparticles by metal overgrowth of the gold nanorod cores in the pores of the silica.[86]

4. Conclusion

As a result of the new properties that anisotropy often introduces to nanoscale systems, it is clear that future applications of nanotechnology will require both the synthetic ability to synthesize a variety of anisotropic nanostructures and an understanding of how the physical and chemical properties are dependent on the nanostructure geometry. Using DDSCAT, we have shown that just as nanorods support transverse and longitudinal LSPRs nanorod matryoshkas are also expected to have plasmon resonances corresponding to their transverse and longitudinal axes. The DDA calculations show that a major advantage of nanorod matryoshkas over nanorods is that while nanorods have only negligible near-field enhancement associated with the transverse LSPR, for nanorod matryoshkas both the transverse and longitudinal plasmons show significant near-field enhancement. The calculations indicate that the magnitude and distribution of the near-field enhancement for the nanorod matryoshka LSPRs depends on the geometry. We have adapted procedures for the syntheses of nanomatryoshkas to produce nanorod matryoshkas with mixed success. While we have been able to coat CTAB-capped gold nanorods in a relatively thin layer of APTMS-functionalized silica and to attach small gold nanoparticles to the silica shell to seed the growth of a gold shell, the growth of a contiguous gold shell has been difficult to achieve with good control over the shell morphology and reproducibility. Some of our difficulties while synthesizing the gold shells, especially their roughness and tendency to aggregate, may be due to the small diameter of our silica-coated gold nanorods. In future work, the synthesis of nanorod matryoshkas may be able to be improved, perhaps by using larger diameter gold nanorods as the cores or through modifications of the gold shell synthesis.

Acknowledgements

We thank the National Science Foundation (CHE-1306596) for funding.

References

1 S. Jiang, Q. Chen, M. Tripathy, E. Luijten, K. S. Schweizer and S. Granick, Janus Particle Synthesis and Assembly, *Adv. Mater.*, 2010, **22**(10), 1060–1071.

2 A. Walther and A. H. E. Müller, Janus Particles: Synthesis, Self-Assembly, Physical Properties, and Applications, *Chem. Rev.*, 2013, **113**(7), 5194–5261.

3 S. Link and M. A. El-Sayed, Spectral Properties and Relaxation Dynamics of Surface Plasmon Electronic Oscillations in Gold and Silver Nanodots and Nanorods, *J. Phys. Chem. B*, 1999, **103**(40), 8410–8426.

4 S. Link, M. B. Mohamed and M. A. El-Sayed, Simulation of the Optical Absorption Spectra of Gold Nanorods as a Function of Their Aspect Ratio and the Effect of the Medium Dielectric Constant, *J. Phys. Chem. B*, 1999, **103**(16), 3073–3077.

5 D. Chen, X. Hou, T. Li, L. Yin, B. Fan, H. Wang, X. Li, H. Xu, H. Lu, R. Zhang, *et al.*, Effects of Morphologies on Acetone-Sensing Properties of Tungsten Trioxide Nanocrystals, *Sens. Actuators, B*, 2011, **153**(2), 373–381.

6 P. K. Jain, K. S. Lee, I. H. El-Sayed and M. A. El-Sayed, Calculated Absorption and Scattering Properties of Gold Nanoparticles of Different Size, Shape, and Composition: Applications in Biological Imaging and Biomedicine, *J. Phys. Chem. B*, 2006, **110**(14), 7238–7248.

7 N. D. Burrows, A. M. Vartanian, N. S. Abadeer, E. M. Grzincic, L. M. Jacob, W. Lin, J. Li, J. M. Dennison, J. G. Hinman and C. J. Murphy, Anisotropic Nanoparticles and Anisotropic Surface Chemistry, *J. Phys. Chem. Lett.*, 2016, **7**(4), 632–641.

8 S. E. Lohse and C. J. Murphy, The Quest for Shape Control: a History of Gold Nanorod Synthesis, *Chem. Mater.*, 2013, **25**(8), 1250–1261.

9 D. V. Talapin, J. H. Nelson, E. V. Shevchenko, S. Aloni, B. Sadtler and A. P. Alivisatos, Seeded Growth of Highly Luminescent CdSe/CdS Nanoheterostructures with Rod and Tetrapod Morphologies, *Nano Lett.*, 2007, **7**(10), 2951–2959.

10 Y. Yin and A. P. Alivisatos, Colloidal Nanocrystal Synthesis and the Organic–Inorganic Interface, *Nature*, 2005, **437**(7059), 664–670.

11 Y. Wang, J. He, C. Liu, W. H. Chong and H. Chen, Thermodynamics Versus Kinetics in Nanosynthesis, *Angew. Chem., Int. Ed.*, 2015, **54**(7), 2022–2051.

12 X. Peng, Mechanisms for the Shape-Control and Shape-Evolution of Colloidal Semiconductor Nanocrystals, *Adv. Mater.*, 2003, **15**(5), 459–463.

13 N. R. Jana, L. Gearheart and C. J. Murphy, Seed-Mediated Growth Approach for Shape-Controlled Synthesis of Spheroidal and Rod-Like Gold Nanoparticles Using a Surfactant Template, *Adv. Mater.*, 2001, **13**, 1389–1393.

14 N. R. Jana, L. Gearheart and C. J. Murphy, Wet Chemical Synthesis of High Aspect Ratio Cylindrical Gold Nanorods, *J. Phys. Chem. B*, 2001, **105**(19), 4065–4067.

15 B. D. Busbee, S. O. Obare and C. J. Murphy, An Improved Synthesis of High-Aspect-Ratio Gold Nanorods, *Adv. Mater.*, 2003, **15**(5), 414–416.

16 B. Nikoobakht and M. A. El-Sayed, Preparation and Growth Mechanism of Gold Nanorods (NRs) Using Seed-Mediated Growth Method, *Chem. Mater.*, 2003, **15**(10), 1957–1962.

17 T. K. Sau and C. J. Murphy, Seeded High Yield Synthesis of Short Au Nanorods in Aqueous Solution, *Langmuir*, 2004, **20**(15), 6414–6420.

18 X. Ye, L. Jin, H. Caglayan, J. Chen, G. Xing, C. Zheng, V. Doan-Nguyen, Y. Kang, N. Engheta, C. R. Kagan, *et al.*, Improved Size-Tunable Synthesis of Monodisperse Gold Nanorods Through the Use of Aromatic Additives, *ACS Nano*, 2012, **6**(3), 2804–2817.

19 X. Ye, C. Zheng, J. Chen, Y. Gao and C. B. Murray, Using Binary Surfactant Mixtures to Simultaneously Improve the Dimensional Tunability and

Monodispersity in the Seeded Growth of Gold Nanorods, *Nano Lett.*, 2013, **13**(2), 765–771.

20 S. E. Lohse, N. D. Burrows, L. Scarabelli, L. M. Liz-Marzán and C. J. Murphy, Anisotropic Noble Metal Nanocrystal Growth: the Role of Halides, *Chem. Mater.*, 2014, **26**(1), 34–43.

21 L. Vigderman and E. R. Zubarev, High-Yield Synthesis of Gold Nanorods with Longitudinal SPR Peak Greater Than 1200 Nm Using Hydroquinone as a Reducing Agent, *Chem. Mater.*, 2013, **25**(8), 1450–1457.

22 Y. Sun, B. Gates, B. Mayers and Y. Xia, Crystalline Silver Nanowires by Soft Solution Processing, *Nano Lett.*, 2002, **2**(2), 165–168.

23 Y. Sun, Y. Yin, B. T. Mayers, T. Herricks and Y. Xia, Uniform Silver Nanowires Synthesis by Reducing $AgNO_3$ with Ethylene Glycol in the Presence of Seeds and Poly(Vinyl Pyrrolidone), *Chem. Mater.*, 2002, **14**(11), 4736–4745.

24 J. Chen, T. Herricks, M. Geissler and Y. Xia, Single-Crystal Nanowires of Platinum Can Be Synthesized by Controlling the Reaction Rate of a Polyol Process, *J. Am. Chem. Soc.*, 2004, **126**(35), 10854–10855.

25 Y. Xiong, H. Cai, B. J. Wiley, J. Wang, M. J. Kim and Y. Xia, Synthesis and Mechanistic Study of Palladium Nanobars and Nanorods, *J. Am. Chem. Soc.*, 2007, **129**(12), 3665–3675.

26 E. Carbó-Argibay, B. Rodríguez-González, I. Pastoriza-Santos, J. Pérez-Juste and L. M. Liz-Marzán, Growth of Pentatwinned Gold Nanorods Into Truncated Decahedra, *Nanoscale*, 2010, **2**(11), 2377–2378.

27 H. Zhang, X. Xia, W. Li, J. Zeng, Y. Dai, D. Yang and Y. Xia, Facile Synthesis of Five-Fold Twinned, Starfish-Like Rhodium Nanocrystals by Eliminating Oxidative Etching with a Chloride-Free Precursor, *Angew. Chem., Int. Ed.*, 2010, **49**(31), 5296–5300.

28 D. V. Talapin, R. Koeppe, S. Götzinger, A. Kornowski, J. M. Lupton, A. L. Rogach, O. Benson, J. Feldmann and H. Weller, Highly Emissive Colloidal CdSe/CdS Heterostructures of Mixed Dimensionality, *Nano Lett.*, 2003, **3**(12), 1677–1681.

29 C. J. DeSantis, A. A. Peverly, D. G. Peters and S. E. Skrabalak, Octopods Versus Concave Nanocrystals: Control of Morphology by Manipulating the Kinetics of Seeded Growth via Co-Reduction, *Nano Lett.*, 2011, **11**(5), 2164–2168.

30 C. J. DeSantis, A. C. Sue, M. M. Bower and S. E. Skrabalak, Seed-Mediated Co-Reduction: a Versatile Route to Architecturally Controlled Bimetallic Nanostructures, *ACS Nano*, 2012, **6**(3), 2617–2628.

31 A. Halder and N. Ravishankar, Ultrafine Single-Crystalline Gold Nanowire Arrays by Oriented Attachment, *Adv. Mater.*, 2007, **19**(14), 1854–1858.

32 Z. Tang, Y. Wang, S. Shanbhag, M. Giersig and N. A. Kotov, Spontaneous Transformation of CdTe Nanoparticles Into Angled Te Nanocrystals: From Particles and Rods to Checkmarks, X-Marks, and Other Unusual Shapes, *J. Am. Chem. Soc.*, 2006, **128**(20), 6730–6736.

33 R. L. Penn and J. F. Banfield, Morphology Development and Crystal Growth in Nanocrystalline Aggregates Under Hydrothermal Conditions: Insights From Titania, *Geochim. Cosmochim. Acta*, 1999, **63**(10), 1549–1557.

34 Z. Tang, N. A. Kotov and M. Giersig, Spontaneous Organization of Single CdTe Nanoparticles Into Luminescent Nanowires, *Science*, 2002, **297**(5579), 237–240.

35 B. Nikoobakht and M. A. El-Sayed, Evidence for Bilayer Assembly of Cationic Surfactants on the Surface of Gold Nanorods, *Langmuir*, 2001, **17**(20), 6368–6374.

36 S. K. Meena and M. Sulpizi, Understanding the Microscopic Origin of Gold Nanoparticle Anisotropic Growth From Molecular Dynamics Simulations, *Langmuir*, 2013, **29**(48), 14954–14961.

37 X. Kou, Z. Sun, Z. Yang, H. Chen and J. Wang, Curvature-Directed Assembly of Gold Nanocubes, Nanobranches, and Nanospheres, *Langmuir*, 2009, **25**(3), 1692–1698.

38 F. Wang, S. Cheng, Z. Bao and J. Wang, Anisotropic Overgrowth of Metal Heterostructures Induced by a Site-Selective Silica Coating, *Angew. Chem., Int. Ed.*, 2013, **52**(39), 10344–10348.

39 N. R. Jana, L. Gearheart, S. O. Obare and C. J. Murphy, Anisotropic Chemical Reactivity of Gold Spheroids and Nanorods, *Langmuir*, 2002, **18**(3), 922–927.

40 K. K. Caswell, J. N. Wilson, U. H. F. Bunz and C. J. Murphy, Preferential End-to-End Assembly of Gold Nanorods by Biotin–Streptavidin Connectors, *J. Am. Chem. Soc.*, 2003, **125**(46), 13914–13915.

41 Z. Nie, D. Fava, E. Kumacheva, S. Zou, G. C. Walker and M. Rubinstein, Self-Assembly of Metal–Polymer Analogues of Amphiphilic Triblock Copolymers, *Nat. Mater.*, 2007, **6**(8), 609–614.

42 B. Wu, D. Liu, S. Mubeen, T. T. Chuong, M. Moskovits and G. D. Stucky, Anisotropic Growth of TiO_2 onto Gold Nanorods for Plasmon-Enhanced Hydrogen Production From Water Reduction, *J. Am. Chem. Soc.*, 2016, **138**(4), 1114–1117.

43 T. Mokari, E. Rothenberg, I. Popov, R. Costi and U. Banin, Selective Growth of Metal Tips Onto Semiconductor Quantum Rods and Tetrapods, *Science*, 2004, **304**(5678), 1787–1790.

44 S. Kudera, L. Carbone, M. F. Casula, R. Cingolani, A. Falqui, E. Snoeck, W. J. Parak and L. Manna, Selective Growth of PbSe on One or Both Tips of Colloidal Semiconductor Nanorods, *Nano Lett.*, 2005, **5**(3), 445–449.

45 Y. Ben-Shahar, F. Scotognella, I. Kriegel, L. Moretti, G. Cerullo, E. Rabani and U. Banin, Optimal Metal Domain Size for Photocatalysis with Hybrid Semiconductor-Metal Nanorods, *Nat. Commun.*, 2016, **7**, 10413.

46 A. M. Alkilany and C. J. Murphy, Toxicity and Cellular Uptake of Gold Nanoparticles: What We Have Learned So Far?, *J. Nanopart. Res.*, 2010, **12**(7), 2313–2333.

47 J. A. Yang, H. T. Phan, S. Vaidya and C. J. Murphy, Nanovacuums: Nanoparticle Uptake and Differential Cellular Migration on a Carpet of Nanoparticles, *Nano Lett.*, 2013, **13**(5), 2295–2302.

48 E. M. Grzincic, J. A. Yang, J. Drnevich, P. Falagan-Lotsch and C. J. Murphy, Global Transcriptomic Analysis of Model Human Cell Lines Exposed to Surface-Modified Gold Nanoparticles: the Effect of Surface Chemistry, *Nanoscale*, 2015, **7**(4), 1349–1362.

49 A. S. D. S. Indrasekara, R. C. Wadams and L. Fabris, Ligand Exchange on Gold Nanorods: Going Back to the Future, *Part. Part. Syst. Charact.*, 2014, **31**(8), 819–838.

50 A. Gole and C. J. Murphy, Polyelectrolyte-Coated Gold Nanorods: Synthesis, Characterization and Immobilization, *Chem. Mater.*, 2005, **17**(6), 1325–1330.

51 S. T. Sivapalan, J. H. Vella, T. K. Yang, M. J. Dalton, R. N. Swiger, J. E. Haley, T. M. Cooper, A. M. Urbas, L.-S. Tan and C. J. Murphy, Plasmonic Enhancement of the Two Photon Absorption Cross Section of an Organic Chromophore Using Polyelectrolyte-Coated Gold Nanorods, *Langmuir*, 2012, **28**(24), 9147–9154.

52 S. Hou, X. Hu, T. Wen, W. Liu and X. Wu, Core–Shell Noble Metal Nanostructures Templated by Gold Nanorods, *Adv. Mater.*, 2013, **25**(28), 3857–3862.

53 Y. Song, G. Liu, X. Dong, J. Wang, W. Yu and J. Li, Yb^{3+}, Er^{3+} Multifunctional Hybrid Nanocomposites with Upconversion Luminescence, Magnetism, and Photothermal Property, *J. Phys. Chem. C*, 2015, **119**(32), 18527–18536.

54 K. Khaletskaya, J. Reboul, M. Meilikhov, M. Nakahama, S. Diring, M. Tsujimoto, S. Isoda, F. Kim, K.-I. Kamei, R. A. Fischer, *et al.*, Integration of Porous Coordination Polymers and Gold Nanorods Into Core–Shell Mesoscopic Composites Toward Light-Induced Molecular Release, *J. Am. Chem. Soc.*, 2013, **135**(30), 10998–11005.

55 L. M. Liz-Marzán, M. Giersig and P. Mulvaney, Synthesis of Nanosized Gold–Silica Core–Shell Particles, *Langmuir*, 1996, **12**(18), 4329–4335.

56 S. O. Obare, N. R. Jana and C. J. Murphy, Preparation of Polystyrene- and Silica-Coated Gold Nanorods and Their Use as Templates for the Synthesis of Hollow Nanotubes, *Nano Lett.*, 2001, **1**(11), 601–603.

57 R. I. Nooney, D. Thirunavukkarasu, Y. Chen, R. Josephs and A. E. Ostafin, Self-Assembly of Mesoporous Nanoscale Silica/Gold Composites, *Langmuir*, 2003, **19**(18), 7628–7637.

58 I. Gorelikov and N. Matsuura, Single-Step Coating of Mesoporous Silica on Cetyltrimethyl Ammonium Bromide-Capped Nanoparticles, *Nano Lett.*, 2008, **8**(1), 369–373.

59 N. S. Abadeer, M. R. Brennan, W. L. Wilson and C. J. Murphy, Distance and Plasmon Wavelength Dependent Fluorescence of Molecules Bound to Silica-Coated Gold Nanorods, *ACS Nano*, 2014, **8**(8), 8392–8406.

60 M. A. Bahri, M. Hoebeke, A. Grammenos, L. Delanaye, N. Vandewalle and A. Seret, Investigation of SDS, DTAB and CTAB Micelle Microviscosities by Electron Spin Resonance, *Colloids Surf., A*, 2006, **290**(1–3), 206–212.

61 C. Radloff and N. J. Halas, Plasmonic Properties of Concentric Nanoshells, *Nano Lett.*, 2004, **4**(7), 1323–1327.

62 C. Ayala-Orozco, J. G. Liu, M. W. Knight, Y. Wang, J. K. Day, P. Nordlander and N. J. Halas, Fluorescence Enhancement of Molecules Inside a Gold Nanomatryoshka, *Nano Lett.*, 2014, **14**(5), 2926–2933.

63 Y. Gao, Y. Li, Y. Wang, Y. Chen, J. Gu, W. Zhao, J. Ding and J. Shi, Controlled Synthesis of Multilayered Gold Nanoshells for Enhanced Photothermal Therapy and SERS Detection, *Small*, 2015, **11**(1), 77–83.

64 L. Lin, H. Gu and J. Ye, Plasmonic Multi-Shell Nanomatryoshka Particles as Highly Tunable SERS Tags with Built-in Reporters, *Chem. Commun.*, 2015, **51**, 17740–17743.

65 C. Ayala-Orozco, C. Urban, M. W. Knight, A. S. Urban, O. Neumann, S. W. Bishnoi, S. Mukherjee, A. M. Goodman, H. Charron, T. Mitchell, *et al.*, Au Nanomatryoshkas as Efficient Near-Infrared Photothermal Transducers for Cancer Treatment: Benchmarking Against Nanoshells, *ACS Nano*, 2014, **8**(6), 6372–6381.

66 H. Xu, Multilayered Metal Core–Shell Nanostructures for Inducing a Large and Tunable Local Optical Field, *Phys. Rev. B: Condens. Matter Mater. Phys.*, 2005, **72**(7), 073405.

67 R. Bardhan, S. Mukherjee, N. A. Mirin, S. D. Levit, P. Nordlander and N. J. Halas, Nanosphere-in-a-Nanoshell: a Simple Nanomatryushka, *J. Phys. Chem. C*, 2010, **114**(16), 7378–7383.

68 A. K. Kodali, X. Llora and R. Bhargava, Optimally Designed Nanolayered Metal-Dielectric Particles as Probes for Massively Multiplexed and Ultrasensitive Molecular Assays Proc, *Proc. Natl. Acad. Sci. U. S. A.*, 2010, **107**, 13620–13625.

69 E. Prodan, C. Radloff, N. J. Halas and P. Nordlander, A Hybridization Model for the Plasmon Response of Complex Nanostructures, *Science*, 2003, **302**(5644), 419–422.

70 X. Xia, Y. Liu, V. Backman and G. A. Ameer, Engineering Sub-100 Nm Multi-Layer Nanoshells, *Nanotechnology*, 2006, **17**(21), 5435–5440.

71 B. J. Jankiewicz, D. Jamiola, J. Choma and M. Jaroniec, Silica-Metal Core–Shell Nanostructures, *Adv. Colloid Interface Sci.*, 2012, **170**(1–2), 28–47.

72 E. Grzincic, R. Teh, R. Wallen, G. McGuire, A. Yella, B. Q. Li and K. Bandyopadhyay, Synthesis of Gold and Palladium Nanoshells by In Situ Generation of Seeds on Silica Nanoparticle Cores, *RSC Adv.*, 2014, **4**(61), 32283–32292.

73 J. Tharion, J. Satija and S. Mukherji, Glucose Mediated Synthesis of Gold Nanoshells: a Facile and Eco-Friendly Approach Conferring High Colloidal Stability, *RSC Adv.*, 2014, **4**(8), 3984–3991.

74 R. Wang, X. Ji, Z. Huang, Y. Xue, D. Wang and W. Yang, Citrate-Regulated Surface Morphology of SiO_2@Au(74) Particles to Control the Surface Plasmonic Properties, *J. Phys. Chem. C*, 2016, **120**(1), 377–385.

75 M. R. Rasch, K. V. Sokolov and B. A. Korgel, Limitations on the Optical Tunability of Small Diameter Gold Nanoshells, *Langmuir*, 2009, **25**(19), 11777–11785.

76 H. Cong, R. Toftegaard, J. Arnbjerg and P. R. Ogilby, Silica-Coated Gold Nanorods with a Gold Overcoat: Controlling Optical Properties by Controlling the Dimensions of a Gold–Silica–Gold Layered Nanoparticle, *Langmuir*, 2010, **26**(6), 4188–4195.

77 D. G. Duff, A. Baiker and P. P. Edwards, A New Hydrosol of Gold Clusters. 1. Formation and Particle Size Variation, *Langmuir*, 1993, **9**(9), 2301–2309.

78 B. T. Draine and P. J. Flatau, Discrete-Dipole Approximation for Scattering Calculations, *J. Opt. Soc. Am. A*, 1994, **11**, 1491–1499.

79 P. B. Johnson and R. W. Christy, Optical Constants of the Noble Metals, *Phys. Rev. B: Solid State*, 1972, **6**(12), 4370–4379.

80 S. Link and M. A. El-Sayed, Shape and Size Dependence of Radiative, Non-Radiative and Photothermal Properties of Gold Nanocrystals, *Int. Rev. Phys. Chem.*, 2000, **19**, 409–453.

81 T. Pham, J. B. Jackson, N. J. Halas and T. R. Lee, Preparation and Characterization of Gold Nanoshells Coated with Self-Assembled Monolayers, *Langmuir*, 2002, **18**(12), 4915–4920.

82 S.-E. Park, M.-Y. Park, P.-K. Han and S.-W. Lee, The Effect of pH-Adjusted Gold Colloids on the Formation of Gold Clusters Over APTMS-Coated Silica Cores, *Bull. Korean Chem. Soc.*, 2006, **27**(9), 1341–1345.

83 Y. Lu, Y. Yin, Z.-Y. Li and Y. Xia, Synthesis and Self-Assembly of Au@SiO(83)$_2$ Core–Shell Colloids, *Nano Lett.*, 2002, **2**(7), 785–788.

84 W. Stöber, A. Fink and E. Bohn, Controlled Growth of Monodisperse Silica Spheres in the Micron Size Range, *J. Colloid Interface Sci.*, 1968, **26**(1), 62–69.

85 W.-C. Wu and J. B. Tracy, Large-Scale Silica Overcoating of Gold Nanorods with Tunable Shell Thicknesses, *Chem. Mater.*, 2015, **27**(8), 2888–2894.

86 M. N. Sanz-Ortiz, K. Sentosun, S. Bals and L. M. Liz-Marzán, Templated Growth of Surface Enhanced Raman Scattering-Active Branched Gold Nanoparticles Within Radial Mesoporous Silica Shells, *ACS Nano*, 2015, **9**, 10489–10497.

Faraday Discussions

PAPER

Natural selection in the colloid world: active chiral spirals

Jie Zhang[a] and Steve Granick*[b]

Received 12th April 2016, Accepted 13th April 2016
DOI: 10.1039/c6fd00077k

We present a model system in which to study natural selection in the colloid world. In the assembly of active Janus particles into rotating pinwheels when mixed with trace amounts of homogeneous colloids in the presence of an AC electric field, broken symmetry in the rotation direction produces spiral, chiral shapes. Locked into a central rotation point by the centre particle, the spiral arms are found to trail rotation of the overall cluster. To achieve a steady state, the spiral arms undergo an evolutionary process to coordinate their motion. Because all the particles as segments of the pinwheel arms are self-propelled, asymmetric arm lengths are tolerated. Reconfiguration of these structures can happen in various ways and various mechanisms of this directed structural change are analyzed in detail. We introduce the concept of VIP (very important particles) to express that sustainability of active structures is most sensitive to only a few particles at strategic locations in the moving self-assembled structures.

"Active matter", by which we mean matter not at thermal equilibrium but instead with mechanisms such that it consumes energy fed into its environment, is of mounting topical interest.[1–4] Nanoparticle and colloidal systems are especially attractive experimental systems as each moving element can with proper design be imaged while simultaneously viewing the collective behavior.[5] Without the averaging that inevitably goes into scattering experiments,[6] and without the coarse-graining that goes into rheology and other ensemble-averaged measurements of system properties,[7] these systems can reveal not only the fine structure but also its larger-scale assembly, provided that the moving elements are large enough to visualize.

Here we are interested in the cases where the moving elements interact with attraction sufficiently strong to produce assembly, but sufficiently weak that the collective assembly has the opportunity to reconfigure in response to the environment. Aspects of traditional self-assembly are involved[8] but are even more complex because the need to respond to the environment can mean that dynamically-assembling systems are not governed primarily by particle–particle

[a]*Department of Materials Science and Engineering, University of Illinois, Urbana, Illinois 61801, USA*
[b]*IBS Centre for Soft and Living Matter, Department of Chemistry, UNIST (Ulsan National Institute of Science and Technology), Ulsan 689-798, South Korea. E-mail: sgranick@ibs.re.kr*

interactions as would be the case in traditional equilibrium self-assembly.[9–11] In this study, we are especially interested in how the system evolves towards a final steady-state. As would be the case for biological evolution, it is natural to anticipate some kind of natural selection of those structures that are most compatible with the environment.

When assemblies of particles move coherently in liquid, their interactions with the environment can generate structures that are chiral, *i.e.* structures that are the same except for different handedness. Examples in the natural world span the range from the mundane, as when water drains in a sink by forming a chiral vortex, to the celestial, as when galaxies rotate clockwise or anti-clockwise. Other examples abound even more in the biological world; for example, lower level organisms such as bacteria often swim in a spiral path as their bodies are never perfectly symmetric.[12] These examples raise interesting puzzles that go beyond simply observing the chirality of the trajectories, puzzles about the relationship between shape and motion, especially when the mounting interest in active material design must ultimately be grounded in coordinating the motion of body parts and the whole body.

Regarding the experimental system, this study goes beyond this laboratory's traditional interest in single-component Janus particles[13–15] and considers instead the new dynamic structures which result when one mixes Janus particles with isotropic particles, both of them in an electric field. Building upon a preliminary study presented elsewhere[16] in which the Janus particles are repulsive to each other, here we consider particles that mutually attract to form active filaments. This is a convenient system in which to explore how structure evolves. Geometrical considerations in this system generate rotating clusters assembled from an isotropic sphere at the core, surrounded by three or four arms composed of strings of self-propelled Janus particles linked head-to-tail. "Evolution" springs from the assembly, disassembly and conformation changes of structures through various "reactions" analyzed below. Long-lived structures are "naturally selected" because their collective motion is better compatible with the motion of the constituent individual particles, which themselves are active. We find that the particles bonded into such long-lived clusters continue to enjoy certain orientational degrees of freedom but not usually so much that this changes the structure or conformation of the moving collective structure.

Experimental

For easiest visualization in an optical microscope, micron-sized particles were selected for study, with an experimental configuration described elsewhere.[16,17] Mixtures of silica spheres (usually 3 or 4 μm diameter) and half-metal coated silica Janus colloidal spheres (3 μm diameter silica coated on one hemispheres with 20 nm of titanium and then 20 nm of SiO_2) are selected with Janus spheres in excess in order to promote the assembly of numerous Janus spheres around a central core. The mixture is suspended in 0.1 mM NaCl aqueous solution and the particle suspension is sandwiched between two conductive transparent ITO coated coverslips and subjected to an AC electric field in the vertical direction. Density mismatch causes the silica spheres to sediment and ensures that the experimental system is 2D. When an electric field is applied, the Janus particles stand up with their Janus interface parallel to the applied field. Observed in a bright field microscope, the metal appears black. Therefore, these Janus particles appear half-black-half-white in the figures described below. Typically, the electric field condition was 300 kHz, 40 V mm^{-1}.

The uncoated silica particles are not propelled by the electric field, but the electric field induces a dipole moment in both the uncoated silica and Janus particles. For the uncoated silica particles, the dipole is induced in the centre, whereas for Janus particles two dipoles are induced, one in each hemisphere of the particle,[18,19] each one shifted from the centre of the sphere.[17] Calculation confirms that the interaction between the metal dipole and silica dipole is attractive[16] and that adjoining Janus particles favor head-to-tail attachment between the metal side and the silica side. Similarly, the metal coated hemispheres of Janus particles are attracted to bare silica particles.

Chiral spirals

The experimental system produces rotating spiral structures in which the core is a silica particle and arms of Janus particles emanate from the core, usually with 3 (for 3 μm cores) or 4 (for 4 μm cores) arms because this is how many arms can nucleate around the core. The arms are found to arrange spaced by equal distance. The structure is not like a rigid pinwheel however: Janus particles within each arm can be perturbed by others and by random fluctuations, so each arm is flexible. At the selected AC frequency it is known that Janus particles are propelled in the direction of the metal coated hemisphere owing to asymmetric induced-charge electroosmotic flow around the two hemispheres each of which has distinct polarizability, as described previously.[20] As the present system is assembled with an excess of Janus particles, some free chains of Janus particles were also observed, unattached to a central core.

As shown in Fig. 1a, the synergistic reorientation of these active building blocks into a spinning structure leads to a self-sustaining rotation that naturally breaks symmetry into spirals of opposite handedness and rotation direction. From a static image, the rotation direction can be predicted from the structure. Each arm is flexible because the head-to-tail arrangement of Janus particles is satisfied over a large degenerate spectrum of possible mutual orientations. Thus, the same connectivity between particles allows numerous conceivable pathways of propelled motion, but only certain arrangements are self-sustaining, *i.e.*, only certain arrangements are compatible with stability of the overall cluster.

Before the AC electric field is applied, there are no dipoles and the mixture of particles is homogeneous. After the AC electric field is switched on, we find that sometimes a disordered cluster assembles in the early stages, and then its structure adapts to become spiral. Alternatively, sometimes we find templated growth in which arms grow regularly from a smaller spiral. The pinwheel structures rotate because of the geometrical arrangement of the Janus particles and although the rotation breaks symmetry, there is no preference for clockwise or counter-clockwise motion. Since each Janus particle in each arm carries its own engine and drives itself, the arms of the pinwheel do not need to be of the same length to take on steady rotation. The particles adapt to the collective structure and to the hydrodynamic forces that the spiral encounters, during rotation, by adapting their orientations depending on where they reside in the cluster.

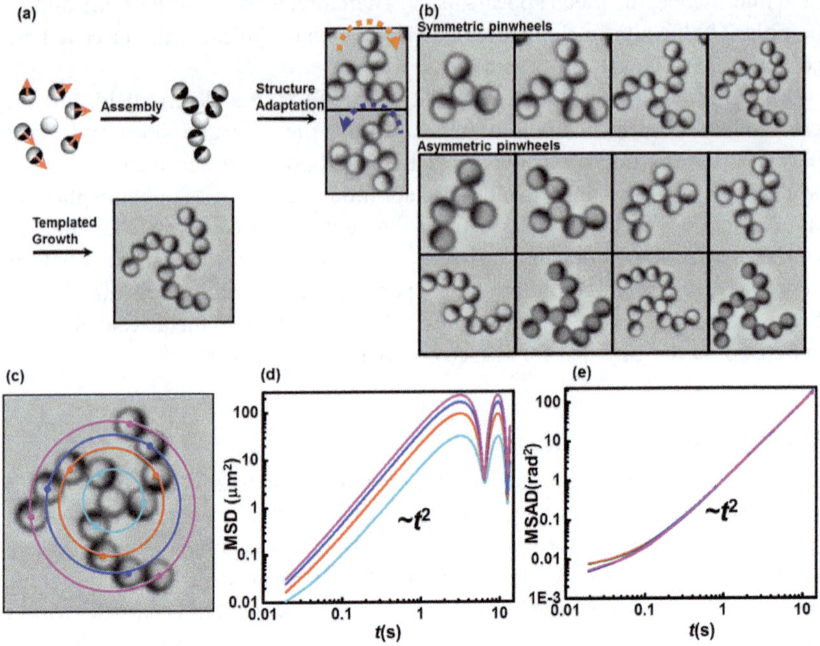

Fig. 1 (a) Schematic showing the growth of chiral clusters from structure adaptation and templated growth. (b) A collection of symmetric and asymmetric chiral pinwheels (only those with clockwise motion are shown). (c) Microscopic image of a three-arm symmetric pinwheel rotating clockwise. Cyan, red, blue and magenta circles label the position of active Janus particles in each layer and their respective trajectories. (d) Mean square displacement of active Janus particles in each layer. Colour coding is the same as in (c). (e) Mean square angular displacement of active Janus particles in each layer. Colour coding is the same as in (c) and (d).

A perfectly symmetric pinwheel structure would have its centre particle fixed in place, but in practice we find they usually jiggle around their starting position, presumably because the arm configurations fluctuate slightly. Similarly, in a perfectly symmetric structure the angular velocity of all participating particles would be the same, and the velocity would depend on distance to the centre. Fig. 1c shows the microscopic image of a pinwheel rotating clockwise with three four-particle-length arms. Cyan, red, blue, and magenta circles with dots label the trajectories and positions of the particles in each layer. All the three particles on the circles trace out identical trajectories and are rotationally the same. Fig. 1d and e show the mean square displacement (MSD) and mean square angular displacement (MSAD) of particles on each layer with the same colour coding as in Fig. 1c. Both the MSD and MSAD scale with time squared, as a result of the constant speed and angular velocity in each layer. To maintain intact structure during rotation, the angular velocity of each layer is the same in Fig. 1e, but the velocity in proportional to the distance of the layer to the centre, as shown in Fig. 1d.

It is striking that arms bend so that their curvature trails the motion of the rotating assembly. In general, arms trail a rotating object when the driving force is applied from the centre and transmitted to the arms, similar to the case if one places a T-shirt on a table and spins the shirt. Despite the fact that in this experimental system each particle is propelled individually, the symmetrical arrangement of arms around a central particle enforces rotation. The conformation and rotation of these pinwheels is equivalent to the case if all the arms were bonded to a centre particle that itself produced the overall rotation. It is pleasing to notice that spiral galaxies similarly display trailing arms with few exceptions.[21]

We can hypothesize speculatively however the different scenario in which the end of each arm would be driven by a tangential external force. This could be produced by a rotating magnetic field if the particles had magnetic susceptibility, as summarized in Fig. 2.

Onset of coherent motion and orientation locking

It is a puzzle why arms lock into distinct mutual curved orientations although they are far away physically. This would be unsurprising if their mutual attachments were permanent and dictated by growth from the centre particle, but in fact the bonding between neighboring particles is fragile and sensitive to orientation of each individual particle.

In Fig. 3a, we show the assembly at 5 successive instants of time up to 3 s, of a simple three-arm pinwheel, each arm containing only one particle, starting from particles that initially are unconnected. To aid visualization, 3 of these particles are labeled with different colours. Fig. 3b shows their time-dependent orientations in the x–y plane. One observes that two particles (blue and magenta) attach to a central particle and start to rotate clockwise: the orientations of these two particles shown on Fig. 3b increase steadily with roughly constant speed until the third (red) particle joins. Orientation of the incoming red particle is nearly constant until it attaches but starting from about 1.4 s, its orientation abruptly changes to a new definite level, under the influence of other free particles nearby, even before attachment to the cluster.

This move happened by chance but was essential to make it possible for the red particle to approach the cluster and become attached to it. Starting from about 2 s, all the three arm particles have joined the cluster. They now move coherently with fixed orientation difference; they rotate together. Also intriguing is to note that participation of the red particle caused the blue and magenta particles to rotate backwards relative to the orientation trajectories at early times, which one observes from the microscopic images and in the unusual turns of the three curves in Fig. 3b at around 2 s elapsed time. It seems that the last member to join played the most important role in determining the configuration of the final cluster.

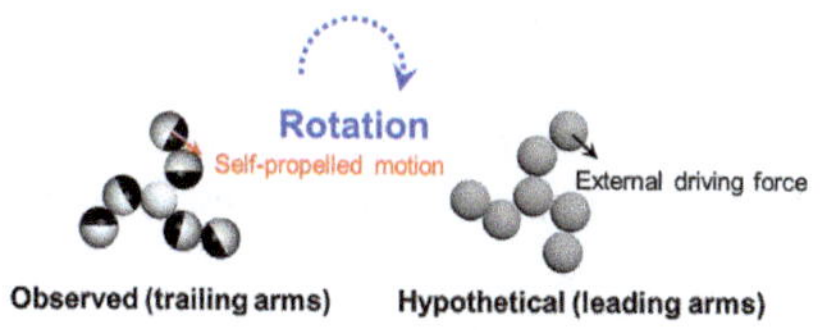

Fig. 2 The trailing (experimentally observed) and leading (hypothesized) arms of a rotating pinwheel.

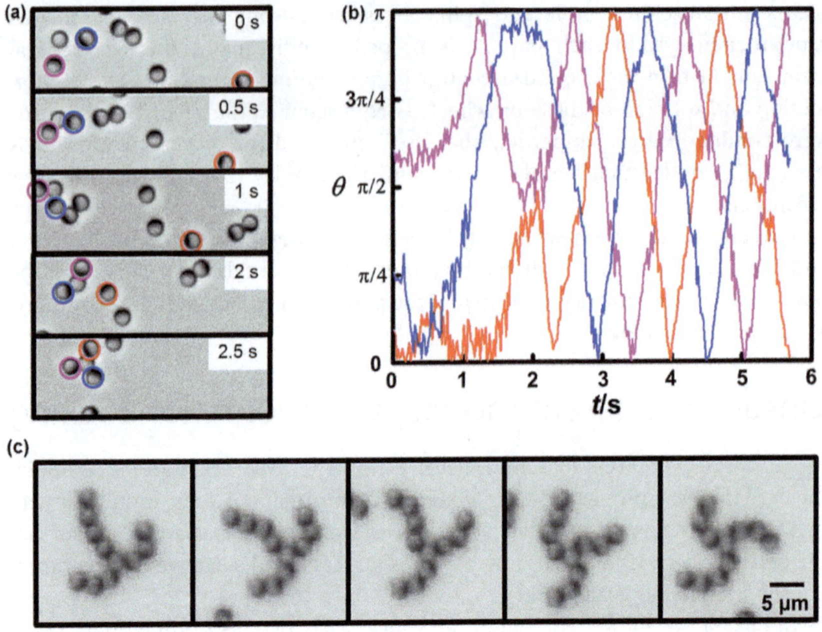

Fig. 3 (a) A time sequence of microscope images showing the simultaneous assembly and coordination of a three-arm pinwheel. Blue, red and magenta circles indicate the identity of participating particles. (b) History of orientation for the three labeled particles in (a). Starting at about 2 s elapsed time, the three active particles coordinate their motion and have fixed orientation differences. Noise of these curves reflects jiggling of particle orientation. (c) The coordination process of a pinwheel with longer arms over two seconds.

Fig. 3c illustrates a pinwheel with three longer arms. The cluster first translates in space because the pinwheel was asymmetric in arm length and arm orientation; motion of the entire cluster is incompatible with that of its individual constituents. This causes the cluster to deform and gradually modulates its shape, tending towards a sustainable shape. In the final state the centre no longer translates apart from fluctuations. It is interesting that before arriving at the final state, the arms seem to be flexible and subject to perturbation, in contrast to the final state in which the arms become reasonably static in shape.

Empirically, we find that the relative orientation of neighboring Janus particles in each arm favors an angle around $\pi/4$, not the value of 0 that might be anticipated solely from considering induced dipole interactions. The observed distribution of orientation angle is plotted in Fig. 4. The discrepancy is believed to be likely to stem from hydrodynamic interactions, but no quantitative explanation is offered at this time. Hydrodynamics probably also promotes the curved shape adopted by rotating filaments, but if we assume the interaction to be local, defects such as the kink in the middle of the free filament in the inset of Fig. 4 cannot be avoided. The arms in rotating pinwheels almost always lack kinks, however.

Kink-free filaments, which are required for the sustainable rotation of a pinwheel, present a result of "evolution" and "natural selection" in this colloidal

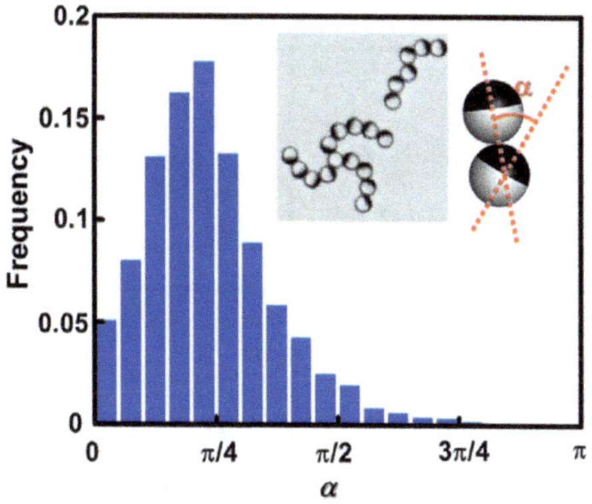

Fig. 4 The distribution of the orientation difference between two adjoining Janus particles, from analyzing chains on rotating pinwheels and also those that form free chains. Inset illustrates the coexistence of free and confined active filaments. The dataset includes about 15 000 pairs of particles over 20 seconds.

world. Here "evolution" means the dynamic assembly and disassembly of all possible structures through the various "reactions" that happen in the system. Long-lived structures are "naturally selected" because the combined collective motion of the clusters is compatible with each active individual particle inside.

Cluster reactions

Imaging the malleability and reconfigurability of this system, we find that this directed dynamics falls into categories that are familiar in organic chemistry: addition, substitution, ring opening and closure, as well as more complex reactions. Fig. 5 shows four typical reactions. In Fig. 5a, the magenta particle approaches the blue particle at the end of an arm on a rotating pinwheel, quickly adjusts its orientation and attaches. Given the requisite distance and mutual orientation between particles, this reaction can also occur internally within a cluster, as illustrated in Fig. 5b, where by chance the blue particle at the end of one arm lags to the point that it "reacts" with the nearby magenta particle on the neighboring arm: the orientation is acceptable and distance sufficiently close, so bonding between blue and magenta particles is created. Ring closure does not necessarily change the rotation of the overall cluster, because each particle is propelled by its own driving force according to its location within the cluster. In Fig. 5c, the cyan particle approaches the blue and magenta particles and inserts between them. Alternatively, this can be interpreted to show that the cyan particle substitutes the magenta particle in bonding with the blue particle. During the transition state the three particles form a triangle and bondings between them shuffle. Fig. 5d illustrates more complex reactions: with the blue particle attacking the magenta particle, the magenta–yellow bonding breaks, which leads to further detachment of the cyan–yellow dimer.

Superficially-similar colloidal "reactions" might be observed in equilibrium self-assembly.[22] The difference here is that the reactions are biased towards the directed evolution of sustainable shapes.

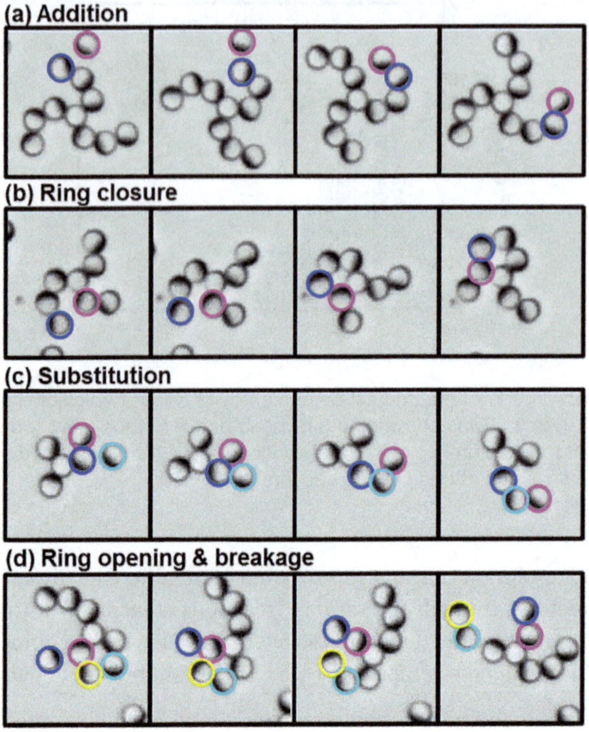

Fig. 5 Four examples of typical cluster "reactions". (a) The free magenta particle attaches to the blue particle on a pinwheel during its steady rotation. (b) The blue particle on one arm attaches to the magenta particle on a neighboring arm of the same pinwheel but the process imposes minimal influence on the rotation of the pinwheel. (c) The cyan particle approaches the bonded blue and magenta particles and inserts between them. (d) The blue particle approaches the magenta particle to open the ring bonding between magenta and yellow particles. The yellow and cyan dimer leaves the cluster due to this unfavorable orientation change.

Chirality switch

We can define a state number to identify the rotation condition and structure of particles and clusters: 0 for the state of random motion without chirality, 1 and −1 for clockwise and counterclockwise rotation, respectively. The natural selection process analyzed in this study concerns how to adapt the initial particle assembly (state 0) into a rotating pinwheel with either the 1 or −1 state. To switch the chirality of a cluster requires the collective change of almost all particle orientations.

To switch the chirality of a cluster which rotates already, one mechanism is through complicated reactions that may involve addition, substitution, ring closure and opening. The connectivity of particles within the cluster need not change, but usually does: this usually involves at least one particle that is not part of the cluster originally. Fig. 6a provides one example in which the reactions reconfigure the cluster and switch the chirality from −1 to 1 without changing the connectivity of the three particles immediately adjacent to the centre particle. The cluster starts with three arms of one-, one- and two-particle length, respectively, rotating counterclockwise, as shown in the first two frames. The chirality switch occurs between frames 2 and 3, when an outside cyan particle approaches randomly nearby, attaches to the red particle, and induces dramatic changes of the orientation of red and magenta particles to favor rotation in the opposite direction. The reaction, however, does not end there. The close distance and compatible orientation of the yellow particle on one arm and the red particle on the neighboring arm drives them to stick and to continue with clockwise rotation. The small size of this ring and the critical orientation difference between neighboring particles results, with further rotation of this cluster, in detachment from the cluster of the cyan–magenta dimer.

The concept of VIP, a very important particle, is illustrated by the development history of this cluster. We can see that the three particles (blue, red, and magenta) that originally form the skeleton of the rotating cluster have always been the VIPs of this particle collection and that the deviation angle of the head of these particles defines the cluster motion. Other particles, whether present as an extension of the blue VIP (yellow particle in frame 1) or as a connection between two VIPs (such as yellow and cyan in frame 5), may bond or debond without affecting significantly the cluster rotation, provided the VIPs do not change their orientations much.

Arms longer than 3 particle diameters can be very flexible as the persistence length of the assembled chains is quite short, as a result of the preferred angle difference between neighboring Janus particles, so the persistence length of assembled chains is quite short. As a result, another way to switch chirality transpires internally but only when the spiral's arm is long enough to be flexible. Although the VIPs immediately adjacent to the centre determine the motion of

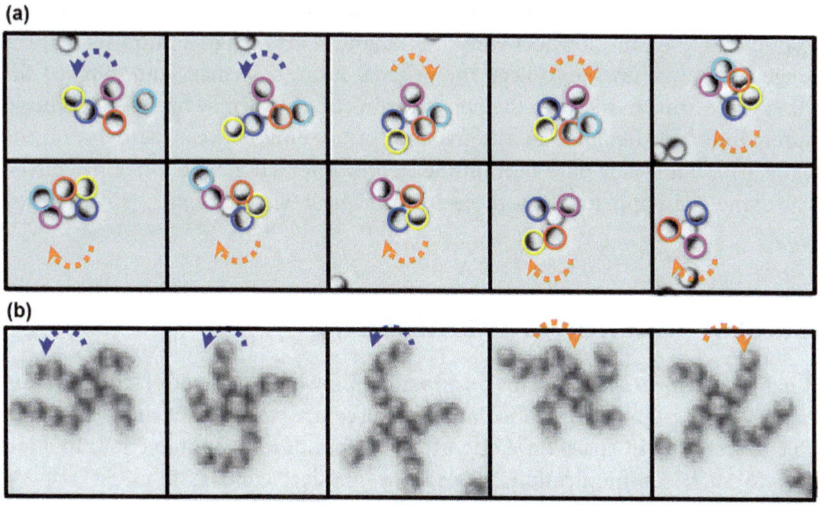

Fig. 6 Two ways to switch chirality. (a) After complicated reactions involving emergence and breakage of bonds, the connectivity of the three particles immediately adjacent to the centre particle stayed the same but the chirality changed in this illustration. (b) With a long enough arm, lack of control of the arm end by the centre particle self-induced the chirality change in this example.

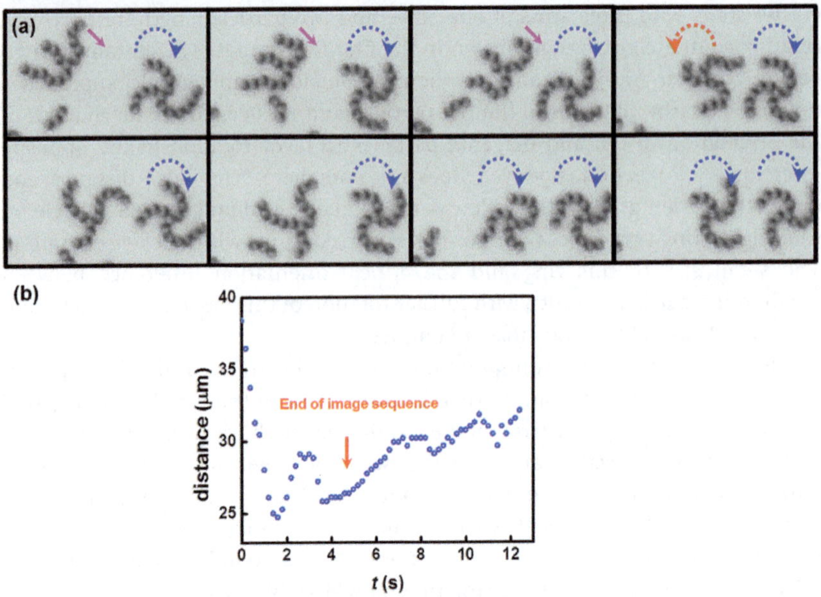

Fig. 7 (a) Time sequence of microscopic images showing the synchronization of two nearby clusters over five seconds. The cluster on the right maintains steady rotation through this time but the cluster on the left starts with directional motion towards the one on the right, goes through the indicated adaptation process, and finishes with rotation in the same direction as the other one. Arrows indicate direction of motion or rotation. (b) Plot of distance between the two cluster centres with time. Red arrow points to where the image sequence ends.

the overall cluster, other particles on a long arm may affect the orientation of VIPs and eventually change the structure and rotation of the cluster. In Fig. 6b, an example of a 4-arm pinwheel with 3, 3, 3, and 5 arm lengths is presented: the longest arm was unable to keep the original form, deformed, and dragged the other three arms to rotate to the opposite direction. This may have been induced purely internally, enabled by the freedom of bonding between particles on the same arm, but it may have been influenced by interaction with a nearby cluster. The same can happen with a three-arm pinwheel when one arm is sufficiently long.

Cluster–cluster interactions

It has been shown that for arms on a rotating pinwheel the rotation coordination is likely to be established *via* mediation by the centre particle. But clusters that are able to translate in space have opportunities for additional interaction. In addition to such communication "cabled" by physical contact, it might also be "wireless" through the longer-range medium of hydrodynamic interactions and induced dipole interactions, each of which could also perturb particle orientations because, as shown in the previous section, small perturbations (internal and external) to the mutual particle orientation within a cluster may result in cascading conformation changes. Physically, one may anticipate intuitively that it would be favorable for nearby clusters to synchronize in the rotation direction that minimizes their mutual influence. Fig. 7a shows one example of synchronization between two nearby clusters, with the left one developing from a translating state. Arrows show direction of translation or rotation for the two clusters. Fig. 7b plots the distance between the two cluster centres with time: it first decreases dramatically due to the translation of the left cluster and then goes up and down as a result of the reconfiguration. Nevertheless, it remains to be determined from a more extensive statistical study how much of conformation change can be attributed strictly to external influence by other clusters, how much of it to purely internal stochastic perturbations.

Conclusion

Natural selection, normally associated with the differential survival advantage of competing organisms, also applies to the world of active matter, where only certain self-assembled motile structures are compatible with long lifetime. This produces biased "evolution" towards those structures. We have studied this problem in a system where each moving element is visualized in real time: a water suspension of half-metal Janus spheres mixed with a minority of silica beads and subjected to AC voltage that produces not only active motion from induced charge electrophoresis but also particle–particle attraction from induced dipole moments. We observe chiral spirals which rotate clockwise or counterclockwise. Mechanisms of spiral formation and of chirality switch are analyzed. Small perturbations to the mutual particle orientation within a given cluster, whether from random internal fluctuations or from interactions with nearby clusters, may result in cascading conformation changes.

Acknowledgements

This work was supported by the New Faculty Fund (1.150068) of UNIST (Ulsan National Institute of Science and Technology). This work was supported at the University of Illinois by the US Department of Energy, Division of Materials Science, under award DE-FG02-07ER46471 through the Frederick Seitz Materials Research Laboratory. At the IBS Centre for Soft and Living Matter, SG acknowledges support by the Institute for Basic Science, project code IBS-R020-D1.

References

1 M. C. Marchetti, J. F. Joanny, S. Ramaswamy, T. B. Liverpool, J. Prost, M. Rao and R. A. Simha, *Rev. Mod. Phys.*, 2013, **85**, 1143.
2 W. Wang, W. T. Duan, S. Ahmed, A. Sen and T. E. Mallouk, *Acc. Chem. Res.*, 2015, **48**, 1938–1946.
3 C. Bechinger, R. Di Leonardo, H. Lowen, C. Reichhardt, G. Volpe and G. Volpe, arXiv:1602.00081, 2016.
4 W. C. K. Poon, arXiv:1306.4799, 2013.
5 W. Wang, W. T. Duan, S. Ahmed, T. E. Mallouk and A. Sen, *Nano Today*, 2013, **8**, 531–554.
6 W. Vanmegen and S. M. Underwood, *Phys. Rev. E: Stat. Phys., Plasmas, Fluids, Relat. Interdiscip. Top.*, 1993, **47**, 248–261.

7 J. C. Crocker, M. T. Valentine, E. R. Weeks, T. Gisler, P. D. Kaplan, A. G. Yodh and D. A. Weitz, *Phys. Rev. Lett.*, 2000, **85**, 888–891.
8 J. Yan, K. Chaudhary, S. C. Bae, J. A. Lewis and S. Granick, *Nat. Commun.*, 2013, **4**, 1516.
9 Q. Chen, S. C. Bae and S. Granick, *Nature*, 2011, **469**, 381–384.
10 Q. Chen, J. K. Whitmer, S. Jiang, S. C. Bae, E. Luijten and S. Granick, *Science*, 2011, **331**, 199–202.
11 J. Zhang, E. Luijten and S. Granick, *Annu. Rev. Phys. Chem.*, 2015, **66**, 581–600.
12 H. S. Jennings, *Am. Nat.*, 1901, **35**, 369–378.
13 L. Hong, A. Cacciuto, E. Luijten and S. Granick, *Nano Lett.*, 2006, **6**, 2510–2514.
14 S. Jiang, J. Yan, J. K. Whitmer, S. M. Anthony, E. Luijten and S. Granick, *Phys. Rev. Lett.*, 2014, **112**, 218301.
15 C. Q. Yu, J. Zhang and S. Granick, *Angew. Chem., Int. Ed.*, 2014, **53**, 4364–4367.
16 A. W. Long, J. Zhang, S. Granick and A. L. Ferguson, *Soft Matter*, 2015, **11**, 8141–8153.
17 J. Zhang, J. Yan and S. Granick, *Angew. Chem., Int. Ed.*, 2016, **55**, 1–5.
18 F. D. Ma, S. J. Wang, D. T. Wu and N. Wu, *Proc. Natl. Acad. Sci. U. S. A.*, 2015, **112**, 6307–6312.
19 F. D. Ma, D. T. Wu and N. Wu, *J. Am. Chem. Soc.*, 2013, **135**, 7839–7842.
20 R. Suzuki, H. R. Jiang and M. Sano, arXiv:1104.5607, 2011.
21 S. Capozziello and A. Lattanzi, *Chirality*, 2006, **18**, 17–23.
22 Y. F. Wang, Y. Wang, D. R. Breed, V. N. Manoharan, L. Feng, A. D. Hollingsworth, M. Weck and D. J. Pine, *Nature*, 2012, **491**, 51–55.

Faraday Discussions

PAPER

Synthesis of Janus plasmonic–magnetic, star–sphere nanoparticles, and their application in SERS detection

Javier Reguera,*[abc] Dorleta Jiménez de Aberasturi,[ac] Naomi Winckelmans,[d] Judith Langer,[ac] Sara Bals[d] and Luis M. Liz-Marzán[abc]

Received 6th February 2016, Accepted 2nd March 2016
DOI: 10.1039/c6fd00012f

Multicomponent nanoparticles are of particular interest due to a unique combination of properties at the nanoscale, which make them suitable for a wide variety of applications. Among them, Janus nanoparticles, presenting two distinct surface regions, can lead to specific interactions with interfaces, biomolecules, membranes *etc.* We report the synthesis of Janus nanoparticles comprising iron oxide nanospheres and gold nanostars, through two consecutive seed-mediated-growth steps. Electron tomography combining HAADF-STEM and EDX mapping has been performed to evaluate the spatial distribution of the two components of the nanoparticle, showing their clear separation in a Janus morphology. Additionally, SERS measurements assisted by magnetic separation were carried out to assess the application of combined plasmonic and magnetic properties for sensing.

1 Introduction

When the dimensions of a material decrease to the nanometric range, unique optical, electronic or magnetic properties may arise, making the resulting nanomaterial useful for a wide set of novel applications. Unlike the bulk material, such properties are dependent on the size and shape of the nanomaterial and can be easily combined through the use of multicomponent nanoparticles. This combination of properties can be exploited to achieve more complex tasks or give rise to new applications. A case of special interest is that of Janus nanoparticles, *i.e.* nanoparticles presenting two chemically different surface domains, which make them particularly interesting in applications where a spatial separation of

[a]*CIC biomaGUNE, Paseo de Miramon 182, 20009 Donostia-San Sebastián, Spain. E-mail: jreguera@cicbiomagune.es*
[b]*Ikerbasque, Basque Foundation for Science, 48013 Bilbao, Spain*
[c]*Biomedical Research Networking Center in Bioengineering Biomaterials and Nanomedicine, Ciber-BBN, Spain*
[d]*EMAT – University of Antwerp, Groenenborgerlaan 171, B-2020 Antwerp, Belgium*

functionalities is required.[1–3] For example, the use of hydrophobic and hydrophilic functionalities generates "nanoamphiphiles" that produce strong adsorption at interfaces and emulsion stabilization,[4,5] creating in some cases thermodynamically stable emulsions.[6,7] Their use has also been proposed in heterogeneous catalysis,[8,9] drug delivery,[10] nanojet motors,[11] antireflecting surfaces,[12] or just as building blocks to form more complex structures such as supracrystals or molecular colloids.[13,14] Despite the high interest in the use of Janus multicomponent nanoparticles, their applicability is still restricted by the limited current synthesis capabilities to obtain nanoparticles with different morphologies, shapes, and materials.

We present a new method that makes use of consecutive seed mediated growth steps to obtain complex nanoparticles with Janus morphology. The resulting nanoparticles comprise two different materials, iron oxide and gold, which present different properties. Iron oxide nanoparticles below a certain size exhibit superparamagnetism,[15] which allows external manipulation by moderate magnetic fields, as well as hyperthermia through alternate magnetic fields.[16] Iron oxide nanoparticles have also been used as contrast agents in magnetic resonance imaging (MRI).[17] Gold nanoparticles, on the other hand, exhibit localized surface plasmon resonances (LSPR), which results in efficient absorption and scattering of light at specific wavelengths that can be tuned through the size and shape of the nanoparticles.[18] This makes them attractive candidates as colorimetric sensors,[19] nanoprobes for surface-enhanced Raman spectroscopy (SERS),[20] labels for imaging,[21] or targeting agents for photothermal treatment of cancer tissue.[22,23] Among the numerous examples of gold nanoparticles, gold nanostars (NSs) are of interest because of their strong light absorption at the near-infrared (NIR) transparency window of biological tissue[24] as well as the high electric field enhancements that can be induced at their tips,[25,26] rendering them highly effective as SERS nanoprobes.[27] Both gold and iron oxide nanoparticles can be easily functionalized, with thiols and amines in the case of gold, and catechols, silanes, diols carboxyls or amines in the case of iron oxide. They are also biocompatible materials so that they are often used in biomedical applications such as drug delivery, biolabelling, hyperthermia treatments, *etc.*

Several approaches have been reported in the literature to produce nanoparticles comprising gold and iron oxide parts, with different morphologies. These include: core–shell spherical nanoparticles,[28] core–shell nanoflowers,[29] core–shell nanostars,[30–32] polymeric assemblies combining iron oxide nanoparticles with nanoparticles or nanostars,[33,34] or porous silica containing iron oxide nanoparticles and gold nanorods.[35] Janus dumbbell-like nanoparticles made of gold and iron oxide have been previously reported, exhibiting magnetic and plasmonic behavior,[36,37] but the gold lobe is typically small (~5–10 nm) and smaller than the iron oxide lobe (~10–20 nm). This results in a weak LSPR band that in general overlaps partially with the scattering by the iron oxide part, thereby hindering applications where a strong plasmonic behavior is necessary.

Here we report the synthesis of Janus iron oxide–gold magnetic nanostars (JMNSs) with a strong plasmonic absorbance, using consecutive steps of seed-mediated-growth starting from gold nanoparticles through dumbbell-like nanoparticles, to JMNSs. UV-Vis spectroscopy and bright field transmission electron microscopy (TEM) were used to evaluate the optical properties and the morphology of the nanoparticles. The spatial distribution of gold and iron oxide in the nanoparticle shows a clear Janus character, as observed in 3D electron tomography combining high angle annular dark field scanning transmission electron microscopy (HAADF-STEM) and energy-dispersive X-ray spectroscopy (EDX) chemical mapping. Finally, as an example of the capabilities of these multifunctional nanoparticles, SERS sensing was performed to detect several dye molecules at low volumes and concentrations. Furthermore, thanks to the magnetic response of these nanoparticles, SERS measurements were performed in magnetically concentrated aggregates producing an effect of amplified SERS detection, which allows detection of much lower concentrations.

2 Results and discussion

Fig. 1a–c shows schematically the different steps performed to obtain Janus nanostars. Steps (a) and (b) were performed following the protocol described by Yu *et al.* for the synthesis of nanodumbbells (NDs).[36] This synthesis can be performed in two different ways. The first method includes two separate steps where gold nanoparticles (Au NPs) are first synthesized with oleylamine and then used as seeds to grow the Fe part by thermal decomposition of $Fe(CO)_5$. The second method, which has been used here, is a one-pot two-step method where $HAuCl_4$ is injected into a solution at 120 °C containing oleylamine, oleic acid, hexadecanodiol, and $Fe(CO)_5$. Au NPs immediately formed after injection; the temperature was then raised up to 300 °C to decompose $Fe(CO)_5$, giving rise to Fe atoms that grow only on one facet of the Au seed. Yu *et al.* described the mechanism by which Fe grows on only one side of Au NPs.[36] As Fe starts to nucleate on Au, the free electrons from Au compensate for the charge induced by the polarized

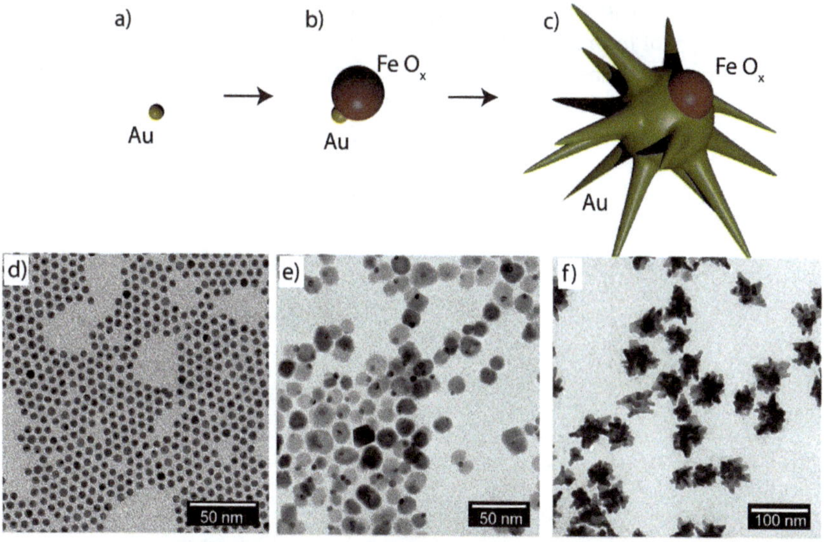

Fig. 1 Scheme of the synthesis of Janus magnetic nanostars through consecutive seed-and-growth-steps. (a) Oleylamine synthesis of Au nanoparticles. (b) Growth of an iron oxide nanoparticle through the decomposition of $Fe(CO_5)$ and subsequent oxidation. (c) Growth of an Au nanostar coated with PVP using the NDs as seeds (d–f) TEM images of different nanoparticles corresponding to the scheme above: (d) Au NPs, (e) NDs, (f) JMNSs.

plane at the interface. As the Au particle has only a limited source of electrons, this compensation makes all other facets of the Au nanoparticle electron deficient and unsuitable for multi-nucleation, resulting in the dumbbell structure. Once the synthesis of the Fe–Au dumbbells is completed, the nanoparticles are exposed to air, oxidizing Fe into a magnetic iron oxide, either as Fe_3O_4 or as γ-Fe_2O_3. We have gone one step further using these ND nanoparticles as seeds for a subsequent seed mediated growth step. With this aim, we performed a ligand exchange of the native capping agent by polymer ligands that allow dispersion in aqueous solution. In this case we used silane-poly(ethyleneglycol) and thiol-poly(*N*-isopropylacrylamide), which bind to the iron oxide and the gold surfaces, respectively. Modifying the polyvinylpyrrolidone (PVP)-assisted Au nanostar synthesis method,[38] the growth of gold nanostars was performed by injection at room temperature of the dumbbell seeds in a pre-reduced DMF solution of $HAuCl_4$. The formation of nanostars was observed by a color change from yellow to blue. Bright field TEM images of NDs and JMNSs are shown in Fig. 1e and f, compared with Au NPs synthesized without the iron oxide component and with sizes ($d = 5.4$ nm) similar to the Au part on the NDs (Fig. 1d). On those TEM images the increase in particle size and the morphological changes can be observed after the consecutive synthesis steps. Along with these morphological changes there is a concomitant change in the absorbance spectrum. Fig. 2 shows the UV-Vis spectra of the three kinds of nanoparticles, JMNSs, NDs and Au NPs. It can be observed that Au NPs feature an LSPR band centered at 520 nm, whereas for NDs the plasmon absorbance red-shifts to ~560 nm, with a lower and broader absorption peak. When the nanostars are grown an intense LSPR band, typical for highly branched morphologies, is observed in the NIR (750–850 nm) which completely damps the iron oxide contribution. This absorbance at the NIR can be highly beneficial in biomedical applications as it is located at the first NIR absorbance window of living tissue and can be used to trigger externally a photothermal effect at the desired region of interest inside the human body.[24]

Contrary to dumbbells (Fig. 1e), the Janus character of the nanostars (Fig. 1f) can be hardly appreciated in bright field TEM images due to the lower contrast of iron oxide as compared to Au and the larger nanostar tips. Thus, confirmation of

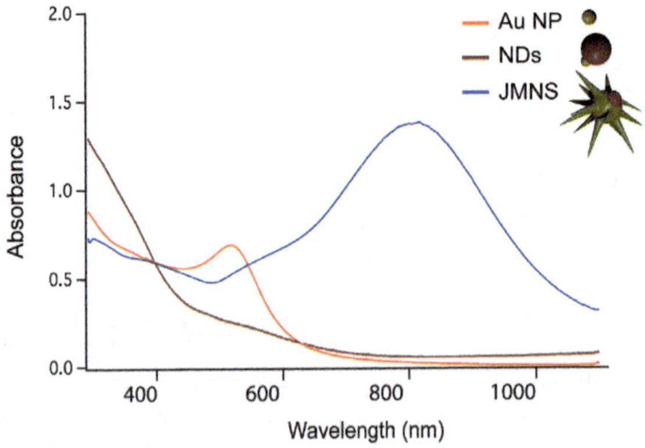

Fig. 2 UV-Vis spectra comparing the absorbance of Au NPs (in red), NDs (in brown), and JMNSs (in blue).

the Janus character required the use of HAADF-STEM (Fig. 3a). In HAADF-STEM mode, the image intensity is determined by the average atomic number Z integrated over the thickness of the sample. Therefore, regions containing Au typically appear with a higher intensity than Fe-rich regions, as indicated in the insets of Fig. 3a. This figure suggests that practically all nanoparticles contain an iron oxide lobe and that this lobe could be exposed on one side of the nanostar. To confirm this observation and to investigate the 3D distribution of the elements, 3D EDX chemical mapping was performed. Therefore, a tilt series of 2D EDX maps was acquired over a tilt range of ±75°. An example of a 2D EDX map for Fe and Au is presented in Fig. 3b, together with an overlay of the maps with an HAADF-STEM image. From these images, a separation of the two components of the nanoparticles can already be distinguished. After using the tilt series of 2D EDX maps

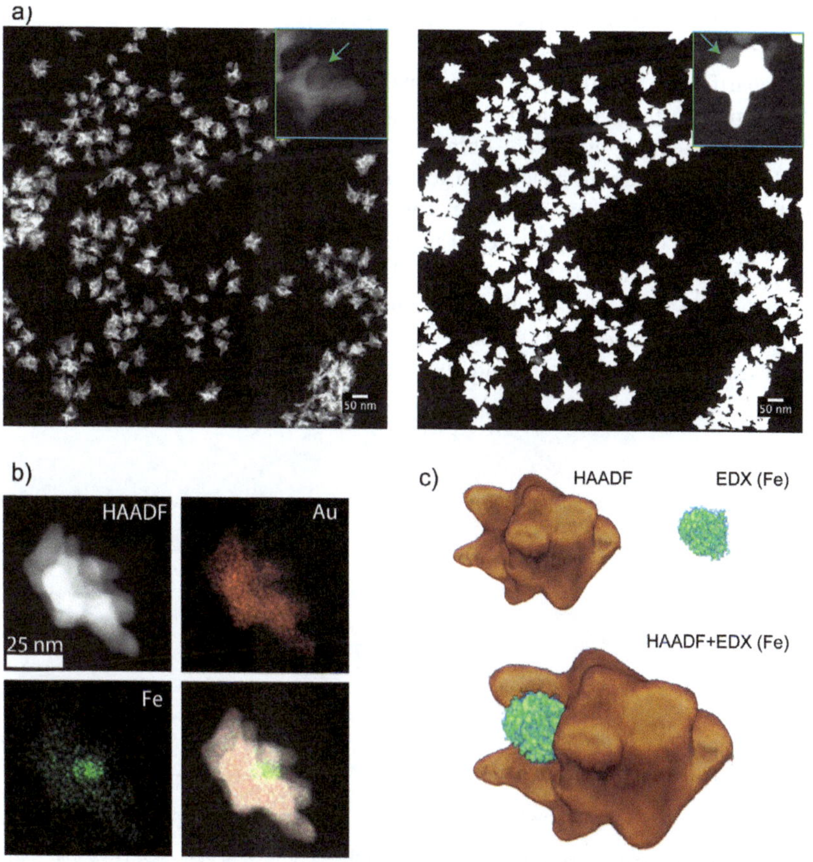

Fig. 3 (a) HAADF-STEM image of JMNSs showing uniformity in size, the inset on the left shows a nanoparticle with a cavity corresponding to the iron oxide component. The image on the right shows the same area but saturating the detector, the inset shows a nanoparticle where the iron oxide part is exposing its surface in a clear Janus configuration. (b) Images of the same nanoparticle using HAADF-STEM, EDX mapping of gold atoms, EDX mapping of Fe atoms and a combination of the three. (c) Tomographic reconstruction of a nanoparticle combining HAADF-STEM tomography and EDX mapping of Fe, at different angles between −75° and 75° at 10° intervals.

as an input for 3D reconstruction, the results presented in Fig. 3c are obtained. The 3D view allows a complete visualization of the nanoparticles, which indicates that the nanostar tips grow towards the opposite and lateral sides of the initial dumbbell nanoparticles, not covering the iron oxide and therefore leaving two different surfaces exposed for further selective functionalization.

A similar analysis was performed on nanoparticles synthesized with different Au-to-seed ratios. Fig. 4a shows the total size of the nanoparticle measured as either average diameter or the maximum tip-to-tip diameter. As expected, an increase of the nanoparticle size is produced when less seeds per gold atom are added to the solution. The synthesis is fairly reproducible as can be seen for the different points corresponding to the same Au–Fe ratios. For intermediate values, two different concentrations of gold were used, indicating that the ratio between the amount of seeds and Au was the main parameter that defines the final particle size. Although the limit of nanoparticle sizes and ratios was not explored in this work, JMNPs were successfully synthesized between 30 and 50 nm. Elemental analysis by inductively coupled plasma mass spectrometry (ICP-MS), showed that the molar ratio between gold and iron varied from [Au]/[Fe] = 74 for the larger nanoparticles to [Au]/[Fe] = 13 for the smaller ones. Electron tomography was also performed on these nanoparticles to evaluate the morphological variation with nanoparticle size (Fig. 4a). The obtained 3D images invariably showed the Janus character of the nanoparticles, with gold branches growing in directions opposite to the iron oxide. Not much difference was observed in tip size and shape for big and small nanoparticles, but a clear increase was observed in the number of tips along with the size of the nanoparticle, *i.e.* bigger nanoparticles are more branched or spiky. Fig. 4b shows the spectra for the three different sizes, displaying the typical optical response of gold nanostars with a main LSPR mode corresponding to the tips around 700–800 nm and a secondary mode (shoulder

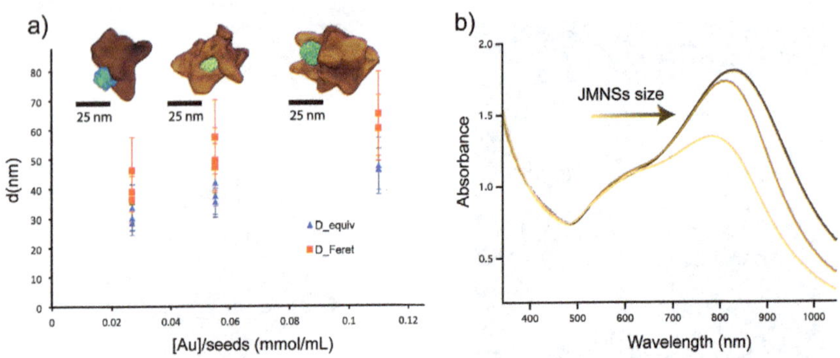

Fig. 4 (a) Variation of JMNSs size depending on the synthesis conditions, *i.e.* varying the ratio between the amount of added ND seeds and $HAuCl_4$ concentration. The size was evaluated as the average diameter by calculating the area from TEM images and assuming a spherical morphology (D_equiv., in blue) and secondly the Feret's maximum diameter (D_Feret, in red), which corresponds to the maximum tip-to-tip diameter. The 3D tomography images corresponding to each size show the presence of nanostar tips in all of them but an increase in the number of tips per nanoparticle. (b) UV-Vis spectra comparing the absorbance of the JMNSs with three different nanoparticle sizes shown in (a).

around 600 nm) corresponding to the central core of the nanoparticles.[38,39] Additionally, the decrease of the size of JMNSs produces a blue-shift of the band corresponding to the main LSPR mode while no big change is observed for the secondary mode. This change in the main mode could be attributed to the lower amount of tips per nanoparticle observed in the tomography images, and a possible variation in the thickness of those tips. These nanoparticles have high colloidal stability, as it has been seen for nanostars stabilized by PVP.[40] In addition, different to the surfactant-free synthesis of nanostars, the PVP stabilization prevents reshaping of the nanostars.[40] After 6 months the JMNSs were still dispersed in water and only a small blue-shift of ~15 nm was observed in the UV-Vis absorption, indicative of a very small rounding of the nanostar tips over time.

To assess the plasmonic and magnetic properties of these nanoparticles in a direct application, SERS was performed for detection of selected molecules. Fig. 5 shows the SERS spectra of an analyte solution containing the model analyte crystal violet (CV) with a concentration of 450 nM (red, upper spectrum). The spectrum clearly shows the characteristic peaks of CV at 1616, 1597, 1293 (assigned as C–C stretching), 1377 (*N*-phenyl-stretching), 1175 (C–H in-plane deformation), 915 and 805 cm^{-1} (C–H out-of-plane deformation) with no external interference.[41] A second spectrum was recorded on the same sample after applying a magnetic field (Fig. 5a, blue line). For this purpose, a small magnet was placed close to the solution for 30 minutes. The nanoparticles magnetically aggregated close to the magnet, thereby increasing the concentration where the laser spot of the Raman spectrometer was focused. Remarkably, we registered SERS signals with an enhancement factor of ~5 times with respect to the signal before magnetic concentration. The measurements produce reproducible results

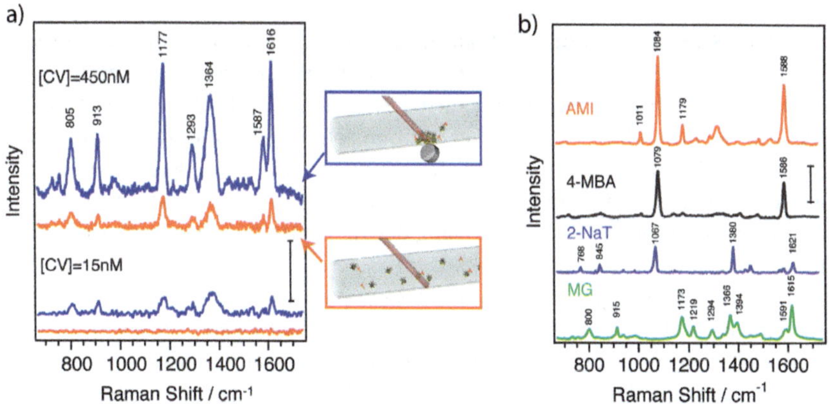

Fig. 5 (a) SERS spectra of crystal violet [CV] containing JMNPs in solution (red) and after magnetic concentration (blue) for two different dye concentrations [CV] = 450 nM (upper spectra) and [CV] = 15 nM (lower spectra). Concentration of JMNSs is $[Au^0]$ = 0.2 and 0.1 mM respectively. The spectra were shifted vertically for clarity with the scale bar corresponding to an intensity of 200 counts. (b) SERS spectra of different molecules after a magnetic aggregation. AMI: acetoamidothiophenol, 4-MBA: 4-mercaptobenzoic acid, 2-NaT: 2-naphthalenethiol, MG: malachite green. Analyte concentration 4.5 μM, $[Au^0]$ = 0.2 mM. AMI, 4-MBA and 2-NaT were measured with 785 nm excitation whereas MG was measured with a 633 nm excitation wavelength. The intensity scale bar corresponds to 500 counts.

for the magnetically aggregated samples down to CV concentrations of 15 nM (Fig. 5a). At such low concentrations no signal can be observed when the measurement is performed in solution without applying a magnetic field (red spectrum). Below 15 nM the spectra are less reproducible due to the low signal-to-noise ratio. The observed increase in signal is most likely due to adsorption of the analytes on the surface of the nanoparticles. When they are in solution, the diffusion time from the analyte position to the surface of a nanoparticle can be very short, making the "capture" of the analyte quite effective. Upon application of a magnetic field, the nanoparticles drag the analytes to a higher concentration area, thereby increasing the measured SERS signal. It should be noted that an optimization of the amount of nanoparticles would be necessary to decrease the detection limit, as too many nanoparticles lead to an increase of the background signal and few nanoparticles decrease the surface available for capturing the analyte.

We additionally carried out SERS measurements with several molecules that have affinity for the gold surface, including molecules containing amino and thiol groups. Fig. 5b shows SERS spectra from solutions containing different molecules at 4.5 μM after magnetic aggregation. At the same concentration, JMNSs yielded clear SERS spectra for all the molecules containing thiols or amines tested here. Furthermore, SERS spectra do not show any meaningful signal from PVP, which was used to stabilize the nanoparticles, or from any other reagent used in the synthesis, resulting in clearly assignable SERS spectra for all the analytes.[34,39,42–46] On the other hand, PVP is a weakly adsorbed polymer, so it can be readily replaced by other molecules or polymers that present a higher affinity for the desired analytes, making the nanoparticles more specific and effective for their detection.

3 Experimental

3.1 Synthesis

3.1.1 Synthesis of dumbbell-like nanoparticles. A solution in 1-octadecane (40 mL) was prepared containing oleic acid (6 mmol), oleylamine (6 mmol) and 1,2-hexadecanodiol (10 mmol) and stirred for 20 min at 120 °C under N_2. $Fe(CO)_5$ (0.3 mL) was then injected and after 3 min a solution containing $HAuCl_4 \cdot 3H_2O$ (0.1 mmol) dissolved in a mixture of oleylamine (0.5 mL) and 1-octadecane (5 mL) was injected and heated up to 300 °C at approximately 1 °C min^{-1}. The solution was left to react for 45 min. After cooling down, the dispersion was exposed to air for 30 min to cause Fe oxidation.

To purify the nanoparticles, 50 mL of isopropanol was added and the solution centrifuged at 4500 RCF for 30 min. The nanoparticles were cleaned two more times after the redispersion with hexane and aggregation with isopropanol. Finally oleylamine (100 μL) was added to store the nanoparticles for long periods of time. The nanoparticles were cleaned again right before functionalization. The final nanoparticle sizes were 5.3 ± 0.8 nm for the Au part and 16.2 ± 2.8 nm for the iron oxide part.

3.1.2 Ligand exchange. Silane-poly(ethylene glycol) (MW = 5 kg mol^{-1}) was purchased from Creative PEGWorks. HS-poly(*N*-isopropylacrylamide) (MW = 12 kg mol^{-1}) was synthesized by RAFT polymerization using ethyl 4-(hydoxymethyl) benzyl carbonotrithioate as RAFT agent and 4,4′-azobis 4(-cyanopentanoic acid) as initiator. 11 mg of nanoparticles was incubated in toluene (20 mL) together with HS-PNIPAM (60 mg, MW 12 kg mol^{-1}), and PEG-silane (110 mg). Acetic acid

(3 μL) was added to induce polycondensation of the silane. The solution was sonicated for 10 min and then left shaking during 2 days. The nanoparticles were purified by transferring into water by adding a small amount of hexane to the solution. The nanoparticles were further purified in water several times using a centrifugation filter (pore size 50 kDa) and finally redispersed in water (10 mL).

3.1.3 Synthesis of Janus nanostars (JMNSs). A solution of $HAuCl_4 \cdot 3H_2O$ (54.6 μL, 50 mM) was added to a solution containing polyvinylpyrrolidone (1 g, MW = 10 kg mol^{-1}) dissolved in DMF (10 mL). The solution was left stirring for 20 min to allow gold salt reduction. The dumbbell nanoparticle solution (25 μL, except those in Fig. 4 described below) was then added and the reaction was left reacting for 1 h, showing a colour change into blue. The concentration of dumbbells and JMNSs was estimated by ICP-MS (see below). The nanoparticles were purified in several centrifugation cycles: the first one at 3420 g for 1 h, 2 more at 2870 g for 1 h and 2 more at 2370 g RCF for 15 min. They were finally redispersed in Milli-Q water.

JMNSs of different sizes. Different nanoparticles were obtained by changing the ratio between Au and the amount of ND seeds. JMNS1 are described above. JMNS2 and JMNS3 have the same [Au] : seed ratio and showed similar sizes.

JMNS1: 25 μL of seeds, 54.6 μL of Au salt (50 mM) in 10 mL of DMF.

JMNS2: 50 μL of seeds, 54.6 μL of Au salt (50 mM) in 10 mL of DMF.

JMNS3: 25 μL of seeds, 27.3 μL of Au salt (50 mM) in 10 mL of DMF.

JMNS4: 50 μL of seeds, 27.3 μL of Au salt (50 mM) in 10 mL of DMF.

3.1.4 Synthesis of Au nanoparticles. To compare the plasmonic behavior of the dumbbells, Au nanoparticles were synthesized with oleylamine to have a similar size. $HAuCl_4 \cdot 3H_2O$ (0.5 g) was dissolved in a mixture of oleylamine (40 mL) and 1,2,3,4-tetrahydronaphthalene (40 mL) at room temperature under gentle N_2 bubbling. Oleylamine (15 mL), 1,2,3,4-tetrahydronaphthalene (15 mL) and a *tert*-butylamineborane complex (2.5 mol) were dissolved and added to the initial solution. The reaction was allowed to proceed for 1 h. Purification was performed by adding isopropanol (320 mL) and centrifuging. Two additional centrifugation steps were performed by redispersing the nanoparticles and precipitating them with isopropanol. Finally, the nanoparticles were dispersed in hexane (20 mL), and oleylamine (200 μL) was added prior to storage. The final nanoparticle size was measured by TEM assuming a spherical shape. Diameter = 5.4 ± 0.4 nm.

3.2 Characterization methods

3.2.1 UV-Vis spectroscopy. Optical extinction spectra were recorded using an Agilent 8453 UV/Vis diode-array spectrophotometer. The resulting spectra were normalized at 400 nm. At this wavelength the absorbance by gold nanoparticles has been shown to be nearly independent of particle size.[47]

3.2.2 ICP-MS. With the aim of obtaining the atomic ratio between metals (Fe and Au) in the JMNs, analysis by inductively coupled plasma mass spectrometry (ICP-MS) (Agilent 7500ce) was performed. An internal calibration was performed using yttrium at a 500 ppb concentration. Measurements were carried out for three different nanoparticle colloids: nanodumbbell seeds (NDs), big Janus magnetic nanostars JMNSs (JMNS1, nanoparticles shown in Fig. 3) and small Janus magnetic nanostars (JMNS4, small nanoparticles shown in Fig. 4). All particles were digested with 30 vol% aqua regia before the ICP analysis.

— NDs: the obtained results yielded 0.29 mg mL^{-1} of Fe and 0.14 mg mL^{-1} of Au.

— JMNS1: 18.24 mg mL^{-1} of Au and 0.07 mg mL^{-1} of Fe.

— JMNS4: 4.03 mg mL^{-1} of Au to 0.09 mg mL^{-1} of Fe.

3.2.3 Bright field transmission electron microscopy (TEM). TEM bright field images were acquired in a JEOL JEM-1400PLUS instrument operating at 120 kV. Sizes of nanoparticles were determined from bright field images with more than 200 nanoparticles. The images were analyzed with the Image J software package. Due to the difficulty of measuring these morphologically complex nanoparticles two different diameters are provided, allowing a comparison between different nanoparticles. An equivalent diameter was obtained by calculating the projected area and assuming a spherical morphology. A second diameter was obtained by calculating the maximum Feret's diameter, *i.e.* the length between two parallel lines that surround the nanoparticle. This diameter corresponds to the maximum tip-to-tip distance of the 2D projected nanoparticle and could thus be slightly underestimated due to the 3D morphology of the nanoparticles.

3.2.4 Electron tomography. HAADF-STEM images, EDX mapping and electron tomography tilt series were acquired using a model 2020 Fischione Instruments tomography holder and a FEI Tecnai Osiris electron microscope operated at 200 kV. A camera length of 87 mm was used. The tilt series of HAADF-STEM and EDX are recorded at angles ranging from +75° to −75° with a 10° increment. For the reconstruction of the series we used the SIRT algorithm, as implemented in the ASTRA toolbox.[48]

3.2.5 Surface enhanced Raman spectroscopy. SERS measurements were performed by means of a confocal Raman microscope (Renishaw inVia) equipped with a 1024 × 512 CCD Peltier-cooled detector using either 633 nm excitation laser with maximal output of 17 mW combined with a 1800 L mm^{-1} diffraction grating or 785 nm excitation laser with maximum output of 270 mW, combined with a 1200 L mm^{-1} diffraction grating. 20 μL of the magnetic Janus particles in aqueous solution ($[Au^0] = 0.2$ mM) were filled in cut Pasteur glass pipettes (1 mm diameter) and mounted on a home-built holder. For measurements in aggregated samples, the edge of a neodymium magnetic disc (1.42 T) was placed below the glass tube and SERS spectra were measured after 1 h. For measurements in non-aggregated samples the same holder without the magnet was used. All SERS spectra were collected using a 10× objective (numerical aperture NA = 0.35) with an integration time of 2 s and a laser power of 5 mW for 633 nm (for CV and MG) accumulating 50 cycles and 16 mW for 785 nm (for AMI, 4-MBA and 2-NaT) accumulating 100 cycles. All spectra presented here are averaged before background correction using the Wire 3.4 software.

4 Conclusions

We have demonstrated in this work that consecutive seed-mediated-growth steps lead to production of complex nanostructures that can be used in sophisticated applications by combining different properties. In this case, nanoparticles were synthesized in three steps, where two seed-mediated-growth steps lead from gold nanoparticles to Janus or dumbbell like nanoparticles, and Janus iron oxide–gold nanosphere–nanostar nanoparticles. These nanoparticles feature superparamagnetic and plasmonic properties and asymmetric activity due to their

Janus character, which are of high potential for diverse applications ranging from interface stabilization of oil–water emulsions with magneto-plasmonic properties to biomedical applications. 3D imaging by tomography based on EDX chemical mapping proved to be an effective method to visualize complex morphologies. The 3D images of these nanoparticles revealed that the growth of the gold tips avoided the surface of the iron oxide nanoparticle, producing Janus nanoparticles with two clearly differentiated surfaces. Moreover, an increase in the size of the gold nanoparticle together with an increase on the number of star tips was produced with the increase of $HAuCl_4$ used on the synthesis of the Janus magnetic nanostars. We finally proved that the plasmonic and magnetic properties of these nanoparticles can be effectively used in a direct application. The particles were used as nanoprobes for SERS measurements in solution thanks to their effective plasmonic properties, without any further surface modification. Furthermore, taking advantage of their magnetic properties, they were magnetically concentrated (magnetic extraction) to produce an additional enhancement of the signal that allowed detection down to 15 nM. Further work would be necessary to functionalize the nanoparticles allowing more specific and effective Raman enhancement of relevant biomolecules. We propose that the Janus character will open new possibilities in applications that require strong interaction with interfaces or in cases where specific functional groups need to be spatially separated, in a similar way as proteins and molecules present different patchy surfaces to achieve their programmed biological functions.

Acknowledgements

This work has been funded by the European Research Council (ERC Advanced Grant #267867, Plasmaquo). N. W. and S. B. acknowledge funding by the European Research Council (ERC Starting Grant #335078, Colouratom).

Notes and references

1 J. Reguera, H. Kim and F. Stellacci, *Chimia*, 2013, **67**, 811–818.
2 D. Rodriguez-Fernandez and L. M. Liz-Marzan, *Part. Part. Syst. Charact.*, 2013, **30**, 46–60.
3 A. Walther and A. H. Muller, *Chem. Rev.*, 2013, **113**, 5194–5261.
4 B. P. Binks and P. D. I. Fletcher, *Langmuir*, 2001, **17**, 4708–4710.
5 A. Kumar, B. J. Park, F. Tu and D. Lee, *Soft Matter*, 2013, **9**, 6604.
6 H. Fan and A. Striolo, *Soft Matter*, 2012, **8**, 9533–9538.
7 R. Aveyard, *Soft Matter*, 2012, **8**, 5233–5240.
8 S. Crossley, J. Faria, M. Shen and D. E. Resasco, *Science*, 2010, **327**, 68–72.
9 J. Faria, M. P. Ruiz and D. E. Resasco, *Adv. Synth. Catal.*, 2010, **352**, 2359–2364.
10 L.-T.-C. Tran, S. Lesieur and V. Faivre, *Expert Opin. Drug Delivery*, 2014, **11**, 1061–1074.
11 G. Zhao, A. Ambrosi and M. Pumera, *Nanoscale*, 2013, **5**, 1319–1324.
12 H. Y. Koo, D. K. Yi, S. J. Yoo and D. Y. Kim, *Adv. Mater.*, 2004, **16**, 274–277.
13 S. C. Glotzer and M. J. Solomon, *Nat. Mater.*, 2007, **6**, 557–562.
14 Q. Chen, J. K. Whitmer, S. Jiang, S. C. Bae, E. Luijten and S. Granick, *Science*, 2011, **331**, 199–202.

15 Y. S. Kang, S. Risbud, J. F. Rabolt and P. Stroeve, *Chem. Mater.*, 1996, **8**, 2209–2211.
16 D. Ho, X. L. Sun and S. H. Sun, *Acc. Chem. Res.*, 2011, **44**, 875–882.
17 U. I. Tromsdorf, O. T. Bruns, S. C. Salmen, U. Beisiegel and H. Weller, *Nano Lett.*, 2009, **9**, 4434–4440.
18 C. L. Nehl and J. H. Hafner, *J. Mater. Chem.*, 2008, **18**, 2415.
19 F. P. Zamborini, L. Bao and R. Dasari, *Anal. Chem.*, 2012, **84**, 541–576.
20 R. A. Alvarez-Puebla and L. M. Liz-Marzan, *Angew. Chem., Int. Ed.*, 2012, **51**, 11214–11223.
21 D. Boyer, P. Tamarat, A. Maali, B. Lounis and M. Orrit, *Science*, 2002, **297**, 1160–1163.
22 S. Jain, D. G. Hirst and J. M. O'Sullivan, *Br. J. Radiol.*, 2012, **85**, 101–113.
23 E. C. Dreaden, A. M. Alkilany, X. Huang, C. J. Murphy and M. A. El-Sayed, *Chem. Soc. Rev.*, 2012, **41**, 2740–2779.
24 R. Weissleder, *Nat. Biotechnol.*, 2001, **19**, 316–317.
25 C. Hrelescu, T. K. Sau, A. L. Rogach, F. Jäckel and J. Feldmann, *Appl. Phys. Lett.*, 2009, **94**, 153113.
26 S. Mazzucco, O. Stéphan, C. Colliex, I. Pastoriza-Santos, L. M. Liz-Marzan, J. Garcia de Abajo and M. Kociak, *Eur. Phys. J.: Appl. Phys.*, 2011, **54**, 33512.
27 C. G. Khoury and T. Vo-Dinh, *J. Phys. Chem. C*, 2008, **112**, 18849–18859.
28 L. Wang, J. Luo, Q. Fan, M. Suzuki, I. S. Suzuki, M. H. Engelhard, Y. Lin, N. Kim, J. Q. Wang and C. J. Zhong, *J. Phys. Chem. B*, 2005, **109**, 21593–21601.
29 Y. Wei, R. Klajn, A. O. Pinchuk and B. A. Grzybowski, *Small*, 2008, **4**, 1635–1639.
30 H. M. Song, Q. S. Wei, Q. K. Ong and A. Wei, *ACS Nano*, 2010, **4**, 5163–5173.
31 P. Quaresma, I. Osorio, G. Doria, P. A. Carvalho, A. Pereira, J. Langer, J. P. Araujo, I. Pastoriza-Santos, L. M. Liz-Marzan, R. Franco, P. V. Baptista and E. Pereira, *RSC Adv.*, 2014, **4**, 3659–3667.
32 J. Li, Y. Hu, J. Yang, P. Wei, W. Sun, M. Shen, G. Zhang and X. Shi, *Biomaterials*, 2015, **38**, 10–21.
33 A. Espinosa, M. Bugnet, G. Radtke, S. Neveu, G. A. Botton, C. Wilhelm and A. Abou-Hassan, *Nanoscale*, 2015, **7**, 18872–18877.
34 A. La Porta, A. Sanchez-Iglesias, T. Altantzis, S. Bals, M. Grzelczak and L. M. Liz-Marzan, *Nanoscale*, 2015, **7**, 10377.
35 P.-J. Chen, Y.-D. Kang, C.-H. Lin, S.-Y. Chen, C.-H. Hsieh, Y.-Y. Chen, C.-W. Chiang, W. Lee, C.-Y. Hsu, L.-D. Liao, C.-T. Fan, M.-L. Li and W.-C. Shyu, *Adv. Mater.*, 2015, **27**, 6488–6495.
36 H. Yu, M. Chen, P. M. Rice, S. X. Wang, R. L. White and S. H. Sun, *Nano Lett.*, 2005, **5**, 379–382.
37 Y. Lee, M. A. Garcia, N. A. Frey Huls and S. Sun, *Angew. Chem., Int. Ed.*, 2010, **49**, 1271–1274.
38 P. Senthil Kumar, I. Pastoriza-Santos, B. Rodriguez-Gonzalez, F. Javier Garcia de Abajo and L. M. Liz-Marzan, *Nanotechnology*, 2008, **19**, 015606.
39 C. G. Khoury and T. Vo-Dinh, *J. Phys. Chem. C*, 2008, **112**, 18849–18859.
40 S. Barbosa, A. Agrawal, L. Rodriguez-Lorenzo, I. Pastoriza-Santos, R. A. Alvarez-Puebla, A. Kornowski, H. Weller and L. M. Liz-Marzan, *Langmuir*, 2010, **26**, 14943–14950.
41 M. V. Cañamares, C. Chenal, R. L. Birke and J. R. Lombardi, *J. Phys. Chem. C*, 2008, **112**, 20295–20300.

42 L. He, N.-J. Kim, H. Li, Z. Hu and M. Lin, *J. Agric. Food Chem.*, 2008, **56**, 9843–9847.
43 R. A. Alvarez-Puebla, D. S. Dos Santos Jr and R. F. Aroca, *Analyst*, 2004, **129**, 1251–1256.
44 A. Michota and J. Bukowska, *J. Raman Spectrosc.*, 2003, **34**, 21–25.
45 Y. Zhang, X. Li, B. Xue, X. Kong, X. Liu, L. Tu and Y. Chang, *Sci. Rep.*, 2015, **5**, 14934.
46 K. M. Balss, T.-C. Kuo and P. W. Bohn, *J. Phys. Chem. B*, 2003, **107**, 994–1000.
47 T. Hendel, M. Wuithschick, F. Kettemann, A. Birnbaum, K. Rademann and J. Polte, *Anal. Chem.*, 2014, **86**, 11115–11124.
48 W. J. Palenstijn, K. J. Batenburg and J. Sijbers, *J. Struct. Biol.*, 2011, **176**, 250–253.

Faraday Discussions

PAPER

Self-folding nanostructures with imprint patterned surfaces (SNIPS)†

Hye Rin Kwag,[a] Jeong-Hyun Cho,[a] Si-Young Park,[a] Jaehyun Park[a] and David H. Gracias*[ab]

Received 22nd February 2016, Accepted 24th February 2016

DOI: 10.1039/c6fd00021e

A significant need in nanotechnology is the development of methods to mass-produce three-dimensional (3D) nanostructures and their ordered assemblies with patterns of functional materials such as metals, ceramics, device grade semiconductors, and polymers. While top-down lithography approaches can enable heterogeneous integration, tunability, and significant material versatility, these methods enable inherently two-dimensional (2D) patterning. Bottom-up approaches enable mass-production of 3D nanostructures and their assemblies but with limited precision, and tunability in surface patterning. Here, we demonstrate a methodology to create Self-folding Nanostructures with Imprint Patterned Surfaces (SNIPS). By a variety of examples, we illustrate that SNIPS, either individually or in ordered arrays, are mass-producible and have significant tunability, material heterogeneity, and patterning precision.

Introduction

Nanostructures that possess morphological and/or functional anisotropy are important building blocks that can display novel properties and function either individually or in ordered assemblies. A number of approaches have been developed to control the morphological anisotropy of nanostructures and their assemblies. These methods include top-down approaches such as advanced photolithography, electron-beam, focused-laser and soft-lithography, and bottom-up methods such as templated growth and directed assembly.[1–24] However, it is still challenging to mass-produce patterned nanostructures and their ordered assemblies in curved or 3D geometries. In this paper, we discuss the development of 3D Self-folding Nanostructures with Imprint Patterned Surfaces (3D SNIPS). SNIPS are fabricated using a combination of highly parallel top-down and bottom-up lithography; and we argue that they provide a means to mass-produce precisely

[a]*Department of Chemical and Biomolecular Engineering, Johns Hopkins University, Baltimore, MD 21218, USA. E-mail: dgracias@jhu.edu*

[b]*Department of Materials Science and Engineering, Johns Hopkins University, Baltimore, MD 21218, USA*

† Electronic supplementary information (ESI) available. See DOI: 10.1039/c6fd00021e

patterned 3D nanoparticles and their ordered assemblies in a highly parallel manner.

In terms of manufacturability, one of the most widely utilized technologies in the commercial fabrication of semiconductor devices is based on top-down lithography with extensive infrastructure, equipment and know-how. Indeed, a very extensive range of lithographic approaches have been developed and refined for use in nanomanufacturing. We investigate the use of two prominent approaches with demonstrated precise nanoscale resolution which are electron-beam lithography (EBL) and nanoimprint lithography (NIL). In EBL, a high energy focused electron beam is rastered over a thin polymer coating (resist) that is sensitive to electron damage *via* polymer chain scission (Fig. 1a). Regions that are exposed to the electron beam undergo polymer chain scission and show enhanced solubility in a developer. EBL has high resolution approaching the atomic scale[25] and excellent tunability such that a variety of 2D patterns can be designed using CAD software and patterns can be readily changed using the software. However, even with advances in multi-beam and high speed rastering,[26] EBL has limited throughput which limits the overall area that can be patterned. In contrast, NIL involves nanoscale patterning using a stamp that is brought into contact with a resist material while increasing the temperature or illuminating the resist with UV light (Fig. 1b). NIL also has extremely high resolution approaching the molecular scale.[27,28] In contrast to EBL, NIL is highly parallel and requires only one master mold to pattern multiple copies. Since large area stamps or rolls can be used in NIL, one can pattern large area substrates with nanoscale features in short time. Even so, it is still expensive to fabricate a single NIL master mold with complex patterns, and achieving alignment registration between sequential NIL overlay patterning steps can be challenging.

While EBL and NIL have exquisite resolution, and despite major advances such as cylindrical roll-to-roll NIL and focused stereoscopic movement of e-beams, both techniques can be used to pattern nanostructures only in an inherently 2D manner. This limitation presents a major challenge in the creation of patterned nanostructures and their ordered assemblies in shapes that deviate significantly from planar geometries. Examples of such shapes include polyhedra, spherical shells (especially their inner surfaces), spirals and helices. Techniques such as 3D stereolithography and 3D printing can be applied to some extent to pattern such structures but they are serial approaches with limited resolution. For example, in 3D stereolithography, laser beams are overlapped at a focal point that can be moved relative to the substrate with 3D stereoscopic control, whereas in 3D printing, nozzles are rastered relative to substrates in a layer by layer fashion; in both cases, stereoscopic or linear rastering involves nanoscale motion using positioning systems which limits throughput. Additionally, there are challenges with material versatility using these approaches; for example, it is challenging to pattern materials such as device grade single crystal silicon (Si) using 3D printing or focused laser beam patterning.

Self-folding is a process that involves bending, curving, folding, or pop-up of structures using a variety of forces derived from capillarity, electrostatic or magnetic interactions, and swelling or shrinking as summarized in previous reviews.[29–38] In this paper, we use three different forces to illustrate the broad processing applicability for SNIPS. We utilize capillary forces derived from the coalescence and reflow of tin (Sn) to form SNIPS with nanoscale dimensions. We

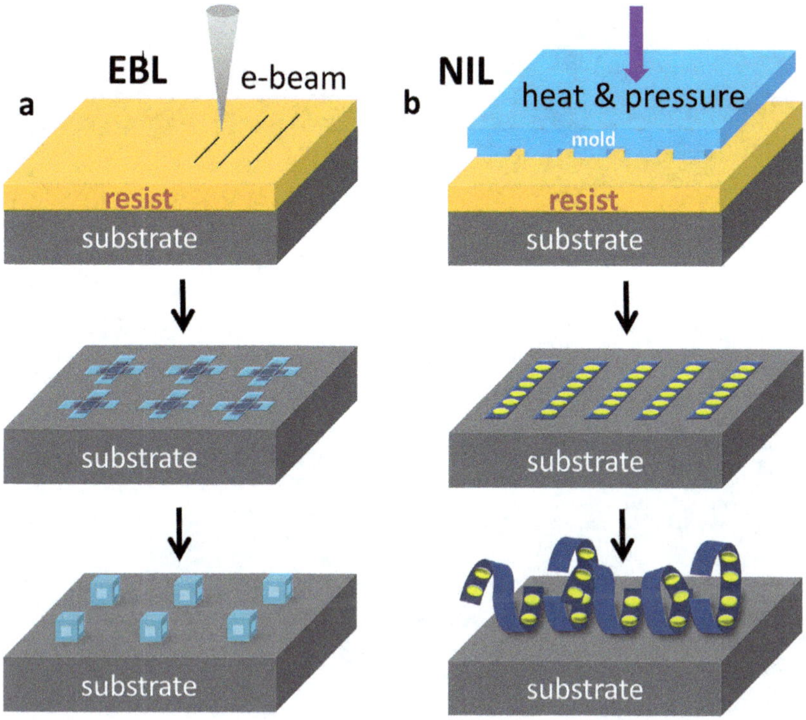

Fig. 1 Schematic illustration of the concept of combining top-down and bottom-up approaches to form (a) EBL patterned 3D nanostructures and (b) NIL patterned SNIPS.

also utilize two other forces, one derived from bilayer strain release and the other from differential swelling to form ordered assemblies of SNIPS in curved and folded geometries. As compared to our previous publications on self-folding nanostructures,[39–41] here, we describe the use of NIL patterning to create a variety of SNIPS with diverse materials and contrast the approach to EBL self-folded nanostructures. Due to the high throughput nature of NIL, we believe that SNIPS represent a new class of mass-producible 3D nanostructures with precise shapes and patterns composed of heterogeneous materials. Importantly, as discussed in this paper, SNIPS and their ordered arrays can be composed of metals, ceramics, polymers and device grade Si.

Experimental

2D pattern transfer by EBL

The patterns formed using EBL were prepared as described previously using an e-beam controlled by a RAITH system.[39] Briefly, PMMA resist was spin coated onto a bare Si wafer and baked. After e-beam exposure, the resist was developed using an MIBK developer.

2D pattern transfer by NIL

A 400 nm thick NXR-1020 thermal resist (Nanonex) was spin coated onto an Si wafer and soft-baked as per the manufacturer's instructions (at 80 °C for 30 min

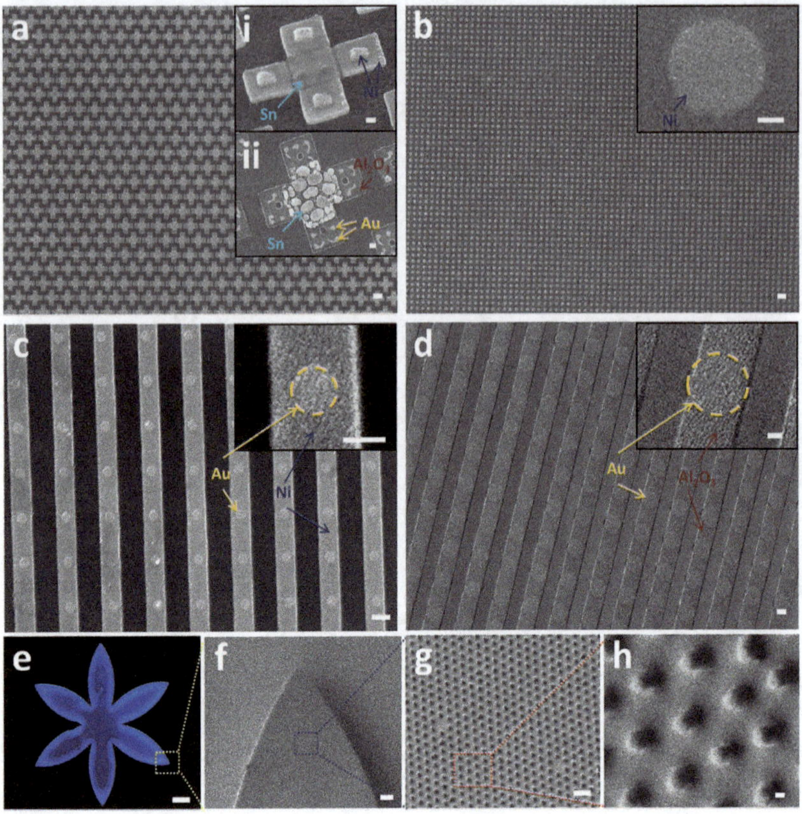

Fig. 2 Top-down patterning of 2D precursors. (a) Ni/Sn cross patterns using multiple layers of EBL, (a-ii) Au patterns on Al_2O_3 panels; (b) circular NIL patterned discs composed of a bilayer of Ni/Sn. Circular Au patterns on bilayer of (c) Ni/Sn and (d) Al_2O_3/Sn strips patterned by NIL. (e) Nanopatterned SU-8 flower (f) zoomed in SEM image of SU-8 flower petal (g and h) zoomed in images of nanobumps on SU-8 flower surface. Scale bars are (a and b) 1 μm and 100 nm for the inset; (c and d) 200 nm and 100 nm for the inset; (e) 100 μm; (f) 10 μm; (g) 1 μm; and (h) 100 nm. Note: the zoom boxes indicate approximate region where the zoomed images are taken and are not drawn in proportion.

in an oven). Periodic circular or linear patterns were imprinted using commercial Si master stamps (LightSmyth Technologies, Inc., see Table S1† for details) and the Nanonex Advanced Nanoimprint Tool NX-2000 or a home-built, bench-top system (described later). After the imprint step, the resist residue was removed by exposing the samples to an O_2 plasma.

EBL and NIL patterns with multiple layers of metals and ceramics

EBL. First, a thin adhesive chromium (Cr) layer (~0.2 nm) was thermally evaporated, and then nickel (Ni) or alumina (Al_2O_3) were deposited by a thermal or e-beam evaporator to a thickness of approximately 40 nm. The resist was dissolved in acetone and a second round of resist was spin coated. The second resist layer was patterned so that the Sn bridged the gap between the Ni panels and

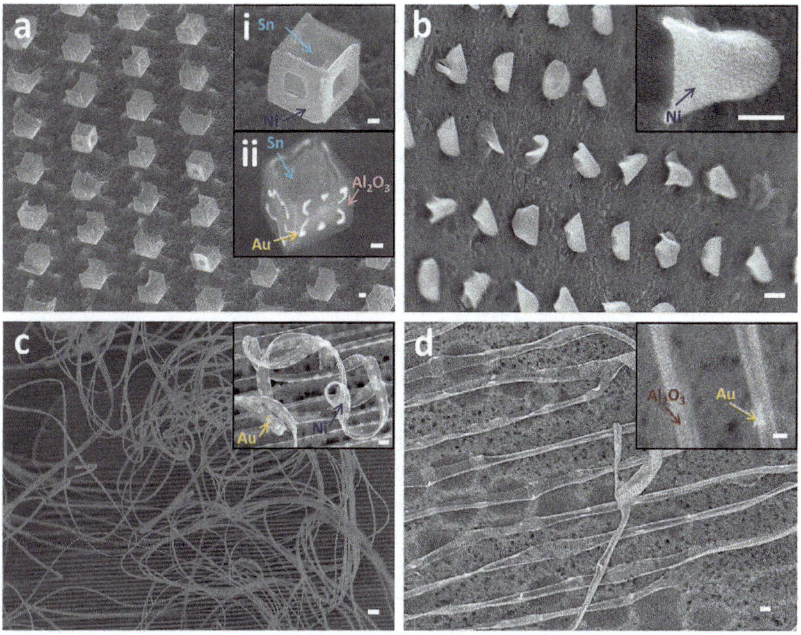

Fig. 3 SEM images of self-folded EBL patterned nanostructures and SNIPS. (a) 500 nm size self-folded EBL patterned cube shaped boxes with patterned surfaces; (b) Ni/Sn taco shaped SNIPS; (c) coiled Ni/Sn strips with circular Au patterns, and (d) tubular Al_2O_3/Sn SNIPS with Au patterns. Scale bars are (a, b and d) 200 nm and 100 nm for the inset (c) 1 μm and 500 nm for the inset.

overlapped the panels at both ends. After Sn deposition of approximately 45 nm, the resist was dissolved in acetone.

NIL. After NIL patterning with circular patterns, an approximately 3 nm gold (Au) thin film was deposited and the NIL resist was dissolved using acetone. A second round of resist was coated and another NIL process was performed using a line pattern mold, and then 10 nm of Ni or 6 nm of Al_2O_3 was deposited. A Sn layer, approximately 2.5 to 5 nm thick, was deposited on top of the Ni or Al_2O_3 layer and the resist was dissolved in acetone.

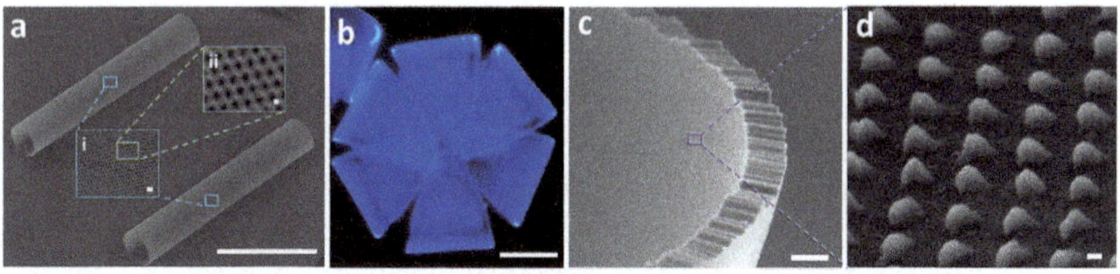

Fig. 4 SNIPS composed of device grade silicon and polymers. (a) Tubular Si_3N_4/Si bilayer SNIPS with 100 nm sized pores, and (b) a representative image of SU-8 (an epoxy polymer) flower shaped SNIPS with nanotopography. (c and d) Zoomed-in image of the curved surface of SU-8 structures showing NIL imprinted nanobumps on the inner surface. Scale bars are (a) 1 mm; (a-i) 1 μm; (a-ii) 200 nm; (b) 100 μm; (c) 10 μm; and (d) 100 nm. Note: the zoom boxes indicate the approximate region where the zoomed images are taken and are not drawn in proportion.

Lift-off and self-folding using forces derived from coalescence and reflow of Sn

The samples were exposed to a CF_4/O_2 plasma to etch the underlying Si (which releases the 2D multilayer nanopatterns from the substrate) and simultaneously drive reflow of Sn and capillary folding as described elsewhere.[39,42] Briefly, the Sn deposited samples were placed in a plasma etcher (Technics Inc., PEII-A) and exposed to CF_4 and O_2 with flow rates of 3.6 sccm and 12 sccm respectively at 25 W RF power.

Fabrication of nanopatterned Si and polymer structures

A silicon nitride (Si_3N_4)/Si bilayer was prepared following the method described elsewhere.[43] Briefly, Si_3N_4 deposited silicon-on-insulator (SOI) wafer was first coated with resist and NIL was performed. After etching away the residual resist, a Cr masking layer was evaporated and the resist was dissolved in acetone. Then the sample was dry etched with SF_6/CF_4 and the Cr and the SiO_2 layer was subsequently etched to release the structures.

For the nanopatterned SU-8 flowers, a polyvinyl alcohol (PVA) solution (1 : 10 in DI water, MW 9000–10 000, Sigma Aldrich) was spin coated on a bare Si wafer (for SNIPS with nanopatterns on the outer surfaces of the flowers) or on a Cr mask (for SNIPS with nanopatterns on the inner surfaces of the flowers) at 1000 rpm and baked at 150 °C for 5–10 min. SU-8 2015 (MicroChem Corp.) was spun on top of the PVA layer at 5000 rpm and was soft baked at 95 °C for 3 min. A Si master mold was attached to a weight of approximately 150 g and placed on top of the SU-8 film at a temperature of 150 °C. We note that the patterns can also be transferred to the SU-8 film by simply pressing down the master mold with the tip of flat metallic wafer forceps. After the imprint step, the sample and the mold were cooled down to room temperature and the mold was detached from the sample. Then the sample was exposed to UV light at around 120 mJ cm^{-2} from either the top of the Si wafer (for SNIPS with nanopatterns on the outer surfaces of the flowers) or from the bottom of the Cr mask (for SNIPS with nanopatterns on the inner surfaces of the flowers) and was post baked at 95 °C for approximately 3–4 min. The micropatterned structures were then developed and put into DI water until release from the substrate. After release, the structures were immersed in acetone overnight and into DI water for the folding step.

Results and discussion

The nanoscale patterns formed by EBL and NIL are realized using resist patterning and additive or subtractive methods (Fig. 1). Consequently, a variety of materials can be patterned. For example, additive methods can be used to enable patterning of metals (evaporation, sputtering, CVD, ALD); semiconductors (CVD), ceramics (evaporation, sputtering, CVD, ALD), polymers (spin coating, casting, atom transfer radical polymerization) and even small molecules (self-assembled monolayers, sublimation). Subtractive methods can also be used for patterning of all materials using dry (plasma, ion etching) or wet etching. Importantly, subtractive etching can be used to pattern high quality device grade materials such as commercial single crystal silicon (*e.g.* SOI), gallium arsenide and piezo-electric ceramics.

In terms of patterning, we encountered the following challenges; EBL was expensive and typical rates for use of instrumentation range from \$50 to \$200 per hour at academic/government user facilities in the US. Hence, due to cost and the serial nature of our e-beam system it was difficult to pattern nanoscale features over millimeter or centimeter areas on the substrate. However, versatility and tunability were high. Consequently, we could readily pattern cross shaped 2D Ni patterns and enable a second sequential layer of patterns of Sn hinges with good registry to the underlying layer (Fig. 2a). In contrast, while NIL can provide multiple replica copies, it also requires a master mold to be fabricated which can be expensive. In order to reduce costs and demonstrate cost effective manufacturing of SNIPS, we utilized low-cost Si stamps (LightSmyth) in this paper (Fig. 2b–d).

Integrating different materials such as Ni, Au, Al_2O_3 and Sn into nanostructures was achieved by multiple layers of EBL. We created a separate CAD design file for each of the different layers; for example, three layers of EBL were used to pattern Au patterns resembling a circuit pattern atop Al_2O_3 (Fig. 2a-ii) panels with Sn hinges for self-folding. Although extremely precise patterning and relatively good registration alignment was possible using EBL, such precision came at low throughput and high cost. In contrast, we could use relatively inexpensive commercial stamps to perform NIL to create bilayer Ni/Sn nano circular discs (Fig. 2b), or even structures requiring two sequential steps of NIL patterning such as Ni/Sn (Fig. 2c) or Al_2O_3/Sn (Fig. 2d) bilayer nanostrips with circular Au patterns. Although the shapes are not as complex as the Au circuit patterns designed for the EBL approach, NIL patterning was achieved at low cost and large numbers of nanostructures could be formed over the centimeter sized stamped region. Alignment was achieved between stamps by careful aligning of the edges of Si master mold. In addition to the patterning of metals and ceramics, we were also able to pattern nanopores in a bilayer composed of single crystal device grade Si on an SOI wafer and a cross-linked epoxy SU-8 (Fig. 2e). Using NIL, a masking layer was patterned on top of the flat bilayer and nanopores were generated by dry etching after which photolithography was further utilized to micropattern the bilayer into rectangular precursors for Si tubes or multi-petal shapes for SU-8 flowers.

NIL often requires relatively expensive commercial imprint tools that can control temperature and pressure in vacuum with high precision. However, in this work we show that in addition to the NIL patterning of substrates with commercial NIL instrumentation shown in Fig. 2b–d, NIL can also be achieved with a low-cost home-built NIL system and we note that the nanoscale patterns in Fig. 2e–h were formed using a bench-top home-built NIL system. Our home-built NIL system is described in Fig. S1a,† where the NIL procedure was achieved using small weights of approximately 150 to 450 grams with a flat surface in ambient air on a hotplate. If higher pressure ($\sim$100 psi) was needed, a commercial heat press (such as Vesta VP3815) was used with the addition of a hotplate (Fig. S1b†). Using the low and high pressure NIL approach in ambient air, we were able to pattern multiple materials including SU-8 and polycaprolactone. A polydimethyl siloxane (PDMS) negative replica of the master mold was also made by molding. The PDMS mold was then used to pattern PMMA at a high temperature in ambient air using capillary force lithography (CFL) (Fig. S2†).[44]

As illustrated in Fig. 2, EBL and NIL enabled patterning with significant versatility, precision and design across material classes but in an inherently 2D layered format. We used selective release and self-folding to transform these 2D patterns on substrates to 3D nanostructures. The resulting SNIPS assembled due to a result of three different forces all of which have been described in detail previously.[39,41,43,45,46] The first was due to the reflow and coalescence of thermally evaporated Sn grains on a semi-wetting Ni or Al_2O_3 substrate (Fig. 3). As previously described,[41,45] the bi-axial strain during Sn reflow can be significant and can result in nanoscale folding of thick panels and nanoscale curving of thin panels. The cruciform shape of Ni panels folded up into cubic box shaped structures with the EBL patterned surfaces taken into the third dimension (Fig. 3a). The methodology has high throughput and we believe that yield was limited by defects in EBL patterning and inhomogeneity of grains, both of which could be improved significantly with improved defect metrology and tighter engineering controls. Importantly, these cubic nanostructures have Au patterns on their surfaces with precise CAD designed shapes whose shape and size could be tuned. We envision that these Au patterns could be functionalized to enable receptor mediated binding of biomolecules and cells in biosensor applications or enable binding sites for bottom-up aggregation of boxes or function as interconnects or electrical bond pads in circuits. The circular bilayer Ni/Sn patterns formed taco shaped SNIPS (Fig. 3b) which is a shape that is difficult to achieve at the nanoscale using current top-down or bottom-up approaches and could be interesting in drug delivery and cell studies. For example, it has been discovered that the phagocytosis process in immune cells is sensitive to the shape and size of the object.[47] Ceramic taco shaped SNIPS with noble metal patterns could also display interesting optical properties. The Ni and Al_2O_3 linear strips with Au circular patterns formed coiled fiber like SNIPS (Fig. 3c) when they folded along the long edge of the strip and tubular SNIPS (Fig. 3d) when they folded along the short edge. When Al_2O_3 of thinner thickness was deposited and etched at the same conditions, it formed tubes with smaller radii of curvature (see Fig. S3†). We observed Au patterned tubular Al_2O_3 SNIPS with diameters as small as 88 nm. We anticipate that such fibrous or tubular metal or ceramic structures with Au patterns could be used to create artificial scaffolds for tissue engineering, plasmonic waveguides, nanomeshes, supported catalysts, and nanofluidic devices.

In addition to metal and ceramic materials, ordered arrays of SNIPS composed of device grade Si or polymers could also be formed (Fig. 4). Self-folding in Si SNIPS was driven by bilayer strain mismatch in Si_3N_4 and Si (Fig. 4a).[33,43,48,49] The tubes roll up with sub-mm scale curvature and well-defined arrays of NIL patterned nanopores on their surfaces; we note that nanopores merely represent a type of pattern and alternate materials such as Au bumps or patterns could easily have been incorporated instead of pores. Such SNIPS represent an assembly of nanostructures in a curved format which is hard to achieve by existing methods. We anticipate utility of these ordered assemblies of SNIPS in microfluidics, catalysis and cell biology. Also, if the two sides of the tubes are sealed and loaded with chemicals, these arrays of SNIPS could be utilized to deliver chemicals with different diffusion rates by controlling the pore size and possibly shapes. Finally, SU-8 flower shaped SNIPS were formed by self-folding based on differential cross-linking.[46] SNIPS in the shape of folded structures were formed using SU-8 thin films patterned with nanoscale bumps by NIL (Fig. 2e–h and

4b-d). It is noteworthy that the flower shaped structures were self-folded in such a way that nanobumps could be patterned either on the inside or outside of the petals, highlighting patterning versatility that would be extremely challenging to achieve by any other method. We anticipate that these structures with nano-topography could be used to trap objects or modulate surface properties for adhesion or cell differentiation.

Conclusions

In summary, in this paper, we have detailed a high-throughput methodology to form SNIPS which are mass-producible, precisely patterned 3D nanostructures formed using NIL and self-folding. We contrast the approach to EBL patterned self-folded structures and highlight versatility in pattern design and material heterogeneity. We have illustrated the need for the development of NIL patterning using low cost molds and bench-top instrumentation. Further refinements are needed to improve alignment accuracy for multilayer heterogeneous integration during NIL as would be required to fabricate more complex patterned structures such as transistors. Also the incorporation of in-line defect metrology, yield analysis, design of experiments and tighter engineering controls would improve yield and reproducibility.

Acknowledgements

We acknowledge funding from the National Science Foundation (Grant: CMMI-1200241). Research performed in part at the NIST Center for Nanoscale Science and Technology.

References

1 S. Park, J. H. Lim, S. W. Chung and C. A. Mirkin, *Science*, 2004, **303**, 348–351.
2 K. L. Ekinci and M. L. Roukes, *Rev. Sci. Instrum.*, 2005, **76**, 061101.
3 B. Tian, X. Zheng, T. J. Kempa, Y. Fang, N. Yu, G. Yu, J. Huang and C. M. Lieber, *Nature*, 2007, **449**, 885–889.
4 A. R. Tao, S. Habas and P. Yang, *Small*, 2008, **4**, 310–325.
5 Y. He, T. Ye, M. Su, C. Zhang, A. E. Ribbe, W. Jiang and C. Mao, *Nature*, 2008, **452**, 198–201.
6 K. J. M. Bishop, C. E. Wilmer, S. Soh and B. A. Grzybowski, *Small*, 2009, **5**, 1600–1630.
7 B. Luk'yanchuk, N. I. Zheludev, S. A. Maier, N. J. Halas, P. Nordlander, H. Giessen and C. T. Chong, *Nat. Mater.*, 2010, **9**, 707–715.
8 H. A. Atwater and A. Polman, *Nat. Mater.*, 2010, **9**, 205–213.
9 E. González, J. Arbiol and V. F. Puntes, *Science*, 2011, **334**, 1377–1380.
10 A. Kuzyk, R. Schreiber, Z. Fan, G. Pardatscher, E. M. Roller, A. Högele, F. C. Simmel, A. O. Govorov and T. Liedl, *Nature*, 2012, **483**, 311–314.
11 C. Auth, C. Allen, A. Blattner, D. Bergstrom, M. Brazier, M. Bost, M. Buehler, V. Chikarmane, T. Ghani, T. Glassman, R. Grover, W. Han, D. Hanken, M. Hattendorf, P. Hentges, R. Heussner, J. Hicks, D. Ingerly, P. Jain, S. Jaloviar, R. James, D. Jones, J. Jopling, S. Joshi, C. Kenyon, H. Liu, R. McFadden, B. McIntyre, J. Neirynck, C. Parker, L. Pipes, I. Post,

S. Pradhan, M. Prince, S. Ramey, T. Reynolds, J. Roesler, J. Sandford, J. Seiple, P. Smith, C. Thomas, D. Towner, T. Troeger, C. Weber, P. Yashar, K. Zawadzki and K. Mistry, *Symp. VLSI Technol., Dig. Tech. Pap.*, 2012, 131–132.

12 P. F. Damasceno, M. Engel and S. C. Glotzer, *Science*, 2012, **337**, 453–457.

13 N. D. Burrows, A. M. Vartanian, N. S. Abadeer, E. M. Grzincic, L. M. Jacob, W. Lin, J. Li, J. M. Dennison, J. G. Hinman and C. J. Murphy, *J. Phys. Chem. Lett.*, 2016, **7**, 632–641.

14 C. R. Martin, *Chem. Mater.*, 1996, **8**, 1739–1746.

15 D. Zhao, J. Feng, Q. Huo, N. Melosh, G. H. Fredrickson, B. F. Chmelka and G. D. Stucky, *Science*, 1998, **279**, 548–552.

16 N. R. Jana, L. Gearheart and C. J. Murphy, *Adv. Mater.*, 2001, **13**, 1389–1393.

17 K. M. Choi and J. A. Rogers, *J. Am. Chem. Soc.*, 2003, **125**, 4060–4061.

18 S. Jeon, V. Malyarchuk, J. A. Rogers and G. P. Wiederrecht, *Opt. Express*, 2006, **14**, 2300–2308.

19 K. T. Nam, D. W. Kim, P. J. Yoo, C. Y. Chiang, N. Meethong, P. T. Hammond, Y. M. Chiang and A. M. Belcher, *Science*, 2006, **312**, 885–888.

20 N. Anscombe, *Nat. Photonics*, 2010, **4**, 22–23.

21 M. R. Jones, K. D. Osberg, R. J. Macfarlane, M. R. Langille and C. A. Mirkin, *Chem. Rev.*, 2011, **111**, 3736–3827.

22 Q. Xu, Y. Lv, C. Dong, T. S. Sreeprased, A. Tian, H. Zhang, Y. Tang, Z. Yu and N. Li, *Nanoscale*, 2015, **7**, 10883–10895.

23 J. R. Tumbleston, D. Shirvanyants, N. Ermoshkin, R. Janusziewicz, A. R. Johnson, D. Kelly, K. Chen, R. Pinschmidt, J. P. Rolland, A. Ermoshkin, E. T. Samulski and J. M. DeSimone, *Science*, 2015, **347**, 1349–1352.

24 J. A. Liddle and G. M. Gallatin, *ACS Nano*, 2016, **10**, 2995–3014.

25 V. R. Manfrinato, L. Zhang, D. Su, H. Duan, R. G. Hobbs, E. A. Stach and K. K. Berggren, *Nano Lett.*, 2013, **13**, 1555–1558.

26 E. Slot, M. J. Wieland, G. de Boer, P. Kruit, G. F. ten Berge, A. M. C. Houkes, R. Jager, T. van de Peut, J. J. M. Peijster, S. W. H. K. Steenbrink, T. F. Teepen, A. H. V. van Veen and B. J. Kampherbeek, *Proc. Soc. Photo-Opt. Instrum. Eng.*, 2008, 6921.

27 S. Y. Chou and P. R. Krauss, *Microelectron. Eng.*, 1997, **35**, 237–240.

28 F. Hua, Y. Sun, A. Gaur, M. A. Meitl, L. Bilhaut, L. Rotkina, J. Wang, P. Geil, M. Shim, J. A. Rogers and A. Shim, *Nano Lett.*, 2004, **4**, 2467–2471.

29 F. Cavallo and M. G. Lagally, *Nano Today*, 2015, **10**, 538–541.

30 D. H. Gracias, *Curr. Opin. Chem. Eng.*, 2013, **2**, 112–119.

31 T. G. Leong, A. M. Zarafshar and D. H. Gracias, *Small*, 2010, **6**, 792–806.

32 X. Li, *J. Phys. D: Appl. Phys.*, 2008, **41**, 193001.

33 Y. Mei, A. A. Solovev, S. Sanchez and O. G. Schmidt, *Chem. Soc. Rev.*, 2011, **40**, 2109–2119.

34 E. A. Peraza-Hernandez, D. J. Hartl, R. J. Malak Jr and D. C. Lagoudas, *Smart Mater. Struct.*, 2014, **23**, 094001.

35 J. S. Randhawa, K. E. Laflin, N. Seelam and D. H. Gracias, *Adv. Funct. Mater.*, 2011, **21**, 2395–2410.

36 J. Rogers, Y. Huang, O. G. Schmidt and D. H. Gracias, *MRS Bull.*, 2016, **41**, 123–129.

37 V. B. Shenoy and D. H. Gracias, *MRS Bull.*, 2012, **37**, 847–854.

38 R. R. A. Syms, E. M. Yeatman, V. M. Bright and G. M. Whitesides, *J. Microelectromech. Syst.*, 2003, **12**, 387–417.

39 J. H. Cho and D. H. Gracias, *Nano Lett.*, 2009, **9**, 4049–4052.
40 J. H. Cho, A. Azam and D. H. Gracias, *Langmuir*, 2010, **26**, 16534–16539.
41 J. H. Cho, T. James and D. H. Gracias, *Adv. Mater.*, 2010, **22**, 2320–2324.
42 J. H. Cho, M. D. Keung, N. Verellen, L. Lagae, V. V. Moshchalkov, P. Van Dorpe and D. H. Gracias, *Small*, 2011, **7**, 1943–1948.
43 J. Park, C. Yoon, Q. Jin, L. Chen and D. H. Gracias, *Proceedings of the 10th IEEE International Conference on Nano/Micro Engineered and Molecular Systems (IEEE-NEMS 2015)*, Xi'an, China, April, 2015, pp. 233–237.
44 K. Y. Suh, Y. S. Kim and H. H. Lee, *Adv. Mater.*, 2001, **13**, 1386–1389.
45 J. H. Cho, D. Datta, S. Y. Park, V. B. Shenoy and D. H. Gracias, *Nano Lett.*, 2010, **10**, 5098–5102.
46 M. Jamal, A. M. Zarafshar and D. H. Gracias, *Nat. Commun.*, 2011, **2**, 527.
47 S. E. A. Gratton, P. A. Ropp, P. D. Pohlhaus, J. C. Luft, V. J. Madden, M. E. Napier and J. M. DeSimone, *Proc. Natl. Acad. Sci. U. S. A.*, 2008, **105**, 11613–11618.
48 V. Y. Prinz, D. Grützmacher, A. Beyer, C. David, B. Ketterer and E. Deckardt, *Nanotechnology*, 2001, **12**, 399–402.
49 L. Zhang, E. Ruh, D. Grützmacher, L. Dong, D. J. Bell, B. J. Nelson and C. Schönenberger, *Nano Lett.*, 2006, **6**, 1311–1317.

Faraday Discussions

PAPER

Organic–inorganic patchy particles as a versatile platform for fluid-in-fluid dispersion stabilisation†

E. Blanco,‡[a] S. K. Smoukov,§[a] O. D. Velev*[a] and K. P. Velikov*[bc]

Received 1st March 2016, Accepted 17th March 2016
DOI: 10.1039/c6fd00036c

We present a new class of organic–inorganic patchy particles for the efficient stabilization of Pickering foams and emulsions. Using solvent-based heterogeneous precipitation, we decorate inorganic silica particles with discrete domains of water insoluble plant protein (zein). By varying the extent of protein coverage on the silica surface, we tune the pH-dependent interactions of the particles and the interfaces. We observe an optimum foam stabilization, which is attributed to the creation of a slightly positive low effective surface potential from positively charged protein patches and the negatively charged silica surface. The effect of surface coverage on foam stability is in line with the predicted low interfacial potential of the patchy particles in water, which determines the energy of particle adsorption. In emulsions, the increase of the protein amount on the silica particles causes a progressive bridging of the oil droplets into a close-packing configuration due to gelation of the protein patches. Protein-based organic–inorganic surface heterogeneous particles represent a new versatile platform for the stabilization of fluid-in-fluid dispersions and as precursors for the assembly of advanced functional materials.

Introduction

Considerable progress has been achieved recently in the formulation of complex colloids with controlled shapes and surface properties (Janus particles, complex-

[a]Department of Chemical & Biomolecular Engineering, North Carolina State University, Raleigh, USA. E-mail: odvelev@ncsu.edu

[b]Unilever R & D Vlaardingen, Olivier van Noortlaan 120, 3133 AT Vlaardingen, The Netherlands. E-mail: krassimir.velikov@unilever.com

[c]Soft Condensed Matter, Debye Institute for Nanomaterials Science, Utrecht University, Princetonplein 5, 3584 CC Utrecht, The Netherlands

† Electronic supplementary information (ESI) available. See DOI: 10.1039/c6fd00036c

‡ Present address: School of Physics and Astronomy, University of Edinburgh, Mayfield Road, Edinburgh EH9 3JZ, UK, E-mail: mblanco@ed.ac.uk

§ Present address: Department of Materials Science and Metallurgy, University of Cambridge, 27 Charles Babbage Rd., Cambridge, CB3 0FS, UK, E-mail: sks46@cam.ac.uk shaped colloids, colloidal molecules, chiral colloids, active colloids, *etc.*). Among these, the formation of particles with chemical anisotropy, patchy colloids, can yield structures with unique functionalities arising from their directional interactions and complex shapes and/or surface properties.[1–10] Despite the broad variety of patchy architectures, great attention is devoted to formulating hybrid bio-inspired colloids in order to create new materials.[7,11] Bio-derived colloids are promising building blocks because of their potential applications in fields such as novel and improved consumer goods, pharmaceuticals, drug delivery or other technological formulations. The conjugation of biomacromolecules to colloidal particles allows the combination of the unusual properties and functionalities of both materials.[12,13] The inclusion of hybrid protein-inorganic patterned particles in a soft material matrix opens a number of options for the convenient control of the properties and stability of multiphase soft matter.

Particle surface properties and shape anisotropy are important parameters affecting the particle-based stabilization of fluid-in-fluid dispersions. The ability of colloidal particles to adsorb at fluid interfaces has been used for the stabilization of soft materials such as foams,[14–16] emulsions,[17–19] bicontinuous gels,[20,21] and dry water.[22] Particle-based or Ramsden–Pickering stabilisation allows immiscible fluids to be structured into emulsions and foams with superior results compared to molecularly dissolved surfactants and proteins.[23,24] Particular interest is focused on Janus particles due their increased energy of adsorption.[25,26] Non-spherical, rod-like particles have been reported as a means of achieving greater stabilization in disperse systems than previously achievable by spherical particles.[27] The use of responsive particles also allows the formation of foams that are responsive to external fields.[28,29] The underlying mechanism of particle stabilisation[30–32] is the creation of a steric barrier and/or bridging[33] at the fluid interfaces, whereas the strength of adsorption depends on the particle contact angle and line tension.[34–36]

The large interest in using natural particle-assisted stabilisation of fluid-in-fluid dispersions is being thwarted by the limited availability of edible, biodegradable materials from sustainable sources.[37] There is a need for the design of novel surface-patterned colloids with tuneable amphiphilic properties. Hence, the trend of the design of novel building blocks based on natural materials contributes to the development of new formulations required for several industrial applications (*e.g.* food, pharmaceutical or agricultural dispersions).[38–41] The range of edible, yet water insoluble, materials is fairly restricted and includes compounds such as cellulose, chitins and water insoluble proteins (*i.e.* prolamins).[38] For this study we chose to use zein, a prolamine rich protein from corn. Zein is an abundant natural material extracted from corn, which is characterized by its poor solubility in water. Nevertheless, zein is soluble in aqueous-ethanol mixtures of specific compositions (achieving a maximum solubility at around 90 vol% ethanol).[42,43] Indeed, the synthesis of zein colloids for emulsion stabilization[44] and oil droplet encapsulation has been demonstrated.[45]

In the present work, a straightforward heterogeneous nucleation method for the production of protein patterned inorganic colloids has been developed. This procedure yields silica particles as the templating core and leads to the heterogeneous nucleation of zein patches on the silica surface. We characterize the resulting composite particles using electron microscopy and electrophoretic mobility measurements. We demonstrate that the new organic–inorganic patchy particles can be used to create very stable foams and emulsions. We relate the effective surface potential, as a result of the partial particle surface coverage, to the pH dependent effective contact angle and hypothesize on the reasons for the high stability observed in these systems.

Experimental

Materials

Zein powder (Z3625, batch# 118K0068), decane (99 wt%), sodium citrate, hydrochloric acid, and sodium hydroxide were obtained from Sigma. Hydrophilic fumed silica (Aerosil® 90, BET = $90 \pm 15\ m^2\ g^{-1}$) was provided by Degussa and colloidal silica particles (100 nm) were from Polyscience. Pure, anhydrous 100 wt% ethanol was obtained through Pharmco-Aaper. Water purified by a MilliQ system was used for all the experiments. All reagents were used without further purification.

Fabrication of zein colloidal particles

The synthesis of the zein particles was done as previously described.[44] In a typical synthesis protocol, 5.0 g of zein was added to an aqueous ethanol mixture (80 vol% EtOH), prepared by mixing 160 mL of anhydrous ethanol with 40 mL of Millipore water. 1.2 L of Millipore water was put into a volumetric cylinder (250 mL) and sheared at 15 000 rpm in a heavy-duty household blender for 2 minutes. The zein solution was added over 2–3 minutes to this cylinder, using a syringe (5 mL). A dispersion with a yellow, milky colour is formed in the 20 vol% EtOH/water medium. The ethanol is removed using rotary evaporation, and the leftover water volume is further reduced to 200 mL to result in a 2.5 wt/vol% aqueous suspension of zein particles. Centrifuging the dispersion for 10 minutes at 1000*g* is done to remove (possible) larger aggregates and improve the stability and monodispersity of the suspension. The final dispersion has a pH of ~4 and is stable at room temperature for at least several weeks.

Zein-coated silica patchy particles

Organic–inorganic patterned colloids were prepared by the heterogeneous nucleation of protein patches on solid surfaces. Stock solutions of zein were prepared by dissolving 0.5 g of zein in 100 mL of an aqueous ethanol mixture (90 vol%). In a typical synthesis, 0.1 g of silica was added to 5 mL of the corresponding zein solution in a 15 mL centrifuge tube. The mixture was sonicated for 5 min. The ethanol concentration was increased to 95 vol% by adding 10 mL of pure (100 wt%) ethanol. The sample was centrifuged at 1300–1500*g* for 5 min. The supernatant was pipetted out. Distilled water was added to obtain a 2 wt/vol% suspension of silica/zein (~5 mL), and the precipitate was redispersed by sonication for 2 min. The zein-to-silica surface ratio (ZSSR, $mg\ m^{-2}$) was calculated as: "mass of zein"/("specific surface area silica" × "mass of silica").

Similarly, we have used 100 nm silica colloidal particles as substrates for zein deposition. In a typical synthesis, 0.1 mL of colloidal silica (3.5 wt%) was added to the zein solution in a 12.5 mL centrifuge tube. The mixture was shaken for 5 min. The ethanol concentration was increased to 95 vol% by adding 14.3 mL of pure

(100 wt%) ethanol. The particles were separated by centrifugation and redispersed in water.

Foam preparation

Foams were prepared by the aeration of a particle suspension (5 mL) in deionized water (95 mL) using a household blender (Oster model 4242). The dispersion was blended for 2 minutes, while the pH was adjusted with hydrochloric acid, and the resulting foam was transferred to a cylindrical glass vessel (250 mL) and left to evolve over time. The foamability was estimated by measuring the initial foam volume right after preparation, $V_f(t = 0)$, while the foam stability was measured by monitoring the foam volume over time, $V_f(t > 0)$. The normalized foam volume was calculated by dividing the initial foam volume by the total volume of the system, V_t.

Emulsion preparation

Decane in water emulsions were prepared in batches of 12 mL according to the following protocol. For the aqueous phase, we used an aliquot of the suspension of patchy particles prepared with different zein/silica ratios. A mixture of different aqueous/oil proportions was sheared at 2000 rpm with a Servodyne mixer head for 2 min and then vigorously shaken by hand.

Surface potential

The ζ-potential of the particles was measured using a Malvern ZetaSizer Nano (Malvern Instruments) at different pH conditions. The pH adjustment was done with HCl or NaOH solution. All measurements were carried out at 25 °C and the results reported are averaged over three readings.

Particle morphology

Transmission electron microscopy (TEM) particle imaging of the zein-patterned silica particles was performed using a JEOL 2000FX TEM at 200 kV, in bright field mode. Regardless of the maintenance of the structural rigidity of the zein protein, glutaraldehyde, a common crosslinking agent was used to crosslink the protein in order to visualize it using TEM. The treatment with the crosslinker was conducted by adding glutaraldehyde (3 vol%) to the solution and stirring the sample with a magnetic bar for a couple of hours at 37 °C. The samples were stained with a 2 wt % phosphotungstic acid solution at pH 7.3 in order to confer more electron density.

Results and discussion

Particle synthesis and characterization

As a result of the heterogeneous precipitation, the surface of the silica particles has been modified by the binding of the zein protein on the silica surface, creating surface anisotropy. In order to provide evidence for the deposition of zein and characterize the surface particles, we have performed TEM on the zein–silica samples. Fig. 1a shows a TEM micrograph of the unmodified fumed silica particles, which appear as fractal-like aggregates. The morphology of the particles

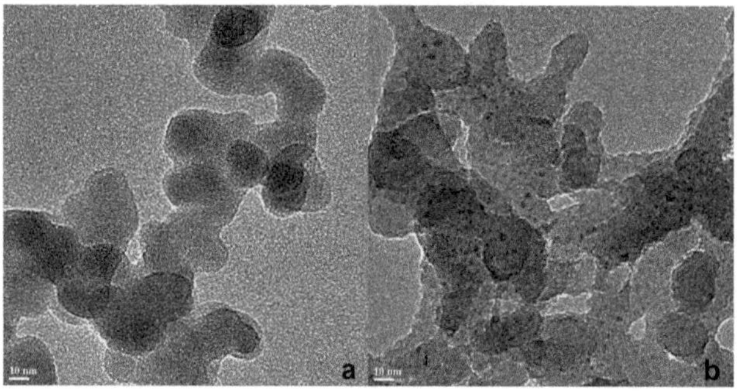

Fig. 1 TEM images of: (a) the fumed silica fractal structure, and (b) fumed silica particles with zein at ZSSR = 3.8 mg m^{-2}. Surface roughness on the scale of less than 10 nm is visible on the surface of the patchy particles in (b). Scale bar = 10 nm.

patterned with zein barely change under this resolution if we compare them with pure silica. The presence of zein however clearly induces smaller scale roughness to the fractal aggregates (see Fig. 1b). This indicates that zein is attached to the silica surface. The ability of zein to adsorb on silica surfaces has been demonstrated previously.[46–48] In order to provide further evidence for the formation of zein patches, we used spherical colloidal silica particles produced by the Stöber method as a model system. Typical TEM micrographs of the organic–inorganic zein–silica patchy colloids are shown in ESI Fig. 1S† (ZSSR = 5.6 mg m^{-2}). The TEM micrographs provide a direct evidence for the ability of zein to form patches on the silica surface. Depending on the conditions used to depose the zein, a complete shell of zein can be formed too. ESI Fig. 1S† shows the TEM micrographs of zein coated silica spheres formed at ZSSR = 62 mg m^{-2}. The presence of a zein shell is evident after partial etching of the silica core by HF (ESI Fig. 1S†).

The zein layering on solid surfaces depends on the lyophobicity of the underlying substrate, as well as the dispersing media.[47] The heterogeneous nucleation process resulted in zein protein deposition on the silica surface. Silica particles and zein were suspended in aqueous ethanol at 90 vol%. Because zein has a high solubility in 90 vol% ethanol, to allow the hydrophobic interaction to take place, zein precipitation in ethanol (95 vol%) was proven to be a simple method to form an amphiphilic surface pattern.

Clearly, depending on the zein amino acid sequence, this protein may adopt distinct spatial conformations and show several functionalities in response to environmental factors such as pH, ionic strength, or the hydrophilic–lipophilic character of the medium.[49] The net charge of the protein directly correlated to the pH of the solution.

After zein nucleation, the fumed silica particle surface is partially covered with zein surface patches. Because of the nature of the nucleation process, the protein groups lead to random patterns and anisotropic coverage. Furthermore, the degree of protein nucleation affects the surface density charge distribution of the resulting particles. The silica surface, which contains mainly silanol groups, is negatively charged over the whole range of pH values, whereas zein has an isoelectric point at pI ≈ 6.5 (Fig. 2). This difference implies that in the pH interval between the two isoelectric points, the interactions between the precipitated zein

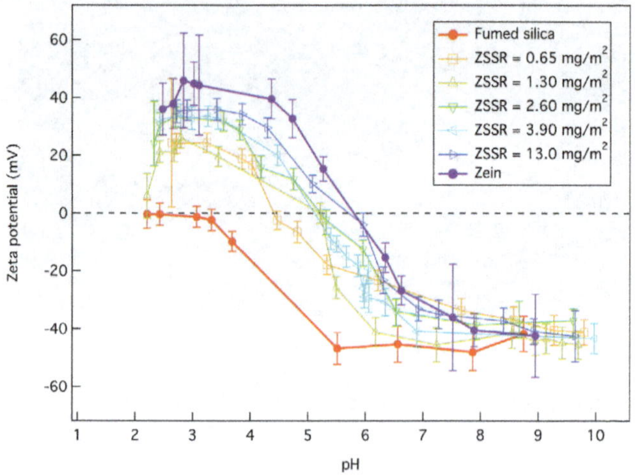

Fig. 2 Zeta potential as a function of pH for silica, zein and zein-coated silica patchy particles at different surface coverages (mg m^{-2}). Deposition of zein gradually shifts the isoelectric point toward the isoelectric point of pure zein particles.

and the silica particles will be attractive. The new anisotropic particles present positively and negatively charged domains on the surface. Both zein and silica domains will be characterized with different pH-dependent surface potentials and contact angles.

The ζ-potential measurements of pure silica, zein and the surface functionalized particles are shown in Fig. 2. The pH-dependent ζ-potential curves of the zein-coated silica are situated in between those of pure zein and the silica colloids, reflecting the gradual change in the surface properties from one dominant species to the other. Furthermore, the composite colloids present a single electrophoretic mobility peak, confirming the existence of a single species of particles with similar surface properties. As a result of the surface modification and the creation of a patchy surface, the mixed particles are also characterized with a different isoelectric point. From the pH dependent ζ-potential curve, we can extract an effective surface coverage x_S which we can use to evaluate what percentage of the original silica surface is covered with zein.

If we assume a simple linear dependence between the surface area of silica covered by zein, x_S, and the surface potential of each material (*i.e.* zein and silica), by fitting the curves for each amount of zein adsorbed one can estimate the surface coverage. On the basis of the electrophoretic mobility data, we estimate that 0.67 mg m^{-2} of zein results in a 40% patchy coverage on the particles after which the coverage increases to 60–70% for a ZSSR of up to 3.9 mg m^{-2}. A nearly full layer of zein on the surface is expected at ZSSR = 13 mg m^{-2}. The pH dependence of the zeta potential of these particles is the closest to that of the pure zein particles (Fig. 2).

The combination of electron microscopy and electrophoretic mobility measurements on the patchy particles presents sufficient evidence for the heterogeneous patterning onto the inorganic colloids. This allows us to propose a model for the patchy zein patterned fumed silica particles (Fig. 3). Depending on

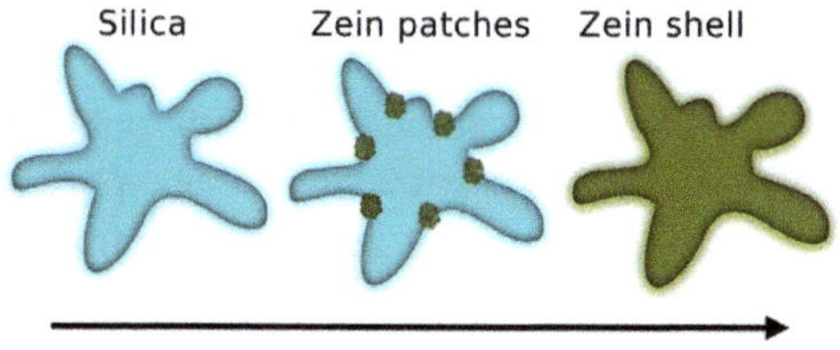

Fig. 3 Schematic representation of the hypothesis for the formation of patches and complete shells of zein, for different levels of deposition, on the surface of irregularly shaped silica particles. The arrow indicates the increase in ZSSR.

the morphology of the silica particles, the ratio between the precipitated zein and the silica, and the available surface areas, spherical or shape anisotropic patchy particles can be prepared. The interaction potential and surface properties of such particles will be difficult to predict.

Foam stabilization

Here we describe the application of these 'inorganic–organic' patchy particles as stabilizers and modifiers of multiphase dispersions. This basic nucleation process promotes the formation of modified silica surfaces by adding protein patches. The key feature of this approach is to change the wetting properties of the particle surfaces. We investigate their behavior at fluid–fluid interfaces and explore the role of chemically anisotropic particles as Pickering-amphiphilic foam stabilizers.

Bare hydrophilic fumed silica particles are not able to stabilize foams effectively,[50] but surface anisotropic particles with different degrees of protein nucleation are. Fig. 4 shows the typical foam time evolution at different pH values (the highest foamed volume was 160 mL). The foam volume decrease with time is

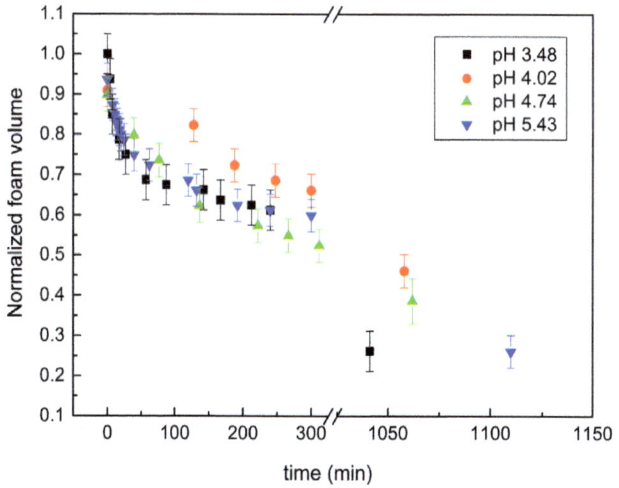

Fig. 4 Time evolution of normalized volumes of foams at different pH values. These foams were prepared using zein-patterned fumed hydrophilic silica particles at 2.7 wt/vol% and ZSSR = 1.3 mg m^{-2}.

slower at pH values close to the isoelectric point of the zein-coated particles (around pH 5).

Indeed, at pH 4.74 we have seen that the foam is still stable after 6 hours; after one day the foam collapsed to a constant cream volume of 62 mL. On the other hand, less stable foams collapsed and partially detached from the cylinder walls already after one day. Images of foams stabilized by heterogeneous patterned colloids at different pH are shown in Supplemental Figure 2S†.

We have observed that the ability of the composite particles to stabilize foams depends strongly on the surface coverage of the zein layer. Interestingly, there is an optimum range of zein coating coverage that results in the largest initial foam volume (Fig. 5). As estimated from ζ-potential measurements this level of 0.67 mg m^{-2} corresponds to approximately 40% patchy coverage of the particles and the level needed for maximum foam stabilization (about ~1.8 mg m^{-2}) corresponds to approximately 60–70% coverage. From the zeta potential we see that the coverage % does not increase linearly, but is almost complete by ZSSR = 13 mg m^{-2}. For a ZSSR of above 5 mg m^{-2}, however, the particle foam stabilization ability falls off down to that of the pure zein particles.

Interpreting the results from these experiments may require knowledge of the molecular packing of zein on the silica surface. The zein protein structure consists of 9 identical α-helix segments. The helices form a plate-like structure, held in place by hydrogen bonds and linked by glutamine loops. The loops form two narrow hydrophilic top and bottom faces, while the two broad helix sides are hydrophobic. Depending on the properties of the underlying substrate and the nature of the solvent, zein adsorbs with different sides facing the substrate surface, and packs differently to minimize unfavourable interactions with the solvent.

In addition to adjusting the hydrophobicity of the individual particles, other effects may be important for the stabilisation of foams. It has been shown, for

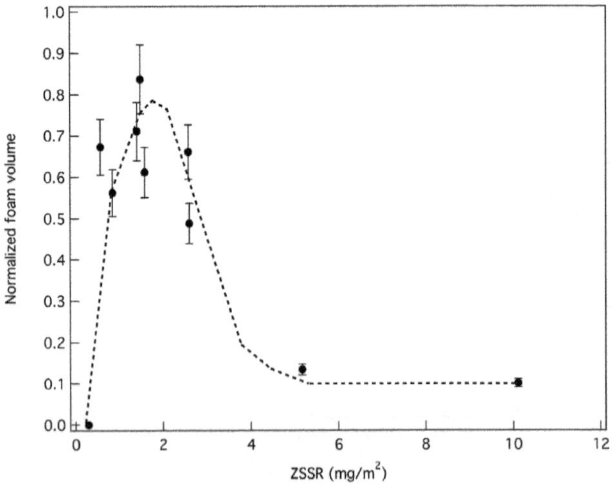

Fig. 5 Normalized foam volume as a function of zein to substrate surface ratio (ZSSR). The particle concentration is 2.7 wt/vol% and the pH ≈ 5. Maximum foam stability is observed at ZSSR ~ 1.8 mg m^{-2}.

example, that elongated, rod-like particles, due to their wood pile-like packing can maintain their initial packing even after complete drying of the foam. Similarly, super stable foams have been achieved with irregularly shaped, modified cellulose particles,[51] not only because they can entangle better with each other than spherical particle stabilizers, but also because the presence of molecular weight fractions with different solubility allows them to stick to each other upon drying of the foam, to form a monolithic foam support structure.

The presence of zein patches on the particle surface can create very heterogeneous surfaces and change the effective contact angle, which enhances the absorption at the fluid interface. Indeed, is not straightforward to have control over the hydrophobic–hydrophilic balance, but it has been found that by using a cationic surfactant (DDAB), modified silica particles present contact angles that vary between 37° (without surfactant) and 75° (highest surfactant concentration).[50] The ability of zein particles to adhere and fuse (which can cause aggregation in dispersions) further can prevent their detachment and stimulate the formation of shell like structures around the air bubble.

The observed high stability of foams with zein-coated particles is interesting from both applied and fundamental points of view. Practically, it is useful to find out that the best stabilization is achieved at pH values near the isoelectric point (pI) of the stabilizing particles, as well as to know that the optimum stabilization is achieved by patchy zein coatings with incomplete coverage, produced from a known range of ZSSR suspensions. In the case of highly irregular particles such as fumed silica particles, the precipitation of zein can take much more complex shapes. Fundamentally, it is interesting to explain the origin of the best foam stabilization with patchy particles, and explain the optimum patchiness found for zein. Notably, because the mode of adsorption of zein depends on the underlying substrate and the zein solvent properties during deposition. A simple assumption that the deposition of zein, which is generally considered more hydrophobic, will increase the particle hydrophobicity and, hence, the ability to stabilise foams is not sufficient, as it cannot explain the observed maximum in foam stabilisation.

Zeta potential measurements provide a potential link between the initial suspension ZSSR and the morphology of the zein layer on the surface of the silica particles, showing that zein completely coats the silica at ZSSR > 13 mg m^{-2}. The presence of such domains not only has implications for the effective contact angle but also for the shape of the contact line. These results suggest that the best stabilization for foams is achieved by heterogeneous patterned particles with an incomplete layer of zein. By modeling the zeta potential of the composite particles as an average of the charges of the exposed areas of pure silica (negatively charged) and pure zein (positively charged), we could find indeed that the behavior of the zeta potential curves changes from that of pure silica near ZSSR = 0 mg m^{-2} to that of pure zein near ZSSR = 13 mg m^{-2}. The best stabilization observed near ZSSR $\sim$ 1.8 mg m^{-2} corresponds to about 60–70% zein surface coverage (Table 1).

Describing the adsorption of shape-anisotropic particles on fluid interfaces is difficult.[52–55] In the case of fumed silica particles, the patchy particles have a complex fractal geometry and the derivation of a link between the energy of adsorption and the contact angle is very difficult. Also the direct determination of the contact angle of such small and irregular particles is very difficult.[56] For this

Table 1 Zeta potential as a function of ZSSR at pH 5

ZSSR (mg m^{-2})	Surface coverage x_S (%)	Zeta potential (mV)
0	0	−46
0.65	40	−10
1.3	55–60	0.4
2.6	~75	10
3.9	~75	11
13	85–95	20
∞	100	25

reason we will present only a qualitative description of the interfacial behaviour of these particles.

In order to establish the origin of the maximum in the foam stability (see Fig. 5), we need to consider the effects of surface patchiness on the energy of particle detachment and the adsorption barrier. If the patches are sufficiently small, we can assume an average contact angle θ_{av}, which we can describe using the model for wetting of patterned surfaces. Nonomura used a similar approach to describe the wettability and the adsorbed position of the particles with extremely rough surfaces using a Cassie–Baxter model to describe the wetting of the patchy surface.[53] They predicted that the adsorbed position and the interfacial energy depend on the interfacial tensions between the solid and liquid phases, the radius of the particle, and the fraction of the particle surface area that is in contact with the external liquid phase.[53] In the case of patchy particles with oppositely charged patches, the origin of the maximum can be related to the effective electroneutrality of the particles. Such electroneutrality is apparent from the electrophoretic measurements (Fig. 2). A similar enhancement of particle stabilization has been observed with mixtures of oppositely charged particles.[57]

It is noticeable that the maximum in foam stability observed by us is not where the ζ-potential is close to zero (*i.e.* the minimum electrostatic interaction). The role of electrostatic interactions is important in determining particle adsorption.[58,59] From our results it appears that low, but positive surface potential is more beneficial for the foam stabilization. It is also known from other experiments that optimal foam stability with electroneutrality is achieved at contact angles less than 90°,[60,61] suggesting that we may not expect a symmetric behavior close to the minimum contact angle, as expected from electroneutral but charged surfaces.

Other possible mechanisms for the increased foam stability may include a decreased adsorption barrier to the fluid–fluid interface due to lower surface charges and hence decreased repulsion.[59] This is directly evident from the surface potential measurements (Fig. 2). Another aspect which can be taken into account is the line tension, which in principle can increase the energy of particle detachment.[62] Although it is not possible for us to establish this contribution, a patchy surface may create an undulating contact line, which should increase the contribution from the line tension. The undulating contact line also changes the lateral capillary forces,[63–65] giving rise to an attraction which can improve the ability of the adsorbed particles to resist lateral displacement. Another important contribution to the improved stability is the hysteresis of the contact line on rough heterogeneous surfaces.[66,67] In the case of patchy particles, the contact angle hysteresis will make particle detachment and rotation on the fluid interface more difficult. The contribution of these effects on the overall foam stability, however, is difficult to estimate.

Emulsion stabilisation

Particle-stabilized emulsions were formed by mechanically shearing a mixture containing the nonpolar oil, decane, and an aqueous dispersion of patchy particles. The appearance of the emulsion at several water fractions is shown in Fig. 6. Remarkably, these novel colloids with random patches stabilize droplets with different size distributions, providing efficient packing and a steric barrier against droplet coalescence. Particles which are heterogeneous in surface charge but electroneutral have been shown to be effective in stabilising emulsions.[57]

Upon analyzing the micrographs we observe a 3D close packed droplet network. The droplet stability against coalescence is related to the particle wetting properties and adsorption at the interface. When the system is loaded with large amounts of zein, the stability and size of the droplets leads to the formation of gelled emulsions. For volume fractions of the aqueous phase $\phi_w = 0.8–0.23$, all emulsions formed were oil-in-water (O/W). For the lowest volume fraction of 0.2, formation of a water-in-oil emulsion (W/O) was observed. Fig. 7 shows the gelled emulsions stabilized by fumed silica particles at different surface coverages. The presence of a gel phase is confirmed by the stress dependence of the storage and loss moduli (ESI Fig. 3S†).

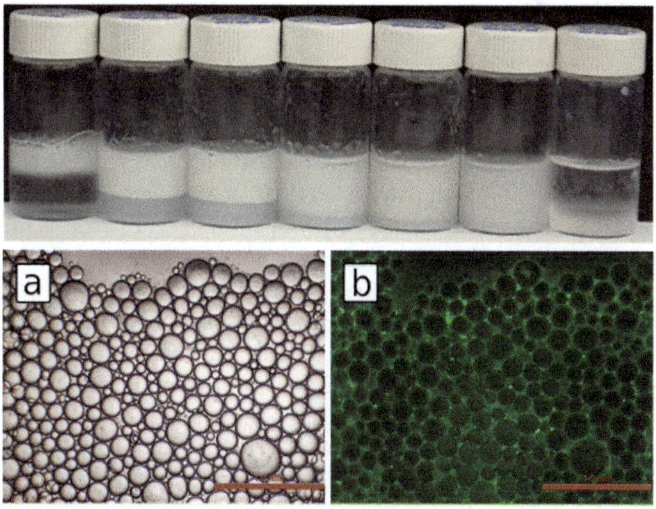

Fig. 6 (Top) Appearance of emulsions stabilized by patchy zein–silica particles (ZSSR = 3 mg m^{-2}). The values of the water fraction are (left to right) $\phi_w = 0.8, 0.5, 0.44, 0.33, 0.28, 0.23$ (all O/W) and 0.2 (W/O). (Bottom) Microscope images of an O/W gel emulsion of close packed droplets $\phi_w = 0.8$: (a) droplet network and (b) fluorescence image where the zein was dyed (FITC). Scale bar 1 mm.

Fig. 7 Gelled emulsions stabilized by zein-patterned fumed silica particles at O/W = 1, pH = 2 and several ZSSR = 0.98, 2.96, 3.7, 4.9, 6.2, 8.6 and 12.3 mg m^{-2} from left to right.

In general, gel emulsions are the result of gel formation in the continuous phase in which the droplets are trapped. Alternatively, gel emulsions are formed due to flocculation of the emulsion droplets. The mechanism of gelation of emulsions stabilized with zein–silica patchy particles is attributed to the low surface potential of the stabilizing particles. When adsorbed, they provide the same low surface potential of the droplet minus their repulsion which allows them to get in close proximity. Once in contact, the protein patches can form gels from the percolated network of the particle stabilized droplets.[68]

One advantage of the extension of patchy particles as stabilizers for high internal phase emulsions is that these particles can be irreversibly adsorbed at the interface of emulsions because of their high energy of attachment, which makes the final emulsions extremely stable, with shelf life stabilities of months.[62]

The field of particle-stabilized fluid-in-fluid dispersions will continue to develop. There is already an active search for surface patterns that allow particles to have optimal surface activity.[26] Further experimental and theoretical studies are necessary to understand the behavior of the surface-active patchy organic–inorganic particles.

Conclusions

Heterogeneous nucleation represents a new approach to fabricate organic–inorganic patchy colloids. Such protein-patterned particles present remarkably tuneable wettability. The role of zein is to modify the wettability of the hydrophilic silica particles, and thus enhance the attachment of patchy particles at the fluid interface. Those colloids, with patches on their surface, can also display amphiphilic properties. The results on the ability of silica or zein particles to stabilize foams suggest that just the pure material colloids, at any pH and ionic strength, are not very efficient foam stabilizers. However, after patterning silica particles with zein, the stabilization capacity remarkably increases. It appears that the optimal stabilization is achieved with incomplete adsorption of zein on the silica area which results in patchy particles. Further layering of the particles results in a full zein shell, which decreases the particle foam stabilization ability to the same level as for pure zein particles.

Acknowledgements

We thank J. de Folter for assistance with zein particle synthesis. This work is partly financially supported by Unilever R & D Vlaardingen, Netherlands, Food & Nutrition Delta 2 program (grant DFN0642300), NanoNextNL and US National

Science Foundation (NSF) Triangle MRSEC on Programmable SoftMatter (DMR-1121107). E. Blanco thanks the Spanish Ministerio de Educación for her post-doctoral fellowship.

Notes and references

1 K. J. Lee, J. Yoon and J. Lahann, Recent advances with anisotropic particles, *Curr. Opin. Colloid Interface Sci.*, 2011, **16**, 195.

2 S. Sacanna and D. J. Pine, Shape-anisotropic colloids: building blocks for complex assemblies, *Curr. Opin. Colloid Interface Sci.*, 2011, **16**, 96.

3 I. C. Pons-Siepermann and S. C. Glotzer, Design of patchy particles using quaternary self-assembled monolayers, *ACS Nano*, 2012, **6**, 3919.

4 G. Van Anders, N. K. Ahmed, R. Smith, M. Engel and S. C. Glotzer, Entropically patchy particles: engineering valence through shape entropy, *ACS Nano*, 2014, **8**, 931.

5 H. Bao, W. Peukert and R. N. K. Taylor, One-pot colloidal synthesis of plasmonic patchy particles, *Adv. Mater.*, 2011, **23**, 2644.

6 L. Feng, R. Dreyfus, R. Sha, N. C. Seeman and P. M. Chaikin, DNA patchy particles, *Adv. Mater.*, 2013, **25**, 2779.

7 F. Sciortino and E. Zaccarelli, Reversible gels of patchy particles, *Curr. Opin. Solid State Mater. Sci.*, 2011, **15**, 246.

8 Z. Zhang and S. C. Glotzer, Self-assembly of patchy particles, *Nano Lett.*, 2004, **4**, 1407.

9 F. Romano, J. Russo and H. Tanaka, Influence of patch-size variability on the crystallization of tetrahedral patchy particles, *Phys. Rev. Lett.*, 2014, **113**, 138303.

10 R. Guo, J. Mao, X. M. Xie and L. T. Yan, Predictive supracolloidal helices from patchy particles, *Sci. Rep.*, 2014, **4**, 7021.

11 A. J. Williamson, A. W. Wilber, J. P. K. Doye and A. A. Louis, Templated self-assembly of patchy particles, *Soft Matter*, 2011, **7**, 3423.

12 G. Singh, S. Pillai, A. Arpanaei and P. Kingshott, Highly ordered mixed protein patterns over large areas from self-assembly of binary colloids, *Adv. Mater.*, 2011, **23**, 1519.

13 J. Sun and B. Bhushan, Hierarchical structure and mechanical properties of nacre: a review, *RSC Adv.*, 2012, **2**, 7617.

14 B. S. Murray and R. Ettelaie, Foam stability: proteins and nanoparticles, *Curr. Opin. Colloid Interface Sci.*, 2004, **9**, 314.

15 A. Stocco, E. Rio, B. P. Binks and D. Langevin, Aqueous foams stabilized solely by particles, *ChemPhysChem*, 2011, **7**, 1260.

16 T. S. Horozov, Foams and foam films stabilised by solid particles, *Curr. Opin. Colloid Interface Sci.*, 2008, **13**, 134.

17 P. S. Clegg, J. W. Tavacoli and P. J. Wilde, One-step production of multiple emulsions: microfluidic, polymer-stabilized and particle-stabilized approaches, *Soft Matter*, 2016, **12**, 998.

18 B. P. Binks and A. T. Tyowua, Oil-in-oil emulsions stabilised solely by solid particles, *Soft Matter*, 2016, **12**, 876.

19 A. Kumar, B. J. Park, F. Tu and D. Lee, Amphiphilic Janus particles at fluid interfaces, *Soft Mater.*, 2013, **9**, 6604.

20 K. Stratford, R. Adhikari, I. Pagonabarraga, J. C. Desplat and M. E. Cates, Colloidal jamming at interfaces: a route to fluid-bicontinuous gets, *Science*, 2005, **309**, 2198.
21 E. M. Herzig, K. A. White, A. B. Schofield, W. C. K. Poon and P. S. Clegg, Bicontinuous emulsions stabilized solely by colloidal particles, *Nat. Mater.*, 2007, **6**, 966.
22 B. P. Binks and R. Murakami, Phase inversion of particle-stabilized materials from foams to dry water, *Nat. Mater.*, 2006, **5**, 865.
23 S. Tcholakova, N. D. Denkov and A. Lips, Comparison of solid particles, globular proteins and surfactants as emulsifiers, *Phys. Chem. Chem. Phys.*, 2008, **10**, 1608.
24 K. P. Velikov and O. D. Velev, *Stabilization of Thin Films, Foams, Emulsions and Bifluid Gels with Surface-Active Solid Particles*, Wiley-VCH Verlag GmbH & Co. KGaA, 2011, pp. 277–306.
25 A. Perro, S. Reculusa, S. Ravaine, E. B. Bourgeat-Lami and E. Duguet, Design and synthesis of Janus micro- and nanoparticles, *J. Mater. Chem.*, 2005, **15**, 3745.
26 H. Rezvantalab and S. Shojaei-Zadeh, Designing patchy particles for optimum interfacial activity, *Phys. Chem. Chem. Phys.*, 2014, **16**, 8283.
27 R. G. Alargova, D. S. Warhadpande, V. N. Paunov and O. D. Velev, Foam superstabilization by polymer microrods, *Langmuir*, 2004, **20**, 10371.
28 S. Lam, E. Blanco, S. K. Smoukov, K. P. Velikov and O. D. Velev, Magnetically Responsive Pickering Foams, *J. Am. Chem. Soc.*, 2011, **133**, 13856.
29 J. Tang, P. J. Quinlan and K. C. Tam, Stimuli-responsive Pickering emulsions: recent advances and potential applications, *Soft Matter*, 2015, **11**, 3512.
30 B. P. Binks, M. Kirkland and J. A. Rodrigues, Origin of stabilisation of aqueous foams in nanoparticle-surfactant mixtures, *Soft Matter*, 2008, **4**, 2373.
31 M. Destribats, S. Gineste, E. Laurichesse, H. Tanner, F. Leal-Calderon, V. Héroguez and V. Schmitt, Pickering emulsions: What are the main parameters determining the emulsion type and interfacial properties?, *Langmuir*, 2014, **30**, 9313.
32 P. M. Kruglyakov, S. I. Elaneva and N. G. Vilkova, About mechanism of foam stabilization by solid particles, *Adv. Colloid Interface Sci.*, 2011, **165**, 108.
33 K. P. Velikov, F. Durst and O. D. Velev, Direct observation of the dynamics of latex particles confined inside thinning water–air films, *Langmuir*, 1998, **14**, 1148.
34 A. Maestro, E. Santini, D. Zabiegaj, S. Llamas, F. Ravera, L. Liggieri, F. Ortega, R. G. Rubio and E. Guzman, Particle and particle–surfactant mixtures at fluid interfaces: assembly, morphology, and rheological description, *Adv. Condens. Matter Phys.*, 2015, **2015**, 917516.
35 A. J. Mendoza, E. Guzmán, F. Martínez-Pedrero, H. Ritacco, R. G. Rubio, F. Ortega, V. M. Starov and R. Miller, Particle laden fluid interfaces: dynamics and interfacial rheology, *Adv. Colloid Interface Sci.*, 2014, **206**, 303.
36 D. J. French, P. Taylor, J. Fowler and P. S. Clegg, Making and breaking bridges in a Pickering emulsion, *J. Colloid Interface Sci.*, 2015, **441**, 30.
37 S. Lam, K. P. Velikov and O. D. Velev, Pickering stabilization of foams and emulsions with particles of biological origin, *Curr. Opin. Colloid Interface Sci.*, 2014, **19**, 490.

38 E. Dickinson, Food emulsions and foams: stabilization by particles, *Curr. Opin. Colloid Interface Sci.*, 2010, **15**, 40.

39 A. R. Patel, C. Remijn, A. I. M. Cabero, P. C. M. Heussen, J. W. M. Seijen ten Hoorn and K. P. Velikov, Novel all-natural microcapsules from gelatin and shellac for biorelated applications, *Adv. Funct. Mater.*, 2013, **23**, 4710.

40 C. C. Berton-Carabin and K. Schroën, Pickering emulsions for food applications: background, trends, and challenges, *Annu. Rev. Food Sci. Technol.*, 2015, **6**, 263.

41 D. Rousseau, Trends in structuring edible emulsions with Pickering fat crystals, *Curr. Opin. Colloid Interface Sci.*, 2013, **18**, 283.

42 J. W. Lawton, Zein: a history of processing and use, *Cereal Chem.*, 2002, **79**, 1.

43 R. Shukla and M. Cheryan, Zein: the industrial protein from corn, *Ind. Crops Prod.*, 2001, **13**, 171.

44 J. W. J. de Folter, M. W. M. van Ruijven and K. P. Velikov, Oil-in-water Pickering emulsions stabilized by colloidal particles from the water-insoluble protein zein, *Soft Matter*, 2012, **8**, 6807.

45 E. Filippidi, A. R. Patel, E. C. M. Bouwens, P. Voudouris and K. P. Velikov, All-Natural Oil-Filled Microcapsules from Water-Insoluble Proteins, *Adv. Funct. Mater.*, 2016, **24**, 5962.

46 F. Dong, G. W. Padua and Y. Wang, Controlled formation of hydrophobic surfaces by self-assembly of an amphiphilic natural protein from aqueous solutions, *Soft Matter*, 2013, **9**, 5933.

47 S. Kim and J. Xu, Aggregate formation of zein and its structural inversion in aqueous ethanol, *J. Cereal Sci.*, 2008, **47**, 1.

48 R. J. Baars, Y. M. van Leeuwen, Y. Hendrix, K. P. Velikov, W. K. Kegel and A. P. Philipse, Morphology-controlled functional colloids by heterocoagulation of zein and nanoparticles, *Colloids Surf., A*, 2015, **483**, 209.

49 Q. Wang, L. L. Yin and G. W. Padua, Effect of hydrophilic and lipophilic compounds on zein microstructures, *Food Biophys.*, 2008, **3**, 174.

50 S. Jiang and S. Granick, Controlling the geometry (Janus balance) of amphiphilic colloidal particles, *Langmuir*, 2008, **24**, 2438.

51 H. A. Wege, S. Kim, V. N. Paunov, Q. X. Zhong and O. D. Velev, Long-term stabilization of foams and emulsions with *in situ* formed microparticles from hydrophobic cellulose, *Langmuir*, 2008, **24**, 9245.

52 Y. Nonomura, S. Komura and K. Tsujii, Adsorption of microstructured particles at liquid–liquid interfaces, *J. Phys. Chem. B*, 2006, **110**, 13124.

53 Y. Nonomura and S. Komura, Surface activity of solid particles with extremely rough surfaces, *J. Colloid Interface Sci.*, 2008, **317**, 501.

54 R. W. Style, L. Isa and E. R. Dufresne, Adsorption of soft particles at fluid interfaces, *Soft Matter*, 2015, **11**, 7412.

55 W. Huang, Q. Lan and Y. Zhang, Colloid particles adsorption and interfacial assembly at the fluids interface, *Progr. Chem.*, 2007, **19**, 212.

56 A. Maestro, E. Guzmán, F. Ortega and R. G. Rubio, Contact angle of micro- and nanoparticles at fluid interfaces, *Curr. Opin. Colloid Interface Sci.*, 2014, **19**, 355.

57 T. Nallamilli, E. Mani and M. G. Basavaraj, A model for the prediction of droplet size in Pickering emulsions stabilized by oppositely charged particles, *Langmuir*, 2014, **30**, 9336.

58 H. Al-Shehri, T. S. Horozov and V. N. Paunov, Adsorption of carboxylic modified latex particles at liquid interfaces studied by the gel trapping technique, *Soft Matter*, 2014, **10**, 6433.
59 V. R. Dugyala, J. S. Muthukuru, E. Mani and M. G. Basavaraj, Role of electrostatic interactions in the adsorption kinetics of nanoparticles at fluid–fluid interfaces, *Phys. Chem. Chem. Phys.*, 2016, **18**, 5499.
60 G. Johansson and R. J. Pugh, The influence of particle size and hydrophobicity on the stability of mineralized froths, *Int. J. Miner. Process.*, 1992, **34**, 1.
61 G. Kaptay and N. Babcsán, *Foam Engineering: Fundamentals and Applications*, Wiley, 2012, ch. 7, pp. 121–143.
62 R. Aveyard, J. H. Clint and T. S. Horozov, Aspects of the stabilisation of emulsions by solid particles: effects of line tension and monolayer curvature energy, *Phys. Chem. Chem. Phys.*, 2003, **5**, 2398.
63 P. A. Kralchevsky, N. D. Denkov and K. D. Danov, Particles with an undulated contact line at a fluid interface: interaction between capillary quadrupoles and rheology of particulate monolayers, *Langmuir*, 2001, **17**, 7694.
64 P. A. Kralchevsky, N. D. Denkov, V. N. Paunov, O. D. Velev, I. B. Ivanov, H. Yoshimura and K. Nagayama, Formation of 2-Dimensional Colloid Crystals in Liquid-Films Under the Action of CapillaryForces, *J. Phys.: Condens. Matter*, 1994, **6**, A395–A402.
65 O. D. Velev, N. D. Denkov, V. N. Paunov, P. A. Kralchevsky and K. Nagayama, Capillary Image Forces .2. Experiment, *J. Colloid Interface Sci.*, 1994, **167**, 66.
66 A. San-Miguel and S. H. Behrens, Influence of nanoscale particle roughness on the stability of Pickering emulsions, *Langmuir*, 2012, **28**, 12038.
67 L. W. Schwartz and S. Garoff, Contact Angle hysteresis on heterogeneous surfaces, *Langmuir*, 1985, **1**, 219.
68 M. N. Lee, H. K. Chan and A. Mohraz, Characteristics of Pickering emulsion gels formed by droplet bridging, *Langmuir*, 2012, **28**, 3085.

Faraday Discussions

PAPER

Programmed assembly of oppositely charged homogeneously decorated and Janus particles†

Alina Kirillova,[ab] Georgi Stoychev[ab] and Alla Synytska*[ab]

Received 29th January 2016, Accepted 11th March 2016
DOI: 10.1039/c6fd00008h

The exploitation of colloidal building blocks with morphological and functional anisotropy facilitates the generation of complex structures with unique properties, which are not exhibited by isotropic particle assemblies. Herein, we demonstrate an easy and scalable bottom-up approach for the programmed assembly of hairy oppositely charged homogeneously decorated and Janus particles based on electrostatic interactions mediated by polyelectrolytes grafted onto their surface. Two different assembly routes are proposed depending on the target structures: raspberry-like/half-raspberry-like or dumbbell-like micro-clusters. Ultimately, stable symmetric and asymmetric micro-structures could be obtained in a well-controlled manner for the homogeneous–homogeneous and homogeneous–Janus particle assemblies, respectively. The spatially separated functionalities of the asymmetric Janus particle-based micro-clusters allow their further assembly into complex hierarchical constructs, which may potentially lead to the design of materials with tailored plasmonics and optical properties.

1. Introduction

Self-assembly is a process where individual building blocks form higher-ordered, organized structures as a consequence of specific, local interactions among the components themselves, in the absence of external guidance.[1] Colloidal particles, which can be synthesized with unprecedented control over their geometry and interactions, may be considered as programmed building blocks for the rational design of functional materials, analogous to atoms being the building blocks for molecules, macromolecules, and crystals.[2] Therefore, colloidal self-assembly represents not only an important model system to study the atomic world,[3] but also a powerful tool for the development of materials with tunable and multi-functional properties.[4–10] Colloidal structures prepared through self-assembly are

[a]*Leibniz Institute of Polymer Research Dresden, Hohe Str. 6, 01069 Dresden, Germany. E-mail: synytska@ipfdd.de*
[b]*Technische Universität Dresden, Fakultät Mathematik und Naturwissenschaften, 01062 Dresden, Germany*
† Electronic supplementary information (ESI) available. See DOI: 10.1039/c6fd00008h

interesting themselves from the theoretical point of view,[11–15] and additionally have a broad variety of potential applications in, for example, photonic/plasmonic devices, photovoltaics, nanoscale electronics, high efficiency energy-conversion/energy-storage, miniature diagnostic systems, drug/gene delivery, and hierarchically structured catalysts.[3,16–18]

The self-assembly process of colloidal particles can be governed by different forces including van der Waals, electrostatic, hydrogen bonding, covalent, coordination, capillary, convective, shear, optical, and electromagnetic forces.[3,19,20] In particular, electrostatic interactions between the particles can be utilized to direct the formation of specially assembled structures.[21–24] The strength of the electrostatic interactions can be tuned continuously by screening the charges and by changing the properties of the medium,[25] and therefore, such interactions can be used to fabricate responsive functional materials.

The aforementioned examples concern the utilization of isotropic colloidal building blocks and their assembly into symmetrical structures. However, the addition of asymmetry at the building block level can facilitate the assembly of such anisotropic building blocks into structures with unique properties, which are not exhibited by isotropic particle assemblies.[11,26–32] For instance, it was suggested that patchy colloids can be used to assemble a diamond lattice of colloidal particles, which provides a larger photonic band gap than many other structures, and thus may be of great use for optoelectronic devices or display applications.[13] Nonetheless, the programmed fabrication of asymmetric structures remains a challenge. Difficulties arise particularly when trying to achieve both spatial and chemical control over the assembly process, which can result in a limited yield, and thus limited applications of the prepared nanomaterials. One direction worth exploring in this context is the creation of a specially designed microscale platform with well-defined spatially separated functionalizations for further use in generating asymmetric assembled structures.

Janus particles (named after the two-faced Roman god) represent a class of multifunctional particles, comprising two different functions on their opposite sides. Their unique anisotropic properties and resemblance to molecular amphiphiles, such as surfactants, phospholipids, and block copolymers, have attracted significant attention in recent years.[33] Due to their multifunctional character, Janus particles have found their applications in a variety of research fields, including stabilization of emulsions,[34–36] chemical catalysis,[37,38] drug delivery,[39] display technology,[40,41] *etc.* Progress has also been made in the area of Janus particle assembly by taking advantage of their asymmetric structure.[34,42–47] However, Janus–Janus particle assembly is typically the focus of the investigations, leaving out the potentially very interesting combination of homogeneous–Janus particles. For instance, a DNA-directed approach was reported for the assembly of asymmetric nanoclusters on a prefabricated Janus particle platform, yielding nanostructures with tunable optical properties.[27] Moreover, the assembly of a binary colloidal system further extends the possibilities to construct complex structures, as colloidal clusters of identical spheres possess limited flexibility in that sense. It was suggested that it is almost impossible to assemble novel colloidal crystals using colloidal clusters of all identical spheres as building blocks.[31,48]

Herein, we report an easy and scalable bottom-up approach for the programmed assembly of binary mixtures of oppositely charged colloidal particles with hairy polymer shells into specially designed target structures: raspberry-like and dumbbell-like micro-clusters. Two different kinds of assemblies are explored in this context: homogeneous–homogeneous particles, and homogeneous–Janus particles, both of which are based on electrostatic interactions mediated by polyelectrolytes on the surface of the particles. The homogeneous particles serve as a platform for generating symmetric structures, while the Janus particles take part in generating asymmetric structures. The obtained results represent a generalized assembly route for the engineering of tunable micro-cluster architectures from the bottom up, which can be further treated as "colloidal molecules" to assemble even more complex hierarchical materials.

2. Experimental section

Materials

Tetraethylorthosilicate (TEOS, Fluka, 99%), ammonium hydroxide (NH_4OH, Acros Organics, 28–30% solution), ethanol abs. (EtOH, VWR, 99.9%), 3-aminopropyltriethoxysilane (APS, ABCR, 97%), α-bromoisobutyryl bromide (Aldrich, 98%), α-bromoisobutyric acid (Aldrich, 98%), anhydrous dichloromethane (Acros Organics, 99.8%), triethylamine (99%, Sigma-Aldrich), fluorescein *o*-acrylate (Aldrich, 97%), rhodamine B isothiocyanate (mixed isomers, BioReagent, Aldrich), copper(II) bromide (Aldrich, 99.999%), tin(II) 2-ethylhexanoate (Aldrich, 95%), tris(2-pyridylmethyl)amine (TPMA, Aldrich, 98%), *N*,*N*,*N*′,*N*′′,*N*′′-pentamethyldiethylenetriamine (PMDTA, Aldrich, 99%), anhydrous *N*,*N*-dimethylformamide (DMF, Sigma-Aldrich, 99.8%), ethyl α-bromoisobutyrate (EBiB, Aldrich, 98%), toluene (Sigma-Aldrich, 99.8%), chloroform (Sigma-Aldrich, 99.5%), hydrochloric acid (Sigma, 36.5–38.0%), methanesulfonic acid (Sigma-Aldrich, 99.5%), sodium hydroxide (pellets, Sigma-Aldrich, 97%), diethyl ether (Aldrich, 99.7%), paraffin wax (mp 53–57 °C, Aldrich), *N*-(3-dimethylaminopropyl)-*N*′-ethylcarbodiimide hydrochloride (EDC, Sigma-Aldrich), *N*-hydroxysuccinimide (NHS, Aldrich, 98%), dichloromethane (Acros Organics, 99.99%), hexane (Sigma-Aldrich, 95%), and carboxy terminated poly(lauryl methacrylate) (PLMA, M_n: 11 000 g mol^{-1}; Polymer Source) were used as received. *tert*-Butyl acrylate (*t*BA, Aldrich, 98%) and 2-(dimethylamino)ethyl methacrylate (DMAEMA, Aldrich, 98%) were passed through basic, neutral, and acidic aluminum oxides prior to the polymerization. Millipore water was obtained from Milli-Q (Millipore, conductivity: 0.055 μS cm^{-1}).

Scanning electron microscopy (SEM)

All scanning electron microscopy (SEM) images were acquired using a NEON 40 EsB CrossBeam scanning electron microscope from Carl Zeiss NTS GmbH, operating at 3 keV in the secondary electron mode. In order to enhance the electron density contrast, samples were coated with platinum (3.5 nm) using a Leica EM SCD500 sputter coater.

Transmission electron microscopy (TEM) and cryo-TEM

Transmission electron microscopy (TEM) and cryo-TEM images were taken with a Libra 120 TEM from Carl Zeiss NTS GmbH equipped with a LaB_6 source. The acceleration voltage was 120 kV, and an energy filter with an energy window of

15 eV was used. Samples for TEM were prepared by immersing a TEM grid into the dispersion with the respective particles for 20 s and removing excess liquid afterwards with filter paper. Gold grids with a carbon film (300 mesh, CF300-Au-50) were used for the analysis (Electron Microscopy Sciences, USA). The PAA-100 and PDMAEMA/NH_2-JP-200 samples for cryo-TEM were prepared as follows: particles were dispersed in water (0.5 mg ml^{-1}) by sonication for 20 min; the dispersion was adjusted to the desired pH value. Prior to the analysis, 3.5 μl of the sample were taken, blotted and vitrified in liquid ethane at −178 °C. Ultimately, an approximately 200 nm thick ice film was examined using TEM.

Dynamic light scattering (DLS) measurements

A Zetasizer Nano ZS (Malvern Instruments, UK) was used for the determination of the particle size (hydrodynamic diameter) using plastic cells for aqueous suspensions. The device is equipped with a 633 nm laser and with non-invasive backscatter (NIBS) technology to increase the particle size sensitivity.

Electrokinetic measurements

The pH-dependent electrokinetic (zeta potential) measurements of the particles in suspension were carried out using a Zetasizer Nano ZS (Malvern Instruments, UK) and an MPT-2 autotitrator. For all the measurements, the particles were suspended in a 10^{-3} M KCl solution in water. The pH of the prepared suspensions was controlled by adding 0.1 M KOH or HCl aqueous solutions. Three measurements were recorded for each sample at each pH value.

Fluorescence microscopy

Fluorescence microscopy images were obtained using an Axio Imager.A1m microscope with a 100× oil immersion objective (Carl Zeiss Microscopy GmbH, Germany) equipped with a mercury lamp. For data acquisition, a standard FITC (exciter: D480/30x; dichroic: 505DCLP; emitter: D535/40m; Chroma Technology Corp., USA), or TRITC (exciter: D540/25x; dichroic: 565DCLP; emitter: D605/55m; Chroma Technology Corp., USA) filter set in conjunction with a Photometrics Cascade II: 512 camera (Visitron Systems GmbH, Germany) and a MetaMorph imaging system (Universal Imaging, USA) was used.

Thermogravimetric analysis (TGA) and gel permeation chromatography (GPC)

Thermogravimetric analysis was performed to measure the polymer layer thickness on the particle surface. All measurements were conducted in the air atmosphere on a TGA Q 5000IR analyzer (TA Instruments, USA). The molecular weight of the bulk polymers obtained after precipitation was determined using GPC (Gradient HPLC HP Series 1100, Agilent Technologies Inc., USA). The thickness of the grafted layer on the SiO_2 particles as well as the grafting density of the attached polymer chains were determined by the equations described elsewhere.[49] The grafting density of the polymer chains on the particle surface was 0.2–0.3 chains per nm^2.

Synthesis of homogeneously decorated hybrid core–shell particles

Synthesis and pre-modification of monodisperse SiO_2 particles. 100–1000 nm silica particles were synthesized using a multistep hydrolysis–condensation procedure of TEOS in an ammonia hydroxide–ethanol solution based on the Stöber approach,[50] and described in ref. 51. In brief, TEOS was added sequentially into a mixture of ethanol and ammonia solution. The particles produced within one step of the synthesis were used as seeds for the next step. Each reaction was carried out by stirring the mixture at 500 rpm overnight at room temperature (starting from the last addition of TEOS). Subsequently, the dispersion with particles of the desired size was separated from the solvent by centrifugation, yielding monodisperse silica spheres. Purified particles were dried in a vacuum oven at 60 °C and then modified with (3-aminopropyl)triethoxysilane (APS) to introduce amino groups onto the surface. This was achieved by stirring the particles for 12 hours in a 5% APS solution in ethanol. The particles were then purified by repeated washing and centrifugation cycles in ethanol, and dried. Afterwards, the ATRP-initiator (α-bromoisobutyryl bromide) was immobilized onto the surface of the dried amino-modified particles. For this purpose, the particles were dispersed in 35 ml of anhydrous dichloromethane, followed by the addition of 1.4 ml of triethylamine and 0.7 ml of α-bromoisobutyryl bromide. The reaction was carried out at room temperature under constant stirring for 2 hours. The modified particles were purified by repeated washing and centrifugation cycles in ethanol, and dried under reduced pressure at 60 °C.

Grafting of PDMAEMA and P*t*BA using surface-initiated ATRP. Poly(2-dimethylaminoethyl methacrylate) (PDMAEMA) was grafted from the 1 μm initiator-modified particles as follows: 3 ml of anhydrous DMF, 48 μl of $CuBr_2$ (0.1 M solution in DMF), 48 μl of PMDTA (0.5 M solution in DMF), 0.15 μl of EBiB, and 3 ml of DMAEMA were added to a test tube containing the initiator-modified silica particles (500 mg) and fluorescein *o*-acrylate (40 mg). The test tube was sealed with a rubber septum and purged with argon. 150 μl of Sn(II) 2-ethylhexanoate were injected. The polymerization was carried out under continuous stirring at 70 °C in a water bath for 2 hours. Particles with the grafted polymer were washed by centrifugation in DMF and ethanol 8 times, and dried under reduced pressure at 60 °C.

A similar procedure was used for the grafting of P*t*BA from the 100, 200, 450, and 600 nm initiator-modified silica particles. Rhodamine B isothiocyanate was added during the preparation of these particles in order to label the resulting PAA-covered particles with a red fluorescent dye. For the polymerization, 7.5 mg of TPMA dissolved in 3 ml of *t*BA, 50 μl of $CuBr_2$ (0.1 M solution in DMF), and 0.15 μl of EBiB were added to the particles. The mixture was sonicated and purged with Ar, followed by the injection of 200 μl of Sn(II) 2-ethylhexanoate. The polymerization was performed under continuous stirring at 115 °C for 2 hours. After the particles were purified by centrifugation and redispersion cycles in toluene and dried, hydrolysis was performed to yield polyacrylic acid (PAA). Briefly, P*t*BA-covered particles were suspended in 5 ml of chloroform in a Teflon centrifuge vial, along with 1 ml of methanesulfonic acid, and the mixture was rapidly stirred for 5 minutes. The modified particles were collected by diluting the mixture with diethyl ether, purified by centrifugation and dried.

Synthesis of hybrid core–shell Janus particles

Preparation of colloidosomes and grafting of the first polymer. Colloidosomes with 1 μm large APS-modified silica spheres were prepared using a wax–water Pickering emulsion approach as described elsewhere.[49,51] The ATRP-initiator (α-bromoisobutyric acid) was then immobilized onto the exposed particle surface.[52] Subsequently, wax was dissolved in hexane, and partly initiator-covered particles were used for polymerization. PDMAEMA was grafted from the modified particle surface in the same manner as described above for the homogeneously decorated particles. Fluorescein *o*-acrylate (2 wt%) was added to the reaction mixture to label the grafted polymer.

Grafting of the second polymer. The "grafting to" approach was utilized to graft the second polymer onto the particles modified in the previous step (PDMAEMA/NH_2-JP). For this purpose, silica particles with the grafted first polymer were dispersed in 20 ml of a 1 wt% carboxy terminated poly(lauryl methacrylate) (PLMA) solution in chloroform, and stirred for 2 hours. Next, the solvent was evaporated and the particles were annealed at 150 °C overnight. The ungrafted polymer chains were removed by multiple redispersion cycles of the particles in appropriate solvents and subsequent centrifugation. As a result, bi-component PDMAEMA/PLMA-JP-1 μm particles were obtained.

Self-assembly experiments

Self-assembly experiments were carried out by mixing different ratios of PAA-covered particles with either PDMAEMA homogeneously decorated particles or PDMAEMA/PLMA Janus particles depending on the target structures: raspberry-like or dumbbell-like micro-clusters. Mass ratios of small-to-big particles used for the preparation of raspberry-like micro-clusters were estimated based on the ratio of the big particle surface area to the area of the small particle projections on the surface of a big particle. The ratios were then varied in order to experimentally obtain the dumbbell-like micro-clusters.

For a typical experiment, 1 mg ml^{-1} dispersions of particles of all sizes were prepared in DI water; the pH was adjusted to 7 in each of them. All the dispersions were kept for an hour in an ultrasonic bath. Then the particles were mixed in the appropriate ratios by taking aliquots from the initial 1 mg ml^{-1} dispersions depending on the target structures. For the raspberry-like micro-clusters, the dispersion with central PDMAEMA homogeneously decorated particles (HP) or PDMAEMA/PLMA Janus particles (JP) was slowly added dropwise to the dispersion of PAA-covered surrounding particles under continuous stirring. *Vice versa*, for the dumbbell-like micro-clusters, the dispersion of PAA-covered particles was slowly added dropwise to the dispersion of PDMAEMA-HP or PDMAEMA/PLMA-JP. The resulting mixtures were sonicated for 10 minutes, and then stirred at 700 rpm for two hours. Samples for SEM imaging were taken at this point. In order to obtain individual micro-clusters, the mixtures were diluted 5 times, sonicated for 10 more minutes and left to settle overnight. Samples for SEM imaging were then taken from the upper layer of the dispersion. A complete overview of the prepared samples using different small-to-big particle ratios is displayed in Table 1. At least five different dispersions were prepared for each of the particle sizes and mass ratios, and analysed by SEM in order to assess the reproducibility of the results.

Table 1 Overview of the particle mixtures prepared for the self-assembly experiments

PAA-covered particle size, nm	Mass ratio PAA : PDMAEMA-HP				Mass ratio PAA : PDMAEMA/PLMA-JP	
100	2 : 1	1 : 1	1 : 2	1 : 20	2 : 1	1 : 20
200	2 : 1	1 : 1	1 : 10	1 : 100	2 : 1	1 : 10
450	4 : 1	2 : 1	1 : 5	1 : 50	4 : 1	1 : 5
600	6 : 1	3 : 1	1 : 3.3	1 : 33	6 : 1	1 : 5

3. Results and discussion

Synthesis and characterization of the core–shell homogeneous and Janus particles

Oppositely charged core–shell particles homogeneously decorated with either negatively charged poly(acrylic acid) (PAA), or positively charged poly(2-dimethylaminoethyl methacrylate) (PDMAEMA) were fabricated by grafting the respective polymers from a silica core. The diameter of the PDMAEMA-decorated particles was 1 μm, while the diameters of the PAA-decorated particles varied from 100 to 600 nm (Table 2). Partly positively charged/amphiphilic 1 μm hybrid hairy Janus particles were synthesized through a combination of the "grafting from" and "grafting to" approaches,[49,51,52] and comprised a silica core with PDMAEMA and poly(lauryl methacrylate) (PLMA) polymer shells on the opposite sides of the core.

The fabricated particles were further characterized using SEM and TEM. TEM images revealed the core–shell structure of the PAA and PDMAEMA homogeneously decorated particles as well as the Janus particles (Fig. S1†). A complete overview of the particle samples with their exact diameters obtained from the SEM images as well as the thicknesses of the grafted polymer shells obtained from the TEM images is displayed in Table 2. Additionally, SEM images of the fabricated Janus particles revealed their Janus character (Fig. S2†). The estimated Janus ratio of the particles is 2 : 1 (PDMAEMA : PLMA).

The zeta potential of the polymer-decorated particles as a function of pH was determined by measuring their electrophoretic mobility (Fig. S3†). The surface charge on the particles functionalized with acidic (in the case of PAA) or basic

Table 2 List of the synthesized core–shell homogeneous and Janus particles, their diameters (*D*), and polymer shell thicknesses (*H*)

Sample ID	D^{core} (SEM), nm	$D^{core+shell}$ (SEM), nm	*H* (TEM), nm
PAA-100	100 ± 10	145 ± 10	9 ± 2
PAA-200	216 ± 12	244 ± 7	10 ± 2
PAA-450	460 ± 15	491 ± 5	14 ± 3
PAA-600	570 ± 22	590 ± 14	12 ± 3
PDMAEMA-1 μm	920 ± 20	950 ± 20	17 ± 5
PDMAEMA/PLMA-JP-1 μm	920 ± 20	945 ± 24	8 ± 2 (PDMAEMA)

(in the case of PDMAEMA) surface groups is highly dependent on the pH of the dispersion. An increase in the negative zeta potential with increasing pH in the case of the PAA-decorated particles is due to the increased dissociation of acidic surface groups (Fig. S3,† red circles); the isoelectric point (IEP) of these particles is at pH < 2. In the case of the basic amino groups on the PDMAEMA-decorated particles, the number of positively charged groups increases with decreasing pH due to their protonation (Fig. S3,† black circles); the IEP of these particles is at pH 9.9. When the PDMAEMA polymer on the PDMAEMA-decorated particles is quaternized (through the addition of iodomethane), the particles become permanently positively charged: their surface charge becomes independent from the pH value of the dispersion (Fig. S3,† green circles). No IEP is observed in this case up to pH 10. The zeta potential curve corresponding to the Janus particles decorated with positively charged hydrophilic PDMAEMA and uncharged hydrophobic PLMA (Fig. S3,† blue circles) is similar to the PDMAEMA-decorated particle curve due to the swelling of PDMAEMA. However, there is a shift in the IEP from pH 9.9 to 8.2 due to the presence of the hydrophobic uncharged PLMA (the IEP value of PLMA-functionalized particles is around pH 4). This indicates that both polymers are present on the Janus particle surface.[52]

Next, the pH-responsive swelling behaviour of the PAA and PDMAEMA polymers grafted onto the silica particles was evaluated using cryo-TEM (Fig. 1). For this purpose, particles with a 100 nm diameter homogeneously decorated with PAA (PAA-100-HP) as well as Janus particles with a 200 nm diameter partly decorated with PDMAEMA (PDMAEMA/NH_2-200-JP) were investigated under different pH values of the media. As expected, the two polymers demonstrated opposite swelling behaviour. The PAA polymer chains on the PAA-100-HP particles are in a highly swollen and stretched out state at pH 10, intermediately swollen at pH 7, and collapsed at pH 2 (Fig. 1a). On the contrary, the PDMAEMA polymer chains are highly swollen and stretched out at pH 2, intermediately swollen at pH 7, and collapsed at pH 10 (Fig. 1b). At pH 7 both of the polymers are charged and swollen at the same time, therefore, this pH value was used during the self-assembly experiments. The thicknesses of the polymer shells at different pH values were evaluated after statistical analysis of the corresponding TEM images. The quantified swelling behaviour of PAA and PDMAEMA as well as their swelling grades under different conditions are given in Table 3.

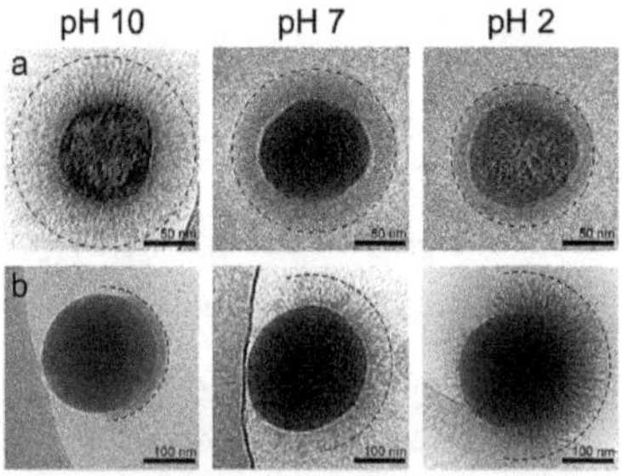

Fig. 1 pH-Responsive behaviour of the grafted polymers evaluated using cryo-TEM images of (a) homogeneous PAA-100-HP particles, and (b) Janus PDMAEMA/NH_2-200-JP particles at different pH values, revealing an opposite swelling behaviour of the PAA and PDMAEMA polymers.

Additionally, the pH-responsive properties of the PAA and PDMAEMA polymers were evaluated using DLS measurements of the corresponding particles (Fig. S4†). The same tendencies in the swelling behaviour as in the cryo-TEM measurements were observed. The hydrodynamic diameter of the 100 nm PAA-decorated particles increases with increasing pH values (Fig. S4a†), while the hydrodynamic diameter of the 1 μm PDMAEMA-decorated particles decreases with increasing pH values (Fig. S4b†).

Self-assembly experiments

We investigated the self-assembly of the synthesized colloidal homogeneous and Janus particles through electrostatic interactions between the charged polyelectrolytes on the surface of the particles. Two main approaches were pursued during the design of the self-assembly experiments depending on the target structures: raspberry-like (or half-raspberry-like in the case of the Janus particles), and dumbbell-like micro-clusters (Fig. 2). For this purpose, we fabricated three types of particles: particles of different sizes homogeneously decorated with PAA, 1 μm particles homogeneously decorated with PDMAEMA, and 1 μm Janus particles decorated with PDMAEMA and PLMA on opposite sides. Large homogeneously decorated particles served as an isotropic platform (especially in approach 1), while the Janus particles served as an anisotropic platform for selective decoration (Fig. 2). In order to demonstrate the generality of our approach, we applied a similar assembly procedure to the mixtures of particles with different size ratios.

First, we explored the self-assembly behaviour of the homogeneously decorated oppositely charged particles by varying the particle mass ratio $M_{PDMAEMA/PAA} = m_{PDMAEMA} : m_{PAA}$, and the particle size ratio $S_{PDMAEMA/PAA} = D_{PDMAEMA} : D_{PAA}$ (Fig. 3). SEM images in the obtained self-assembly diagram demonstrate a clear tendency of the particles to form raspberry-like structures at higher PAA particle concentrations. As expected, with an increase in the PAA-decorated particle diameter, the number of PAA-decorated particles on each large PDMAEMA-decorated particle decreases. With a decrease in the PAA-decorated particle concentration, a tendency to form dumbbell-like structures takes place. Using the more concentrated as-prepared samples, we could obtain networks of the micro-clusters formed during self-assembly (Fig. 3, low magnification images), whereas upon diluting the samples and leaving the mixtures to settle overnight, individual micro-clusters can be obtained (Fig. 3, high magnification images).

Table 3 pH-Responsive behaviour of the grafted PAA and PDMAEMA polymers evaluated using cryo-TEM measurements

	$H^{cryo\text{-}TEM}$, nm			
System/swelling grade	Dry state	pH 10	pH 7	pH 2
PAA-100-HP	9 ± 2	47 ± 4	27 ± 4	15 ± 2
Swelling grade PAA	—	5.0	3.0	1.7
PDMAEMA/NH_2-200-JP	10 ± 3	14 ± 3	42 ± 8	80 ± 10
Swelling grade PDMAEMA	—	1.4	4.0	8.0

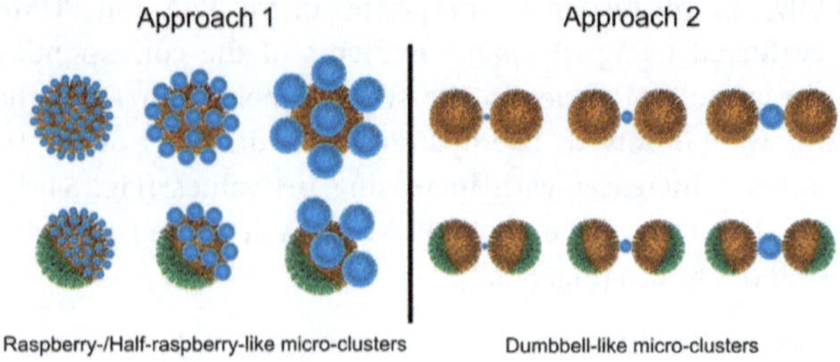

Fig. 2 Proposed schemes for the programmed assembly of homogeneously decorated and Janus core–shell particles in dispersions.

Next, we investigated the self-assembly of homogeneously decorated particles with Janus particles, again, by varying the particle mass ratio $M_{JP/PAA} = m_{JP} : m_{PAA}$, and the particle size ratio $S_{JP/PAA} = D_{JP} : D_{PAA}$ (Fig. 4). The spatially separated regions on a single Janus particle have different chemical properties and surface charges, thus enabling us to assemble asymmetric structures. As a result, half-raspberry-like micro-clusters were formed at higher PAA-decorated particle concentrations, and dumbbell-like micro-clusters were formed at higher Janus particle concentrations (Fig. 4).

Furthermore, we calculated the maximum number of small PAA-decorated particles on a single large (homogeneously decorated or Janus) particle surface. For this purpose, we used the effective diameter of the particles, which was

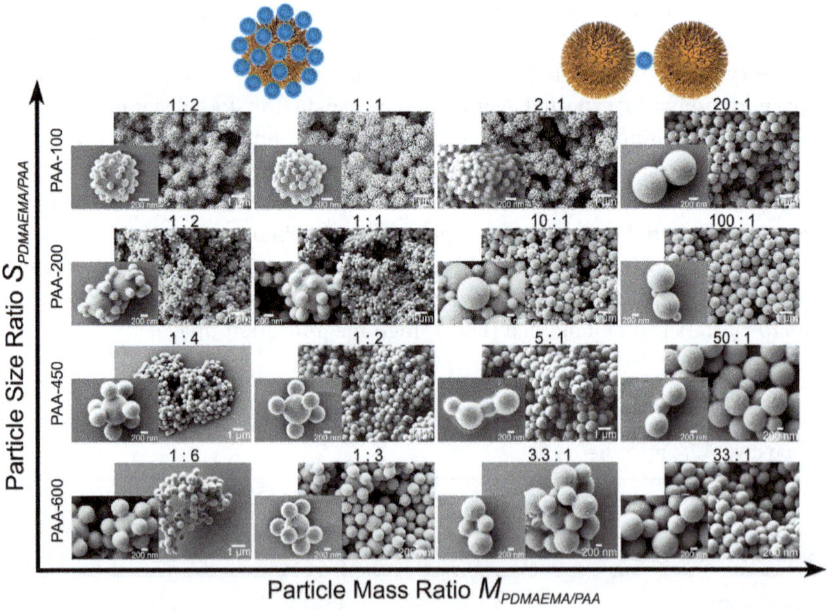

Fig. 3 Diagram with representative SEM images of the structures formed during the self-assembly of particles homogeneously decorated with PAA and PDMAEMA; the particle size ratio $S_{PDMAEMA/PAA}$, and particle mass ratio $M_{PDMAEMA/PAA}$ were varied.

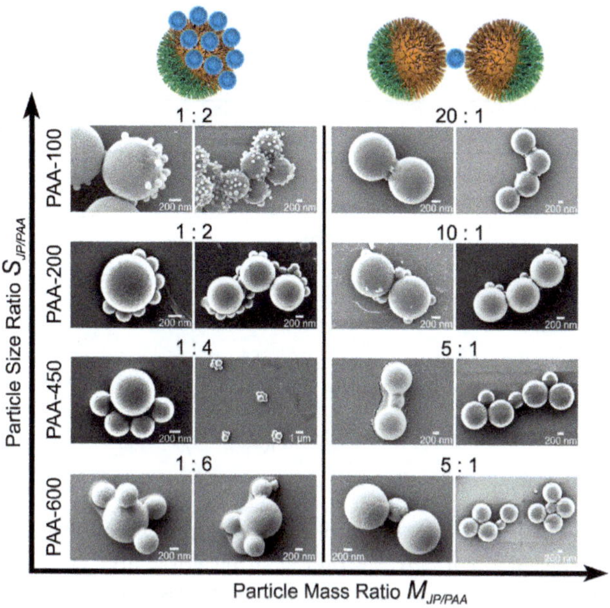

Fig. 4 Diagram with representative SEM images of the micro-clusters formed during the self-assembly of particles homogeneously decorated with PAA and PDMAEMA/PLMA-JP-1 µm; the particle size ratio $S_{JP/PAA}$, and particle mass ratio $M_{JP/PAA}$ were varied.

estimated based on the polymer shell thickness (H) in the dry state and the swelling grade of the respective polymer at pH 7. The results for all the particles are displayed in Table 4. The maximum number of small PAA-covered particles on a single large particle (N) was calculated as previously described in ref. 27, using eqn (1) for the homogeneously decorated large particle (HP), and eqn (2) for the Janus particle (JP), where 2/3 of the particle surface are covered with PDMAEMA (Table 5). Herein, R is the effective radius of the large particle (homogeneously decorated or Janus), whereas r is the effective radius of a PAA-decorated small particle.

$$N = \frac{2}{1 - \frac{\sqrt{R^2 + 2Rr}}{R + r}} \tag{1}$$

Table 4 Particle effective diameters based on the estimated polymer shell thickness in the dry state ($H^{\text{dry state}}$, TEM) and in the swollen state at pH 7 ($H^{\text{swollen state}}$, cryo-TEM)

System	D^{core}, nm	$H^{\text{dry state}}$, nm	$H^{\text{swollen state}}$, nm	$D^{\text{effective}}$, nm
PAA-100	100	9	27	154
PAA-200	216	10	30	276
PAA-450	460	14	42	544
PAA-600	570	12	36	642
PDMAEMA-1 µm	920	17	68	1056
PDMAEMA/PLMA-JP-1 µm	920	8	32	984

Table 5 Theoretical maximum number (N^{theory}) and the experimentally obtained average number ($N^{experiment}$) of the small spheres on a single 1 μm large sphere

PAA system	N^{theory} PDMAEMA-HP	N^{theory} PDMAEMA/PLMA-JP	$N^{experiment}$ PDMAEMA/PLMA-JP
PAA-100	246	145	36 ± 13
PAA-200	92	55	7 ± 2
PAA-450	36	20	4 ± 1
PAA-600	27	16	2 ± 1

$$N = \frac{4}{3\left(1 - \frac{\sqrt{R^2 + 2Rr}}{R + r}\right)} \quad (2)$$

We investigated the surface assembly degree by comparing the experimentally obtained number of small particles on the surface of a large Janus particle with the maximum theoretical number (Table 5, Fig. 5). The same trend exists in the experimental results as in the theoretical ones: with increasing the PAA-covered particle size, their number on the surface of a large Janus particle decreases (Fig. 5). However, in the experiments much lower numbers of PAA-covered particles were attached to the PDMAEMA-covered parts of the Janus particles. This can be explained by the fact that PAA-decorated particles randomly attach to the PDMAEMA-covered parts on the Janus particle surface. Unlike in a hexagonal close packing model, such random attachment of the smaller particles leads to large empty spaces between the particles. This is further facilitated by the similar negative charges present on the PAA-covered particles, which act as repulsive forces. Therefore, the particles cannot come near enough to each other to form a close packing model. Statistical analysis of the number of PAA-decorated particles on the surface of a single Janus particle is given in Fig. 6 based on five independent experiments for each particle size ratio. Depending on the

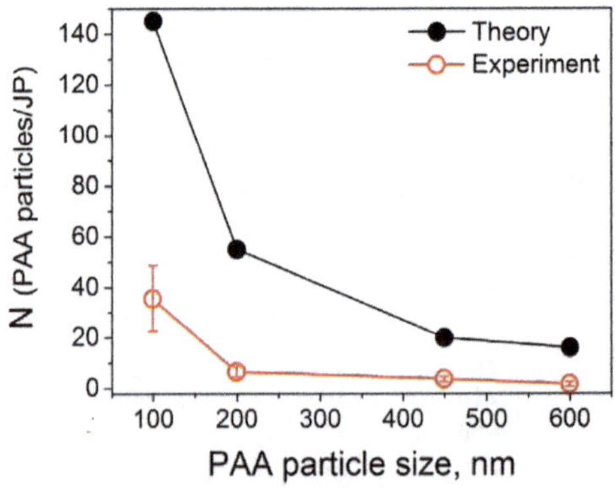

Fig. 5 Comparison of the theoretical and experimental numbers (N) of PAA-decorated particles on a single Janus particle depending on the PAA-decorated particle size.

PAA-covered particle size, different numbers of particles are preferentially attached to the Janus particle. A narrow number distribution in the experimental results indicates that the clusters are easily formed in a controlled manner as well as showing very good reproducibility of the results (Fig. 6).

In order to visualize the micro-clusters in dispersion and to exclude the "drying effect" of the SEM imaging, we stained the PAA-covered particles with a red fluorescent dye (rhodamine), and the PDMAEMA shell of the Janus particles with a green fluorescent dye (fluorescein). Three representative examples of the micro-clusters formed in dispersion from PAA-450 homogeneously decorated particles and PDMAEMA/PLMA-JP-1 μm Janus particles at a 4 : 1 ratio are displayed in Fig. 7. Using a standard TRITC filter for imaging, we can see the small PAA-decorated particles, while using a FITC filter the large Janus particles are seen. By combining the two colours, we can observe red-orange spots of PAA-covered particles surrounding the green Janus particle, and forming half-raspberry-like micro-clusters like the ones observed using SEM.

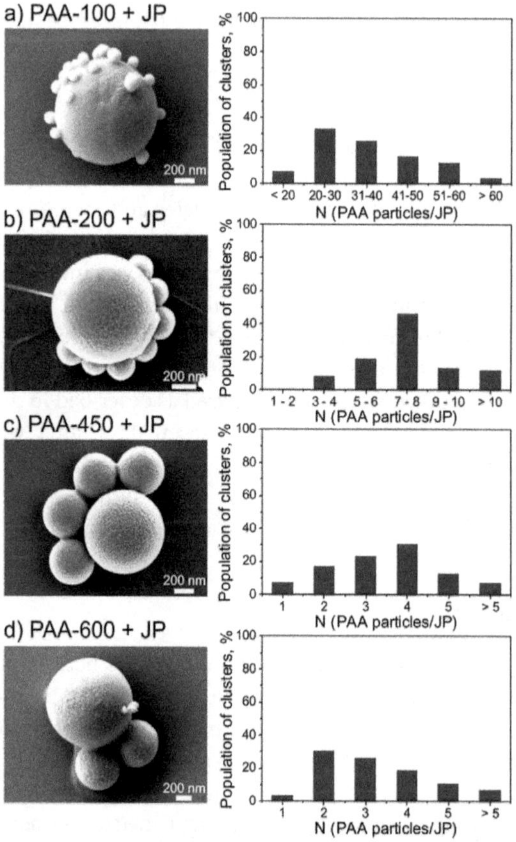

Fig. 6 Representative SEM images of the micro-clusters formed from PAA-decorated particles and PDMAEMA/PLMA-JP, and statistical analysis of the number of PAA-decorated particles on a single Janus particle depending on the PAA-decorated particle size: (a) 100 nm; (b) 200 nm; (c) 450 nm; (d) 600 nm. 200–250 clusters were counted for each particle size ratio (a–d), based on the results from five independent experiments.

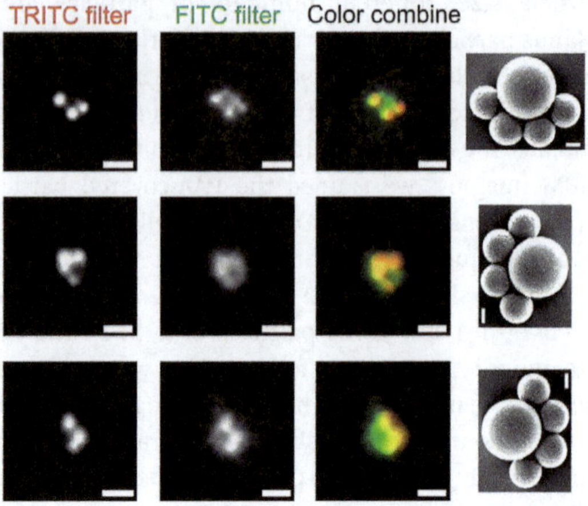

Fig. 7 Fluorescence microscopy images obtained with different filters, and with the colours combined, of the micro-clusters formed in dispersion from PAA-450 homogeneously decorated particles and PDMAEMA/PLMA-JP-1 μm Janus particles (4 : 1). PAA particles adsorb onto one hemisphere of the JPs in the same manner as in the SEM images in the dry state (on the right). Scale bars: 1 μm in the fluorescence microscopy images; 200 nm in the SEM images.

Furthermore, we used cryo-SEM in order to visualize the dumbbell-like structures formed in dispersion from PAA-450 homogeneously decorated particles and PDMAEMA/PLMA-JP-1 μm Janus particles at a 1 : 5 ratio (Fig. S5†). The same micro-clusters as in the dry state were observed in the frozen state (Fig. 4). Additionally, the stability of the assembled structures was investigated by imaging the same dispersions using SEM directly after preparation, and after one week (Fig. S6†). PAA-450 and PDMAEMA/PLMA-JP-1 μm assemblies were chosen as the representative ones for the experiment – both raspberry-like and dumbbell-like structures. The initial dispersions were prepared as described in the Experimental section, and were then left untouched for seven days, after which they were sonicated for 30 minutes. The same micro-clusters were observed in dispersions after 1 week and after additional sonication, which indicates that the structures are perfectly stable in dispersions (Fig. S6,† right panel).

4. Conclusions

Conclusively, we demonstrated an easy and adjustable bottom-up approach for the programmed assembly of hairy oppositely charged colloidal particles: homogeneously decorated particles and Janus particles. Electrostatic interactions between the particles were mediated by polyelectrolytes on their surface, which also introduced pH-responsive properties into the system. We designed two different assembly routes depending on the target structures: raspberry-like (half-raspberry-like) or dumbbell-like micro-clusters. Both homogeneous–homogeneous and homogeneous–Janus particle assemblies were explored in dispersions and in the dry state. Ultimately, stable symmetric and asymmetric micro-structures could be obtained in a well-controlled manner. Moreover, both kinds of particles (homogeneous and Janus) can be easily prepared on a large scale, thus opening more perspectives for the applications of the assembled micro-clusters. The obtained results therefore represent a generalized assembly route for the large-scale engineering of tunable micro-cluster architectures using the bottom-up approach. Furthermore, the spatially separated functionalities of the asymmetric Janus particle-based micro-clusters allow their further assembly into complex hierarchical constructs, which may potentially lead to the design of materials with tailored plasmonics and optical properties.

Acknowledgements

Deutsche Forschungsgemeinschaft (DFG, Grant SY 125/4-1) and the Leibniz Institute of Polymer Research Dresden (IPF) are acknowledged for financial support. The authors thank Dr Ivan Raguzin for the drawing of the particle 3D pictures.

References

1 S. A. van der Meulen, G. Helms and M. Dogterom, *J. Phys.: Condens. Matter*, 2015, **27**, 233101.
2 Y. Wang, Y. Wang, D. R. Breed, V. N. Manoharan, L. Feng, A. D. Hollingsworth, M. Weck and D. J. Pine, *Nature*, 2012, **491**, 51–55.
3 F. Li, D. P. Josephson and A. Stein, *Angew. Chem.*, 2011, **50**, 360–388.
4 M. H. Lash, J. C. Jordan, L. C. Blevins, M. V. Fedorchak, S. R. Little and J. J. McCarthy, *Angew. Chem.*, 2015, **54**, 5854–5858.
5 H. Y. Koo, D. K. Yi, S. J. Yoo and D. Y. Kim, *Adv. Mater.*, 2004, **16**, 274–277.
6 Y. Cheng, M. Wang, G. Borghs and H. Chen, *Langmuir*, 2011, **27**, 7884–7891.
7 C. M. Liddell and C. J. Summers, *Adv. Mater.*, 2003, **15**, 1715–1719.
8 V. Kitaev and G. A. Ozin, *Adv. Mater.*, 2003, **15**, 75–78.
9 N. Vogel, M. Retsch, C.-A. Fustin, A. del Campo and U. Jonas, *Chem. Rev.*, 2015, **115**, 6265–6311.
10 Y. Wang, Y. Wang, X. Zheng, É. Ducrot, M.-G. Lee, G.-R. Yi, M. Weck and D. J. Pine, *J. Am. Chem. Soc.*, 2015, **137**, 10760–10766.
11 D. Ortiz, K. L. Kohlstedt, T. D. Nguyen and S. C. Glotzer, *Soft Matter*, 2014, **10**, 3541–3552.
12 E. Sanz, M. E. Leunissen, A. Fortini, A. van Blaaderen and M. Dijkstra, *J. Phys. Chem. B*, 2008, **112**, 10861–10872.
13 Z. Zhang, A. S. Keys, T. Chen and S. C. Glotzer, *Langmuir*, 2005, **21**, 11547–11551.
14 Z. Zhang and S. C. Glotzer, *Nano Lett.*, 2004, **4**, 1407–1413.
15 S. Ni, J. Leemann, H. Wolf and L. Isa, *Faraday Discuss.*, 2015, **181**, 225–242.
16 M. Karg, T. A. F. König, M. Retsch, C. Stelling, P. M. Reichstein, T. Honold, M. Thelakkat and A. Fery, *Mater. Today*, 2015, **18**, 185–205.
17 G. A. Ozin, K. Hou, B. V. Lotsch, L. Cademartiri, D. P. Puzzo, F. Scotognella, A. Ghadimi and J. Thomson, *Mater. Today*, 2009, **12**, 12–23.
18 M. Tebbe, M. Mayer, B. A. Glatz, C. Hanske, P. T. Probst, M. B. Muller, M. Karg, M. Chanana, T. A. F. Konig, C. Kuttner and A. Fery, *Faraday Discuss.*, 2015, **181**, 243–260.
19 Z. Mao, H. Xu and D. Wang, *Adv. Funct. Mater.*, 2010, **20**, 1053–1074.
20 Y. Min, M. Akbulut, K. Kristiansen, Y. Golan and J. Israelachvili, *Nat. Mater.*, 2008, **7**, 527–538.

21 M. E. Leunissen, C. G. Christova, A. P. Hynninen, C. P. Royall, A. I. Campbell, A. Imhof, M. Dijkstra, R. van Roij and A. van Blaaderen, *Nature*, 2005, **437**, 235–240.
22 E. V. Shevchenko, D. V. Talapin, N. A. Kotov, S. O'Brien and C. B. Murray, *Nature*, 2006, **439**, 55–59.
23 B. Bharti, G. H. Findenegg and O. D. Velev, *Langmuir*, 2014, **30**, 6577–6587.
24 M. Rasa, A. P. Philipse and J. D. Meeldijk, *J. Colloid Interface Sci.*, 2004, **278**, 115–125.
25 E. Spruijt, H. E. Bakker, T. E. Kodger, J. Sprakel, M. A. Cohen Stuart and J. van der Gucht, *Soft Matter*, 2011, 7, 8281–8290.
26 A. H. Groschel and A. H. E. Muller, *Nanoscale*, 2015, **7**, 11841–11876.
27 H. Xing, Z. Wang, Z. Xu, N. Y. Wong, Y. Xiang, G. L. Liu and Y. Lu, *ACS Nano*, 2012, **6**, 802–809.
28 S. C. Glotzer and M. J. Solomon, *Nat. Mater.*, 2007, **6**, 557–562.
29 G. R. Yi, D. J. Pine and S. Sacanna, *J. Phys.: Condens. Matter*, 2013, **25**, 193101.
30 A. Perro, E. Duguet, O. Lambert, J. C. Taveau, E. Bourgeat-Lami and S. Ravaine, *Angew. Chem.*, 2009, **48**, 361–365.
31 Y.-S. Cho, G.-R. Yi, J.-M. Lim, S.-H. Kim, V. N. Manoharan, D. J. Pine and S.-M. Yang, *J. Am. Chem. Soc.*, 2005, **127**, 15968–15975.
32 P. Datskos, D. A. Cullen and J. Sharma, *Angew. Chem., Int. Ed.*, 2015, **54**, 9011–9015.
33 A. Walther and A. H. E. Müller, *Chem. Rev.*, 2013, **113**, 5194–5261.
34 S. Jiang, Q. Chen, M. Tripathy, E. Luijten, K. S. Schweizer and S. Granick, *Adv. Mater.*, 2010, **22**, 1060–1071.
35 B. P. Binks and P. D. I. Fletcher, *Langmuir*, 2001, **17**, 4708–4710.
36 A. Walther, M. Hoffmann and A. H. Muller, *Angew. Chem.*, 2008, **47**, 711–714.
37 J. Faria, M. P. Ruiz and D. E. Resasco, *Adv. Synth. Catal.*, 2010, **352**, 2359–2364.
38 A. Kirillova, C. Schliebe, G. Stoychev, A. Jakob, H. Lang and A. Synytska, *ACS Appl. Mater. Interfaces*, 2015, **7**, 21218–21225.
39 S. Hwang and J. Lahann, *Macromol. Rapid Commun.*, 2012, **33**, 1178–1183.
40 J. M. Crowley, N. K. Sheridon and L. Romano, *J. Electrost.*, 2002, **55**, 247–259.
41 T. Nisisako, T. Torii, T. Takahashi and Y. Takizawa, *Adv. Mater.*, 2006, **18**, 1152–1156.
42 Q. Chen, S. C. Bae and S. Granick, *Nature*, 2011, **469**, 381–384.
43 Q. Chen, J. K. Whitmer, S. Jiang, S. C. Bae, E. Luijten and S. Granick, *Science*, 2011, **331**, 199–202.
44 Q. Chen, S. C. Bae and S. Granick, *J. Am. Chem. Soc.*, 2012, **134**, 11080–11083.
45 L. Hong, A. Cacciuto, E. Luijten and S. Granick, *Nano Lett.*, 2006, **6**, 2510–2514.
46 A. A. Shah, B. Schultz, W. Zhang, S. C. Glotzer and M. J. Solomon, *Nat. Mater.*, 2015, **14**, 117–124.
47 M. S. Fernandez, V. R. Misko and F. M. Peeters, *Phys. Rev. E: Stat., Nonlinear, Soft Matter Phys.*, 2015, **92**, 042309.
48 Y. Xia, B. Gates and Z. Y. Li, *Adv. Mater.*, 2001, **13**, 409–413.
49 S. Berger, A. Synytska, L. Ionov, K.-J. Eichhorn and M. Stamm, *Macromolecules*, 2008, **41**, 9669–9676.
50 W. Stöber, A. Fink and E. Bohn, *J. Colloid Interface Sci.*, 1968, **26**, 62–69.
51 S. Berger, L. Ionov and A. Synytska, *Adv. Funct. Mater.*, 2011, **21**, 2338–2344.
52 A. Kirillova, G. Stoychev, L. Ionov and A. Synytska, *Langmuir*, 2014, **30**, 12765–12774.

Faraday Discussions

PAPER

Templated growth of gold satellites on dimpled silica cores†

C. Chomette,[a] E. Duguet,*[a] S. Mornet,[a] E. Yammine,[a] V. N. Manoharan,[bc] N. B. Schade,[c] C. Hubert,[d] S. Ravaine,*[d] A. Perro[e] and M. Tréguer-Delapierre*[a]

Received 23rd February 2016, Accepted 29th February 2016
DOI: 10.1039/c6fd00022c

We synthesize robust clusters of gold satellites positioned with tetrahedral symmetry on the surface of a patchy silica core by adsorption and growth of gold on the patches. First we conduct emulsion polymerization of styrene in the presence of 52 nm silica seeds whose surface has been modified with methacryloxymethyltriethoxysilane (MMS). We derive four-dimple particles from the resulting silica/polystyrene tetrapods. Polystyrene chains are covalently bound to the silica surface within the dimples due to the MMS grafts and they may be thiolated to induce adsorption of 12 nm gold particles. Using chloroauric acid, ascorbic acid and sodium citrate at room temperature, we grow gold from these 12 nm seeds without detachment from or deformation of the dimpled silica surface. We obtain gold satellites of tunable diameter up to 140 nm.

1 Introduction

Colloidal particles with an anisotropic shape or chemical functionality are building blocks for the directed assembly of new materials.[1–4] For instance, particles with chemical anisotropy can assemble into unusual lattices[5] and dimpled particles can bind to spherical particles in a lock-and-key scheme due to depletion forces.[6] Both types of anisotropic particles can be used to assemble clusters of controlled size as well.[6,7]

Gold patches are of special interest because of their intrinsic optical properties and the ease with which their surfaces can be chemically modified. One way to

[a]CNRS, Univ. Bordeaux, ICMCB, UPR 9048, F-33600 Pessac, France. E-mail: etienne.duguet@icmcb.cnrs.fr
[b]Harvard Univ., School of Engineering and Applied Sciences, Cambridge, MA 02138, USA
[c]Harvard Univ., Department of Physics, Cambridge, MA 02138, USA
[d]CNRS, Univ. Bordeaux, CRPP, UPR 8641, F-33600 Pessac, France
[e]Univ. Bordeaux, CNRS, ISM, UMR 5255, F-33405 Talence, France

† Electronic supplementary information (ESI) available: TEM images of silica/polystyrene tetrapods from silica seeds previously surface-modified with acetoxypropyltrimethoxysilane; pictures of the nitrobenzyl pyridine assay on dimpled silica particles before and after the chloromethylation reaction. See DOI: 10.1039/c6fd00022c

prepare particles with gold patches is through physical vapor deposition of gold onto a monolayer or multilayer of spherical particles,[5,8,9] but this technique suffers from limited versatility and scalability.[5] An alternative method is to seed the growth of gold patches using small silver patches in a surface-conformal strategy.[10] Such a strategy would be quite easily up-scaled, but the size, number, and position of the gold patches then remain uncontrolled, because of the random positioning of the silver precursor patches.[11] One may instead nucleate and grow gold on isotropic particles, but this yields homogeneous gold coatings[12,13] or gold patches of uncontrolled size and position.[14] Therefore synthesis routes that permit control over gold patches in colloidal batch quantities have remained elusive.

One simple route would be to start with particles with highly controlled, symmetric patches, and then to modify them with gold. However, there are only a few established protocols for the synthesis of particles with controlled numbers and arrangements of patches or dimples of any sort, *e.g.* with linear, triangular or tetrahedral symmetry. Based on the steric hindrance phenomenon, they generally involve the confinement of a controlled number of spherical particles (which become the patches, patch precursors or dimple precursors) in a droplet, through sticky protrusions or at the surface of another colloid. For instance, Kim *et al.* stabilized oil-in-water emulsions with negatively-charged polystyrene (PS) particles and then photopolymerized the oil to obtain patchy microparticles.[15] Conversely, Kraft *et al.* prepared polymeric particles bearing liquid protrusions and then polymerized the liquid.[16] Wang and colleagues exploited the direct confinement by droplets of particles with specific surface functionalities.[7,17]

We recently reported the synthesis of silica particles with dimples arranged symmetrically, such that the number (1, 2, 3, 4, 6 or 12) and depth of dimples may be tuned.[18,19] Anchored PS chains reside at the bottom of each dimple and may be chemically modified to be sticky in specific assembly conditions.

In the present paper, we describe the use of these dimpled silica particles to adsorb and grow gold nanoparticles as metallic patches. Each dimple ultimately serves as a cradle for a single gold particle, and thus the dimple arrangement determines the final number, size, and positions of the gold satellites. We first explain some new insights regarding features of the residual PS chains in the dimples, and then we describe how we chemically modify them with thiol groups such that they interact favorably with gold nanoparticles. We next report the adsorption of citrate-stabilized gold nanoparticles within the dimples and their subsequent growth according to three different synthesis recipes that we investigated. Here we focus on particles with four dimples, but the same strategy could be used with any other number of dimples.

2 Materials and methods

2.1 Materials

We used styrene (Sigma-Aldrich, 99%), methacryloxymethyltriethoxysilane (MMS, ABCR, 98%), acetoxypropyltrimethoxysilane (AMS, ABCR), sodium persulfate (Sigma-Aldrich, 99%), Synperonic® NP30 (Aldrich), sodium dodecylsulfate (SDS, Sigma-Aldrich, >90%), tetraethoxysilane (TEOS, Sigma-Aldrich, 99%), ammonia (28–30% in water, SDS), $KAuCl_4$ (Sigma-Aldrich, 98%), $HAuCl_4 \cdot 3H_2O$ (Sigma-Aldrich, 99.9%), trisodium citrate (Sigma-Aldrich, 99%), $SnCl_4$ (Sigma-Aldrich,

>99%), hydrochloric acid (37%, Sigma-Aldrich) and ammonia (28–30% in water, SDS), NaSH (Alfa-Aesar, 98%), $NaBH_4$ (Sigma-Aldrich, 96%), cetyltrimethylammonium bromide (CTAB, Sigma-Aldrich, 99%), and ascorbic acid (Sigma-Aldrich, 99%) as we received them. We obtained deionized water with a resistivity of 18.2 Ω cm at 25 °C from a Milli-Q system (Millipore). We purchased tetrahydrofuran (THF), dimethylformamide (DMF), chloroform, and absolute ethanol from Sigma-Aldrich. Lastly, we prepared butyl chloromethyl ether according to a recipe already published.[20]

2.2 Synthesis of dimpled silica particles

We prepared batches of tetrapods, consisting of a central silica core and four PS satellite nodules each, by seeded-growth emulsion polymerization of styrene, according to a procedure we published previously.[21–23] For a typical batch, we used 52 nm silica particles whose surface had been modified with MMS (0.5 molecule per nm^2) as seeds (18×10^{15} particles per L) for the aqueous emulsion polymerization of styrene (100 g L^{-1}) at 70 °C for 6 h in the presence of sodium persulfate (0.5 g L^{-1}) and a surfactant mixture (3 g L^{-1}) of Synperonic® NP30 and SDS (5 wt%). The monomer-to-polymer conversion efficiency was 78%. Transmission electron microscopy revealed that the resulting batch consisted of tetrapods (73%), tripods (16%), hexapods (6%), mono- or bipods (4%) and pentapods (1%).

We regrew the silica cores to create dimples using a method that we had also reported previously.[18,19] We prepared a mixture of 450 mL ethanol and 35 mL ammonia (1 M) and added first 10 mL of the polymerization medium containing 3.6×10^{14} tetrapods and then 4.8 mL of TEOS at a rate of 1 mL h^{-1}. We dissolved the PS satellite nodules by performing three centrifugation/redispersion cycles in THF, which leaves the dimples in place. The final batch contains 110 nm silica particles bearing four (74%), three (18%), six (5%), one/two (2%) or five (1%) dimples. In parallel experiments, we performed the synthesis using AMS instead of MMS for the preliminary surface modification of the silica seeds.

2.3 Characterization of the residual PS macromolecules at the bottom of the dimples

We collected, dried, and weighed the first THF supernatant containing the dissolved PS chains, and then dissolved it again in THF at a mass concentration of 1 mg mL^{-1} for size exclusion chromatography (SEC) experiments (UV detector at 260 nm and 0.2 vol% of trichlorobenzene as standard).

2.4 Synthesis of citrate-stabilized gold nanoparticles

Our recipe was inspired by a previously reported technique.[24] In a 1 L round bottom flask maintained at a temperature of 100 °C, we sequentially added 700 mL water, 314 mg $KAuCl_4$ diluted in 50 mL water, and 1.53 g trisodium citrate diluted in 75 mL water. The solution changed from yellow to colourless, bluish gray, and then red. We let the dispersion react for 15–20 min at 100 °C. TEM images showed that the average diameter of the gold nanoparticles was 12 nm.

2.5 Thiolation of the residual PS macromolecules at the bottom of the dimples

We have already reported on the chloromethylation stage of the dimples' surface elsewhere.[19] Briefly, after transferring the as-prepared dimpled silica particles in chloroform, we added butyl chloromethyl ether in chloroform (3 M; 5 mL) and 0.3 mL $SnCl_4$. We set the temperature to 45 °C and then aged the mixture overnight. Finally we washed the dispersion with three cycles of centrifugation/redispersion in HCl solution (4 wt% in water) and then in water/ethanol (50/50 wt %) before redispersing in 20 mL of DMF. We performed the thiolation stage after redispersion of 10^{13} chloromethylated patchy particles in ethanol. We added 2 g NaSH in 10 mL of ethanol and let the system react overnight under stirring. Finally, we washed the particles 5 times by centrifugation at 10 000*g* for 15 min in ultrapure ethanol. At this time, the sample could be stored in ethanol. Before use, we transferred the particles into water, and then reduced the potential disulfide bonds by adding a few drops of an aqueous solution of 1 M $NaBH_4$. Finally we washed the dispersion by centrifugation and stored it in degassed water. To confirm the presence of thiol groups, we incubated the particles with citrate-stabilized gold nanoparticles in water in a gold-to-silica particle ratio of 400 : 1. After overnight incubation over stirring, we removed the excess gold nanoparticles by centrifugation using ultrapure water at 5000*g* for 10 min and then we adjusted the concentration of the patchy silica particles to 9×10^{11} particles per mL.

2.6 Seeded-growth of gold nanoparticles within the dimples

We performed experiments with the goal of enlarging the diameter of the gold seeds, assuming that there is one gold seed per dimple. We investigate three different well-known seeded-growth recipes in solution.

Recipe #1. We took 20 mL of the thiolated patchy particle dispersion, seeded with citrate-stabilized gold nanoparticles, and added it to 250 mL of an aqueous solution containing CTAB (0.015 M), ascorbic acid (1 mM) and $HAuCl_4$ (0.5 mM) at 35 °C for one hour. The resulting solution contained a small amount of free gold nanoparticles, which we could remove by centrifugation.

Recipe #2. We took 15 mL of the thiolated patchy particle dispersion seeded with citrate-stabilized gold nanoparticles, centrifuged it, dispersed it in 15 mL sodium citrate solution (2.2 mM), and transferred it into a 20 mL flask. We heated the dispersion to 90 °C and then added 0.2 mL of $HAuCl_4$ (25 mM) in two equal fractions at 30 minute intervals. We assumed the gold nanoparticle concentration was roughly 3×10^{12} particles per mL. After 30 min, we removed 5.5 mL of the reacting volume from the flask and added 5.3 mL ultrapure water with 0.2 mL citrate sodium solution (60 mM). After temperature stabilization (5 min), we considered the growth cycle of the process completed. Finally, to stop the process, we allowed the solution to cool down to room temperature after completion of the last cycle. By repeating this process, we grew the gold particles within each dimple through up to 9 iterations.

Recipe #3. We took 1 mL of the thiolated patchy particle dispersion seeded with citrate-stabilized gold nanoparticles, and placed it in a three-neck 20 mL round bottom flask. The volume of the suspension contained between 10^{10} and 10^{12} gold seeds. To this we added a 10 mL aliquot of precursor solution A, containing $HAuCl_4 \cdot 3H_2O$, and 10 mL of reducing solution A′, containing trisodium citrate and ascorbic acid. We added these solutions separately at room temperature through syringes *via* a peristaltic pump under vigorous stirring over a period of about 45 min (13.3 mL h^{-1}). Immediately after this was complete, we brought the mixture to boiling and maintained this temperature for about 30 min. Finally, we allowed the solution to cool down.

We prepared solution A by diluting a $HAuCl_4 \cdot 3H_2O$ stock solution (0.2% w/v) to 10 mL. Similarly, we prepared solution A′ by diluting a mixture of an ascorbic acid solution (1% w/v) and a trisodium-citrate solution (1% w/v) to 10 mL as well. The relative volume ratios of all three stock solutions were always 8 : 4 : 1. We estimated the volume of the gold precursor solution $V_{\text{Au solution}}$ required to grow the seeds from an initial diameter D_{initial} to a final diameter D_{final} by using the following equation:

$$V_{\text{Au solution}} = \rho_{\text{Au}} N_{\text{Au}} \pi (D_{\text{final}}{}^3 - D_{\text{initial}}{}^3)/6 C_{\text{Au solution}} M_{\text{Au}}$$

where ρ_{Au} is the density of gold, M_{Au} the gold molar mass, and N_{Au} the number of gold nanoparticles in the considered volume.

We diluted the solution to obtain a final volume of 10 mL before loading it into the syringe and initiating the growth process.

3 Results and discussion

Our multistep synthesis route to dimpled silica particles is summarized in Scheme 1 and was reported previously.[18,19,21–23] Transmission electron microscopy (TEM) revealed that our silica/PS tetrapod mixture contained 73% tetrapods with byproducts ranging from monopods to hexapods. After regrowing the silica and dissolving the polystyrene to develop dimples on the particles, 74% of the contents were 110 nm silica particles bearing four dimples with a depth of about 27 nm. The byproducts are dimpled silica particles derived from the other multipods in the same proportions.

We previously showed that PS bumps remain at the bottom of the dimples (Fig. 1a and b), and we postulated that these grafted PS macromolecules result from the copolymerization of styrene with MMS.[18,19] To test this hypothesis, we performed the same synthesis of PS/silica tetrapods and then dimpled particles, but we replaced MMS with acetoxypropyltrimethoxysilane (AMS). The acetoxy end of AMS is saturated and therefore cannot polymerize according to a chain-growth mechanism. The resulting tetrapods were not as well-defined (see ESI Fig. S1†)

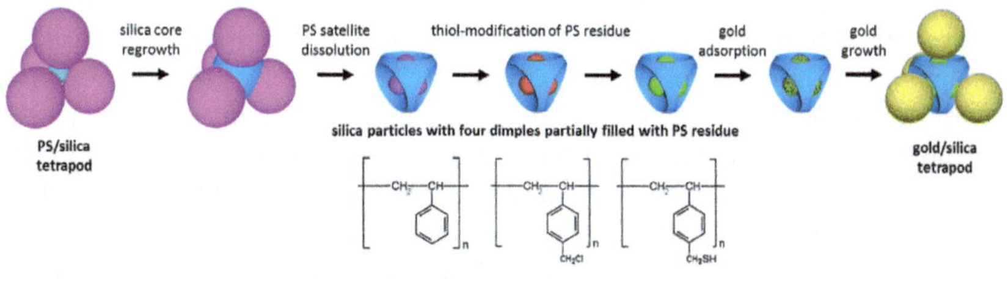

Scheme 1 Main steps of the synthesis of gold/silica tetrapods from PS/silica tetrapods *via* seeded growth of gold within the dimples.

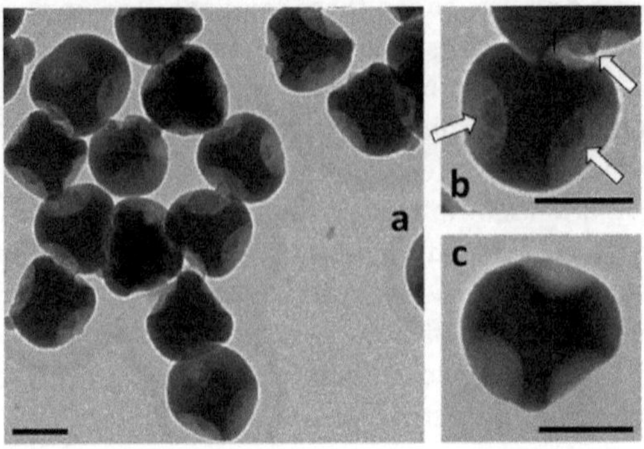

Fig. 1 Transmission electron micrographs of silica particles with four dimples each, obtained from MMS-modified silica seeds (a and b) and from AMS-modified silica seeds (c). White arrows identify polystyrene bumps at the bottom of the dimples. Scale bars are 100 nm.

and, as we have already reported,[25] we found that they fall down onto TEM grids more frequently. Nevertheless, in the TEM image of a dimpled silica particle derived from one of these tetrapods, the PS bumps did not appear within the dimples (Fig. 1c). This confirms the origin of the grafted PS macromolecules and the role of MMS at their anchoring points.

The length of the grafted macromolecules had been studied on alumina or silica/PS or poly(ethyl)acrylate core–shell particles obtained in similar emulsion polymerization conditions, using silanes as surface coupling agents.[26–28] The average molar mass and molar mass distribution of the grafted chains were often reported higher than those of the free macromolecules. In some cases, the authors pointed to siloxane bonding between the grafted molecules, suggesting that the MMS grafts serve as cross-linking groups. They assumed that the average molar masses of the grafted and non-grafted chains would be the same otherwise. Using our SEC experiments performed on the non-grafted PS chains recovered at the stage of their dissolution in THF, we found an average molar mass of 540 000 g mol^{-1} (number-averaged) or 980 000 g mol^{-1} (weight-averaged). These values are comparable to those typically obtained from emulsion polymerization. It means that these chains are long and can readily serve as extended anchoring arms for interacting with gold satellites.

However, since our goal was to develop covalent coupling with gold satellites, we could not use the PS bumps in their original state. Instead, we modified the PS macromolecules with thiol groups, which are well-known for their straightforward coupling to gold surface atoms. We used a two-step scheme involving chloromethylation of PS and then nucleophilic substitution of the chloromethyl groups with sodium hydrosulfide in ethanol (Scheme 1). We performed the first step in chloroform using a non-commercial chloromethylating agent prepared from butyl alcohol, paraformaldehyde and gaseous HCl at low temperature.[20] To verify successful completion of this stage, we performed a qualitative colorimetric test with nitrobenzyl pyridine (see ESI Fig. S2†).[29] We then carried out the thiol-modification in ethanol, even though chloromethylated PS is probably only partially soluble in ethanol. We indirectly verified the success of this stage by mixing the resulting colloids with citrate-stabilized 12 nm gold particles. The excess of gold particles was removed by centrifugation. The resulting particles were redispersed in water and observed under TEM (Fig. 2). We used citrate-stabilized gold nanoparticles because citrate binds weakly to gold, and thus thiol groups readily displace citrate ligands. We found that the gold particles are attached only at the thiolated PS bumps within the dimples (Fig. 2). In control experiments (not shown) with chloromethylated PS bumps without thiol-modification, we saw no gold nanoparticle adsorption.

To take advantage of this specific adsorption site for gold nanoparticles within the dimples, we investigated methods for promoting and directing subsequent gold growth in order to obtain larger gold satellites within each dimple. The main challenges were: (1) preserving the gold/thiol bonds; (2) restricting gold growth to the dimple area to preserve the tetrahedral symmetry of the nanostructure; and (3) ultimately obtaining gold/silica tetrapods (Scheme 1). The seeds act as efficient redox catalysts for metal ion reduction, which occurs selectively at their surface, as the energetic barrier for heterogeneous nucleation is much lower than that for homogeneous nucleation. In the literature, many seed-mediated recipes based on the use of citrate-capped gold seeds and chloroauric acid as a gold source have been reported. Three of them are particularly relevant here since they can yield gold particles with diameters of up to 200 nm.[30–32] We investigated each of them to assess their applicability for gold growth from the 12 nm gold seeds within our dimples. In all cases, for the sake of simplicity, we assumed one gold seed per dimple in our colloids, on average.

The first recipe uses ascorbic acid as a reducing agent and cetyltrimethylammonium bromide (CTAB) as a surfactant. This approach produces quasi-spherical gold particles up to 180 nm in diameter with a narrow size distribution.[30] Unfortunately, in our experiments with citrate-capped gold particles anchored to the dimple surface, the procedure yielded quite complex and aggregated objects. Not only did most of the gold structures exhibit highly non-spherical shapes, but also they detached from the dimple areas (Fig. 3). We assume this failure is due to CTAB detrimentally affecting the gold–thiol bonding.

The second recipe, performed at 90 °C in the presence of sodium citrate as a reducing and capping agent, generally yields uniform spherical gold particles

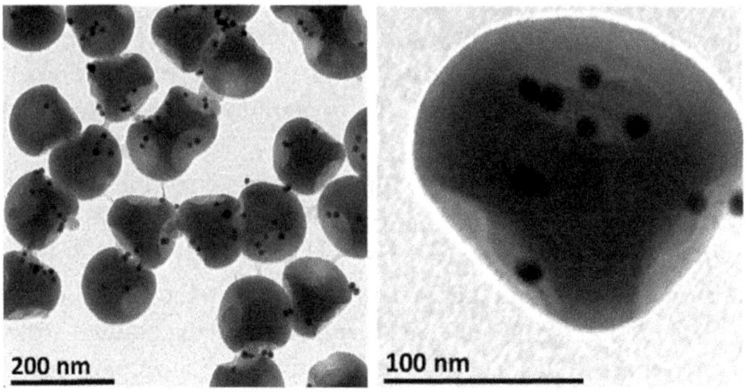

Fig. 2 Transmission electron micrographs of the dimpled silica particles after thiol-modification of the PS chains at the bottom of the dimples and incubation with citrate-stabilized gold nanoparticles.

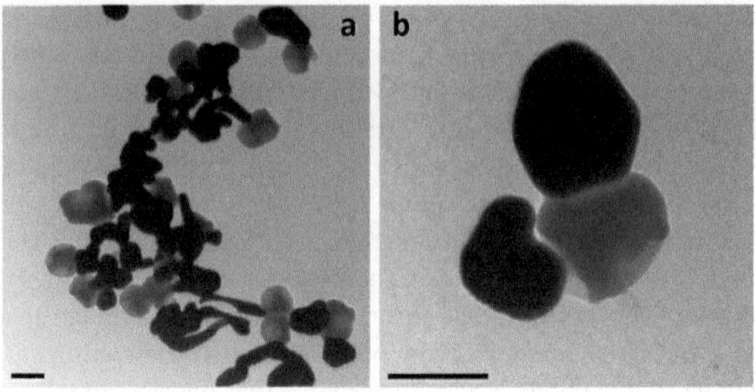

Fig. 3 Transmission electron micrographs of gold/silica nanostructures obtained after gold regrowth performed in the presence of ascorbic acid and CTAB. Scale bars are 100 nm.

with diameters of tens of nanometers.[31] We implemented this multi-step procedure for seeded growth on the surface of our particles' dimples. Starting from 12 nm particles and repeating the process up to 9 times, we found that this approach can produce gold satellites with diameters of about 90 nm or even larger. In contrast, the previous recipe yields gold particles that remain attached to the dimples and that are relatively spherical (Fig. 4). However, the silica cores show damage from the successive treatments. After the ninth step, it becomes difficult to identify the core morphology. We assume the silica cores experience gradual etching due to the presence of sodium ions (counter-ions of the citrate precursor) in the reaction medium. This hypothesis is consistent with other experimental[33] and theoretical[34] studies in the literature. Moreover, the reduction process occurred at 90 °C, which has been reported as harsh conditions for silica.[35]

The UV-visible spectra of these plasmonic nanoclusters agree with these overall observations. The surface plasmon resonance (SPR) band redshifts with increasing particle size. It peaks at 524 nm after one regrowth step, 531 nm after three, 538 nm after six, and 548 nm after nine. We also observe a large SPR broadening and increased scattering during the cyclic growth process. The gradual dissolution of the silica cores led to uncontrolled distances between the gold satellites. This in turn gives rise to uneven optical coupling between the gold particles, which broadens the absorption band.

Overall, we found that this second recipe is sufficient for enlarging the gold seeds while preserving their positions within the dimples. However, the silica particles do not fare well through the many iterations necessary for obtaining large gold satellites.

The third recipe uses milder reduction conditions (ascorbic acid and sodium citrate at room temperature) and is reportedly capable of producing free gold particles with diameters in the range of 15–300 nm.[32] When we applied this recipe to our system, we found encouraging results as shown in the series of TEM images in Fig. 5. The reduction of the metallic precursor led to the gradual growth of the gold satellites onto the silica surface, and the dimpled silica particles were only slightly damaged. The gold nanoparticles remained strongly anchored to the dimples during the growth iterations and the largest satellites reached sizes of 120–140 nm after three cycles. We observe single gold satellites in a majority of

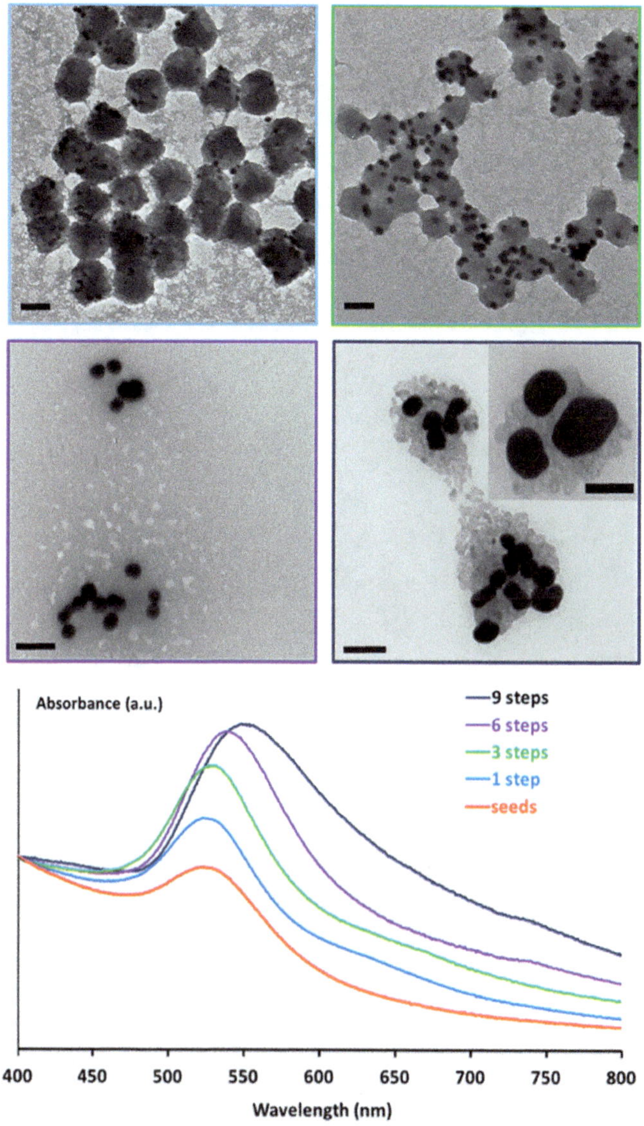

Fig. 4 Transmission electron micrographs and corresponding UV-visible spectra (normalized at 400 nm) of gold/silica nanostructures obtained after 1, 3, 6, and 9 growth iterations performed in the presence of sodium citrate at 90 °C. Scale bars are 100 nm.

the dimples, particularly on silica cores with a tetrahedral arrangement of dimples. The gold satellites are more oblate than spherical. These silica/gold nanostructures are stable for more than one month, but they sediment quickly due to their high mass.

We measured the optical properties of these core/satellites clusters during the growth process, as shown in Fig. 5. As expected, the electric dipole surface plasmon resonance redshifted with increasing gold satellite size. We found increased scattering with successive growth iterations as well.

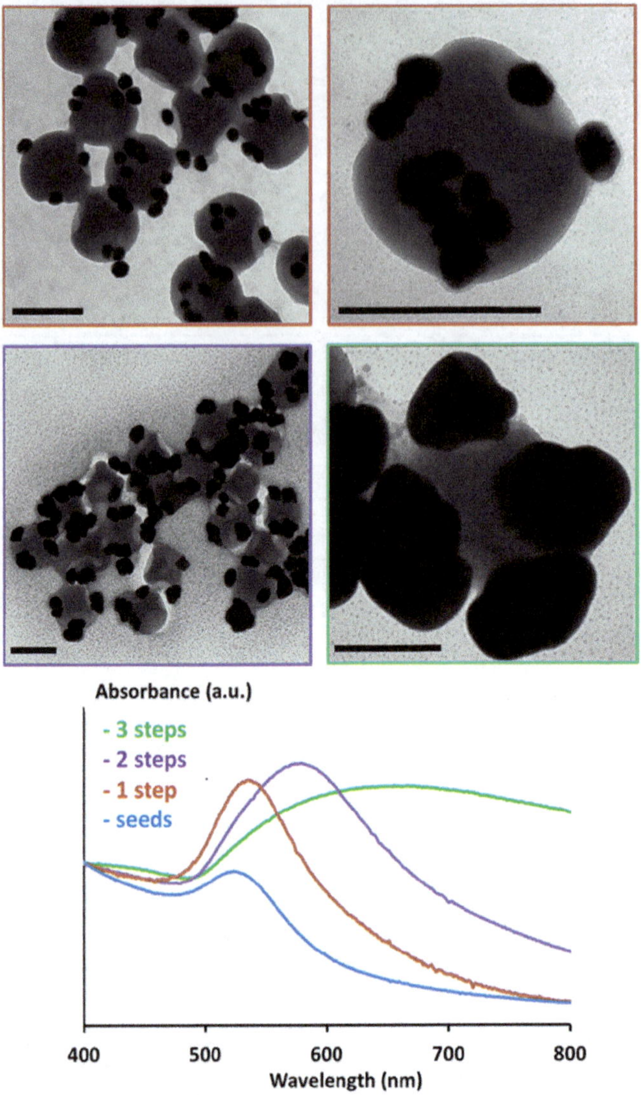

Fig. 5 Transmission electron micrographs and corresponding UV-visible spectra (normalized at 400 nm) of gold/silica nanoclusters obtained after 1, 2, or 3 growth iterations performed in the presence of sodium citrate and ascorbic acid at room temperature. Scale bars are 100 nm.

4 Conclusion

We have shown that the PS macromolecules are grafted at the bottom of the dimples in our silica particles due to the copolymerization of styrene and MMS-derived grafts. These macromolecules are long and easily thiol-modified, which makes them suitable adsorption sites for citrate-stabilized gold nanoparticles. Our efforts to grow larger gold satellites, using the adsorbed nanoparticles as seeds, led to moderate success. Two key factors arise for minimizing polydispersity and controlling the gold satellite morphologies and positions: (1) the applied mixture of ascorbic acid and sodium citrate as reducing and stabilizing agents; and (2) the process of slowly adding separate reducing and precursor solutions at room temperature, as demonstrated in the third recipe. These mild reaction conditions preserve the gold satellite positions within the dimples as well as the geometry of the dimpled silica cores. This procedure yields well-defined gold patches with diameters up to 140 nm, although they are not perfectly spherical. The tetrahedral symmetry of the dimpled structure is preserved throughout the growth of the gold satellites.

The surface functionality of these structures means that they have potential for further chemical modification and may find use in colloidal self-assembly schemes. Moreover, these gold clusters could be useful in their own right as elements of optical metafluids due to their tetrahedral symmetry and plasmonic properties.[36]

Acknowledgements

TEM experiments were performed at the Plateforme de Caractérisation des Matériaux and Bordeaux Imaging Center of the University of Bordeaux. This work was supported by the LabEx AMADEus (ANR-10-LABX-42) and IdEx Bordeaux (ANR-10-IDEX-03-02), *i.e.* the Investissements d'Avenir programme of the French government managed by the Agence Nationale de la Recherche.

References

1 A. Van Blaaderen, *Nature*, 2006, **439**, 545.
2 F. Li, D. P. Josephson and A. Stein, *Angew. Chem., Int. Ed.*, 2011, **50**, 360.
3 S. Sacanna, D. J. Pine and G.-R. Yi, *Soft Matter*, 2013, **9**, 8096.
4 E. Duguet, C. Hubert, C. Chomette, A. Perro and S. Ravaine, *C. R. Chim.*, 2016, **19**, 173.
5 Q. Chen, S. C. Bae and S. Granick, *Nature*, 2011, **469**, 381.
6 S. Sacanna, W. T. M. Irvine, P. M. Chaikin and D. J. Pine, *Nature*, 2010, **464**, 575.
7 Y. Wang, Y. Wang, D. R. Breed, V. N. Manoharan, L. Fen, A. D. Hollingsworth, M. Weck and D. J. Pine, *Nature*, 2012, **491**, 51.
8 G. Zhang, D. Wang and H. Möhwald, *Nano Lett.*, 2005, **5**, 143.
9 A. B. Pawar and I. Kretzschmar, *Langmuir*, 2008, **24**, 355; A. B. Pawar and I. Kretzschmar, *Langmuir*, 2009, **25**, 9057.
10 H. Bao, B. Butz, Z. Zhou, E. Spiecker, M. Hartmann and R. N. Klupp Taylor, *Langmuir*, 2012, **28**, 8971.
11 R. N. Klupp Taylor, H. X. Bao, C. T. Tian, S. Vasylyev and W. Peukert, *Langmuir*, 2010, **26**, 13564.
12 S. J. Oldenburg, R. D. Averitt, S. L. Westcott and N. J. Halas, *Chem. Phys. Lett.*, 1998, **288**, 243.
13 J. Yang, Y. Li, L. Zu, L. Tong, G. Liu, Y. Qin and D. Shi, *ACS Appl. Mater. Interfaces*, 2015, **7**, 8200.
14 H. Bao, T. Bihr, A.-S. Smith and R. N. Klupp Taylor, *Nanoscale*, 2014, **6**, 3954.
15 S. H. Kim, G. R. Yi, K. H. Kim and S. M. Yang, *Langmuir*, 2008, **24**, 2365.

16 D. J. Kraft, J. Hilhorst, M. A. P. Heinen, M. J. Hoogenraad, B. Luigjes and W. K. Kegel, *J. Phys. Chem. B*, 2011, **115**, 7175.
17 Y. Wang, Y. Wang, X. Zheng, G. R. Yi, S. Sacanna, D. J. Pine and M. Weck, *J. Am. Chem. Soc.*, 2014, **136**, 6866.
18 A. Désert, C. Hubert, Z. Fu, L. Moulet, J. Majimel, P. Barboteau, A. Thill, M. Lansalot, E. Bourgeat-Lami, E. Duguet and S. Ravaine, *Angew. Chem., Int. Ed.*, 2013, **52**, 11068.
19 C. Hubert, C. Chomette, A. Désert, M. Sun, M. Treguer-Delapierre, S. Mornet, A. Perro, E. Duguet and S. Ravaine, *Faraday Discuss.*, 2015, **181**, 139.
20 A. Warshawsky and A. Deshe, *J. Polym. Sci., Polym. Chem. Ed.*, 1985, **23**, 1839.
21 A. Perro, E. Duguet, O. Lambert, J.-C. Taveau, E. Bourgeat-Lami and S. Ravaine, *Angew. Chem.*, 2009, **121**, 367.
22 A. Désert, I. Chaduc, S. Fouilloux, J.-C. Taveau, O. Lambert, M. Lansalot, E. Bourgeat-Lami, A. Thill, O. Spalla, S. Ravaine and E. Duguet, *Polym. Chem.*, 2012, **3**, 1130.
23 A. Désert, J. Morele, J. C. Taveau, O. Lambert, M. Lansalot, E. Bourgeat-Lami, A. Thill, O. Spalla, L. Belloni, S. Ravaine and E. Duguet, *Nanoscale*, 2016, **8**, 5454.
24 A. Wang, H. P. Ng, Y. Xu, Y. Li, Y. Zheng, J. Yu, F. Han, F. Peng and L. Fu, *J. Nanomater.*, 2014, **2014**, 1.
25 A. Perro, D. Nguyen, S. Ravaine, E. Bourgeat-Lami, O. Lambert, J. C. Taveau and E. Duguet, *J. Mater. Chem.*, 2009, **19**, 4225.
26 D. Yin, Q. Zhang, H. Zhang and C. Yin, *J. Polym. Res.*, 2010, **17**, 689.
27 A. Guyot and P. Espiard, *Polymer*, 1995, **36**, 4391.
28 Z. Zeng, J. Yu and Z.-X. Guo, *Macromol. Chem. Phys.*, 2005, **206**, 1558.
29 F. Galindo, B. Altava, M. I. Burguete, R. Gavara and S. V. Luis, *J. Comb. Chem.*, 2004, **6**, 859.
30 J. Rodríguez-Fernández, J. Pérez-Juste, F. J. García De Abajo and L. M. Liz-Marzán, *Langmuir*, 2006, **22**, 7007.
31 N. G. Bastus, J. Comenge and V. Puntes, *Langmuir*, 2011, **27**, 11098.
32 C. Ziegler and A. Eychmüller, *J. Phys. Chem. C*, 2011, **115**, 4502.
33 E. Mahon, D. R. Hristov and K. A. Dawson, *Chem. Commun.*, 2012, **48**, 7970.
34 P. M. Dove, N. Han, A. F. Wallace and J. J. De Yoreo, *Proc. Natl. Acad. Sci. U. S. A.*, 2008, **105**, 9903.
35 Y. J. Wong, L. Zhu, W. S. Teo, Y. W. Tan, Y. Yang, C. Wang and H. Chen, *J. Am. Chem. Soc.*, 2011, **133**, 11422.
36 Y. A. Urzhumov, G. Shvets, J. Fan, F. Capasso, D. Brandl and P. Nordlander, *Opt. Express*, 2007, **15**, 14129.

Faraday Discussions

DISCUSSIONS

Janus and patchy nanoparticles: general discussion

Alberto Striolo, Jongwook Kim, Luis Liz-Marzán, Luciano Tadiello, Matthias Pauly, Catherine Murphy, Anna Roig, David Gracias, Younan Xia, Javier Reguera, Axel Mueller, Kevin Critchley, Mathias Brust, Leonardo Scarabelli, Martin Mayer, Matthias Thiele, Martin Buzza, András Deák, Ana Maria Bago Rodriguez, Christian Kuttner, Heiko Wolf, Euan Kay, Antonio Stocco, David Portehault, Hedi Mattoussi, Kelley Heatley, Eugenia Kumacheva, Guillermo González, Christoph Hanske, Wenming Tong, Muhammad Nawaz Tahir, Benjamin Abécassis, Steve Granick, Etienne Duguet, Alla Synytska and Krassimir Velikov

DOI: 10.1039/C6FD90048H

Wenming Tong opened discussion of the introductory lecture by Catherine Murphy: If the silica shell is thin enough, would there be a strong plasmon coupling?

Catherine Murphy replied: Yes!

Luis Liz-Marzán commented: If you intend to use multishell nanoparticles as SERS tags, the separation between the Au core and Au shell must be very small, meaning that your silica shells may not be suitable. An alternative is the use of Raman-active molecules to separate the core and the shell. Some work has been reported in this direction by us[1] and others and it seems to work fine.

1 Y. Wang, K. Sentosun, A. Li, M. Coronado-Puchau, A. Sánchez-Iglesias, S. Li, X. Su, S. Bals and L. M. Liz-Marzán, *Chem. Mater.*, 2015, **27**, 8032–8040.

Catherine Murphy responded: Intuitively it does seem that having no dielectric space layers between the metal nanostructures would be best. That is why the paper by Bhargava *et al.* mentioned in my lecture is so interesting, they calculate that, counterintuitively, a spacer layer gives you more of a boost than you realize, even though you are turning off chemical enhancement mechanisms to a large extent.[1] But since we cannot readily make these structures, molecular spacer layers are probably the best right now in practice.

1 A. K. Kodali, X. Llora, and R. Bhargava, *Proc. Natl. Acad. Sci. U. S. A.*, 2010, **107**, 13620–13625.

Eugenia Kumacheva asked: Are there any interference effects that depend on the thickness of the ligand shell? By tuning the shell thickness you can get both constructive and destructive interference.

Catherine Murphy responded: If you mean the optical interference effects of a multilayered shell that contains insulator and metal layers, then yes, we do expect that, per the paper by Bhargava *et al.* and references therein.[1] Other papers use Peter Nordlander and Naomi Halas' hybridization model of plasmonic interactions to assess the optical properties of multilayered nanoshells; see for example **ref. 2 and 3** below. If there is a paper that does the exact calculations you are thinking of, I have yet to find it!

1 A. K. Kodali, X. Llora, and R. Bhargava, *Proc. Natl. Acad. Sci. U. S. A.*, 2010, **107**, 13620–13625.
2 C. Liu and B. Q. Li, *J. Phys. Chem. C*, 2011, **115**, 5323–5333.
3 S. Gao, P. Li and F. Li, *Appl. Phys. Lett.*, 2013, **102**, 123107.

Christian Kuttner remarked: For core–shell particles, there are different design concepts like a closed metal shell[1] or a layer of densely packed particles. Could you comment on the particular benefits of these designs in regard to the possible applications of these particle types?

1 M. Tebbe, C. Kuttner, M. Mayer, M. Maennel, N. Pazos-Perez, T. König and A. Fery, *J. Phys. Chem. C*, 2015, **119**, 9513–9523.

Catherine Murphy replied: In our case we really want a spacer layer in there to separate the metals from each other, in order to get all the neat plasmonic field enhancements that other groups have calculated. Luis Liz-Marzán did mention that there is work with only a molecule spacing two metal layers, which seems incredible but also seems to work. But for us, we actually need 5–50 nm type spacers for constructive and destructive interference sorts of phenomena to happen. You are right, there is a lot of good work on metal-on-metal nanoparticles with all kinds of interesting strain/epitaxy interfaces.

Leonardo Scarabelli commented: During your talk you mentioned that you encountered some problems trying to achieve a smooth second gold layer deposited on the silica. Have you thought about exploiting a thermal effect? The silica layer should protect the internal rod from reshaping, while the gold particles deposited on the outside layer may melt together, resulting in a uniform coverage of the entire surface.

Catherine Murphy replied: Yes, we have played a little bit with heating the solutions up. We are always worried about particles aggregating though, so it is a delicate balance. But probably we should be more systematic and push this further.

Axel Mueller opened discussion of the paper by Steve Granick: The study of non-equilibrium is extremely interesting. How do you control the rotational speed and arm-length? What are the effects of the viscosity, field strength and frequency?

Steve Granick responded: Your interesting question is about system-specific details. As we describe in the paper, the parameters are many and among the multitude of combinatorial possibilities this scouting, exploratory study selected just a few of them. In this study our main aim was to identify and articulate new problems.

Alberto Striolo commented: In the description of the guiding principles for the dynamic out-of-equilibrium assembly of particles you discuss what happens when the Janus particles are all of the same type; the results are fascinating. In addition, in the paper you presented here, you have mixtures of Janus and isotropic particles. It appears that the dynamic features you reported in this paper are due to the diversity of the system composition.

The question focuses on the importance of having a mixture to observe rich behaviour: can you comment on the effect of system composition on the results? Have you had the chance to study multi-component systems? From a different perspective, and perhaps as a joke, does this teach us anything about the role of diversity in society?

Steve Granick responded: As a great English poet once wrote, "Variety is the very spice of life, that gives it all its flavor." (William Cowper, 1731–1800).

Antonio Stocco said: Dipolar and hydrodynamic interactions are expected to be different between two active Janus particles and between a bare particle and a Janus particle. If θ is the angle between the Janus axis and the direction connecting the centers of mass of the two particles, have you observed a difference between the angles of the Janus–Janus particle and Janus–bare particle systems? Recently, Spagnolie *et al.*[1] described the hydrodynamic entrapment and scattering of an active particle near a spherical obstacle (or a passive particle), can one compare your experimental angle for the Janus–bare particle system with the angle described by Spagnolie *et al.*?

1 S. E. Spagnolie, G. R. Moreno-Flores, D. Bartolo and E. Lauga, *Soft Matter*, 2015, **11**, 3396–3411.

Steve Granick replied: It would indeed be interesting to consider the system to which you refer. It was not the goal of our study, however.

Kelley Heatley asked: In your paper you state the electric field conditions are 300 kHz and 40 V mm^{-1}. Typically at these field conditions dipolar interactions dominate and many groups see colloidal particles stacking in the direction of the applied field. Do you ever see stacking or can you comment on why this does not occur?

Steve Granick responded: You are correct of course. In other studies we have studied the stacking effects explicitly.[1] In this study we designed the system to be two-dimensional in order to facilitate the quantitative analysis presented in this paper.

1 N. Yanai, M. Sindoro, J. Yan and S. Granick, *J. Am. Chem. Soc.*, 2013, **135**, 34–37.

Mathias Brust said: Could you please comment on the energy conversion processes in your system? Does the applied field supply energy for the motion of particles, or is this entirely Brownian motion and the field merely induces oscillating dipole moments in the Janus particles?

Steve Granick answered: As you correctly guess, the applied field supplies energy; as in other active matter systems, motion consumes energy that is fed externally into the system. The mechanism of translation is discussed in our paper.

David Portehault asked: Using an electric field as a stimulus provides access to complex functions, such as exponential decays or square waves for instance. Do you think that other dynamic patterns of assembly or transient states might emerge by using such functions?

Steve Granick replied: You are correct that the potential diversity of the electric field waveform is essentially unlimited! Such complexity may introduce interesting second-order effects, in my opinion, but is unlikely to modify qualitatively the notions of natural selection introduced in this paper.

Eugenia Kumacheva said: How can you design particles to control the formation of dipoles? Are there any design rules?

Steve Granick responded: The design rules for the approach taken here were described in our recent paper, see **ref. 1** below. Briefly, the two Janus hemispheres have differential responses to AC electric fields.

1 J. Yan, M. Han, J. Zhang, C. Xu, E. Luijten and S. Granick, *Nat. Mater.*, 2016, DOI: 10.1038/nmat4696.

Mathias Brust commented: The individual beads you address with the AC field are small enough to undergo Brownian motion but large enough to respond to the external stimulus by polarisation. To what extent are these phenomena scalable? Can you go into the molecular size regime and still have an external handle on the constituents?

Steve Granick answered: We are engaged in the sorts of efforts you describe and I am hopeful that such phenomena may scale down to nm-sized particles, but it is too soon to know conclusively.

Christian Kuttner remarked: Aiming at the rational design of suitable building blocks for self-assembled structures, the ratio of attractive and repulsive interactions is related to the dipole moments of the two hemispheres. Could the Janus balance be tuned to serve as a control parameter for the head-to-tail attraction?

Steve Granick replied: Indeed this is a possibility but I am not sure of the need to engage in this approach as attraction can already be modulated using the approach described in our contributed paper.

Axel Mueller asked: Would you be able to scale the size of the particles down to the range below 100 nm? Would Brownian motion disturb the observed effects?

Steve Granick answered: You are correct that Brownian motion would render the system response more complex, but I am hopeful that the general idea will scale down to nm-sized particles. Studies in this direction are under way in our laboratory.

Axel Mueller commented: The study of 20 nm particles will be challenging due to Brownian motion. Also it will not be easy to take a movie of particles of that size.

Steve Granick responded: Indeed. The advances being made using modern liquid-phase TEM methods do however promise to provide hard data at this size regime. In spite of the fact that the technique requires further development, especially to fully understand when sample damage becomes a serious concern and when it does not, I find the possibilities opened by this technique very exciting.

Heiko Wolf asked: Do you have strategies to lock the spiral geometries at a certain stage and permanently link connected particles? Such larger clusters may be interesting building blocks in further assemblies.

Steve Granick replied: Indeed! We are pursuing this direction of investigation. However, user-friendly chemical reactions that are compatible with the other conditions of self-assembly are less trivial to implement than one might wish.

Hedi Mattoussi opened discussion of the paper by Javier Reguera: During the growth of the star-shaped hybrid nanostructures that you have discussed in your presentation, your structures seem to consistently show metal stars only on one "half" of the staring magnetic nanoparticles. Do you know why? Are there specific local initiation spots on the iron oxide nanoparticles you start with? Are there instances when the metal covers the full surface of the starting magnetic nanoparticles?

Javier Reguera replied: The Janus character is due to two causes. First, it is the use of seeds that are already heterodimers, *i.e.* they already show a small gold nanoparticle on one side. Second, it is the tendency of the newly created gold (after the reduction to Au(0)) to grow on the gold part instead of the iron oxide one, which is energetically more favorable. In the case of these nanoparticles, the nanoparticles always showed Janus morphology even for a high gold to iron oxide ratio. It never covers the whole iron oxide nanoparticle; however, it is possible to obtain nanostars with an iron oxide core if the initial seed is a core–shell iron oxide–gold nanoparticle.

Hedi Mattoussi commented: Can the two metal portions of the hybrid nanostructures be coated with the same ligand? In other words, when you want to surface modify the hybrid structure, do you have to use a mixture of two ligands with distinct coordination to satisfy the different affinity of the magnetic and metallic surfaces to specific anchoring groups?

Javier Reguera responded: The fact that both parts can be independently functionalized is a clear advantage of these nanoparticles. In the case that one wants to cover them with the same ligand, several strategies could be used. For example, they could be coated with amine terminated ligands that bind to the two materials, with polymers or block copolymers that wrap around the nanoparticles, or with pairs of molecules that are similar except for the anchoring groups. Nevertheless, in that case it is probably better to have a core–shell structure instead of Janus.

David Portehault remarked: The size distribution has been measured from regular 2D TEM images. Do you think it would be feasible to obtain a more accurate description of the system using tomography? How long does one 3D reconstruction take?

Javier Reguera replied: Indeed, it is always better to perform size measurements in 3D. Depending on the complexity of the structures, such investigations may be very time demanding, not only the acquisition, but also certainly the segmentation afterwards (can even be several weeks). On the other hand, there are recent advances that will overcome this in the near future.[1]

1 E. Bladt, D. M. Pelt, S. Bals and K. Joost Batenburg, *Ultramicroscopy*, 2015, **158**, 81–88.

Luis Liz-Marzán remarked: For the synthesis of your initial dumbbells you generate iron particles, which nucleate on pre-formed gold nuclei, then you oxidise iron into iron oxide. Why does iron nucleate on gold and why does it not detach after oxidation?

Javier Reguera responded: I completely agree with you. The growth of iron is through the decomposition of $Fe(CO)_5$, so it grows as metallic iron on the gold seed and it is oxidized after it is exposed to air. The change from iron to iron oxide should produce stress at the gold-oxide interface. We have seen that, despite this stress, the binding is very stable and the nanoparticles resist quite well prolonged sonication in a sonication bath or even with sonication probes. This stress could however change the properties of the nanoparticles and it would be very interesting to study.

Martin Mayer asked: How does the high refractive index of the iron oxide section influence the plasmonic properties of the Janus particle? Does it even render the plasmonic properties anisotropic?

Javier Reguera responded: This is a very good point. The iron oxide certainly adds anisotropy to the refractive index surrounding the nanoparticle, which should create anisotropy in the plasmonic modes. This has never been measured for this kind of nanoparticle, but this optical anisotropy would definitely be a very important property and would be very interesting to measure.

Kevin Critchley commented: The absorption band in the UV-Vis spectrum is broad and I am interested in the origin of this. Would each individual particle be expected to have a broad plasmon band or would you expect that the individual particles have a relatively narrow band and therefore the resultant absorption spectrum reflects the heterogeneity within the sample? For example, the implication of the latter explanation would mean that only a fraction of the particles would be active during NIR laser-based photothermal treatment.

Javier Reguera replied: The plasmonic behavior of the nanoparticles is produced through the hybridization of the core and tip modes where the core serves as a nanoantenna dramatically increasing the excitation cross section of the tip plasmons.[1] As there are many tips of different shapes and orientations (many plasmonic modes) the combined plasmonic peak becomes relatively broad for every nanoparticle. Moreover, the width of the plasmonic band highly depends on the morphology of the nanoparticle, for example, on the size (related to the number of peaks) or the sharpness of the tips, as has been reported in **ref. 2** below. The polydispersity is also expected to play a role in the width of the plasmonic bands as a decrease in size tends to blueshift the plasmonic bands. From that point of view, an improvement in the nanoparticle uniformity through better synthesis control or a size separation step would improve the plasmonic behavior of the nanoparticles.

1 F. Hao, C. L. Nehl , J. H. Hafner and P. Nordlander, *Nano Lett.*, 2007, **7**, 729–732.
2 S. Barbosa, A. Agrawal, L. Rodríguez-Lorenzo, I. Pastoriza-Santos, R. A. Alvarez-Puebla, A. Kornowski, H. Weller and L. M. Liz-Marzán, *Langmuir*, 2010, **26**, 14943–14950.

Anna Roig commented: How can you explain the star shape (with spikes) of the gold nanoparticles?

Javier Reguera responded: This is a surfactant directed growth of nanoparticles. There are several surfactants described in the literature that can achieve an anisotropic growth and PVP is one of them. The shape of the final nanoparticles synthesized with PVP depends on the synthesis conditions (solvent, PVP concentration, *etc.*). We have followed the procedure described by Kumar *et al.* where a very high quantity of PVP induces the branching of gold with single crystalline tips giving rise to a gold nanostar.[1]

1 P. S. Kumar, I. Pastoriza-Santos, B. Rodríguez-González, F. J. García de Abajo and L. M Liz-Marzán, *Nanotechnology*, 2008, **19**, 015606.

Jongwook Kim remarked: Gold nanorods typically lose their shape and optical anisotropies under irradiation with NIR lasers. Do you see the same changes after irradiation or a certain time after synthesis?

Javier Reguera replied: Laser irradiation experiments with these nanoparticles have not been done yet, but this is an important point for applications where long and intense exposure to irradiation is needed, like in photothermal treatments for example. For these applications the photostability (*i.e.* the change in the morphology and wavelength of maximum absorption with illumination) would need to be studied. For the SERS experiments explored in this work, no effect was observed probably due to the fact that the illumination occurred for very short times and in a small region.

Regarding the time after synthesis, the nanoparticles are highly stable. After 6 months in solution no appreciable difference could be observed using TEM and only a small blue shift was observed in the spectrum (~15 nm) probably due to the rounding of the nanoparticle tips. This was produced, however, for highly cleaned nanoparticles (five centrifugation cycles). A longer time stability is expected for nanoparticles containing more PVP or other stabilizing ligands.

Leonardo Scarabelli asked: Can you comment on the possible applications of the system you designed? Do you think there are any synergistic effects arising from the double nature of your nanostructures (both plasmonic and magnetic)?

Javier Reguera responded: These nanoparticles are very versatile since they combine superparamagnetic and plasmonic properties which, as has been demonstrated in separate nanomaterials, have given rise to a large number of applications. We demonstrated their combined use, for example, as a very effective SERS substrate, but it is very easy to think of applications in optics (responsiveness, magnetic self-assembly, *etc.*) or in biomedicine (combination of photo- and magneto-hyperthermia, magnetic manipulation, imaging, *etc.*). I believe, however, that the main difference to other magnetoplasmonic nanoparticles is the Janus character which could direct the interaction of these nanoparticles or could give rise to unique anisotropic properties. This Janus magnetoplasmonic character has never been exploited and could, in the near future, give rise to new, unique, and exciting applications.

Christian Kuttner commented: Fig. 5 shows the application of Janus particles for surface-enhanced Raman spectroscopy. The SERS measurements were assisted by magnetic separation, *i.e.* the crowding of particles. Does the enhancement arise from an increased concentration of analyte or can it also be partly attributed to electromagnetic interparticle coupling (formation of hot-spots) between the particles?

Javier Reguera answered: Different to spherical nanoparticles, nanostars already have hot spots at their tips. With the aggregation of spherical nanoparticles hot spots are generated at the contact points resulting in an increase of the SERS signal. On the contrary, in the case of gold nanostars, this increase in the signal due to aggregation is normally not observed. This is probably due to the fact that no new hot spots are generated or that some hot spots decrease due to the coupling of nanoparticle plasmonic modes to compensate the generation of new ones. We believe that our magnetoplasmonic nanoparticles should behave in a similar way to conventional nanostars, so the enhancement in the signal is most probably due to the capture of the analyte while the nanoparticles are in suspension, and this increase in concentration results in the nanoparticles aggregating after the magnetic field is applied.

It is worth mentioning that having the nanoparticles in solution, instead of having an initial aggregate, is highly beneficial as it decreases the diffusion distance between the analyte and the nanoparticle (SERS substrate) improving the coupling between both. After this fast capture, the magnetic concentration mechanism is a very simple step to concentrate both the nanoparticles and analyte.

Euan Kay asked: Do you have any insight into the morphology of the aggregates when the magnetic field is applied and how this is affected by the Janus nature of the nanoparticles? For example, does the gold region on one nanoparticle orient towards the gold region on a neighbouring nanoparticle?

Javier Reguera responded: These are a completely new kind of nanoparticle and an extensive characterization has not been done yet. Certainly, it would be very interesting to know if the magnetic field affects the assembly of the nanoparticles that are constrained by their nanostar "backpack". This assembly could certainly affect the plasmonic behavior and efficiency as a SERS substrate.

It has been shown that magnetic nanoparticles tend to form long filaments, and for the cases where Janus morphology was studied the filaments showed an inner core formed of the magnetic parts.[1] It still needs to be studied whether this filament configuration would be produced for the Janus nanostar–sphere nanoparticles and if the filaments would remain in the final aggregate. Certainly it is worth studying this.

1 L. Zhang, F. Zhang, W.-F. Dong, J.-F. Song, Q.-S. Huo and H.-B. Sun, *Chem. Commun.*, 2011, **47**, 1225–1227.

Wenming Tong commented: Is it possible to correlate the analyte concentration and SERS signals?

Javier Reguera answered: Yes, it is possible. This would require calibration in the medium where the analyte is going to be measured, and most probably should be calibrated for every batch of nanoparticles. In this case we have not performed this calibration but a clear increase in the intensity of the Raman signals was observed with the increase in concentration.

Muhammad Nawaz Tahir asked: Is it possible to grow gold stars on the Au Domain of the Au–Fe_2O_3 Janus nanoparticles or does it depend on the size of the Au domain, and why do gold stars not grow on the iron oxide domain?

Javier Reguera answered: Yes, it should grow on the Au domain independently of the size. For example, in our case the nanostars grew on a gold domain of about 5–6 nm, but the standard synthesis of gold nanostars is performed with 13–14 nm seeds. Other seed sizes should be possible too. We believe that it is energetically more favorable for Au to grow on the Au rather than on the iron oxide surface when both are available at the same time and the growth is slow enough.

Axel Mueller opened discussion of the paper by David Gracias: Ionov and Stamm at Dresden use polymer bilayers for self-folding. What is the difference to your approach?

David Gracias replied: Yes, there are many research groups (>20) who study self-folding. Regarding polymer bilayers, to my knowledge the first polymer bilayer self-folding paper was by Hu *et al.*[1] Our paper is not focused on the mechanism of folding. Both folding mechanisms involving the use of capillary forces with reflowing tin and differential swelling of a polymer using a gradient (both very different from the polymer bilayer work) were invented and have been published by our lab in previous work, and some of these papers have been cited in the references. The work presented in this paper is unique in the following ways: a) we use highly parallel and cost-effective nanoimprint lithography (NIL) to pattern precursors for folding with different heterogeneous materials. e.g. gold (metal) and Al_2O_3 (dielectric) as a first step towards creating functional electronic or plasmonic devices with well defined heterogeneous material patterns on the surfaces of the nanostructures; b) by combining highly parallel and cost effective NIL and highly parallel and cost-effective self-folding, the SNIPS platform shows how one could mass produce precisely patterned 3D nanostructures as well as macrostructures with nanoscale patterns. This work is significant from a nano-manufacturing perspective since such high value nano- and macro-devices cannot be produced in a cost-effective manner at the present time; c) as we have shown here and in previous papers, capillary forces are very strong at the nanoscale and can be used to curve patterned templates with very small 10 nm scale radii. In comparison, although there are isolated reports of folding bilayers with very small radii, it is generally very challenging to reproducibly curve polymer bilayers with nanoscale radii due to the inability to deposit very thin polymer films with an appropriate modulus and the swelling ratio mismatch needed to achieve such small radii. Further, the functionality of many practical electronic devices necessitates the use of not just polymers but also metals, device grade semi-conductors and inorganics.

1 Z. Hu, X. Zhang and Y. Li, *Science*, 1995, **269**, 525–527.

Eugenia Kumacheva asked: Can you form different structures from the same building block?

David Gracias answered: Yes, it is possible to create different folding outcomes from the same building block (or net). We recently published a paper where we describe the formation of "isomer" analogs during self-folding.[1]

1 S. Pandey, D. Johnson, R. Kaplan, J. Klobusicky, G. Menon and D. H. Gracias, *PLoS One*, 2014, **9**, e108960.

Eugenia Kumacheva continued: Is the statistical distribution of angles acceptable?

David Gracias responded: We utilize several methods of self-folding (strain, surface tension or capillary forces and swelling). We have previously described the use of so called "locking hinges" which correct errors and self-seal the structures.[1] By using such liquid hinges there are a large number of perfect structures that are formed with very little statistical distribution of angles. In the absence of such error correcting mechanisms, the process is sensitive to engineering parameters and a larger statistical distribution would likely result which may or may not be important, depending on the application. For example, in a gripping device, the exact angles that the fingers move is not that critical as long as the device can hold on to an object. In other cases, such as aggregative self-assembly, a distribution of angles in polyhedra may not be acceptable.

1 T. G. Leong, P. Lester, T. Koh, E. Call and D. H. Gracias, *Langmuir*, 2007, **23**, 8747–8751.

Luis Liz-Marzán commented: What happens to metals and metal oxides when they fold? Are cracks formed? Also, if you intend to integrate transistors and other elements into a 100 nm object, how stable will the nanostructure be?

David Gracias replied: Our self-folding is driven by capillary forces or mismatched strain. Since 100 nm sized transistor elements are routinely used in planar VLSI, we do not anticipate any major issues other than those already known for planar VLSI such as gate leakage, interdiffusion, electromigration and short channel effects. Other than that our process uses plasma etching and thermal annealing (below the annealing temperatures used in traditional VLSI) so we do not anticipate any major issues. Nevertheless, the long term reliability of our 3D structures have not been studied and will be a priority once the fabrication has been put into commercial practice.

Mathias Brust opened a general discussion of the papers by Steve Granick and David Gracias: Profs. Granick and Gracias have shown very different approaches to ordered processes/objects that are in an intermediate size regime that is perhaps difficult to access through diffusion or stirring while the components are still too small to be handled efficiently for top down assembly. Whereas Granick emphasises the fundamental nature of his work, Gracias has mentioned that he can envisage that components for future computers could be made by his approach. I wonder if the authors agree with me that there might be an opportunity for using a "Granick type methodology" to extend self-assembly from the molecular to the mesoscale, which may ultimately lead to the 3D assembly of complex functional structures from mesoscale devices such as transistors made in a way similar to that discussed by Gracias in his paper.

David Gracias responded: That is a very interesting point and certainly researchers would like to utilize non-equilibrium self-assembly to create adaptive computational or smart devices with functionalities similar to that in living systems. I envision that this will be possible over the next few decades by merging concepts from robotics, colloidal assembly, computer science, equilibrium/non-equilibrium molecular and mesoscale self-assembly, VLSI and self-folding.

Steve Granick responded: It is an intriguing vision that you describe. Ultimately the sort of thing you describe is, I think, a quest to which many of us in this field would subscribe. The devil is in the details.

Christian Kuttner commented: As a more general prospect, one might ask what the future of device fabrication might look like? In particular the combination of soft and hard materials, like conducting/semiconducting particles and soft functional frameworks, is a challenge but might open new avenues for advanced devices. Scalable techniques like self-folding nanostructures with imprint patterned surfaces (SNIPS) show that the future of device fabrication could lie in the development of large-scale compatible techniques for the mass production of 3D nanostructures. What is your opinion toward future device fabrication?

David Gracias answered: It is likely that no one method will dominate future device fabrication, there is room for several methods. SNIPS leverage top-down VLSI and bottom-up strain engineered self-folding to create 3D structures that are precisely patterned. In this regard it is a unique approach for creating curved and polyhedral patterned nanostructures, but may not be necessary if unpatterned polyhedral structures are fabricated.

Steve Granick responded: Indeed this can be said to be a goal for many involved in justifying the utility of self-assembly approaches. In order for the potential of this vision to be realized, it will be necessary to find ways to avoid the incidence of errors in self-assembly. Without wishing to overstate the problem, it is a fact that the current top-down approaches do not permit failure; an aeroplane must perform flawlessly without crashing, for example. We have not yet achieved this level of infallible control.

Catherine Murphy asked: Can you tell us the status of the 3D printing of nanoparticles/nanoparticle composites?

David Gracias replied: The 3D printing of nanoparticles has advanced significantly and it is easy to 3D print nanoparticles provided that the correct surface chemistry, solubility, viscosity of the solvent, *etc.* is worked out. However, as compared to our SNIPS approach, we note that it is (a) challenging to 3D print device grade single crystal silicon and compound semiconductor piezoelectrics in an integrated device with nanoscale features/resolution; (b) if such devices are 3D printed using nanoparticles, then when sintering *etc.* is used to fuse the nanoparticles to make 3D structures by 3D printing, the properties of those materials may be different from high quality materials. The attractive feature of SNIPS is that we can leverage the infrastructure and know how from 2.5 D VLSI integration and create curved and folded 3D devices in a highly parallel and cost effective manner.

Luis Liz-Marzán continued discussion of the paper by Steve Granick: Do you see a potential functionality of the materials you have presented? You have shown strings; chiral arrangements. Have you identified any properties of these assemblies that can provide them with useful functionality?

Steve Granick responded: The functionality in which I believe fervently, it is the value of basic science. Basic understanding leads inexorably to better functionality in technology, provided that one does not seek results that are too easy, too cheap or too obvious. My research is currently carried out at the Institute of Basic Science in South Korea, whose belief is that well-selected basic science is the best possible investment in the potential game-changing technologies of the future. If you are curious to learn more, please check out our website: softmatt.ibs.re.kr.

David Gracias remarked: To date there have been a few studies on model systems of the non-equilibrium assembly of interacting objects such as bubble rafts, magnetic spinners and your system. What have these studies taught us about life or living systems? If they haven't already, could you conjecture on how they might teach us something fundamental about living systems or would they all just be system specific?

Steve Granick responded: Your interesting question is on the philosophical side, isn't it? You are asking: how does science make progress, by what measures can we know whether science makes progress, and also you ask how, from measurements which necessarily are about specific systems, to infer generalizations of wide applicability. In my personal opinion, it would be hubris to imagine that from the studies of our model systems we learn about how a rabbit develops or how Beethoven composed great music, for example. However, I would insist on the usefulness of seeking general principles. In the specific paper that I presented at this conference, we sought to elucidate, in a system selected for the reasons described in the paper, how observations of self-assembly in our non-equilibrium conditions could be integrated into the more general considerations of natural selection.

Younan Xia continued the general discussion of the papers by Steve Granick and David Gracias: Can your assembly processes be conducted in the gas phase or under vacuum rather than in the liquid phase?

David Gracias replied: Yes, our tin based self-folding process is carried out in a plasma etcher which is under moderate vacuum. At small size scales, since shear forces in liquid can cause a large resistance to folding, it may be advantageous to perform self-folding in the gas phase or under vacuum. One mechanism for the arc discharge formation of CNTs is that small graphene sheets roll up in the gas phase which suggests also that gas/vacuum based self-folding is an attractive methodology.

Steve Granick responded: Your question raises the interesting issue of how to manipulate the particle–particle interactions which for self-assembly are needed. They must be sufficiently strong to keep the particles together but not so strong that the particles become frozen and stick however they encounter each other; they must be not too weak, not too strong, but "just right". In the liquid phase this is relatively easy to implement. Under vacuum or in the gas phase, to accomplish this goal would be a more challenging task but surely it could be accomplished if one had a reason to tackle the problem.

Benjamin Abécassis addressed Steve Granick: In the case of the swarms of active particles, such phenomena have already been observed in other types of particles (non-Janus). How specific is the Janus character in your case? Do they have to be Janus or do they form swarms just because they are "swimming"?

Steve Granick replied: In a recent paper we described the rules that govern when our particles form swarms or other collective structures.[1] It is definitely not as simple as you suggest.

1 J. Yan, M. Han, J. Zhang, C. Xu, E. Luijten and S. Granick, *Nat. Mater.*, 2016, DOI: 10.1038/nmat4696.

Hedi Mattoussi continued discussion of the introductory lecture by Catherine Murphy: The cyclic voltammetry (CV) profiles collected from dopamine containing compounds tend to exhibit a pH-dependent oxidation potential. When you showed the CV data from the solution containing dopamine and other compounds (including CTAB) I saw no changes in the oxidation potential with pH; is this significant?

Catherine Murphy answered: We did CV of hydroquinone, not dopamine. We only ran the CV at the pH at which we do the gold nanorod reaction, which is rather acidic. I too would suspect a pH dependence for the CV but we did not investigate that. The CV profiles are with and without CTAB at the same pH.

Hedi Mattoussi continued discussion of the paper by Javier Reguera: Can you grow other star metals on the magnetic substrates, such as silver? Are your star nanostructures similar to those reported by Teri Odom at Northwestern University? Their stars are not grown on other nanostructured substrates.

Javier Reguera replied: The nanoparticles presented here are the first ones obtained with this morphology, but we are working to obtain nanoparticles with other materials too. In principle, dumbbell-like nanoparticles have been obtained with many metal–metal oxide materials (see for example **ref. 1** below). The strategy that we have shown here is quite general and it should be applicable to other different metals, like silver nanostars for example. In addition, using this strategy of consecutive seed-mediated growth steps it is also easy to think about more complex structures with more than one layer of materials.

The nanostars synthesized here have a different synthesis protocol to the ones used by Teri Odom, which are produced using a seedless synthesis. In our case a seed-mediated growth process is required so the nanostar grows on the heterodimer seed and gives rise to the Janus morphology. It is possible, however, that the formation of nanostars in HEPES buffer, like the synthesis you mention, in the presence of the dumbbells can also give rise to Janus nanoparticles, but this has not been tried yet.

1 M. R. Buck, J. F. Bondiand R. E. Schaak, *Nat. Chem.*, 2012, **4**, 37–44.

Hedi Mattoussi asked: The absorption spectra do not show any characteristic shifts as are often reported for Au nanorods and nanoshells, for example. Is that due to the rather broad range of the individual rod aspect ratios and diameters of those stars you showed in the presentation and report?

Javier Reguera replied: In our case we do observe a redshift of the plasmon absorbance when we obtain bigger nanoparticles, as is shown in Fig. 4 and has been previously reported.[1] This shift is obviously smaller that the ones you would obtain, for example, by changing the length of a nanorod. What we have seen in our nanoparticles is that when they get bigger in size the number of tips tends to increase much more than the sharpness does, which would explain the relatively small redshift.

1 P. S. Kumar, I. Pastoriza-Santos, B. Rodríguez-González, F. J. García de Abajo and L. M Liz-Marzán, *Nanotechnology*, 2008, **19**, 015606.

Axel Mueller opened discussion of the paper by Alla Synytska: Are your structures kinetically determined, or are they already in equilibrium?

Alla Synytska responded: The particle aggregates do not change with time. However, it is very difficult to say if they are in global or in local thermodynamic minimum or their reorganization is slow.

Martin Buzza asked: One potential problem with mixing oppositely charged particles is that instead of forming clusters you can form a gel. To what extent can you prevent the individual clusters from sticking to each other in your method?

Alla Synytska replied: The formation of a gel depends on the ratio between the particles. An excess of one sort of particle results in the formation of agglomerates with either positive or negative surface charge (depending on which kind of the particles are in excess). These agglomerates repel each other and do not form a gel.

Alberto Striolo said: The self-assembly experiments you discussed in your paper were performed at pH 7. However, you showed that the "corona" polymer is strongly responsive to pH. I was wondering whether you have attempted to change the pH, *e.g.* to 2 or to 11, after the assemblies have been created? If you have not conducted these experiments, what would you expect to see if you were to conduct these experiments?

Alla Synytska responded: We have not performed this experiment yet. However, I would expect that the behavior of the particle aggregates at pH 2 and pH 11 will be similar to the behavior of the corresponding interpolyelectrolyte complex (PDMAEMA-PAA) — the aggregates will disassemble. PAA is uncharged at pH 2 and its interactions with PDMAEMA will be guided by the formation of hydrogen bonds and hydrophobic interactions. Similar behavior will be observed at pH 11 when PDMAEMA is uncharged.

Mathias Brust asked: When decorating a larger sphere with many smaller ones, what rationale, if any, guided your choice of the size ratio between the two types of spheres?

Alla Synytska responded: We varied the size ratio in order to investigate the effect of this parameter on the structure of the assembled aggregates. The Stoeber method allows the synthesis of SiO_2 particles with sizes between *ca.* 100 nm and 1 μm and we varied the size of the particles in this range.

Javier Reguera remarked: The satellite particles seem to have a clear separation among them and a homogenous distribution through the surface of the core particles or the Janus domain. This is probably due to electrostatic or other kind of repulsions. Could you comment on the maximum density of satellite particles that can be achieved with this method and what strategies could be used to improve the system?

Alla Synytska replied: The origin of the separation of satellite particles is electrostatic and steric repulsions (the polymer layers are swollen) between the polymer coronas. Since the SEM images were recorded in a dry state, this gap appears. The density can be increased by reducing the thickness of the polymer corona, which can be achieved by decreasing the molecular weight of the polymer, reducing the grafting density and reducing the swelling degree through the pH and salt concentration.

Christoph Hanske said: How stable are the core particles in the presence of salt? Is it possible to control their coverage with satellite particles by varying the ionic strength in the adsorption step?

Alla Synytska answered: The assembly process depends on the properties of the polymer coronas and interactions between them, which depend on the pH and ionic strength. An increase in the salt concentration must lead to the weakening of the electrostatic interactions between the polyelectrolytes and disassembly of agglomerates.

Christoph Hanske commented: How much can you decrease the size of the core particles in this assembly process?

Alla Synytska responded: The Stoeber method allows for the synthesis of SiO_2 particles with sizes between *ca.* 100 nm and 1 μm and we varied the size of the particles in this range.

David Portehault asked: The size distribution is relatively broad. Are there ways to improve it during the synthesis or through sorting after the synthesis?

Alla Synytska answered: Shear forces can be used to control the size and size distribution of the agglomerates.

Leonardo Scarabelli said: Would it be possible to reiterate the self-assembly process, attaching something else to your satellite structure? The presence of a large charged core should allow you to have fine control over the attachment of the third layer (which would present the same charge as the core) to the oppositely charged satellites. Moreover, given that your approach is completely reversible, if the third layer is designed to switch to neutral at a different pH compared to the bigger core, you should be able to selectively reverse each step of the self-assembly process.

Alla Synytska responded: The adsorption of a third layer of particles onto the satellite particles is possible. It will work then as a layer-by-layer approach. The global neutralization of the charge by adding oppositely charged particles is not possible because the neutralization occurs locally at the points of contact between the particles.

András Deák commented: It looks like the smaller particles around the bigger ones are located at the substrate level. Are capillary forces responsible for this finding and to what extent might they influence the results?

Alla Synytska replied: Clearly, the formation of aggregates is due to the electrostatic attraction between the oppositely charged polyelectrolyte-modified particles. The influence of capillary forces must be negligible because the surface tension between the swollen polyelectrolytes and solvent (water) will be very low because of their similarity.

Luciano Tadiello commented: Can you discuss the mechanical stability of your self-assembled soft particle system when subjected to shearing forces, for example in laminar flow? Do you think that controlling the shearing forces may give you a tool for controlling the structural changes within your system made of relatively soft particles? See for example **ref. 1** below, where it is possible to evolve the particle size distribution in a relatively simple system (PS latex).

1 G. Barthelmes, S. E. Pratsinis and H. Buggisch, *Chem. Eng. Sci.*, 2003, **58**, 2893–2902.

Alla Synytska answered: The particle agglomerates are stabilized by electrostatic attractions and interchain entanglement between the polymer coronas. While we did not perform any quantitative measurements of the stability, the agglomerates appear rather stable as they do not break down during ultrasonication for half an hour. I think that the use of only very large shear forces can lead to the disassembly of agglomerates and will allow for the control of their size.

Christian Kuttner asked: The presented microclusters exhibit a distinct disorder of the satellite particles at the core–particle surface, supposedly owing to particle jamming, see also Höller *et al.*[1] Does the assembly follow a random sequential adsorption mechanism with a hit-and-stick character? How could the order and coverage of the satellite particles be increased?

1 R. P. M. Höller, M. Dulle, S. Thomä, M. Mayer, A. M. Steiner, S. Förster, A. Fery, C. Kuttner and M. Chanana, *ACS Nano*, 2016, **10**, 5740–5750.

Alla Synytska responded: The adsorption mechanism has a hit-and-stick character. The distance between the satellite particles depends on the thickness of the polymer shell and strength of the electrostatic repulsion between the satellite particles. The density can be increased by reducing the thickness of the grafted polymer layers, which can be achieved by decreasing the molecular weight of the polymer, reducing the grafting density and reducing the swelling degree through the pH and salt concentration.

Christian Kuttner continued: This comment aims at the possibilities to control the adsorption of particles. Have you considered using grafting for the polymer modification of the core particles? The use of polymers with an internal hierarchy, like block copolymers, could allow for the formation of a hemisphere with a distinct patchiness, as shown on microparticles[1] and flat surfaces.[2] This could allow for the selective adsorption of satellite particles and possibly even evoke symmetric order.

1 C. Kuttner, P. C. Maier, C. Kunert, H. Schlaad and A. Fery, *Langmuir*, 2013, **29**, 16119–16126.
2 C. Kuttner, A. Hanisch, H. Schmalz, M. Eder, H. Schlaad, I. Burgert and A. Fery, *ACS Appl. Mater. Interfaces*, 2013, **5**, 2469–2478.

Alla Synytska replied: The use of block copolymer brushes will lead to the formation of clusters (patches) on the surface of the particle. The effect of these clusters on the character of the self-assembly of the particles will be noticeable when the size of the satellite particles is comparable to the size of the cluster. This can allow for the selective adsorption of satellite particles.

Axel Mueller commented: What is the influence of the quaternization of PDMAEMA?

Alla Synytska responded: We used non-quaternized PDMAEMA, which is a weak polyelectrolyte and its charge depends on the pH. The quaternization of PDMAEMA to PDMAEMAq changes its properties and it becomes a strong polyelectrolyte, whose charge is independent of pH. The assembly of the particles will be similar to the behavior of the corresponding polyelectrolyte complex quat-PDMAEMA-PAA. The aggregates will be formed at a pH > pKa (PAA) when both polyelectrolytes are charged. Thereby, one can control and program the assembly between the quaternized PDMAEMA-modified and PAA-modified particles in a larger pH range compare to that for the non-quaternized PDMAEMA. At very low pH values, between pH 2 and pKa (PAA), the PAA chains are almost uncharged and the assembly will be determined by hydrophobic interactions and hydrogen bonds. In this case, the assembly between quaternized PDMAEMA-modified and PAA-modified particles into clusters will be non-specific and therefore hard to control and to program.

Axel Mueller commented: What is the typical degree of polymerization or length of a polymer chain? Can you vary this length?

Alla Synytska replied: The molecular weight of the polymers is *ca.* 100 kDa, which corresponds to a polymerization degree of 600. The polymerization degree can be controlled, for example, by varying the polymerization time and concentration of the monomer.

Younan Xia opened discussion of the paper by Etienne Duguet: The synthesis is very interesting, but it seems to involve too many steps. If the yield of each step is 70%, the total yield will only be 20% after four steps.

Etienne Duguet responded: In such a multistage synthesis of clusters two types of yield shall be considered. The morphology yield is the fraction of clusters that present the expected morphology at the end of the process. We demonstrated that it depends essentially on the first stage, *i.e.* the seeded-growth emulsion polymerization leading to silica/polystyrene clusters. If the fraction of tetrapods in the cluster mixture is 70%, the fraction of tetrapod-like silica/gold clusters at the end of the process will be very close to 70%. The amount yield, *i.e.* the more conventional definition of yield, depends on the number of stages and more particularly on the number of clusters that may be lost through the washing/purification steps (essentially centrifugation/redispersion cycles or dialysis) that are performed after each stage. It depends on the cluster amount and may be estimated to 95% at each stage, leading to a total yield between 50 and 70% depending on the number of gold regrowth/etching cycles.

Axel Mueller commented: The formation of a diamond lattice is the holy grail of colloidal self-assembly. Can you reach this either with polystyrene or with gold on your tetrapods?

Etienne Duguet responded: Tetrapod-like silica/polystyrene or silica/gold clusters are not good building blocks as demonstrated using digital experiments because of the satellite–satellite interactions that cannot be easily made directional. Nevertheless, silica particles with four sticky dimples may be good candidates to obtain a diamond lattice. If the original optical properties such as a full photonic bandgap are expected, three extra conditions shall be considered: 1) the diamond lattice needs to be open enough. Therefore, spacers such as spherical colloids will allow for the connection of two dimples and control of the distance between them. They could be envisioned as colloidal bond pairs mimicking the covalent electronic doublet in the structure of conventional diamond; 2) the contrast of refractive indices between the building blocks and the matrix will be maximal. Therefore, it does not allow any chemical composition; 3) metal such as gold colloids are not good candidates because of the inherent losses due to their high absorption cross-section.

Axel Mueller asked: To improve the synthesis, could you use an amine-containing polymer like 2-vinyl pyridine, then load $HAuCl_4$ and make gold seeds through reduction?

Etienne Duguet answered: Because 2-vinylpyridine may be polymerized using conventional emulsion polymerization, it could be a good idea to use it instead of styrene and therefore free ourselves of the polystyrene modification to make it aminated.

Christoph Hanske remarked: In this work you attached very small gold nanospheres to your tetrapods to overgrow them further with gold. In some images one can see several gold seeds in a single dimple. Would it be feasible to start with larger spheres to improve the homogeneity of the final clusters?

Etienne Duguet answered: It would be possible but the larger the gold seeds the more difficult their attachment to the dimples.

Matthias Thiele asked: The nanoparticles are huge (200 nm). Is it possible to make them smaller? Is it further possible to make the silica cores smaller?

Etienne Duguet responded: We have shown that it is possible to get good morphology yields of the polystyrene/silica tetrapods from 44 nm silica seeds.[1] By limiting the silica regrowth to a thickness of 10 nm, dimpled silica particles with a diameter of 65 nm could be surely obtained.

1 A. Désert, J. Morele, J.-C. Taveau, O. Lambert, M. Lansalot, E. Bourgeat-Lami, A. Thill, O. Spalla, L. Belloni, S. Ravaine and E. Duguet, *Nanoscale*, 2016, **8**, 5454–5469.

Matthias Thiele continued: Is it possible to attach a small dye inside the silica to achieve a plasmonic coupling with the gold?

Etienne Duguet answered: Because the silica seeds are prepared independently, they may be loaded with any other small particles or molecules. In particular, any fluorescent silica seed may be used for the emulsion polymerization stage provided that the MMS coupling agent is grafted with the right density on its surface. Therefore, within the resulting patchy particles, the fluorescent core could achieve a plasmonic coupling with the gold patches.

Guillermo González enquired: Would it be possible to attach gold nanospheres of 100 nm to the silica core in order to obtain clusters of gold satellites with tetrahedral symmetry in one step and high yield?

Etienne Duguet answered: We attempted such a direct assembly, but it failed very probably because of the huge density difference between gold and silica which would be detrimental to high collision probability.

Catherine Murphy addressed Etienne Duguet and David Gracias: Based on the previous session regarding 2D to 3D assembly, it seems that there might be a system in which gold particles are put on a surface that then pops into a tetrahedral shape? Has this been demonstrated?

David Gracias replied: Placing gold particles on a surface and then folding has been done. For two recent papers see **ref. 1 and 2** below, one on e-beam patterned gold and the other on gold nanoparticle deposition.

1 J.-H. Cho, M. D. Keung, N. Verellen, L. Lagae, V. V. Moshchalkov, P. Van Dorpe and D. H. Gracias, *Small*, 2011, 7, 1943–1948.
2 K. J. Si, D. Sikdar, Y. Chen, F. Eftekhari, Z. Xu, Y. Tang, W. Xiong, P. Guo, S. Zhang, Y. Lu, Q. Bao, W. Zhu, M. Premaratne and W. Cheng, *ACS Nano*, 2014, **8**, 11086–11093.

Etienne Duguet answered: As far as we know, this process has not been demonstrated yet.

Martin Buzza asked: Instead of trying to make a diamond lattice with the tetrapods, could you co-assemble tetrahedral cups and spheres at high concentration?

Etienne Duguet answered: This suggested route is the most promising.

Alberto Striolo addressed Etienne Duguet and Javier Reguera: Earlier Dr. Reguera showed us that pyramidal gold nano-structures can be grown on Fe particles (DOI: 10.1039/C6FD00012F). I was wondering whether the tetrapods you have described as seeds in your paper could be used as "seeds" for growing pyramidal gold particles? I would expect these structures would be much more regular than those presented by Dr. Reguera, because the tetrapods provide a naturally well-controlled substrate. Could you comment on whether changing the surfactants would be sufficient for growing the structures I described, and whether you see any use for such particles?

Javier Reguera answered: In my opinion, Prof. Duguet already demonstrated in his paper that the small gold nanoparticles attached to the tetrapods can act as seeds for the further growth of gold. This is very similar to our system with multicomponent nanoparticles, in our case iron oxide–gold, where only the gold part behaves as seeds during the gold growth. On the other hand, the formation of anisotropic gold shapes, like our branched gold nanostars, is directed by the surfactants used during the synthesis, so it seems very reasonable to think that this system could also have similar gold shapes to our Janus magnetic nanostars. In this case the nanoparticle would be composed of the central silica part and have four nanostars attached symmetrically to it. However, the nanostars still would have irregular gold branches as this is inherent to the nanostar synthesis.

I think they could become very interesting nanoparticles and deserve being synthesized. The resulting nanoparticle formed from four symmetrically arranged nanostars could be an excellent substrate for optical applications like SERS. The central silica part could help with the stability of the nanoparticles while the gold nanostars could perform the plasmonic functionality.

Etienne Duguet responded: Using our silica/gold tetrapods could be a way to control the direction of the star branches, but it seems more speculative to envision that the number of branches would be restricted to four. Because our patchy particles are larger than the seeds used in the paper by Javier Reguera, the final morphology would be more readily expected to be four half-stars located at the bottom of the four silica dimples.

Matthias Pauly commented: A tetrapod with 4 different substituents is chiral if the 4 satellites are different. This must be the case as the 4 gold nanoparticles are at least slightly different in size. Have you measured the chiroptical properties of the individual tetrapods? Do you see any way to control the chirality of such a structure, for instance by grafting 4 different nanoparticles (different in size and/or composition)? Is there a way to avoid a racemic mixture and favor one "enantiomer"?

Etienne Duguet answered: We did not measure the chiroptical properties of the individual clusters described in this paper and obtained by site-specific nucleation and growth. We do not see how this pathway would provide homogeneous batches of chiral clusters. As suggested, such chiral clusters would be more easily obtained through an assembly route. Such experiments are in progress for getting racemic mixtures! Favoring one colloidal enantiomer remains a true bottleneck.

Christian Kuttner asked: Have you considered adsorbing hydrogel-encapsulated gold nanoparticles into the silica templates? Hydrogels, *e.g.* PNIPAM, could facilitate the integration of gold particles into the template related to the soft nature of the swollen shell.[1] By using a hydrogel with residual vinyl functions, thiol-ene "click" chemistry might serve for the covalent attachment onto the thiol-modified PS residue inside the template.[2] In addition, hydrogel shells have been shown to be permeable for gold ions, which allows for post-modification like the overgrowth of the gold particles to the desired satellite sizes, inside the template structure.[3]

1 C. Kuttner, M. Chanana, M. Karg and A. Fery, Macromolecular decoration of nanoparticles for guiding self-assembly in 2D and 3D, in *Macromolecular Self-Assembly*, eds. L. Billon and

O. V. Borisov, John Wiley & Sons, Inc., 2016, pp. 159–192, DOI: 10.1002/9781118887813.ch6.
2 C. Kuttner, P. C. Maier, C. Kunert, H. Schlaad and A. Fery, *Langmuir*, 2013, **29**, 16119–16126.
3 M. B. Müller, C. Kuttner, T. König, V. V. Tsukruk, S. Förster, M. Karg and A. Fery, *ACS Nano*, 2014, **8**, 9410–9421.

Etienne Duguet responded: The alternative of creating new clusters through the assembly of dimpled silica particles on one hand and spherical colloids on the other was demonstrated by us, but has not yet been published. We successfully used covalent bond formation such as amidation or Huisgen cycloaddition. Hydrogels could actually be promising alternatives providing simultaneously sticky components and cradles for gold regrowth.

David Portehault asked: How much would the assembly, for instance into a diamond structure, be tolerant to size distribution?

Etienne Duguet responded: The formation of any crystalline structure, in particular the diamond-like one, requires a high monodispersity of the size and shape of the building blocks. More precisely, simulations showed that any deviation of the size larger than 2 to 5% is crippling to the colloidal crystallization for the formation of long range structures.

Benjamin Abécassis asked: The ultimate goal of the synthesis and self-assembly of these complex nanoparticles is to make a crystal. What volume fraction would you need? Is it reasonable with the current synthesis scheme?

Etienne Duguet answered: Making non close-packed crystals is indeed one of the ultimate goals of the patchy particles design. Please, have a look at the digital experiments of Smallenburg and Sciortino considering four patches with different patch-to-particle diameter ratios.[1] Our synthesis scheme is suitable for getting patchy particles at the gram scale. The main issue concerns the purity/polydispersity of the batches as already discussed.

1 F. Smallenburg and F. Sciortino, *Nat. Phys.*, 2013, **9**, 554–558.

Anna Roig remarked: Do you have control of the dimple size? Can you use polystyrene spheres of different sizes to control the dimple size?

Etienne Duguet answered: The dimple size can be controlled. Their diameter depends on the diameter of the polystyrene spheres and the extent of the silica core regrowth; their depth depends essentially on the extent of the silica core regrowth.

Christoph Hanske asked: Have you considered incorporating gold particles of different morphologies, for example thin end-functionalized nanorods?

Etienne Duguet responded: We attempted such an experiment with small nanorods but we found several of them attached within the same dimple. For attaching one nanorod per dimple it is mandatory to match the size of the sticky patch and the gold rod. Large gold nanorods would be more difficult to attach because of their high density.

Wenming Tong continued discussion of the paper by Alla Synytska: For the dumbbell shape assembly, is it possible to assemble the small spheres linearly in between the two large spheres? If yes, how long can the linear chain be in terms of particle numbers?

Alla Synytska replied: Linear structures will not be possible in our case. Addition of satellite particles to our Janus particle results in head-to-head sticking of the Janus particles. Linear structures are formed in the case of head-to-tail sticking, which can be achieved when satellite particles of one sort stick to one side of the Janus particles and satellite particles of another sort stick to the other side of the Janus particles.

Ana Maria Bago Rodriguez communicated with Krassimir Velikov:† You have studied both the foam and emulsion stability by varying the zein-to-silica surface ratio (ZSSR). For the foam study at pH 5 you found an optimum range of zein coating coverage (from 1 to 3 mg m^{-2}, approximately) that results in the largest initial foam volume. Are the most stable emulsions also obtained in this ZSSR range?

Krassimir Velikov communicated in response: Because of the different nature of emulsions (*i.e.* having a fixed volume fraction, the opposite to foams where the volume fraction of air dispersed in the liquid can change), a direct comparison with the emulsion stability as a function of the ZSSR could not be done in our experiment.

Ana Maria Bago Rodriguez communicated: In the experimental procedure for the preparation of the emulsions, is there any reason why after the mixing at 2000 rpm with a Servodyne mixer head for 2 min you have then shaken them vigorously by hand?

Krassimir Velikov communicated in response: The shaking by hand was performed to ensure that the sample was sufficiently homogeneous prior to transferring to a glass vial for storage.

† The paper by Krassimir Velikov was not presented at the meeting and questions were submitted by email after the meeting.

Faraday Discussions

PAPER

Optical anisotropy and sign reversal in layer-by-layer assembled films from chiral nanoparticles†

Zhumei Liang,[abc] Kalil Bernardino,[d] Jishu Han,[ab] Yunlong Zhou,[ab] Kai Sun,[e] André F. de Moura*[d] and Nicholas A. Kotov*[abcef]

Received 31st March 2016, Accepted 27th April 2016
DOI: 10.1039/c6fd00064a

Chiral anisotropy and related optical effects at the nanoscale represent some of the most dynamic areas of nanomaterials today. Translation of optical activity of chiral semiconductor and metallic nanoparticles (NPs) into optoelectronic devices requires preparation of thin films from chiral NPs on both flat and curved surfaces. In this paper we demonstrate that chiral NP films can be made *via* layer-by-layer assembly (LBL) using negatively charged chiral CdS NPs, stabilized by D- and L-cysteine and positively charged polyelectrolytes, as building blocks. LBL coatings from NPs combine simplicity of preparation and strong optical activity. Circular extinction measurements using circular dichroism instruments indicate that the film possess four chiroptical bands at 280, 320, 350, and 390 nm. The latter two bands at 390 and 350 nm are associated with the band gap transitions (chiral excitons), while the former two are attributed to transitions involving surface ligands. When NPs are assembled in LBL films, the rotatory activity and the sign for circular extinction associated with the electronic transition in the inorganic core of the NPs is conserved. However, this is not true for circular extinction bands at short wavelengths: the sign of the rotatory optical activity is reversed. This effect is attributed to the change of the conformation of surface ligands in the polyelectrolyte matrix, which was confirmed both by semi-empirical and density functional (DFT) quantum mechanical calculations. Circular dichroism spectra calculated using a DFT algorithm closely match the experimental spectra of CdS NPs. These findings indicate that the spectroscopic methods sensitive to chirality of the surface ligands can be used to investigate fine structural changes in the surface layer of

[a]*Department of Chemical Engineering, University of Michigan, Ann Arbor, Michigan, 48109, USA. E-mail: kotov@umich.edu*

[b]*Biointerfaces Institute, University of Michigan, Ann Arbor, Michigan, 48109, USA*

[c]*School of Chemistry and Chemical Engineering, State Key Laboratory Metal Matrix Composites, Shanghai Jiao Tong University, 800 Dongchuan Road, Shanghai, 200240, China*

[d]*Department of Chemistry, Federal University of São Carlos, São Carlos, 13565-905, Brazil. E-mail: moura@ufscar.br*

[e]*Department of Materials Science and Engineering, University of Michigan, Ann Arbor, Michigan 48109, USA*

[f]*Michigan Center for Integrative Research in Critical Care, Ann Arbor, MI, USA*

† Electronic supplementary information (ESI) available. See DOI: 10.1039/c6fd00064a

nanocolloids. Strong rotatory optical activity of nanostructured semiconductor films opens the possibilities for new polarization-based optical devices.

Introduction

Geometrical objects with non-superimposable mirror images are defined as chiral. Understanding, controlling and utilizing the subtle anisotropy of chiral molecular, nano-, and mesoscale interactions was foundational for advances in molecular biology, drug discovery, liquid crystal electronics, and other essential areas of science and technology. Over the last decade, there has been an emergence of scientific interest in the optical, chemical, and biological properties of chiral inorganic nanoparticles (NPs).[1–5] In general terms, the chiroptical activity of individual NPs can be assigned to the deviations of their shapes from a perfect sphere or rod with correspondingly spherical or cylindrical symmetry, respectively, to those lacking central symmetry. Due to the complexity of the atomic geometry of the NP cores and the surface layers with a lot of degrees of freedom and the possibility to deviate from perfection, one can safely state that most of the individual NPs, are in fact chiral, although typical dispersions are racemic unless a deliberate chiral bias was applied during their preparation. Considering the size, molecular mass, and surface chemistry of many water-soluble NPs, one could also see some similarity with nanoscale biomacromolecules, for instance, globular proteins.[6] The chirality of the latter in a variety of different asymmetric reactions and processes is well-known and is widely utilized both in the lab and by nature.

The syntheses of chiral NPs have been developed extensively and a variety of nanocrystalline dispersions can be obtained in liquid media.[1,7–11] The use of chiral stabilizing surface ligands is a relatively universal method for their synthesis and was implemented for synthesizing CdS,[9] CdSe,[12] and CdTe.[10] The chiroptical bands measured using conventional circular dichroism spectrometers were found for both NP semiconductor band-gap states[13,14] and plasmonic transitions.[7,8,15,16] Nanoscale assemblies of NPs[15] offer the possibility to amplify rotatory optical activity using geometrical factors[17–20] and field-enhancement effects.[21–25]

Besides being fundamentally interesting from chemical, optical, and biological standpoints, and potentially from the perspective of classical geometry and mathematics, NP chirality is a technologically relevant property and can be successfully used in bioanalytics.[15,18,21,26,27] Atomic chirality of the NP surface is the primary influential factor in the interactions of NPs with each other as well as in biological systems.[11,14]

Nanoscale chirality of the asymmetric NP cores with tetrahedral shapes reminiscent of similar geometries[10] and familiar from organic chemistry are known to affect the assembly patterns of large ensembles of NPs.[28] Strong polarization rotation stemming from the chirality at nanometer and submicron scales characteristic for individual NP and their higher order superstructures, respectively, represent a great resource for optical devices.[29,30] Chiral NPs and their assemblies can also provide new pathways for heterogeneous chiral catalysis,[31,32] separation,[33] biosensing,[34] and optical devices.[35–37]

From these standpoints, it will be useful to have a proven and universal tool for the preparation of coatings from optically active NPs. This process will complement the synthetic methods for preparing dispersions of chiral NPs as well as lithographically defined chiral plasmonic surfaces.[38–41]

In this work, we demonstrate that the chirality of individual CdS NPs can be transferred to thin films using layer-by-layer (LBL) assembly.[42–45] Their universality, transparency, and exceptional mechanics[46,47] make them valuable for further research on polarization optics based on NP composites.[48,49] LBL assembly makes it possible to "design" the spectrum of rotatory optical activity by varying the peak intensity and combining NPs[44] with other chiral compounds, such as polyelectrolytes,[50] cellulose,[51] and proteins.[52] The anisotropy factor of the chiral thin films made in this study is comparable or higher than that of dyes[53] and DNA,[54] and cholesteric liquid crystal.[55] Unlike previous studies of chiroptically active thin films involving NPs,[56,57] the comparison between rotatory optical activity in dispersions and in LBL films allows one to gain insights about the possible origin of specific chiroptical bands of NPs and how they are affected by the media around the particles. Application of quantum mechanical methods for description of optical activity makes possible assignment of specific bands and identification of molecular level changes in the chiral shell from organic surface ligands coating the NPs. Such knowledge is essential for fundamental understanding of NP interactions[58] and their biological functions.[59,60]

Results and discussion

Film deposition and structure

Negatively charged CdS NPs were synthesized using chiral L-cysteine and D-cysteine as stabilizers and will be referred to as CdS-L and CdS-D, respectively. Note that these notations are used to specify the method of NP preparation and the type of the surface ligands used for their synthesis but not the chiral symmetry of the NPs core or its organic shell. The chirality of NPs encompasses both molecular and nanometer scales, and thus, the description of the geometry of NP stereoisomers must account for their multidimensional chirality.

The NPs were assembled with cationic poly(diallyldimethylammonium) chloride (PDDA) in a conventional LBL manner; the corresponding films are denoted as D_n and L_n, respectively where n represents the number of deposition cycles. Quartz substrates were initially coated with a single layer of PDDA (0.5 cycle) and then immersed into the aqueous solution of CdS NPs dispersion. After 1 min of adsorption time they were taken out and rinsed with water, followed by drying for 1 min. This process represents one deposition cycle resulting in a deposition of a single PDDA/NP bilayer.

The tests of ellipsometric thickness of the LBL film coatings made after 5, 10, 15, and 20 cycles showed that they increased linearly with the increase of cycles (see ESI†). After one cycle of PDDA/CdS NPs deposition, obvious NPs can be seen on the AFM images (Fig. 1a and d). Upon subsequent deposition, the increasing amount of NPs is apparent (Fig. 1b and e). Importantly, no striation or other preferential alignment of NPs within the plane of the substrate can be seen. They would indicate that linear dichroism (LD) and linear birefringence (LB) may be associated with optical properties of these films. Having uniform distribution of NPs increases the probability of optical homogeneity of the coatings. Observing the section (Fig. 2a–d) and surface images (Fig. 2e and f), the thicknesses of the films linearly increased with increasing numbers of deposition cycles, which is

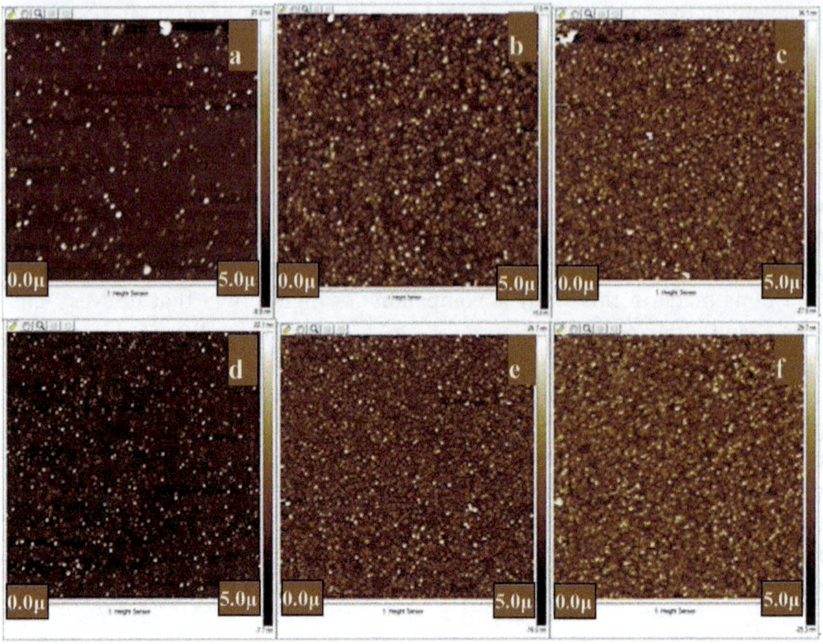

Fig. 1 AFM images of D_1 (a), D_2 (b), D_5 (c) and L_1 (d), L_2 (e), L_5 (f); vertical color bars are (a) 21 nm; (b) 17 nm; (c) 36 nm; (d) 22 nm; (e) 26 nm; (f) 26 nm.

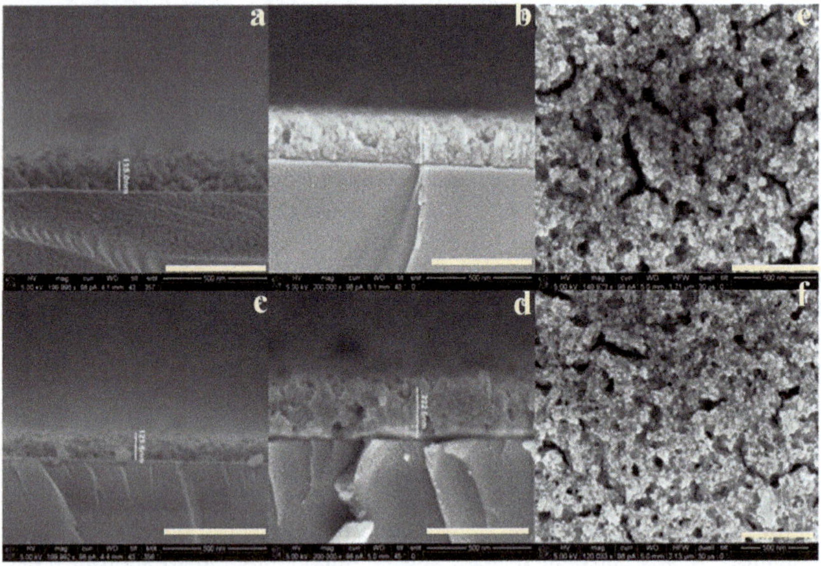

Fig. 2 SEM images of D_{15} (a), D_{20} ((b), (e)) and L_{15} (c), L_{20} ((d), (f)). All scale bars are 500 nm.

technologically convenient. After 20 deposition cycles, a dense layer formed on the substrate (Fig. 2e and f), and completely transparent LBL-prepared PDDA/CdS films were obtained.

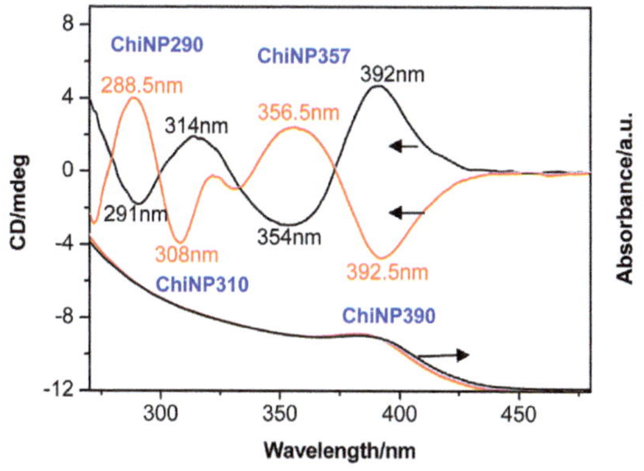

Fig. 3 UV-Vis (lower lines) and corresponding CD (upper lines) spectra of CdS-D (black) and CdS-L (red) NPs.

Chiroptical activity of the LBL films

Similar to other reports on CdS NPs decorated with chiral ligands,[9,61] CdS-D and CdS-L in dispersion possess four positive and negative circular dichroism (CD) bands from 270 to 450 nm (Fig. 3), that could be denoted using their spectral locations as ChiNP290, ChiNP310, ChiNP357, and ChiNP390. Free D- and L-cysteine molecules do not absorb light at these wavelengths.[62] The ChiNP390 band can be assigned to the bandgap transition based on the match with the peak in the absorption spectra, however the remaining bands – ChiNP290, ChiNP310, and ChiNP357 – are not resolved in the absorption spectra (Fig. 3). They are "invisible" in the absorption spectra of the dispersions because of broadening due to size inhomogeneity or other reasons. They are likely to correspond to the electronic transition of higher energy associated with either excitonic or surface states such as those formed by hybridization of the NP state with the molecular states of cysteine.[14,63]

Let us look now at the optical properties of the corresponding LBL films. As expected, absorbance of the films increased with increased numbers of deposition cycles (Fig. 4a and b). The emission spectra of all the films show broad bands between 450 and 700 nm with the maximum wavelength in the range of 540–560 nm; under UV light, both D- and L-films gave bright green emission.

The intensity of the rotatory optical activity is varied by changing the number of film bilayers (Fig. 4c). Importantly, the spectra obtained using standard CD spectrometers for solid thin films may include other anisotropic components common for light–matter interactions, such as optical activity associated with dichroic and birefringent effects in complex optical media.[64] For instance, small variations of dipping angle during LBL deposition may lead to additional components' polarization rotation originating from LD and LB, despite the fact that the distribution of NPs in the polymer matrix is uniform as can be seen in Fig. 2. Although we did not see any variations of CD spectra when carrying out simple tests meant to identify some of these effects,[65] we shall nevertheless be cautious and refer to optical data in thin films as circular extinction (CE).

D_n and L_n films displayed increased CE responses with increasing number of LBL deposition cycles. Taking the interval from 15 to 20 deposition cycles as an example, the degree of rotations per one millimeter of film thickness was

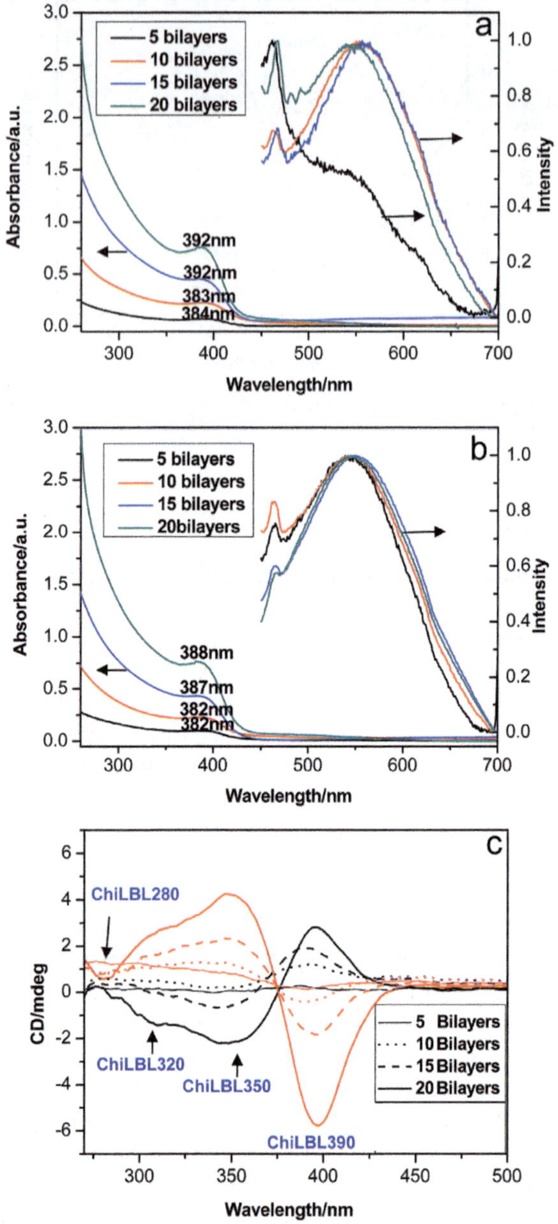

Fig. 4 ((a), (b)) UV-Vis and luminescence spectra for different numbers of bilayers of D_n (a) and L_n (b) films. (c) Circular extinction spectra for different numbers of film bilayers; D_n and L_n films are denoted with black and red colors, respectively.

calculated, assuming that most of the polarization rotation originated from the chirality of individual NPs. These calculations are carried out to evaluate the upper bounds of chiroptical anisotropy of these films (Table 1). The thickness measurements were obtained using an ellipsometer (Table 1 in ESI†), and was confirmed by scanning electron microscope (SEM) images (Fig. 2a–d). For ChiLBL390, the rotation of the film increased from 2.48 (D_{15}) to 5.02 (D_{20}) mdeg

Table 1 CE data and thickness of LBL films at 15 and 20 deposition cycles

Film/deposition cycles	Thickness/nm	CE (mdeg)/ wavelength (nm)		Rotation/mdeg
D_{15}	230	−0.68/343	1.79/396	2.45
D_{20}	338	−2.21/343	2.80/396	5.02
L_{15}	270	2.35/347	−1.79/397	4.14
L_{20}	374	4.29/347	−5.73/397	10.02

and the thicknesses increased from 230 (D_{15}) to 338 (D_{20}) nm. The degree of rotation of the D_n film was 23.5 deg per mm. Similarly, the degree of rotation of L_n film was 56.6 deg per mm. For comparison, cholesteric liquid crystals display an optical rotation power of 10^3–10^5 deg per mm.[55]

According to Lambert–Beer's law, $\Delta\varepsilon = \frac{\Delta A}{cb}$, $\Delta A = \frac{\theta\ (\text{mdeg})}{33\ 000}$, and c is the concentration of CdS NPs in a LBL film, which can be calculated by the thermogravimetric analysis (TGA) data of the film, b is the path length of light, (Table 1). $\Delta\varepsilon$ of D_{15}, D_{20}, L_{15}, and L_{20} at different wavelengths were obtained (Table 2 in ESI†). Compared with some organic and bio-organic molecules of similar molecular mass (Table 3 in ESI†), D_n and L_n have similar $\Delta\varepsilon$ values.[53]

Optical activity at different wavelengths

Similarly to free NPs, we observed four different bands in the CE spectra of corresponding LBL films: ChiLBL280, ChiLBL320, ChiLBL350, and ChiLBL390. CE peaks demonstrated a slight red shift as the number of layers increased (Fig. 4a and b), as expected from absorbance in densely packed arrays of NPs. The ChiLBL390 band corresponds to the bandgap transition and is the same as seen in individual CdS NPs. Note that the other three bands – ChiLBL280, ChiLBL320, ChiLBL350 – retain approximately the same positions as the bands in CdS NP dispersions. It is instructive to follow the signs of the CD/CE bands for NPs with the same chiral organic shell, for instance, CdS-L. Starting from the longer wavelengths, the sequence of CD band signs for NP dispersions in Fig. 3 is negative (ChiNP390), positive (ChiNP357), negative (ChiNP310), and positive (ChiNP290). At the same time, the sequence of CE band signs in LBL films from CdS-L in Fig. 4 in the same order is different, namely, negative (ChiLBL390), positive (ChiLBL350), positive (ChiLBL320), and negative (ChiLBL280). In other words, the signs of the CE bands close to the bandgap remain the same as CD bands for NP dispersions, while the signs of the CE bands at shorter wavelengths are opposite in LBL films compared to CD bands of NPs in dispersion. The similar sequence of signs for polarization rotation is observed for CdS-D NPs with the difference that all of the bands are mirrored with respect to abscissa in the CE spectrum.

The reversal of the sign for rotatory optical activity when NPs become incorporated into the thin composite film is unexpected and, therefore, worth additional attention. Considering the relative strength and similarity of the band position for NP in dispersion and NPs in the film, one has reason to suppose the band changes sign due to the change of chiral geometry associated with this band.

First let us briefly discuss the origin of chirality in NPs to better understand the potential reasons for the sign reversal. It can originate from multiple sources and the two that received most of the experimental support are the chiral arrangements of the surface ligands on the NP surface[66,67] and asymmetry of the inorganic core.[1,11,28,68] These are non-mutually exclusive reasons for NP chirality. The chiral distribution of atoms in the CdS core can originate from the influence of cysteine ligands on the atoms in the surface of the NP core.[1]

The CD/CE band at 390 nm coincides well with the first excitonic transition of CdS NPs visible absorption spectrum. This transition is associated purely with the NP core and does not include any influence of the surface ligand. The conservation of the sign of 390 nm rotatory optical activity when comparing the NPs in solution and LBL films is consistent with such an explanation because the atomic distribution of atoms is unlikely to change as a result of the soft matrix around it.

The same explanation could potentially be valid in case of the band at 350 nm; it could be assigned to the next allowed (or 'bright') excitonic transition in CdS NPs.[69,70] However, it does not apply to the bands at 310 and 290 nm. The fact that these bands change their chirality when the NPs' surroundings change is indicative that the chiral organization of atoms is affected by the media. They could be associated with the transitions involving surface ligands and can be generically described as "surface states". Using semi-empirical calculations with the PM6 algorithm, we found that the electronic transitions involving the cysteine surface ligands are indeed possible (Fig. 5). The calculations were carried out for hypothetical NP geometries with nearly perfect tetrahedrons as well as their truncated versions. The possibility of some portion of electron density to be located on the ligands was also demonstrated for high energy transitions in gold clusters.[1] Therefore, the change of sign for optical anisotropy for NPs in aqueous dispersions and in polymer matrices may originate from the change of geometry of the surface ligands.

Further insight in the role played by the ligands on each of the CD bands may be obtained from time-dependent density functional theory (TD-DFT) calculations. The calculated CD spectrum for the CdS-D NP (Fig. 6) based on 16 representative transitions along the spectral region of interest was in excellent agreement with experimental data (Fig. 3); both the position of the bands and their signs matched the CD spectra. Although the number of the atoms in the computational NP is smaller than those in dispersions, so as to become computationally tractable, the agreement between calculated and experimental CD spectra demonstrated that the main features of a typical CdS-D NP were

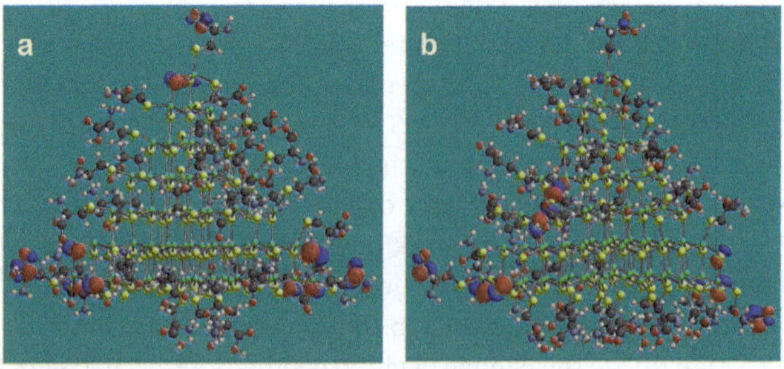

Fig. 5 (a) LUMO 1 and (b) LUMO 2 for CdS-D calculated using semi-empirical PM6 algorithm in SPARTAN computational environment.

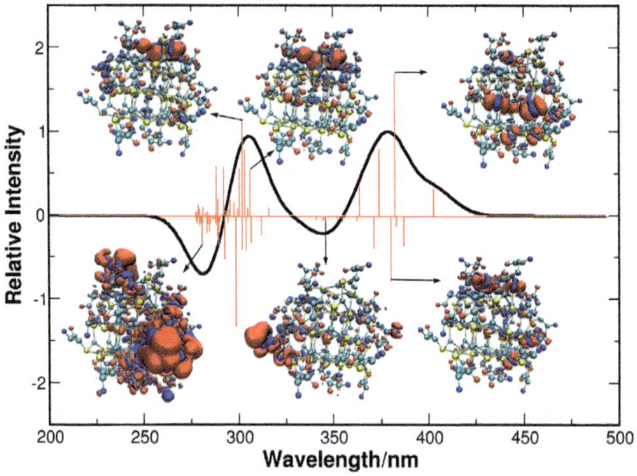

Fig. 6 TD-DFT CD spectrum for CdS-D. Vertical red lines stand for the calculated transitions, whereas the black curve stands for the convoluted spectrum (Gaussian line-shape with half-width of 12 nm). The convoluted curve was normalized to have unity intensity for its most intense band. The charge carriers for six selected states are depicted, using red surfaces for the holes and blue surfaces for the electrons of each transition. All surfaces were plotted with an isovalue of 0.0005 e $bohr^{-3}$ and the hydrogen atoms of cysteine ligands were removed for better visualization.

reproduced by the computational model. Also important was that findings of TD-DFT methods confirmed those in the simpler semi-empirical model in Fig. 5 that surface atoms of the NP core and surface ligands are essential for understanding the CD/CE spectra in the UV-Vis region. The advantage of the semi-empirical PM6-based model is the ability to calculate NPs with a much larger number of atoms than TD-DFT models.

As a general trend, the bands associated with high energy transitions above 390 nm were located mainly on NP atoms close to the surface cysteine molecules (Fig. 6 and ESI†). The isovalue which was chosen to render the surfaces of electronic density differences was large enough to allow the visualization of only the largest contributions to the overall electronic density reorganization taking place upon excitation.

The TD-DFT models showed that, in addition to the bandgap transition associated with the NP core as a whole, the ChiNP390 chiroptical band is likely to have a contribution from reorganization of the electronic density of a few sulfur atoms, always of the p → s type (Fig. 6 and ESI†), with only minor contributions from cysteine molecules. We expect that the valence band–conduction band transition will dominate optical absorption for these NPs.

The ChiNP357 band may also involve charge reorganization of cysteine molecules but the optical absorption is likely to be still dominated by the exciton transition. The contributions from cysteine molecules bonded to the truncated vertexes of NPs, and contributions from higher level bandgap transitions and from ligands on the edges and faces of the NPs should be considered.

The transition densities for the ChiNP310 band displayed a large contribution from transitions located on sulfur atoms with a significant contribution of density reorganization from cysteine molecules. Finally, the ChiNP290 band displayed the most complex transition densities, with larger contributions from the carboxyl groups of cysteine molecules. The transition densities also suggest some degree of charge transfer between cysteine molecules, with some molecules losing electronic density while others gained electronic density.

The TD-DFT data confirmed that conformational change in the ligands associated with the transition from the aqueous dispersion to LBL film must result in the change of the chiroptical signature of the NPs. One can notice an additional component of chirality in NPs. If the NP core was a perfect tetrahedron, all the vertexes would be identical and should have equivalent contributions in the transition densities for each band. However, the CdS core became distorted and the four vertexes were not equivalent, which was previously observed in the small gold clusters.[1,8,66,71] In order to assess these subtle structural differences, the Cd–S bonds were classified by their length as either shorter than 0.26 nm or longer than 0.26 nm (Fig. 7). The first vertex in Fig. 7 (S atoms in red) had the larger number of Cd–S bonds shorter than 0.26 nm as compared to the other vertexes, and the nearby bonds in the faces and edges also tended to be shorter when close to this vertex. It is well-known from quantum mechanics that more confined systems should display more energetic transitions. Shorter bond lengths between the same pair of atoms can be associated with increased confinement effects for highly localized transitions as those we have just described. In other words, quantum confinement itself has become chiral, leading to dissymmetric contributions from each corner to the observed bands, and, therefore, to chiral excitons. Even considering that all the electrons should be regarded as belonging to the whole NP, it may be expected that regions with shorter bond lengths (and larger

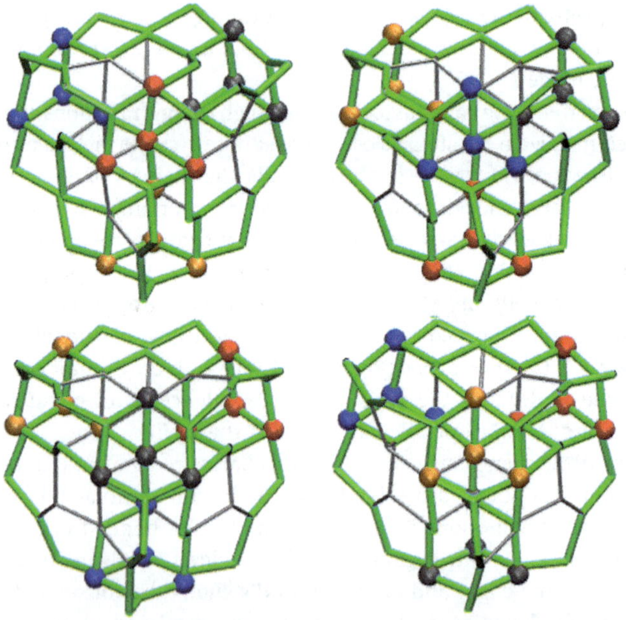

Fig. 7 Representation of the vertexes of CdS NP core, where the four S atoms defining a vertex are highlighted with a different color for each vertex. Thick green lines: Cd–S bonds with length shorter than 0.26 nm. Thin gray lines: Cd–S bonds with bond length longer than 0.26 nm.

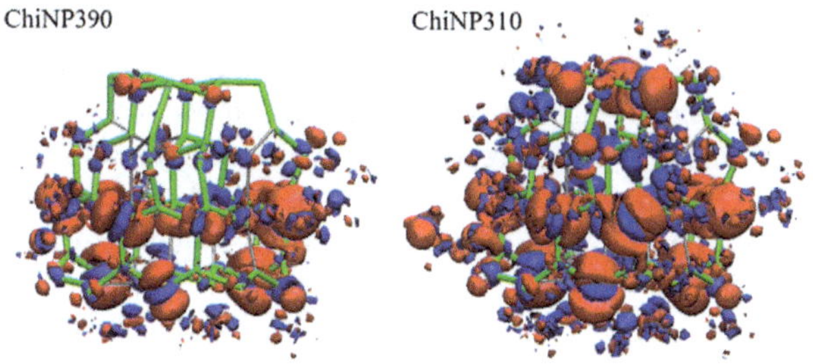

Fig. 8 Superposition of all relevant transition orbitals for the ChiNP390 and ChiNP310 bands. Only the CdS core is shown for clarity, using the same representation of Fig. 7. Individual transitions comprising these representations may be found in the ESI.†

confinement effects) should contribute less to low-energy transitions. Indeed, if we superimpose the transition densities for all relevant transitions comprising a band, it becomes clear that the denser vertex (with S atoms depicted in red in Fig. 7) had almost no change in the electron density for the transitions comprising the ChiNP390 band, whereas significant contributions from this vertex were observed for the higher-energy ChiNP310 band (Fig. 8).

In summary, thin films from chiral NPs can be made using LBL assembly. These films displayed rotatory optical activity with optical anisotropy similar to that of organic molecules and bio-molecules. The relatively low energy circular extinction bands were attributed to electronic transitions in the slightly twisted crystal lattice. The change of molecular environment of NPs from dispersion to LBL films has little effect on their spectral position and chiroptical polarity. The sign reversal of circular extinction bands at high energy when NPs assembled into LBL films was attributed to the change of conformation of surface ligands. Although more detailed study of this effect is needed, perhaps including *ab initio* level quantum mechanical calculations capable of more accurate attribution of specific absorption bands, it is clear that the chiroptical spectroscopy toolbox can be used for studies of fine structural effects in NPs because the transitioned invisible bands in UV-Vis and luminescence spectra due to broadening become more apparent in optical rotation spectra.

Methods

Materials

Medium molecular weight (PDDA) was purchased from Aldrich and diluted to 1 wt% in deionized water. D- and L-cysteine hydrochloride monohydrate, $Cd(ClO_4)_2\cdot 6H_2O$, and thioacetamide were all received from Aldrich and used as received.

Synthesis of D- and L-cysteine stabilized CdS NPs (CdS-D and CdS-L)

The procedure for the synthesis of CdS NPs was adapted from a procedure published before.[9] Briefly, $Cd(ClO_4)_2\cdot 6H_2O$ (1.2 mmol) and D- or L-cysteine (2.4 mmol) were dissolved in deionized water (60 mL), followed by adjusting the pH value to 10 with 1 M NaOH. The solution was placed in a three-necked flask with refluxing and de-aeration by nitrogen (N_2) gas for 0.5 h. Then, thioacetamide (0.6 mmol) was added to the above N_2-saturated aqueous solution. The solution was de-aerated by N_2 for another 0.5 h, then was heated to 100 °C in an oil bath for 6 h.

Preparation of LBL films

The LBL films were prepared by the traditional dip-coating technique. A piranha-washed quartz (or silicon) substrate was first soaked in 1 wt% PDDA solution for 1 min. It was then rinsed under a stream of deionized water for 30 s and dried with compressed air. The substrate was then soaked in the CdS NPs solution for 1 min before being rinsed and dried again. The four-step sequence was repeated for each of the cycles.

Instrumentation

The growth of the film was evaluated by light transmission measurements obtained using an Agilent 8453E UV-visible spectrometer. Photoluminescence (PL) spectra were obtained on a Jobin Yvon Horiba FluoroMax-3 instrument. Circular dichroism and circular extinction measurements were performed using a J-715 spectropolarimeter. The thickness of the film was obtained using a M-44IR spectroscopic ellipsometer (J.A. Woollam Co. Inc.). Transmission electron microscopy (TEM) observation was performed using a JEOL 3011 high-resolution electron microscope, and TEM specimens were prepared by dropping a cleaned aqueous solution of CdS NPs on an ultrathin carbon film on a holey carbon support film, 400 mesh grid supplied by Ted Pella, Inc. The specimens were left to dry at room temperature. Scanning electron microscopy (SEM) images were obtained with an FEI Nova 200 Nanolab. Atomic force microscopy (AFM) imagery was performed using a Nanoscope III (Digital Instruments/Veeco Metrology Group).

Computational studies

A tetrahedral NP was drawn with stoichiometry $Cd_{31}S_{40}$, with (111) faces terminated in sulfur atoms and truncated corners. Three D-cysteine molecules were added to each face, along with two extra ligands at each edge, resulting in an organic layer comprising 24 D-cysteine molecules. The cysteine molecules were treated as neutral amino acids ($-NH_2$ and –COOH groups). Geometry optimization was carried out with the PM7 Hamiltonian[72] as implemented in the MOPAC2012 (ref. 73 and 74) software, using the COSMO solvation model for water.[75] TD-DFT calculations were performed with the Becke's three-parameters exchange functional and the Lee–Yang–Par correlation functional (the so-called B3LYP hybrid functional),[76] using the LANL2DZ-ECP basis set.[77,78] Altogether, 80 excited singlet states were computed, using an active space for the Davidson procedure in the range of Kohn–Sham orbitals from 850 to 875 (HOMO was orbital number 867). Solvent effects were included using the PCM water model[79] with UFF atomic radii scaled by 1.5 (so as to avoid artificial solvent cavities inside the NP). All TD-DFT calculations were performed with the Gaussian 09 program (revision C.01).[80]

Acknowledgements

The central part of this work was supported by the NSF project "*Energy- and Cost-Efficient Manufacturing Employing Nanoparticles*" NSF 1463474. Financial support from FAPESP (2013/07296-2). K. B. thanks CNPq for a scholarship and A. F. M. thanks MEC/PET for a fellowship. Partial support of this work was also made by the Center for Photonic and Multiscale Nanomaterials (C-PHOM) funded by the National Science Foundation (NSF) Materials Research Science and Engineering Center program DMR 1120923 as well as NSF projects 1403777, 1411014, 1463474, 1538180 and Department of Army W911NF-10-1-0518 Reconfigurable Matter from Programmable Colloids. The authors thank the University of Michigan's EMAL for its assistance with electron microscopy, and for the NSF grant #DMR-9871177 for funding of the JEOL 2010F analytical electron microscope used in this work.

References

1 C. Noguez and I. L. Garzón, Optically active metal nanoparticles, *Chem. Soc. Rev.*, 2009, **38**, 757–771.
2 A. Guerrero-Martínez, J. L. Alonso-Gómez, B. Auguié, M. M. Cid and L. M. Liz-Marzán, From individual to collective chirality in metal nanoparticles, *Nano Today*, 2011, **6**, 381–400.
3 Y. Mastai, Enantioselective crystallization on nanochiral surfaces, *Chem. Soc. Rev.*, 2009, **38**, 772–780.
4 S. Srivastava, A. Santos, K. Critchley, K.-S. Kim, P. Podsiadlo, K. Sun, J. Lee, C. Xu, G. D. Lilly, S. C. Glotzer and N. A. Kotov, Light-controlled self-assembly of semiconductor nanoparticles into twisted ribbons, *Science*, 2010, **327**, 1355–1359.
5 Y. Xia, Y. Zhou and Z. Tang, Chiral inorganic nanoparticles: Origin, optical properties and bioapplications, *Nanoscale*, 2011, **3**, 1374–1382.
6 N. A. Kotov, Inorganic nanoparticles as protein mimics, *Science*, 2010, **330**, 188–189.
7 C. Gautier and T. Bürgi, Chiral *N*-isobutyryl-cysteine protected gold nanoparticles: Preparation, size selection, and optical activity in the UV-vis and infrared, *J. Am. Chem. Soc.*, 2006, **128**, 11079–11087.
8 T. G. Schaaff and R. L. Whetten, Giant gold–glutathione cluster compounds: Intense optical activity in metal-based transitions, *J. Phys. Chem. B*, 2000, **104**, 2630–2641.
9 M. P. Moloney, Y. K. Gun'ko and J. M. Kelly, Chiral highly luminescent CdS quantum dots, *Chem. Commun.*, 2007, 3900.
10 Y. Zhou, M. Yang, K. Sun, Z. Tang and N. A. Kotov, Similar topological origin of chiral centers in organic and nanoscale inorganic structures: Effect of stabilizer chirality on optical isomerism and growth of CdTe nanocrystals, *J. Am. Chem. Soc.*, 2010, **132**, 6006–6013.
11 J. E. Govan, E. Jan, A. Querejeta, N. A. Kotov and Y. K. Gun'ko, Chiral luminescent CdS nano-tetrapods, *Chem. Commun.*, 2010, **46**, 6072–6074.
12 S. A. Gallagher, M. P. Moloney, M. Wojdyla, S. J. Quinn, J. M. Kelly and Y. K. Guni'ko, Synthesis and spectroscopic studies of chiral CdSe quantum dots, *J. Mater. Chem.*, 2010, **20**, 8350.
13 S. D. Elliott, M. P. Moloney and Y. K. Gun'ko, Chiral shells and achiral cores in CdS quantum dots, *Nano Lett.*, 2008, **8**, 2452–2457.
14 N. Suzuki, Y. Wang, P. Elvati, Z.-B. Qu, K. Kim, S. Jiang, E. Baumeister, J. Lee, B. Yeom, J. H. Bahng, *et al.*, Chiral graphene quantum dots, *ACS Nano*, 2016, **10**, 1744–1755.
15 W. Chen, A. Bian, A. Agarwal, L. Liu, H. Shen, L. Wang, C. Xu and N. A. Kotov, Nanoparticle superstructures made by polymerase chain reaction: Collective

interactions of nanoparticles and a new principle for chiral materials, *Nano Lett.*, 2009, **9**, 2153–2159.

16 G. Shemer, O. Krichevski, G. Markovich, T. Molotsky, I. Lubitz and A. B. Kotlyar, Chirality of silver nanoparticles synthesized on DNA, *J. Am. Chem. Soc.*, 2006, **128**, 11006–11007.

17 A. Kuzyk, R. Schreiber, Z. Fan, G. Pardatscher, E.-M. Roller, A. Högele, F. C. Simmel, A. O. Govorov and T. Liedl, DNA-based self-assembly of chiral plasmonic nanostructures with tailored optical response, *Nature*, 2012, **483**, 311–314.

18 X. Wu, L. Xu, L. Liu, W. Ma, H. Yin, H. Kuang, L. Wang, C. Xu and N. A. Kotov, Unexpected chirality of nanoparticle dimers and ultrasensitive chiroplasmonic bioanalysis, *J. Am. Chem. Soc.*, 2013, **135**, 18629–18636.

19 W. Ma, H. Kuang, L. Wang, L. Xu, W.-S. Chang, H. Zhang, M. Sun, Y. Zhu, Y. Zhao, L. Liu, *et al.*, Chiral plasmonics of self-assembled nanorod dimers, *Sci. Rep.*, 2013, **3**, 1934.

20 B. Auguié, J. L. Alonso-Gómez, A. Guerrero-Martínez and L. M. Liz-Marzán, Fingers crossed: Optical activity of a chiral dimer of plasmonic nanorods, *J. Phys. Chem. Lett.*, 2011, **2**, 846–851.

21 E. Hendry, T. Carpy, J. Johnston, M. Popland, R. V. Mikhaylovskiy, A. J. Lapthorn, S. M. Kelly, L. D. Barron, N. Gadegaard and M. Kadodwala, Ultrasensitive detection and characterization of biomolecules using superchiral fields, *Nat. Nanotechnol.*, 2010, **5**, 783–787.

22 I. Lieberman, G. Shemer, T. Fried, E. M. Kosower and G. Markovich, Plasmon-resonance-enhanced absorption and circular dichroism, *Angew. Chem., Int. Ed.*, 2008, **47**, 4855–4857.

23 Z. Zhu, W. Liu, Z. Li, B. Han, Y. Zhou, Y. Gao and Z. Tang, Manipulation of collective optical activity in one-dimensional plasmonic assembly, *ACS Nano*, 2012, **6**, 2326–2332.

24 A. O. Govorov, Z. Fan, P. Hernandez, J. M. Slocik and R. R. Naik, Theory of circular dichroism of nanomaterials comprising chiral molecules and nanocrystals: Plasmon enhancement, dipole interactions, and dielectric effects, *Nano Lett.*, 2010, **10**, 1374–1382.

25 M. Chekini, R. Filter, J. Bierwagen, A. Cunningham, C. Rockstuhl and T. Bürgi, Fluorescence enhancement in large-scale self-assembled gold nanoparticle double arrays, *J. Appl. Phys.*, 2015, **118**, 233107.

26 L. Wang, L. Xu, H. Kuang, C. Xu and N. A. Kotov, Dynamic nanoparticle assemblies, *Acc. Chem. Res.*, 2012, **45**, 1916–1926.

27 Y. Zhao, L. Xu, W. Ma, L. Wang, H. Kuang, C. Xu and N. A. Kotov, Shell-engineered chiroplasmonic assemblies of nanoparticles for zeptomolar DNA detection, *Nano Lett.*, 2014, **14**, 3908–3913.

28 J. Yeom, B. Yeom, H. Chan, K. W. Smith, S. Dominguez-Medina, J. H. Bahng, G. Zhao, W.-S. Chang, S.-J. Chang, A. Chuvilin, *et al.*, Chiral templating of self-assembling nanostructures by circularly polarized light, *Nat. Mater.*, 2014, **14**, 66–72.

29 W. Li, Z. J. Coppens, L. V. Besteiro, W. Wang, A. O. Govorov and J. Valentine, Circularly polarized light detection with hot electrons in chiral plasmonic metamaterials, *Nat. Commun.*, 2015, **6**, 8379.

30 N. Yu and F. Capasso, Flat optics with designer metasurfaces, *Nat. Mater.*, 2014, **13**, 139–150.

31 T. Yasukawa, H. Miyamura and S. Kobayashi, Chiral metal nanoparticle-catalyzed asymmetric C–C bond formation reactions, *Chem. Soc. Rev.*, 2014, **43**, 1450–1461.
32 T. Hu, B. P. Isaacoff, J. H. Bahng, C. Hao, Y. Zhou, J. Zhu, X. Li, Z. Wang, S. Liu, C. Xu, *et al.*, Self-organization of plasmonic and excitonic nanoparticles into resonant chiral supraparticle assemblies, *Nano Lett.*, 2014, **14**, 6799–6810.
33 N. Shukla, M. A. Bartel and A. J. Gellman, Enantioselective separation on chiral Au nanoparticles, *J. Am. Chem. Soc.*, 2010, **132**, 8575–8580.
34 S. Li, L. Xu, W. Ma, X. Wu, M. Sun, H. Kuang, L. Wang, N. A. Kotov and C. Xu, Dual-mode ultrasensitive quantification of microRNA in living cells by chiroplasmonic nanopyramids self-assembled from gold and upconversion nanoparticles, *J. Am. Chem. Soc.*, 2016, **138**, 306–312.
35 A. Baev, M. Samoc, P. N. Prasad, M. Krykunov and J. Autschbach, A quantum chemical approach to the design of chiral negative index materials, *Opt. Express*, 2007, **15**, 5730.
36 N. Yu, F. Aieta, P. Genevet, M. A. Kats, Z. Gaburro and F. Capasso, A broadband, background-free quarter-wave plate based on plasmonic metasurfaces, *Nano Lett.*, 2012, **12**, 6328–6333.
37 J. Zhang, W. Feng, H. Zhang, Z. Wang, H. A. Calcaterra, B. Yeom, P. A. Hu and N. A. Kotov, Multiscale deformations lead to high toughness and circularly polarized emission in helical nacre-like fibres, *Nat. Commun.*, 2016, **7**, 10701.
38 M. Hentschel, M. Schäferling, T. Weiss, N. Liu and H. Giessen, Three-dimensional chiral plasmonic oligomers, *Nano Lett.*, 2012, **12**, 2542–2547.
39 K. Jefimovs, N. Saito, Y. Ino, T. Vallius, P. Vahimaa, J. Turunen, R. Shimano, M. Kauranen, Y. Svirko and M. Kuwata-Gonokami, Optical activity in chiral gold nanogratings, *Microelectron. Eng.*, 2005, **78–79**, 448–451.
40 R. Ogier, Y. Fang, M. Svedendahl, P. Johansson and M. Käll, Macroscopic layers of chiral plasmonic nanoparticle oligomers from colloidal lithography, *ACS Photonics*, 2014, **1**, 1074–1081.
41 A. V. Kildishev, A. Boltasseva and V. M. Shalaev, Planar photonics with metasurfaces, *Science*, 2013, **339**, 1232009.
42 G. Decher, Fuzzy nanoassemblies: toward layered polymeric multicomposites, *Science*, 1997, **277**, 1232–1237.
43 P. Podsiadlo, A. K. Kaushik, E. M. Arruda, A. M. Waas, B. S. Shim, J. Xu, H. Nandivada, B. G. Pumplin, J. Lahann, A. Ramamoorthy, *et al.*, Ultrastrong and stiff layered polymer nanocomposites, *Science*, 2007, **318**, 80–83.
44 N. A. Kotov, I. Dékány and J. H. Fendler, Layer-by-layer self-assembly of polyelectrolyte-semiconductor nanoparticle composite films, *J. Phys. Chem.*, 1995, **99**, 13065–13069.
45 C. Jiang, S. Markutsya, Y. Pikus and V. V. Tsukruk, Freely suspended nanocomposite membranes as highly sensitive sensors, *Nat. Mater.*, 2004, **3**, 721–728.
46 Z. Tang, N. A. Kotov, S. Magonov and B. Ozturk, Nanostructured artificial nacre, *Nat. Mater.*, 2003, **2**, 413–418.
47 A. A. Mamedov and N. A. Kotov, Free-standing layer-by-layer assembled films of magnetite nanoparticles, *Langmuir*, 2000, **16**, 5530–5533.
48 Y. Kim, J. Zhu, B. Yeom, M. Di Prima, X. Su, J.-G. Kim, S. J. Yoo, C. Uher and N. A. Kotov, Stretchable nanoparticle conductors with self-organized conductive pathways, *Nature*, 2013, **500**, 59–63.

49 L. Xu, W. Ma, L. Wang, C. Xu, H. Kuang and N. A. Kotov, Nanoparticle assemblies: Dimensional transformation of nanomaterials and scalability, *Chem. Soc. Rev.*, 2013, **42**, 3114–3126.

50 Y. Lvov, S. Yamada and T. Kunitake, Non-linear optical effects in layer-by-layer alternate films of polycations and an azobenzene-containing polyanion, *Thin Solid Films*, 1997, **300**, 107–112.

51 M. Agarwal, Q. Xing, B. S. Shim, N. Kotov, K. Varahramyan and Y. Lvov, Conductive paper from lignocellulose wood microfibers coated with a nanocomposite of carbon nanotubes and conductive polymers, *Nanotechnology*, 2009, **20**, 215602.

52 J. J. Richardson, M. Bjornmalm and F. Caruso, Technology-driven layer-by-layer assembly of nanofilms, *Science*, 2015, **348**, 6233, DOI: 10.1126/science.aaa2491.

53 N. Berova, L. Di Bari and G. Pescitelli, Application of electronic circular dichroism in configurational and conformational analysis of organic compounds, *Chem. Soc. Rev.*, 2007, **36**, 914–931.

54 V. Rizzo and J. A. Schellman, Matrix-method calculation of linear and circular dichroism spectra of nucleic acids and polynucleotides, *Biopolymers*, 1984, **23**, 435–470.

55 S. Furumi, Recent progress in chiral photonic band-gap liquid crystals for laser applications, *Chem. Rev.*, 2010, **10**, 394–408.

56 H. S. Oh, S. Liu, H. Jee, A. Baev, M. T. Swihart and P. N. Prasad, Chiral poly(fluorene-*alt*-benzothiadiazole) (PFBT) and nanocomposites with gold nanoparticles: Plasmonically and structurally enhanced chirality, *J. Am. Chem. Soc.*, 2010, **132**, 17346–17348.

57 X. Zhu and M. Liu, Layer-by-layer assembly of a new L-glutamic acid based polymer with PAH: Formation of hollow sphere during deposition, dye loading and chiroptical switch, *Soft Matter*, 2011, **7**, 11447.

58 C. A. Silvera Batista, R. G. Larson and N. A. Kotov, Nonadditivity of nanoparticle interactions, *Science*, 2015, **350**, 1242477.

59 P. D. Pino, B. Pelaz, Q. Zhang, P. Maffre, G. U. Nienhaus and W. J. Parak, Protein corona formation around nanoparticles – from the past to the future, *Mater. Horiz.*, 2014, **1**, 301–313.

60 R. Huang, R. P. Carney, F. Stellacci and B. L. T. Lau, Protein–nanoparticle interactions: The effects of surface compositional and structural heterogeneity are scale dependent, *Nanoscale*, 2013, **5**, 6928–6935.

61 M. P. Moloney, J. Govan, A. Loudon, M. Mukhina and Y. K. Gun'ko, Preparation of chiral quantum dots, *Nat. Protoc.*, 2015, **10**, 558–573.

62 T. Ralph, M. Hitchman, J. Millington and F. Walsh, Evaluation of a reactor model and cathode materials for batch electrolysis of L-cystine hydrochloride, *J. Electroanal. Chem.*, 1999, **462**, 97–110.

63 Y. Zhou, Z. Zhu, W. Huang, W. Liu, S. Wu, X. Liu, Y. Gao, W. Zhang and Z. Tang, Optical coupling between chiral biomolecules and semiconductor nanoparticles: Size-dependent circular dichroism absorption, *Angew. Chem., Int. Ed.*, 2011, **50**, 11456–11459.

64 Y. Kim, B. Yeom, O. Arteaga, S. Jo Yoo, S.-G. Lee, J.-G. Kim and N. A. Kotov, Reconfigurable chiroptical nanocomposites with chirality transfer from the macro- to the nanoscale, *Nat. Mater.*, 2016, **15**, 461–468.

65 B. Yeom, H. Zhang, H. Zhang, J. I. Park, K. Kim, A. O. Govorov and N. A. Kotov, Chiral plasmonic nanostructures on achiral nanopillars, *Nano Lett.*, 2013, **13**, 5277–5283.

66 S. Knoppe and T. Bürgi, Chirality in thiolate-protected gold clusters, *Acc. Chem. Res.*, 2014, **47**, 1318–1326.

67 C. Zeng, Y. Chen, K. Kirschbaum, K. Appavoo, M. Y. Sfeir and R. Jin, Structural patterns at all scales in a nonmetallic chiral $Au_{133}(SR)_{52}$ nanoparticle, *Sci. Adv.*, 2015, **1**, e1500045.

68 M. V. Mukhina, V. G. Maslov, A. V. Baranov, A. V. Fedorov, A. O. Orlova, F. Purcell-Milton, J. Govan and Y. K. Gun'ko, Intrinsic chirality of CdSe/ZnS quantum dots and quantum rods, *Nano Lett.*, 2015, **15**, 2844–2851.

69 P. A. Kurian, C. Vijayan, K. Sathiyamoorthy, C. S. Suchand Sandeep and R. Philip, Excitonic transitions and off-resonant optical limiting in CdS quantum dots stabilized in a synthetic glue matrix, *Nanoscale Res. Lett.*, 2007, **2**, 561–568.

70 D. O. Demchenko and L.-W. Wang, Optical transitions and nature of Stokes shift in spherical CdS quantum dots, *Phys. Rev. B: Condens. Matter Mater. Phys.*, 2006, **73**, 155326.

71 O. Lopez-Acevedo, H. Tsunoyama, T. Tsukuda, H. Häkkinen and C. M. Aikens, Chirality and electronic structure of the thiolate-protected Au_{38} nanocluster, *J. Am. Chem. Soc.*, 2010, **132**, 8210–8218.

72 J. J. P. Stewart, Optimization of parameters for semiempirical methods VI: More modifications to the NDDO approximations and re-optimization of parameters, *J. Mol. Model.*, 2013, **19**, 1–32.

73 J. J. P. Stewart, Application of localized molecular orbitals to the solution of semiempirical self-consistent field equations, *Int. J. Quantum Chem.*, 1996, **58**, 133–146.

74 J. D. C. Maia, G. A. Urquiza Carvalho, C. P. Mangueira, S. R. Santana, L. A. F. Cabral and G. B. Rocha, GPU linear algebra libraries and GPGPU programming for accelerating MOPAC semiempirical quantum chemistry calculations, *J. Chem. Theory Comput.*, 2012, **8**, 3072–3081.

75 A. Klamt and G. Schuurmann, COSMO: A new approach to dielectric screening in solvents with explicit expressions for the screening energy and its gradient, *J. Chem. Soc., Perkin Trans. 2*, 1993, 799.

76 A. D. Becke, Density-functional thermochemistry. III. The role of exact exchange, *J. Chem. Phys.*, 1993, **98**, 5648.

77 P. J. Hay and W. R. Wadt, *Ab initio* effective core potentials for molecular calculations. Potentials for the transition metal atoms Sc to Hg, *J. Chem. Phys.*, 1985, **82**, 270.

78 W. R. Wadt and P. J. Hay, *Ab initio* effective core potentials for molecular calculations. potentials for main group elements Na to Bi, *J. Chem. Phys.*, 1985, **82**, 284.

79 J. Tomasi, B. Mennucci and R. Cammi, Quantum mechanical continuum solvation models, *Chem. Rev.*, 2005, **105**, 2999–3093.

80 M. J. Frisch, G. W. Trucks, H. B. Schlegel, G. E. Scuseria, M. A. Robb, J. R. Cheeseman, G. Scalmani, V. Barone, B. Mennucci, G. A. Petersson, H. Nakatsuji, M. Caricato, X. Li, H. P. Hratchian, A. F. Izmaylov, J. Bloino, G. Zheng, J. L. Sonnenberg, M. Hada, M. Ehara, K. Toyota, R. Fukuda, J. Haseg and D. F., *Gaussian 09, Revision B.01*, Gaussian Inc., Wallingford, CT, 2010.

Faraday Discussions

PAPER

Template-assisted colloidal self-assembly of macroscopic magnetic metasurfaces†

Martin Mayer,[ac] Moritz Tebbe,[b] Christian Kuttner,[ac] Max J. Schnepf,[a] Tobias A. F. König*[acd] and Andreas Fery*[ace]

Received 12th February 2016, Accepted 22nd February 2016
DOI: 10.1039/c6fd00013d

We demonstrate a template-assisted colloidal self-assembly approach for magnetic metasurfaces on macroscopic areas. The choice of anisotropic colloidal particle geometry, assembly pattern and metallic film is based on rational design criteria, taking advantage of mirror-charge effects for gold nanorods placed on gold film. Monodisperse gold nanorods prepared utilizing wet-chemistry are arranged with high precision on wrinkled templates to form linear array-type assemblies and subsequently transferred to a thin gold film. Due to the obtained particle-to-film distance of 1.1 nm, the plasmonic mode of the nanorod is able to couple efficiently with the supporting metallic film, giving rise to a magnetic mode in the visible spectrum (721 nm). Conventional UV-vis-NIR measurements in close correlation with electromagnetic simulations provide evidence for the presence of a magnetic resonance on the macroscopic area. The herein presented scalable lithography-free fabrication process paves the road towards colloidal functional metasurfaces with an optical response in the effective magnetic permeability.

Introduction

Recent interest in designing metamaterials focuses on the step from 3D bulk materials to 2D metasurfaces, which feature similar optical features.[1–6] Metasurfaces are optical functional interfaces based on regular patterns of subwavelength repeating metaatoms (unit cells). The controlled arrangement of metaatoms allows the engineering of novel optical devices based on controlled

[a]*Leibniz-Institut für Polymerforschung Dresden e.V., Institute of Physical Chemistry and Polymer Physics, Hohe Str. 6, 01069 Dresden, Germany. E-mail: koenig@ipfdd.de; fery@ipfdd.de*

[b]*Department of Physical Chemistry II, University of Bayreuth, Universitätsstr. 30, 95440 Bayreuth, Germany*

[c]*Cluster of Excellence Centre for Advancing Electronics Dresden (CFAED), Technische Universität Dresden, 01062 Dresden, Germany*

[d]*Technische Universität Dresden, Fakultät Mathematik und Naturwissenschaften, 01062 Dresden, Germany*

[e]*Department of Physical Chemistry of Polymeric Materials, Technische Universität Dresden, Hohe Str. 6, 01069 Dresden, Germany*

† Electronic supplementary information (ESI) available: Additional simulations, experimental details, optical data and effective optical parameters are provided. See DOI: 10.1039/c6fd00013d

reflectance,[7,8] efficient energy transport and wave guiding,[9,10] as well as perfect lensing[11] and optical magnetism.[12] For many of these applications, the main challenge is the control of the magnetic resonances in the visible optical range. The lack of magnetic response in natural optical materials[13] was first overcome by artificial magnetic resonances obtained for split-ring resonators in the micro-wave regime.[3] However for split-ring resonators, the limited scalability and saturation effects[14] deny their application in the visible optical range.

Consequently, the fabrication of nanostructures with magnetic resonances remains a challenge especially on macroscopic areas. The design of such magnetic metasurfaces needs to be rationally linked to scalable fabrication techniques of hierarchical assemblies in *quasi* 2D-resonators.[5] The incident electric-field component induces oscillating charges within the resonators and, hence, gives rise to an effective magnetic moment; this excitation is commonly referred to as a magnetic resonance.[15–17] To achieve these effects, metasurfaces have to be built with nanometre precision.[18,19] Common fabrication techniques rely on lithographic methods, like e-beam lithography or ion-beam milling, which have several drawbacks, such as losses due to polycrystallinity and lack in resolution and scalability.[20–24] The colloidal bottom-up assembly approach, which exploits the advantages of noble metal nanoparticles synthesized using wet-chemistry (*e.g.*, single-crystallinity, scalability), can provide alternatives to lithographic top-down techniques.[25–27] However, those approaches often rely on random particle deposition, which strongly limits the complexity of the prepared unit cells and the homogeneity of the obtained metasurfaces.[28–30] To boost the complexity of the fabricated metasurfaces, plasmonic nanoparticles are coupled to metallic films to form mirror charges in the emerging cavities.[8,10,31–33] Template-assisted convective self-assembly allows the design of unit cells with high complexity and in a large variety by surface-alignment of nanoparticles in well-defined orientations. Such assemblies are only limited by the scalability and spatial resolution of the directing template.[34–36]

In this work, we use a lithography-free colloidal bottom-up approach to fabricate a macroscopic magnetic metasurface. Plasmonic nanoparticles in close contact with a metallic film induce electric current loops due to mirrored charges at the substrate, which give rise to a magnetic mode. We provide theoretical explanations for the experimentally observed magnetic mode through electromagnetic modelling and conventional UV-vis-NIR spectroscopy. Anisotropic nanoparticles are used, due to the following reasons: the short axis of the nanorods is used for magnetic mode excitation and their long axes provide control over their orientation in template-assisted self-assembly. For the building block, we used tailored wet-chemical synthesis to obtain monodisperse, gold nanorods with defined crystalline facets (flat faces). Furthermore, due to a defined spacer material we obtain a reliable particle-to-film distance. Using colloidal self-assembly, we were able to fabricate a macroscopic magnetic metasurface, which shows a clear signature of an effective magnetic permeability in the visible wavelength range. With this colloidal approach, we aim for a reliable method to make the unique manipulation of light on the macroscale possible.

Results and discussion

Design of the metasurface unit cell

One of the central prerequisites for obtaining a magnetic resonance is the induction of a circular oscillating current flow within the unit cell. This can be achieved by the efficient coupling of the surface plasmon with the electric component of the incident light.[15,37] Isolated nanoparticles in solution inherently exhibit electric modes in the form of localized surface plasmon resonances, as indicated for an exemplary dipolar-isolated mode (EM) by its surface charges (Fig. 1A, electric mode).[38] Breaking the symmetry is achieved by coupling a plasmonic nanoparticle to a substrate. The emerging modes depend strongly on the

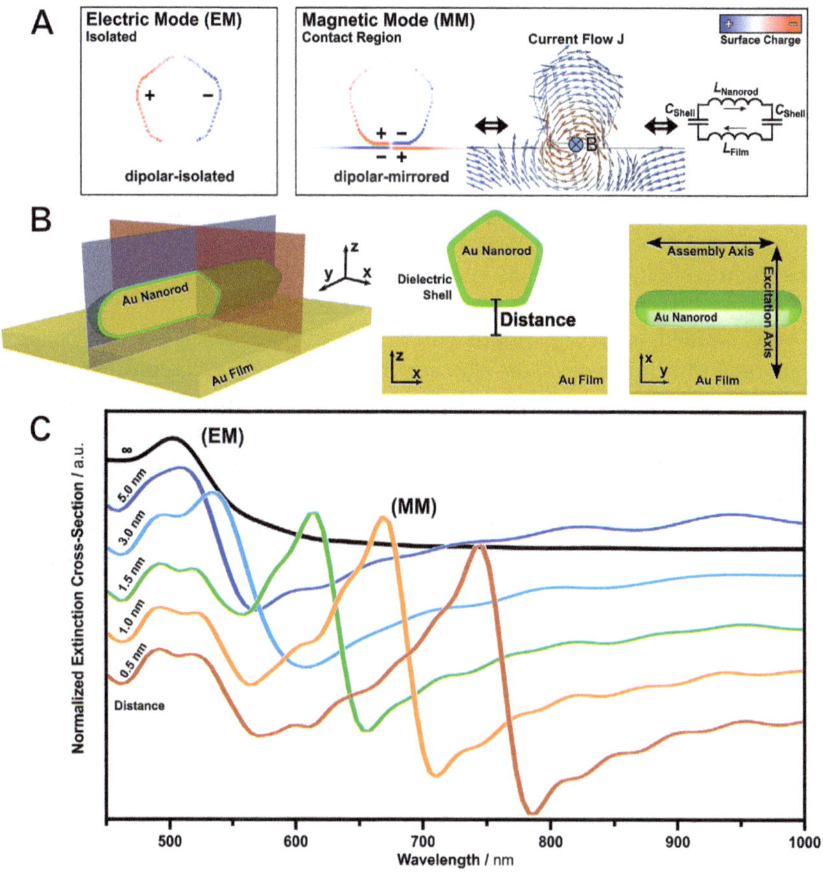

Fig. 1 Evolution of magnetic modes – from electric to magnetic resonance through metallic film-coupled plasmonic nanoparticles. (A) Surface charge pattern of an electric mode (EM), magnetic mode (MM), characteristic electric current loop (flow J) and representative inductor–capacitor (LC) model for one unit cell. (B) Schematic representation of the unit cell: gold nanorods (160 nm × 30 nm) on a 40 nm thin gold film. The short axis of the nanorod is used for the excitation by linear polarized light (transversal mode) and the long axis is used for directing the assembly process. (C) Calculated extinction cross-section at different particle-to-film distances, normalized against the bare gold film.

substrate material, nanoparticle geometry and particle-to-film distance. In contrast to nanospheres, nanoparticles with flat faces are advantageous because of their larger contact area with the substrate, which results in narrow resonances and enhanced excitation.[39] The interaction of a nanoparticle with a dielectric substrate results in hybridized modes.[40] On the contrary, when a particle is coupled to a metallic substrate, the dipolar mode induces a mirror charge pattern at the surface of the metallic film (Fig. 1A, magnetic mode).[41–43] These mirror charges may induce an electric current loop (current density J), as calculated by electromagnetic simulations. According to Ampère's circuital law, a magnetic moment is excited.

Recently, Wang *et al.* have numerically demonstrated that magnetic modes can be represented by an inductor–capacitor (LC) model (Fig. 1A).[8] For film-coupled nanoparticles, the metal components act as inductors L, whereas the dielectric insulator acts as a capacitor C. Thus, the resonance wavelengths $\lambda_{(\mathrm{MM})}$ of the dipolar-mirrored mode (MM) can be expressed using the following equation:

$$\lambda_{(\mathrm{MM})} = 2\pi c\sqrt{\left(L_{\mathrm{nanorod}} + L_{\mathrm{film}}\right)C_{\mathrm{shell}}/2} \quad (1)$$

However, an analytical calculation by means of eqn (1) is not straightforward, since the inductance and the capacitance depend on the curvature and geometry of the nanoparticle. Nevertheless, the model provides an intuitive description of the counter-oscillating currents in the unit cell. Consequently, a metasurface can be designed based on film-coupled nanoparticles.

We decided to design our magnetic metasurface based on film-coupled nanorods utilizing their short axes for magnetic mode excitation (transversal mode) and their long axes for providing orientational order upon self-assembly (Fig. 1B). The transversal magnetic mode is excited if the electric field component is planar with the cross-section of the unit cell (xz-plane, red). In contrast, the longitudinal magnetic mode, which could be excited with the electric field parallel to the long axis (yz-plane, blue), is significantly shifted outside the frequency range of visible light and will not be discussed further. Fig. 1B shows a schematic depiction of the proposed metasurface unit cell consisting of a nanoparticle as a building block positioned over a metallic film. The capping agent of the nanoparticle serves as a thin dielectric shell. For observation of the magnetic mode with conventional spectroscopy methods, an array of well-aligned anisotropic nanoparticles is required. For colloidal metasurface design, 1D periodic arrangements are ideal as they provide collective alignment of nanorods along their long axes. At the same time, a distance of at least five nanorod radii between the lines will suppress grating effects.[44] Tip-to-tip coupling within nanorod lines is decoupled from transversal excitation and does not interfere with the induction of the magnetic resonance.

Comprehensive numerical simulations of this unit cell clearly reveal that the spectral position of the magnetic mode can be tailored by tuning the following four parameters: (I) the dimensions and shape of the plasmonic nanoparticle; (II) the refractive index of the dielectric spacer (*e.g.*, capping agent); (III) the material of the plasmonic nanoparticle and (IV) the coupling distance between the nanoparticle surface and the substrate. These four parameters need to be precisely balanced to provide a strong magnetic resonance within the visible spectrum. However, tuning of the first three parameters is limited by: (I) the dimension of the nanorods (approximately 30 nm in width);[45] (II) the index of refraction of the spacer material (roughly from 1.4 to 1.5 for polymeric materials);[46] and (III) the limited choice in plasmonic metals. Silver as well as gold both possess plasmonic properties in the optical range. However, even though silver features lower damping, offering narrower line-widths and higher intensities,[47,48] it is vulnerable against sulfidation and oxidation in the air.[49] Consequently, we choose gold in the present study. In contrast, the particle-to-film separation can be controlled by the capping agent and an additional spacer layer at the substrate, *e.g.*, by layer-by-layer techniques.[50] In combination, the cavity size can be tailored with nanometre precision. Taking the limited range of optimization of the nanorod morphology and the index of refraction of the building blocks into account, we will focus on the size of the nanoparticle-film cavity in order to further optimise the metasurface design.

Fig. 1C shows the extinction cross-section σ_{ext} of a nanorod at varying distances from a metallic film. The extinction spectra were normalized against the transmission (T) of the bare gold film ($\sigma_{ext} = -\log 1/T$, transparent for green light only). Owing to the transmission characteristics of gold, the extinction cross-sections show a pseudo-resonance below the interband gap of gold (<512 nm) and a pronounced increase in extinction towards lower energy. The isolated particle monomer, *i.e.* at infinite separation, shows a single plasmonic resonance (electric mode, EM). This transversal excitation corresponds to the electric mode as discussed above. Approaching the surface, the dipolar electric mode is converted to a magnetic dipolar-mirrored mode (magnetic mode, MM) and a higher-order magnetic mode. This quadrupolar-mirrored mode is further discussed in the ESI, Fig. S1.† The magnetic resonance shifts towards lower energy in comparison to the isolated electric mode. This characteristic red-shift is also known as a plasmon ruler.[51] The spectral position of the magnetic mode is particularly sensitive to distance changes even in the angstrom range. Thus, the resonance frequency may be tuned over the entire optical range. To avoid effects like quantum tunnelling, a minimum distance of 0.4 nm should be sustained.[52,53] These separations lie within the range of conventional capping agents.[54] Next, we will address the fabrication of suitable nanoparticles as building blocks.

Synthesis of well-defined building blocks

Gold nanorods possess intrinsic anisotropy in both their geometrical and optical properties. Thus, nanorods, as well as their top-down lithographic equivalents, are common building blocks in metamaterial fabrication, *e.g.*, as in the well-known double-bar setup.[15,18,55] The fabrication of hierarchical unit cells of high precision requires building blocks of well-defined size and shape (Fig. 2). High structural integrity is critical for the performance of the final system, because disorder and polydispersity in size and shape yield additional undesired resonance broadening. By using wet-chemical bottom-up synthesis, a variety of monodisperse building blocks is accessible. In particular, the control over nanoparticle size, shape, and crystallinity allows for tailored morphological and optical properties.[56,57] The penta-twinned nanorods used were prepared by seed-mediated growth based on a modified protocol by Pérez-Juste *et al.*[45,58] combined with a supplemental purification step employing depletion forces.[59,60]

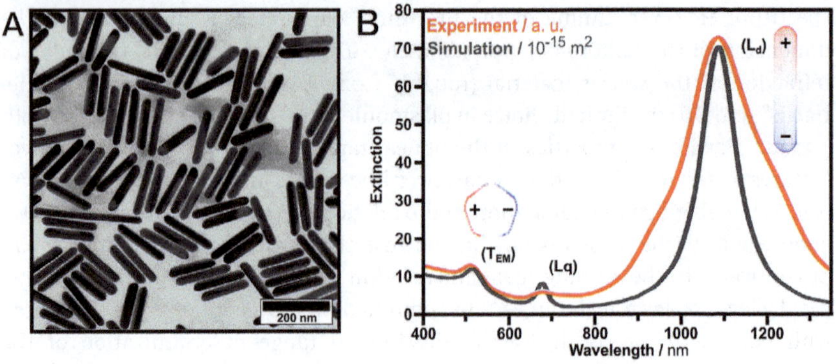

Fig. 2 Gold nanorods in solution – nature of the electric modes. (A) TEM image of the purified gold nanorods with mean dimensions of 161.0 ± 14.8 nm in length and 30.2 ± 3.7 nm in width. (B) Experimental (red) and simulated (black) extinction spectra. The gold nanorods exhibit three electric plasmon resonances, which can be attributed to dipolar transversal (T_{EM}, 512 nm), dipolar longitudinal (L_d, 1087 nm) and quadrupolar longitudinal (L_q, 680 nm) modes. The corresponding surface charges of the dipolar excitations are shown as insets.

With this method, penta-twinned gold nanorods can be produced which exhibit larger thicknesses and higher aspect ratios compared to the more commonly used single-crystalline gold nanorods.[61] Additionally, their pentagonal cross-section and their flat sides are beneficial for substrate-supported effects due to the increased interaction areas. However, owing to the use of twinned seeds only 20% of the resulting nanoparticles are nanorods with the majority being spherical and 2D by-products. These by-products need to be removed as they strongly impede the bottom-up self-assembly of hierarchical unit cells. Consequently, a supplemental purification step is required to yield narrow distributions. Efficient separation can be achieved by exploiting depletion forces with additional surfactants acting as depletants, as published recently.[59,60] Here, the effective physical parameter is the interaction area and, thus, the morphology of the nanoparticles in solution. To avoid time-consuming screening for an adequate depletant (*i.e.*, surfactant) concentration, we measured a calibration curve which allows the purification of penta-twinned gold nanorods depending on their longitudinal localized surface plasmon resonance (see ESI Fig. S2†). This purification protocol enables almost quantitative removal of spherical by-products and other morphologies. A second purification step is performed to remove longer and thinner nanorods to selectively tune and enhance the size-distribution. Fig. 2 shows a TEM image (A) and extinction spectra (B) of the synthesized and purified nanorod sample. The nanorods exhibited mean dimensions of 161.0 ± 14.8 nm in length and 30.2 ± 3.7 nm in width (for as-synthesized TEM images, UV-vis spectra and histograms of the size distribution, see ESI Fig. S2†). Consequently, the herein presented 3-step protocol is an ideal means to precisely tailor the aspect ratio of nanorods. The dipolar electric characters of the transversal (T_{EM}) and the predominant longitudinal plasmon resonance (L_d) are depicted by their surface charges in Fig. 2 (insets).

For the successful template-assisted convective assembly the wetting behaviour of the nanorods is critical. As shown recently by our group, surface chemistry providing enhanced wettability can be adjusted using surface functionalisation with a hydrophilic protein shell.[21,55] Consequently, concentrated highly stable solutions of gold nanorods with bovine serum albumin (BSA)[21] as dielectric shells were obtained for subsequent assembly. In the dry state, the BSA layer exhibits a shell thickness below 2 nm as we have shown in previous work. Consequently, these nanoparticles match the requirements for reliable template-assisted colloidal self-assembly. In the following, we focus on the self-assembly of the nanorods into array-type structures.

Fabrication of film-coupled nanorod arrays as magnetic unit cells

Controlled colloidal self-assembly of the nanoscale building blocks into the proposed magnetic unit cell is an important step towards a macroscopic magnetic metasurface (Fig. 3). Recently, we demonstrated the precise alignment of nanorods into well-defined 1D particle lines using template-assisted convective self-assembly employing wrinkled elastomeric substrates.[55] In the first fabrication step, the nanorods are assembled in the form of tip-to-tip arrangements within the periodic structure of the wrinkle. In order to achieve periodic macroscopic alignment in single particle lines the templates need to be homogeneous on square-centimetre scales. The dimensions of its features are crucial and need to precisely match the nanorod dimensions (*i.e.*, width). The employed wrinkled templates are fabricated by plasma treatment of a macroscopic uniaxially stretched polydimethylsiloxane (PDMS) elastomer. The plasma treatment introduces a thin oxidized film at the surface. Upon strain release, periodic wrinkles are formed due to the mechanical mismatch between the substrate and the thin film.[62] The template periodicity is critical for controlling the period of the alignment. To avoid any grating effects the periodicity was chosen to be sufficiently large (see discussion above for more details). Thus, periodicity and amplitude were tuned to values of 210 nm and 40 nm, respectively.[62] Subsequently, the amplitude was reduced to 20 nm by additional plasma post-treatment without altering the periodicity to match the dimensions of the employed gold nanorods.[55] At the same time, the surface is hydrophilized sufficiently to enable subsequent nanoparticle assembly from aqueous solution (see plasma post-treatment calibration in the ESI Fig. S3†).

Nanorod assembly into single particle lines was achieved by dip coating. The unidirectional laminar flow promotes the orientation of the anisotropic particles along the major (long) axis. Fig. 3A displays schematically the self-assembly process. Macroscopic templates are withdrawn from a highly concentrated nanoparticle solution (2 mg mL^{-1}). During withdrawal, the nanorods are aligned into the channels by capillary forces within the confined meniscus (see Fig. 3B). Due to the confinement of the wrinkle structure, we are able to align the building blocks for a further transfer step.

In the second fabrication step, we use the wet-contact print method to transfer the pre-aligned anisotropic nanorods onto the gold film. The building blocks were transferred to a hydrophilized gold film while maintaining their orientational order. Utilizing template stripping, a surface roughness of the gold film below 1 nm was achieved. Consequently, a uniform contact between the nanorods and the supporting gold film is provided. Quantitative transfer can be achieved by wet-transfer printing of the prefilled templates and subsequent lift-off of the emptied

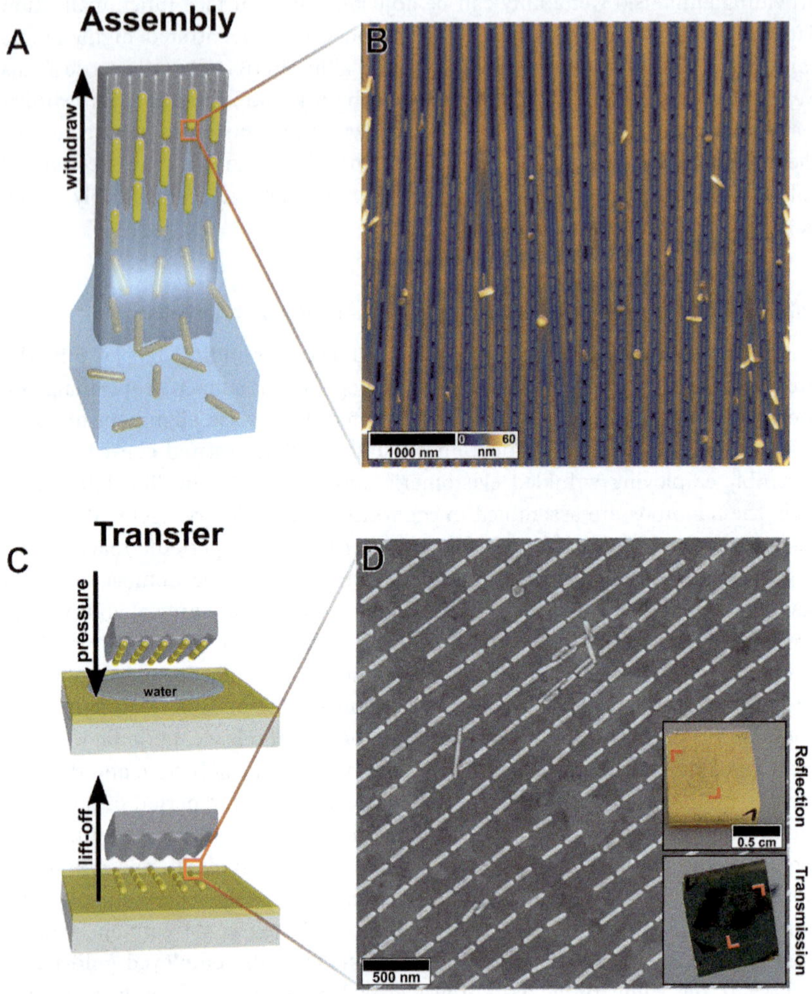

Fig. 3 Macroscopic template-assisted self-assembly using mechanical instabilities. (A) Schematic template-assisted self-assembly process by dip coating the wrinkled templates in a highly concentrated gold nanorod solution. (B) AFM image of the self-assembled array of gold nanorods in the wrinkle template. (C) Schematic transfer process using wet-contact printing and subsequent lift-off of the emptied template. (D) SEM image of the assembled structures transferred to a smooth gold film and photographs of the macroscopic substrate in reflection and transmission (upper and lower inset, respectively).

template as schematically depicted in Fig. 3C. Contrary to the transfer to other surface materials,[55] an additional adhesion promoter is not required. The combination of van der Waals interactions between the gold nanoparticles and gold film and the gold-selective binding groups of the protein shell (*e.g.*, amine, thiol) is sufficient to guarantee a complete transfer. Thus, sufficient particle–substrate spacing is given solely by the thickness of the dried shell of the stabilizing agent at the nanorod surface. The successful transfer covers an area of 5 mm × 5 mm as evidenced by SEM micrographs and macroscopic photographs (see Fig. 3D).

The quality of the assembled structures was evaluated using statistical analysis of the position and orientation of the particles with high magnification SEM images (360 individual nanorods, for further information see ESI Fig. S4†). The tip-to-tip particle spacing was determined to be 15.6 ± 5.8 nm within the lines and 209 ± 6 nm between neighbouring lines, in-line with the template periodicity. The high structural precision of the assembly was evidenced by the averaged deviation θ to the principle alignment direction $\langle\sigma_\theta\rangle$ being only 2.69° corresponding to a 2D-order parameter $\langle S_{2D}\rangle$ of 0.99 ($\langle S_{2D}\rangle = \langle 2\cos^2(\theta) - 1\rangle$). This large-yield and scalable method enabled us to fabricate a metasurface array of film-coupled nanorods on the macroscale. The following section will elucidate its optical characteristics.

Optical characterisation of film-coupled nanorod arrays

The optical properties of the macroscopic assemblies were investigated using conventional far-field UV-vis-NIR measurements. Since the magnetic resonance is dominated by absorption, spectroscopy was performed in the transmission geometry.[63] In order to allow spectroscopy in transmission, the thickness of the gold film was set to 34 nm (Fig. 3D, insets). In the visible wavelength range, the magnetic mode is excited only if the polarized light is in plane with the cross-section of the nanorod (transversal mode). Experimentally, this excitation was

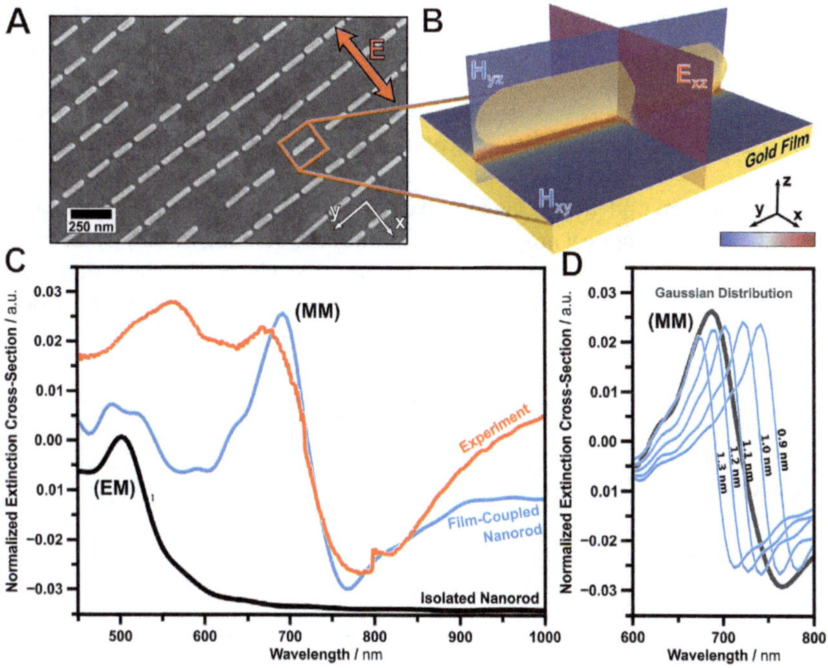

Fig. 4 Macroscopic magnetic metasurfaces. (A) SEM image of the template-assisted self-assembled array of film-coupled nanorods. (B) Simulated magnetic ($|H|^2/|H_0|^2$, coloured scale range from 1 to 4) and electric field intensity ($|E|^2/|E_0|^2$, coloured scale range from 1 to 100) of the unit cell. (C) Experimentally observed UV-vis-NIR (red) and modelled (blue) extinction spectra of the metasurface. Modelled transversal extinction spectrum of an isolated nanorod (black, not to scale). (D) Modelled extinction spectra of the magnetic mode at various particle-to-film distances (0.9 nm to 1.3 nm) and the averaged result.

performed on a spot size of 4 mm × 4 mm using polarized spectroscopy measurements. Fig. 4 shows the extinction spectrum normalized against the bare gold film.

Experimentally we observe the clear spectral signature of a magnetic mode (MM) with its peak intensity at 675 nm and point of inflexion at 721 nm (Fig. 4C). In addition, the quadrupolar-mirrored mode at 556 nm is present. The experimental observation is in excellent agreement with the modelling results of a film-coupled nanorod array (blue). The extinction cross-section of an isolated nanorod in air (transversal excitation, black) indicates the position of the initial electric mode (EM). From the position of the magnetic mode (MM), a mean particle-to-film distance of 1.1 nm was extracted. This distance is still above the quantum tunnelling limit of 0.4 nm.[53] The small deviation between the experiment and the model can be attributed to the high sensitivity towards particle-to-film distances. This influence is determined in Fig. 4D for various distances from 0.9 nm to 1.3 nm. This further emphasises the high sensitivity of the mode towards distance variations. Consequently, macroscopic magnetic substrates may be tuned over the entire visible spectrum.

Effective magnetic properties of the metasurface

To quantify the magnitude of the magnetic response of the metasurface, we modelled the effective optical properties using the *S*-parameter retrieval method established by D. R. Smith *et al.*[64] This method is in close analogy to the inductor–capacitor (*LC*) model (see eqn (1) and Fig. 5A). For this purpose, the anisotropic optical response must be calculated in the forward and backward direction (*S*-parameters). The model was built based on identical parameters (*i.e.*, nanorod dimensions, shell and film thickness, particle-to-film distance) to those described above.

Before we can understand the effective magnetic permeability of the unit cell, we will discuss the optical properties of the following building blocks: (I) the bare gold film and (II) the gold nanorod monomer (ESI Fig. S5†). These steps are important to prove the stability of the *S*-parameter method in this frequency

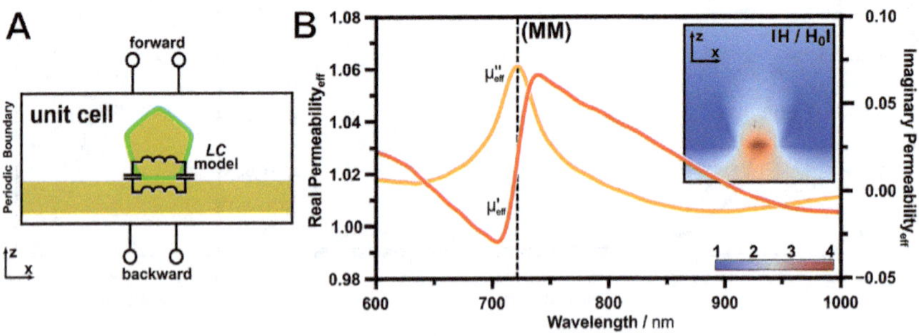

Fig. 5 Effective magnetic properties of film-coupled nanorods. (A) Schematic representation of the *S*-parameter retrieval method and analogy to the inductor–capacitor (*LC*) model. The unit cell is treated as an effective medium, since only the *S*-parameters are evaluated in the forward and backward direction. (B) Calculated magnetic permeability for the film-coupled nanorod revealing the presence of a magnetic resonance and related magnetic field (inset).

range. (I) The calculated effective electric permittivity of the bare gold film (40 nm thickness) agrees well with literature values (see Methods part). (II) The optical properties of the individual gold nanorod monomer are of particular interest for colloidal optical metamaterial design, because the bulk real part of the permittivity of gold is always negative. Once the metallic particle is surrounded by dielectric medium, the real part of the permittivity turns positive, which is in good agreement with the observations described in literature.[65]

Adding a film to the nanorod will result in a clear response in the complex magnetic permeability (Fig. 5B). The imaginary part of the permeability reveals the energetic location of the magnetic resonance through its characteristic peak at the point of inflexion of the extinction, whereas the real part shows the typical slope at the same energy. This relation is defined generally as the Kramers–Kronig relation, which indicates the relation between the absorption and dispersion. At the same energy, an electric response was also observed in the permittivity (ESI Fig. S5†). Consequently, the observed resonances show a mode characteristic for an electric and a magnetic resonance. This is not surprising due to the strong electromagnetic coupling of the plasmonic particle to the metallic film. We want to advise that the retrieved values below the interband gap of gold (<515 nm) should be taken with caution. In this short wavelength regime the permittivity is in the non-radiative limit for a surface plasmon resonance, which renders the response of the particle to a highly damped mode.[66] However, the signature of the magnetic mode is clearly revealed by the magnetic permeability. Consequently, the herein fabricated system represents a magnetic metasurface, which is capable of tuning the effective permeability by the induction of a magnetic moment.

Conclusion

Within this study, we developed a scalable colloidal self-assembly approach towards magnetic metasurfaces using anisotropic metal nanoparticles coupled to a supporting noble metal film. The assembly was based on a rational design strategy, which used the short axis for magnetic mode excitation and the long axis for the template-assisted self-assembly process. In particular, our approach utilizes mirror charges induced by the gold nanorods into the gold film to excite a magnetic resonance. With this approach, we were able to identify a suitable unit cell geometry: (I) anisotropic building blocks with flat faces to ensure a large contact area to the supporting substrate; (II) a suitable spacer material for a well-defined nanoscale particle-to-film distance and (III) sufficient line-to-line spacing to neglect grating effects. The building blocks are assembled through a template-assisted colloidal self-assembly technique resulting in a well-defined orientation and scalability. Applying nanoparticles with a thin homogenous dielectric shell enabled us to realize a mean coupling distance of 1.1 nm. The 2D-order parameter was determined to be close to unity (0.99) clearly proving the high potential of the employed convective self-assembly technique. Conventional UV-vis-NIR spectroscopy in close agreement with electromagnetic simulations was used to reveal the signature of the optical properties of the fabricated magnetic metasurface. The main target optical feature, the magnetic resonance, was identified at a wavelength of 721 nm. Retrieval of the effective optical properties (*S*-parameter method) clearly identified its presence and the magnetic mode was found in absorption as well as dispersion.

Achieving optical metasurface features on macroscopic areas requires assembly concepts that are robust in the sense that local variations are suppressed or can be tolerated. In the present study, we achieve robustness in several ways: first, the use of nanorods allows for control over orientational order by exploiting their intrinsic anisotropy (anisotropy in geometrical and optical properties), which is demonstrated by the high 2D-order parameter. In addition, line spacings (structure periodicity) that are much larger than the relevant particle length scale (in this case the particle width) grant robustness by suppressing undesired coupling between lines. Finally, the use of a defined nanoparticle coating as a spacer layer results in uniform particle-to-film distances.

In terms of functionality, these aspects allow for the extension of this concept to a functional magnetic metasurface. In particular, the high sensitivity of the film-coupled concept (sharp drop of the magnetic resonance) makes it interesting for sensing applications. For example, stimulus responsive polymers could be used as spacer layers and thus could translate configurational changes into an optically detectable response. Thus, the proposed approach can be generalized towards the colloidal fabrication of functional optical metasurfaces on a macroscopic area.

Methods

Materials

Ammonia hydroxide solution (NH_3, 25%), ascorbic acid (AA, $C_6H_8O_6$, >99%), benzyldimethylhexadecylammonium chloride (BDAC), bovine serum albumin (BSA, 98%), hydrogen peroxide (H_2O_2, 30%), hydrogen tetrachloroaurate ($HAuCl_4$, >99.9%) and sodium borohydride ($NaBH_4$, 99%) were purchased from Sigma Aldrich. 1 M sodium hydroxide (NaOH) solutions and trisodium citrate ($Na_3C_6H_5O_7$, >99%) were supplied by Grüssing. Hexadecyltrimethylammonium bromide (CTAB, 99%) was received from Merck KGaA. Hexadecyltrimethylammonium chloride (CTAC, 99%) was supplied by Molekula. Sylgard 184 PDMS elastomer kits were purchased from Dow Corning. NOA81 UV Curing Optical Adhesive was supplied by Norland Products. Gold films were evaporated on RCA-SC1 cleaned silicon wafers. All chemicals and solvents were used as received. MilliQ-grade water (18.2 M cm, pH 8) was used as the solvent in all preparations.

Penta-twinned gold nanorod synthesis

The penta-twinned gold nanorod synthesis protocol was adopted from Pérez-Juste *et al.* with some recently published optimisations.[45,58]

3.5 nm sized citrate-capped gold seeds. Briefly, 20 mL of a 0.125 mM $HAuCl_4$, 0.25 mM trisodium citrate aqueous solution was prepared and stirred for 10 min at room temperature. Next, 600 μL of a freshly prepared 0.1 M $NaBH_4$ solution was added quickly under vigorous stirring. After 2 min, the stirring rate was reduced and the seeds were aged for 40 min under slow stirring at room temperature. To ensure complete removal of excessive $NaBH_4$, the solution was stirred at 40–45 °C for another 10 min.

5.5 nm sized CTAB-capped gold seeds. 8 mL of a growth solution consisting of 40 mM CTAB and 0.125 mM $HAuCl_4$ was prepared. The Au(III) was reduced to Au(I) utilizing 20 μL of a 0.1 M AA solution (f.c. 0.25 mM). Then, 1333 μL of the citrate-capped gold seeds were added quickly and the solution was mixed thoroughly for 30 s using Vortex-Finger treatment. The 5.5 nm sized CTAB-capped gold seeds were aged for 3 h at 23 °C prior to the final overgrowth step.

Penta-twinned gold nanorod growth. 4 L of a growth solution containing 8 mM CTAB and 0.125 mM $HAuCl_4$ was prepared. The solution was kept at 20 °C for at least 30 min. Next, 12.5 mL of a 0.1 M AA solution (f.c. 0.313 mM) was added and gently stirred leading to the solution becoming clear within minutes. Finally, 6.67 mL of 5.5 nm sized CTAB-capped seeds were added quickly and mixed thoroughly. For the nanorod growth, the solution was kept at 20 °C overnight.

Nanorod purification

Nanorod purification was carried out in a manner analogous to the purification protocol by Scarabelli *et al.*[60] For penta-twinned gold nanorods of an average length and width of about 160 nm and 30 nm, respectively, the final surfactant (CTAC) concentration, determined by the calibration curve in ESI Fig. S2,† was set to 70 mM with a 25 wt% CTAC solution leading to selective flocculation and sedimentation overnight. The supernatant was discarded and the sediment containing the nanorods was redispersed in 12 mL of a 10 mM BDAC solution for the next purification step. The longer gold nanorods were slowly sedimented in a 10 mM BDAC solution over the course of 5 days. The targeted gold nanorods in the supernatant were collected and the longer nanorods in the precipitate were discarded.

Ligand exchange and functionalization

For functionalization with BSA, the BDAC concentration in 6 mL of a penta-twinned gold nanorod solution was set to 0.5 mM (correlating to the critical micelle concentration, CMC, of BDAC) by repeated centrifugation at 1400 RCF for 30 min. Directly before functionalization, the surfactant concentration was reduced below the CMC to 0.25 mM by dilution. The nanorod solution was added quickly to a 30 mL BSA solution (pH 8, 10 mg mL^{-1}) containing 0.02 wt% trisodium citrate under sonication. After 20 min of sonication, the solution was centrifuged at 1500 RCF for 45 min and the concentrated nanorods were redispersed in one drop of NaOH solution at pH 11 and diluted with 30 mL of BSA solution (pH 10, 1 mg mL^{-1}) containing 0.02 wt% trisodium citrate. Finally, after incubation overnight at 4 °C, the BSA-functionalized gold nanorods were washed 3 times at 1500 RCF for 45 min with NaOH solution at a pH value of 10.5 to remove excess protein. Subsequently the solution was concentrated to a gold concentration of 2 mg mL^{-1}.

Wrinkle preparation for templates

Wrinkled templates with a wavelength and amplitude of approximately 210 nm and 30 nm, respectively, were produced as described elsewhere.[55] Briefly, PDMS was prepared by casting 25 g of the mixture of cross-linker and pre-polymer (1 : 5 w/w) in a balanced polystyrene dish (10 cm × 10 cm). First, the PDMS was pre-cross-linked at room temperature for 24 h and afterwards cross-linked at 80 °C for 4 h. The final PDMS was cut into 1 cm × 4.5 cm strips and individually stretched above the critical strain in a home-built stretching apparatus. Next, the strips were plasma-treated for 1200 s in an oxygen atmosphere at 0.6–1.2 mbar and relaxed slowly after cooling to room temperature inducing wrinkle formation. The wrinkled PDMS substrates were cut into 1 cm × 0.7 cm templates. To adjust the amplitude to the desired value of 15–20 nm the substrate was post-treated with oxygen plasma at 0.6 mbar with a power of 100 W for 540 s (see calibration curve in ESI Fig. S3†). Finally, the templates were covalently fixed on plasma-activated glass strips (6 mm × 76 mm) with the wrinkles oriented parallel to the long axis of the glass slides, which later act as a fixture for the subsequent assembly process by dip-coating.

Substrates

Smooth gold film substrates were produced by template stripping. Therefore, a 34.19 ± 0.02 nm thick gold film (see also in ESI Fig. S6 and Table S1†) was evaporated on a RCA SC-1 cleaned silicon wafer without adhesive agent. Afterwards, 1.5 cm × 1.5 cm glass slides were placed on top with 3 μL of NOA81 UV Curing Optical Adhesive and cured with an approximated energy exposure of 2 J cm^{-2} from a hand-held UV light source followed by 12 h final curing at 50 °C. Finally, the glass slides were carefully stripped using a scalpel with the smooth gold film on top. The resulting gold films exhibited a root-mean-square roughness (r_{RMS}) of 0.38 nm and were spectrally transparent within an optical window between 400 nm and 750 nm.

Convective self-assembly and transfer

Template-assisted convective self-assembly of gold nanorods was performed by dip-coating as published recently by Tebbe *et al.*[55] Therefore, the wrinkled templates fixed on the glass strip were clamped into the dip-coater and moved into 1 cm × 1 cm glass cuvettes, which act as containers. Afterwards, 600 μL of the gold nanorod solution (2 mg mL^{-1}, pH 10.5, BSA-stabilized) was poured slowly into the container. Assembly is performed by withdrawing the template with a commercially available dip-coater (DC/D/LM system, KSV Instruments, Germany) at a speed of 10 μm min^{-1}. Transfer to the gold substrate was adopted from Hanske *et al.*[21] Briefly, the gold substrate was hydrophilized with oxygen plasma for 7 min (0.2 mbar, 100 Ws). The resulting hydrophilic gold substrate was wetted with 6 μL of a pH 10.5 NaOH solution and the nanorod-filled wrinkled template was pressed onto the substrate with a pressure of 140 kPa. After 5 h, the water was dried and the transfer was complete. Finally, the empty template was carefully lifted off.

Modelling and numerical simulation

A commercial-grade simulator from Lumerical Solutions, Inc. (FDTD Solutions, Version 8.11.442 on Linux x86_64) based on the finite-difference time-domain (FDTD) method was used to perform the calculations. The penta-twinned gold nanorods were modelled with a pentagonal cross section in water (constant index of refraction of 1.333) and a rounded cap to represent their true structure. Nanorod dimensions were set according to the TEM analysis (see size distributions in ESI Fig. S2D and E,† 161 nm × 30 nm) and the rods were covered with a homogeneous 3 nm layer of BSA with a fixed index of refraction of 1.48. The edge rounding and tip length were rendered to agree with the experimental UV-vis-NIR measurements of the purified nanorod solutions (10% and 11 nm, respectively). Assemblies were modelled with optimised gold nanorods on a gold substrate in air (constant index of refraction of 1). To emulate the dried state on the gold substrate, the thickness of the BSA layer was reduced to 1.5 nm. Considering the smoothness of the gold substrate and the good agreement with the experiment, the optical roughness of the film was neglected. On the basis of the experimental data of Johnson and Christy an approximating polynomial function for the optical constants of gold is used by the FDTD software (6 coefficients, 1 imaginary weight: 0.241 RMS error).[67] Zero-conformal-variant mesh refinement with a mesh size of 0.5 nm was used for the best representation of the rounded structure and in the cavity the mesh size was reduced to 0.1 nm. All simulations reached the automatic shut-off level of 10^{-6} before reaching 300 fs of simulation time. The simulation space was restricted by perfectly matched layers for isolated nanorods as boundary conditions and by periodic boundaries for the unit cell setup. The fields were normalized against the unit cell without the nanorod but with the gold film as reference. Therefore, the presented data include only the contribution of the nanorod whereas the contribution of the gold film serves as the background. To calculate the effective optical properties of film-coupled nanorods, *S*-parameter extraction analysis from Lumerical Solutions Inc. was used. To retrieve the optical constants for the inhomogeneous asymmetric structure, we calculated the $S11$ and $S22$ parameters by simulating in the forward and backward source direction, respectively, at an automatic shut-off level of 10^{-7}.

Characterisation

Optical characterization of the solutions was performed in a 1 mm pathway cuvette without dilution using a Cary 5000 spectrometer (Agilent, USA) using the Cary Universal Measurement Accessory (UMA), which allowed the measurement of aqueous nanoparticle solutions in the range of 200 nm to 1850 nm. The gold substrates were measured in the transmission geometry on the Cary 5000 spectrometer (Agilent, USA) with an attached Cary Universal Measurement Accessory (UMA) at a spot size of 4 mm × 4 mm with linearly polarized light. The background was recorded on the same substrate prior tot he transfer of the gold nanorods. The spot size was fixed to the same value for UV-vis and NIR detectors resulting in slit widths of 1 nm and 4 nm, respectively.

The gold substrate was characterized using variable-angle spectroscopic ellipsometry at incident angles from 45° to 85° in 5° steps in a wavelength range from 193 nm to 1690 nm using a RC2 (J.A. Woollam Co., USA) and was fitted using the CompleteEASE software package (see ESI Fig. S6 and Table S1† for fitting parameters).

AFM measurements of all substrates (wrinkled templates, gold films, assemblies in the wrinkled templates and nanorods transferred to the gold films) were performed using a Nanoscope Dimension V (Bruker, USA), operated in the Tapping Mode™. Al-coated silicon cantilevers (OTESPA, Bruker) with a stiffness of typically 35–47.2 N m^{-1} and typical resonance frequencies of 300 kHz were utilized.

TEM measurements of the gold nanoparticles at all relevant stages of the synthesis were performed using a Zeiss 922 OMEGA EFTEM at a voltage of 120 kV. Zero-loss filtered images were recorded using a bottom mounted Ultrascan 1000

(Gatan) CCD camera system. Gatan Digital Micrograph 3.9 for GMS 1.4 software was used for image acquisition. TEM samples were concentrated and the surfactant concentration (if present) was set to the corresponding critical micelle concentration. Droplets of 5 μL were dried on a Quantifoil 300 mesh copper grid with carbon films placed on a hydrophobic film to trap the droplet during drying. For the determination of the mean particle dimensions 250 particles per sample were measured.

SEM measurements were recorded using a LEO 1530 FE-SEM (Zeiss, Germany) with in-lens and SE2 detectors using an acceleration voltage of 2–3 kV.

Acknowledgements

This study was funded by the European Research Council under grant ERC-2012-StG 306686 (METAMECH: Template-assisted assembly of METAmaterials using MECHanical instabilities). M. T. was supported by the Elite Network Bavaria within the framework of the Elite Study Program "Macromolecular Science" and funded *via* a grant for PhD candidates according to Bavarian elite promotion act (BayEFG).

References

1 A. V. Kildishev, A. Boltasseva and V. M. Shalaev, *Science*, 2013, **339**, 1289.
2 J. B. Pendry, *Phys. Rev. Lett.*, 2000, **85**, 3966–3969.
3 J. B. Pendry, A. J. Holden, D. J. Robbins and W. J. Stewart, *IEEE Trans. Microwave Theory Tech.*, 1999, **47**, 2075–2084.
4 D. R. Smith, W. J. Padilla, D. C. Vier, S. C. Nemat-Nasser and S. Schultz, *Phys. Rev. Lett.*, 2000, **84**, 4184–4187.
5 N. Yu and F. Capasso, *Nat. Mater.*, 2014, **13**, 139–150.
6 S. P. Burgos, R. de Waele, A. Polman and H. A. Atwater, *Nat. Mater.*, 2010, **9**, 407–412.
7 G. M. Akselrod, T. Ming, C. Argyropoulos, T. B. Hoang, Y. Lin, X. Ling, D. R. Smith, J. Kong and M. H. Mikkelsen, *Nano Lett.*, 2015, **15**, 3578–3584.
8 H. Wang, K. O'Dea and L. Wang, *Opt. Lett.*, 2014, **39**, 1457–1460.
9 A. Rose, T. B. Hoang, F. McGuire, J. J. Mock, C. Ciracì, D. R. Smith and M. H. Mikkelsen, *Nano Lett.*, 2014, **14**, 4797–4802.
10 J. B. Lassiter, F. McGuire, J. J. Mock, C. Ciracì, R. T. Hill, B. J. Wiley, A. Chilkoti and D. R. Smith, *Nano Lett.*, 2013, **13**, 5866–5872.
11 S. Kawata, A. Ono and P. Verma, *Nat. Photonics*, 2008, **2**, 438–442.
12 M. Kataja, T. K. Hakala, A. Julku, M. J. Huttunen, S. van Dijken and P. Torma, *Nat. Commun.*, 2015, **6**, 1–8.
13 A. Alù and N. Engheta, *Opt. Express*, 2009, **17**, 5723–5730.
14 J. Zhou, T. Koschny, M. Kafesaki, E. N. Economou, J. B. Pendry and C. M. Soukoulis, *Phys. Rev. Lett.*, 2005, **95**, 223902.
15 G. Dolling, C. Enkrich, M. Wegener, J. F. Zhou, C. M. Soukoulis and S. Linden, *Opt. Lett.*, 2005, **30**, 3198–3200.
16 C. M. Soukoulis and M. Wegener, *Nat. Photonics*, 2011, **5**, 523–530.
17 D. Lin, P. Fan, E. Hasman and M. L. Brongersma, *Science*, 2014, **345**, 298–302.
18 Z.-G. Dong, H. Liu, M.-X. Xu, T. Li, S.-M. Wang, S.-N. Zhu and X. Zhang, *Opt. Express*, 2010, **18**, 18229–18234.

19 N. I. Zheludev, *Science*, 2010, **328**, 582–583.
20 N. Meinzer, W. L. Barnes and I. R. Hooper, *Nat. Photonics*, 2014, **8**, 889–898.
21 C. Hanske, M. Tebbe, C. Kuttner, V. Bieber, V. V. Tsukruk, M. Chanana, T. A. F. König and A. Fery, *Nano Lett.*, 2014, **14**, 6863–6871.
22 H. Ditlbacher, A. Hohenau, D. Wagner, U. Kreibig, M. Rogers, F. Hofer, F. R. Aussenegg and J. R. Krenn, *Phys. Rev. Lett.*, 2005, **95**, 257403.
23 B. J. Wiley, D. J. Lipomi, J. Bao, F. Capasso and G. M. Whitesides, *Nano Lett.*, 2008, **8**, 3023–3028.
24 G. V. Naik, V. M. Shalaev and A. Boltasseva, *Adv. Mater.*, 2013, **25**, 3264–3294.
25 J.-S. Huang, V. Callegari, P. Geisler, C. Brüning, J. Kern, J. C. Prangsma, X. Wu, T. Feichtner, J. Ziegler, P. Weinmann, M. Kamp, A. Forchel, P. Biagioni, U. Sennhauser and B. Hecht, *Nat. Commun.*, 2010, **1**, 150.
26 Y. Cui, M. T. Björk, J. A. Liddle, C. Sönnichsen, B. Boussert and A. P. Alivisatos, *Nano Lett.*, 2004, **4**, 1093–1098.
27 Z. Li, S. Butun and K. Aydin, *ACS Photonics*, 2014, **1**, 228–234.
28 M. J. Rozin, D. A. Rosen, T. J. Dill and A. R. Tao, *Nat. Commun.*, 2015, **6**, 1–7.
29 W. Chen, M. Tymchenko, P. Gopalan, X. Ye, Y. Wu, M. Zhang, C. B. Murray, A. Alu and C. R. Kagan, *Nano Lett.*, 2015, **15**, 5254–5260.
30 G. M. Akselrod, J. Huang, T. B. Hoang, P. T. Bowen, L. Su, D. R. Smith and M. H. Mikkelsen, *Adv. Mater.*, 2015, **27**, 8028–8034.
31 H. Chen, T. Ming, S. Zhang, Z. Jin, B. Yang and J. Wang, *ACS Nano*, 2011, **5**, 4865–4877.
32 H. Chen, L. Shao, T. Ming, K. C. Woo, Y. C. Man, J. Wang and H.-Q. Lin, *ACS Nano*, 2011, **5**, 6754–6763.
33 T. Maurer, P.-M. Adam and G. Lévêque, *Nanophotonics*, 2015, **4**, 363–382.
34 A. Rey, G. Billardon, E. Lortscher, K. Moth-Poulsen, N. Stuhr-Hansen, H. Wolf, T. Bjornholm, A. Stemmer and H. Riel, *Nanoscale*, 2013, **5**, 8680–8688.
35 S. Ni, J. Leemann, H. Wolf and L. Isa, *Faraday Discuss.*, 2015, **181**, 225–242.
36 X. Zhou, Y. Zhou, J. C. Ku, C. Zhang and C. A. Mirkin, *ACS Nano*, 2014, **8**, 1511–1516.
37 A. Vallecchi, M. Albani and F. Capolino, *Opt. Express*, 2011, **19**, 2754–2772.
38 C. F. Bohren and D. R. Huffman, in *Absorption and Scattering of Light by Small Particles*, Wiley-VCH Verlag GmbH, 2007, ch. 12, pp. 325–380, DOI: 10.1002/9783527618156.
39 T. König, R. Kodiyath, Z. A. Combs, M. A. Mahmoud, M. A. El-Sayed and V. V. Tsukruk, *Part. Part. Syst. Charact.*, 2014, **31**, 274–283.
40 S. Zhang, K. Bao, N. J. Halas, H. Xu and P. Nordlander, *Nano Lett.*, 2011, **11**, 1657–1663.
41 G. Lévêque and O. J. F. Martin, *Opt. Express*, 2006, **14**, 9971–9981.
42 A. Pors and S. I. Bozhevolnyi, *Opt. Express*, 2013, **21**, 27438–27451.
43 D. E. Gómez, Z. Q. Teo, M. Altissimo, T. J. Davis, S. Earl and A. Roberts, *Nano Lett.*, 2013, **13**, 3722–3728.
44 M. B. Müller, C. Kuttner, T. A. F. König, V. V. Tsukruk, S. Förster, M. Karg and A. Fery, *ACS Nano*, 2014, **8**, 9410–9421.
45 M. Mayer, L. Scarabelli, K. March, T. Altantzis, M. Tebbe, M. Kociak, S. Bals, F. J. García de Abajo, A. Fery and L. M. Liz-Marzán, *Nano Lett.*, 2015, **15**, 5427–5437.
46 J. E. Mark, *Polymer Data Handbook*, Oxford University Press, 2009.

47 N. Dmitruk, S. Malynych, I. Moroz and V. Kurlyak, *Semicond. Phys., Quantum Electron. Optoelectron.*, 2010, **13**, 369–373.
48 J. Becker, I. Zins, A. Jakab, Y. Khalavka, O. Schubert and C. Sönnichsen, *Nano Lett.*, 2008, **8**, 1719–1723.
49 J. L. Elechiguerra, L. Larios-Lopez, C. Liu, D. Garcia-Gutierrez, A. Camacho-Bragado and M. J. Yacaman, *Chem. Mater.*, 2005, **17**, 6042–6052.
50 G. Decher, *Science*, 1997, **277**, 1232–1237.
51 P. K. Jain, W. Huang and M. A. El-Sayed, *Nano Lett.*, 2007, **7**, 2080–2088.
52 L. Yang, H. Wang, Y. Fang and Z. Li, *ACS Nano*, 2016, **10**, 1580–1588.
53 J. A. Scholl, A. Garcia-Etxarri, G. Aguirregabiria, R. Esteban, T. C. Narayan, A. L. Koh, J. Aizpurua and J. A. Dionne, *ACS Nano*, 2016, **10**, 1346–1354.
54 C. Kuttner, M. Chanana, M. Karg and A. Fery, in *Macromolecular Self-Assembly*, ed. L. Billon, O. V. Borisov, Wiley-VCH, 2016.
55 M. Tebbe, M. Mayer, B. A. Glatz, C. Hanske, P. T. Probst, M. B. Muller, M. Karg, M. Chanana, T. A. F. Konig, C. Kuttner and A. Fery, *Faraday Discuss.*, 2015, **181**, 243–260.
56 E. C. Dreaden, A. M. Alkilany, X. Huang, C. J. Murphy and M. A. El-Sayed, *Chem. Soc. Rev.*, 2012, **41**, 2740–2779.
57 Y. Xia, Y. Xiong, B. Lim and S. E. Skrabalak, *Angew. Chem., Int. Ed.*, 2009, **48**, 60–103.
58 J. Pérez-Juste, L. M. Liz-Marzán, S. Carnie, D. Y. C. Chan and P. Mulvaney, *Adv. Funct. Mater.*, 2004, **14**, 571–579.
59 K. Park, H. Koerner and R. A. Vaia, *Nano Lett.*, 2010, **10**, 1433–1439.
60 L. Scarabelli, M. Coronado-Puchau, J. J. Giner-Casares, J. Langer and L. M. Liz-Marzán, *ACS Nano*, 2014, **8**, 5833–5842.
61 L. Scarabelli, A. Sánchez-Iglesias, J. Pérez-Juste and L. M. Liz-Marzán, *J. Phys. Chem. Lett.*, 2015, **6**, 4270–4279.
62 B. A. Glatz, M. Tebbe, B. Kaoui, R. Aichele, C. Kuttner, A. E. Schedl, H.-W. Schmidt, W. Zimmermann and A. Fery, *Soft Matter*, 2015, **11**, 3332–3339.
63 U. K. Chettiar, A. V. Kildishev, T. A. Klar and V. M. Shalaev, *Opt. Express*, 2006, **14**, 7872–7877.
64 D. R. Smith, D. C. Vier, T. Koschny and C. M. Soukoulis, *Phys. Rev. E: Stat., Nonlinear, Soft Matter Phys.*, 2005, **71**, 036617.
65 C. Rockstuhl, F. Lederer, C. Etrich, T. Pertsch and T. Scharf, *Phys. Rev. Lett.*, 2007, **99**, 017401.
66 M. Tebbe, C. Kuttner, M. Mayer, M. Maennel, N. Pazos-Perez, T. A. F. König and A. Fery, *J. Phys. Chem. C*, 2015, **119**, 9513–9523.
67 P. B. Johnson and R. W. Christy, *Phys. Rev. B: Solid State*, 1972, **6**, 4370–4379.

Faraday Discussions

PAPER

Studying nanoparticles' 3D shape by aspect maps: Determination of the morphology of bacterial magnetic nanoparticles

D. Peddis,*[ak] G. Muscas,[ab] R. Mathieu,[c] P. Anil Kumar,[j] G. Varvaro,[a] G. Singh,[d] I. Orue,[i] D. Gil-Carton,[f] L. Marcano,[g] A. Muela[eh] and M. L. Fdez-Gubieda[eg]

Received 25th March 2016, Accepted 29th April 2016

DOI: 10.1039/c6fd00059b

Magnetic nanoparticles (MNPs) are widely investigated due to their potential use in various applications, ranging from electronics to biomedical devices. The magnetic properties of MNPs are strongly dependent on their size and shape (*i.e.*, morphology), thus appropriate tools to investigate their morphology are fundamental to understand the physics of these systems. Recently a new approach to study nanoparticle morphology by Transmission Electron Microscopy (TEM) analysis has been proposed, introducing the so-called Aspect Maps (AMs). In this paper, a further evolution of the AM method is presented, allowing determination of the nanoparticles' 3D shape by TEM image. As a case study, this paper will focus on magnetite nanoparticles (Fe_3O_4), with a mean size of ~45 nm extracted from *Magnetospirillum gryphiswaldense* magnetostatic bacteria (MTB). The proposed approach gives a complete description of the nanoparticles' morphology, allowing estimation of an average geometrical size and shape. In addition, preliminary investigation of the magnetic properties of MTB nanoparticles was performed, giving some insight into interparticle interactions and on the reversal mechanism of the magnetization.

[a]*Istituto di Struttura della Materia – CNR, 00015 Monterotondo Scalo, RM, Italy. E-mail: davide.peddis@ism.cnr.it*

[b]*Department of Physics and Astronomy, Uppsala University, Box 516, SE-751 20 Uppsala, Sweden*

[c]*Department of Engineering Sciences, Uppsala University, Box 534, SE-751 21 Uppsala, Sweden*

[d]*Department of Materials Science and Engineering, Norwegian University of Science and Technology, Trondheim-7491, Norway*

[e]*BCMaterials, Building no 500, Technological Park of Biscay, Derio, Spain*

[f]*Structural Biology Unit, CIC BioGUNE, CIBERehd, Derio, Spain*

[g]*Department Electricidad y Electronica, University of Basque Country, UPV-EHU, Spain*

[h]*Department Inmunología, Microbiología y Parasitología, University of Basque Country, UPV-EHU, Spain*

[i]*SGIker Medidas Magnéticas, University of Basque Country, UPV-EHU, Spain*

[j]*Faculty of Physics and CENIDE, University of Duisburg-Essen, Lotharstr. 1, 47048 Duisburg, Germany*

[k]*The Vinca Institute of Nuclear Sciences, University of Belgrade, P.O. Box 522, 11001 Belgrade, Serbia*

1. Introduction

The energy of a magnetic particle (MNP) is generally dependent on the magnetization direction and, for uniaxial anisotropy and an easy axis aligned with the direction of external field (H), it can be written as[1]

$$E_a = K_a V \sin^2\theta - M_s V \mu_0 H \cos\theta \tag{1}$$

where K_a is the anisotropy constant, V the particle volume, M_s the saturation magnetization, and θ the angle between magnetization vector and the easy direction. According to the Néel–Brown theory,[2] above a certain temperature (the blocking temperature, T_B) and on a certain time scale, the particle moment can produce a thermally activated transition. This behaviour is analogous to paramagnetism but with a different time and magnetization scale, and for this reason it is called "superparamagnetism". Below T_B, the particle's moment is blocked and unable to rotate over the barrier in the time of a measurement. Tuning the magnetic properties of nanoparticles (*e.g.*, T_B, magnetic anisotropy) means controlling the energy profile of the "single" nanoparticle. Eqn. (1) shows that E_a is proportional to V and K_a, where the latter is mainly dependent on the structure and chemical composition of the material (*i.e.*, magnetocrystalline anisotropy)[3] and the surface to volume ratio (*i.e.*, surface anisotropy). Also the magnetostatic energy can give an important contribution to the anisotropy. For a particle with finite magnetization and non-spherical shape, the magnetostatic energy will be larger for some orientations of the magnetic moments than for others. Thus, the shape determines the magnitude of magnetostatic energy and this type of anisotropy is often known as shape anisotropy.[4] Particle's shape has been identified as a major tool for controlling MNPs' ability to self-assemble in different morphologies under external magnetic field[5,6] as well as engineering the magnetic properties of nanomaterials, with additional attractive feature of displaying different properties in different directions.[7,8] The magnetic behaviour of a nanoparticle assembly can also be strongly affected by interparticle interactions that lead to an increase in the energy barrier.[9–12] The magnetic interactions can be due to dipolar and exchange coupling among nanoparticle surface atoms and they play a fundamental role in the physics of these systems. It should be underlined that for several applications (*e.g.*, biomedical, energy, *etc.*), concentrated assemblies of individually responding magnetic entities are required. This last requirement is difficult to be satisfied as interparticle dipole–dipole interactions are strongly enhanced by both an increase in the size and concentration of MNPs.[9]

Starting from this landscape, it is clear that the morphology of the particles and their arrangements (*i.e.*, interparticle interactions) are crucial issues to understand the physics of these systems. In this view, advancements in the analysis of both size and shape (*i.e.*, morphology) of nanoparticles becomes a key issue to design new nanostructured materials for specific application.[13–15] Recently a new statistical approach to investigate the morphological features of magnetic nanoparticles has been proposed to follow the growth of nanoparticles by thermal decomposition of a metal–organic precursor in a high boiling solvent.[16] This statistical approach is based on the correlation of particle size and shape creating the so-called "aspect maps" (AMs). In the present paper, AMs have been applied to magnetite nanoparticles (Fe_3O_4) extracted from the *Magnetospirillum gryphiswaldense* magnetostatic bacteria (MTB). The magnetosomes present high structural and chemical quality, giving them very good magnetic properties for some biomedical applications (*i.e.*, magnetic hyperthermia, drug delivery). AMs give a complete description of nanoparticle morphology, and an improvement of this method is proposed in order to obtain 3D information from TEM images. A preliminary discussion of magnetic properties will be also given, with particular attention to interparticle interactions and the magnetization reversal mechanism of nanoparticles both outside (O-MTB) and inside (I-MTB) the bacteria.

2. Experimental

2.1 Synthesis

Magnetospirillum gryphiswaldense was cultured in a medium containing 100 μM Fe(III)-citrate as described elsewhere.[17] The magnetosomes were extracted from bacteria according to the following protocol: cells were collected by centrifugation at 8000 × *g* and were lysed using a French-press (1.4 kbar). The lysate was centrifuged at 800 × *g* to remove cell debris and the magnetosomes were isolated from the supernatant by magnetic separation. Finally, the magnetosomes were dispersed in ultra-pure water by sonication for 30 minutes. For magnetic measurements, both the bacterial cells and the magnetosomes were lyophilized.

2.2 Samples characterization

DC magnetization measurements were performed using a Quantum Design – Superconducting Quantum Interference Device (SQUID) magnetometer equipped with a superconducting magnet ($H_{max} = \pm 5.5$ T). To avoid any displacement of the nanoparticles during the measurements, the samples, in the form of powders, were immobilized in an epoxy resin.

Electron microscopy was performed on magnetosomes adsorbed onto glow-discharged 200-mesh carbon-coated copper grids. The 2D TEM images were obtained with a Philips CM120 Biofilter electron microscope at an accelerating voltage of 100 kV. In order to obtain three-dimensional characterization of the magnetosomes, a series of different electron tomograms was obtained on a JEM-2200FS/CR (JEOL, Ltd.) field emission gun electron microscope, operating at 200 kV. The magnetosomes were dropped in grids including 10 nm gold nanoparticles (as fiducial markers) for alignment processing. A single-axis tomographic tilt series, at a nominal magnification of 30.000×, producing a final pixel size of 0.71 nm per pixel, were collected on an UltraScan 4000 CCD camera (Gatan Inc.), over a tilt range of ±64 with 2° increments and at underfocus values ranging from 5 to 8 μm, using the semiautomatic data acquisition software SerialEM2. For alignment and 3D reconstruction of the tilted series, we used IMOD and/or Tomo3D software.[18,19] 3D reconstruction was carried out by weight back-projection. No contrast transfer function (CTF) correction was applied, thus limiting our reconstructions to the first zero of the CTF.

3. Results and discussion

Magnetospirillum gryphiswaldense bacteria are well known to produce magnetite nanoparticles with a cubo-octahedral shape.[17,20] In order to better understand the magnetic proprieties of such bio-mineralized nanoparticles and their magnetic collective behavior, a detailed morphological analysis is required.

From several TEM images, ~360 particles have been examined. For each particle the profile contour has been defined using a semiautomatic analysis tool provided by FIJI software,[21] then several morphological parameters such as minimum and maximum diameter, perimeter, and area of projected image have been evaluated. The average diameter has been calculated as the mean value between width and height. Finally, the frequency counts of particle size has been fitted with a log-normal function estimating average diameter, $\langle D_{\mathrm{TEM}} \rangle$, and standard deviation, σ.[22–25]

To estimate the polydispersity of the sample with respect to the average particle size, an empirical parameter has been defined as:

$$\mathrm{PD} = 100 \times \frac{\sigma}{D_{\mathrm{TEM}}} \tag{2}$$

The circularity (C) has been used to describe the particle shape:

$$C = 4\pi \frac{A}{P^2} \tag{3}$$

where A and P are the particle's area and perimeter respectively.[16] Circularity is widely used to define the deviation from a perfect spherical shape, which is represented by a value of 1, while lower values describe square, triangular or more complex shapes. TEM images show the particles' homogenous size, evidencing squared shapes (Fig. 1a–c). Some twinned particles, evidenced with red arrows in Fig. 1c, are also present. The average particle size obtained by the fit is 43.0(1) nm, with a sigma value of 5.6(1) nm and a polydispersity (PD) of 13% (Fig. 1c). The sample exhibits a narrow shape distribution (Fig. 1d), with a mode value of

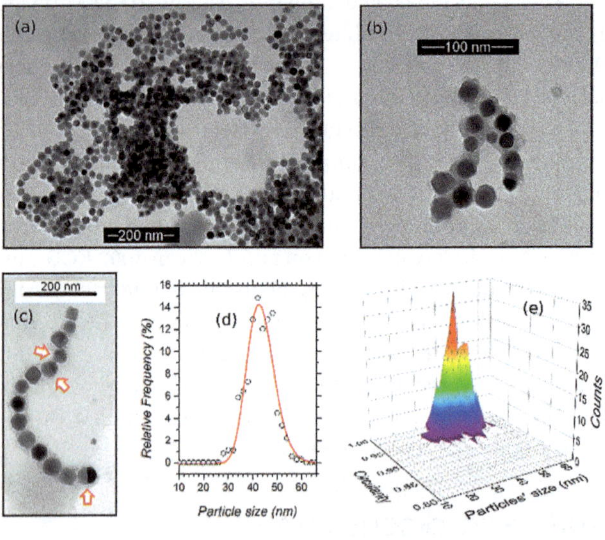

Fig. 1 TEM images of nanoparticles in panel (a)–(c), where in the last one red arrows indicate twinned particles. Particle size distributions with the corresponding aspect map are reported in panel (d) and (e), respectively.

0.97(2), which usually represents a good approximation of spherical particles. On the other hand, a clear square shape is visible to the naked eye. For a deeper and more accurate analysis of sub-distributions, an aspect map was extracted (Fig. 1e). The aspect map gives a complete description of a nanoparticle's morphology, correlating for each one size and shape, identifying a point on a surface, which represents the parameter called "aspect". This allows a simpler and faster identification of each particle, distinguishing two of them not only for a difference in size, but also for a difference in shape. As in the case of a particle system, where the difference in diameter is too low to be able to clearly distinguish two groups, but the difference in volume is large enough to clarify the presence of a bimodal distribution, in the same way the "aspect" may show different groups of particles otherwise not clearly identifiable.[16] AM shows a double peak, probably due to the relatively low number of particles analyzed, with a main mode at $\langle D \rangle = 42(1)$ nm and $C = 0.96(2)$ with a second sub-peak at $\langle D \rangle = 47(1)$ nm and $C = 0.96(2)$; no evidence of smaller particle sub-distribution has been detected.

3.1 Nanoparticle's 3D shape investigation

The average circularity presents a value larger than expected, closer to that of a sphere, despite evident squared shapes are shown by TEM images. It is known that *Magnetospirillum gryphiswaldense* biomineralizes cubo-octahedra magnetite;[20] for this reason a more detailed analysis was carried out. TEM images are the projection of the 3D shape onto the plane, thus it was assumed that each particle sits on the TEM grid on a flat face, and that is not possible to sit on an edge or a corner. For a cube, all six faces have same area and the same projection is obtained whatever face the solid is set on; the same results are produced for a pure octahedron. The cube–octahedron is obtained from a mixture of a cube (C) and an octahedron (O) in different proportions. Starting from a pure cube with a ratio $O : C = 0.0$, different shapes have been simulated in steps of 0.1, up to a pure octahedron with ratio 1.0. The mixture of the two shapes produces square and triangular faces, with a different area according to the ratio O : C (Fig. 2).

For a cube-octahedron two projected images are possible, one when the solid sits on a triangular-like face and the other when sitting on a square-like face. Thus for each O : C ratio it is possible to identify two possible projected figures for isolated particles, corresponding to specific values of circularity. We have considered the probability of a solid to sit on a square or a triangular face proportional to the % area of each one, thus for each O : C ratio, a statistical average value of circularity has been defined.

Fig. 3b presents for each O : C ratio a value of circularity corresponding to the solid sitting on the triangular or on the square face, and the average value weighted for the probability of the two options. The solid with an O : C ratio of 0.8 is the one showing an average C value of 0.94(2), the most compatible with the average circularity of the sample estimated in 0.96(2) (Fig. 3a), thus it is the best candidate to represent the average shape of the nanoparticles in the sample.

In order to confirm our hypothesis, three dimensional characterization of the MTB nanoparticles has been carried out. To assess the 3D structure of these nanoparticles, we performed electron tomography on the isolated magnetosomes. Interestingly, some magnetosomes displayed planar faces corresponding to different O : C ratios. Some isolated magnetosomes have polygonal features

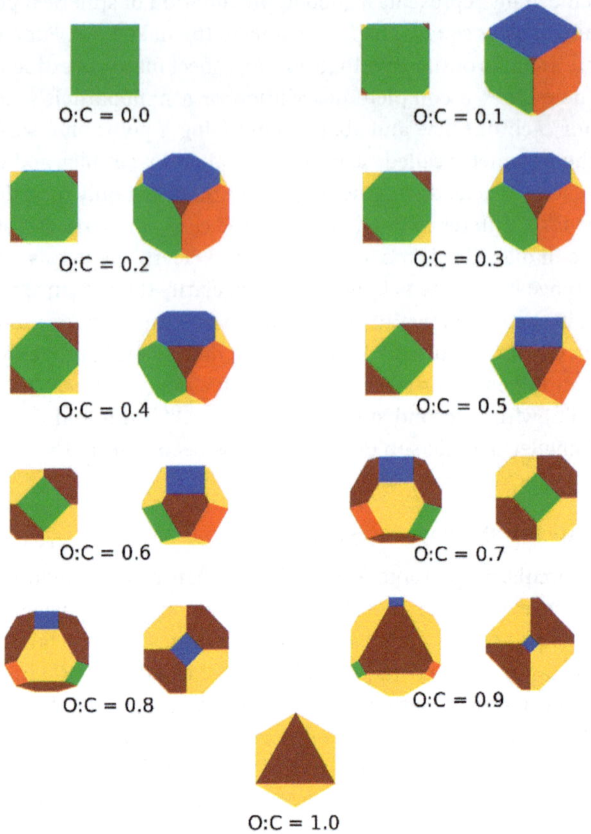

Fig. 2 Projected images for the two possible sitting configurations for several O : C ratios.

with planar facets (40 nm in length) and formed angles between 120 and 130° (Fig. 4a). This observation confirmed the presence of different cube-octahedral conformations with a major population similar to the model with an O : C ratio of about 0.8.

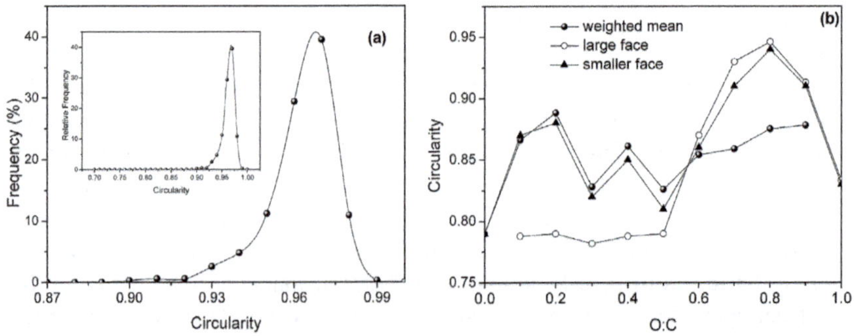

Fig. 3 Panel (a) reports the distribution of particles' circularity measured by TEM images; panel (b) shows the circularity values for particles sitting on the small face (triangles), the large one (circles) and the statistical weighted average value (squares).

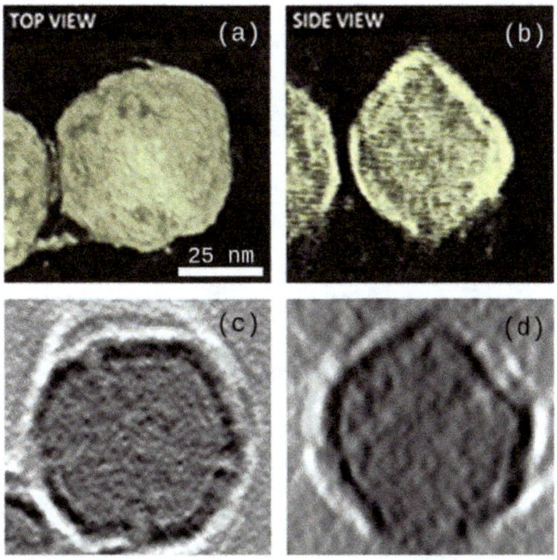

Fig. 4 (a, top left) Top view (*xz*) of the reconstructed tomogram represented as iso-surface. Scale bar represents 25 nm. (a, top right). Side view (*xy*) of the reconstructed tomogram. Tomograms were displayed in Chimera.[5] (b) Tomographic central *xy* (c) and *zy* (d) sections of a representative magnetosome nanoparticle with O : C ratio ~0.8.

3.2 Magnetic properties

A preliminary magnetic investigation was carried on magnetic MTB nanoparticles both inside (I-MTB) and outside (O-MTB) the bacteria. The O-MTB consists of randomly oriented magnetic nanoparticles coated by a lipid bilayer membrane (thickness ~ 2 nm)[17] while the I-MTB sample consists of small randomly oriented nanoparticle chains inside the whole cell. Previous magnetic relaxation measurements suggest that such particles' chains behave as very large magnetic dipoles with the intra-chain magnetic moments aligned along their long axis.[26]

Fig. 5 (upper part) shows Zero Field Cooled (ZFC, full circles) and Field Cooled (empty circles) magnetization curves for I-MTB (a) and O-MTB (b) samples. ZFC and FC magnetization measurements were carried out by cooling the sample from room temperature to 5 K in a zero magnetic field; then, a static magnetic field of 5 mT was applied. M_{ZFC} was measured during the warming up from 5 K to 300 K, whereas M_{FC} was recorded during the following cooling. M_{ZFC} of both I-MTB and O-MTB samples shows an evident kink around 106 K that can be clearly ascribed to the first-order Verwey phase transition, which can be considered as a finger print of magnetite.[27] This transition is due to an ordering of the random distribution of ferrous and ferric ions at the interstitial sites with octahedral (O_h) and tetrahedral (T_d) symmetry.[28] Below the so-called Verwey temperature (T_V), an abrupt increase of resistivity is observed, as the result of the localization of interacting electrons at particular positions ($T > T_V$, $(Fe^{3+})T_d\,[Fe^{2.5+}]O_h$; $T < T_V$, $(Fe^{3+})T_d\,[Fe^{3+}Fe^{2+}]O_h$). This electronic charge ordering is related to a change in the structure symmetry: Fe^{3+} and Fe^{2+} located at interstitial sites with octahedral symmetry arrange themselves on alternate (001) planes, leading to an orthorhombic distortion.[29] It is worth noting that O-MTB shows a slightly less evident magnetic signature of Verwey transition with respect to the I-MTB sample, with a wider magnetization peak. As discussed by Prozorov and co-workers,[30] the

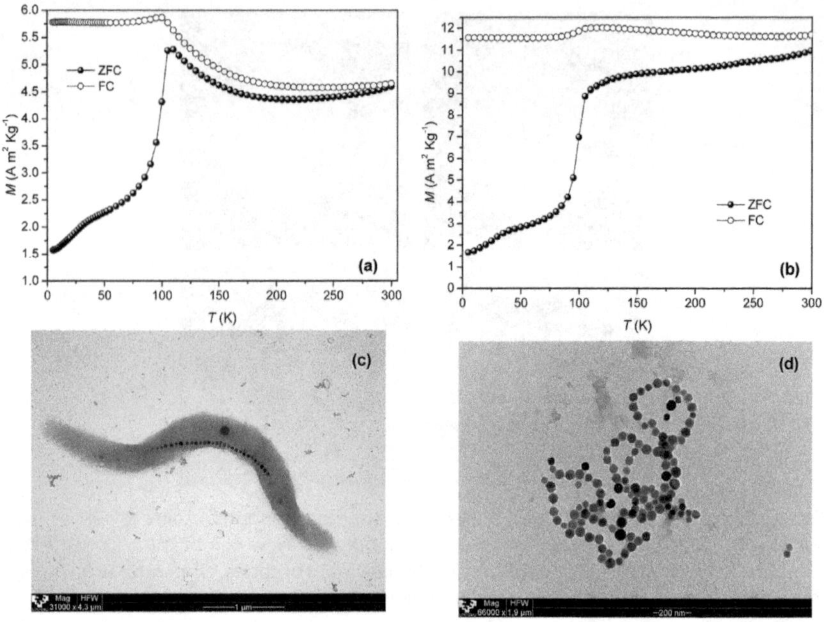

Fig. 5 ZFC (full circles) and FC (empty circles) for I-MTB (a), O-MTB (b); TEM images for I-MTB (c) and O-MTB (d).

difference between I-MTB and O-MTB can be ascribed to the different "organization of nanoparticles". Indeed, in randomly organized nanoparticles (O-MTB, Fig. 5d), magnetic fluctuations are strong enough to overcome the Néel barrier and reorientation cannot be considered because the nanoparticle moments are continually fluctuating. As a result, a less evident and slightly wider peak is observed due to the finite distribution of particle barriers and directions. On the other hand nanoparticles organized in chains (I-MTB, Fig. 5c) act as long dipoles with enhanced effective anisotropy along the chain, and thermal fluctuations are insufficient to overcome this barrier.

Both samples show clear irreversibility between FC and ZFC in all the explored temperature range, suggesting that T_B is higher than 300 K. The low-temperature-independent behaviour of M_{FC} qualitatively indicates the presence of magnetic interparticle interactions, leading to a collective behaviour that strongly influences the magnetization dynamic at low temperatures.[31]

Field-dependent magnetization curves recorded at 5 K (not reported here) indicate for both samples a tendency towards uniaxial anisotropy although a slight increase of reduced remanence magnetization is observed in the I-MTB sample (M_r/M_s O-MTB ≅ 0.44; M_r/M_s I-MTB ≅ 0.55); the coercivity is larger for MNPs inside the bacteria ($\mu_0 H_C$, I-MTB ≅ 60 mT; $\mu_0 H_C$, O-MTB ≅ 40 mT).

To investigate thoroughly the role of interparticle interactions and the effect of nanoparticle arrangement on the magnetization reversal, remanence magnetization studies were carried out at 5 K. Both Isothermal Remanent Magnetization (IRM) and Direct Current Demagnetization (DCD) remanent curves were measured. The starting point for an IRM experiment is a totally de-magnetized sample cooled in a zero magnetic field. In the present case, a small external field was applied at a fixed temperature and, after 10 s, it was switched off and the remanence was measured (M_{IRM}). The process was repeated, increasing the field until the sample reached saturation. A DCD experiment is basically similar to an IRM one, but the sample is initially in a saturated state. An external field of −5 T was applied for 10 s, then a positive reverse field (H_{rev}) was applied and, after 10 s, it was switched off and the remanent magnetization (M_{DCD}) was measured.[32,33] This sequence was repeated increasing the field until saturation in the opposite direction was reached. As an example, Fig. 6a reports the IRM and DCD curves for the O-MTB sample.

By differentiating the two remanence curves with respect to the field, it is possible to obtain the irreversible susceptibility χ^{irr} that represents a map of the switching field distribution (SFD) in both the magnetizing (DCD) and demagnetizing mode (IRM). The DCD curves and their derivatives for both the samples are reported in Fig. 6b and c, respectively. SFD depends on both the intrinsic distribution of the energy barrier (related to a dispersion of particle volume, shape and anisotropy) and the easy axis spatial spread, being also strongly influenced by interparticle interactions.

It should be mentioned that for an ideal system consisting of non-interacting single-domain uniaxial nanoparticles, the two differentiated curves should

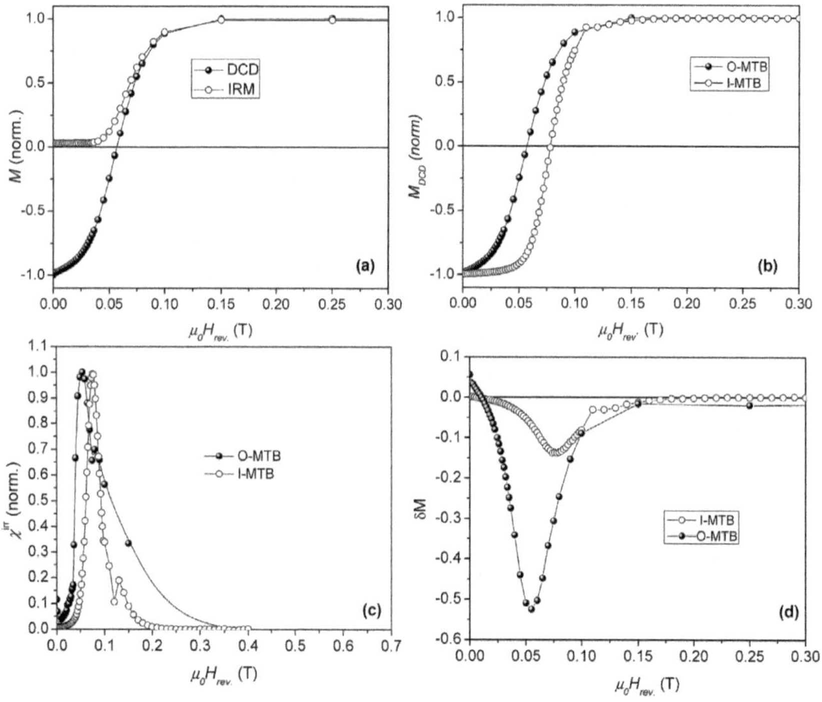

Fig. 6 (a) Normalized M_{DCD} and M_{IRM} recorded at 5 K for the O-MTB sample; (b) normalized M_{DCD} recorded at 5 K for O-MTB (full circles) and I-MTB samples (empty circles); (c) irreversible susceptibility obtained by M_{DCD} for O-MTB (full circles) and I-MTB samples (empty circles); (d) δM plots obtained by the eqn (5) for O-MTB (full circles) and I-MTB samples (empty circles).

overlap with a factor of 2 in the height of the $dM_{r,IRM}/dH$ relative to the $dM_{r,DCD}/dH$ curve as derived from the Wohlfarth relation[34]

$$M_{DCD} = 1 - 2M_{IRM} \tag{4}$$

In a real system, any deviation from this relationship can be attributed to the effect of interactions as proposed by Kelly[35] who expressed the parameter δM as

$$\delta M = M_{DCD} - [1 - 2M_{IRM}] \tag{5}$$

This parameter, which is zero in absence of interactions, is therefore a direct measure of deviations from the non-interacting case. Positive values of δM are due to interactions promoting the magnetized state, whereas negative values of δM are due to demagnetizing interactions.[36] The δM plots of the two samples are reported in Fig. 6d, indicating the prevalence of dipolar interactions as expected for nanoparticles coated by an organic membrane. In a sample of randomly distributed nanoparticles with average magnetic moment μ and average inter-particle distance d, the energy due to dipole–dipole interactions for one particle is on the order of:

$$E_{dip} \cong \frac{\mu_0}{4\pi}\frac{\mu^2}{d^3} \tag{6}$$

The lower value of remanent coercivity (H_{cr}), defined as the reverse field giving zero irreversible magnetization (Fig. 6b), observed in the O-MTB sample ($\mu_0 H_C$, I-MTB $\cong$ 80 mT; $\mu_0 H_C$,O-MTB $\cong$ 60 mT) suggests that dipolar interactions promote a highly demagnetized state for the whole ensemble, in agreement to that indicated by the δM plot, where a much stronger negative deviation is evident with respect to the I-MTB sample. It is worth mentioning that the observed change of the H_{cr} follows that of H_C.

Following eqn. (6) and being μ equal in both the samples, the lower negative dipolar interactions in the I-MTB sample can be either due to a larger distance among nanoparticles in different bacteria or to the presence of intra-bacterial particle chains with a parallel alignment of particle moments[37–39] that would result in a positive contribution to the δM plot. Moreover, this particle arrangement could also lead to a reduction of the overall randomness of particle easy axis orientation in the I-MTB sample, thus explaining the lower width of the SFD, with respect to what is observed in the O-MTB sample, where no specific geometrical arrangement is present. This is also in agreement with a small increase of M_r/M_S on the O-MTB sample. It should be underlined that these effects have been already observed in several systems of magnetic nanoparticles that were artificially synthesized.[3,40,41]

4. Conclusions

Recently a new statistical approach, based on the correlation of particle size and shape creating the so-called "aspect maps" (AMs), has been proposed to investigate the morphological features of magnetic nanoparticles.[16] In the present paper an improvement of the AM method has been applied on magnetite nanoparticles

(Fe_3O_4) extracted from *Magnetospirillum gryphiswaldense* allowing to obtain 3D information from TEM images. The obtained results have been verified by tomography measurements, showing a very good agreement with experimental results. Preliminary magnetic measurements clearly indicate that depending on the geometrical arrangement, identical nanoparticles can show significantly different magnetic properties in terms of magnetic interactions, coercivity, remanence coercivity and switching field distribution. Relaxation measurements are in progress to understand deeply the reversal mechanism of magnetization in these systems. In particular comparison with nanoparticles with the same diameters and different shapes (*e.g.* spherical, cubic) will be developed in order to investigate the effect of morphology on interparticle interactions and the reversal of magnetization.

Acknowledgements

The Research Council of Norway is acknowledged for the support to the Norwegian Micro- and Nano-Fabrication Facility, NorFab. Thanks to the Swedish Research Council (VR).

Notes and references

1 M. El-Hilo, K. O'Grady and R. W. Chantrell, *J. Magn. Magn. Mater.*, 1992, **117**, 21–28.
2 L. Néel, *Ann. Geophys.*, 1949, **5**, 99–136.
3 D. Peddis, C. Cannas, A. Musinu, A. Ardu, F. Orru, D. Fiorani, S. Laureti, D. Rinaldi, G. Muscas, G. Concas and G. Piccaluga, *Chem. Mater.*, 2013, **25**, 2005–2013.
4 J. Crangle, *Solid State Magnetism*, London, 1991.
5 G. Singh, H. Chan, T. Udayabhaskararao, E. Gelman, D. Peddis, A. Baskin, G. Leitus, P. Kral and R. Klajn, *Faraday Discuss.*, 2015, **181**, 403–421.
6 G. Singh, H. Chan, A. Baskin, E. Gelman, N. Repnin, P. Král and R. Klajn, *Science*, 2014, **345**(6201), 1149–1153.
7 L. M. L. Marzan, *J. Mater. Chem.*, 2006, **16**, 3891–3892.
8 G. R. Patzke, F. Krumeich and R. Nesper, *Angew. Chem., Int. Ed.*, 2002, **41**, 2446–2461.
9 D. Parker, V. Dupuis, F. Ladieu, J. P. Bouchaud, E. Dubois, R. Perzynski and E. Vincent, *Phys. Rev. B: Condens. Matter Mater. Phys.*, 2008, **77**, 104428–104429.
10 L. Suber and D. Peddis, in *Nanomaterials for Life Science*, 2009, vol. 4.
11 J. Alonso, M. L. Fdez-Gubieda, J. M. Barandiarán, A. Svalov, L. Fernández Barquín, D. Alba Venero and I. Orue, *Phys. Rev. B: Condens. Matter Mater. Phys.*, 2010, **82**, 054406.
12 A. M. Huízar-Félix, D. Muñoz, I. Orue, C. Magén, A. Ibarra, J. M. Barandiarán, A. Muela and M. L. Fdez-Gubieda, *Appl. Phys. Lett.*, 2016, **108**, 063109.
13 M. Gonzales-Weimuller, M. Zeisberger and K. M. Krishnan, *J. Magn. Magn. Mater.*, 2009, **321**, 1947–1950.
14 R. Hergt, S. Dutz and M. Röder, *J. Phys.: Condens. Matter*, 2008, **20**, 385214.
15 M. Artus, L. Ben Tahar, F. Herbst, L. Smiri, F. Villain, N. Yaacoub, J.-M. Grenèche, S. Ammar and F. Fiévet, *J. Phys.: Condens. Matter*, 2011, **23**, 506001.

16 G. Muscas, G. Singh, W. R. Glomm, R. Mathieu, P. A. Kumar, G. Concas, E. Agostinelli and D. Peddis, *Chem. Mater.*, 2015, **27**, 1982–1990.

17 M. L. Fdez-Gubieda, A. Muela, J. Alonso, A. García-Prieto, L. Olivi, R. Fernández-Pacheco and J. Barandiarán, *ACS Nano*, 2013, **7**, 3297–3305.

18 J. R. Kremer, D. N. Mastronarde and J. R. McIntosh, *J. Struct. Biol.*, 1996, **116**, 71–76.

19 E. F. Pettersen, T. D. Goddard, C. C. Huang, G. S. Couch, D. M. Greenblatt, E. C. Meng and T. E. Ferrin, *J. Comput. Chem.*, 2004, **25**, 1605–1612.

20 C. T. Lefèvre and D. A. Bazylinski, *Microbiol. Mol. Biol. Rev.*, 2013, **77**, 497–526.

21 J. Schindelin, I. Arganda-Carreras, E. Frise, V. Kaynig, M. Longair, T. Pietzsch, S. Preibisch, C. Rueden, S. Saalfeld, B. Schmid, J.-Y. Tinevez, D. J. White, V. Hartenstein, K. Eliceiri, P. Tomancak and A. Cardona, *Nat. Mater.*, 2012, **9**, 676–682.

22 E. Limpert, W. a. Stahel and M. Abbt, *BioScience*, 2001, **51**, 341.

23 L. B. Kiss, J. Söderlund, G. a. Niklasson and C. G. Granqvist, *Nanotechnology*, 1999, **10**, 25–28.

24 C. G. Granqvist and R. A. Buhrman, *J. Appl. Phys.*, 1976, **47**, 2200–2219.

25 A. T. Ngo, P. Bonville and M. P. Pileni, *Eur. Phys. J. B*, 1999, **9**, 583–592.

26 R. Prozorov, Y. Yeshurun, T. Prozorov and A. Gedanken, *Phys. Rev. B: Condens. Matter*, 1999, **59**, 6956–6965.

27 E. J. W. Verwey, *Nature*, 1939, **144**, 327.

28 V. Dediu, E. Arisi, I. Bergenti, A. Riminucci, M. Solzi, C. Pernechele and M. Natali, *J. Magn. Magn. Mater.*, 2007, **316**, e721–e723.

29 G. Muscas, G. Concas, C. Cannas, A. Musinu, A. Ardu, F. Orrù, D. Fiorani, S. Laureti, D. Rinaldi, G. Piccaluga, *et al.*, *J. Phys. Chem. C*, 2013, **117**, 23378–23384.

30 R. Prozorov, T. Prozorov, S. K. Mallapragada, B. Narasimhan, T. J. Williams and D. a. Bazylinski, *Phys. Rev. B: Condens. Matter Mater. Phys.*, 2007, **76**, 1–10.

31 D. Peddis, C. Cannas, A. Musinu and G. Piccaluga, *J. Phys. Chem. C*, 2008, **112**, 5141–5147.

32 D. Peddis, C. Cannas, A. Musinu and G. Piccaluga, *Chem.–Eur. J.*, 2009, **15**, 7822–7829.

33 K. O'Grady and R. W. Chantrelll, *Int. Workshop Stud. Magn. Prop. Fine Part. Relevance Mater. Sci.*, 1991, 93–102.

34 E. P. Wohlfarth, *J. Appl. Phys.*, 1958, **29**, 595–596.

35 P. E. Kelly, K. O. Grady, P. I. Mayo and R. W. Chantrell, *IEEE Trans. Magn.*, 1989, **25**, 3881–3883.

36 D. Peddis, P. E. Jonsson, G. Varvaro and S. Laureti, in *Nanomagnetism: Fundamentals and Applications*, ed. C. Binns, Elsevier B.V, Oxford, UK, 2014, pp. 129–189.

37 S. Mørup, M. F. Hansen and C. Frandsen, *Beilstein J. Nanotechnol.*, 2010, **1**, 182–190.

38 J. M. Thomas, E. T. Simpson, T. Kasama and R. E. Dunin-Borkowski, *Acc. Chem. Res.*, 2008, **41**, 665–674.

39 R. E. Dunin-Borkowski, T. Kasama, A. Wei, S. L. Tripp, M. J. Hytch, E. Snoeck, R. J. Harrison and A. Putnis, *Microsc. Res. Tech.*, 2004, **64**, 390–402.

40 M. P. Pileni, *J. Phys. D: Appl. Phys.*, 2008, **41**, 134002.

41 Y. Lalatonne, L. Motte, V. Russier, a. T. Ngo, P. Bonville and M. P. Pileni, *J. Phys. Chem. B*, 2004, **108**, 1848–1854.

Faraday Discussions

PAPER

Linear assembly of patchy and non-patchy nanoparticles

Rachelle M. Choueiri,†[a] Elizabeth Galati,†[a] Anna Klinkova,[a] Héloïse Thérien-Aubin[a] and Eugenia Kumacheva*[abc]

Received 19th March 2016, Accepted 19th April 2016
DOI: 10.1039/c6fd00057f

Linear assemblies of nanoparticles show promising applications due to their collective electronic, optical and magnetic properties. Rational design and controllable organization of nanoparticles in one-dimensional structures can strongly benefit from the marked similarity between conventional step-growth polymerization reactions and directional step-wise assembly of nanoparticles in linear chains. Here we show different aspects of the "polymerization" approach to the solution-based self-assembly of polymer-functionalized metal nanoparticles with different chemical compositions, shapes and dimensions. The self-assembly was triggered by inducing solvophobic attraction between polymer ligands, due to the change in solvent quality. We show that both anisotropic (patchy) nanoparticles and nanoparticles uniformly capped with polymer molecules can self-assemble in linear chains. We explore the control of chain length, morphology, and composition, discuss the ability to form isotropic and hierarchical structures and show the properties and potential applications of linear assemblies of plasmonic nanoparticles.

Introduction

Linear, one-dimensional assemblies of metal nanoparticles (NPs) have promising applications in the sensing of chemical and biological species and in the fabrication of waveguides and antennas.[1] The collective optical properties of NP assemblies depend on their aggregation number, the distance between the NPs in the chain, NP collinearity and chain rigidity, to name just a few important characteristics.[2,3] For example, the structure and composition of NP chains influence the spatial distribution of charge polarization, as well as the splitting and position

[a]Department of Chemistry, University of Toronto, 80 Saint George Street, Toronto, Ontario M5S 3H6, Canada. E-mail: ekumache@chem.utoronto.ca

[b]Department of Chemical Engineering and Applied Chemistry, University of Toronto, 200 College Street, Toronto, Ontario M5S 3E5, Canada

[c]The Institute of Biomaterials and Biomedical Engineering, University of Toronto, 4 Taddle Creek Road, Toronto, Ontario M5S 3G9, Canada

† R. M. Choueiri and E. Galati contributed equally to this work.

of the plasmon resonance peak. Specific design rules have to be applied to the fabrication of such assemblies, in order to achieve their best performance.

One of the promising approaches to the rational design of self-assembled NP structures relies on a remarkable resemblance between linear polymer molecules and one-dimensional NP assemblies, in which NP building blocks act as repeat units linked with physical or chemical bonds. For example, for the chain of rod-like NPs shown schematically in Fig. 1, it is possible to define the degree of polymerization, x (equal to 3), the bond length and the bond angle.

Furthermore, building on this analogy, it would be possible to make the correlation between the mean aggregation number of the NP chains and their number average or weight average degree of polymerization, as well as polydispersity index, PDI[4] as

$$\overline{X}_{\mathrm{n}} = \frac{\sum n_x x}{\sum n_x},\ \overline{X}_{\mathrm{w}} = \frac{\sum n_x x^2}{\sum n_x x},\ \mathrm{PDI} = \frac{\overline{X}_{\mathrm{w}}}{\overline{X}_{\mathrm{n}}}, \tag{1}$$

where n_x is the number of chains containing x NPs. Based on the similarity between polymer molecules and linear assemblies of NPs, we introduce the concept of nanoscopic monomers (*nanomers*) forming a colloidal analogue of a molecular polymer, that is, a *nanopolymer*.

The analogy between molecular polymerization reactions and linear self-assembly of NPs can benefit both areas of research. Direct visualization of nanostructures formed in different stages of self-assembly and analysis of their shape, composition and size distribution offers the capability to examine the effect of reactivity of nanomers and explore the role of side reactions on nanopolymer structure and composition. Such analysis would enable validation of theoretical models developed for synthetic polymer chemistry. Furthermore, nanoscience-specific techniques, *e.g.*, the measurement of the red-shift of surface plasmon resonance of metal NPs or small-angle X-ray scattering can be used for the characterization of colloidal polymerization. On the other hand, strategies developed in synthetic polymer chemistry, *e.g.*, the copolymerization of comonomers with distinct properties, the synthesis of specific isomers, and control of polymer molecular weight by using "chain stoppers", offer a platform for the self-assembly of nanopolymers with well-defined structure and properties.

Here we discuss different aspects of one-dimensional self-assembly of polymer-functionalized metal NPs with different compositions, shapes and dimensions. We show that both patchy and non-patchy NPs coated with polymer ligands can assemble in nanopolymer chains, with a striking similarity between molecular step-growth polymerization and nanoparticle assembly. We also show an example of a higher-level organization of NP assemblies in complex, hierarchical

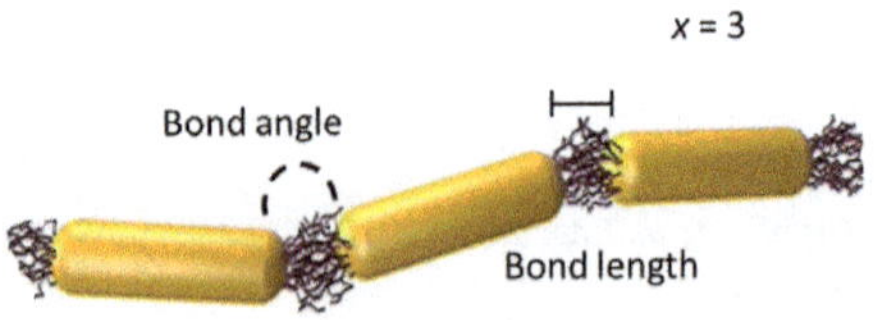

Fig. 1 A chain of three polymer-functionalized nanorod repeat units forming bonds with a particular angle and length.

structures. Finally, we demonstrate optical properties of linear assemblies of plasmonic NPs, such as their surface enhanced Raman scattering (SERS), as well as new properties of the chains of face-to-face assembled metal nanocubes, due to the formation of a Fabry–Perot resonator structure.[5]

Experimental

Gold, silver and palladium NPs were synthesized using metal precursor salts as described elsewhere.[6–12] The syntheses were carried out in water, with cetyltrimethylammonium bromide (CTAB) used as a stabilizer. To exchange the CTAB ligands with polymer molecules, a concentrated NP solution in water was added to the solution of thiol-terminated polystyrene (PS) or thiol-terminated poly(styrene-*co*-isoprene) in tetrahydrofuran. The resulting solution was sonicated for 30 min and equilibrated for at least 24 h. The synthesis of two-component heterostructured iron oxide–gold NPs is reported elsewhere.[13] For these NPs, after the replacement of CTAB ligands with PS molecules on the surface of the gold constituent, an additional ligand exchange step was carried out to replace oleic acid with dopamine-terminated polyethylene glycol on the iron oxide surface.

After completion of the ligand exchange procedure, the NPs were transferred into dimethylformamide (DMF), a good solvent for both CTAB and PS. Linear NP self-assembly was triggered by reducing the quality of solvent for the polymer ligands. A solvent mixture containing DMF and water was added to the NP solution to reach a pre-determined concentration of water, C_w, in the solution. When thiol-terminated poly(styrene-*co*-isoprene) was used as a ligand, after a particular self-assembly time, the NP solution in the DMF–water mixture in the presence of azobisisobutyronitrile (AIBN) was subjected to 5–25-min-long ultra-violet irradiation using a UV-A lamp (Hönle, UVAPrint 40C, $\lambda = 365$ nm, intensity 30 mW cm^{-2}), to achieve permanent crosslinking of the self-assembled NP chains.

The evolution of the self-assembled structures was followed by monitoring the shift in the extinction peak of the NPs using a Varian Cary 5000 spectrometer. Raman scattering measurements were carried out using a Raman spectrometer (DeltaNu Advantage Raman Series with a 785 nm laser line). Characterization of NP dimensions and electrokinetic potential were carried on a Zetasizer Nano instrument (Malvern) using a 633 nm HeNe laser. Imaging of the self-assembled structures was carried out by transmission electron microscopy (TEM) and scanning electron microscopy (SEM) using a Hitachi H-7000 transmission electron microscope and a Hitachi HD-2000 scanning transmission electron microscope, respectively. A droplet of the NP solution was deposited on the carbon-coated grid, and after ~60 s the solvent was removed with a tissue.

Results and discussion

Linear assembly of anisotropic (patchy) nanoparticles

End-functionalization of gold spherocylindrical and arrow-headed nanorods (both denoted as NRs) with PS ligands was achieved by fine-tuning the concentration of PS in the solution in tetrahydrofuran in the ligand exchange procedure, so that CTAB was replaced only at the NR edges due to its weaker attachment to the 111 crystalline plane at the NR ends.[14] The end-attachment of PS endowed patchy NRs with two reactive sites (Fig. 2A). To induce linear NR self-assembly, the

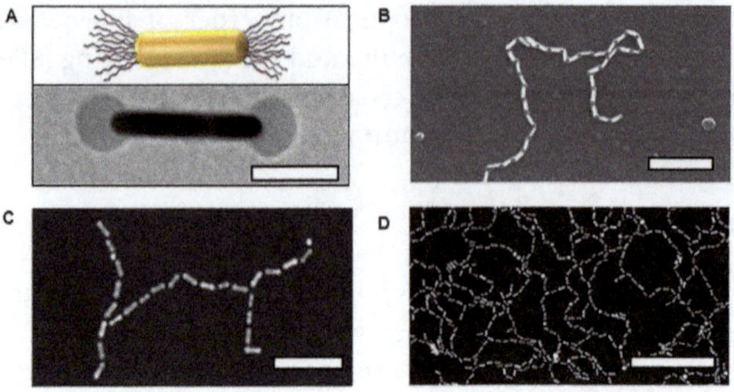

Fig. 2 (A) Schematic (top) and TEM image (bottom) of a gold NR end-tethered with PS ligands. Scale bar is 100 nm. (B) SEM image of a linear chain of gold NRs end-tethered with PS ligands. Scale bar is 500 nm. (C) Dark-field TEM image of PS-end-tethered gold arrow-headed NRs self-assembled into a branched polymer (H-polymer). Scale bar 200 nm. SEM image of the self-assembled network of gold NRs end-terminated with PS molecules. Scale bar is 500 nm. In (B–D) the self-assembly occurred in the DMF–water solution at $C_w = 15$ vol%.

reactivity of the patches was activated by introducing water to the NR solution in DMF, thereby reducing solvent quality for the hydrophobic PS molecules. To reduce the free surface energy of the system, the PS ligands screened unfavorable interactions with the poor solvent by PS–PS association, thus forming a physical bond between the NR ends. This process resulted in the formation of linear NR chains shown in Fig. 2B. The self-assembly resembled molecular polymerization, with individual NRs acting as monomers (or *nanomers*) forming a colloidal analogue of a molecular polymer (*nanopolymer*). Similar to molecular polymers, we observed the formation of other nanopolymer structures such as branched H-polymers (Fig. 2C) and networks (Fig. 2D).

The evolution of the nanopolymers in the course of self-assembly was characterized by statistical analysis of the self-assembled structures. By analogy with molecular polymerization, the average number of NRs in the chains was defined as the number average degree of polymerization, $\bar{X}_n$.

The self-assembly kinetics and statistics of the distribution of the degree of polymerization of the nanopolymer followed closely those of a conventional molecular step-growth polymerization.[15] The value of $\bar{X}_n$ of the nanopolymer increased linearly with the self-assembly time t as

$$\bar{X}_n = 2[M]_0 k_p t + 1, \quad (2)$$

where $[M]_0$ is the initial NR concentration and k_p is the polymerization rate constant. The relation $\bar{X}_n \sim t$ was characteristic of reaction-controlled step growth polymerization, in which the reactivity of monomer functional groups does not depend on the polymer's degree of polymerization (Flory's assumption).[4] The rate of assembly characterized as $d\bar{X}_n/dt$ linearly increased with nanomer concentration (Fig. 3A). Furthermore, the variation in polydispersity index, PDI, of the nanopolymers was similar to the change in PDI of step-growth polymers, which increase as[4,16] $\text{PDI} = 2 - 1/\bar{X}_n$ and approach 2 at high conversion (Fig. 3B).

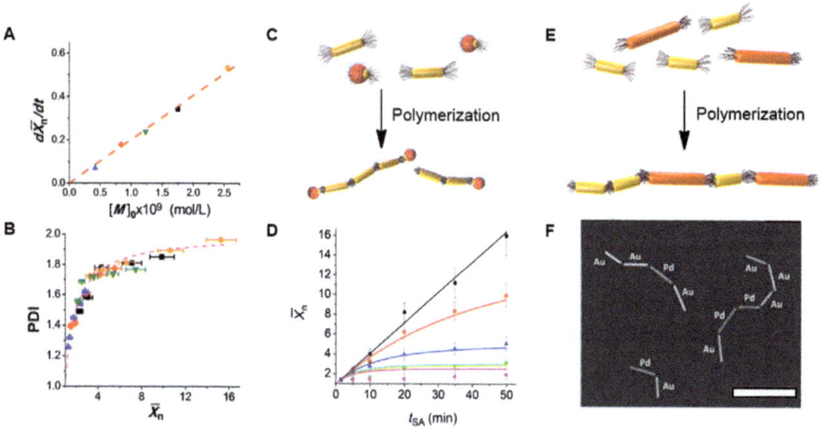

Fig. 3 (A and B) Polymerization of NRs at initial NR concentration, $[M]_0$, of 0.42×10^{-9} (solid blue triangles), 0.84×10^{-9} (solid red diamonds), 1.24×10^{-9} (solid green inverted triangles), 1.76×10^{-9} (solid black squares), and 2.56×10^{-9} (solid orange circles) (mol L^{-1}). (A) Variation in the rate of NR polymerization, $d\bar{X}_n/dt$, plotted as a function of NR concentration. (B) Variation in polydispersity index, PDI, in the course of NR polymerization. The dashed line shows the theoretically expected variation PDI $= (2 - 1/\bar{X}_n)$. (A) and (B) are adapted from ref. 15. (C) Schematic of NR polymerization in the presence of colloidal chain stoppers. The yellow and red colours correspond to the gold and iron oxide constituents of the colloidal chain stoppers. (D) Variation in $\bar{X}_n$ of the NR chains with self-assembly time at different chain stopper concentrations. (C) and (D) are adapted from ref. 17. (E) Schematic of copolymerization of distinct nanomers shown with different colours; (F) dark field TEM image of chains of copolymerized palladium and gold NRs. Scale bar is 500 nm. (F) is adapted from ref. 20.

By analogy with molecular polymers, we estimated the mean contour and persistence lengths of the nanopolymers formed by the NRs, which at varying solvent quality were up to 800 and 250 nm, respectively. We note that these characteristics should be treated with caution, since they were determined by analyzing NR images and may be affected by drying NR solutions.

We further explored the similarity between the self-assembly of patchy NRs and polymer synthesis by using the strategies used in molecular step-growth polymerization, namely, control of polymer molecular weight by using monofunctional chain stoppers and copolymerization of chemically distinct nanomers.

Control of nanopolymer chain length

To control the value of $\bar{X}_n$ in conventional step-growth polymerization, the stoichiometry between bifunctional monomers is broken by adding a small amount of monofunctional comonomers which "zip" growing polymer chains.[4] We explored this approach for nanopolymers by introducing in the solution of self-assembling bifunctional patchy NRs a small quantity of "monofunctional" nanospheres (NSs), as illustrated in Fig. 3C.[17] The colloidal analogues of monofunctional chain stoppers (CSs) were heterostructured NSs, which contained two constituents: a reactive gold constituent capped with PS ligands and an iron oxide constituent capped with dopamine-tethered polyethylene glycol. The ratio of the dimensions of the gold and iron oxide components of the NSs was carefully tuned to ensure size complementarity between the end of the gold NR nanomer and the reactive PS-capped gold domain of the CS (polyethylene glycol end-tethered to the non-reactive iron oxide domain acted as a stabilizer). A barrier preventing further NR or NS attachment to the "zipped" chain end was provided by a sufficiently large size of the iron oxide domain.

The heterostructured CSs were mixed with the NRs in DMF in the ratio $0 \leq [\mathrm{CS}]/[\mathrm{NR}] \leq 1.55$, where [CS] and [NR] were the molar concentrations of the CSs and the NRs, respectively. When the quality of the solvent for the PS ligands was reduced by adding water to DMF to achieve $C_\mathrm{w} = 15$ vol%, the two-patch NRs assembled in linear chains, while the single-patch patch NSs attached to chain end(s) and in this manner, "zipped" them to suppress further NR self-assembly. The processes of self-assembly and chain zipping, as well as side reactions of CSs were visualized at different stages of the self-assembly process by using electron microscopy. Imaging of NPs was complemented by the analysis of the shift in surface plasmon resonance of the NRs. In this manner, we were able to ascertain the prevalence of side reactions of the monofunctional CSs such as their dimerization or binding to junctions between adjacent NRs in the chains. Based on the kinetic and statistical analysis of chain growth, we developed a model of the kinetics of the colloidal polymerization in the presence of a particular concentration of CSs. In particular, *via* statistical analysis of the TEM images we accounted quantitatively for the contribution of side reactions in the course of NR chain growth. The efficiency, E, of the CS attachment to the end of the chain (the desired "zipping" step) enabled the prediction of $\bar{X}_\mathrm{n}$ of the nanopolymer as

$$\overline{X}_\mathrm{n} = \frac{1}{1 - P_\mathrm{NR} + \frac{1}{2}EP_\mathrm{CS}\frac{[\mathrm{CS}]}{[\mathrm{NR}]}}, \tag{3}$$

where P_NR is the conversion of NRs into nanopolymers in the course of self-assembly, and P_CS the conversion of CSs (defined as the fraction of NR ends and CSs, respectively, that has reacted at a particular self-assembly time). The ability to visualize and quantify the effect of side reactions highlighted the value of the colloidal approach to polymerization.

The ratio [CS]/[NR] affected the length of the NR chains: for a particular self-assembly time, the value of $\bar{X}_\mathrm{n}$ of the nanopolymer reduced with increasing concentration of monofunctional CSs (Fig. 3D). The co-assembly of CSs with the NR chains also resulted in a deviation from the linear dependence of $\bar{X}_\mathrm{n}$ *vs.* t, which was characteristic of the NR self-assembly in the absence of CSs.

By varying the ratio [CS]/[NR] at the beginning of the self-assembly process, a particular value of $\bar{X}_\mathrm{n}$ of the nanopolymer could be achieved and effectively controlled by adjusting the initial concentration of colloidal CSs, in striking similarity with molecular step-growth polymerization.

Step-growth copolymerization

Future applications of self-assembled nanostructures require NP ensembles with a high degree of complexity, which may be achieved by heteroassembly of NPs with different dimensions, shapes, and chemical compositions. For example, new optical modes, such as Fano resonance are expected for heterostructures of NPs.[18,19] Since linear self-assembly of NPs is analogous to a polymerization reaction, we explored the step-growth polymerization-based approach for the assembly of *nanocopolymers*,[20] in which distinct NP repeat units were assembled in linear chains, as shown in the scheme in Fig. 3E. Experimentally, this approach was explored for the self-assembly of PS end-tethered gold and palladium NR co-nanomers, and for the self-assembly of gold NRs with difference lengths in the DMF–water mixture. A study of the co-assembly of gold NRs with different lengths revealed that the polymerization rate reduced with increasing NR length, due to the combined effect of reduced NR mobility and increased excluded volume. The copolymerization of gold NRs with different lengths was examined to determine the composition and microstructure of the resulting copolymer chains. When the NRs with different length were co-assembled, the rate of copolymerization was related to the rate of the homopolymerization as

$$k_{\text{copol}} = k_{11}f^2 + k_{22}(1-f)^2 + 2k_{12}f(1-f) \tag{4}$$

where $f = [\text{NR}_1]_0/([\text{NR}_1]_0 + [\text{NR}_2]_0)$, and $[\text{NR}_1]_0$ and $[\text{NR}_1]_0$ are the molar concentrations of short and long NRs, respectively, at the beginning of the co-assembly process. The constants k_{11}, k_{22} and k_{12} are the rate constants of the assembly of two short NRs, two long NRs and a short and a long NRs, respectively.

The structure of the nanocopolymer was found to be random. The microhomogeneity factor (H) of the nanocopolymer was determined by analyzing the distribution of dyads of two NP types in the nanocopolymer as

$$H = \frac{P_{\text{NP}_1-\text{NP}_2}}{P_{\text{NP}_1-\text{NP}_2} + 2P_{\text{NP}_1-\text{NP}_1}} + \frac{P_{\text{NP}_1-\text{NP}_2}}{P_{\text{NP}_1-\text{NP}_2} + 2P_{\text{NP}_2-\text{NP}_2}}, \tag{5}$$

where P is the fraction of dyads of a particular type. For molecular copolymers, the value of H is 0, 1 and 2 for alternating, random and block copolymers, respectively. TEM analysis of the structure of the copolymers of gold NRs with different lengths revealed the values of H varying from 0.94 to 1.01, typical of a random copolymer.[4]

Following the same strategy, we co-assembled (copolymerized) gold and palladium NRs with lengths of ~180 and ~270 nm, respectively, that were introduced in the solution at concentrations of 43 and 8.6 nM, respectively. The coassembly yielded a random copolymer (shown in Fig. 3F) with a microhomogeneity factor $H = 1.13$.

Photocrosslinking of nanopolymers

Photocrosslinking of the polymer ligands bridging NRs in the nanopolymer was used to gain control over the properties of the self-assembled chains.[21] The NRs end-terminated with poly(styrene-*co*-isoprene) (PS-*co*-PI) were dispersed in DMF containing an initiator azobisisobutyronitrile (AIBN). The addition of water to this mixed solution led to a reduction of the quality of the solvent for both PS-*co*-PI[22] and AIBN, which resulted in two effects: the self-assembly of the NRs in linear chains and the partition of the hydrophobic initiator in the PS-*co*-PI ligands bridging the NRs (Fig. 4A). After a particular self-assembly time (determining the value of $\bar{X}_n$), the crosslinking reaction was triggered by illuminating the NR chains with ultra-violet (UV) light. The UV-irradiation of the AIBN molecules created free

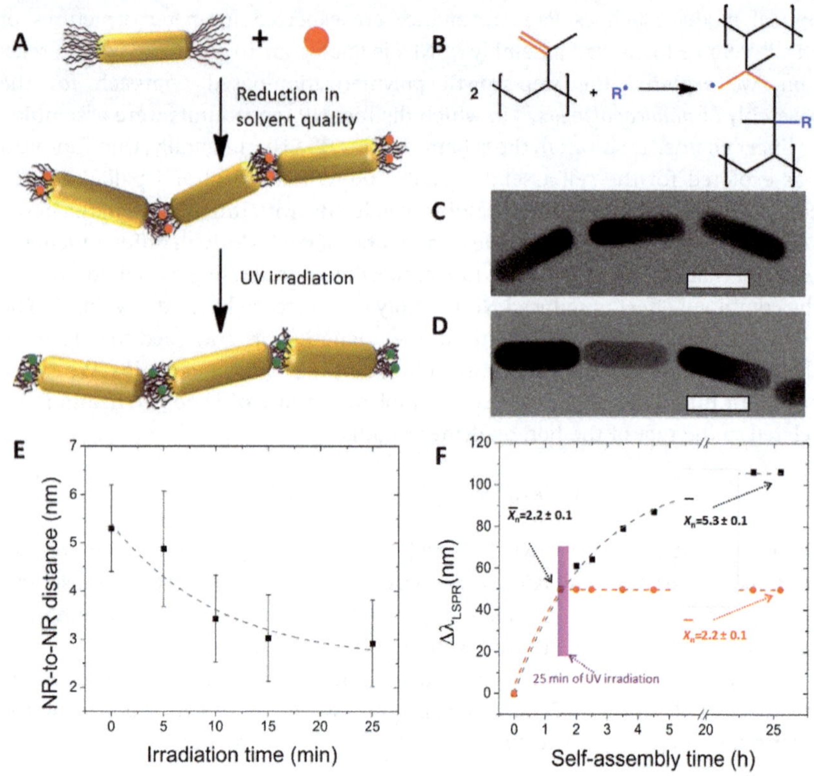

Fig. 4 Photocrosslinking of self-assembled NR chains. (A) Schematic of chain crosslinking. Top: The NRs end-functionalized with poly(styrene-*co*-isoprene) ligands are mixed with a photoinitiator, AIBN (red dot), in the solution in DMF; middle: NR self-assembly is triggered by adding of water to the NR solution in DMF. The hydrophobic AIBN molecules partition in the copolymer between adjacent NRs in the chains; bottom: chain photocrosslinking is initiated by the irradiation at 365 nm to crosslink the 3,4-isoprene units in the copolymer. (B) Chemical reaction taking place in the copolymer during photocrosslinking. A fragment of the NR chain before (C) and after (D) photocrosslinking. The scale bars are 25 nm. (E) Effect of the photocrosslinking time on the average distance between two adjacent NRs in the chain. (F) Effect of photocrosslinking on the NR self-assembly, presented as a temporal shift of the surface plasmon resonance with time. After 1.5 h assembly (at $\bar{X}_n = 2.2$), the NRs were irradiated for 25 min at 365 nm. The NR chain growth was frozen at $\bar{X}_n = 2.2$ (red circles), while without irradiation the self-assembly continued and the value of $\bar{X}_n$ increased. (A, C, D, E and F) are reproduced from ref. 21.

radicals which initiated the crosslinking of the 3,4-isoprene units in adjacent PS-*co*-PI molecules (Fig. 4B).

One of the important outcomes of the NR chain photocrosslinking was a decrease in the volume of the copolymer ligands bridging NR ends in the chain. This effect resulted in a reduction of the gap separating two adjacent NRs (Fig. 4C–E). Approximately 55% reduction in the inter-rod distance was achieved after 25 min of UV-irradiation, in comparison with the non-crosslinked nanopolymer. This effect has important implications for potential applications of NR ensembles in surface-enhanced Raman scattering (SERS)-based sensing, since the enhancement of the SERS signal increases with a reducing distance between plasmonic NPs.[21,23] Furthermore, the crosslinking of the copolymer also resulted in the quenching of the growth of the nanopolymer. This effect was caused by the crosslinking of the terminal copolymer ligands at the NR chain ends, thus suppressing the interpenetration of the copolymer molecules and preventing further addition of nanomers to the chain. This effect is explicitly shown in Fig. 4F. After the value of $\bar{X}_n$ of 2.2 was reached (corresponding to a particular shift in the surface plasmon resonance of the NR chains) and the polymer was crosslinked, no further increase in $\bar{X}_n$ occurred within 25 h, nor variation in the NR chain optical properties, while without irradiation the NR chains continued to grow reaching a value of $\bar{X}_n$ of 5.3 after 24 h.

Finally, the photocrosslinking of the copolymer ligands led to an increase of collinearity of adjacent NRs in the chains. Crosslinked NR chains had a higher fraction of bond angles exceeding 120°, compared to the non-crosslinked NR chains. A higher rigidity of the photocrosslinked NR chains was attributed to a reduced conformational entropy due to the reduction in the volume of the polymer ligands connecting the NRs in the chains.

Linear assembly of isotropic (non-patchy) nanoparticles

We found that surface or shape anisotropy was not required for directional linear self-assembly of polymer-functionalized NPs, in comparison with linear assembly of bifunctional "patchy" NPs.[24] In a poor solvent, NPs uniformly coated with PS ligands also assembled in nanopolymers. This effect was attributed to directional, anisotropic forces governing linear NP assembly and was examined for gold NSs and silver and palladium nanocubes (NCs), all coated with PS ligands.

Gold nanospheres. A cartoon of non-patchy gold NS uniformly surface-functionalized with PS ligands is shown in Fig. 5A. Gold NSs with a mean diameter of 20 nm were capped with PS ligands of different molecular weights and dispersed in DMF. The self-assembly was triggered by reducing solvent quality for the polymer by adding water to the NS solution in DMF. Using two variables, namely, the quality of the solvent (by changing water concentration, C_w, in the mixed solution from 2.5 to 15 vol%) and by varying PS molecular weight (from 5000 to 50 000 g mol^{-1}), we induced NS assembly in linear and globular structures (Fig. 5B–D).[25] The effect of solvent quality and polymer molecular weight on the resulting structure of NS assemblies is summarized in a phase-like diagram (Fig. 5G), which shows globular (G), chain (C), a mixed (G + C) and individual NS (I) regimes. The formation of globules was favored for the NSs capped with high molecular weight PS ligands at low water content, whereas for higher C_w values (dependent on polymer molecular weight), linear NS assemblies were the dominant species. At $C_w \geq 15$ vol%, regardless of the polymer molecular weight, the NSs assembled only in a linear manner.

The schematics of two limiting configurations formed by three adjacent NSs in the chains and globules are shown in Fig. 5E and H, respectively. The angle θ between the lines connecting the center of the 2nd NS with the centers of adjacent 1st and 3rd NSs was 60° and 180° for the globular and linear assembly, and thus determined the type of structure formed.

The formation of a particular structure (globular *vs.* linear) was governed by the interplay of forces leading to the minimum total free energy of the system,

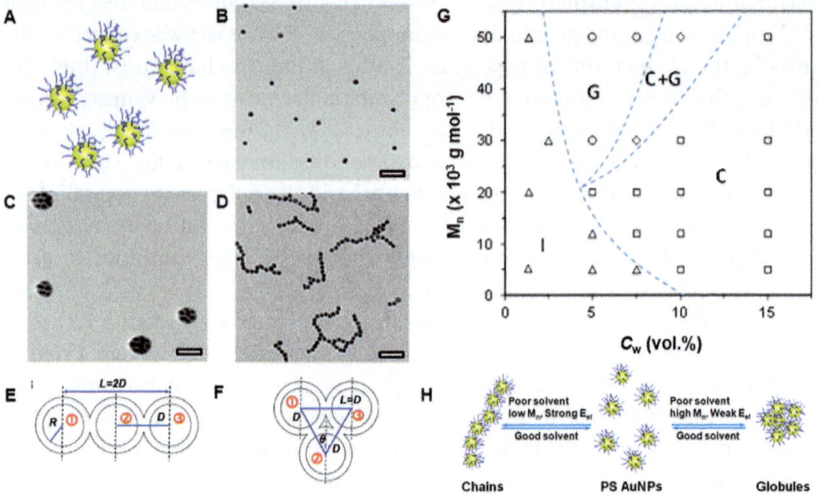

Fig. 5 (A) Schematic of individual gold NSs functionalized with PS ligands. (B) TEM image of individual PS-50k-functionalized gold NSs deposited from DMF. (C) TEM image of globules formed by gold NSs functionalized with PS-50k after 1 h assembly in the DMF–water solution at $C_W = 5\%$. (D) TEM image of chains of gold NSs functionalized with PS-5k after 1 h assembly in the DMF–water solution at $C_W = 15\%$. (G) Phase diagram of self-assembly of gold NSs. Schematics of three assembled gold NSs with 180° (E) and 60° (F) the center of the second NS with the centers of adjacent 1st and 3rd NSs (H) schematic of the factors determining NS self-assembly in linear and globular structures. The grafting density of PS-50k on the NS surface was ~0.08 chains per nm^2 and the mean square end-to-end distance for PS-50k was ~220 nm^2. Adapted from ref. 25.

$$\Delta E_\mathrm{t} = \Delta E_\mathrm{el} + \Delta E_\mathrm{h}, \tag{6}$$

where ΔE_el is the change in energy of repulsive electrostatic interactions and ΔE_h is the change in energy representing a combined effect of hydrophobic/poor solvency attraction forces. Short-range van der Waals interactions were ignored.

The value of ΔE_h was calculated as

$$\Delta E_\mathrm{h} = \Delta E\gamma \times \Delta A, \tag{7}$$

where ΔE_st is the reduction in the surface energy when two NSs form contact, γ is the interfacial tension between the collapsed PS ligands and the solvent, and ΔA is the change in area of the interface when two NSs associate in a poor solvent.

The value of ΔE_el for two associating NSs was calculated as[26]

$$\Delta E_\mathrm{el} = \frac{\pi^2}{2}\frac{kT}{l_\mathrm{B}} R \ln\left(\frac{2D - 2R}{L - 2R}\right) \tag{8}$$

where k is Boltzmann's constant, l_B is the Bjerrum length, R is the NS radius, D is the interparticle spacing and L is the distance between the first and third NS (Fig. 5E and F).

The variation in ΔE_t was evaluated for the assembly of NSs stabilized with PS molecules with a molecular weight of 50 000 g mol^{-1} at C_W of 5 and 15 vol%, at which the NPs assembled into the globules and the chains, respectively. Indeed, in agreement with experimental results, for C_w = 5 vol%, due to the stronger contribution of the surface energy, the globular NS assembly (θ = 60°) had a minimum in ΔE_t and was thus favored. At C_w = 15 vol%, the minimum value of ΔE_t was achieved at θ = 180° (corresponding to the NS chains), due to the stronger and larger contribution of electrostatic repulsion.

We found that reduction in electrostatic repulsion achieved by addition of salts or tetrahydrofuran (a solvent with a lower dielectric constant than DMF) to the NS solution in the DMF/water mixture, resulted in a chain-to-globule transition. In addition, variation in the molecular weight of the polymer ligands could be used to control the formation of a particular structure. The utilization of longer polymer ligands resulted in a greater screening of the NS charge and the propensity of NSs to form globular structures increased even at high solvent polarity. Fig. 5H summarizes the role of different factors on the ability to form linear nanopolymers or globular structures from non-patchy NSs.

Silver nanocubes. Silver NCs capped with PS ligands also assembled in linear chains in the DMF–water mixture under suitable solvent quality and polarity.[5] We varied the flexibility of these NC chains, as well as the shape and volume of junctions formed by PS between adjacent NCs in the chains by varying the ratio between the length of the PS ligands and the NC size. Chains of NSs coated with PS molecules were used as a control system.

For PS-coated NCs, it was expected that due to the NC geometry, their face-to-face assembly is preferred, due to the maximum reduction in solvent-exposed PS surface area. For this binding modality, the formation of rigid chains was expected. We found, however, that PS ligands with a high molecular weight screened the NC geometry, thus yielding fewer face-to-face NC contacts. Consequently, chains with a larger number of kinks were formed, resembling assembly of PS-coated NSs. Fig. 6 shows the chains formed by PS-capped silver NSs (Fig. 6A)

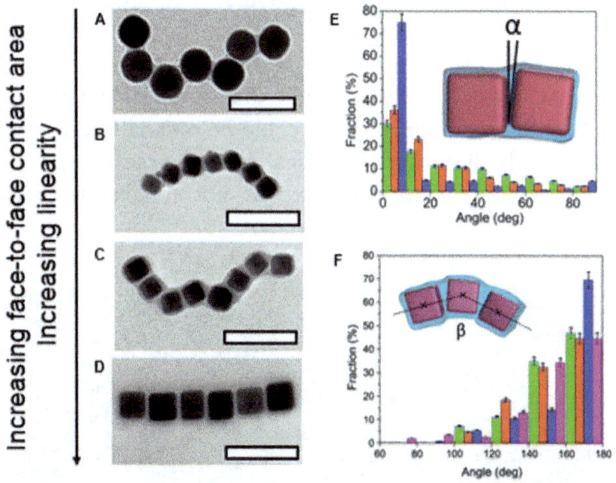

Fig. 6 (A) Chain of silver NSs; (B, C and D) representative TEM images of silver NCs assembled in nanopolymer structures. Scale bars are 100 nm. (E) Fraction of α angles for 25 (green), 35 (red) and 45 nm (blue) silver NCs. (F) Fraction of β angles for 25 (green), 35 (red) and 45 nm (blue) silver NCs and silver NSs (pink). Adapted from ref. 5.

and the chains of NCs at a reducing ratio of the PS length-to-NC size (top-to-bottom, Fig. 6B–D). When the ratio of polymer length-to-NC size reduced, the fraction of face-to-face NC contacts increased, the NC chains became more rigid and exhibited a smaller number of kinks (Fig. 6E and F). Quantification of the number of face-to-face contacts was carried out using a MATLAB code that analysed the TEM images of the self-assembled chains and measured the angles formed by two or three adjacent NPs in the chains (Fig. 6E and F, insets). The statistical analysis of the angles formed by two or three adjacent NCs confirmed quantitatively the trend of increased NC collinearity with increasing ratio of NC size-to-polymer length.

Palladium nanocubes. The self-assembly of uniformly functionalized NPs was further studied for palladium NCs capped with PS molecules with a molecular weight of 5000 and 50 000 g mol^{-1} (PS-5k and PS-50k, respectively).

The morphology of the self-assembled structures in the DMF/water mixture at $C_w = 20$ vol% depended on the length of the polymer ligands (controlled by its molecular weight), in qualitative agreement with the results obtained for assembly of gold NSs and silver NCs. Fig. 7 shows that for the same self-assembly time and solvent quality, the structures formed by palladium NCs functionalized with PS-5k exhibited a stronger trend to form linear structures than those formed by the NCs capped with PS-50k. The latter had a larger aggregation number due to the stronger solvophobic attraction between the longer polymer ligands. In addition, the assemblies formed by PS-50k-capped palladium NCs had fewer face-to-face contacts between adjacent NCs. The latter effect stemmed from the screening of the cubic shape of the NCs with longer polymer ligands. In contrast, palladium NCs capped with PS-5k showed the propensity to form chain-like and open network assemblies.

Hierarchical assembly of nanoparticles

An extension of the anisotropic assembly of isotropically PS functionalized gold NSs was the self-assembly of networks of globules composed of these NSs. Given a sufficiently long incubation time of 24–48 h in the DMF–water mixture, the globules of with an average diameter of approximately 85 nm (Fig. 8A) assembled into chains and networks (Fig. 8B). We note that in the chains, the globules did not merge to form larger entities, which could be rationalized by the enthalpic penalty involved in the breaking of physical bonds to maximize PS–PS interactions. In addition, although the solvent was not sufficiently polar to cause

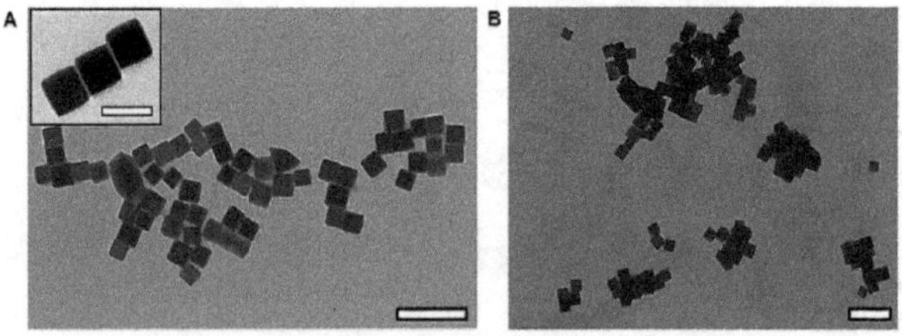

Fig. 7 TEM images of nanostructures formed by palladium NCs capped with (A) PS-5k and (B) PS-50k in the DMF/water mixture at $C_w = 20$ vol%. Scale bars are 200 nm. Inset in (A) shows an enlarged fragment of the NC chain. Scale bar is 10 nm.

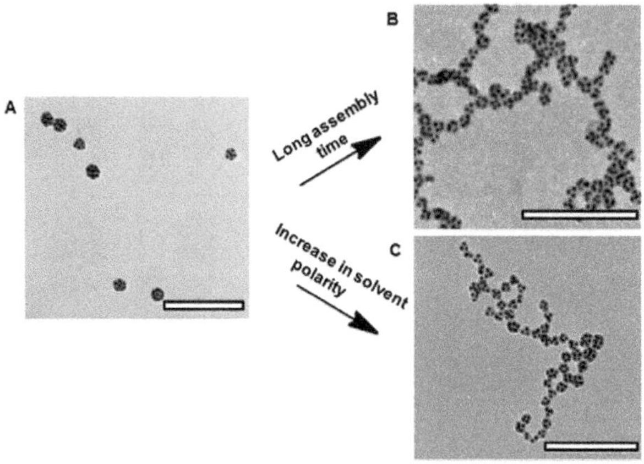

Fig. 8 (A) Individual globules formed by gold NSs isotropically functionalized with PS-50k in the DMF–water solution at $C_W = 7.5\%$ after 1 h self-assembly. Networks of globules formed from the globules after 24 h assembly (B) and after increasing C_W to 15 vol% at the self-assembly time of 2 h.

anisotropic assembly of individual NSs, linear and network diffusion-controlled assembly of the globules was favored due to the larger overall charge of the NS globules, in comparison with individual NSs.[27]

Furthermore, reducing solvent quality by increasing C_W from 5 to 15 vol% resulted in qualitatively similar self-assembly of chains and networks of globules at a reduced time of 2 h (Fig. 8C), due to the more efficient bonding of the NS globules.

Plasmonic properties of nanopolymers

Nanopolymers display directional coupling of optical, electronic, and magnetic properties as a result of interactions between the surface plasmons, excitons, or magnetic moments, respectively, of individual NPs. The formation of a periodic array of hot spots in the gaps between the NPs assembled in chains is important for SERS applications of plasmonic nanopolymers, as they provide significant signal enhancement of up to 10^{15} orders of magnitude.[28] Thus, the use of hot spots enables the observation of single-molecule SERS and sets the limit for the spatial resolution in scanning probe microscopy with high chemical specificity (tip-enhanced Raman spectroscopy).[29]

Dynamic linear self-assembly of metal NPs provides the capability to explore the role of order in NP assemblies on their SERS properties. While a direct qualitative correlation exists between extinction and SERS of individual metal NPs, it deteriorates dramatically in strongly coupled systems with a large distribution of spatially localized resonances. At the same time, there is a direct correlation between SERS enhancement and the dynamic evolution of electric field enhancement in the hot spots, as assessed by solving numerically Maxwell's equations by iteration over time.[23]

We assembled nanopolymers from gold NRs end-tethered with PS ligands and found that for the nanopolymers formed by gold NRs exhibited a non-linear increase with increasing $\bar{X}_n$. Fig. 9B shows the variation in the normalized

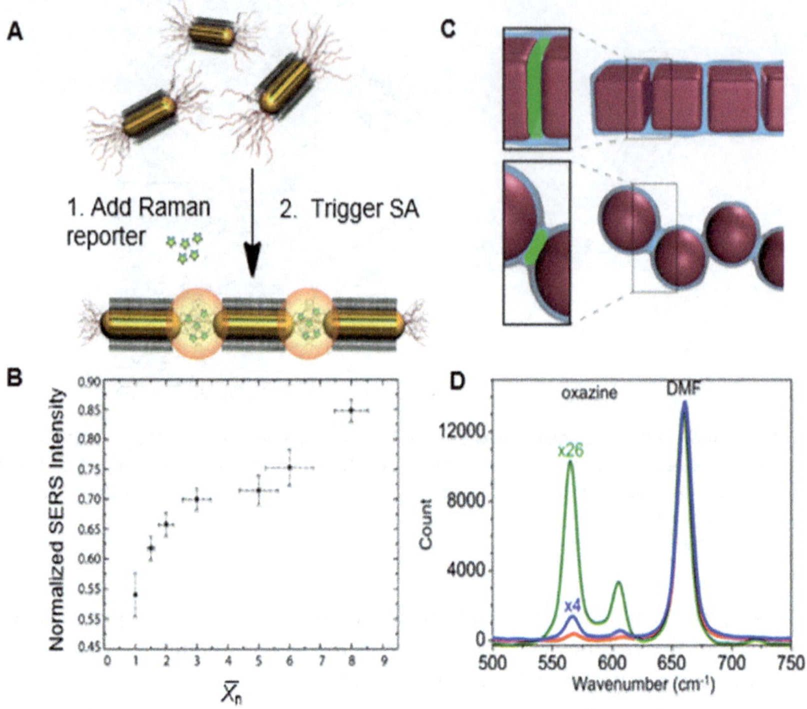

Fig. 9 (A) Schematic representation of the Raman reporter partitioning in the gaps between linearly self-assembled gold NRs. (B) Normalized SERS intensity with respect to $\bar{X}_n$ of gold NR chains. Adapted from ref. 22. (C) Illustration of the hot spots localized between adjacent NCs and adjacent NSs in the nanopolymers. (D) SERS profiles of oxazine solution in DMF (red trace), and oxazine partitioned in the PS junctions binding adjacent NSs (blue trace) in the chains and NCs (green traces) in the nanopolymer. Adapted from ref. 5.

intensities of the Raman peak of a reporter molecule (oxazine) uptaken by the chains of gold NRs and localized in the inter-NR gaps filled with PS (Fig. 9A). The collinearity of the NR chains had a prodigious effect on the electric field, and therefore resulted in SERS enhancement: for bent chains, there was a significant reduction in extinction and SERS enhancement, compared to the co-linear NR assemblies.

Furthermore, for plasmonic nanopolymers formed by face-to-face assembled silver NCs (shown in Fig. 6), we found that the geometry of hot spots played an important role in associated SERS enhancement factors (Fig. 9C). In comparison with chains of silver NSs, nanopolymers formed by face-to-face bonded NCs showed a higher SERS intensity (Fig. 9D). This effect was attributed to the larger volume of hot spots and a nearly uniform electric field in the gaps between collinear NCs, as well as new coupling modes in the plasmonic polymer formed by the NCs. A pair of face-to-face assembled NCs formed a Fabry–Perot resonator[30] with a resonance wavelength determined by the NC dimensions and the size of the inter-NC gap. When Fabry–Perot resonance corresponded to the excitation wavelength, strong enhancement in electric field and SERS was observed.[5] Thus, the optical properties of plasmonic polymers can be programmed by using NP building blocks with specific shapes in sub-wavelength sensing and imaging enhancement.

Conclusions

We have shown that linear nanopolymer chains can be assembled from nanomers with surface and shape anisotropy. This process follows the kinetics and statistics of molecular step-growth polymerization. Following this analogy, we used the strategies of molecular polymerization such as the utilization of chain stoppers and copolymerization to precisely tune the morphology of the nanopolymers. Photocrosslinking of the polymer ligands binding NPs in the nanopolymers enabled control over interparticle distance, chain rigidity and the degree of polymerization of the nanopolymer. We have also demonstrated that surface and shape anisotropy of NPs is not a requirement for assembling one-dimensional NP structures, since anisotropy may arise from the competition of nanoscale forces. Isotropically functionalized metal NSs and NCs self-assembled into chains when polymer ligand and self-assembly medium characteristics were judiciously selected. We also demonstrate that the generated self-assembled structures may form hierarchical self-assemblies, including chain networks and globular networks. Lastly, the effect of self-assembled structure morphology with respect to SERS enhancement was explored yielding important insight into the effects of hot spot geometry and nanopolymer length on plasmonic performance. These findings pave the way for the systematic design of self-assembled nanostructures in the quest for their applications or in the exploration of new self-assembly parameter spaces.

Acknowledgements

We thank the Connaught Foundation and the Natural Sciences and Engineering Research Council of Canada (Discovery and Canada Research Chair programs) for financial support of this work. R. M. C. acknowledges Natural Sciences and Engineering Research Council of Canada for a PGS-D scholarship. A. K. acknowledges the Connaught fund. The authors are grateful to Ilya Gourevich for help with imaging experiments.

Notes and references

1 A. Klinkova, R. M. Choueiri and E. Kumacheva, *Chem. Soc. Rev.*, 2014, **43**, 3976–3991.

2 P. K. Jain, K. S. Lee, I. H. El-Sayed and M. A. El-Sayed, *J. Phys. Chem. B*, 2006, **110**, 7238–7248.

3 L. S. Slaughter, B. A. Willingham, W.-S. Chang, M. H. Chester, N. Ogden and S. Link, *Nano Lett.*, 2012, **12**, 3967–3972.

4 G. Odian, *Principles of Polymerization*, John Wiley & Sons, Inc., Hoboken, NJ, USA, 2004.

5 A. Klinkova, H. Thérien-Aubin, A. Ahmed, D. Nykypanchuk, R. M. Choueiri, B. Gagnon, A. Muntyanu, O. Gang, G. C. Walker and E. Kumacheva, *Nano Lett.*, 2014, **14**, 6314–6321.

6 B. Nikoobakht and M. A. El-Sayed, *Chem. Mater.*, 2003, **15**, 1957–1962.
7 B. D. Busbee, S. O. Obare and C. J. Murphy, *Adv. Mater.*, 2003, **15**, 414–416.
8 Y. Xiang, X. Wu, D. Liu, L. Feng, K. Zhang, W. Chu, W. Zhou and S. Xie, *J. Phys. Chem. C*, 2008, **112**, 3203–3208.
9 Y.-H. Chen, H.-H. Hung and M. H. Huang, *J. Am. Chem. Soc.*, 2009, **131**, 9114–9121.
10 N. R. Jana, L. Gearheart and C. J. Murphy, *Langmuir*, 2001, **17**, 6782–6786.
11 F. Lu, Y. Tian, M. Liu, D. Su, H. Zhang, A. O. Govorov and O. Gang, *Nano Lett.*, 2013, **13**, 3145–3151.
12 W. Niu, Z.-Y. Li, L. Shi, X. Liu, H. Li, S. Han, J. Chen and G. Xu, *Cryst. Growth Des.*, 2008, **8**, 4440–4444.
13 H. Yu, C. Min, P. M. Rice, S. X. Wang, R. L. White and S. Sun, *Nano Lett.*, 2005, **5**, 379–382.
14 Z. Nie, D. Fava, E. Kumacheva, S. Zou, G. C. Walker and M. Rubinstein, *Nat. Mater.*, 2007, **6**, 609–614.
15 K. Liu, Z. Nie, N. Zhao, W. Li, M. Rubinstein and E. Kumacheva, *Science*, 2010, **329**, 197–200.
16 W. H. Carothers, *Trans. Faraday Soc.*, 1936, **32**, 39–49.
17 A. Klinkova, H. Thérien-Aubin, R. M. Choueiri, M. Rubinstein and E. Kumacheva, *Proc. Natl. Acad. Sci. U. S. A.*, 2013, **110**, 18775–18779.
18 L. Hu, Y. Huang, L. Fang, G. Chen, H. Wei and Y. Fang, *Sci. Rep.*, 2015, **5**, 16069.
19 K. C. Woo, L. Shao, H. Chen, Y. Liang, J. Wang and H.-Q. Lin, *ACS Nano*, 2011, **5**, 5976–5986.
20 K. Liu, A. Lukach, K. Sugikawa, L. Chung, J. Vickery, H. Therien-Aubin, B. Yang, M. Rubinstein and E. Kumacheva, *Angew. Chem., Int. Ed.*, 2014, **53**, 2648–2653.
21 A. Lukach, K. Liu, H. Therien-Aubin and E. Kumacheva, *J. Am. Chem. Soc.*, 2012, **134**, 18853–18859.
22 A. Stewart, A. Lee, S. Ip, E. Kumacheva and G. A. Walker, *ACS Nano*, 2014, **8**, 5462–5467.
23 A. Lee, G. F. S. Andrade, A. Ahmed, M. L. Souza, N. Coombs, E. Tumarkin, K. Liu, R. Gordon, A. G. Brolo and E. Kumacheva, *J. Am. Chem. Soc.*, 2011, **133**, 7563–7570.
24 G. A. DeVries, *et al.*, *Science*, 2007, **315**, 358–361.
25 R. Choueiri, A. Klinkova, H. Thérien-Aubin, M. Rubinstein and E. Kumacheva, *J. Am. Chem. Soc.*, 2013, **135**, 10262–10265.
26 E. B. Zhulina, A. B. Boulakh and O. V. Borisov, *Z. Phys. Chem.*, 2012, **226**, 625–643.
27 M. Yang, G. Chen, Y. Zhao, G. Silber, Y. Wang, S. Xing, Y. Han and H. Chen, *Phys. Chem. Chem. Phys.*, 2010, **12**, 11850–11860.
28 D. J. Anderson and M. Moskovits, *J. Phys. Chem. B*, 2006, **110**, 13722–13727.
29 R. C. Maher, in *Raman Spectroscopy for Nanomaterials Characterization*, Springer Berlin Heidelberg, Berlin, Heidelberg, 2012, pp. 215–260.
30 S. A. Maier, *Plasmonics Fundamentals and Applications*, Springer, New York, 2007.

Faraday Discussions

PAPER

Gold nanorod synthesis catalysed by Au clusters†

Yasser A. Attia,[ab] Carlos Vázquez-Vázquez,[a] M. Carmen Blanco,[a] David Buceta[a] and M. Arturo López-Quintela*[a]

Received 15th February 2016, Accepted 8th March 2016
DOI: 10.1039/c6fd00015k

Gold nanorods have been successfully synthesized by the seed mediated method using Au clusters. This synthesis does not require silver ions to obtain large amounts of Au nanorods and has good control over their aspect ratio. Au clusters are produced with the same recipe as for Au seeds, but using shorter reaction times. This very simple scheme confirms the important catalytic influence of clusters in the anisotropic growth control.

Introduction

The different plasmon absorption properties of Au nanostructures, ranging from visible to near infrared, in the last few years have been exploited for developing a large number of important applications, such as sensing, drug delivery, optics, *etc.*[1–12] Due to this importance several methods have been proposed so far for the synthesis of Au nanostructures.[13–20] In spite of the large efforts devoted for more than one decade to the seeded growth synthesis of Au nanostructures, and in particular of Au nanorods (NRs), since its introduction at the beginning of the new century[21–23] a complete understanding of the involved mechanisms is still lacking.[24,25] Particularly, the influence of silver ions is still very controversial.[26] Although it is possible to obtain Au NRs in the absence of silver ions, its presence was shown to greatly improve the amount of Au NRs and to control their aspect ratio.[27] Recently, we have shown[28] that Ag_3 clusters (localized at the tips of the Au NRs) catalyse the anisotropic growth of Au NRs thus introducing a new parameter in this very complex reaction, which was formerly totally overlooked. We could demonstrate that such small Ag clusters are formed in the experimental conditions used for the seeded growth in the presence of Ag ions. It is now well established that small clusters display excellent catalytic activities, mainly in redox reactions. Thus, for example, Corma *et al.* described exceptional catalytic

[a]*Instituto de Investigacións Tecnolóxicas & Facultade de Química, Departamento de Química Física, Universidade de Santiago de Compostela, E-15782 Santiago de Compostela, Spain. E-mail: malopez.quintela@usc.es*

[b]*National Institute of Laser Enhanced Sciences, Cairo University, 12631, Egypt*

† Electronic supplementary information (ESI) available. See DOI: 10.1039/c6fd00015k

activities of small Au clusters (in the range 5 to 10 atoms) for the aerobic oxidation of thiophenol, comparable to sulfhydryl oxidases;[29] Lee *et al.* reported the high activity of Au clusters of similar sizes for propene epoxidation,[30] and also the activity of positively charged Ag clusters (2 to 4 atoms) for the same reaction;[31] Zhu *et al.*[32–34] reported the selective catalytic oxidation of styrene by different monodispersed, thiolate-protected Au clusters (25, 38 and 144 atoms), demonstrating that the catalytic activity increases when the cluster size decreases. Leelavathi *et al.*[35] reported the catalytic activity of thiolate-protected Ag clusters (7 and 8 atoms) for *p*-nitrophenol reduction with $NaBH_4$; and Vilar-Vidal *et al.* reported the high catalytic activity of Cu clusters of 5 and 13 atoms (but the inactivity of clusters with 20 atoms) for the reduction of methylene blue with hydrazine.[36] Due to the well-demonstrated, high catalytic activities displayed by metal clusters of different nature, we hypothesized that other clusters, different to Ag_3 could also be used for the anisotropic growth of Au NRs. Moreover, due to the fact that clusters are almost always formed before the formation of nuclei,[37] one would expect that the catalytic Ag clusters (generated in the presence of Ag ions) could be replaced by Au clusters (because of their expected similar catalytic properties[38]), which could be generated by the same reaction used for the preparation of Au seeds, without using Ag ions.

We will show here that, indeed, this is the case: using Au clusters one can get the same level of control and tunability of the gold NRs formation, as when Ag ions are present in the growth solution. Moreover, Au clusters can be very easily prepared following the same recipe as to produce Au seeds, but allowing the reaction to proceed only until shorter reaction times than for the formation of seeds.

Experimental

Materials

Hexadecyltrimethylammonium bromide (CTAB, 98%, Sigma-Aldrich, Fluka), sodium borohydride ($NaBH_4$, 95%, Riedel-de Haen AG), hydrogen tetrachloroauratetrihydrate ($HAuCl_4 \cdot 3H_2O$, 99.999%, Aldrich), silver nitrate ($AgNO_3$, 99%, Aldrich), and L-ascorbic acid (99%, Aldrich) were used as supplied. Deionized water was used in all the experiments. All glassware was cleaned with aqua regia and rinsed with deionized water prior to the experiments.

Synthesis methods

Seed solution. 5 ml of 0.2 M CTAB solution was mixed with 2.5 ml of 5×10^{-4} M $HAuCl_4$. While stirring the solution, 30 μl of ice-cold freshly prepared 0.1 M $NaBH_4$ was added, which results in the formation of a solution with a brownish yellow colour. Vigorous stirring of the seed solution was continued for 2 minutes. The seeds were used after 10 min of preparation.

Cluster solution. 5 ml of 0.2 M CTAB solution was mixed with 5 ml of 5×10^{-4} M $HAuCl_4$. While stirring the solution, 60 μl of ice-cold freshly prepared 0.1 M $NaBH_4$ was added, which results in the formation of a solution with a brownish yellow colour. Vigorous stirring of the cluster solution was continued and used after 10 s, 30 s, 1 min and 5 min of preparation.

Growth solution with Ag ions. 2.5 ml of 0.2 M CTAB solution were added to 100 μl of 4×10^{-3} M $AgNO_3$ solution at 25 °C. To this solution, 2.5 ml of 10^{-3} M

$HAuCl_4$ were added and, after gentle mixing of the solution, 35 μl of 0.0788 M ascorbic acid (freshly prepared) were added. The growth solution colour changed from dark yellow to colourless. For comparison experiments, 6 μl of the seed solution were added. The colour of the solution gradually changed within 10–20 min.

Growth solution without Ag ions. 5 ml of the gold cluster solution were added to 2.5 ml of deionized water at 25 °C. To this solution, 2.5 ml of 10^{-3} M $HAuCl_4$ were added and, after gentle mixing of the solution, 35 μl of 0.0788 M ascorbic acid (freshly prepared) were added. The colour of growth solution changed from pale yellow to colourless within a few seconds. For comparison experiments, 6 μl of the seed solution were added. The colour of the solution gradually changed within 10–20 min.

Characterization

The atomic force microscopy (AFM) measurements were conducted under normal ambient conditions using a XE-100 instrument (Park Systems) in non-contact mode. The AFM probe was a non-contact silicon cantilever with high resonant frequency (325 kHz) and backside aluminium reflex coating (ACTA, supplied by Park Systems). For AFM imaging, a drop of a diluted sample was deposited onto a freshly cleaved mica sheet (SPI Supplies, Grade V-1 Muscovite), which was thoroughly washed with Milli-Q water and dried under nitrogen gas flow.

UV-Vis spectra were measured with a Thermo Evolution 300 UV-Visible spectrophotometer using 1 cm path length Hellma quartz cuvettes.

Transmission Electron Microscopy (TEM) images were obtained with a JEOL JEM-1011 microscope working at 100 kV. A drop from a diluted sample solution was deposited onto an amorphous carbon film on 400 mesh copper grids and left to evaporate at room temperature.

ESI mass spectra experiments were carried out on a Synapt G2 traveling wave ion-mobility spectrometer (TWIMS) time-of-flight (TOF) mass spectrometer (Waters, Manchester, UK), equipped with a standard electrospray ionization (ESI) source and modified to allow for ion/molecule reactions, as previously described.[39] The cluster solution was measured after precipitating the CTAB in excess by rapidly cooling down the reaction with ice followed by filtration. The sample was ionized by ESI using a sample pump rate of 14 μl min^{-1}, a source temperature of 100 °C, and a cone voltage of 40 V. As blank, 5 ml of 0.2 M CTAB solution was mixed with 5 ml of 5×10^{-4} M $HAuCl_4$.

Results and discussion

It is now recognized that stable clusters can be formed prior to the formation of the first stable nuclei.[37] Such clusters are kinetically trapped during the first steps of the reaction, and this can be conveniently used for the synthesis of clusters, either using strong binding ligands or using slow reaction rates. In the seed-mediated growth of Au NRs, the preparation of Au seeds is carried out aging the solution for a particular reaction time (see Scheme 1). Therefore, playing with times shorter than that particular reaction time, one could trap the formation of small Au clusters, which could be used for catalysing the anisotropic growth of Au, as shown in Scheme 1.

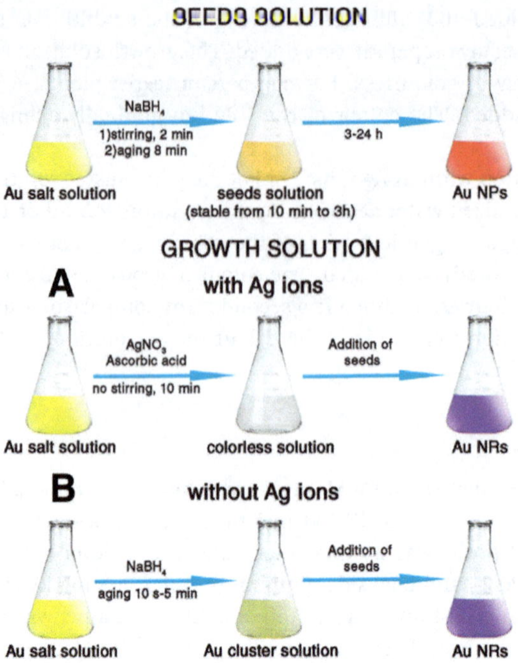

Scheme 1 Scheme of the common method for the production of Au NRs using silver ions (A) and the method here proposed using Au clusters (B).

The UV-Vis spectrum of a seed solution left for reaction during 5 min can be observed in Fig. 1A, showing that samples do not display any surface plasmon resonance (SPR) band. For comparison purposes, in the same Figure one can see the UV-Vis spectrum of a similar seed solution after a reaction time of 10 min (needed for the formation of seeds), showing the incipient appearance of the SPR band; and also after 24 h clearly showing the presence of the SPR band of Au nanoparticles. Fig. 1C shows the AFM picture of the seed solution after 5 min of reaction. The estimated size of the formed species is 0.4 ± 0.2 nm. This size is smaller than the seeds size (1.5 ± 0.2 nm, see Fig. S1†). Therefore, the very small size of the formed species, and the absence of plasmon band points to the formation of small Au clusters with less than 12 atoms.[40]

The presence of such clusters can also be detected because of their luminescence properties,[41] as it is observed in Fig. 1B. Assuming the simple spherical jellium model, the number of atoms in the cluster can be calculated by the simple expression $N = (E_F/E_g)^3$, where E_F and E_g represent the Fermi level (5.5 eV for bulk Au) and the HOMO–LUMO energy bandgap, respectively. The value of the bandgap can be estimated using the emission peak energy (412 nm, *i.e.* 3.0 eV), from which $N = 6.2$ can be obtained. Therefore, we can assume that Au_6 clusters can be kinetically trapped during the first 5 min of the seed solution reaction.

We tried further the cluster characterization by ESI-TOF mass spectrometry, which is a very good technique normally used for the identification of clusters. However, the presence of the surfactant, CTAB (even after purification), makes the unambiguous detection of the clusters extremely difficult, as it can be seen in

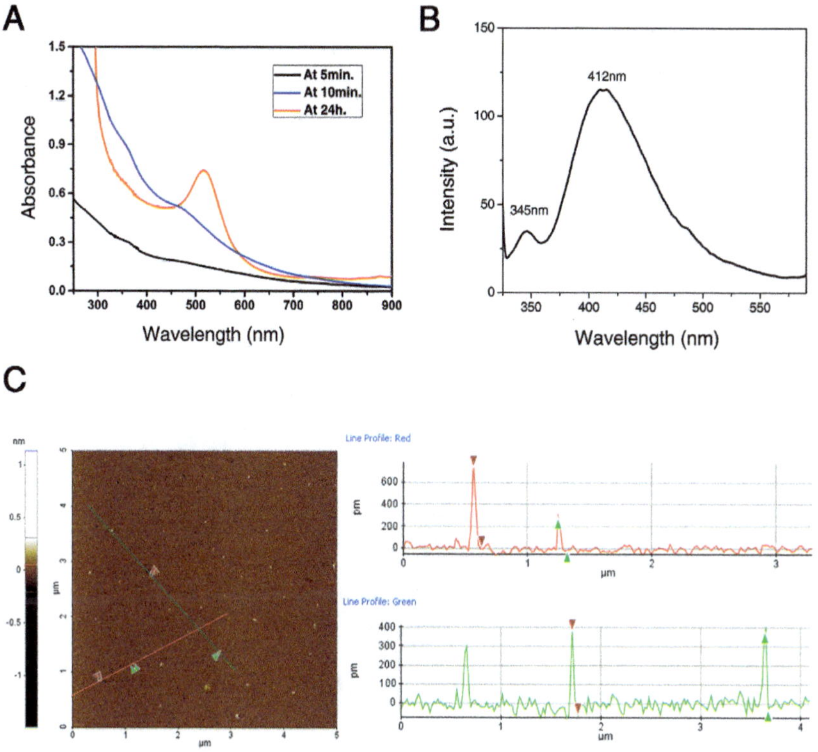

Fig. 1 UV-Vis spectra of seed solutions at 5 min, 10 min and 24 h (A), luminescence spectra (B) and AFM picture (C) of seed solutions after 5 min of reaction time showing the presence of Au clusters.

Fig. S2,† where the spectrum shows only the presence of CTAB. However, by careful comparison with the blank spectrum (in three consecutive experiments), one can just see a peak that can be associated with $[Au_6(CTA)_2B_6H_{9-11}]$ (as it can be seen in Fig. S3†). We recognize this alone cannot be taken as an strong proof of the presence of such clusters, since its signal is very small. However, the results nicely agree and support the previously discussed spectrometry results.

Fig. 2 shows the TEM micrographs of different samples obtained by adding different Au cluster solutions to an Au seeds solution. These different Au cluster solutions were obtained at different reduction times before the beginning of the seed formation (<10 min, see ESI†). It can be observed the preferential formation of Au NRs with high aspect ratios using cluster solutions aged for ≈1 min (Fig. 2C). The Au NRs yield is ≈95%, with ≈85% of large aspect ratio and 6% of lower aspect ratio (see Table 1). The observed selectivity and aspect ratio are higher than those previously reported using Ag ions (see *e.g.* ref. 28 for comparison purposes).

Therefore, the anisotropic particle growth can be explained as being induced by the catalytic properties of the kinetically trapped Au_6 clusters formed at short reaction times of the seeds solution. According to the mechanism proposed before by Attia *et al.*[28] the optimum Au NRs growth will occur when the [seeds] ≈

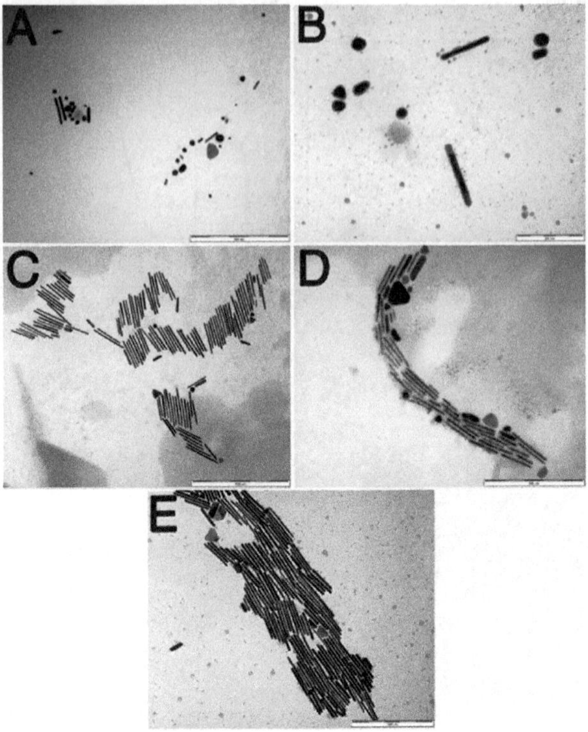

Fig. 2 TEM pictures of Au NRs prepared with Au clusters obtained after different reaction times of the seed solution: (A) 10 s; (B) 30 s; (C) 1 min; (D) 5 min; (E) 5 min using a double concentration of seeds.

[clusters]. From these results, it seems that this optimal condition occurs after a reduction time of $\approx$1 min.

Luminescent properties (as well as AFM) reveal that the aging time of the seeds solution (in the time frame 1–5 min) does not appreciably change the cluster size. Therefore, during this time, there is mainly an increase in the cluster concentration without an appreciable change in the cluster size. This fact explains that, when the concentration of seeds is increased, the reaction time needed to attain the optimum conditions also increases, as it is experimentally observed in Fig. 2E,

Table 1 Yield of different nanostructures produced by the Au-cluster method using a seed solution at different reduction times. 2A–E refer to the corresponding pictures in Fig. 2

	Aging time				
Nanostructures	10 s (2A)	30 s (2B)	1 min (2C)	5 min (2D)	5 min (2E)
Long rods	8%	15%	89%	34%	90%
Short rods	20%	7%	6%	35%	6%
Other structures	72%	78%	4%	30%	3%

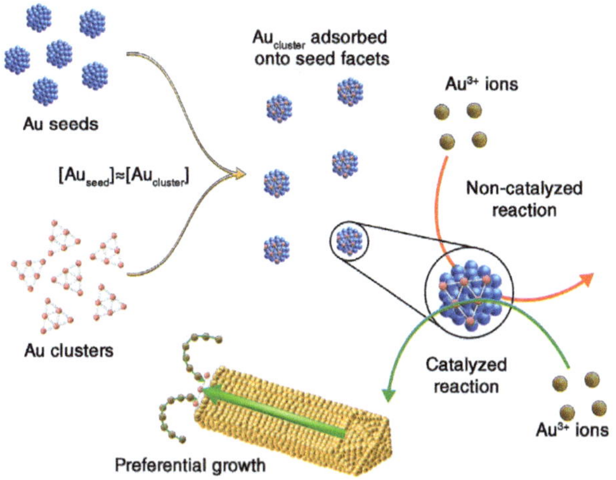

Scheme 2 Proposed mechanism for Au NR formation catalysed by Au clusters.

showing that 5 min are now needed to attain approximately a double concentration of catalytic clusters than in 1 min.

The mechanism for the Au nanorod formation catalysed with Au clusters can be explained in a similar way to that previously reported for Ag clusters. When the cluster/seeds ratio is ≈ 1, Au clusters (adsorbed onto a particular facet of the Au seeds) break the growth symmetry, causing a particular facet to grow much faster than the others, by catalysing the reduction in the facet they are adsorbed. This leads to the preferential formation of Au nanorods as can be seen in Scheme 2.

Conclusions

A new method to direct the formation of Au NRs is reported, consisting of the mixing two Au seed solutions at different reduction times: one large (10 min < time < 3 hours) for the formation of Au seeds; and the other short (<5 min) for the formation of catalytic Au_6 clusters. Au clusters could then be adsorbed onto particular crystal planes of the Au seeds, catalysing the anisotropic formation of Au NRs, as it was previously postulated.[4] With this simple method, without using Ag ions, Au NRs with high yield and large aspect ratio can be obtained. Therefore, we confirm here that the presence of catalytic clusters in the reaction of the nanoparticle formation (a largely hidden parameter), has to be also taken into account to explain the complex mechanisms involved in these very complicated reactions.

Acknowledgements

This work was supported by Obra Social "la Caixa" (OSLC-2012-007), European Commission through FEDER, FP7 and H2020 programs (0681_InveNNta_1_E; FutureNanoNeeds, FP7-Grant 604602, H2020-INSPIRED N. 646155); MINECO, Spain (MAT2012-36754-C02-01, MAT2015-67458-P); Xunta de Galicia, Spain (GRC2013-044, FEDER Funds). D. B. thanks the postdoctoral grant from Xunta de Galicia, Spain (POS-A/2013/018).

Notes and references

1 K. M. Mayer and J. H. Hafner, *Chem. Rev.*, 2011, **111**, 3828–3857.

2 E. C. Dreaden, M. a. Mackey, X. Huang, B. Kang and M. a. El-Sayed, *Chem. Soc. Rev.*, 2011, **40**, 3391–3404.

3 P. Ghosh, G. Han, M. De, C. K. Kim and V. M. Rotello, *Adv. Drug Delivery Rev.*, 2008, **60**, 1307–1315.

4 G. V. Hartland, *Chem. Rev.*, 2011, **111**, 3858–3887.

5 K. E. Sapsford, K. M. Tyner, B. J. Dair, J. R. Deschamps and I. L. Medintz, *Anal. Chem.*, 2011, **83**, 4453–4488.

6 M. Grzelczak, J. Vermant, E. M. Furst and L. M. Liz-Marzán, *ACS Nano*, 2010, **4**, 3591–3605.

7 A. M. Alkilany and C. J. Murphy, *J. Nanopart. Res.*, 2010, **12**, 2313–2333.

8 C. J. Murphy, T. K. Sau, A. M. Gole, C. J. Orendorff, J. Gao, L. Gou, S. E. Hunyadi and T. Li, *J. Phys. Chem. B*, 2005, **109**, 13857–13870.

9 W. Shi, H. Zeng, Y. Sahoo, T. Y. Ohulchanskyy, Y. Ding, Z. L. Wang, M. Swihart and P. N. Prasad, *Nano Lett.*, 2006, **6**, 875–881.

10 C. Sonnichsen and a P. Alivisatos, *Biophys. J.*, 2005, **88**, 364A–365A.

11 W. Huang, W. Qian and M. El-Sayed, *J. Phys. Chem. B*, 2005, **109**, 18881–18888.

12 A. G. Tkachenko, H. Xie, D. Coleman, W. Glomm, J. Ryan, M. F. Anderson, S. Franzen and D. L. Feldheim, *J. Am. Chem. Soc.*, 2003, **125**, 4700–4701.

13 J. Zhang, M. R. Langille, M. L. Personick, K. Zhang, S. Li and C. A. Mirkin, *J. Am. Chem. Soc.*, 2010, **132**, 14012–14014.

14 H. M. Chen and R. S. Liu, *J. Phys. Chem. C*, 2011, **115**, 3513–3527.

15 J. E. Millstone, S. Park, K. L. Shuford, L. Qin, G. C. Schatz and C. A. Mirkin, *J. Am. Chem. Soc.*, 2005, **127**, 5312–5313.

16 T. Ming, W. Feng, Q. Tang, F. Wang, L. Sun, J. Wang and C. Yan, *J. Am. Chem. Soc.*, 2009, **131**, 16350–16351.

17 M. L. Personick, M. R. Langille, J. Zhang, N. Harris, G. C. Schatz and C. a. Mirkin, *J. Am. Chem. Soc.*, 2011, **133**, 6170–6173.

18 Y. Ma, Q. Kuang, Z. Jiang, Z. Xie, R. Huang and L. Zheng, *Angew. Chem., Int. Ed.*, 2008, **47**, 8901–8904.

19 Q. Xu, R. M. Rioux and G. M. Whitesides, *ACS Nano*, 2007, **1**, 215–227.

20 H. P. Liang, L. J. Wan, C. L. Bai and L. Jiang, *J. Phys. Chem. B*, 2005, **109**, 7795–7800.

21 B. Nikoobakht and M. A. El-Sayed, *Chem. Mater.*, 2003, **15**, 1957–1962.

22 B. Nikoobakht and M. A. El-Sayed, *Langmuir*, 2001, **17**, 6368–6374.

23 N. R. Jana, L. Gearheart and C. J. Murphy, *Chem. Mater.*, 2001, **13**, 2313–2322.

24 S. E. Lohse and C. J. Murphy, *Chem. Mater.*, 2013, **25**, 1250–1261.

25 L. Scarabelli, M. Grzelczak and L. M. Liz-Marzán, *Chem. Mater.*, 2013, **25**, 4232–4238.

26 M. Grzelczak, J. Pérez-Juste, P. Mulvaney and L. M. Liz-Marzán, *Chem. Soc. Rev.*, 2008, **37**, 1783–1791.

27 N. R. Jana, L. Gearheart and C. J. Murphy, *Adv. Mater.*, 2001, **13**, 1389–1393.

28 Y. A. Attia, D. Buceta, C. Blanco-Varela, M. B. Mohamed, G. Barone and M. A. López-Quintela, *J. Am. Chem. Soc.*, 2014, **136**, 1182–1185.

29 A. Corma, P. Concepción, M. Boronat, M. J. Sabater, J. Navas, M. J. Yacaman, E. Larios, A. Posadas, M. A. López-Quintela, D. Buceta, E. Mendoza, G. Guilera and A. Mayoral, *Nat. Chem.*, 2013, **5**, 775–781.

30 S. Lee, L. M. Molina, M. Lopez, J. A. Alonso, B. Hammer, B. Lee, S. Seifert, R. E. Winans, J. W. Elam, M. J. Pellin and S. Vajda, *Angew. Chem., Int. Ed.*, 2009, **48**, 1467–1471.

31 Y. Lei, F. Mehmood, S. Lee, J. Greeley, B. Lee, S. Seifert, R. E. Winans, J. W. Elam, R. J. Meyer, P. C. Redfern, D. Teschner, R. Schlögl, M. J. Pellin, L. A. Curtiss and S. Vajda, *Science*, 2010, **328**, 224–228.

32 Y. Zhu, H. Qian and R. Jin, *J. Mater. Chem.*, 2011, **21**, 6793–6799.

33 Y. Zhu, H. Qian, M. Zhu and R. Jin, *Adv. Mater.*, 2010, **22**, 1915–1920.

34 Y. Zhu, H. Qian, B. A. Drake and R. Jin, *Angew. Chem., Int. Ed.*, 2010, **49**, 1295–1298.

35 A. Leelavathi, T. U. Bhaskara Rao and T. Pradeep, *Nanoscale Res. Lett.*, 2011, **6**, 132.

36 N. Vilar-Vidal, J. Rivas and M. A. López-Quintela, *ACS Catal.*, 2012, **2**, 1693–1697.

37 Y. Piñeiro, D. Buceta, J. Calvo, S. Huseyinova, M. Cuerva, Á. Pérez, B. Domínguez and M. A. López-quintela, *Adv. Colloid Interface Sci.*, 2015, **215**, 1–7.

38 Y. Piñeiro, J. Rivas and M. A. Lopez-Quintela, in, *Colloidal Foundations of Nanoscience*, ed. D. Berti and G. Palazzo, Elsevier, Amsterdam, 2014, pp. 81–105.

39 N. J. Rijs, T. Weiske, M. Schlangen and H. Schwarz, *Chem. Phys. Lett.*, 2014, **608**, 408–424.

40 D. Buceta, M. C. Blanco, M. A. López-Quintela and M. B. Vukmirovic, *J. Electrochem. Soc.*, 2014, **161**, D3113–D3115.

41 B. Santiago-González and M. A. López-Quintela, in, *Functional Nanometer-Sized Clusters of Transition Metals: Synthesis, Properties and Applications*, ed. W. Chen and S. Chen, The Royal Society of Chemistry, Cambridge, 2014, pp. 25–50.

Faraday Discussions

PAPER

Gold nanotriangles decorated with superparamagnetic iron oxide nanoparticles: a compositional and microstructural study

J. A. Hachtel,[ab] S. Yu,[c] A. R. Lupini,[ab] S. T. Pantelides,[abd] M. Gich,[c] A. Laromaine[c] and A. Roig*[c]

Received 29th February 2016, Accepted 11th March 2016

DOI: 10.1039/c6fd00028b

The combination of iron oxide and gold in a single nanoparticle results in both magnetic and plasmonic properties that can stimulate novel applications in bio-sensing, medical imaging, or therapeutics. Microwave assisted heating allows the fabrication of multi-component, multi-functional nanostructures by promoting selective heating at desired sites. Recently, we reported a microwave-assisted polyol route yielding gold nanotriangles decorated with iron oxide nanoparticles. Here, we present an in-depth microstructural and compositional characterization of the system using scanning transmission electron microscopy (STEM) and electron energy loss spectroscopy (EELS). A method to remove the iron oxide nanoparticles from the gold nanocrystals and some insights on crystal nucleation and growth mechanisms are also provided.

Introduction

Colloidal synthesis of multi-material nanocrystals with selectable morphologies, beyond core–shell and dimer geometries, is progressing quickly in nanocrystal research.[1–3] Novel uses of multi-material anisotropic nanocrystals are expected in fields ranging from nanomedicine to photocatalysis.[4–6] Specifically, the combination of gold and iron oxide in a single nanoparticle (NP) exhibiting both magnetic and plasmonic properties has been proposed for applications related to bio-sensing, medical imaging or therapeutics.[7–10] Superparamagnetic iron oxide nanoparticles (SPIONs) are biocompatible and extensively validated for uses in nanomedicine, such as magnetic resonance imaging,[11] drug delivery,[12] cell labeling,[13] bio-separation[14] and hyperthermia.[15] On the other hand, the optical

[a]*Department of Physics and Astronomy, Vanderbilt University, Nashville, TN 37235, USA*

[b]*Materials Science and Technology Division, Oak Ridge National Laboratory, Oak Ridge, TN 37831, USA*

[c]*Institut de Ciència de Materials de Barcelona, ICMAB-CSIC, Campus UAB, Bellaterra, Catalunya, E-08193 Spain. E-mail: roig@icmab.es*

[d]*Department of Electrical Engineering and Computer Science, Vanderbilt University, Nashville, TN 37235, USA*

properties and the localized surface plasmon resonance (LSPR) of gold nanoparticles (Au NPs) have also motivated their biomedical uses in imaging,[16] sensing,[17] drug and gene delivery[18] as well as in photothermal therapies.[19]

The shape[20] and size[21] of the Au NPs are crucial in determining the LSPR wavelength. Au NPs with anisotropic shapes displaying an absorption band in the near-infrared region (NIR) are appealing for photothermal therapies because NIR wavelengths can penetrate the biological tissues more efficiently.[22] Moreover, gold nanotriangles (Au NT) are highly effective when used in surface-enhanced Raman spectroscopy for biosensing[23] and photothermal therapy.[24,25] Thus, fabrication methods for creating multi-component structures consisting of Au NPs with anisotropic planar shapes (*e.g.* triangles and hexagons) and a magnetic component are gathering increased attention.

Most synthesis methods for gold nanoparticles with anisotropic shapes are based on a multi-step seed-mediated approach using a cationic surfactant, most often cetyltrimethylammonium bromide (CTAB).[23,26,27] The polyol process with ethylene glycol (EG) is another synthesis route used to generate seeds with different defects, such as multiple twin defects or stacking faults, yielding noble metal nanoparticles in the shape of dodecahedra or icosahedra (in the case of twins) and thin plates (in the case of stacking faults). The polyol route was initially employed for the fabrication of silver nanoparticles of different shapes and later the process was expanded to other systems, such as gold, platinum, rhodium and palladium nanoparticles.[28–30]

Recently, we reported a microwave-assisted polyol route yielding monodisperse Au NTs, decorated with a thin layer of SPIONs. These multi-material nanotriangles are readily dispersible in water, can self-assemble forming a monolayer at the liquid–air interface, display an excellent magnetic response, and exhibit a LSPR band in the NIR region.[31] Here, we present an in-depth, nanoscale structural and compositional characterization of the system using scanning transmission electron microscopy (STEM) with electron energy loss spectroscopy (EELS) and provide insights on the early stages of crystal nucleation and growth. We present a technique to remove the SPIONs from the Au NPs post-synthesis, determine the phase of iron oxide present in the SPIONs, and find some suggestions that SPIONs might preferentially bond to different gold surfaces.

Experimental

Materials

Iron(III) acetylacetonate ($Fe(acac)_3 \geq 97.0\%$), polyvinylpyrrolidone (PVP, average molecular weight: 10 000 g mol^{-1}), and hydrogen tetrachloroaurate trihydrate ($HAuCl_4 \cdot 3H_2O \geq 99.9\%$) were purchased from Sigma-Aldrich. Anhydrous benzyl alcohol (99%) was from Scharlau and ethylene glycol (EG $\geq$ 99%), HCl (37%) and acetone were bought from Panreac. All materials were used as-received without further purification. Milli-Q water (MQ-H_2O) was used in all experiments.

Synthesis of PVP coated SPIONs

SPIONs with a PVP surface coating (PVP–SPIONs) were synthesized following our microwave (MW) synthesis protocol.[32,33] Briefly, 680 mg of PVP were dissolved in

4 mL of anhydrous benzyl alcohol using continuous sonication. Then, 12.36 mg of $Fe(acac)_3$ (0.035 mmol) were mixed with the above prepared solution to give a homogenous solution of a dark red color. The solution was heated for 5 min at 60 °C and 10 min at 180 °C (reactor power at 250 W). PVP–SPIONs (5 nm in diameter) were collected by adding 30 mL of acetone, centrifuging at 1300*g* for 30 min and redispersing in 4 mL of EG to be further used.

Synthesis of hybrid Au–SPION nanoparticles

25 mg of PVP were added to the as-obtained PVP–SPION dispersions in EG (4 mL) and sonicated to obtain a homogenous mixture where 40 μL of 100 mM $HAuCl_4$ were added. The molar ratio of free PVP to $HAuCl_4$ is 0.625 : 1 and that of SPIONs to $HAuCl_4$ is 1.3 : 1. The solution was heated at 120 °C for 10 min in the microwave reactor (reactor power at 250 W). The as-obtained Au–SPIONs were washed twice with acetone to remove the free PVP and non-attached SPIONs, collected by centrifugation at 1300*g* for 2 min and redispersed in 2 mL of MQ-H_2O.

Removal of SPIONs from Au NT surfaces

SPIONs were removed from the Au NP surface using diluted HCl (1 : 10 dilution, pH ≈ 3). Au–SPION composites were concentrated by centrifugation at 3000 rpm for 15 min, and after decanting the supernatant the concentrated composites were redispersed in 0.5 mL of Milli-Q water. 1 mL of diluted HCl was then added dropwise to the system and incubated overnight. The acidic solution was replaced by fresh Milli-Q water and the system was used for further characterization.

Material characterization

Hydrodynamic sizes were determined using a Zetasizer Nano ZS (Malvern) with a He/Ne 633 nm laser at an angle = 173° at 25 °C. Ultraviolet-visible-near infrared (UV-Vis-NIR) spectra were collected on a Cary-5000 UV-Vis-NIR spectrophotometer. Superconductive Quantum Interference Device (SQUID) measurements were recorded on a magnetometer (Quantum Design MPMS5XL). The magnetization curves of the PVP–SPIONs and the hybrid Au NT–SPIONs were recorded at 5 K as a function of the applied magnetic field. The morphologies of the Au–SPIONs were analysed in a JEOL JEM-1210 electron microscope operating at 120 kV. High-resolution transmission electron microscopy (HRTEM) was performed using a JEOL 2010F HRTEM STEM and EELS was performed using a Nion UltraSTEM operating at 200 kV with a Gatan Enfinium ER EEL spectrometer.

Results and discussion

In our previous report on the synthesis of planar gold nanotriangles and nanohexagons (Au NT and NH) decorated with SPIONs, TEM analysis showed that 75% of the observed gold morphologies consisted of NT (60%) and NH (15%), and the remaining nanoparticles were a mixture of platonic structures (23%) and smaller rounded particles (2%).[31] The report also demonstrated that this synthetic approach offers a straightforward and fast route to produce gold planar heterostructures with a 10 wt% magnetic fraction and with the LSPR band in the NIR region.[31] These heterostructures are formed by maghemite nanoparticles (γ-Fe_2O_3) that are superparamagnetic at room temperature and that can self-assemble as a monolayer in liquid–air interfaces. Fig. 1 contains a schematic representation of the synthesis and the physical and chemical characterization of the as-obtained nanostructures.

Bright field (BF) STEM images of a Au NT–SPION and a Au NH–SPION are shown in Fig. 2. It can be seen that the Au NTs are of well-defined equilateral shape with their surface decorated by a monolayer of SPIONs as shown in Fig. 2a and b. The equivalent images of a SPION decorated Au NH are shown in Fig. 2c and d. The resolution of the BF images in Fig. 2b and d are high enough to see the lattice planes of the SPIONs, and the insets contain the fast Fourier transform (FFT) of their respective images. In the FFTs, each spot indicates a different crystal plane, and a circular ring pattern is an indication of many non-aligned crystal grains/nanoparticles. The circular FFT pattern is observed in both the NT and NH, indicating that the SPIONs are randomly oriented with respect to the Au NT, and no epitaxial relationship between the two exists.

To investigate the composition of the nanotriangles, EEL spectrum imaging (SI) was performed and then analysed using multiple-linear least squares (MLLS) fitting. Fig. 3 displays the composition map of a Au NT using MLLS-EELS mapping. Fig. 3a shows a high-angle annular dark field (HAADF) image of the nanoparticle, and Fig. 3b shows the elemental composition, where red is iron, blue is gold, and yellow is carbon. Fig. 3c–e show each of the individual compositional maps respectively. Fig. 3f contains the reference spectra used for MLLS fitting. The fit region is from 45 eV to 65 eV due to similar positions of the bulk

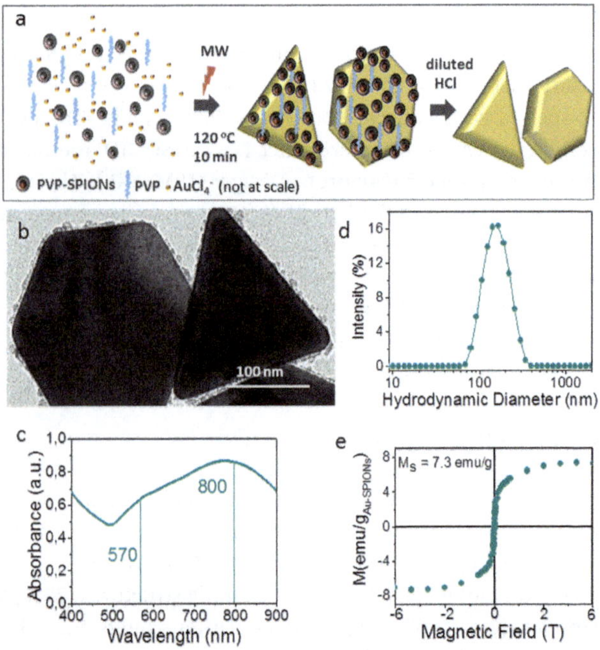

Fig. 1 (a) Schematic representation of the synthesis route. (b) HRTEM image of a Au NH–SPION and a Au NT–SPION. Characterization of the heterostructures: (c) UV-Vis-NIR spectrum. (d) Hydrodynamic diameter of the Au–SPIONs measured using DLS. (e) Magnetization curve up to 6 T at 5 K.

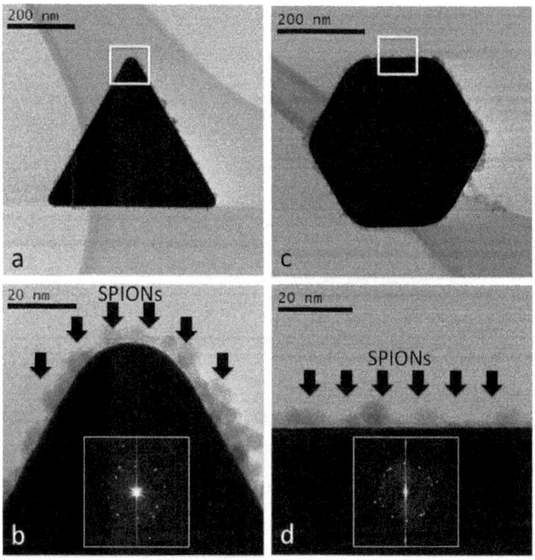

Fig. 2 (a) Bright-field (BF) scanning transmission electron microscopy (STEM) image of a single Au NT–SPION nanostructure. (b) BF-STEM image of the NT tip outlined in (a) showing the crystallinity of the SPIONs. The inset contains an fast Fourier transform (FFT) of the BF image and shows FFT spots arranged in a circular pattern indicating that the SPIONs are randomly oriented with respect to the NT. (c and d) BF-STEM images of a Au NH–SPION structure and the top edge containing the SPIONs. In the inset of (d) the same circular FFT pattern is visible, indicating that no epitaxial relationship between the Au and SPIONs occurs in either geometry.

plasmons for iron and carbon in the 15 to 40 eV band. Due to the presence of the prominent Fe-M edge at 54 eV and the significantly different low-loss character of gold, the 45–65 eV range presents three completely different spectra that are suitable for an MLLS fit.

It is important to note that both the gold and iron maps show high intensities along the edges and weak/no intensity in the bulk. This is due to the thickness of the gold in the bulk of the nanoparticle, which scatters a large portion of the beam away from the EELS detector. As a result, in thick samples of strongly scattering materials such as gold, the EELS intensity is reduced. Even planar structures, such as the NHs and NTs have a three-dimensional character, which can become quite thick. EELS log-ratio thickness measurements show that the NT are 138 nm thick. Such thicknesses are sufficient to result in reduced signal in the nanoparticle across the entire EEL spectrum.

Due to the fact that the faceted edges of the NT taper to a point at the edges, the elemental composition of the Au NT–SPIONs can be observed and analysed at thinner regions of thickness. Fig. 3g shows a magnified view of the edge of the structure outlined in Fig. 3a. Three regions are indicated: the SPIONs, the faceted edge, and the bulk of the NT. Fig. 3h shows the low loss EEL spectrum from each of those three regions. Points 1 and 3 show the characteristic low-loss spectra of iron-oxide and gold respectively; however, at point 2 we see a low-loss spectrum that appears to be a linear combination of the iron and gold spectra. This indicates that even where the iron EELS signal cannot be detected due to scattering from the thick gold, there are SPIONs present on all facets of the NT.

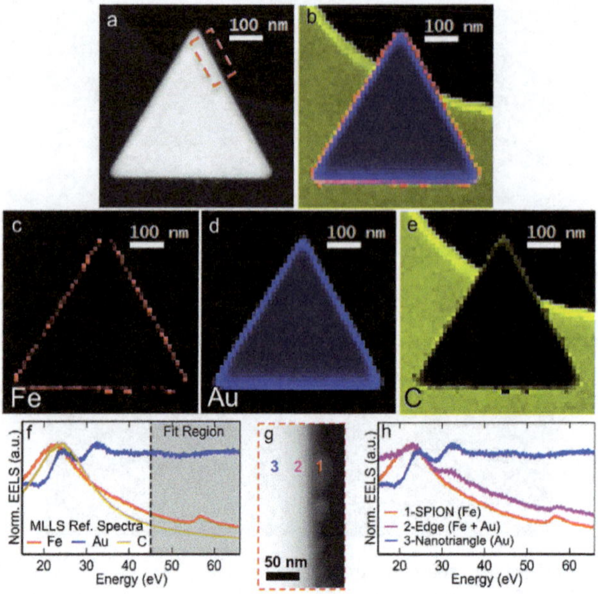

Fig. 3 (a) High angle annular dark field (HAADF) image of a Au NT–SPION composite. (b) The elemental map of the Au NT–SPION composed of the EEL spectrum image (SI) maps of Fe, Au, and C (shown in (c–e) respectively) determined through multiple linear least squares (MLLS) fitting. (f) The MLLS reference spectra used for the fitting. The fit region is 45 eV through to 65 eV to avoid the similar bulk plasmons of Fe and C. (g) Close up of the outlined region in (a), showing the three regions: the Fe in SPIONs, the faceted edge of the structure, and the thick bulk of the NT. (h) Spectra from each of the three regions highlighted in (g), showing that the signal from the faceted edge is a combination of Fe and Au and that the coverage of the nanoparticles is uniform across all the facets.

To assess crystallinity in thin structures, STEM can be used with a defocused probe to generate a 'Ronchigram'. In the Ronchigram, diffraction in the sample results in a Kikuchi pattern.[34] Fig. 4a shows a HAADF image of a Au NH–SPION nanostructure. The sample is tilted so that the (111) zone axis is parallel to the beam, and the Kikuchi pattern for the (111) zone axis is shown to be uninterrupted across the entire NH in Fig. 4b, indicating that the entire structure is a single crystal. For the thicker structures where the Kikuchi pattern is not easily observed, crystallinity can be determined through convergent beam electron diffraction (CBED). Fig. 4c shows a HAADF image of a thicker structure where no strong Kikuchi pattern is observed. Fig. 4d and e show CBED patterns from the gold bulk and the thin carbon supporting film. A 16×16 region including the NT is scanned, and the HAADF intensity along with CBED patterns are collected at each position. Fig. 4f–h show the resulting maps, where Fig. 4f shows the HAADF map, Fig. 4g shows the map from the amorphous carbon diffraction ring seen in Fig. 4e (green squares), and Fig. 4h shows the most intense diffraction spots from the gold (111) pattern (red squares). From the maps we can see that the same (111) CBED pattern is present at all points in the structure and no shift in the diffraction pattern is observed, indicating that the entire NT is a single crystal.

SPIONs were easily dissolved from the Au NP surface using a diluted HCl aqueous solution (1 : 10 dilution, pH $\approx$ 3). Fig. 5 shows Au NT–SPIONs before and

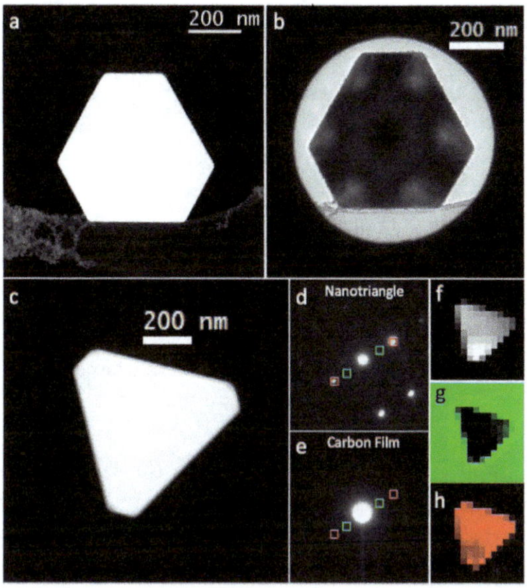

Fig. 4 (a) HAADF image of a Au NH–SPION structure. (b) STEM Ronchigram of the NH aligned with the (111) crystal axis parallel to the beam. The (111) Kikuchi diffraction pattern can be observed to be uninterrupted across the entire structure, indicating that the NH is a single crystal. (c) HAADF image of a Au NT, too thick to observe the Kikuchi pattern. (d and e) Convergent beam electron diffraction (CBED) pattern from the Au NT (d) and the support C film (e) for the Au NT of (c). (f–h) 16 × 16 maps of the nanotriangle are made using the HAADF intensity (f), the intensity at the amorphous C diffraction ring (g) (green squares in (d) and (e)), and (h) the Au (111) diffraction spots (red squares). The entire NT shows the same (111) diffraction pattern, indicating that is a single crystal.

after the treatment. Fig. 5a shows a NT with the SPIONs still adhered to the surface, while Fig. 5b shows a NT without any SPIONs. The SPIONs are visible in the HAADF image presented in Fig. 5a, but none are observed in Fig. 5b. To further demonstrate the absence of iron from the treated NTs, the EELS core-loss data from the Fe-L_3 edge was collected from a large region at the bottom edge of both NTs. In the NT–SPIONs the Fe-L_3 edge is easily observed, but is completely absent from the NT post-treatment with HCl, indicating that the SPIONs have been removed. The ability to dissolve the SPIONs post synthesis opens the door to investigate to which extent the plasmonic modes of the single platelet are affected by the dielectric media modification induced by the SPIONs.

Next, insights into the gold nanocrystal nucleation and growth mechanisms are provided. In the polyol synthesis, the heating of $HAuCl_4$ in EG in the presence of PVP leads to the reduction of Au(III) to Au(0), where EG serves as the solvent and reducing agent and PVP as the stabilizer. In our route, PVP–SPIONs are also placed into the solution. They are negatively charged ($Z_{potential} = -30$ mV) exposing the OH^- group of the PVP. Under the effect of microwave radiation, the polar OH^- rich zones of the SPION surface likely serve as hot spots and thus preferential sites where the gold nucleates, similarly as we reported for the *in situ* synthesis of other inorganic nanoparticles on hydroxyl rich substrates.[35,36] As a result, gold seeds nucleate in the presence of SPIONs and are attached to them

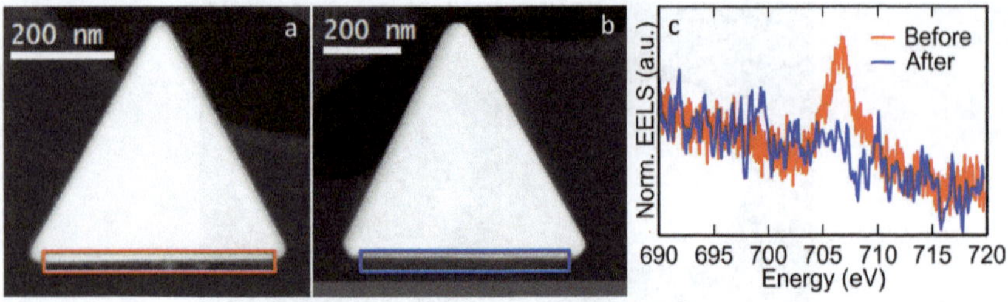

Fig. 5 HAADF images of different Au NT–SPIONs from before (a) and after (b) the SPION removal treatment. EEL spectra are collected and summed over the outlined region in both images and plotted in (c). In the 'before' EELs the Fe L_3-peak is clearly visible, but it is absent in the 'after', indicating that Fe is not present in this sample.

via PVP. Those initial gold seeds will grow from the later reduced gold atoms, resulting in gold particles decorated by a thin layer of SPIONs. However, an explanation why triangular nanocrystals are the most common geometry is still missing. The shape outcome is determined by thermodynamic parameters (*e.g.*, reduction potential and surface capping) and kinetic parameters (*e.g.*, concentration, mass transport, temperature, and the involvement of foreign species) which are intimately and intricately entangled with each other.[37,38] As pointed out recently in a review by Xia *et al.*,[39] it is extremely difficult to track alterations to the nuclei at initial crystallization stages due to the lack of experimental tools capable of identifying and monitoring these structures during growth. Nevertheless, the prevalence of planar structures in our case is probably a consequence of a slow initial reduction rate,[40] maybe due to the rather low set-up temperature of 120 °C. However, this argument is not entirely convincing since higher local temperatures may be present in the MW route. Another possibility is that, as an oxidative etchant, the Fe(II)/Fe(III) pair is kinetically controlling the morphology of the metal nanocrystals by retarding the rate of atom supply from the gold precursor.[41]

In order to gain further insights into the SPIONs' role in the growth kinetics, we started by assessing the SPIONs' coverage on the different facets of the gold structures semi-quantitatively. It is expected that during crystal growth, free PVP and PVP–SPIONs will selectively adsorb on the (111) facets rather than the (100) facets.[39] In the NH, the crystal facets enclosing the hexagon are the (111) and (100) families in an alternating sequence. Fig. 6 depicts how the preferential bonding of PVP–SPIONs to the (111) facets is empirically determined. The system is imaged in two tilt orientations, termed max and min tilt. In each orientation the sample is tilted to one extreme, which causes one specific facet of the NH to be parallel to the beam, allowing that facet to be imaged effectively. Since the facets are alternating, if a (100) facet is aligned with the beam at max tilt (as seen in Fig. 6a), then a (111) facet is aligned with the beam at min tilt (Fig. 6b). The tilt is performed such that two congruent facets of the nanoparticle are able to be seen at both max and min tilt, and due to the alternating pattern of the (100) and (111) facets, this results in the imaging of two different (100) facets (max-right, min-left) and two different (111) facets (max-left, min-right). HAADF images for a NH at max and min tilt are shown in Fig. 6c and d respectively, and indeed it can be seen that at max tilt, the left side has a high coverage while the right side is sparse, and at min tilt the inverse is true. To help visualize the coverage, line profiles from the max

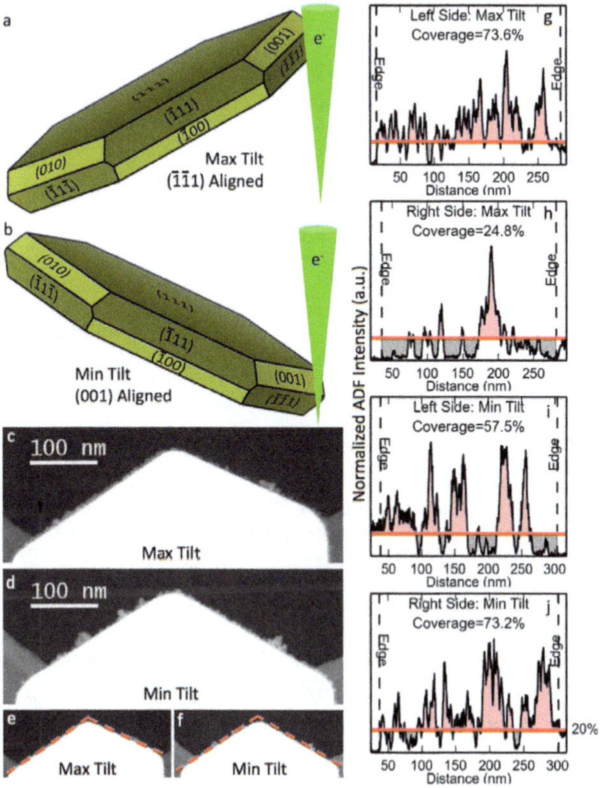

Fig. 6 Preferential bonding of the SPIONs to a specific facet is determined through tilting the sample. (a) and (b) show two orientations of the NH with respect to the beam, termed max and min tilt respectively, which align alternating facets with the electron beam. HAADF images of the max (c) and min (d) tilts show that on alternating facets there are varying concentrations of the SPIONs. Line profiles along the edges of the two tilts, ((e)-max and (f)-min) are plotted and the percent coverage is calculated by seeing where along the line profile the HAADF intensity is greater than a 20% threshold (red-line). (g–j) show the line profiles on the two different edges of the NH at the two tilts, and it can be seen that at max-left (g) and min-right (j) the coverage is significantly higher than max-right (h) and min-left (i). Due to the alternating nature of the facets, this demonstrates that SPIONs preferentially bond to one facet of the NH.

and min tilt (shown in Fig. 6e and f respectively) are performed, and the coverage is analysed.

The coverage is calculated as the percentage of the line profile with a HAADF intensity above a threshold; 20% of the maximum HAADF intensity is chosen qualitatively as giving the best representation of the nanoparticle coverage. In Fig. 6g–j the four line profiles are plotted, showing high coverage at max-left and min-right, and low coverage at max-right and min-left. It is worth noting that in the min-left profile (Fig. 6i) the coverage is artificially increased due to a large portion of the supporting carbon film that overlaps with the NH edge. Determining which crystal axis corresponds to the high-concentration facets is not possible from the STEM data, as the tilt limits of the STEM holders are $\sim 20^\circ$, which is only enough to preferentially align a facet, but not to tilt into a crystal

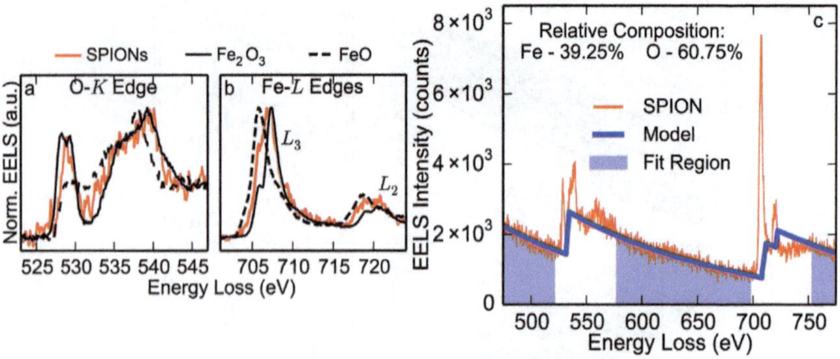

Fig. 7 (a) O-K edge EELS. The two reference samples (Fe_2O_3-solid, FeO-dashed) show two distinct behaviours in the relative intensity of the pre-peak at 529 eV; the SPIONs (red) match the Fe_2O_3. (b) Fe-L edge EELS. A 2 eV shift (708 eV-Fe_2O_3, 706 eV-FeO) is observed in the peak intensity of the L edge, and the SPIONs share the position of the Fe_2O_3 peak at 708 eV. (c) Relative composition of O and Fe. By fitting X-ray photo-absorption cross sections to the tails of the O and Fe edges, the intensities of the two edges can be directly compared and the relative composition calculated. The result is a 1.548 ratio between O and Fe, further confirming the SPIONs are in the Fe_2O_3 phase.

zone access normal to the facet. As a result, the data is not sufficient to absolutely confirm that SPIONs selectively bond to the ⟨111⟩ plane, but it does provide direct experimental evidence of preferential bonding on different gold facets in the Au NH–SPIONs.[39]

Finally, in Fig. 7, EELS fine structure along with quantitative analysis of the relative composition of iron and oxygen are used to determine the phase of the iron oxide present in the SPIONs. Fig. 7a and b show the O-K and Fe-L edges taken from three different samples: a reference sample of Fe_2O_3 (with 3+ valence), a reference sample of FeO (with 2+ valence), and finally the SPIONs. In Fig. 7a the O-K edges from the three samples are shown. The pre-peak at 529 eV shows that there is a clear distinction between Fe_2O_3 and FeO, and more importantly, that the pre-peak in the SPIONs is almost an exact match of the Fe_2O_3. Additionally, in the Fe-L edge EELS shown in Fig. 7b, it can be seen that the peak energy of Fe_2O_3 and FeO differ significantly (2+ at 706 eV, 3+ at 708 eV), and that the SPION peak clearly aligns with the Fe_2O_3 peak at 708 eV. Moreover, it is also observed that the FeO L_2 edge has only a single peak at 718 eV while both the SPIONs and the Fe_2O_3 have two peaks at 719 eV and 721 eV. There are some dissimilarities between the Fe_2O_3 reference sample and the SPIONs, namely a stronger pre-peak on the L_3 edge and a larger area beneath the L_2 edge. The likely reason for these differences is sample damage from the electron beam.

However, other iron-oxide phases besides Fe_2O_3 and FeO exist and in order to further investigate the SPION composition, we quantify the relative amounts of iron and oxygen in the SPIONs. The relative concentration is determined by fitting and then subtracting the power-law EELS background and then integrating the intensity of the tails of the oxygen and iron EELS edges.[42] The near-edge fine structure varies significantly across different compounds, so these spectral regions are excluded from the integration. X-ray photo-absorption (XRPA) cross sections are known to be excellent analogues of the corresponding EELS cross sections, and the intensity of the EELS signal is determined by fitting these XRPA cross sections to the power-law tails of the EELS edges.[43] A quantitative model of the EELS signal is formed and plotted against the experimental EELS signal in Fig. 7c, where it can be seen that the quantitative model matches well with the power-law tails of the edges and ignores the variations due to fine-structure. From the quantitative model a relative composition of 39.25% Fe and 60.75% O is determined, which corresponds to the composition of Fe_2O_3 and is in agreement with the fine-structure analysis in Fig. 7a and b. These results indicate that the Fe retains its Fe(III) character and does not reduce to Fe(II).

Conclusions

Here, we presented the nanoscale structural and compositional characterization of gold planar nanostructures (triangles and hexagons) decorated with superparamagnetic maghemite nanoparticles. The magnetic gold planar structures are synthesized through 10 minutes of microwave heating, are dispersible in water, display the LSPR band in the NIR region, and can have the magnetic nanoparticles easily removed from the gold surface using a diluted acidic solution. High spatial-resolution STEM analysis shows that the gold structures are all single crystals and that the SPIONs adhere across the entire gold nanostructures with no epitaxial relationship. However, the different gold facets are shown to interact differently with the PVP polymer, changing the coverage of the SPIONs on the different facets. As a result of the differences in facet coverage, the SPIONs could be further exploited as a probe to follow initial seed formations through the nanocrystal growth process, which are difficult to observe directly. Finally, the phase of the iron oxide in the SPIONs was determined through STEM-EELS to be Fe_2O_3, ruling out that the SPIONs undergo a reduction from Fe(III) to Fe(II) during the synthesis of the gold planar structures.

Acknowledgements

This research was partially funded by the Spanish Ministry of Economy and Competitiveness in co-funding with European Social Funds through the MAT2012-35324 project, the Severo Ochoa Programme for Centers of Excellence in R&D (SEV-2015-0496) and the Ramón y Cajal program (AL, RyC-2010-06082) and the Generalitat de Catalunya (2014SGR213). Authors also acknowledge support of the Chinese Scholarship Council (SMY, 201206150053) and the following projects: COST Action MP1202, and the U.S. Department of Energy grant DE-FG02-09ER46554 (JAH, STP). The authors thank Dr. Gabriel Veith of Oak Ridge National Laboratory for providing the iron-oxide reference samples. Work at Oak Ridge National Laboratory was supported by the Department of Energy, Office of Science, Basic Energy Sciences, Materials Sciences and Engineering Division.

References

1 L. Carbone and P. D. Cozzoli, *Nano Today*, 2010, **5**, 449.
2 C. de Mello Donegá, *Chem. Soc. Rev.*, 2011, **40**, 1512.
3 M. R. Buck and R. E. Schaak, *Angew. Chem., Int. Ed.*, 2013, **52**, 6154.

4 W. Shi, X. Liu, C. Wei, Z. J. Xu, S. Siong, W. Sim, L. Liu and C. Xu, *Nanoscale*, 2015, 7, 17249.
5 B. Wu, D. Liu, S. Mubeen, T. T. Chuong, M. Moskovits and G. D. Stucky, *J. Am. Chem. Soc.*, 2016, **138**, 1114.
6 S. Bai, L. Yang, C. Wang, Y. Lin, J. Lu, J. Jiang and Y. Xiong, *Angew. Chem., Int. Ed.*, 2015, **54**, 14810.
7 J. Lee, J. Yang, H. Ko, S. Oh, J. Kang, J. Son, K. Lee, S. W. Lee, H. G. Yoon, J. S. Suh, Y. M. Huh and S. Haam, *Adv. Funct. Mater.*, 2008, **18**, 258.
8 C. Xu, B. Wang and S. Sun, *J. Am. Chem. Soc.*, 2009, **131**, 4216.
9 C. Wang and J. Irudayaraj, *Small*, 2010, **6**, 283.
10 F. Mohammad, G. Balaji, A. Weber, R. M. Uppu and C. S. S. R. Kumar, *J. Phys. Chem. C*, 2010, **114**, 19194.
11 E. Taboada, R. Solanas, E. Rodríguez, R. Weissleder and A. Roig, *Adv. Funct. Mater.*, 2009, **19**, 2319.
12 S. Guo, D. Li, L. Zhang, J. Li and E. Wang, *Biomaterials*, 2009, **30**, 1881.
13 E. Carenza, V. Barceló, A. Morancho, J. Montaner, A. Rosell and A. Roig, *Acta Biomater.*, 2014, **10**, 3775.
14 C.-J. Yu, C.-Y. Lin, C.-H. Liu, T.-L. Cheng and W.-L. Tseng, *Biosens. Bioelectron.*, 2010, **26**, 913.
15 R. Di Corato, A. Espinosa, L. Lartigue, M. Tharaud, S. Chat, T. Pellegrino, C. Ménager, F. Gazeau and C. Wilhelm, *Biomaterials*, 2014, **35**, 6400.
16 G. Luke, D. Yeager and S. Emelianov, *Ann. Biomed. Eng.*, 2012, **40**, 422.
17 L. Xu, H. Kuang, C. Xu, W. Ma, L. Wang and N. A. Kotov, *J. Am. Chem. Soc.*, 2011, **134**, 1699.
18 B. Duncan, C. Kim and V. M. Rotello, *J. Controlled Release*, 2010, **148**, 122.
19 L. R. Hirsch, R. J. Stafford, J. A. Bankson, S. R. Sershen, B. Rivera, R. E. Price, J. D. Hazle, N. J. Halas and J. L. West, *Proc. Natl. Acad. Sci. U. S. A.*, 2003, **100**, 13549.
20 N. G. Bastús, J. Comenge and V. F. Puntes, *Langmuir*, 2011, **27**, 11098.
21 J. S. DuChene, W. Niu, J. M. Abendroth, Q. Sun, W. Zhao, F. Huo and W. D. Wei, *Chem. Mater.*, 2012, **25**, 1392.
22 M. Smith, M. C. Mancini and S. Nie, *Nat. Nanotechnol.*, 2009, **4**, 710.
23 L. Scarabelli, M. Coronado-Puchau, J. J. Giner-Casares, J. Langer and L. M. Liz-Marzán, *ACS Nano*, 2014, **8**, 5833.
24 B. Pelaz, V. Grazu, A. Ibarra, C. Magen, P. del Pino and J. M. de la Fuente, *Langmuir*, 2012, **28**, 8965.
25 M. Perez-Hernandez, P. del Pino, S. G. Mitchell, M. Moros, G. Stepien, B. Pelaz, W. J. Parak, E. M. Galvez, J. Pardo and J. M. de La Fuente, *ACS Nano*, 2015, **9**(1), 52.
26 T. K. Sau and C. J. Murphy, *J. Am. Chem. Soc.*, 2004, **126**, 8648.
27 J. E. Millstone, G. S. Métraux and C. A. Mirkin, *Adv. Funct. Mater.*, 2006, **16**, 1209.
28 B. Wiley, Y. Sun, B. Mayers and Y. Xia, *Chem.–Eur. J.*, 2005, **11**, 454.
29 I. Washio, Y. Xiong, Y. Yin and Y. Xia, *Adv. Mater.*, 2006, **18**, 1745.
30 Y. Sun and Y. Xia, *Science*, 2002, **298**, 2176.
31 S.-M. Yu, J. A. Hachtel, M. F. Chisholm, S. T. Pantelides, A. Laromaine and A. Roig, *Nanoscale*, 2015, **7**, 14039.
32 O. Pascu, E. Carenza, M. Gich, S. Estradé, F. Peiró, G. Herranz and A. Roig, *J. Mater. Chem. C*, 2012, **116**, 15108.

33 S.-M. Yu, A. Laromaine and A. Roig, *J. Nanopart. Res.*, 2014, **16**, 1.

34 D. B. Williams and C. B. Carter, *Transmission Electron Microscopy*, Springer, USA, 1996.

35 O. Pascu, J. M. Caicedo, M. Lopez-Garcia, V. Canalejas, A. Blanco, C. Lopez, J. Arbiol, J. Fontcuberta, A. Roig and G. Herranz, *Nanoscale*, 2011, **3**, 4811.

36 M. Zeng, A. Laromaine, W. Feng, P. A. Levkin and A. Roig, *J. Mater. Chem. C*, 2014, **2**, 6312.

37 N. Malikova, I. Pastoriza-Santos, M. Schierhorn, N. A. Kotov and L. M. Liz-Marzán, *Langmuir*, 2002, **18**, 3694.

38 Y. Xiong, I. Washio, J. Chen, M. Sadilek and Y. Xia, *Angew. Chem., Int. Ed.*, 2007, **46**, 4917.

39 Y. Xia, X. Xia, H.-C. Peng and J. Liu, *J. Am. Chem. Soc.*, 2015, **137**, 7947.

40 Y. Wang, H.-C. Peng, J. Liu, C. Z. Huang and Y. Xia, *Nano Lett.*, 2015, **15**, 1445–1450.

41 J. Chen, T. Herricks and Y. Xia, *Angew. Chem., Int. Ed.*, 2005, **44**, 2589.

42 R. F. Egerton, *Rep. Prog. Phys.*, 2009, **72**, 016502.

43 R. F. Egerton, *Ultramicroscopy*, 1993, **50**, 13–28.

Faraday Discussions

DISCUSSIONS

Anisotropic nanoparticles: general discussion

Andrea Castelli, Alberto Striolo, Anna Roig, Catherine Murphy, Javier Reguera, Luis Liz-Marzán, Axel Mueller, Kevin Critchley, Yu Zhou, Mathias Brust, Antoine Thill, Leonardo Scarabelli, Luciano Tadiello, Tobias A. F. König, Beate Reiser, M. Arturo López-Quintela, Martin Buzza, András Deák, Christian Kuttner, Estefania Gonzalez Solveyra, Lucia Pasquato, David Portehault, Hedi Mattoussi, Nicholas A. Kotov, Eugenia Kumacheva, Kelley Heatley, Julian Bergueiro, Guillermo González, Wenming Tong, Muhammad Nawaz Tahir, Benjamin Abécassis, Oscar Rojas-Carrillo, Younan Xia, Martin Mayer and Davide Peddis

DOI: 10.1039/c6fd90049f

Alberto Striolo opened a general discussion of the paper by Nicholas Kotov: In your paper you describe how by changing the properties and shapes of the nanoparticles you can have very pronounced chiral optical properties. As such, I was wondering if you could comment on how robust the layer-by-layer system you presented in your paper is with respect to poly-dispersity in both size and shape of the nanoparticles. Also, is it possible that the polymers affect the system properties as well?

Nicholas Kotov responded: Layer-by-layer assembly is attractive for many nanoparticles precisely because of the robustness of the process and robustness of the resulting materials. It is certainly possible that polymers affect the behavior of the coatings, which is an advantage of the layer-by-layer composites because it adds additional degrees of freedom to the system.

Tobias König then spoke: I would like to ask you about the reliability of the measured chirality effect. You used a commercially available J715 spectropolarimeter from JASCO. Commercial CD spectrometers have the reputation of not always making correct measurements.[1] Furthermore, your signal to noise ratio is rather low, which has been also observed at a low degree of rotation. You also indicate that cholesteric liquid crystals display a three to five order of magnitude higher optical rotation power. A clear trend of the CD effect has been proved by adding further bilayers to the structure. To be on the safe side, the CD rotation should be observed in two directions: first, CD transmission starting from the front side; and second, CD transmission starting from the back side. As result, the circular extinction spectra should be inverted. Did you measure the CD extinction in both ways?

1 Oriol Arteaga Barriel: http://www.mmpolarimetry.com/circular-dichroism/

Nicholas Kotov answered: We agree that in the case of solid films on a substrate, care needs to be exercised when attributing reading in CD instruments to circular dichroism. We have extensively discussed this problem of optically active films in previous publications.[1,2] One of them was written together with the author of the reference that was cited by you, Dr. Oriol Arteaga Barriel. Mueller matrix polarimetry was applied to determine the different components of the optical activity. Similarly to previous studies, for the films we described in this publication, we indeed applied the front–back test that you described when reporting the optical activity data. Please also note that we use the term Circular Dichroism only when we refer to optical activity in dispersions of CdS nanoparticles. When we describe the optical activity of LBL films made from them, we use the term circular extinction, which covers all potential sources of polarization rotation, similarly to that used in ref. 2. Although the contributions of the linear effects here might be small, one still needs to be careful with this attribution even when the back–front test gives the same spectrum.

1 B. Yeom, H. Zhang, H. Zhang, J. I. Park, K. Kim, A. O. Govorov and N. A. Kotov, Chiral plasmonic nanostructures on achiral nanopillars, *Nano Lett.*, 2013, **13**(11), 5277–5283.
2 Y. Kim, B. Yeom, O. Arteaga Barriel, S. Y. Yoo, S. G. Lee, J-G. Kim and N. A. Kotov, Reconfigurable chiroptical nanocomposites with chirality transfer from the macro- to the nanoscale, *Nat. Mater.*, 2016, **15**(4), 461–468.

Benjamin Abécassis asked: You have showed dipolar moments in gold nanorods, do you have any idea of the value of the dipole?

Nicholas Kotov replied: In this presentation we showed that the gold nanorods have surface potential anisotropy. Please note that this is somewhat different than the dipole in chemical molecules. The dipole moment assumes the existence of an electrical field inside the volume occupied by the molecule. For gold nanorods there is no electrical field inside (!) the gold core. This anisotropy is confined to the thin surface layer. Perhaps one can calculate some effective dipole moment, but its value could be misleadingly large.

Eugenia Kumacheva enquired: You use polystyrene, polyacrylates and polybutadiene. Can you say a few words about the general design rules and the dependence on certain parameters, to help other groups to make similar structures with different polymers? What are the challenges?

Nicholas Kotov answered: The design rules for these films are the same as for standard LBL films that were described in the paper by Gero Decher.[1] Their specificity in case of LBL films for nanoparticles is given in my paper.[2] When one would use intrinsically chiral NPs with chiral ligands and an inorganic core (as we use in this study), the higher volumetric fraction of the NPs in the composite film will give higher optical activity.

Since I know you have been working in field of NPs and polymer films, I am positive you might be aware of this point. Addressing an extension of your question, one might want to ask the question, what are the design rules for nanoparticle composite films with chiroptical activity? Now we arrive in virgin territory, which still needs to be explored. Based on the data that our group, your group, Decher's group and others obtained, I would put forward the rule that in order to obtain high circular polarization activity in NP films, one needs to organize them not only in the plane of the substrate but also along the out-of-plane axis.

1 G. Decher, Fuzzy nanoassemblies: Toward layered polymeric multicomposites, *Science*, 1997, **277**(5330), 1232–1237.
2 N. A. Kotov, Ordered layered assemblies of nanoparticles, *MRS Bull.*, 2001, **26**(12), 992–997.

Eugenia Kumacheva then asked: Do you have information about the distance between the nanoparticles in the films?

Nicholas Kotov replied: In my estimation, the distance between NPs in these films is several NP diameters.

Lucia Pasquato said: You study quite small clusters, is there an upper limit above which you don't observe any more optical activity? Is there any correlation between the optical activity and the number of the stereogenic centers of the organic coating? In other words, from small nanoparticles to large ones is necessary to increase the size of the optically active chiral ligand in order to have a measurable optical activity of the nanoparticle?

Nicholas Kotov responded: The optical activity of interest here originates from chiral excitons. They represent the chirality of the semiconductor core of the nanoparticles. Our current hypothesis is that it is the result of the staple-like attachment of the cysteine ligands to the CdS surface. Another possibility is the increase of chiral dislocations identified by the recent work of Yura Gun'ko for CdSe nanorods.[1]

If our hypothesis is correct, then the atomic chirality of the inorganic core should increase when the presence of chiral organic ligands on the surface is increased. I agree that this effect should be dependent on the size of the inorganic component of the nanoparticles. What we learned so far is that the presence of chiral organic ligands on the NP surface allows to make not only small chiral nanoparticles but also nanohelices with a submicron scale pitch length.[2] This occurs because, besides the effects of surface ligands on the core, there are also effects of surface ligands on the self-organization phenomena determined by their chirality.[3]

1 A. S. Baimuratov, I. D. Rukhlenko, Y. K. Gun'ko, A. V. Baranov and A. V. Fedorov, Dislocation-induced chirality of semiconductor nanocrystals, *Nano Lett.*, 2015, **15**, 1710.
2 J. Yeom, B. Yeom, H. Chan, K. W. Smith, S. Dominguez-Medina, J. Hwan Bahng, G. Zhao, W. S. Chang, S. J. Chang, A. Chuvilin, D. Melnikau, A. L. Rogach, P. Zhang, S. Link, P. Král and N. A. Kotov, Chiral templating of self-assembling nanostructures by circularly polarized light, *Nat. Mater.*, 2015, **14**, 66–72.
3 Y. Zhou, R. Marson, G. van Anders, J. Zhu, G. X. Ma, P. Ercius, K. Sun, B. J. Yeom, S. C. Glotzer and N. A. Kotov, Biomimetic hierarchical assembly of helical supraparticles from chiral nanoparticles, *ACS Nano*, 2016, **10**(3), 3248–3256.

Christian Kuttner spoke: This question is related to Prof. Kumacheva's earlier question, as she pointed out that you could use Müller matrix ellipsometry to measure the complex optical properties. Chirooptical activity (*e.g.*, circular dichroism) may also arise from a combination of linear effects, like linear dichroism and linear birefringence. Have you looked deeper into these linear effects?

Nicholas Kotov answered: Yes we have. They are described in the paper related to LBL films of gold nanoparticles published earlier this year.[1]

1 Y. Kim, B. Yeom, O. Arteaga Barriel, S. Y. Yoo, S. G. Lee, J-G. Kim and N. A. Kotov, Reconfigurable chiroptical nanocomposites with chirality transfer from the macro- to the nanoscale, *Nat. Mater.*, 2016, **15**(4), 461–468.

Christian Kuttner then asked: Concerning the intensity of the rotatory optical activity, it is mentioned in your paper that the dipping angle during LBL deposition may lead to additional contributions, for example by changes in linear dichroism and linear birefringence. How does the dipping angle affect the morphology within the layers and to what extent can this influence the chirooptical properties of the material?

Nicholas Kotov responded: We speculate that it might affect it, but we did not observe this effect for CdS nanoparticle films. One still needs to be careful, and anticipate that for some other nanoparticles it could. The mechanism of this effect is preferential alignment of nanoparticles along the dipping direction.

Kevin Critchley asked: It is proposed that some of the bands in your circular dichroism (CD) spectrum could be associated with excitation bands. The assignment for the first excitation fits well. However, I suggest performing the same CD measurement on several nanoparticle sizes; the excitation bands would shift with size due to quantum confinement, and thus would enable greater confidence in the band assignment. Would this be a useful experiment?

Nicholas Kotov replied: Absolutely, yes. This would be an excellent experimental series to carry out for dispersed nanoparticles. In this case the CD "wave" corresponding to chiral excitons should move as the size is changed. At the same time, the bands localized at the surface should remain the same or experience little change.

Tobias König spoke: In your publication you studied the CD absorption details. Including the emission, what is the application of your CD absorption molecules?

Nicholas Kotov answered: Applications of chiral nanoparticles and circularly polarized emission are, indeed, the questions that we are working on now. Some of the applications in bioanalytical fields for the detection of peptides, DNA and small molecules have been shown in the past, for instance in ref. 1 and 2. Applications in polarization optoelectronics have been addressed in the recent 2016 *Nature Materials* paper, I cited in a response above. Other groups have also demonstrated applications in chiral photonics and broadband polarization devices.

Regarding circularly polarized luminescence (CPL), chiral excitons, in principle, may lead to emission of circularly polarized photons. This is not always the case, however. So far, we observed only weakly polarized emission and the reasons for this are being established. I am not ready to speculate on this topic yet. There are multiple quantum mechanical reasons for and against a high degree of polarization in CPL for chiral semiconductor nanoparticles. One, possibly obvious, mechanism of how the rotational polarization can be lost between absorption and emission is thermalization to lower energy states.

1 Y. Zhao, L. Xu, W. Ma, L. Wang, H. Kuang, C, Xu and N. A. Kotov, Shell-engineered chiroplasmonic assemblies of nanoparticles for zeptomolar DNA detection, *Nano Lett.*, 2014, **14**(7) 3908–3913.
2 X. Wu, L. Xu, L. Liu, W. Ma, H. Yin, H. Kuang, L. Wang, C. Xu and N. A. Kotov, Unexpected chirality of nanoparticle dimers and ultrasensitive chiroplasmonic bioanalysis, *J. Am. Chem. Soc.*, 2014, **135**(49), 18629–18636.

Alberto Striolo opened a general discussion of the paper by Tobias König: You showed a very interesting and powerful technique to prepare assemblies of nanoparticles. I think one of the first steps is the preparation of a 'wrinkled' substrate. Can you tell us what is the minimal distance between the valleys of the wrinkles you can achieve?

Tobias König replied: The minimal corrugation length for our substrate/film bi-layer system is now 170 nm,[1] however with other substrate/film combinations periodicities down to 2 nm are achievable.[2] Most of those systems are, however, neither free of lateral cracks in the surface nor they are tunable in terms of their hydrophilicty – both very important factors to grant uniform, large-area particle assemblies.

1 M. Tebbe *et al.*, *Faraday Discuss.*, 2015, **181**, 243.
2 P. Rofouie *et al.*, *J. Chem. Phys.*, 2015, **143**(11), 114701.

Martin Buzza remarked: In terms of plasmonic properties, how important is it to control the lateral and longitudinal distance between rods?

Tobias König responded: From a plasmonic point of view, it is very important. In our work we focused on the transverse mode only, because this mode is pronounced in the visible wavelength range. The longitudinal mode is neglected because it is located in the infrared region. Including the longitudinal coupling it is red-shifted even further. Finally, I'd like to point out that we used the longitudinal axes for assembly purposes only.

Luciano Tadiello remarked: Which is the most important parameter that you control to achieve the best alignment and least amount of defects? Do defects stem from the wet printing technique or rather from particle size distribution?

Tobias König replied: Thank you for your question; please refer to the answer I gave to Alberto Striolo, above.

Javier Reguera addressed Tobias König and Martin Mayer: Your system is very sensitive to the separation between the nanorod and the gold surface, however you use BSA as stabilizer that could, in principle, create inhomogeneities (anisotropy) on the surface coating even if this is a 'purely monodisperse polymer'. I wonder if you could use linear molecules with an end-anchoring point, such as thiolated self-assembled monolayers, to achieve a more perfect and controllable organic shell around the nanoparticle?

Could you also comment on the big tip-to-tip distance in comparison with the nanoparticle–substrate distance?

Martin Mayer responded: We have previously shown that BSA allows a homogeneous and complete surface coverage while maintaining high colloidal stability and simultaneously completely substitutes the CTAB bilayer.[1] Especially, the latter is crucial for our assembly method, because a negative surface charge and a contact angle close to zero is required.[2,3] I totally agree that thiolated self-assembled monolayers would also allow for a homogeneous surface coverage. However, the need for a negative charge inhibits a direct exchange of CTAB with low-molecular ligands due to the required transition of the point of zero charge. At this point the low electrostatic stabilisation may result in agglomeration or particle loss. Proteins and also polyelectrolytes can directly overcompensate the initial charge, therewith restoring the negative net charge required for colloidal stability.

The tip-to-tip distance and particle-to-film distance emerge from different effects and are set at different stages of the sample fabrication. The particle-to-film distance is defined during transfer by the thickness of the dried protein shell and the strong van der Waals force between the gold film and the gold particle. This results in a distance as small as 1.1 nm.

The tip-to-tip distance is regulated during self-assembly by the equilibrium of attractive capillary forces, attractive van der Waals forces, and repulsive forces owing to electrostatics and steric hindrance of the protein, but also adhesion between the particle and template. If you compare different particle morphologies and dimensions this distance can vary essentially, *e.g.* spherical particle tend towards a close-packed assembly, while smaller nanorods form an intermediate distance compared to the herein reported distance of 15.6 nm, see C. Hanske *et al.* for spherical and M. Tebbe *et al.* for smaller rod-like particles.[2,3]

1 M. Tebbe, C. Kuttner, M. Männel, A. Fery, and M. Chanana, Colloidally stable and surfactant-free protein-coated gold nanorods in biological media, *ACS Appl. Mater. Interfaces*, 2015, **7**, 5984–5991.

2 M. Tebbe, M. Mayer, B. A. Glatz, C. Hanske, P. T. Probst, M. B. Müller, M. Karg, M. Chanana, T. A. F. König, C. Kuttner, and A. Fery, Optically anisotropic substrates *via* wrinkle-assisted convective assembly of gold nanorods on macroscopic areas, *Faraday Discuss.*, 2015, **181**, 243–260.

3 C. Hanske, M. Tebbe, C. Kuttner, V. Bieber, V. V. Tsukruk, M. Chanana, T. A. F. König, and A. Fery, Strongly coupled plasmonic modes on macroscopic areas *via* template-assisted colloidal self-assembly, *Nano Lett.*, 2014, **14**, 6863–6871.

Estefania Gonzalez Solveyra asked: Since the position of the magnetic mode is highly sensitive to the particle-to-film distance, have you considered changing the surface density of the BSA on the surface or using polymers of different lengths (or molecular weight) as a way to modify such distance and translate that into an optical response? In the same line, do you think that the system would be sensitive to detect changes in spacer length due to responses to changes in the environment (such as changes in pH, presence of ions, *etc.*) if using responsive polymers to modify the surface of the Au-NRs? Could you tune the chemistry of the polymer or the spacer you are using and translate it into an optical response?

Tobias König responded: We agree and we have unpublished results on tuning the particle-to-film distance. Indeed the results show interesting plasmonic properties. As indicated in Fig. 4d in our manuscript, we have pointed out a high sensitivity of 150 nm per particle-to-film distance changes [*Faraday Discuss.*, 2016, DOI: 10.1039/c6fd00013d].

András Deák asked: What is the origin of the negative extinction cross-sections observed for both experimental and simulated values in your Fig. 4c and 4d at certain wavelengths?

Tobias König replied: This negative extinction value results from the normalization procedure to the gold film. In the case in Fig. 4c we normalized the transmission of the metasurface (T) to the transmission of the gold film (T_0). The normalized transmission ($\mathrm{d}\alpha$) we calculated with the following relation: $\mathrm{d}\alpha = -\log(T/T_0)$. More details can be found in our electronic supporting information. Furthermore, by normalization to the reflection, only positive extinction cross-section will be obtained. We chose the normalization in respect to transmission to highlight the sensing application, where a sharp flank at the resonant wavelength is necessary for the feedback lock.

András Deák asked: How is the change in the magnitude of the calculated magnetic permeability (in your Fig. 5) related to the extinction peak intensity associated with this mode (strength of this mode)?

Tobias König responded: We found an optimal magnetic permeability response between the 1.0 and 1.5 nm particle-to-film distance. In this particle-to-film distance range the extinction cross-section is also strongest (Fig. 2c in our paper). Outside this range the magnetic permeability is decreasing.

Catherine Murphy asked: When you said you used BSA as the surface coating on the gold nanorods to transfer them between substrates, I immediately thought of how BSA is 'sticky'. It is well-known to protein chemists that BSA will adsorb to 'any' surface. So, is this why you use BSA for the surface functionalization, or are there other parameters or proteins that work?

Martin Mayer answered: In order to achieve a complete transfer, wettability of the substrate plays a key role. Thus, the target substrate needs to be more hydrophilic than the template, which results in the required 'stickiness'. However, if the hydrophobicity contrast is not granted, an adhesive layer, like PEI, can promote the transfer here; for the targeted gold substrate this was not required. See also C. Hanske *et al.* for the case of spherical and M. Tebbe *et al.* for rod-shaped particles.[1,2] Apart from protein, the transfer is also feasible for other surface modifications, as shown by M. B. Müller *et al.* for pNIPAM-encapsulated particles.[3,4]

1 C. Hanske, M. Tebbe, C. Kuttner, V. Bieber, V. V. Tsukruk, M. Chanana, T. A. F. König, and A. Fery, Strongly coupled plasmonic modes on macroscopic areas *via* template-assisted colloidal self-assembly, *Nano Lett.*, 2014, **14**, 6863–6871.
2 M. Tebbe, M. Mayer, B. A. Glatz, C. Hanske, P. T. Probst, M. B. Müller, M. Karg, M. Chanana, T. A. F. König, C. Kuttner, and A. Fery, Optically anisotropic substrates *via* wrinkle-assisted convective assembly of gold nanorods on macroscopic areas, *Faraday Discuss.*, 2015, **181**, 243–260.
3 M. B. Müller, M. Karg, A. Fortini, T. Hellweg, and A. Fery, Wrinkle-assisted linear assembly of hard-core/soft-shell particles: Impact of the soft shell on the local structure, *Nanoscale*, 2012, **4**, 2491–2499.
4 M. B. Müller, M. Tebbe, D. V. Andreeva, M. Karg, R. A. Alvarez Puebla, N. Pazos-Perez, M. Karg, and A. Fery, Large-area organization of pNIPAM-coated nanostars as SERS platforms for polycyclic aromatic hydrocarbons sensing in gas phase, *Langmuir*, 2012, **28**, 9168–9173.

Christian Kuttner remarked: This comment follows on from the earlier question by Dr Luciano Tadiello on how the best alignment and least amount of defects can be obtained. Structural defects and misalignment may occur at different steps of the sample preparation but can be avoided if the relevant preconditions are met:

(I) The nanoparticles' size and aspect ratio (in particular their width) need to be narrowly dispersed, which can be controlled by the wet-chemical synthesis protocol and additional purification processes.[1,2] In addition, the particles' surface chemistry needs to be adjusted for an adequate surface charge (to match the template) and to provide for a high colloidal stability and thus high particle concentration; a prerequisite to achieve high filling of the template structure.[3]

(II) In order to achieve highly periodic particle arrays homogeneously distributed on a large area, the template has to comply with two requirements: Nanoscale confinement is required, for which the structure of the template (*i.e.*, periodicity and amplitude of wrinkles) and the nanoparticle dimensions need to be matched. Second, the template should provide homogeneous, ideally complete, wetting. In the case of wrinkled PDMS templates, this requires suppression of the formation of cracks in the hydrophilic surface layer to prevent contact line pinning, which is possible by appropriate substrate fabrication conditions.[4]

(III) The latter is essential for the assembly process in the template. We used dip-coating to align gold nanorods utilizing the convective flow resulting from the moving contact line along the template.[5]. The attractive capillary forces promote nanorod deposition in close-packed assemblies, which is required for high filling rates.

(IV) Likewise, homogenous, ideally complete, wetting is required for the wet-printing transfer to the target substrate in order to avoid poor coverage and inhomogeneity. Defects related to incomplete transfer are most likely due to poor substrate wettability, for example by contamination with surfactant and contact line pinning, which needs to be avoided. In addition, the transfer can be assisted by an additional adhesion layer on the target substrate (see also the response of Martin Mayer, above).

In conclusion, the design of appropriate nanoscale confinement is essential and needs to be addressed first. The other parameters are of a physiochemical nature and need to be optimized accordingly.

1 M. Tebbe, M. Mayer, B. A. Glatz, C. Hanske, P. T. Probst, M. B. Müller, M. Karg, M. Chanana, T. A. F. König, C. Kuttner, and A. Fery, Optically anisotropic substrates *via* wrinkle-assisted convective assembly of gold nanorods on macroscopic areas, *Faraday Discuss.*, 2015, **181**, 243–260.

2 M. Mayer, L. Scarabelli, K. March, T. Altantzis, M. Tebbe, M. Kociak, S. Bals, F. J. Garcia de Abajo, A. Fery and L. M. Liz-Marzan, Controlled living nanowire growth: Precise control over the morphology and optical properties of AgAuAg bimetallic nanowires, *Nano Lett.*, 2015, **15**, 5427–5437.
3 M. Tebbe, C. Kuttner, M. Männel, A. Fery, and M. Chanana, Colloidally stable and surfactant-free protein-coated gold nanorods in biological media, *ACS Appl. Mater. Interfaces*, 2015, **7**, 5984–5991.
4 B. A. Glatz, M. Tebbe, B. Kaoui, R. Aichele, C. Kuttner, A. E. Schedl, H.-W. Schmidt, W. Zimmermann and A. Fery, Hierarchical line-defect patterns in wrinkled surfaces, *Soft Matter*, 2015, **11**, 3332–3339.
5 A. Rey, G. Billardon, E. Lörtscher, K. Moth-Poulsen, N. Stuhr-Hansen, H. Wolf, T. Björnholm, A. Stemmer and H. Riel, Deterministic assembly of linear gold nanorod chains as a platform for nanoscale applications, *Nanoscale*, 2013, **5**, 8680.

Eugenia Kumacheva asked: Using the deposition method that you presented, you can bring together different types of nanoparticles in the grooves and subsequently, explore coupling of their properties. Have you tried this approach?

Tobias König responded: We agree and we have unpublished results on this. What we have done in earlier publications is to fabricate single/dimer/tetramer chains with isotropic particles and anisotropic particles, which shows hybridized plasmonic optical properties.[1,2]

1 C. Hanske, M. Tebbe, C. Kuttner, V. Bieber, V. V. Tsukruk, M. Chanana, T. A. F. König, and A. Fery, Strongly coupled plasmonic modes on macroscopic areas *via* template-assisted colloidal self-assembly, *Nano Lett.*, 2014, **14**, 6863–6871.
2 M. Tebbe, M. Mayer, B. A. Glatz, C. Hanske, P. T. Probst, M. B. Müller, M. Karg, M. Chanana, T. A. F. König, C. Kuttner, and A. Fery, Optically anisotropic substrates *via* wrinkle-assisted convective assembly of gold nanorods on macroscopic areas, *Faraday Discuss.*, 2015, **181**, 243–260.

Eugenia Kumacheva commented: Stephan Link from Rice University uses more complicated nanofabrication methods but essentially, forms similar arrays of plasmonic nanoparticles. You may wish to take a look at his single particle spectroscopy experiments. Did you try to use single particle spectroscopy to distinguish coupling of plasmonic properties with the substrate and between the nanoparticles?

Tobias König answered: We do not need to use single particle spectroscopy. We'd like to clarify this with important issue with our unique selling points: (i) we use single-crystalline and monodisperse building blocks with strong extinction cross-section (sizes about 30-100 nm) and (ii) large-area self-assembly method with a high order parameter on the centimeter scale. (iii) We use conventional UV-vis-IR spectroscopy for optical characterization of the building blocks in the non-coupling regime, which show (iv) high agreement with electromagnetic simulations.[1–4] This optical response results in much better signal-to-noise ratio in comparison to single spectroscopy methods.

1 M. Tebbe, C. Kuttner, M. Mayer, M. Maennel, N. Pazos-Perez, T. A. König and A. Fery, 2015. Silver-overgrowth-induced changes in intrinsic optical properties of gold nanorods: from noninvasive monitoring of growth kinetics to tailoring internal mirror charges, *J. Phys. Chem. C*, 2015, **119**, 9513–9523.
2 C. Hanske, M. Tebbe, C. Kuttner, V. Bieber, V. V. Tsukruk, M. Chanana, T. A. F. König, and A. Fery, Strongly coupled plasmonic modes on macroscopic areas *via* template-assisted colloidal self-assembly, *Nano Lett.*, 2014, **14**, 6863–6871.

3 M. Tebbe, M. Mayer, B. A. Glatz, C. Hanske, P. T. Probst, M. B. Müller, M. Karg, M. Chanana, T. A. F. König, C. Kuttner, and A. Fery, Optically anisotropic substrates *via* wrinkle-assisted convective assembly of gold nanorods on macroscopic areas, *Faraday Discuss.*, 2015, **181**, 243–260.
4 M. B. Müller, C. Kuttner, T. A. F. König, V. V. Tsukruk, S. Förster, M. Karg, and A. Fery, Plasmonic library based on substrate-supported gradiential plasmonic arrays, *ACS Nano*, 2014, **8**, 9410–9421.

Wenming Tong commented: Is there any specific reason for using penta-twinned gold nanorods instead of single crystal gold nanorods?

Tobias König responded: We chose penta-twinned gold nanorods due to their flat faces, which increase the contact area to the gold substrate and consequently increase the magnetic response. For the best of our knowledge a specific nanorod diameter of 30 nm and the monodispersity below 10% is not achievable with a single crystal nanorod synthesis.[1]

1 M. Mayer, L. Scarabelli, K. March, T. Altantzis, M. Tebbe, M. Kociak, S. Bals, F. J. Garcia de Abajo, A. Fery and L. M. Liz-Marzan, Controlled living nanowire growth: Precise control over the morphology and optical properties of AgAuAg bimetallic nanowires, *Nano Lett.*, 2015, **15**, 5427–5437.

Tobias König opened a general discussion of the paper by Davide Peddis: In our publication, we study the response of the magnetic dipole moment as effective magnetic permeability (modelling only). In your publication, you are using a Quantum Design Superconducting Quantum Interference Device (SQUID) to measure the magnetic response. Is it possible to measure our particle-to-film induced magnetic dipole moment with your device?

Davide Peddis answered: In principle yes. Of course some preliminary tests are necessary.

M. Arturo López-Quintela enquired: How do you take into account the different degree of crystallinity to extract the influence on the shape of the materials?

Davide Peddis responded: This method has been applied for the first time to magnetite nanoparticles produced by magnetotactic bacteria because this material has excellent morpho-structural features, with a very high degree of crystallinity. How to take into account different degrees of crystallinity is definitely still an open question for us.

David Portehault asked: To discriminate cubes from cube–octahedra, you are making the assumption that the cubes should lie on their faces and never on their edges or corners. However, in case of aggregated particles, there is a non-negligible probability that indeed, cubes could as well be positioned on these edges or corners. Hence, could you comment on this assumption?

Davide Peddis replied: Of course in order to apply our method, magnetic nanoparticles must be well dispersed and the absence of aggregation is required

Andrea Castelli said: You show that the 3D shape distribution of convex nanoparticles can be calculated starting from TEM images employing the circularity of the nanoparticles' projection. Could this method work also for concave and branched nanoparticles?

Davide Peddis responded: This method is relatively new and we are testing in it in several systems, including convex particles. Hopefully we will have results very soon

Alberto Striolo opened a general discussion of the paper by Eugenia Kumacheva: When I was studying polymers I was fascinated by the stimuli-responsive ones, such as, for example, PNIPAM. Looking at the structures you are producing, I wonder whether you will be able to build stimuli-responsive complexes, whose structure would change when you change the temperature, pH, salt concentration, *etc.*, but reversibly.

Eugenia Kumacheva responded: This is an excellent question. One of the ways to make our plasmonic colloidal polymers stimulus-responsive is to use stimulus-responsive polymer ligands linking nanoparticles in the chains. We indeed used temperature-responsive poly(*N*-isopropylacrylamide) (PNIPAM) to assemble gold nanorods in chains. Upon irradiation of the nanorods end-tethered with PNIMAM at the wavelength corresponding to the surface plasmon resonance of the nanorods, they absorbed light energy and subsequently, released it as heat through non-radiative energy transfer. This heat was transferred to PNIPAM which, under elevated temperature, became hydrophobic. In order to avoid unfavorable interactions with the solvent (water), PNIMAM ligands associated and the nanorods formed chains. Upon cooling, the chains dissociated in individual nanorods.[1]

1 D. Fava, M. A. Winnik and E. Kumacheva, Photothermally-triggered self-assembly of gold nanorods, *Chem. Commun.*, 2009, 2571–2573.

Catherine Murphy commented: I noticed in your theoretical section that the bond angle between 'monomers' was a parameter. Does that mean you can control the bond angle between the rods? How so?

Eugenia Kumacheva responded: There are several ways to control bond angles. For example, for flat (planar) interacting nanoparticle facets, the most favorable configuration providing the maximum screening of the ligands from the unfavorable solvent is a face-to-face assembly, in which the bond angle is equal to zero. This was observed for the silver and gold nanocubes. For the arrow-headed nanorods, a facet-to-facet assembly resulted in either a 90 or 180 degree angle.

For spherocylindrical nanorods or for spherical nanoparticles, bond angles can be tuned by varying the 'rigidity' of the bond, *e.g.*, by using rigid and/or short ligands or by crosslinking polymer ligands between the nanoparticles after the self-assembly took place.

At the moment, we do not pursue a detailed exploration of this parameter mostly, because we see a 2D projection of the nanoparticle chains on the surface, which may not correspond to the real conformation of chains in 3D space.

Catherine Murphy continued: So the end facets of the nanorods, even though they are covered with polymer, still help direct the organization of the nanorod assembly?

Eugenia Kumacheva responded: Yes, this is correct. In addition to my previous answer, I should stress that long polymer ligands can 'screen' the nanoparticle shape governing facet-to-facet interactions. Thus the ratio between the length of the ligand and nanoparticle size should not be too large. We showed this effect for nanocubes with different dimensions, which were capped with polystyrene molecules of a specific length.[1] Large nanocubes coated with short polystyrene ligands formed rigid chains with a bond angle close to zero.

1 Klinkova *et al.*, *Nano Lett.*, 2014, **14**, 6314–6321.

David Portehault asked: To follow up on the analogy with molecular polymerisation, do you think that living, or at least controlled, polymerisation would be accessible? You would need, for instance, a chain stopper that is able to adsorb reversibly at the end of the chains to put them into a 'dormant' state. What possibilities do you have to reach such systems?

Eugenia Kumacheva responded: You are absolutely right; however we draw the analogy with step-growth, rather than chain polymerization. We are working on this problem.

Javier Reguera commented: The use of nanopolymers, instead of molecular polymers, certainly can confer new properties, such as plasmonics, to the polymer. It is still unknown if these nanopolymers can have other properties of polymers. From your point of view, would this nanopolymer exhibit thermodynamic properties such the glass transition or the viscoelasticity of thermoplastics?

Eugenia Kumacheva responded: This is a great question. We did not explore in detail the properties of nanopolymers that originate from the polymer ligands. I think they should exhibit the glass transition temperature that corresponds to the glass transition temperature of the polymer molecules. We measured T_g and found that it was well above room temperature. Certainly, the polymer molecules confined between adjacent nanoparticles should exhibit viscoelasticity.

Axel Mueller added: I doubt that it will be so easy to perform a living polymerisation of the nanorods, similar to ATRP. So far we have been dealing with a step-growth polymerisation, where oligomers can add to each other, and we first would have to convert the mechanism to chain-growth, where we add on monomer at a time.

Axel Mueller then enquired: This question is related to copolymerisation of Au/PD or long/short Au nanorods: My understanding from the mechanism is that the driving force is mostly hydrophobic interactions of polystyrene endcaps and a minor contribution of electrostatic forces. If it is more hydrophobic interactions, it wouldn't matter if you have palladium or gold, short or long, thus the reactivity ratio will be near to unity. Do you have any data on the reactivity ratios or rate constants?

Eugenia Kumacheva responded: This is correct. It is the nature of polymer ligands (and their interactions with the solvent) that determines the reactivity ratio. Having said that, I should stress that this is true only for nanoparticles with the same shape and dimensions. For example, shorter nanorods will exhibit faster self-assembly (a higher rate constant) than longer nanorods (a smaller rate constant), due to a larger excluded volume of the latter. Thus the reactivity of these nanoparticles will be different.

Axel Mueller asked: With respect to silver cubes: why don't they form 2D and 3D structures?

Eugenia Kumacheva replied: It is the balance between the solvophobic (hydrophobic) forces and electrostatic repulsion that gives the lowest surface energy of the system at a specific solvent composition. We showed this effect for nanospheres and nanocubes. When electrostatic repulsion is suppressed by adding a salt or a co-solvent with a low dielectric constant, linear chains transform into 3D clusters.

Mathias Brust enquired: Have you tried to envisage a 'Lego' type approach, *i.e.* two different sides using non-covalent interactions, electrostatic/pi, having a plus and minus side as a way to assemble; would you have more control over structures that you make?

Eugenia Kumacheva answered: This is an excellent point. We have tried to use electrostatic forces. The level of control of the self-assembly process was similar to that achieved with solvophobic forces using polymer ligands in a poor solvent. In principle, any forces leading to non-covalent bonding of nanoparticles would work in the self-assembly process, and unless one needs alternating 'nanomers' in the nanopolymer chains, the polymer approach would work.

Beate Reiser asked: I am curious why you chose iron oxide containing chain stoppers in the step-growth polymerization approach. Has it been the aim to exploit their magnetic properties, for instance to use the variable mobility in an electric field as a method to separate the nanopolymers by number of nanomers in a chain?

Eugenia Kumacheva responded: The requirements of the reactive and non-reactive patches of chain stopper nanoparticles (NPs) were as follows: (i) the ligands capping two NP patches had to be distinct in terms of their solubility in the solvent and sufficiently short to avoid the screening of the counterpart patch; (ii) the size of reactive and nonreactive NP patches had to be sufficient to cap the nanorod end and to avoid the attachment of multiple NPs.

A previously reported protocol for the synthesis of gold-iron oxide NPs satisfied requirement (ii). To address requirement (i), we attached thiol-terminated polystyrene to the gold patch and dopamine-terminated PEG to the iron oxide patch. We indeed attempted to use the magnetic properties of the terminal chain stopper NPs to move or separate the chains, however due to their small size, we needed a very strong magnetic field, which was not available in our experiments.

Benjamin Abécassis remarked: Your aggregation/polymerisation numbers are low in comparison to other polymers, why is that? Do you have ideas on how to make it higher?

Eugenia Kumacheva answered: The degree of polymerization of the nanorod chains can be very high. Occasionally, we observed chains with a degree of polymerization of 100 or so. The problem is technical: due to the high density of gold, these chains are heavy. They precipitate from the solution and are difficult to characterize.

Luciano Tadiello remarked: Suppose you are comparing two linear arrays of nanoparticles, same surface chemistry and same number of particle per array, but one array is made of spherical and the other of cubic nanoparticles. Which one of the two arrays will be subjected most likely to chain branching? On one hand, I expect spherical particles have less contact points (less strong interaction), but cubic particles create more steric hindrance and do not allow much branching around the linear chain.

Eugenia Kumacheva answered: I think that branching will be favored for chains formed by cubic nanoparticles, as it provides a stronger reduction in energy per facet-to-facet contact. Importantly, small nanocubes coated with long polymer ligands behave as spherical nanoparticles and in such case, the chain propensity for branching will be similar for the two types of nanoparticles.

Martin Buzza asked: Is it possible to weaken the hydrophobic attraction between nanoparticles so that the association between the nanoparticles becomes reversible and we get into the 'living polymer' regime?

Eugenia Kumacheva answered: In principle, this is possible. We did not pursue this direction, as we plan to use a different route to the 'living' nature of the chains.

Kelley Heatley commented: You stated you tried different polymers aside from polystyrene. Are you able to control the rigidity or flexibility of your chains depending on the polymer used to connect them?

Eugenia Kumacheva responded: The most rigid polymer ligand was poly(vinyl carbazole). Qualitatively, we indeed observed a higher chain rigidity (gold nanorod co-linerarity) than for the chains formed by polystyrene-tethered nanorods, although the solubilities of these polymers in the DMF/water solvent were similar.

Quantitatively, we did not characterize these structures.

Antoine Thill asked: In your paper and presentation you mentioned a possible chain to globule transition when salt is added to the suspension (or when THF is added). Is it possible to reverse this process (by dialysis for example)? Behind this question is of course the idea of growing faster longer chains in presence of salt.

Eugenia Kumacheva responded: In principle, this is possible but due to the multiple nanoparticle interactions in a globule, the transition from a globule to a chain requires more energy than the chain-to-globule transition and can be achieved with sonication.

Alberto Striolo opened a general discussion of the paper by M. Arturo López-Quintela: Yesterday Professor Murphy told us that it is not possible to produce gold nanorods of various aspect ratios. I was wondering whether using the process you described today it is possible to produce reliably gold nanorods of any desired aspect ratio.

M. Arturo López-Quintela answered: We have seen that the aspect ratio (a.r.) of gold nano-rods (GNRs) synthesized using external-produced, catalytic Ag clusters is larger than the one using Ag ions (see, *e.g.* Fig. 1 in ref. 1, below), but we didn't investigate further this matter with GNRs. Instead, using that approach (helped by the growth inhibition of adsorbed polymers) we were able to synthesize silver nanowires (SNWs) of a.r. ~1000 (20 nm × 20 μm), which is a technology transferred to the company NANOGAP (www.nanogap.es), because of the large interest in such SNW products for the preparation of high-quality transparent conductive films.

1 Y. A. Attia, D. Buceta, C. Blanco-Varela, M. B. Mohamed, G. Barone and M. Arturo López-Quintela, Structure-directing and high-efficiency photocatalytic hydrogen production by Ag clusters, *J. Am. Chem. Soc.*, 2014, **136**, 1182–1185.

Catherine Murphy said: I have a philosophical question about your use of the term 'catalyst'. A classical catalyst, of course, speeds up a reaction and participates in it, but is not consumed during the course of the reaction. Are these clusters, then, classic catalysts? Do the clusters become part of the rod?

M. Arturo López-Quintela answered: The clusters are TRUE classical catalysts. They are not consumed during the formation of the rod. If they were consumed and became part of the rod, they would not exist at the end of the reaction at the tips of the rods, so that one would not observe all the photocatalytic activities that are observed at the end of the reaction; as we have shown during the presentation, and which are described in detail in ref. 1.

1 Y. A. Attia, D. Buceta, C. Blanco-Varela, M. B. Mohamed, G. Barone and M. Arturo López-Quintela, Structure-directing and high-efficiency photocatalytic hydrogen production by Ag clusters, *J. Am. Chem. Soc.*, 2014, **136**, 1182–1185.

Luis Liz-Marzán said: From the TEM pictures and from your schematic drawings of the geometry of the nanorods, you are obtaining pentatwinned rods. You are however comparing the synthesis with that assisted by silver ions, from which single crystal rods are obtained. Do you know how twinning forms when you carry out the synthesis using Ag clusters?

M. Arturo López-Quintela answered: Sorry for the confusion, but the schematic drawing in Scheme 2 of our paper was intended only to show that the adsorption of one catalytic Au cluster on one face of the crystalline seed can break the symmetrical growth, inducing the formation of gold nanorods. We didn't, however, carry out HR-TEM measurements to investigate the crystal symmetry of the obtained nanorods. At the moment, we only know that using catalytic Ag clusters we obtain the same crystal structure than using silver ions,[1] but we don't know if Au clusters display similar properties to Ag clusters in the preferential growth of gold nanorods.

1 Y. A. Attia, D. Buceta, C. Blanco-Varela, M. B. Mohamed, G. Barone and M. Arturo López-Quintela, Structure-directing and high-efficiency photocatalytic hydrogen production by Ag clusters, *J. Am. Chem. Soc.*, 2014, **136**, 1182–1185.

David Portehault asked: You show that the clusters persist all along the nanorod growth. Are you able to detect their optical properties during and after the synthesis?

M. Arturo López-Quintela responded: We were not able to detect their optical properties during the synthesis, (neither Au clusters, nor the Ag clusters used in our previous paper[1]) because they are in a very small concentration. We were, however, able to detect them by other techniques, such as EXAFS-XANES (see *e.g.* ref. 2). We could optically identify the formation of Au clusters in a solution before the formation of seeds, which was then used for the catalytic anisotropic growth of gold nanorods.

1 Y. A. Attia, D. Buceta, C. Blanco-Varela, M. B. Mohamed, G. Barone and M. Arturo López-Quintela, Structure-directing and high-efficiency photocatalytic hydrogen production by Ag clusters, *J. Am. Chem. Soc.*, 2014, **136**, 1182–1185.
2 Y. A. Attia, D. Buceta, F. G. Requejo, L. J. Giovanetti and M. A. López-Quintela, Photostability of gold nanoparticles with different shapes: the role of Ag clusters, *Nanoscale*, 2015, 7, 11273.

Eugenia Kumacheva said: How do you know the number of Au ions in the cluster? Such clusters have a fluorescence that is dependent on the cluster size. You could possibly use it to find out the number of ions in the cluster?

M. Arturo López-Quintela responded: This question was already discussed earlier. Just to mention an important point: the clusters discussed here were clusters of atoms (joined by covalent bonds, as it was shown in many theoretical papers dealing with atom clusters[1]). We never mentioned the formation of clusters of ions. However, for the clusters of atoms that we discussed, we have found that there is a very good correlation between the emission energy and the number of atoms (for a detailed discussion see *e.g.* ref. 2).

1 Y. Piñeiro *et al.*, The emergence of quantum confinement in atomic quantum clusters, in *Colloidal Foundations of Nanoscience*, ed. D. Berti and G. Palazzo, ch. IV, pp. 81–105, Elsevier, Amsterdam, 2014.
2 B. Santiago-González *et al.* New strategies and synthetic routes to synthesize fluorescent atomic quantum clusters, in *Functional Nanometer-Sized Clusters of Transition Metals: Synthesis, Properties and Applications*, ed. Wei Chen and Shaowei Chen, RSC Smart Materials no.7, ch. III, pp. 25–50, Royal Society of Chemistry, UK, 2014.

Leonardo Scarabelli commented: In your manuscript you describe the gold cluster as a shape catalyst. Do you observe also a catalytic effect in the reduction of the gold precursor onto the growing particles? In other terms, have you studied the kinetics of reduction? Have you observed any trend in changing the amount of gold clusters added to the growth solution?

M. Arturo López-Quintela answered: It is now well-recognized that small metal clusters are very active as catalysts (for a general discussion see e.g. ref. 1). In particular, Au clusters (with a number of atoms in the range 5 to 10) show exceptional catalytic activities in redox reactions, as *e.g.* the oxidation of thiols,[2] and also Ag clusters of even smaller sizes (Ag_3) display catalytic activities for the reduction of gold ions.[3] Therefore, it is expected that Au clusters of similar sizes catalyze also the reduction of gold ions. What we show in our manuscript is that Au clusters (Au_6) can be formed in large amounts during the process of the formation of Au seeds. Therefore, we used such clusters to catalyze the reaction to grow the Au seeds. Because clusters are easily adsorbed onto particular planes of the Au seeds, if one uses a 1 : 1 ratio of the concentration of seeds to the concentration of catalytic clusters, one can favor the growth of the crystal in that particular plane, giving the formation of nanorods. If one increases further the concentration of clusters, the catalytic growth of different planes can be driven, and other structures, like triangles *etc.*, can be obtained as we have shown before with catalytic Ag_3 clusters.[3] Briefly, the gold clusters are redox catalysts, which have an influence on the nanoparticle shape because of their preferential adsorption onto specific crystal planes. The existence of such clusters at the planes they catalyze, at the end of the formation of a nanostructure, can be checked by different photocatalytic experiments showing that the photocorrosion takes place at the planes they were assumed to catalyze when the nanostructure is formed (see *e.g.* ref. 4).

1 N. Vilar-Vidal, J. Rivas, M.A. López-Quintela, Metal(0) Clusters in Catalysis, in *Functional Nanometer-Sized Clusters of Transition Metals: Synthesis, Properties and Applications*, ed. Wei Chen, Shaowei Chen, RSC Smart Materials No.7, ch. VIII, pp. 226–260, Royal Society of Chemistry, UK, 2014.

2 A. Corma, P. Concepcion, M. Boronat, M. J. Sabater, J. Navas, M. J. Yacaman, E. Larios, A. Posadas, M. A. Lopez-Quintela, D. Buceta, E. Mendoza, G. Guilera and A. Mayoral, *Nat. Chem.*, 2013, **5**, 775–781.

3 Y. A. Attia, D. Buceta, C. Blanco-Varela, M. B. Mohamed, G. Barone and M. Arturo López-Quintela, Structure-directing and high-efficiency photocatalytic hydrogen production by Ag clusters, *J. Am. Chem. Soc.*, 2014, **136**, 1182–1185.

4 Y. A. Attia, D. Buceta, F. G. Requejo, L. J. Giovánetti and M. A. López-Quintela, Photostability of gold nanoparticles with different shapes: the role of Ag clusters, *Nanoscale*, 2015, 7, 11273.

Guillermo González commented: Au clusters can be used for the synthesis of gold nanorods. However, are the Au clusters limited to the preparation of gold nanorods? Can they be also exploited for the synthesis of other anisotropic gold nanoparticles, such gold nanotriangles?

M. Arturo López-Quintela answered: We have not performed further such experiments because this has already been demonstrated previously for Ag clusters.[1] Using a 1 : 1 ratio of clusters/seeds (like now for Au clusters) one obtains nanorods because only the growth of one plane is catalyzed. By increasing the ratio to 3 : 1 one obtains triangles, and increasing further to 9 : 1 one obtains multipods.

1 Y. A. Attia, D. Buceta, C. Blanco-Varela, M. B. Mohamed, G. Barone and M. Arturo López-Quintela, Structure-directing and high-efficiency photocatalytic hydrogen production by Ag clusters, *J. Am. Chem. Soc.*, 2014, **136**, 1182–1185.

Alberto Striolo asked: I think in your presentation you briefly mentioned that you can use the gold nanorods to conduct reactions in the presence of ethanol. Could you explain what reaction you were referring to? Also, I was wondering whether you can comment on where the preferential active sites for this reaction are located on the gold nanorod?

M. Arturo López-Quintela answered: 1. The reaction we conducted was the photooxidation of ethanol. The purpose of this was to prove that the photocorrosion of gold nanorods (GNRs) was due to the Ag clusters physisorbed on the tips of the GNRs. Ethanol was used as a hole scavenger, *i.e.* a molecule which is easy to oxidize (in comparison to gold). We observed that, indeed, ethanol is photooxidized by the strong oxidizing holes generated by very mild UV irradiation absorbed by the 'semiconductor' Ag clusters. The presence of ethanol prevented in this way the photocorrosion of GNRs. 2. The preferential active sites for the photocatalytic experiments mentioned above are the tips of the GNRs. This confirms that, indeed, Ag clusters are formed during the synthesis of GNRs, and contribute to the anisotropic growth (due to their high catalytic activity for redox reactions, as used for the formation of GNRs) because they have preferential adsorption to the crystal planes of the tips of the GNRs.

Hedi Mattoussi enquired: The Au_6 clusters you showed seem to have a rather low emission intensity; are you sure that this emission is generated from the quantum confinement effects or from the ligands?

M. Arturo López-Quintela answered: The low emission intensity of Au_6 clusters is due to the absence of strong binding ligands, which would avoid the luminescence quenching. The observed catalytic properties of such clusters for the redox reaction associated with the gold anisotropic growth indicate again that the recombination of photo-produced charges is not favored. Moreover, the absence of luminescence in the starting solution, and also on the seed solution, both of which should not contain clusters, confirms that the origin of the luminescence comes from the formed clusters. The presence of such clusters in solution was further confirmed by other techniques (such as mass spectrometry, *etc.*), as we show in our paper. Moreover, such luminescent properties nicely agree with those expected for gold clusters of such size (see detailed discussion in our paper).

Hedi Mattoussi remarked: Have you also looked at other types of clusters? They will have better defined absorption and, potentially, emission features? Are these real nanoscale size clusters or more metal-complexes?

M. Arturo López-Quintela replied: Yes, we have used Ag_3 clusters before for the synthesis of gold nanorods.[1] Because of this, we hypothesized that Au clusters could be used for the same purpose (due to the similarities of the catalytic properties observed for small Au and Ag clusters – see references in our answers to other questions). And, indeed we checked that this is the case, as we show in our paper.

As we answered in another question, we were not interested in the emission properties of clusters, but in their catalytic properties, which are much higher in unprotected clusters. We showed in our paper (and other previous papers – see references in our paper) that the clusters are formed by metal(0) atoms, and not by metal ions, as it is the case of metal–organic complexes. XPS, EXAFS, XANES, among other techniques, were used to prove the difference between those species.

1 Y. A. Attia, D. Buceta, C. Blanco-Varela, M. B. Mohamed, G. Barone and M. Arturo López-Quintela, Structure-directing and high-efficiency photocatalytic hydrogen production by Ag clusters, *J. Am. Chem. Soc.*, 2014, **136**, 1182–1185.

Yu Zhou opened a general discussion of the paper by Anna Roig: In your hybrid nanoparticle system, can you reduce the size of the triangular nanoparticle, and how?

Anna Roig responded: Yes, the size of the triangular particles can be reduced by decreasing the molar ratio [SPIONs]/[Gold precursor], although by reducing the size a decrease in the percentage of nanotriangles (NT) is observed.

As an example:

—for [SPION]/[gold precursor] = 0,7, the mean edge size of the NT is 187 nm, while NT are 55% of the shape population.

—for [SPION]/[gold precursor] = 0,13, the mean edge size of the NT is 110 nm, while NT are 30% of the shape population.

Yu Zhou commented: How stable are the oxide nanoparticles attached to the nanotriangles? When you apply a laser to the surface, do you expect them to be released from the surface? This is good for drug delivery but not so good for hypothermia. Can you comment on this?

Anna Roig replied: Question 1: The iron oxide nanoparticles are strongly attached to the nanotriangles' surfaces. When using an external magnet to attract the particles in one spot, the liquid phase is completely clear, meaning that all gold particles are decorated by magnetic particles and that these are firmly attached. Moreover, EELS composition analysis of the sample indicates a continuous carbon layer over the triangles due to the PVP. PVP is the polymer used to anchor the iron oxide to the nanotriangles.

Question 2: Further experiments will be required to answer this question. My guess is that it will very much depend on the power of the laser used. Too strong a power may melt the PVP and release the particles. Indeed, this scenario would be beneficial as a drug delivery strategy but it would not work for hyperthermia treatments.

Oscar Rojas-Carrillo asked: I am wondering if you tried to decorate your particles with another metal like platinum or palladium?

Anna Roig replied: We have succeeded in decorate the metal gold triangles with TiO_2 nanoparticles. We have not tried to decorate gold particles with metallic particles.

Hedi Mattoussi said: In your reaction, which molecules absorb the microwaves to produce the heating required for your growth to progress? If you are relying on the PEG compound, does the ligand/compound have enough absorption cross-section to produce the necessary heat at micromolar concentrations?

Anna Roig answered: In our reaction, the polar solvent, ethylene glycol, is the heat generating agent. It does so in a volumetric homogeneous way inside the reaction vessel. In addition the hydroxyls of the PVP generate hot spots, and we believe serve as nucleation sites for the gold. This is the reason why gold nucleates on the SPIONs (which are PVP coated) and the gold crystals are always 'breaded' with SPIONs.

For microwave reactions one needs either a polar solvent with sizeable $\tan\delta = \varepsilon''/\varepsilon'$ or ionic charges to be able to generate heat.

Luis Liz-Marzán asked: The spectra from your nanoplatelets in solution is quite broad. Is this due to polydispersity in size or to the different plasmon resonances of triangles and hexagons or other shapes? Can you compare the spectra of different shapes using single particle EELS as you have shown for triangles?

Anna Roig replied: Question 1: We believe that the rather broad absorption peak is due to both factors; diversity of gold nanoparticles morphologies and size polydispersity within each morphology. A 15% size polydispersity has been computed for nanotriangles, which are the most abundant shape (60%).

Question 2: Please see Fig. 1 and Fig. 2 here for preliminary unpublished results of the plasmon resonance modes of two single particles, a triangle and what we assume is a dodecahedron with a pentagonal 2D projection.

Alberto Striolo commented: In the images you presented the SPIONs seem to be almost homogeneously distributed on the surface of the gold particles.

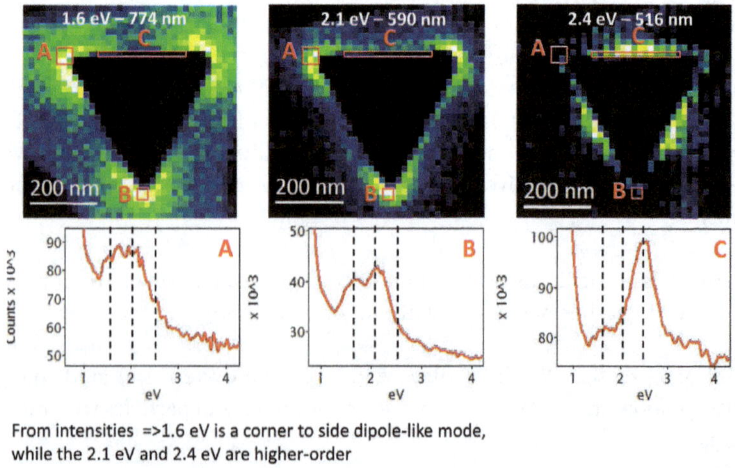

Fig. 1 Plasmon modes in a single nanotube.

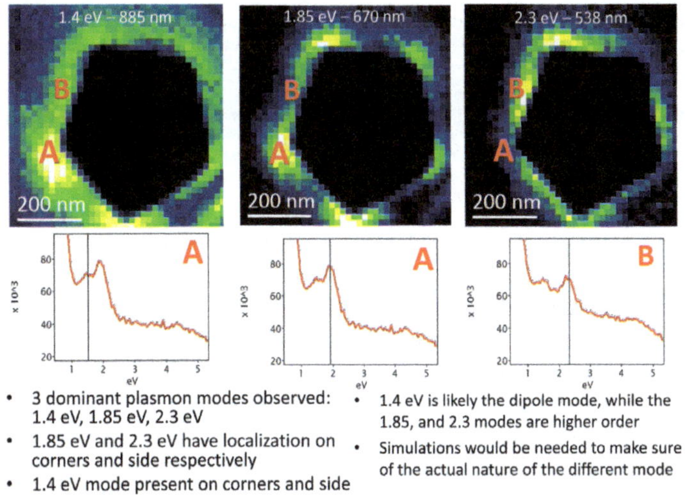

Fig. 2 Pentagons' plasmonic response

However, in some cases it seems that SPIONs adsorb on SPIONs. Is this a possibility, or was I looking at an aberration of the images?

Anna Roig responded: From the analysis of a large number of images on these systems we can say that SPIONs are homogeneously distributed on all gold surfaces and for all gold nanoparticle shapes. We have never observed gold nanoparticles 'naked', without SPION coverage. Equally, we have always observed that the gold particles are covered by a monolayer of SPIONs. Nevertheless in most of our syntheses we worked with an excess of SPIONs and those unattached nanoparticles are visible in the TEM images. Sample purification can be undertaken to remove the excess of SPIONs.

Guillermo González commented: The use of SPIONs as seeds for the synthesis of magnetic gold nanotriangles is quite attractive. How would the concentration of SPIONs affect the size and shape of the nanoparticles? Would be possible to use this method for the synthesis of gold nanoparticles of different shape?

Anna Roig answered: Indeed, concentration of SPIONs can be used to tune the size of nanotriangles. We have observed that by decreasing the concentration of SPIONs we are able to decrease the edge length of the nanotriangles, in our case from 240 nm down to 80 nm. Unfortunately by doing so the percentage yield of nanotriangles also decreases from over 50% yield to approximately 30% yield. I believe that by fine-tuning all other reaction conditions one could be able to control the size and shape of gold nanocrystals with the concentration of SPIONs. Even so, we have observed that the amount of free PVP used for the gold nanocrystals synthesis is the parameter that has the most influence on the yield of nanotriangles, not their size.

Oscar Rojas-Carrillo enquired: Your Fig. 1, in the UV vis spectrum, shows two maximum peaks, why is a maximum peak at 570 nm shown, considering that only triangular particles are present?

Anna Roig responded: The peak at 570 nm corresponds to the rounded smaller gold particles.

Anna Roig then opened a general discussion of the paper by Eugenia Kumacheva: When mixing rods of different materials (*i.e.*, Au, Pd) how can you achieve alternation of materials inside the chain?

Eugenia Kumacheva answered: This brings the concept of copolymerization, which is extensively used for copolymers formed by monomer MOLECULES. As long as we can show that the self-assembly of nanoparticles follows the kinetics and statistics of molecular copolymerization, in the next step, we can explore the possibility to form nanoparticle heterochains. This has been done for gold nanorods with different lengths and for the co-assembly of gold and palladium nanorods. The rate constants of homo-assembly and hetero-assembly and the concentration of each population of nanorods determined the chain composition and nanorod sequences in the chains. In this manner, we made random copolymers. Notably, we also used the approaches used in conventional molecular copolymerization to assemble block copolymers. With regards to alternating copolymer chains, the approach is to use different ligands on two populations of 'nanomers'. The ligands should show specific recognition for each other. This approach was demonstrated a while ago by Murphy. Alternatively, hybridization of DNA attached to two populations of nanoparticles would lead to alternating copolymers.

Mathias Brust asked: Your growing chains rapidly become quite large and will no longer diffuse very much. To what extent do these effects slow down the polymerisation process, and what can be done to accelerate it?

Eugenia Kumacheva responded: Actually, the polymerization process is terminated by chain precipitation, due to the high density of nanoparticles. Thus we do not suffer too much from the slow diffusion of long chains. Tentatively, for gold nanorods with a length of 4050 nm, we noticed that the deviation from the kinetics and statistics of molecular polymerization occurs at a degree of polymerization exceeding 20. To accelerate the polymerization process several approaches can be used such as an increase in the nanoparticle concentration, the addition of salts, or an increase in water concentration (thus further reducing the quality of the solvent).

Hedi Mattoussi asked: When you make these 'polymer rods', have you explored the effects of changing the molecular weight of the polymer? Does changing the polymer molecular weight affect the size and structure of the polymer-rod assemblies?

Eugenia Kumacheva replied: The change in the molecular weight of polymer ligands has a great effect of the morphology of the nanorod assemblies. The variation in the molecular weight of the polymer molecules and their relative location with respect to the long side of the nanorods results in two competing modes of nanorod assembly: the side-by-side and the end-to-end assembly. The resulting structures include bundles, chains, rings, and bundled chains of nanorods. These results have been reported in the paper published as ref. 1.

1 Z. Nie, D. Fava, M. Rubinstein and E. Kumacheva, "Supramolecular" assembly of gold nanorods end-terminated with polymer "pom-poms": effect of pom-pom structure on the association modes, *J. Am. Chem. Soc.*, 2008, **130**, 3683–3689.

Muhammad Nawaz Tahir remarked: (1) When we have polymer based nano-assemblies of inorganic nanoparticles, do these polymers show synergistic effects to enhance the properties through electronic coupling between polymer and different inorganic domains, or these polymers insulate the interface? (2) It is also possible to get inorganic nanoparticle assemblies through epitaxially intergrown domains? Which one do you think will be better for devices: (a) polymer mediated self-assembled structures; or (b) epitaxially grown superparticles?

Eugenia Kumacheva replied: This nanoassemblies may have a synergistic effect, *e.g.*, plasmonic–excitonic coupling arising from plasmonic properties of the nanoparticles and the electronic properties of polymer-separating nanoparticles. We did not investigate these properties. I think it depends on device function; it is hard to say at this stage.

Christian Kuttner remarked: For many applications it is beneficial to have small particle spacings. Could you comment on how small such spacings can be achieved while maintaining the stability of the linear chain assembly? What is the lower limit of coating material required for particle assembly in solution?

Eugenia Kumacheva replied: The distance between the nanorods is determined by: (i) the molecular weight of the polymer ligand; and (ii) the quality of the solvent.

We found that changing the molecular weight of the ligand is more efficient, and for the polystyrene ligands with a molecular weight of 2000 g mol^{-1} we achieved a distance between the nanorod ends as small as 1–2 nm. A further reduction in the molecular weight of polystyrene ligands suppressed nanorod assembly, due to the low enthalpy gain in interactions of polymer ligands.

Christian Kuttner remarked: How does the grafting density of such polymeric coatings affect the robustness of the assembly? Is it already clear whether chain density or chain length is more important?

Eugenia Kumacheva responded: We did not explore the effect of grafting density on the self-assembly. For polystyrene molecules with a molecular weight of 50 000 g mol^{-1}, the grafting density was approximately 0.022 $\pm$ 0.0006 chains nm^{-2}.

The effect of the molecular weight of the polymer ligands on nanorod self-assembly was studied. There is an optimized range of molecular weights that results in a well-defined linear nanorod assembly. The utilization of too short polymer ligands does not lead to self-assembly, while too long polymer ligands may cause side-by-side assembly of gold nanorods (see *e.g.*, ref. 1).

1 Z. Nie, D. Fava, M. Rubinstein and E. Kumacheva, "Supramolecular" assembly of gold nanorods end-terminated with polymer "pom-poms": effect of pom-pom structure on the association modes, *J. Am. Chem. Soc.*, 2008, **130**, 3683–3689.

Axel Mueller asked: Coming back to an earlier question on branching: If you mix nanorods and spherical nanoparticles, would this not be a perfect way to form branch points?

Eugenia Kumacheva responded: This is correct. I should also mention that in addition to nanoparticle attachment to the chain end, it may attach to the junction between the two adjacent nanorods, thus leading to a branching point. Nanorods can also attach to the junctions, but for small nanospheres the steric hindrance is weaker.

Axel Mueller added: This leads to 3D structures, which are difficult to characterize; maybe tomography would help?

Julian Bergueiro enquired: Which other supramolecular interactions could work instead of the hydrophobic interaction?

Would you need a multivalent effect to achieve enough interaction?

Eugenia Kumacheva responded: Any type of forces that act in supramolecular chemistry – hydrophobic forces, hydrogen bonding, electrostatic forces, biospecific forces – would work. Obviously, for 'nanomers' that are significantly larger than molecular monomers, these forces should be sufficiently strong. Multivalency is not mandatory: each nanoparticle can carry many molecules that interact with each other. Polymer ligands are beneficial, since they have many monomer–monomer interactions.

Tobias König said: Following on from Axel Mueller's question, from a plasmonic point of view, it is more interesting to attach the spherical nanoparticle at the long face of the nanorod. This interaction will result in so-called plasmonic hybridisation, which has been shown by Shao[1]

1 L. Shao *et al.*, *Nano Lett.*, 2012, **12**, 1424–1430.

Eugenia Kumacheva replied: I agree. We have unpublished results on selective attachment of spherical nanoparticles to the long side or ends of nanorods. They indeed show interesting plasmonic properties.

Alberto Striolo addressed Eugenia Kumacheva, M. Arturo López-Quintela and Anna Roig: Yesterday we heard a short comment on the mass production of nanostructured systems, including nanoparticles, nano-rods, *etc.* Can you comment on the possibility of producing large quantities of the materials you have described today?

M. Arturo López-Quintela replied: I mentioned in my answer to the Alberto Striolo, above, that we already transferred the technology to the company NANOGAP for the mass production of silver nanowires. We also transferred the technology for the mass-production of luminescent gold clusters with giant Stokes-shifts, and with tuned excitations from the UV until the near IR. Such clusters are of interest for applications in different fields, such as biosensors, spotting counterfeiting,*etc.*

Anna Roig then responded: The microwave synthesis proposed for our multimaterial system, consisting of gold nanotriangles decorated with iron oxide nanoparticles, is undertaken in a batch mode. Batch mode production is always more difficult to scale up than other continuous fabrication processes. Up to now, we have not studied the possibility of scaling up the production of the multimaterial presented in our paper. However, in a previous work, we demonstrated that by using a very specific microwave oven (from Milestone) that includes 15 identical reaction vessels we were able to multiply the production of iron oxide nanoparticles by 100 as compared with our initial synthesis. We could obtain up to 3 g per batch of iron oxide nanoparticles as a stable colloidal dispersion using a 10 min synthesis route.

1 L. Gonzalez-Moragas, S. M. Yu, N. Murillo-Cremaes, A. Laromaine, A. Roig, Scale-up synthesis of iron oxide nanoparticles by microwave-assisted thermal decomposition, *Chem. Eng. J.*, 2015, **281**, 87–95 (DOI:10.1016/j.cej.2015.06.066).

Eugenia Kumacheva also replied: The issue of mass production of nanostructured systems, including nanoparticles, is very important. In my opinion, mass production should not compromise nanoparticles' size and shape distribution. Unfortunately, this is not always the case for nanoparticles purchased from companies claiming high nanoparticle quality. In my laboratory, for applications that require a lot of nanoparticles, I encourage students to scale up nanoparticle synthesis. This is generally done using a 'numbering out' method, that is, by using parallel synthesis in many small reactors. Recently, we reported a significantly increased nanoparticle yield using a "suspension-based" method.[1]

Another elegant method would be to use continuous microfluidic synthesis in multiple parallel microfluidic reactors (again, a 'numbering out' method). I have great faith in this method!

1 A. Klinkova, E. M. Larin, E. Prince, E. H. Sargent and E. Kumacheva, Large-scale synthesis of metal nanocrystals in aqueous suspensions, *Chem. Mater.*, 2016, **28**, 3196–3202.

Catherine Murphy also added: Just a comment on scaling up: It is possible for many inorganic nanoparticles now, to purchase them through Strem or Sigma-Aldrich. This is true of gold nanorods, silver nanowires, and others. Maybe Younan Xia can comment further, but flow synthesis, as opposed to batch synthesis, can work in many cases. For us, the gold nanorods we make in flow are not as nice as the ones in batch, but for some applications that might not matter too much. The advantage of flow is that you just keep adding feed solutions and then the reactions just keep going, so you can make liters and liters.

Younan Xia then commented: For most types of nanostructures, it has been very difficult to scale the production volume using batch reactors without compromising the quality due to the high sensitivity of nucleation and growth toward experimental conditions. One solution to this issue is to disperse the reaction solution into a train of small droplets in continuous flow. This way, one can obtain samples even more uniform than what has been obtained in batch reactors.

Anna Roig continued: Flow synthesis is possible with microwave ovens but is not as studied as the batch mode. Regarding the batch mode, a particular advantage of microwave nanoparticle synthetic routes is the volumetric heating mode that facilitates the scaling up of a given process.

Hedi Mattoussi concluded: I would second Catherine Murphy's description. Continuous flow growth has been adapted by chemical engineers to prepare larger amount of nanocrystals (including luminescent quantum dots). There are still hurdles in terms of reproducing exactly the same NR aspect ratio or the same nanocrystal emission wavelength every time, but small changes in these parameters can usually be tolerated in most applications.

Faraday Discussions

PAPER

Novel stabilisation of emulsions by soft particles: polyelectrolyte complexes†

Ana Maria Bago Rodriguez,[a] Bernard P. Binks*[a] and Tomoko Sekine[b]

Received 1st February 2016, Accepted 17th February 2016
DOI: 10.1039/c6fd00011h

We put forward the concept of a novel particle stabiliser of oil–water emulsions, being the polyelectrolyte complex (PEC) formed between oppositely charged water-soluble polymers in cases where either polymer alone is incapable of stabilising an emulsion. Using poly(4-styrene sulfonate) sodium salt, PSSNa and poly(diallyldimethylammonium chloride), PDADMAC, of low polydispersity and similar molecular mass, we correlate the behaviour of their mixtures in water with that of emulsions after addition of oil. In aqueous mixtures, spherical particles of diameters between 100 and 150 nm are formed through electrostatic interactions between charged polymer chains. Around equal mole fractions of the two polymers, the zeta potential of the particles reverses in sign and emulsions of oil-in-water (o/w) for a range of oils can be prepared which are the most stable to coalescence and creaming. The effects of PEC concentration and the oil : water ratio have been examined. All emulsions are o/w and stability is achieved by close-packed particle layers at drop interfaces and particle aggregation in the continuous phase. Increasing the salt concentration initially causes destabilisation of the aqueous particle dispersion due to particle aggregation followed by dissolution of particles at high concentrations; the corresponding emulsions change from being stable to completely unstable and are then re-stabilised due to adsorption of uncharged individual polymer molecules.

Introduction

An emulsion can be defined as a heterogeneous system of two immiscible liquid phases in which one of the phases (dispersed phase) is dispersed in the other (continuous phase) as drops of microscopic or colloidal size.[1,2] Emulsions are thermodynamically unstable and there is a tendency with time for them to revert to the separated liquid phases. Therefore, in order to protect the formed drops from coalescence, a surface-active material or emulsifier should be added rendering the emulsions kinetically stable.[1]

[a]Department of Chemistry, University of Hull, Hull, HU6 7RX, UK. E-mail: b.p.binks@hull.ac.uk
[b]Shiseido Global Innovation Center, 2-2-1 Hayabuchi, Tsuzuki-ku, Yokohama 224-8558, Japan

† Electronic supplementary information (ESI) available. See DOI: 10.1039/c6fd00011h

One way to stabilise emulsions is with surfactant molecules. The mechanism of stabilisation involves the reduction of the interfacial tension between two immiscible phases by adsorption at the oil–water interface.[3] Emulsions can also be stabilised by surface-active polymers. Here, polymer molecules form structured interfacial films that prevent the coalescence of drops.[4] Proteins and polysaccharides are natural polymers of great interest in the food industry.[5] Due to their balanced hydrophilic/lipophilic structure, proteins place themselves at the boundary between an oil and a water phase, thereby contributing to the suppression of the interfacial tension.[6] Their amphiphilic nature, together with the formation of a viscoelastic film at the interface, provides electrostatic and steric stabilisation of emulsions.[7] Many proteins such as caseins,[8–10] whey proteins,[11,12] ovalbumins[13] and bovine serum albumin[14] have been known for decades as emulsifiers. On the other hand, water-soluble polysaccharides are not considered true emulsifiers by colloid scientists as they do not absorb at liquid interfaces.[15–17] Only certain hydrocolloids, such as gum Arabic, are known to exhibit emulsification properties.[15] Their surface activity is related to hydrophobic proteinaceous moieties bonded to the polysaccharide backbone.[16] Garti and Leser also demonstrated however that certain hydrophilic polysaccharides can display emulsification properties even after the removal of protein residues.[16]

A third type of emulsion stabiliser is solid particles. Particle-stabilised emulsions, whose discovery was attributed to Pickering in 1907,[18] have gained special attention due to the enhanced stability to coalescence compared to the common surfactant-stabilised emulsions.[19] Moreover, as certain particles are biologically compatible and environmentally friendly, their use has spread rapidly in the food, biomedicine, pharmaceutical and cosmetic industries. The stabilisation mechanism is based on the strategic location of the solid particles at the interface. This leads to the formation of a rigid barrier that prevents or inhibits coalescence. Such an effect is explained by the partial wettability of the particles at the oil–water interface as described by Finkle *et al.*[20] The stabilisation is linked to the contact angle that the particle makes with the oil–water interface.[21] Emulsion stabilisation by hard or non-deformable particles has been extensively studied and many examples are available in the literature spanning inorganic materials such as silica,[22–24] metal,[25,26] carbon[27,28] and clay[29–31] particles. On the contrary, reports on emulsions stabilised by soft or deformable particles like microgels are relatively recent. Microgels can be defined as cross-linked polymer particles that are swollen by a solvent.[32] The principles of Pickering emulsions cannot be applied entirely to microgel-stabilised emulsions. Due to their softness, microgels behave differently compared with rigid particles as they become deformed at the oil–water interface.[33] This deformability is crucial in understanding the emulsion stability in such systems.[34] Richtering and co-workers have evaluated such behaviour using their own synthesized microgel of poly(*N*-isopropylacrylamide)-*co*-methacrylic acid (PNIPAM-*co*-MAA).[34–37] Destribats *et al.* have worked with a whey protein microgel[38] as well as microgels made of PNIPAM of variable cross-linking degrees.[39–41] Microgel synthesis is however quite involved and some of the monomers are expensive.

The use of mixtures of particles of opposite charge to form aggregates of low overall charge has been shown to be an effective way of preparing surface-active particles *in situ*, capable of stabilising emulsions.[29,42–44] In these cases, the separate particles of negative or positive charge were too hydrophilic to enable emulsion stabilisation. With this in mind, we had the idea of investigating if mixtures of water-soluble polymers of opposite charge could likewise act as stabilisers of oil–water emulsions, wherein systems containing either polymer alone were ineffective. A polyelectrolyte (PEL, also called polyion) can be defined as a polymer consisting of a macromolecule bearing numerous ionisable groups (either cationic or anionic) and low molecular weight counterions to ensure electroneutrality.[45] Due to the charges along the polymer chain, they are hydrophilic and water-soluble. This feature enables them to form complexes with oppositely charged polyelectrolytes, leading to the formation of the so-called polyelectrolyte complex (PEC) or polyion complex (PIC). Their formation is mainly due to electrostatic interactions between the charged domains of two oppositely charged polyelectrolytes.[46–48] Besides this, additional inter-macromolecular interactions such as van der Waals, hydrogen bonding, hydrophobic and dipole interactions are involved in the formation of the complex but are not the driving force for complexation.[46,47] It is well-known that PEC formation can be considered an entropy driven process as it involves the release of counterions initially bound to the ionic groups of the polymer chain leading to an increase of the system entropy.

PECs in solution can show three different kinds of behaviour in terms of solubility depending on the ionic groups and the differences in molar mass of the two polymers.[46,49] Soluble complexes are achieved by mixtures in a non-stoichiometric ratio of polyelectrolytes containing weak ionic groups and a large difference in molar mass. The formation and stability of these aggregates are dependent on the presence and concentration of soluble salts, on the charge ratio of the two polyions and on the pH of the solution. By altering the above conditions, water-soluble PECs can form stable macroscopic homogeneous systems which coexist in solution with insoluble, colloidally stable PECs, or aggregate and precipitate. Secondly, complex formation between polyelectrolytes with strong ionic groups and/or high and similar molecular weight, usually results in highly aggregated, macroscopic heterogeneous systems which tend to flocculate. However, this aggregation process can be stopped at a colloidal level by working at low or moderate ionic strengths, in extremely diluted solutions and in non-stoichiometric charge ratio conditions. Finally, when concentrated solutions of polyelectrolytes of high and similar molecular weight are mixed near stoichiometric ratios, a two-phase system is formed composed of a liquid phase and a PEC aggregate-rich phase. However, at sufficiently low concentrations, soluble complexes can be obtained even when the charges are neutralized.

The aim of this work is therefore to explore if PECs formed in aqueous solution from near-monodisperse anionic and cationic polymers of similar molar mass attain sufficient surface activity to adsorb at the oil–water interface of emulsion drops after addition of oil. To the best of our knowledge, this is the first such use of PECs as soft interfacial particles enabling emulsion stabilisation. We do however acknowledge the literature on protein–polysaccharide mixtures, whose interactions have been studied since 1896 following the pioneering work of Beijerinck.[50] Complex formation is mainly due to electrostatic interactions between oppositely charged domains of each individual biopolymer.[51] The complex at the emulsion droplet surface can enhance emulsion stability compared with emulsions of protein alone by the formation of a thick layer around the droplets that improves the steric stabilisation. The presence of the polysaccharide during the emulsification process leads to a reduction of the droplet size, which causes a decrease in the rate of creaming.[52] Despite the many examples of emulsions stabilised by protein–polysaccharide mixtures in the literature,[53–63] all of them refer to systems in which the protein alone is an emulsifier, unlike in the systems reported here. We first examine the behaviour of aqueous mixtures of an anionic and a cationic polyelectrolyte and characterise the size and charge of the PECs formed. Emulsions are then prepared from the aqueous polymer mixtures and oil, and their stability, drop sizes and arrangement of PEC particles around drops is evaluated. The effects of PEL and PEC concentration, salt concentration, oil volume fraction and oil type are investigated. Air–water and oil–water interfacial tensions of planar interfaces are also monitored.

Experimental

Materials

Poly(4-styrene sulfonate) sodium salt, PSSNa, and poly(diallyldimethylammonium chloride), PDADMAC were purchased from Polymer Standard Services (PSS, Mainz) and were used as received. The chemical structures as well as some other properties are shown in Table 1. Two pure samples of PSSNa of different average molecular weight differing by a factor of 6.1 were chosen. The degree of sulfonation of PSSNa is greater than 90% in both cases (from the certificate of analysis). The second polyelectrolyte, PDADMAC, has a molecular weight centred at 160 kDa. For the emulsion preparation, six different oils were selected (Table 1). They include non-polar alkanes, aromatic oils and a silicone oil. Prior to use, all the oils were passed twice through a basic alumina column (particle size: 0.06–0.20 mm, Merck kGaA) to remove polar impurities. Water was first passed through a reverse osmosis unit and then a Milli-Q reagent water system (Millipore). After treatment, its surface tension measured with a Krüss K11 tensiometer and Wilhelmy plate was 72.0 mN m^{-1} at 25 °C. Sodium chloride, NaCl (Fisher Chemical, 99.9%), hydrochloric acid, HCl (Fisher Chemical, 37%) and sodium hydroxide, NaOH (Fisher Scientific, >97%) were used as received.

Methods

(a) **Preparation and characterisation of PEC aqueous dispersions.** Individual polyelectrolyte solutions of different concentrations (0.01–1 g L^{-1}) were prepared by weighing the corresponding amount of each PEL and dissolving them in Milli-Q water. PEL solutions were prepared at their natural pH. The pH of both the individual PEL solutions (1 g L^{-1} PSSNa 6.32; 1 g L^{-1} PDADMAC 5.44) and the PEC solution (prepared from the 1 g L^{-1} individual PEL solutions 5.71) was measured with a pH meter (3510, Jenway). All solutions were kept at room temperature, in the dark and sealed with parafilm to avoid evaporation.

Aqueous PEC dispersions of different mole fractions of PSSNa (x_{PSSNa}) were obtained by mixing known volumes of each individual polyelectrolyte solution of a fixed concentration with a magnetic stirrer (VWR VMS-C7, stirrer speed = 4) at room temperature. All solutions were prepared in 14 mL screw-cap glass vials. Due to the remarkable influence of the mixing procedure on the particle size,[64–67] we decided to work under fixed conditions in order to obtain reproducible results. Therefore, the order of addition was set as follows: water, PSSNa solution, PDADMAC solution. Water and PSSNa solution were added in one shot additions

Table 1 Molecular characteristics of polyelectrolytes employed in PEC formation and chemical structure, source, purity and density (20 °C) of all the oils used

Polyelectrolyte	Repeat unit	Molar mass per charged unit/g mol^{-1}	$M_w{}^a$/Da	$M_n{}^b$/Da	$M_p{}^c$/Da	PDI[d] (M_w/M_n)
PSSNa	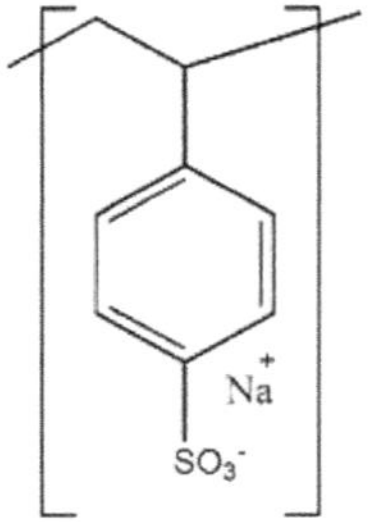	207.20 207.20	148 000 976 000	— —	152 000 976 000	<1.20 <1.20
PDADMAC	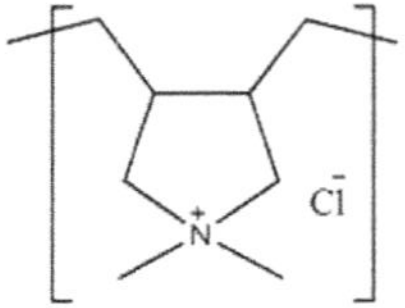	161.67	160 000	101 000	—	1.58

Name	Structure	Supplier	Purity/%	Density/ g cm^{-3}
n-Dodecane		Alfa Aesar	>99	0.796

Table 1 (Contd.)

Name	Structure	Supplier	Purity/%	Density/ g cm^{-3}
Squalane	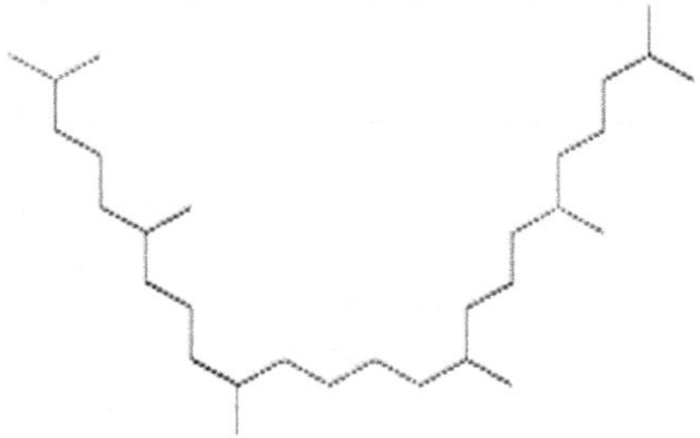	Aldrich	≥99	0.818
Isopropyl myristate	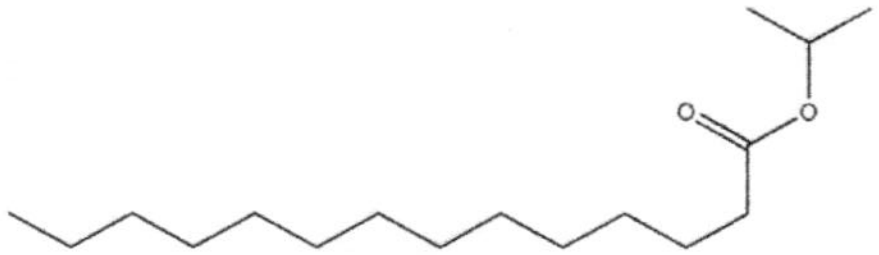	Aldrich	>98	0.859
Toluene	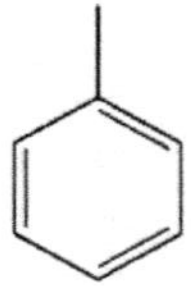	Analar Normapur	100	0.866

Table 1 (*Contd.*)

Name	Structure	Supplier	Purity/%	Density/ g cm^{-3}
50 cS polydimethyl siloxane (PDMS)	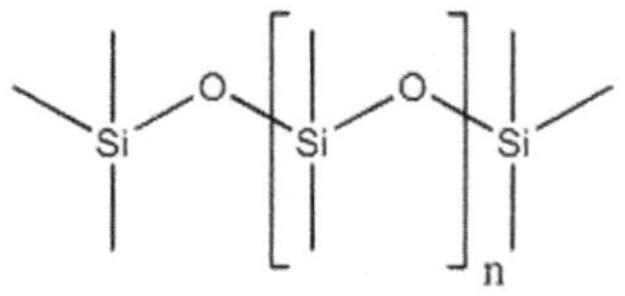	Dow Corning	100	0.964
Paraffin oil	Mixture of hydrocarbons from petroleum	Sigma-Aldrich	—	0.869

[a] M_w: weight average molecular weight. [b] M_n: number average molecular weight. [c] M_p: molar mass at the peak maximum. [d] PDI: polydispersity index.

with a micropipette. PDADMAC solution was then added to the above mixture whilst stirring. The addition of PDADMAC was done sequentially every min up to the total desired volume with a micropipette. This was done to allow the added polyelectrolyte to interact with the oppositely charged species present in the vial. For PEC aqueous dispersions with $x_{PSSNa} < 0.3$, each addition was of 2000 μL, for x_{PSSNa} between 0.3 and 0.5 the addition was of 1000 μL, for x_{PSSNa} between 0.5 and 0.8 it was 500 μL and for $x_{PSSNa} > 0.8$ it was 200 μL. With this protocol, the total mixing time for all the samples was around 4 min. After the complete addition of PDADMAC solution, mixing was kept at the same speed for an additional minute.

For the study of the influence of the salt concentration on the stability of complexes, PEC aqueous dispersions were prepared as above. Immediately after preparation, stipulated amounts of solid NaCl were added into each dispersion in order to obtain a NaCl concentration range from 0 to 5 M. To evaluate the effect of pH on the stability of complexes, individual PEL solutions were prepared in 0.01 M NaOH (pH = 12) and in 0.01 M HCl (pH = 2). After that, PEC preparation was performed as described above.

Dynamic light scattering (DLS) was used to determine the size and the size distribution (polydispersity index, PDI) of aqueous PEC dispersions prepared at natural pH with no added electrolyte. Both parameters were derived from the cumulant analysis.[68] Measurements were carried out at 25 °C using a Zetasizer Nanoseries NanoZS (ZEN3600, Malvern Instruments). The instrument was equipped with a 4 mW He–Ne laser beam as a light source, operating at $\lambda = 633$ nm under a scattering angle of 173°. To run the measurement, samples were placed in a plastic disposal cuvette of 1 cm path length. The PDI is a dimensionless value defined by the software as $2c/b^2$, where b is the second order cumulant and c is the coefficient of the squared term of the correlation function. The results of both parameters are given as the average of three measurements of 60 s each. The relative standard deviation was lower than 2% and 10% for the size and PDI, respectively. The DLS technique was limited to aqueous PEC dispersions prepared from individual PEL solutions with a concentration < 1 g L^{-1}. The zeta potential was measured at 25 °C by a Zetasizer Nanoseries NanoZS (Malvern Instruments) equipped with a 4 mW He–Ne laser beam operating at $\lambda = 633$ nm. Measurements were made by introducing a universal dip cell (ZEN1002, Malvern Instruments) inside a plastic disposal cuvette. The Smoluchowski

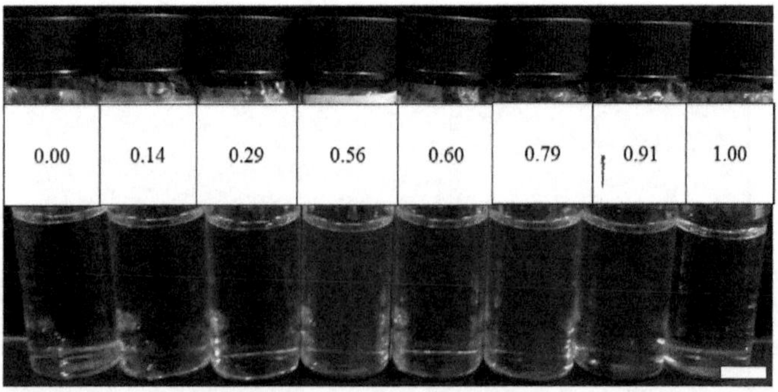

Fig. 1 Appearance of freshly prepared aqueous PEC dispersions prepared from the 0.1 g L^{-1} individual PEL solutions at different x_{PSSNa} (given). Scale bar = 1 cm.

approximation[69] was used to convert measured electrophoretic mobilities to zeta potentials. Each value was averaged from three parallel measurements of 12 runs each. The refractive indices of water and aqueous PEC dispersions were obtained using a refractometer (M46 313, Hilger) at 25 °C and were 1.333 and 1.380, respectively. Transmittance measurements on aqueous PEC dispersions were carried out with a UV-Vis spectrophotometer (Perkin Elmer Lambda 25) at λ = 400, 500 and 700 nm. At these wavelengths none of the polyelectrolytes absorb. Samples were placed in a quartz cuvette of 1 cm path length with Milli-Q water being used as a reference. Transmittance (%) values are given as the average of three measurements.

Scanning electron microscopy (SEM) was used to obtain micrographs of the particles. One drop of the aqueous PEC dispersion was applied to a carbon disc and left to evaporate at room temperature for 3 days. Once dried, the disc was coated with a thermally evaporated carbon film (10 nm thick) using an Edwards high vacuum coating unit. Micrographs were taken with a Zeiss EVO 60 scanning

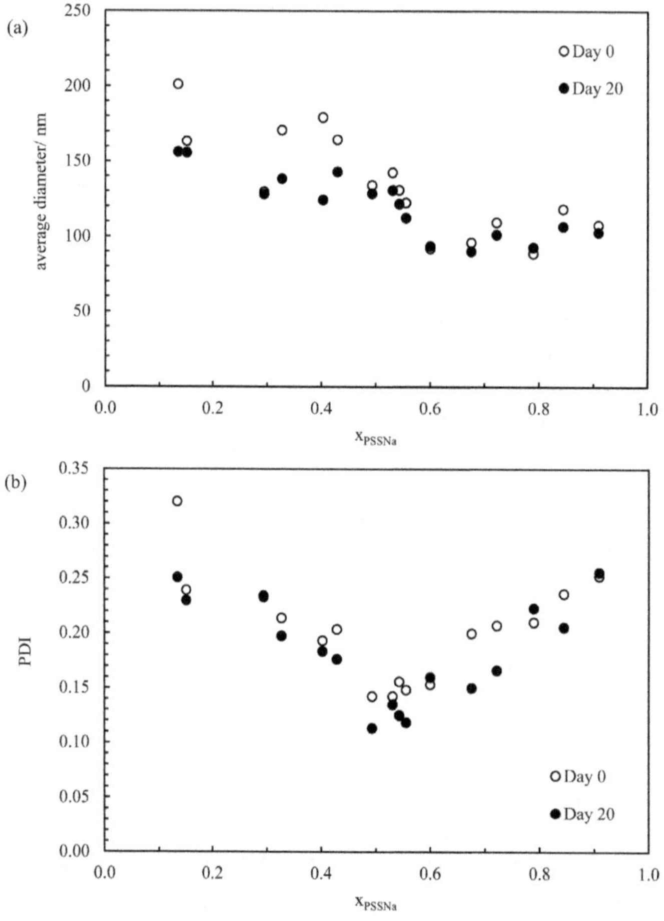

Fig. 2 (a) Average particle diameter and (b) PDI in size for aqueous PEC dispersions prepared from 0.1 g L^{-1} individual PEL solutions *versus* x_{PSSNa} for freshly prepared dispersions (unfilled circles) and 20 days after preparation (filled circles).

electron microscope at a voltage of 20 kV and a probe current of 70 pA. The average particle diameter was calculated from at least fifty individual entities on SEM images with ImageJ 1.47v software. Energy dispersive X-ray spectra (EDX) were also acquired.

(b) **Preparation and characterisation of emulsions.** The density of the oils was determined using a density meter (DMA 35N, Anton Paar) at 20 °C. Emulsions containing an aqueous PEC dispersion and *n*-dodecane were prepared in 14 mL screw-cap glass vials. The mixture was emulsified with an Ultra-Turrax homogenizer (IKA® T25 digital) with a dispersing element of 8 mm (stator diameter). Mixing was for 2 min at a constant speed of 13 000 rpm. Immediately after preparation, the emulsion type was inferred from the drop test, by checking the miscibility of a drop of emulsion with both water and oil. Photographs of the obtained emulsions were taken just after preparation and over time to evaluate their stability. Optical micrographs of the freshly prepared emulsions were obtained on a dimple glass slide (Fisher Scientific) with a cover slip (Scientific Laboratory Supplies LTD) using an Olympus BX-51 microscope fitted with a DP50 digital camera. The mean droplet diameter of the emulsion was calculated from at least fifty individual droplets with ImageJ 1.47v. The fraction of water (f_w) released five days after preparation was measured as a height. However, due to the very low volume of oil separated, for the oil fraction (f_o) the oil was removed carefully from above the emulsion with a Pasteur pipette and weighed.

Selected stable emulsions were imaged with cryo-SEM. A small amount of the emulsion was mounted on an aluminium sample holder (diameter ~ 10 mm) with a Pasteur pipette or a spatula depending on its viscosity. The sample was plunged into liquid nitrogen slush at a temperature of −210 °C. The frozen sample was placed inside the cryo preparation chamber (PP3010T, Quorum Technologies Ltd.) and was fractured with a sharp knife at −140 °C under high vacuum. After that, it was coated with platinum to a thickness of ~2 nm. The sample was then transferred to a Zeiss EVO 60 SEM chamber for imaging at a voltage of 15 kV and a probe current of 30 pA at −140 °C. In some cases, in order to obtain a clearer image of the sample, sublimation of the surface water (ice) was performed. This step was done at −75 °C for 10 min.

Different sets of emulsions were systematically prepared by varying one of the following parameters: x_{PSSNa}, concentration of the starting PEL solutions and oil volume fraction. Moreover, the effect of salt addition before or after emulsification was also studied. In the first case, the aqueous PEC dispersion contained NaCl before addition of oil, whereas in the second case, 1 mL of a 5 M NaCl solution was added to a pre-formed emulsion. In addition to this, selected emulsions containing different oils were also prepared.

(c) **Surface and interfacial tension.** The air–water surface tension of aqueous PEC dispersions was measured with a Krüss K11 tensiometer using a Pt Wilhelmy plate at 25 °C. Surface tensions are given as the average of three measurements. After each measurement, the plate was rinsed with ethanol and heated to glowing in a blue Bunsen flame. The interfacial tension between *n*-dodecane and an aqueous PEC dispersion was measured with the Krüss K11 tensiometer and the du Noüy ring method (Pt–Ir) at 25 °C. The density of the two phases were measured and inputted in the software. The applied correction method was that of Harkins and Jordan.[70] The results of 3 separate measurements were averaged. All measurements were carried out at room temperature unless stated otherwise.

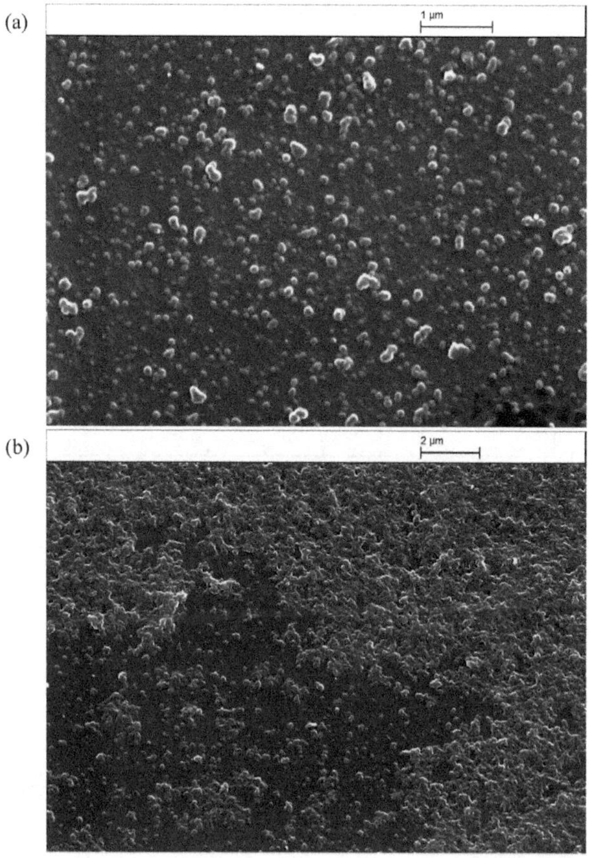

Fig. 3 SEM images of aqueous PEC dispersion ($x_{PSSNa} = 0.56$) prepared from (a) 0.1 g L^{-1} and (b) 0.5 g L^{-1} individual PEL solutions.

Results and discussion

Characterisation of aqueous dispersions of PEC

Before studying the behaviour of emulsions stabilized by polyelectrolyte complexes, a comprehensive study on PEC characterisation at low concentrations was carried out. To gain a deep insight into their formation and properties, aqueous PEC dispersions were first characterised in terms of size and zeta potential along with microscopy. The appearance of aqueous PEC dispersions prepared from 0.1 g L^{-1} individual PEL solutions at different x_{PSSNa} is shown in Fig. 1, where x refers to mole fraction using the values of M_w given in Table 1 (148 kDa PSSNa and 160 kDa PDADMAC). When either PDADMAC or PSSNa are in excess (low and high x_{PSSNa}, respectively), solutions are transparent. However, around the point where all the charges are expected to be neutralised, $x_{PSSNa} = 0.56$, the solutions appear slightly opalescent.

Fig. 2(a) shows the average diameter of the PEC particles as a function of x_{PSSNa} of the above dispersions. The results for freshly prepared dispersions and for the same 20 days after preparation are included. As seen, particles seem to shrink

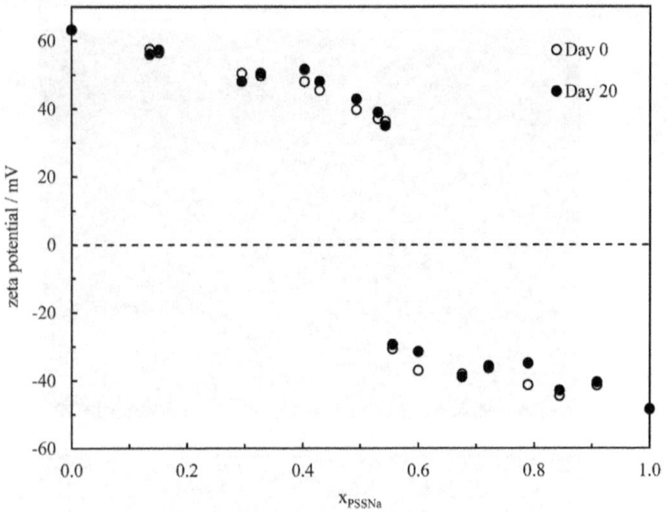

Fig. 4 Variation of zeta potential with x_{PSSNa} for aqueous PEC dispersions prepared from 0.1 g L^{-1} individual PEL solutions for freshly prepared dispersions (unfilled circles) and 20 days after preparation (filled circles).

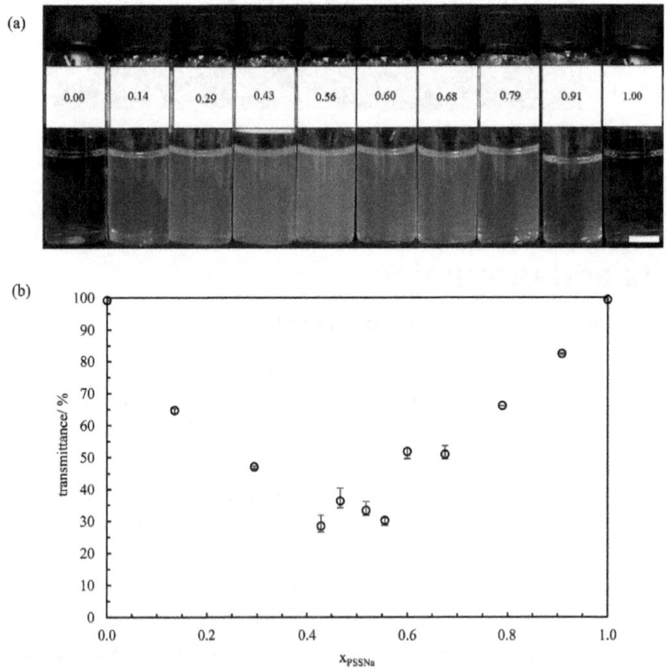

Fig. 5 (a) Appearance of freshly prepared aqueous PEC dispersions prepared from 0.5 g L^{-1} individual PEL solutions at different x_{PSSNa} (given). Scale bar = 1 cm. (b) Transmittance at $\lambda = 400$ nm for the above.

with time when PDADMAC is in excess. It is worth noting that the values for the individual polyelectrolyte solutions were not measurable with dynamic light scattering. Owing to the long-ranged electrostatic repulsion between charged monomers within the chain, polyelectrolytes adopt a more expanded, rigid rod-like conformation.[71,72] However, when PEC particles are formed, the individual polyelectrolytes do not adopt this conformation any more as they interact with each other through electrostatic forces. As shown, particles are spherical of diameter between *ca.* 100 and 200 nm depending on x_{PSSNa}. On the basis of these results, PEC particles obtained when PDADMAC is in excess (low x_{PSSNa}) seem to be slightly larger and, as soon as the mole fraction of the anionic polyelectrolyte increases, the particles become smaller. When PSSNa is in excess ($x_{PSSNa} > 0.5$) the particle diameter remains constant at around 100 nm. These results are in agreement with those obtained by Mende *et al.*[67] with the same polyelectrolyte system (PSSNa-70 kDa and 1000 kDa, PDADMAC-5 kDa and 290 kDa) who found that the particle size does not vary significantly when PSSNa is in excess. The PSSNa–PDADMAC system was also studied by Dautzenberg (PSSNa-66 kDa, PDADMAC-250 kDa).[73] He reported that the sizes of PEC particles seem to

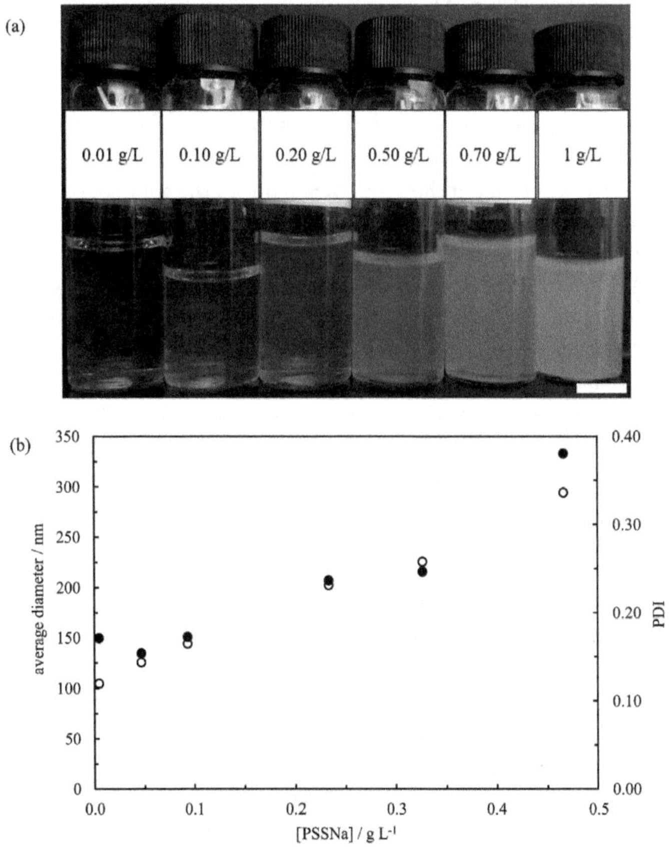

Fig. 6 (a) Appearance of aqueous PEC dispersions ($x_{PSSNa} \approx 0.54$) obtained from the individual PEL solutions of the given concentrations. Scale bar = 1 cm. (b) Average particle diameter (unfilled circles) and PDI (filled circles) *versus* [PSSNa] for the above.

increase as long as the proportion of PDADMAC in the mixture increases, again consistent with our results. Nevertheless, in the literature there is not full agreement about the dependence of the particle size on the mixing ratio. Depending on the polyelectrolytes used and their concentrations, different tendencies have been reported. While in some cases[67,74,75] particles with the smallest size are obtained around the point of charge neutralization and the size tends to increase as soon as one of the components is in excess (displaying a minimum), other authors[76,77] show that particles are smaller when one of the components is in excess and the largest size is obtained at the point where all the charges are compensated.

The formation of more compact structures for $x_{PSSNa} > 0.5$ could be related to the experimental procedure. As described earlier, PDADMAC solution was added in different amounts into PSSNa solutions. Therefore, on the left hand side of the plot, the major component was added to the minor while for $x_{PSSNa} > 0.5$, the minor component was added into the major. According to Müller,[64] for the minor-to-major scenario, more equilibrated smaller PEC particles were achieved as the charge sign is never reversed. Applied to our system, for $x_{PSSNa} > 0.5$, a small volume of PDADMAC solution is dosed into a big volume of PSSNa solution. Consequently, the particles are electrostatically stabilised by the excess like-charged component. However, for the major-to-minor case, immediately after exceeding the critical 1 : 1 stoichiometry the excess oppositely charged component can "cross-link" the secondary particles to form colloidal networks with lower structural density. Similar trends reported by other authors support this hypothesis.[65,66] The corresponding plot of the PDI of particles *versus* x_{PSSNa} is given in Fig. 2(b). Interestingly, the polydispersity in size is least when the proportion of both polyelectrolytes is about the same. When one of the components is in excess, the PDI increases revealing a minimum. This is consistent with the results obtained earlier.[67,73,77]

Fig. 3(a) shows an SEM image of an aqueous PEC dispersion prepared from the 0.1 g L^{-1} individual PEL solutions at $x_{PSSNa} = 0.56$. In general, particles can be considered quasi-spherical despite some aggregation also being detected. The average particle diameter measured from the micrograph of approximately 100 nm is in good agreement with the one acquired by light scattering. The EDX spectra of the particles verified the presence of sulphur (data not shown). The zeta potentials were determined for freshly prepared dispersions and for those 20 days after preparation. As shown in Fig. 4, when PDADMAC is in excess ($x_{PSSNa} < 0.54$), the zeta potential is positive since the positively charged polyelectrolyte in excess surrounds the particles. This value decreases slightly as the PDADMAC proportion decreases. On the other hand, when PSSNa is in excess, the zeta potential is negative and increases slightly in magnitude with increasing x_{PSSNa}. These results are in line with those obtained by other authors.[75,77,78] There is a dramatic change in both the sign and value of the zeta potential at x_{PSSNa} between 0.54 and 0.56. This region refers to complete charge neutralisation and fits with the PDI plot, where the particles are least polydisperse.

The same experimental procedure was applied for aqueous PEC dispersions prepared from the 0.5 g L^{-1} individual PEL solutions at different x_{PSSNa} to corroborate the above tendencies. As depicted in Fig. 5(a), dispersions become noticeably more turbid as soon as x_{PSSNa} approaches 0.56. The transmittance of these dispersions is given in Fig. 5(b). The increase in turbidity around the charge

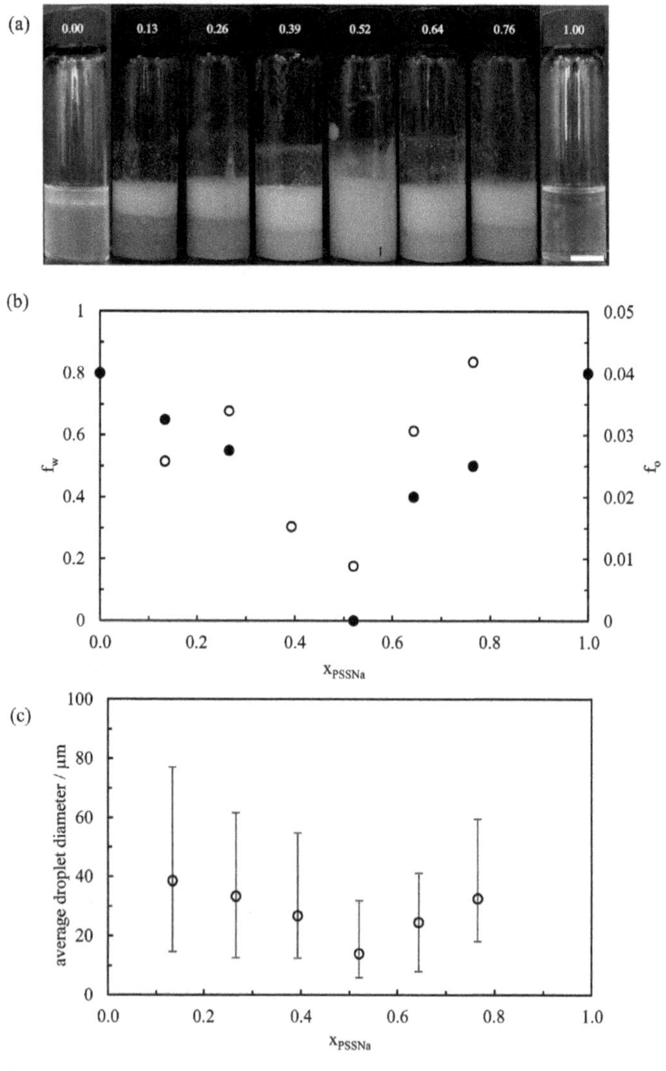

Fig. 7 (a) Appearance of dodecane-in-water emulsions ($\phi_o = 0.2$) stabilised by PEC particles prepared from 20 g L^{-1} individual PEL solutions at different x_{PSSNa} given. Scale bar = 1 cm. (b) Fraction of water (filled points) and fraction of oil (unfilled points) released from emulsions after 1 week *versus* x_{PSSNa}. (c) Average droplet diameter of above emulsions after preparation *versus* x_{PSSNa}.

neutralisation point may be attributed to an increase in the overall particle concentration. The possibility of particle aggregation was discounted from the light scattering measurements as the smallest particles were obtained when both charges were neutralised (Fig. S1(a)†). For this set of PEC dispersions prepared at a higher concentration, profiles for the average particle diameter, PDI in particle size and zeta potential (Fig. S1 and S2†) were entirely consistent with the ones obtained at the low concentration range. The SEM image of the PEC dispersion ($x_{PSSNa} = 0.56$) from the 0.5 g L^{-1} individual PEL solutions is shown in Fig. 3(b). Here, the concentration of particles is seen to be higher than in the previous case.

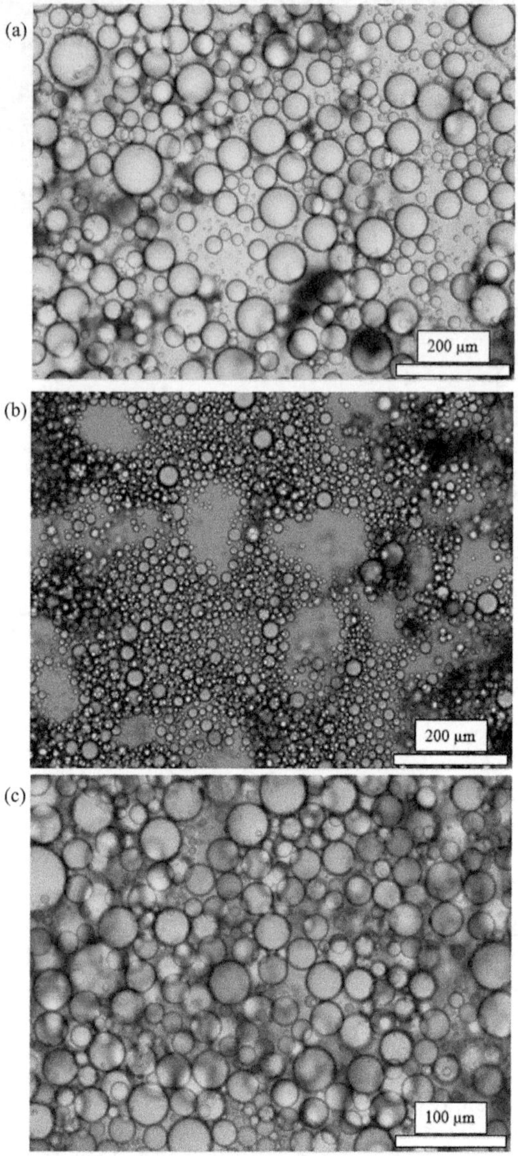

Fig. 8 Optical microscope images of selected emulsions in Fig. 7 at x_{PSSNa} of (a) 0.13, (b) 0.52 and (c) 0.64. The bare patches in (b) correspond to the glass slide.

It is likely however that the film of aggregated particles observed is most likely formed during the slow evaporation process during sample preparation. Finally, to evaluate the influence of the PEL concentration on the structural parameters of the obtained particles, solutions of higher concentrations were prepared at a constant $x_{PSSNa} \approx 0.54$. As shown in Fig. 6(a), the higher the initial PEL concentration, the more turbid the dispersion (Fig. S3(a)†). From the dynamic light scattering measurements, Fig. 6(b), it can be concluded that particles seem to aggregate as both the average diameter and the PDI increase with the polyelectrolyte concentration. This can be gleaned from the results of Starchenko *et al.*[79] for the same polyelectrolyte system. By increasing the PEL concentration, and consequently the concentration of primary particles, it accelerates the aggregation and increases the size of the secondary particles.[79,80] The zeta potential of the particles appears to increase slightly with particle concentration until it reaches a plateau around −38 mV (Fig. S3(b)†).

In order to judge the influence of the polymer molecular weight on the structural parameters of the formed PECs, a set of aqueous PEC dispersions at different values of x_{PSSNa} were prepared from PELs of different molecular weights: PDADMAC (160 kDa) and PSSNa (976 kDa). As shown in Fig. S4,† no remarkable differences in either the size or PDI are observed in comparison to the previous results where both polymers were of comparable M_w, although the value of x_{PSSNa} at neutralisation is considerably lower as expected. The average diameter for the PEC particles with $x_{PSSNa} > 0.25$ (anionic polyelectrolyte in excess) is approximately 25 nm larger than that for particles prepared from PEL of similar M_w. Generally, small complexes are formed by small polyelectrolytes, while larger

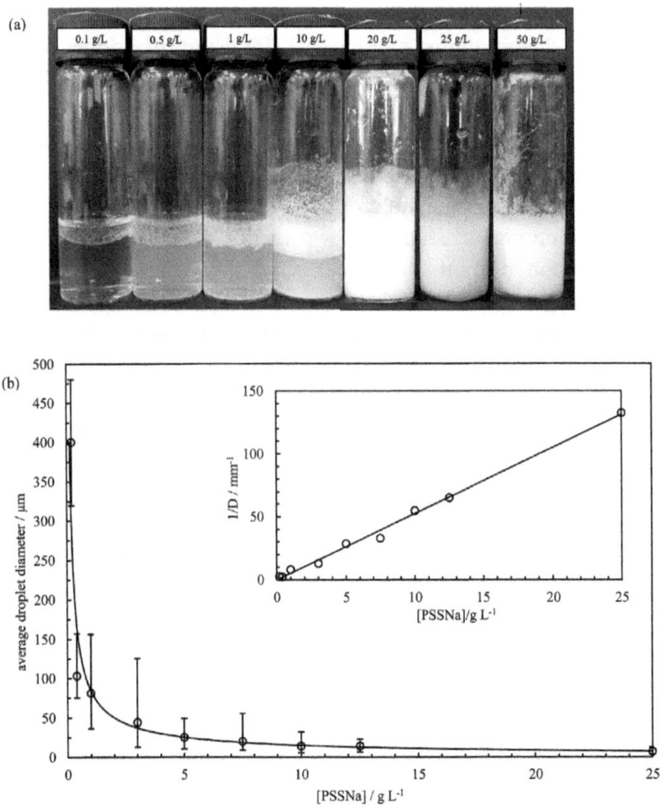

Fig. 9 (a) Appearance of dodecane-in-water emulsions ($\phi_o = 0.2$) stabilised by PEC particles ($x_{PSSNa} \approx 0.50$) prepared from the individual PEL solutions of the given concentrations immediately after preparation. (b) Average droplet diameter *versus* initial [PSSNa] in water for the above emulsions. Inset – inverse of average droplet diameter as a function of initial [PSSNa] in water.

polyelectrolytes give larger complexes.[65,66,81] Ankerfors *et al.*[82] related this tendency to the diffusion-controlled formation of "pre-complexes". While low M_w polymers form stable complexes quickly, larger polyelectrolytes are more prone to aggregation due to longer diffusion times. However, the results presented in this study do not follow the above general tendency, also reported in ref. 80 and 83 for the same pair of polyelectrolytes. Dautzenberg[83] attributed this surprising result to the kinetics of the process of PEC formation that prevails and suppresses the effect of the molecular weight on the resulting structures. For the zeta potential data given in Fig. S5(a) ,† slightly higher negative values were achieved for $x_{PSSNa} >$ 0.25 compared to the results from polymer mixtures of the same M_w. This can be attributed to the increase in the amount of negatively charged groups on each PSSNa chain.

Oil–water emulsions prepared from polymer mixtures

(a) **Systems without salt.** Our interest here is whether the PEC particles prepared in water are surface-active enough to adsorb to an oil–water interface created on emulsifying the aqueous phase with a non-polar alkane. We evaluate the effect of three parameters on the emulsion behaviour. Two of them are related to characteristics of the aqueous PEC dispersion (x_{PSSNa} and PEL concentration) whilst the other is the oil volume fraction (ϕ_o). Emulsions of *n*-dodecane ($\phi_o = 0.2$) and an aqueous PEC dispersion from the 20 g L^{-1} individual PEL solutions were prepared for different values of x_{PSSNa} from 0 to 1. From Fig. 7(a), it can be seen that oil-in-water (o/w) emulsions prepared with the polyelectrolytes alone (x_{PSSNa} = 0 and 1) are extremely unstable and phase separate completely immediately after preparation; *i.e.* the polymers are not surface-active. This finding contrasts that in the field of emulsions stabilized by protein–polysaccharide complexes, for example, where one or both components stabilize the emulsion alone.[53–63] However, once PECs are present in the aqueous phase, long-term stable o/w emulsions are achieved, implying that PEC particles are the stabilizing emulsifier. These emulsions cream with time with water separating below the cream, apart from the emulsion prepared close to charge neutralization ($x_{PSSNa} = 0.52$) which is stable. The fractions of water (due to creaming) and oil (due to coalescence) released one week after preparation are plotted in Fig. 7(b). In both cases, a minimum is achieved for x_{PSSNa} around 0.5, and the extent of coalescence is extremely small (<0.05) for all the polymer mixtures. The average droplet size of the emulsions immediately after preparation is given in Fig. 7(c) and passes through a shallow minimum value of *ca.* 15 μm for x_{PSSNa} around 0.5, consistent with the inhibition of creaming at this condition. Optical micrographs of some of the emulsions are shown in Fig. 8 where all droplets are spherical. It thus appears that, by reducing the overall charge of PEC particles in mixtures of polymers, their hydrophobicity is increased to such an extent that they prefer to adsorb to the oil–water interface during mixing. This argument was used to explain the stabilization of emulsions at optimum ratios of anionic and cationic solid particles.[42,44]

Depending on the initial emulsifier concentration, two main régimes are distinguished with respect to emulsion formation in particle-stabilised emulsions.[84] When the system is emulsified at low concentration of stabilizer, known as the emulsifier-poor régime, droplets are partially covered with the emulsifier. Therefore, once the agitation is stopped, droplets coalesce to a limited extent. The

(a)
(b)
(c)

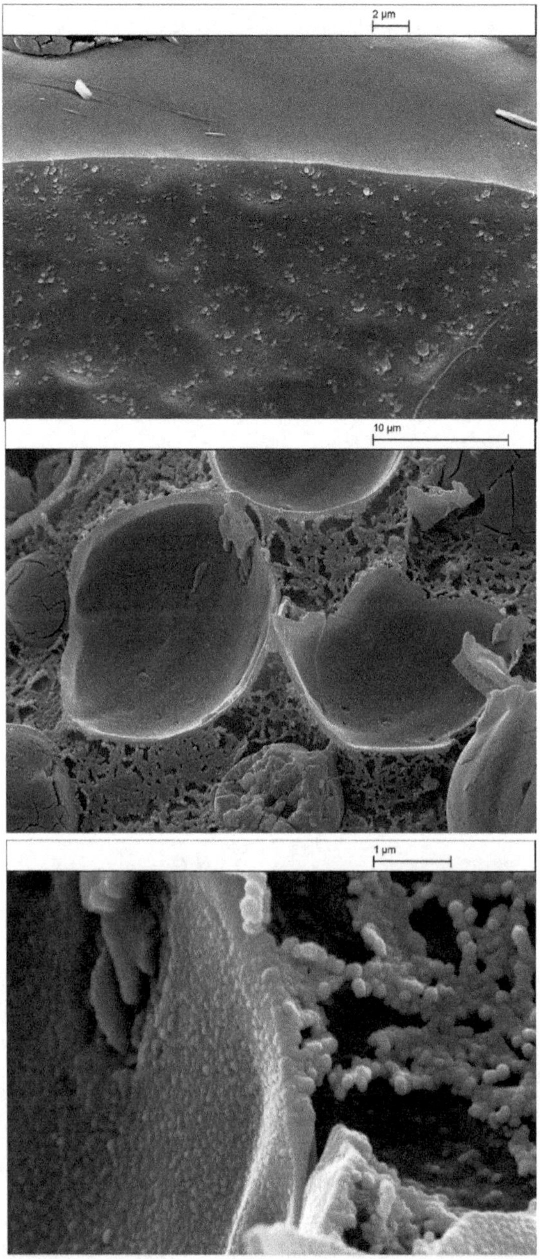

Fig. 10 Cryo-SEM images of selected emulsions in Fig. 9 for (a) 1 g L^{-1} PEL, $x_{PSSNa} = 0.47$ and (b), (c) 25 g L^{-1} PEL, $x_{PSSNa} = 0.52$.

degree of interface coverage by particles increases leading to a reduction of the total interfacial area between oil and water. This prevents further coalescence events.[85] In this régime, the mean drop size decreases with increasing emulsifier concentration. In contrast, at high emulsifier concentrations (emulsifier-rich régime) the average drop size is practically independent of the stabilizer

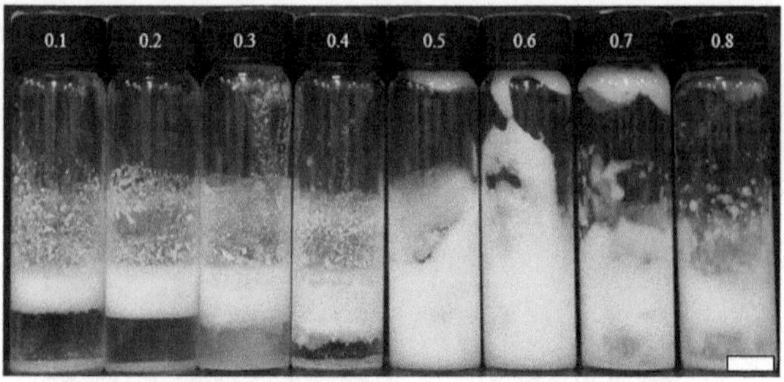

Fig. 11 Appearance of dodecane-in-water emulsions immediately after preparation for different ϕ_o values (given). [PSSNa] in each emulsion is 8 g L^{-1} and x_{PSSNa} is 0.54. Scale bar = 1 cm.

concentration and is mainly governed by the process of drop fragmentation. For emulsions experiencing limited coalescence, the following equation can be written, considering complete adsorption of spherical particles at the interface,[86]

$$\frac{1}{D_{[3,2]}} = \frac{1 - \phi_o}{6\phi_o \tau} c \tag{1}$$

where ϕ_o is the volume fraction of the dispersed phase, c is the initial emulsifier concentration, τ is the stabilizer adsorption density and $D_{[3,2]}$ is the Sauter mean diameter obtained from

$$D_{[3,2]} = \frac{\sum_i N_i D_i^3}{\sum_i N_i D_i^2} \tag{2}$$

(a) (b)

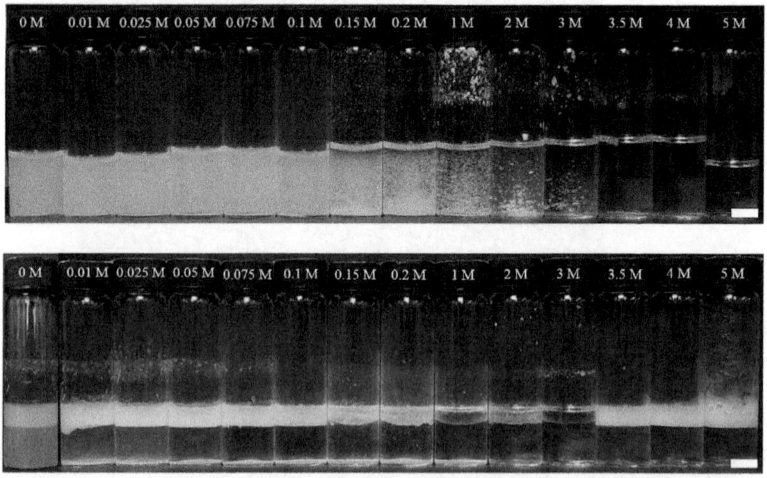

Fig. 12 (a) Appearance of aqueous PEC dispersions prepared from 1 g L^{-1} individual PEL solutions and $x_{PSSNa} = 0.52$ at different concentrations of NaCl (given) after preparation. (b) Appearance of dodecane–water emulsions ($\phi_o = 0.2$) prepared from the dispersions in (a). Scale bar = 1 cm.

where N_i is the total number of droplets with diameter D_i.

By plotting the inverse of the mean drop diameter *versus* the concentration of the emulsifier, a straight line is expected to be obtained. From the slope, the stabilizer adsorption density can be calculated. For that purpose, a series of emulsions of *n*-dodecane-in-water ($\phi_o = 0.2$) were prepared from an aqueous PEC dispersion at $x_{PSSNa} \approx 0.5$ from the individual PEL solutions at different concentrations. The appearance of selected emulsions can be seen in Fig. 9(a). At low concentrations (<0.1 g L^{-1}) no stable emulsions were achieved. However, as soon as the starting PEL concentration increased, emulsions stable to coalescence were formed. For the emulsions prepared with a [PEL] $\geq$ 20 g L^{-1} the emulsions also showed no sign of creaming and their viscosity increased considerably. The droplet sizes of the stable emulsions were determined directly from optical micrographs and are plotted as a function of the polyelectrolyte concentration in the aqueous phase in Fig. 9(b). In the inset, the plot corresponding to eqn (1) is also given. The linear relationship confirms that these emulsions experience limited coalescence in the emulsifier-poor régime. For the explored concentration range, the transition towards the emulsifier-rich régime was not reached and droplets with average diameter as low as 8 μm are formed. From the slope of the graph, the calculated adsorption density by particles is 0.13 g m^{-2}. Without knowing the particle density, it is difficult to compare this value with that expected for close packing of monodisperse particles. An alternative form of eqn (1) given in ref. 85 is based on the number of particles used to estimate the surface coverage. However, this cannot be calculated at present and attempts to determine it by evaporating an aqueous PEC dispersion and imaging it with SEM proved difficult as NaCl crystals masked some of the particles.

Cryo-SEM analysis was attempted with several o/w emulsions around the composition $x_{PSSNa} \approx 0.5$. At a relatively low initial PEL concentration, Fig. 10(a) shows part of a dodecane drop partly covered by discrete PEC particles. The frozen oil constituting the drop has been removed allowing visualization of the interface. The particle diameter estimated from the image (230 nm) is comparable to that measured for particles in water with dynamic light scattering (250 nm). In Fig. 10(b) and (c), emulsion droplets obtained from the PEC solution prepared from 25 g L^{-1} PEL solutions are shown at two magnifications. In this case, the frozen water has been sublimed to better visualize the particles. PEC particles are seen to be densely packed at the oil–water interface and excess particles form a network in the continuous aqueous phase. This dual location of particles explains the excellent stability of emulsions to both coalescence and creaming. A preliminary experiment aimed at establishing the percentage of particles adsorbed yielded the following: starting with 9.6 g L^{-1} PEC in water, the resolved aqueous phase after creaming of the o/w emulsion contained 5.4 g L^{-1}, *i.e.* 56% was non-adsorbed at this relatively high concentration. Closer inspection of Fig. 10(c) and S6† may indicate that more than one particle layer exists at the droplet interface; however it may be that these thicker layers are those responsible for the bridging between neighboring droplets.

In order to establish whether catastrophic phase inversion is achievable in this system, the influence of the oil volume fraction was studied by fixing both the polyelectrolyte concentration in the final emulsion (8 g L^{-1}) and the value of x_{PSSNa} (0.54). The appearance of the emulsions at ϕ_o between 0.1 and 0.8 is shown in Fig. 11. Upon increasing ϕ_o, all emulsions are o/w in which the volume of

(a)

(b)

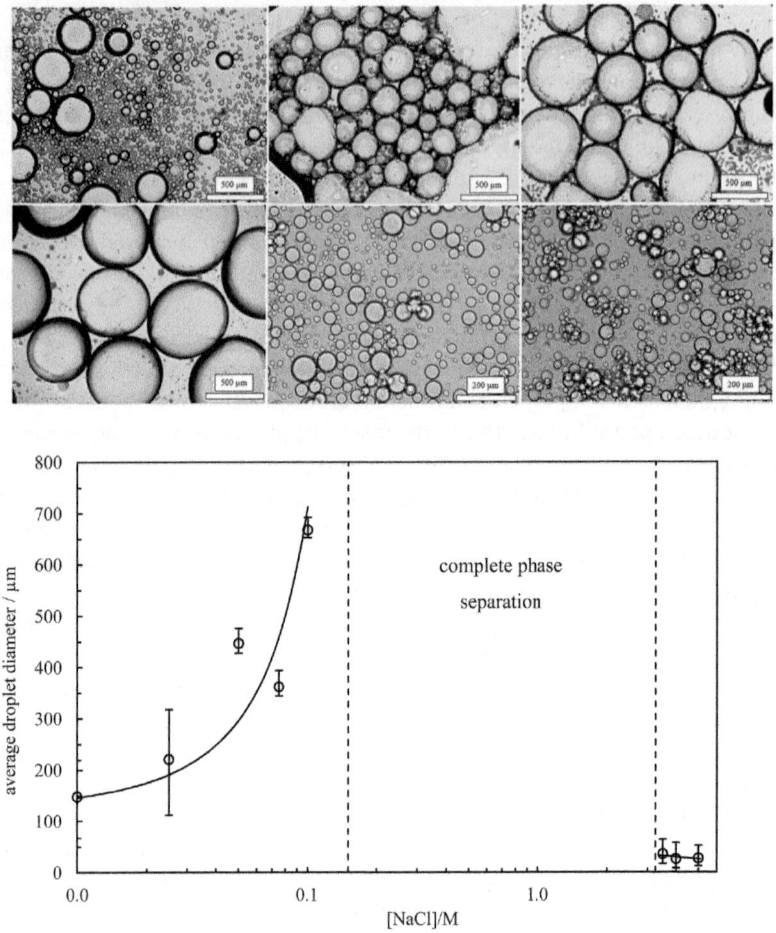

Fig. 13 (a) Optical microscope images of dodecane-in-water emulsions ($\phi_o = 0.2$) stabilised by PEC particles prepared from 1 g L^{-1} individual PEL solutions at different [NaCl] concentrations (from left to right and up to down): 0, 0.025, 0.05, 0.1, 4 and 5 M. (b) Average droplet diameter *versus* [NaCl] for emulsions in (a).

emulsion prepared also increases until all the oil and water mixture becomes emulsified at $\phi_o \geq 0.5$. Emulsions at higher oil fractions were noticeably more viscous but completely stable for at least 6 months. The average droplet diameter determined from optical micrographs decreased from 133 ± 32 μm at $\phi_o = 0.1$ to 10 ± 4 μm at $\phi_o = 0.6$. For this polymer mixture and oil chosen, catastrophic inversion cannot be achieved. The preferred emulsion at $\phi_o = 0.5$ being o/w implies that the PEC particles so formed are only partially hydrophobic.

(b) Effect of salt concentration. One important parameter that influences PEC formation and final structure is the concentration of salt as its ions interact with the charges on the oppositely charged polyelectrolytes. For this reason, its influence on both aqueous PEC dispersions and emulsion stability was evaluated. Salt was added after PEC formation. The appearance of aqueous PEC dispersions prepared at different NaCl concentrations (from 0 to 5 M) from the 1 g L^{-1} PELs solutions at $x_{PSSNa} = 0.52$ is shown in Fig. 12(a). Three different regions can be distinguished. Between 0 and 0.1 M NaCl, turbid stable dispersions are obtained. The transmittance at 700 nm decreased from 39% (no added salt) to 13% (in 0.1 M NaCl). The average diameter of the PEC particles was 294 nm without salt and 873 nm in the presence of 0.1 M NaCl, with a broad size distribution in both cases (PDI > 0.3). Moreover, after allowing the dispersions to stand for a couple of days, precipitation of white PEC aggregates was visible at the bottom of the vessel. Taken together, the evidence suggests that the colloidally stable particles without salt begin to aggregate due to the screening effect of the counterions (Na^+ and Cl^-). The second region, between 0.15 M and 3.0 M NaCl, is where flocs of PEC aggregates of several hundred microns appear. Within this range, the floc concentration decreases with salt concentration. Finally, for aqueous PEC dispersions prepared at [NaCl] $\geq$ 3.5 M, they were completely transparent. However, immediately after salt addition, white flocs were obtained which rapidly dissolved with time. The absence of colloidal particles in these solutions was confirmed through dynamic light scattering. At such high ionic strength conditions, electrostatic interaction between the PELs no longer exists because of the high screening effect of the counterions. Therefore, the PEC particles dissolve liberating the initial soluble chains of both polyelectrolytes which remain unchanged for over 3 months. Our findings are in good agreement with those of a very recent study by Zhang *et al.*[87] for the same pair of polyelectrolytes but of higher polydispersity. Without added salt, neutral small size primary particles surrounded by excess polyelectrolyte were formed (stable PEC). At intermediate salt concentration, small counterions screened the PECs leading to aggregation in large secondary particles that eventually precipitated (unstable PEC). At even higher salt concentrations, dissolution of PECs into the individual chains occurred.

From these dispersions/solutions, emulsions containing *n*-dodecane ($\phi_o = 0.2$) were prepared by the standard procedure. Their appearance immediately after homogenisation is shown in Fig. 12(b). Three different scenarios can be differentiated which correlate with the three regions described above for aqueous PEC dispersions. Emulsions prepared with a salt content between 0 M and 0.1 M were o/w and stable to coalescence. Those prepared at [NaCl] between 0.15 M and 3.0 M exhibited complete phase separation as if no polymer mixture was present. Emulsions prepared at salt concentrations greater than 3.5 M were o/w and stable again. We thus see that emulsions stabilised by PEC particles are stimuli-responsive being destabilised and subsequently re-stabilised by salt addition and their behaviour is closely linked to the properties of the pre-cursor aqueous dispersion/solution. The average droplet diameter was measured from optical micrographs for the stable emulsions. By increasing the salt concentration from 0 M to 0.1 M, the droplet size increased progressively and at very high ionic strengths the droplet size was relatively small as shown in the optical micrographs of Fig. 13(a). We plot the average droplet diameter as a function of NaCl concentration in Fig. 13(b). The increase of the droplet diameter in the first region is related to the aggregation process of particles in water before emulsification. Moreover, the overall PEC concentration decreases due to this aggregation process. In the second region, aggregation levels are exceptionally high. The inability to stabilise any emulsion may be due to a change in the hydrophobicity of the particles following salt addition at this level or to the fact that very large

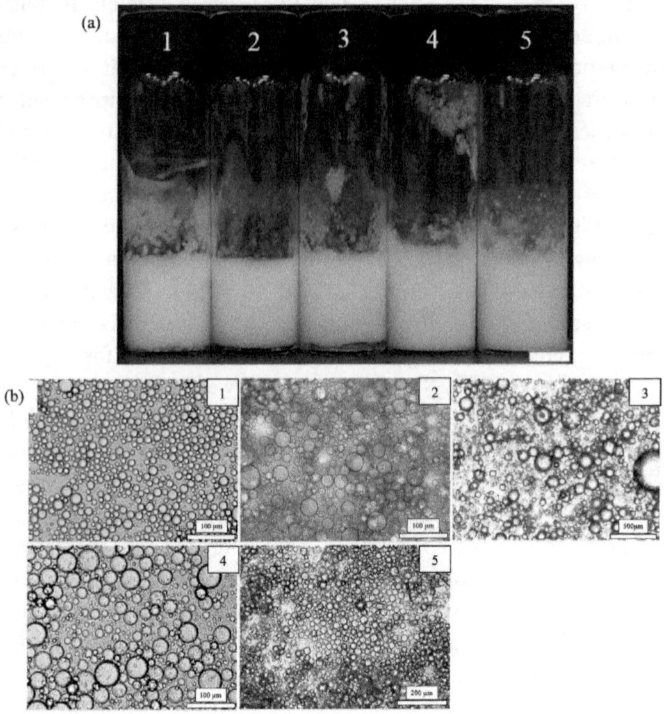

Fig. 14 (a) Appearance of o/w emulsions of an aqueous PEC dispersion (50 g L^{-1} individual PEL solutions) and different oils ($\phi_o = 0.2$) three months after preparation. (1) Squalane, (2) isopropyl myristate, (3) toluene, (4) 50 cS PDMS and (5) liquid paraffin. (b) Corresponding optical micrographs for the above emulsions.

aggregates are easily dislodged from droplet interfaces if they initially adsorb. In the third region, we unexpectedly find that stable emulsions appear again, this time stabilised by individual polyelectrolyte molecules. The average droplet diameter for these emulsions (17 μm) was even lower than the one achieved for the initial emulsion with no added salt stabilised by PEC particles (103 μm). Moreover, during emulsification, a volume of foam was created which was not observed for any of the other emulsions. This suggests that the polyelectrolyte molecules have become surface-active in this high concentration of salt. As a result, we decided to investigate briefly the behaviour of the polyelectrolytes alone dissolved in 5 M NaCl. In Fig. S7(a),† the appearance of 1 g L^{-1} individual PEL solutions in 5 M NaCl after hand-shaking for 30 s is shown for both polymers. A compact foam was only obtained for PSSNa, implying this polyelectrolyte becomes surface active at the air–water interface. Similarly, the emulsification of both polyelectrolyte solutions with dodecane ($\phi_o = 0.2$) was attempted (Fig. S7(b)†). During homogenisation, air bubbles were formed in both cases but were smaller and more compact in the case of PSSNa. For both polymers, a stable o/w emulsion below the foam was formed, confirming their surface activity when alone at the oil–water interface. There are literature reports indicating that polyelectrolytes at high ionic strength behave as neutral polymers due to the neutralisation of the ionized groups,[88,89] but we are unaware of studies demonstrating the stabilisation of emulsions in this case.

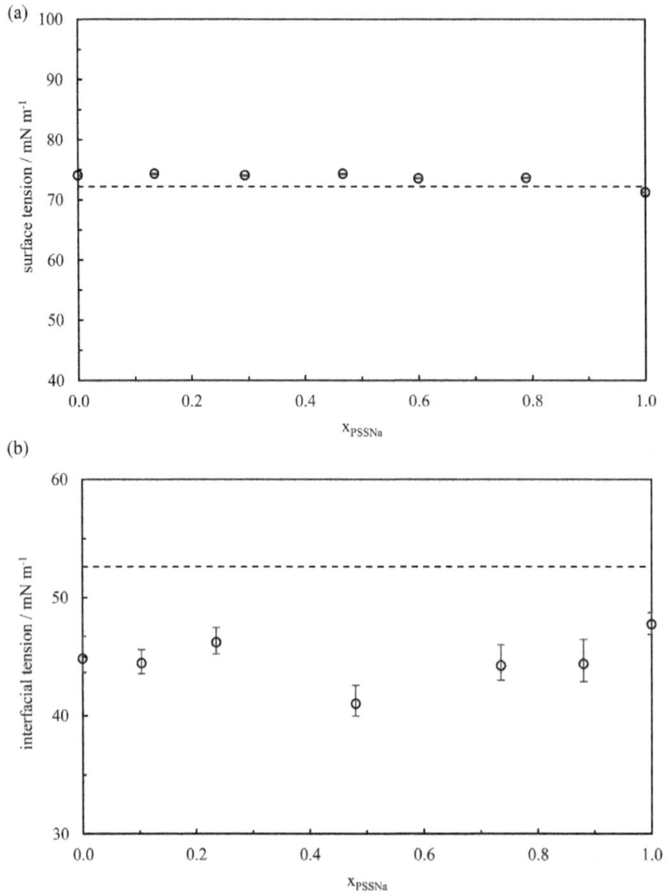

Fig. 15 (a) Air–water surface tension *versus* x_{PSSNa} for PEC aqueous dispersions prepared from 0.5 g L^{-1} individual PEL solutions; dashed line is bare surface. (b) Dodecane–water interfacial tension *versus* x_{PSSNa} for PEC dispersions prepared from 1 g L^{-1} individual PEL solutions; dashed line is bare interface.

The influence of salt on an already prepared o/w emulsion was also evaluated by adding 1 mL of 5 M NaCl solution to 5 mL of emulsion resulting in an overall salt concentration in the emulsion of 1 M. This final concentration was chosen as complete phase separation was obtained when salt was added before emulsification. Surprisingly, in this case no phase separation was achieved and the o/w emulsion was stable, although with a larger droplet size (163 μm) compared with the emulsion prepared with no added electrolyte (103 μm). The effect of salt is thus not as pronounced if added after emulsification. Here, salt is not only added after PEC formation (compared to earlier) but also after their adsorption to the oil–water interface. It is conceivable that salt ions dissolved outside of droplets may not access all parts of interfacially bound particles but the origin of the difference needs to be probed in a future study.

The influence of pH on both aqueous PEC dispersions and emulsions was also evaluated (not shown). Both PSSNa and PDADMAC are strong polyelectrolytes dissociating completely in the entire pH range. The influence of pH would be negligible for each polyelectrolyte alone. However, its influence is relevant to complex formation as the ion concentration increases due to addition of HCl or NaOH used to prepare the acid and basic solutions. The behaviour was evaluated at pH 2 and 12 for aqueous PEC dispersions prepared from 1 g L^{-1} individual PEL solutions. Their appearance was similar to those obtained from the polyelectrolyte mixture at natural pH (Fig. 6(a)). The transmittance decreased from 39.0% for the PEC prepared at natural pH to 26.4% and 32.9% for the PEC dispersions prepared at pH 2 and 12, respectively. Therefore, the same effect as the one obtained after salt addition was noticed. Emulsions prepared from these dispersions and dodecane ($\phi_o = 0.2$) either completely phase separated (pH = 2) or contained much larger droplets (660 μm, pH = 12) compared with that prepared at natural pH (103 μm). More work is required to understand the origin of these effects.

(c) Variation of oil type. The study until now has been with a non-polar alkane which has no groups to modify the surface properties of PEC particles. In order to establish how generic this novel emulsion stabilisation mechanism is, we prepared emulsions from aqueous PEC dispersions from 50 g L^{-1} individual PEL solutions and different oils. Their appearance and optical microscope images are given in Fig. 14 for a long chain alkane, an ester, an aromatic oil, a silicone oil and a commercial paraffin. In all cases, o/w emulsions stable to coalescence were obtained. The average drop diameter was approx. 20 μm for all oils except toluene for which it was ≈60 μm. We also verified that emulsions with the different oils containing each polyelectrolyte alone (25 g L^{-1}) exhibited complete phase separation after mixing, confirming that the PELs are not surface-active alone.

Behaviour at air–water and oil–water planar interfaces

Finally, in the light of emulsion stabilisation by PEC particles above, we have investigated their surface activity at planar air–water and oil–water interfaces *via* interfacial tension measurements. Aqueous PEC dispersions varying in x_{PSSNa} from 0 to 1 obtained from 0.5 g L^{-1} PEL solutions were prepared. As shown in Fig. 15(a), the air–water surface tension of all the dispersions and solutions, including the ones in which the polyelectrolytes alone are present, is 72.0 ± 0.4 mN m^{-1} representing no reduction compared to that for pure water. The surface tension for an aqueous PEC dispersion at higher concentration (1 g L^{-1}, x_{PSSNa} = 0.53) was also 72.0 mN m^{-1}. The PELs and PECs do not therefore lower the tension of the air–water surface, as shown earlier for sodium polyacrylate.[90] Likewise, the dodecane–water interfacial tension was measured for aqueous PEC dispersions prepared from 1 g L^{-1} PEL solutions at different x_{PSSNa}, Fig. 15(b). In the absence of polymer, the interfacial tension of 52.6 ± 0.2 mN m^{-1} is in agreement with the value reported in ref. 91 of 52.5 mN m^{-1} confirming the absence of surface-active impurities. Here however, all polymer-containing dispersions/solutions exhibit a lower tension than the bare interface and the tension passes through a shallow minimum at intermediate values of x_{PSSNa}. The maximum lowering however is only *ca.* 10 mN m^{-1} compared with low molar mass surfactant systems which typically lower interfacial tensions to *ca.* 1–5 mN m^{-1} above their critical micelle concentration.[92] As with emulsions stabilised by hard particles, the reduction of the interfacial tension is not the operative mechanism. We are not in a position however to say what level of tension reduction is required for the classical picture of emulsion formation as observed with surfactants. Further, these may not be equilibrium tensions as an energy barrier to adsorption of PEC particles may not be surmounted in the absence of stirring. The findings are in line with those of several authors interested in adsorption of particles from bulk water. Drelich *et al.*[93] found that the paraffin oil–water tension in the presence of hydrophobic fumed silica particles was not significantly different from the value obtained without particles. Similar results were obtained by Vignati *et al.*[94] with silanized silica particles at isooctane or octanol–water interfaces. Even though particles strongly adhered to the interface of emulsion drops, no reduction of the interfacial tension was detected. Aveyard *et al.*[95] found that polystyrene latex particles spontaneously adhere to the octane–water interface, lowering its tension by approximately 4 mN m^{-1}. Importantly, only when *spread* monolayers of particles are compressed and particles in close proximity exhibit mutual repulsion, does the interfacial tension reach very low values (<2 mN m^{-1}). Thus, the modest lowering mentioned above for *adsorbed* films does not imply that PEC particles are surface-inactive at the oil–water interface.

Conclusions

We demonstrate the use of polyelectrolyte complexes as novel oil–water emulsion stabilisers using relatively monodisperse polyelectrolytes of PSSNa and PDADMAC of similar molecular mass. We confirm that the anionic and cationic polyelectrolytes alone are not emulsifiers as complete phase separation occurred immediately after mixing. The polyelectrolyte complex formed in mixtures through electrostatic interactions yields particles of between 100 and 150 nm in diameter depending on the ratio of polymers. Oil-in-water emulsions for a range of oils at different oil : water ratios can be prepared exhibiting exceptional resistance to coalescence. The most stable emulsion to both creaming and coalescence is formed at around equal mole fraction of the two polymers, and it possesses the smallest drop size. The average drop size could be tuned depending on the initial polyelectrolyte concentrations. Using cryo-SEM, we observe that particles are located at the oil–water interface of emulsion drops. Their distribution is not uniform at low polyelectrolyte concentrations, whereas at high concentrations a close-packed layer of particles covers the interfaces and excess particles aggregate in water enhancing emulsion stability. The addition of increasing concentrations of salt to aqueous particle dispersions causes a transition from stable dispersions to aggregated and unstable dispersions and finally to dissolution of the particles of complex yielding solutions of the individual polymer molecules. The corresponding emulsions are initially destabilized completely at intermediate salt concentrations but, at high salt concentrations, emulsions are re-stabilised in this case by adsorbed polymer molecules.

We suggest a number of strategies as future directions in this emerging area of soft matter science: (i) *influence of type of electrolyte* – it has been shown that PEC formation in water is influenced by the radius and valence of the ions in added salt,[96] but nothing is known in emulsions stabilized by such complexes; (ii) *variation of polymer charge density* – what changes occur in mixing a strong and a weak polyelectrolyte instead of two strong polyelectrolytes as reported here?

Detailed studies on this are now under way in our laboratory; (iii) *investigate the morphology of adsorbed PEC particles* – using novel electron microscope techniques, can we ascertain whether PEC particles adsorb to droplet interfaces as spherical entities or deform once adsorbed akin to certain microgel particles?[41]; (iv) *increase the hydrophobicity of PEC particles* – by careful selection of the precursor polymers, can we design PEC particles of relatively high hydrophobicity and hence explore their potential as stabilizers of water-in-oil emulsions?; (v) *simulations* – PEC formation in water has been widely studied using Monte Carlo simulations,[97–99] including the effects of adding salt.[100] Can these simulations be extended to the behaviour of PEC particles at fluid interfaces?

Acknowledgements

We are grateful to Shiseido (Japan) for a fully funded PhD studentship award to AMBR and to Mr A. Sinclair (University of Hull) for the SEM images and EDX spectra.

References

1 *Modern Aspects of Emulsion Science*, ed. B. P. Binks, Royal Society of Chemistry, Cambridge, 1998, pp. 1–48.
2 I. D. Morrison and S. Ross, *Colloidal Dispersions. Suspensions, Emulsions, and Foams*, Wiley, New York, 2002.
3 P. Becher, *Encyclopedia of Emulsion Technology*, Marcel Dekker, New York, 1996, vol. 4, pp. 1–37.
4 C. Wollenweber, A. V. Makievski, R. Miller and R. Daniels, *Colloids Surf., A*, 2000, **172**, 91–101.
5 I. A. Nnanna and L. J. Xia, Protein-based Surfactants. Synthesis, Physicochemical Properties and Applications, *Surfactant Science Series*, Marcel Dekker, New York, 2011, vol. 101.
6 R. S. H. Lam and M. T. Nickerson, *Food Chem.*, 2013, **141**, 975–984.
7 C. J. Beverung, C. J. Radke and H. W. Blanch, *Biophys. Chem.*, 1999, **81**, 59–80.
8 M. Hu, D. J. McClements and E. A. Decker, *J. Agric. Food Chem.*, 2003, **51**, 1696–1700.
9 D. M. Mulvihill and P. C. Murphy, *Int. Dairy J.*, 1991, **1**, 13–37.
10 D. G. Dalgleish, *Colloids Surf., B*, 1993, **1**, 1–8.
11 J. Li, Y. Cheng, P. Wang, W. Zhao, L. Yin and M. Saito, *Food Hydrocolloids*, 2012, **26**, 448–455.
12 A. Ye and H. Singh, *Food Hydrocolloids*, 2006, **20**, 269–276.
13 Y. Mine, T. Noutomi and N. Haga, *J. Agric. Food Chem.*, 1991, **39**, 443–446.
14 C. Castelain and C. Genot, *J. Agric. Food Chem.*, 1996, **44**, 1635–1640.
15 N. Garti, *J. Dispersion Sci. Technol.*, 1999, **20**, 327–355.
16 N. Garti and M. E. Leser, *Polym. Adv. Technol.*, 2001, **12**, 123–135.
17 E. Dickinson, in *Gums and Stabilisers for the Food Industry*, ed. G. O. Phillips and P. A. Williams, Royal Society of Chemistry, 1988, vol. 4, pp. 249–263.
18 S. U. Pickering, *J. Chem. Soc. Trans.*, 1907, **91**, 2001–2021.
19 B. P. Binks and S. O. Lumsdon, *Langmuir*, 2000, **16**, 8622–8631.
20 P. Finkle, H. D. Draper and J. H. Hildebrand, *J. Am. Chem. Soc.*, 1923, **45**, 2780–2788.

21 J. H. Schulman and J. Leja, *Trans. Faraday Soc.*, 1954, **50**, 598–605.

22 P. S. Clegg, E. M. Herzig, A. B. Schofield, T. S. Horozov, B. P. Binks, M. E. Cates and W. C. K. Poon, *J. Phys.: Condens. Matter*, 2005, **17**, S3433–S3438.

23 N. Yan, M. R. Gray and J. H. Masliyah, *Colloids Surf., A*, 2001, **193**, 97–107.

24 B. Braisch, K. Köhler, H. P. Schuchmann and B. Wolf, *Chem. Eng. Technol.*, 2009, **32**, 1107–1112.

25 J. Zhou, X. Qiao, B. P. Binks, K. Sun, M. Bai, Y. Li and Y. Liu, *Langmuir*, 2011, **27**, 3308–3316.

26 H. Duan, D. Wang, N. S. Sobal, M. Giersig, D. G. Kurth and H. Möhwald, *Nano Lett.*, 2005, **5**, 949–952.

27 J. Kim, L. J. Cote, F. Kim, W. Yuan, K. R. Shull and J. Huang, *J. Am. Chem. Soc.*, 2010, **132**, 8180–8186.

28 H. Katepalli, V. T. John and A. Bose, *Langmuir*, 2013, **29**, 6790–6797.

29 S. Abend, N. Bonnke, U. Gutschner and G. Lagaly, *Colloid Polym. Sci.*, 1998, **276**, 730–737.

30 B. P. Binks, J. H. Clint and C. P. Whitby, *Langmuir*, 2005, **21**, 5307–5316.

31 J. Dong, A. J. Worthen, L. M. Foster, Y. Chen, K. A. Cornell, S. L. Bryant, T. M. Truskett, C. W. Bielawski and K. P. Johnston, *ACS Appl. Mater. Interfaces*, 2014, **6**, 11502–11513.

32 A. Fernandez-Nieves, H. Wyssm, J. Mattsson and D. A. Weitz, *Microgel Suspensions: Fundamentals and Applications*, Wiley-VCH, New York, 2011, pp. 3–4.

33 R. W. Style, L. Isa and E. R. Dufresne, *Soft Matter*, 2015, **11**, 7412–7419.

34 W. Richtering, *Langmuir*, 2012, **28**, 17218–17229.

35 B. Brugger, B. A. Rosen and W. Richtering, *Langmuir*, 2008, **24**, 12202–12208.

36 B. Brugger and W. Richtering, *Langmuir*, 2008, **24**, 7769–7777.

37 B. Brugger, S. Rütten, K. Phan, M. Möller and W. Richtering, *Angew. Chem., Int. Ed.*, 2009, **48**, 3978–3981.

38 M. Destribats, M. Rouvet, C. Gehin-Delval, C. Schmitt and B. P. Binks, *Soft Matter*, 2014, **10**, 6941–6954.

39 M. Destribats, M. Wolfs, F. Pinaud, V. Lapeyre, E. Sellier, V. Schmitt and V. Ravaine, *Langmuir*, 2013, **29**, 12367–12374.

40 M. Destribats, V. Lapere, E. Sellier, F. Leal-Calderon, V. Schmitt and V. Ravaine, *Langmuir*, 2011, **27**, 14096–14107.

41 M. Destribats, M. Eyharts, V. Lapeyre, E. Sellier, I. Varga, V. Ravaine and V. Schmitt, *Langmuir*, 2014, **30**, 1768–1777.

42 B. P. Binks, W. Liu and J. A. Rodrigues, *Langmuir*, 2008, **24**, 4443–4446.

43 T. Nallamilli, E. Mani and M. G. Basavaraj, *Langmuir*, 2014, **30**, 9336–9345.

44 T. Nallamilli, B. P. Binks, E. Mani and M. G. Basavaraj, *Langmuir*, 2015, **31**, 11200–11208.

45 H. Dautzenberg and B. Philipp, *Polyelectrolytes: Formation, Application and Characterisation*, Hanser Publishers, New York, 1994, pp. 1–45.

46 J. Koetz and S. Kosmella, *Polyelectrolytes and Nanoparticles*, 2007, Springer-Verlag, Berlin, pp. 1–4, 36–45.

47 A. V. Il'ina and V. P. Varlamov, *Appl. Biochem. Microbiol.*, 2005, **41**, 5–11.

48 H. V. Sæther, H. K. Holme, G. Maurstad, O. Smidsrød and B. T. Stokke, *Carbohydr. Polym.*, 2008, **74**, 813–821.

49 P. Coimbra, P. Ferreira, P. Alves and M. H. Gil, in *Carbohydrates Applications in Medicine*, ed. M. H. Gil, Research Signpost, Kerala, 2014, pp.1–29.

50 M. W. Beijerinck, *Zentbl. Bakteriolog. P.*, 1896, **2**, 697–699.

51 A. K. Ghosh and P. Bandyopadhyay, in *The Complex World of Polysaccharides*, ed. D. N. Karunaratne, InTech, 2012, pp. 395–408.

52 A. Benichou, A. Aserin and N. Garti, *J. Dispersion Sci. Technol.*, 2002, **23**, 93–123.

53 R. Pongsawatmanit, T. Harnsilawat and D. J. McClements, *Colloids Surf., A*, 2006, **287**, 59–67.

54 D. Guzey and D. J. McClements, *J. Agric. Food Chem.*, 2007, **55**, 475–485.

55 L. Jourdain, M. E. Leser, C. Schmitt, M. Michel and E. Dickinson, *Food Hydrocolloids*, 2008, **22**, 647–659.

56 R. Lutz, A. Aserina, L. Wicker and N. Garti, *Colloids Surf., B*, 2009, **72**, 121–127.

57 S. Padala, P. A. Williams and G. O. Phillips, *J. Agric. Food Chem.*, 2009, **57**, 4964–4973.

58 M. Klein, A. Aserin, I. Svitov and N. Garti, *Colloids Surf., B*, 2010, **77**, 75–81.

59 Y. Li, M. Hub, H. Xiao, Y. Dua, E. A. Decker and D. J. McClements, *Eur. J. Pharm. Biopharm.*, 2010, **76**, 38–47.

60 A. K. Stone and M. T. Nickerson, *Food Hydrocolloids*, 2012, **27**, 271–277.

61 K. G. Zinoviadou, E. Scholten, T. Moschakis and C. G. Biliaderis, *Int. Dairy J.*, 2012, **26**, 94–101.

62 B. Yin, W. Deng, K. Xu, L. Huang and P. Yao, *J. Colloid Interface Sci.*, 2012, **380**, 51–59.

63 F. A. Perrechil and R. L. Cunha, *Food Hydrocolloids*, 2013, **30**, 606–613.

64 M. Müller, *Adv. Polym. Sci.*, 2014, **256**, 197–260.

65 C. Schatz, A. Domard, C. Viton, C. Pichot and T. Delair, *Biomacromolecules*, 2004, **5**, 1882–1892.

66 C. Schatz, J. M. Lucas, C. Viton, A. Domard, C. Pichot and T. Delair, *Langmuir*, 2004, **20**, 7766–7778.

67 M. Mende, H. Buchhammer, S. Schwarz, G. Petzold and W. Jaeger, *Macromol. Symp.*, 2004, **211**, 121–133.

68 D. E. Koppel, *J. Chem. Phys.*, 1972, **57**, 4814–4820.

69 M. Smoluchowski, *Handbuch der Electrizität und des Magnetismus*, ed. L. Graetz and F. Auerbach, J. A. Barth, Germany, 1921, vol. 2, p. 366.

70 W. D. Harkins and H. F. Jordan, *J. Am. Chem. Soc.*, 1930, **52**, 1751–1772.

71 S. Förster, M. Schmidt and M. Antonietti, *Polymer*, 1990, **31**, 781–792.

72 E. Josef and H. Bianco-Peled, *Soft Matter*, 2012, **8**, 9156–9165.

73 H. Dautzenberg, *Macromolecules*, 1997, **30**, 7810–7815.

74 M. Sotiropoulou, C. Cincu, G. Bokias and G. Staikos, *Polymer*, 2004, **45**, 1563–1568.

75 A. Matralis, M. Sotiropoulou, G. Bokias and G. Staikos, *Macromol. Chem. Phys.*, 2006, **207**, 1018–1025.

76 L. Gärdlund, L. Wågberg and M. Norgren, *J. Colloid Interface Sci.*, 2007, **312**, 237–246.

77 M. Mende, G. Petzold and H. Buchhammer, *Colloid Polym. Sci.*, 2002, **280**, 342–351.

78 D. Le Cerf, A. S. Pepin, P. M. Niang, M. Cristea, C. Karakasyan-Dia and L. Picton, *Carbohydr. Polym.*, 2014, **113**, 217–224.

79 V. Starchenko, M. Müller and N. Lebovka, *J. Phys. Chem. B*, 2012, **116**, 14961–14967.

80 V. Starchenko, M. Müller and N. Lebovka, *J. Phys. Chem. C*, 2008, **112**, 8863–8869.
81 H. V. Sæther, H. Holme, G. Maurstad, O. Smidsrød and B. Stokke, *Carbohydr. Polym.*, 2008, **74**, 813–821.
82 C. Ankerfors, S. Ondaral, L. Wågberg and L. Ödberg, *J. Colloid Interface Sci.*, 2010, **351**, 88–95.
83 H. Dautzenberg, in *Physical Chemistry of Polyelectrolytes, Surfactant Science Series*, ed. T. Radeva, Marcel Dekker, New York, 2001, vol. 99, pp. 743–792.
84 S. Tcholakova, N. D. Denkov and A. Lips, *Phys. Chem. Chem. Phys.*, 2008, **10**, 1608–1627.
85 S. Arditty, C. P. Whitby, B. P. Binks, V. Schmitt and F. Leal-Calderon, *Eur. Phys. J. E*, 2003, **11**, 273–281.
86 M. Destribats, M. Rouvet, C. Gehin-Delval, C. Schmitt and B. P. Binks, *Soft Matter*, 2014, **10**, 6941–6954.
87 Y. Zhang, E. Yildirim, H. S. Antila, L. D. Valenzuela, M. Sammalkorpi and J. L. Lutkenhaus, *Soft Matter*, 2015, **11**, 7392–7401.
88 P. Walstra, *Physical Chemistry of Foods*, Marcel Dekker, New York, 2001.
89 S. Dumitiu, *Polysaccharides. Structural Diversity and Functional Versatility*, Marcel Dekker, New York, 2nd edn, 2004.
90 T. Okubo, *J. Colloid Interface Sci.*, 1988, **125**, 386–398.
91 S. Zeppieri, J. Rodríguez and A. L. López de Ramos, *J. Chem. Eng. Data*, 2001, **46**, 1086–1088.
92 R. Aveyard, B. P. Binks and J. Mead, *J. Chem. Soc., Faraday Trans. 1*, 1987, **83**, 2347–2357.
93 A. Drelich, F. Gomez, D. Clausse and I. Pezron, *Colloids Surf., A*, 2010, **365**, 171–177.
94 E. Vignati, R. Piazza and T. P. Lockhart, *Langmuir*, 2003, **19**, 6650–6656.
95 R. Aveyard, J. H. Clint, D. Nees and N. Quirke, *Langmuir*, 2000, **16**, 8820–8828.
96 S. Dragan and M. Cristea, *Eur. Polym. J.*, 2001, **37**, 1571–1575.
97 C. F. Narambuena, E. P. M. Leiva, M. Chávez-Páez and E. Pérez, *Polymer*, 2010, **51**, 3293–3302.
98 V. Starchenko, M. Müller and N. Lebovka, *J. Phys. Chem. C*, 2008, **112**, 8863–8869.
99 J. Rydén, M. Ullner and P. Linse, *J. Chem. Phys.*, 2005, **123**, 034909.
100 Y. Hayashi, M. Ullner and P. Linse, *J. Phys. Chem. B*, 2003, **107**, 8198–8207.

Faraday Discussions

PAPER

Numerical analysis of Pickering emulsion stability: insights from ABMD simulations†

François Sicard* and Alberto Striolo

Received 16th March 2016, Accepted 29th March 2016
DOI: 10.1039/c6fd00055j

The issue of the stability of Pickering emulsions is tackled at a mesoscopic level using dissipative particle dynamics simulations within the Adiabatic Biased Molecular Dynamics framework. We consider the early stage of the coalescence process between two spherical water droplets in a decane solvent. The droplets are stabilized by Janus nanoparticles of different shapes (spherical and ellipsoidal) with different three-phase contact angles. Given a sufficiently dense layer of particles on the droplets, we show that the stabilization mechanism strongly depends on the collision speed. This is consistent with a coalescence mechanism governed by the rheology of the interfacial region. When the system is forced to coalesce *sufficiently slowly*, we investigate at a mesoscopic level how the ability of the nanoparticles to stabilize Pickering emulsions is discriminated by nanoparticle mobility and the associated caging effect. These properties are both related to the interparticle interaction and the hydrodynamic resistance in the liquid film between the approaching interfaces.

1 Introduction

Pickering emulsions,[1] *i.e.* particle-stabilized emulsions, have been used in various applications including biofuel processing,[2] encapsulation for drug delivery,[3] and food preservation.[4,5] When nanoparticles (NPs) are used as emulsifiers, their characteristics might have critical effects on properties such as interfacial tension,[6] droplet size, and emulsion stability.[7–13] Although the factors affecting Pickering emulsion stability are still a key topic of research, the density of NPs on the droplet surface is known to be among the most important parameters.[14] Indeed, the interfacial tension of the droplet depends strongly on the NP coverage,[14] which also depends on the NP affinity to the interface, *i.e.*, the NP desorption energy. When the NP density on the droplet is large enough, the three-phase contact angle discriminates between NPs that are effective at stabilizing the emulsion and those that are not.[15,16] It is known that particles, once adsorbed, are rather difficult to displace from the interface. It is this property that makes

Department of Chemical Engineering, University College London, Torrington Place, London WC1E 7JE, UK. E-mail: francois.sicard@free.fr

† Electronic supplementary information (ESI) available. See DOI: 10.1039/c6fd00055j

particles such good colloidal stabilizers for emulsions and foams. Besides the structure of NP monolayers, the NP diffusion is also believed to affect the system stability. Janus nanoparticles are believed to be more effective than homogeneous NPs in the stabilization of emulsions.[15] Indeed, Janus NPs typically have high adsorption energies, and therefore are expected to pack densely at an interface.

In our present work, we study the early stage of the coalescence process between two spherical water droplets in a decane solvent. The droplets are stabilized with various Janus NP types. We implement Adiabatic Biased Molecular Dynamics (ABMD) simulations,[17–19] where the system is forced to coalesce *adiabatically*. We show that, for a sufficiently dense layer of particles adsorbed on water droplets, the stabilization mechanism can strongly depend on the collision speed achieved during the merging process. For collision velocities sufficiently slow compared to the rate of drainage of the thin liquid film trapped between the coalescing emulsion droplets, the ability of the NPs to stabilize Pickering emulsions is dictated by NP mobility and the associated caging effect. This coalescence process is compared with simulations at sufficiently high collision speed, which suggest the droplets' faculty to undergo large elastic deformation before coalescing is the stabilization mechanism.

2 Methods and algorithms

2.1 MD simulation

The Dissipative Particle Dynamics (DPD) simulation method[20] was implemented within the simulation package LAMMPS.[21] The procedure and the parameterisation details are fully described in prior work.[14,22] The system simulated here is composed of water, oil (decane), and NPs. One "water bead" (w) represents 5 water molecules. One decane molecule is modelled as two "oil beads" (o) connected by one harmonic spring of length $0.72R_c$ and spring constant $350k_BT/R_c$,[23] where R_c is the DPD cutoff distance. The size of the simulation box is $L_x \times L_y \times L_z \equiv 72 \times 72 \times 140R_c^3$, where L_i is the box length along the ith direction. Periodic boundary conditions are applied in all three directions. The NPs are modelled as hollow rigid ellipsoids and contain polar (p) and nonpolar (ap) DPD beads on their surface. One DPD bead was placed at the NP center for convenience, as described elsewhere.[14,22] Hollow models have been used in the literature to simulate NPs, and hollow NPs can also be synthesized experimentally.[24] We considered spherical and ellipsoidal NPs with shapes defined by the equation $x^2/b^2 + y^2/b^2 + z^2/c^2 = 1$, where x, y, and z are Cartesian coordinates and c and b are the semi-axes of the ellipsoid. When $b = c$, spherical NPs are obtained. When $c < b$, the ellipsoid is oblate; when $b < c$, it is prolate. All NPs simulated had the same volume, $4/3\pi a_0^3$, where a_0 is the radius of the equivalent sphere. We imposed $a_0 = 2R_c \approx 1.5$ nm. All types of beads in our simulations have a reduced mass of 1. The total number of beads on one NP surface changes with the aspect ratio c/b. This allows us to keep the surface bead density sufficiently high to prevent other DPD beads (either decane or water) from penetrating the NPs (which would be unphysical), as already explained elsewhere.[14] To differentiate every NP, we report the nonpolar fraction N of the NP surface beads. For example, $N = 70$ indicates that 70% of the beads on the NP surface are nonpolar. The interaction parameters shown in Table 1 are used herein. These parameters were adjusted to reproduce selected atomistic simulation results, as explained in prior work.[22] By tuning the interaction parameters between polar or nonpolar NP beads and the water and decane beads present in our system, it is possible to quantify the effect of surface chemistry on the structure and dynamics of NPs at water–oil interfaces. All simulations were carried out in the NVE ensemble.[22] The scaled temperature was 1, equivalent to ambient temperature. The DPD time scale can be gauged by matching the self-diffusion of water. As demonstrated by Groot and Rabone,[23] the time constant of the simulation can be calculated as $\tau = \frac{N_m D_{sim} R_c^2}{D_{water}}$, where τ is the DPD time constant, N_m is the number of water molecules incorporated in a simulation bead, D_{sim} is the simulated water self-diffusion coefficient, and D_{water} is the experimental water self-diffusion coefficient. When the water–water interaction parameter $a_{w\text{-}w} = 131.5 k_B T/R_c$ (*cf.* Table 1), we obtained $D_{sim} = 0.0063 R_c^2/\tau$. Since for experimental water $D_{water} = 2.43 \times 10^{-3}$ cm^2 s^{-1},[25] we finally obtain $\tau = 7.6$ ps. In the simulations discussed herein, the size of the droplet was fixed. At the beginning of each simulation, the solvent (oil) beads were distributed within the simulation box forming a cubic lattice. Two water droplets of radius $\approx 20R_c$ were generated by replacing the oil beads with water beads within the volume of two spherical surfaces separated by a distance of $50R_c$.

A number of spherical or ellipsoidal NPs were placed randomly at the water–decane interface with their polar (nonpolar) part in the water (oil) phase to reach the desired water–decane interfacial area per NP. Following previous work,[26] the NPs considered in this study are oblate with an aspect ratio $c/b = 1$ when spherical and $c/b = 2$ when ellipsoidal. Note that nonspherical NPs have been considered as it has been reported they can be more efficient in stabilizing emulsions than spherical NPs. Indeed, it has been reported that the minimum amount of particle needed to stabilize an emulsion decreases as the particle aspect ratio increases.[27,28] The initial configuration obtained was simulated for 10^6 timesteps in order to relax the density of the system and the contact angle of the nanoparticles on the droplet. The system pressure and the three-phase contact angles did not change notably after 5000 simulation steps.

2.2 Droplet characteristics

It is well established that the stability of Pickering emulsions is chiefly determined by the size, topology, and concentration of particles at the interface, as well as their wetting properties.[14,15,26,29,30] In the following we consider a system made by two identical spherical water droplets immersed in oil, and stabilized by a sufficiently dense layer of NPs. By a *sufficiently dense* layer of NPs, we mean the NP concentration must be dense enough that the liquid film drainage time, *i.e.*, the time interval from droplet contact to fusion, is significant. This can be tackled in the first instance by considering *steered* simulations with a range of pulling rates representative of shear rates that can be reached in conventional mechanical

Table 1 DPD interaction parameters a_{ij} expressed in $k_B T/R_c$ units. Symbols w, o, ap, and p stand for water beads, oil beads, NP nonpolar beads, and NP polar beads, respectively

	w	o	ap	p
w	131.5	198.5	178.5	110
o		131.5	161.5	218.5
ap			450	670
p				450

Table 2 Interfacial area per NP ($\mathcal{A}_{NP}$), contact angle (θ_C), average orientation angle (θ_a), radius of gyration (R_G), asphericity (A_s), and relative shape anisotropy (A_n), obtained from simulating the various NPs at the decane–water interface. In all cases, the water droplets were simulated in a decane solvent. The statistical errors are estimated as one standard deviation from the average contact angles obtained for equilibrated trajectories

		144 NPs	160 NPs	165 NPs	170 NPs
S-55JP	$\mathcal{A}_{NP}(R_c^2)$	30.84 ± 0.06	27.61 ± 0.06	26.75 ± 0.08	25.96 ± 0.13
	θ_C	$91.4° \pm 2.1°$	$91.3° \pm 2.1°$	$91.6° \pm 2.2°$	$91.6° \pm 2.3°$
	$R_G(R_c)$	13.951 ± 0.003	13.824 ± 0.003	13.777 ± 0.003	15.562 ± 0.004
	$A_s(R_c)$	1.54 ± 0.28	1.68 ± 0.24	1.70 ± 0.26	2.81 ± 0.16
	A_n	$(2.4 \pm 1.6) \times 10^{-4}$	$(2.13 \pm 1.29) \times 10^{-4}$	$(7.18 \pm 3.64) \times 10^{-4}$	$(12.6 \pm 2.8) \times 10^{-4}$
S-60JP	$\mathcal{A}_{NP}(R_c^2)$	31.11 ± 0.07	27.82 ± 0.06	26.89 ± 0.07	26.10 ± 0.10
	θ_C	$95.2° \pm 2.4°$	$95.4° \pm 2.3°$	$95.3° \pm 2.6°$	$95.3° \pm 2.5°$
	$R_G(R_c)$	13.937 ± 0.003	13.810 ± 0.003	13.761 ± 0.003	13.728 ± 0.003
	$A_s(R_c)$	1.53 ± 0.28	1.54 ± 0.29	1.65 ± 0.24	2.35 ± 0.18
	A_n	$(2.42 \pm 1.49) \times 10^{-4}$	$(2.71 \pm 1.75) \times 10^{-4}$	$(2.93 \pm 1.34) \times 10^{-4}$	$(12.7 \pm 3.2) \times 10^{-4}$
S-70JP	$\mathcal{A}_{NP}(R_c^2)$	31.44 ± 0.08	28.14 ± 0.05	27.21 ± 0.05	26.29 ± 0.06
	θ_C	$100.6° \pm 2.8°$	$100.5° \pm 2.8°$	$100.6° \pm 3.2°$	$101.0° \pm 3.2°$
	$R_G(R_c)$	13.921 ± 0.003	13.806 ± 0.003	13.771 ± 0.003	13.703 ± 0.004
	$A_s(R_c)$	1.53 ± 0.25	1.54 ± 0.25	1.90 ± 0.25	1.85 ± 0.25
	A_n	$(2.17 \pm 1.06) \times 10^{-4}$	$(2.70 \pm 1.42) \times 10^{-4}$	$(5.42 \pm 2.46) \times 10^{-4}$	$(6.18 \pm 3.28) \times 10^{-4}$
E-70JP	$\mathcal{A}_{NP}(R_c^2)$	31.07 ± 0.11	27.67 ± 0.09	26.75 ± 0.09	25.93 ± 0.09
	θ_a	$61.2° \pm 7.6°$	$56.7° \pm 10.1°$	$52.8° \pm 10.8°$	$51.6° \pm 11.3°$
	$R_G(R_c)$	14.035 ± 0.003	13.805 ± 0.004	15.812 ± 0.004	13.649 ± 0.004
	$A_s(R_c)$	1.61 ± 0.30	1.97 ± 0.33	1.97 ± 0.33	1.69 ± 0.27
	A_n	$(3.02 \pm 1.92) \times 10^{-4}$	$(5.92 \pm 3.27) \times 10^{-4}$	$(3.77 \pm 2.06) \times 10^{-4}$	$(3.86 \pm 2.19) \times 10^{-4}$

Table 3 Water droplet radius, R_{drop}, expressed in R_c units, for various NP types and concentrations. The statistical errors are estimated as one standard deviation from the average obtained for equilibrated trajectories. R_{drop} is the observable that allows us to fix the target value S_{target} for the coordinate used in ABMD simulations

	144 NPs	160 NPs	165 NPs	170 NPs
S-55JP	18.8 ± 0.2	18.8 ± 0.3	18.7 ± 0.4	18.7 ± 0.6
S-60JP	18.9 ± 0.3	18.8 ± 0.3	18.8 ± 0.3	18.8 ± 0.5
S-70JP	19.0 ± 0.3	18.9 ± 0.2	18.9 ± 0.3	18.9 ± 0.3
E-70JP	18.9 ± 0.4	18.8 ± 0.4	18.7 ± 0.4	18.7 ± 0.4

emulsification[31] or ultrasonication processes.[32] In the following, we consider a limited range of NP concentrations and types (*i.e.* contact angles) as well as different shapes (spherical and ellipsoidal), as initial conditions. Table 2 summarizes the thermodynamic (three-phase contact angle, NP orientation angle) and geometrical properties[33] (radius of gyration, asphericity, and relative shape anisotropy of the

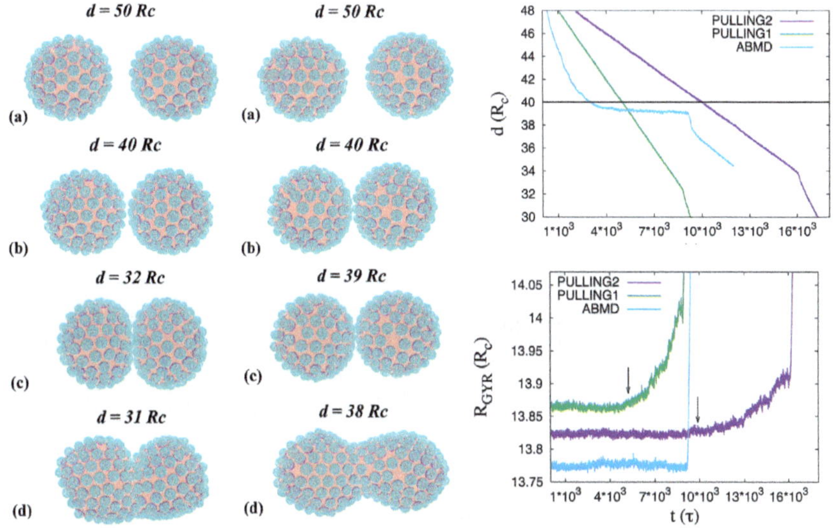

Fig. 1 (Left panel) Sequence of simulation snapshots representing typical coalescence processes of two water droplets stabilized with 160 spherical nanoparticles of type 55JP. Cyan and purple spheres represent polar and apolar beads, respectively. The distance between the two droplets is measured between their centers of mass and decreases from top to bottom. The left (right) column represents the sequence of a pulling (ABMD) simulation where (a) is at the initial distance, (b) is the first contact between the two droplets, (c) is at the coalescence threshold, and (d) is the merging of the two droplets. (Right panel) Temporal evolution of the distance between the centers of mass of the two droplets, d, (top) and the radius of gyration, R_{GYR}, of one incoming droplet (bottom) during pulling and ABMD simulations. The temporal evolutions of R_{GYR} are shifted arbitrarily along the ordinate axis for clarity. The two arrows in the bottom panel indicate the departure from the initial plateau of R_{GYR} indicating the beginning of the shape alteration of the droplets. This corresponds to the distance $d = 40R_c$ (black plain line) at which the two coalescing droplets interact through NP–NP interactions. We use the notation: Pulling1 ($v_P = 0.001R_c$/time step, $k_P = 5000k_BT/R_c^2$), Pulling2 ($v_P = 0.0005R_c$/time step, $k_P = 5000k_BT/R_c^2$), and ABMD ($\alpha = 1000k_BT/R_c^2$, $S_{target} = 39R_c$).

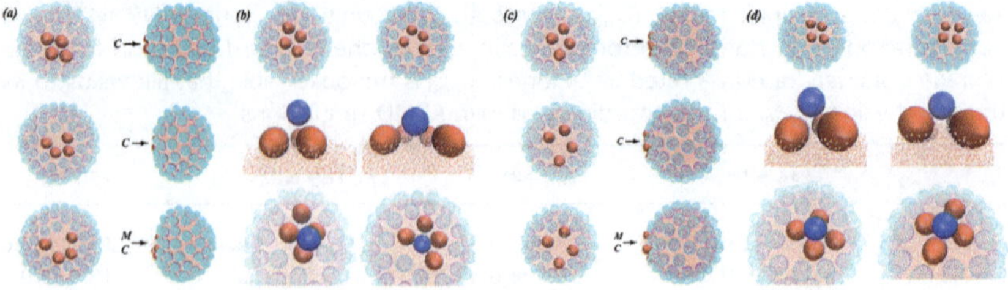

Fig. 2 Sequence of simulation snapshots representing typical coalescence processes of two water droplets stabilized with 160 spherical nanoparticles of type 55JP during pulling (a and b) and ABMD simulations (c and d). Cyan and purple spheres represent polar and apolar beads, respectively. (M) and (C) represent contact and merging points between the incoming NP (blue sphere) and the set of NPs surrounding the neighbour droplet that defines the interaction area (red spheres). The contact point (C) corresponds to the first NP–NP interaction between the incoming NP and its neighbour droplet. The merging point (M) is where the water bridge is formed between the two droplets. The interaction area is delimited by the set of NPs (red spheres) on the approaching droplet that defines the first neighbouring shell, *i.e.*, the distance before merging between the center of the incoming (blue) NP and the centers of each red NP is less than $7R_c$. In the configuration considered here, the merging point is located in the interaction area, but this is not always the case (see Fig. 4).

droplet) for different NP types and concentrations that we consider throughout this study.

Three-phase contact angle. To estimate the three-phase contact angle on the droplets we calculate the fraction of the spherical NP surface area that is wetted by water, and we obtain θ_C as

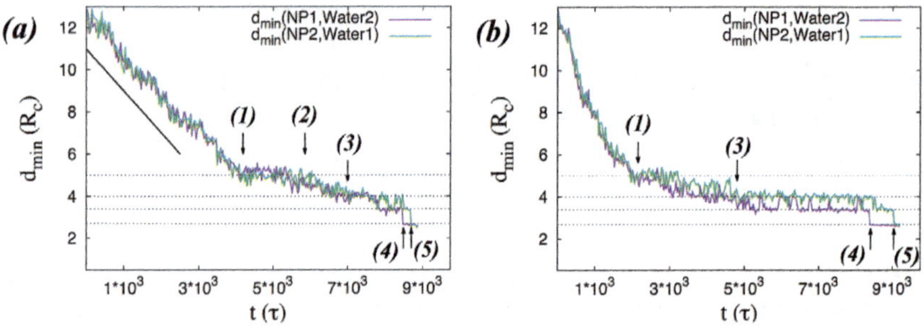

Fig. 3 Temporal evolution of the minimal distance d_{min} between the center of the incoming droplet (NP1 or NP2) and the water molecules from the other droplet (Water2 or Water1, respectively). Two water droplets stabilized with 160 spherical nanoparticles of type 55JP in a decane solvent are considered. During the pulling simulation (panel a) the distance decreases consistently with the pulling simulation rate (black plain line). The distance d_{min} reaches a plateau (1) when the incoming NP interacts with a NP belonging to the second droplet. The incoming NP then interacts with the solvent shell surrounding the droplet (2) and the water molecules (3). The two droplets coalesce when a bridge is formed between two NPs from the two different droplets (4 and 5). Considering the ABMD simulation (panel b), the distance d_{min} does not show the NP–NP interaction plateau (1). In this system the incoming NP interacts with the solvent shell surrounding the droplet and the water molecules (3). As in the pulling simulation, the two droplets coalesce when a bridge is formed and facilitated by two NPs from the two different droplets (4 and 5).

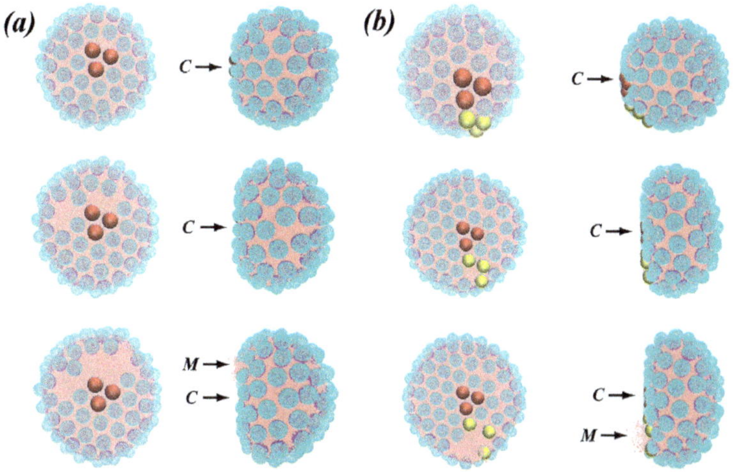

Fig. 4 Sequence of simulation snapshots representing typical coalescence processes of two water droplets stabilized with 160 spherical NPs (panel a) and 165 spherical NPs (panel b) of type 60JP during pulling simulations. Cyan and purple spheres represent polar and apolar beads, respectively. (M) and (C) represent contact and merging points, respectively, between one NP on the incoming droplet (blue sphere) and the interaction area delimited by the NPs highlighted in red on the second droplet (see main text and Fig. 2 for definitions). In panel (a), the droplet deforms generating a rather large interfacial area not covered by NPs. This interfacial area favours the coalescence. In panel (b), the increase of the NP density prevents the occurrence of the large NP-free interfacial area. One NP on the incoming droplet finds itself in contact with the water molecules from the other droplet (yellow spheres), favouring coalescence. This NP, which effectively initiates coalescence, is different from the NP that induced the compression of the second droplet when the two droplets initially came into contact.

$$\theta_C = 180 - \arccos\left(1 - \frac{2A_w}{4\pi R^2}\right), \tag{1}$$

where A_w is the area of the NP surface that is wetted by water and R is the radius of the NP. The ratio $A_w/4\pi R^2$ is obtained by dividing the number of NP surface beads (ap or p) which are wetted by water, by the total number of beads on the NP surface (192 for spherical NPs). One surface bead is wet by water if a water bead is the solvent bead nearest to it. One standard deviation from the average is used to estimate the statistical uncertainty.

Orientation angle. To characterize the ellipsoidal NP structure on a droplet, we focus on the orientation of their longest axes (the c axes) with respect to the radial directions from the center of the droplet.[26] The orientation angle θ_a is defined as

$$\theta_a = \arccos(\mathbf{u}_i \cdot \mathbf{R}_i), \tag{2}$$

where $\mathbf{u}_i$ is the unit vector along the NP c axis and $\mathbf{R}_i$ is the unit vector representing the radial direction from the center of the droplet. When $\theta_a = 0(\pi/2)$, the corresponding NP is parallel (perpendicular) to the droplet radial direction, and therefore perpendicular (parallel) to the local liquid–liquid interface. Note that we define the angle θ_a so that it is never larger than $\pi/2$. One standard deviation from the average is used to estimate the statistical uncertainty.

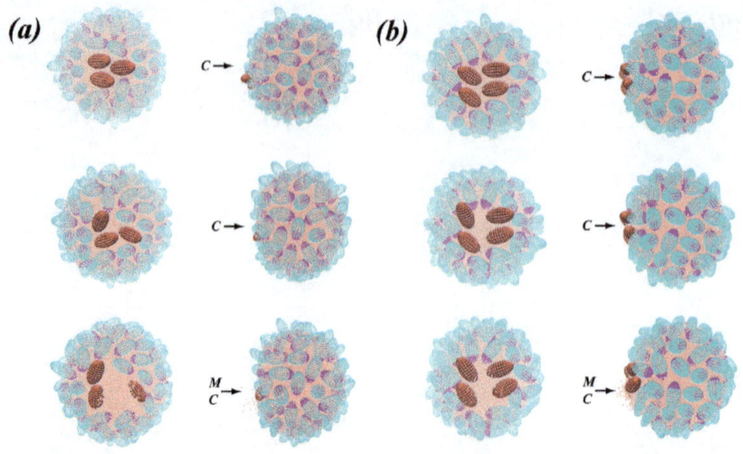

Fig. 5 Sequence of simulation snapshots representing typical coalescence processes of two water droplets stabilized with 160 ellipsoidal nanoparticles of type 70JP during pulling (panel a) and ABMD simulations (panel b). Cyan and purple spheres represent polar and apolar beads, respectively. M and C represent contact and merging points between one NP on the incoming droplet and a set of NPs on the second droplet highlighted in red. These red spheres define the interaction area (see main text and Fig. 2 for definitions).

Radius of gyration, asphericity and relative shape anisotropy. The description of the geometrical properties of complex systems by generalized parameters such as the radius of gyration or principal components of the gyration tensor has a long history in macromolecular chemistry and biophysics.[33,34] Indeed, such descriptors allow an evaluation of the overall shape of a system (spherical, oblate, or prolate) and reveal its symmetry. Considering, *e.g.*, the following definition for the gyration tensor,

$$\mathscr{T}_{\mathrm{GYR}} = \frac{1}{N}\begin{bmatrix} \sum x_i^2 & \sum x_i y_i & \sum x_i z_i \\ \sum x_i y_i & \sum y_i^2 & \sum y_i z_i \\ \sum x_i z_i & \sum y_i z_i & \sum z_i^2 \end{bmatrix}, \tag{3}$$

where the summation is performed over N atoms and the coordinates x, y, and z are related to the geometrical center of the atoms, one can define a reference frame where $\mathscr{T}_{\mathrm{GYR}}$ can be diagonalized:

$$\mathscr{T}_{\mathrm{GYR}}^{\mathrm{diag}} = \begin{bmatrix} S_1 & 0 & 0 \\ 0 & S_2 & 0 \\ 0 & 0 & S_3 \end{bmatrix}. \tag{4}$$

In this format we obey the convention of indexing the eigenvalues according to their magnitude. We thus define the radius of gyration $R_{\mathrm{GYR}}{}^2 \equiv S_1{}^2 + S_2{}^2 + S_3{}^2$, the asphericity $A_\mathrm{s} \equiv S_1 - \frac{1}{2}(S_2 + S_3)$, which measures the deviation from the spherical symmetry, and the relative shape anisotropy $A_\mathrm{n}{}^2 \equiv 1 - 3\frac{S_1S_2 + S_1S_3 + S_2S_3}{(S_1 + S_2 + S_3)^2}$, which reflects both the symmetry and dimensionality of a system.[33]

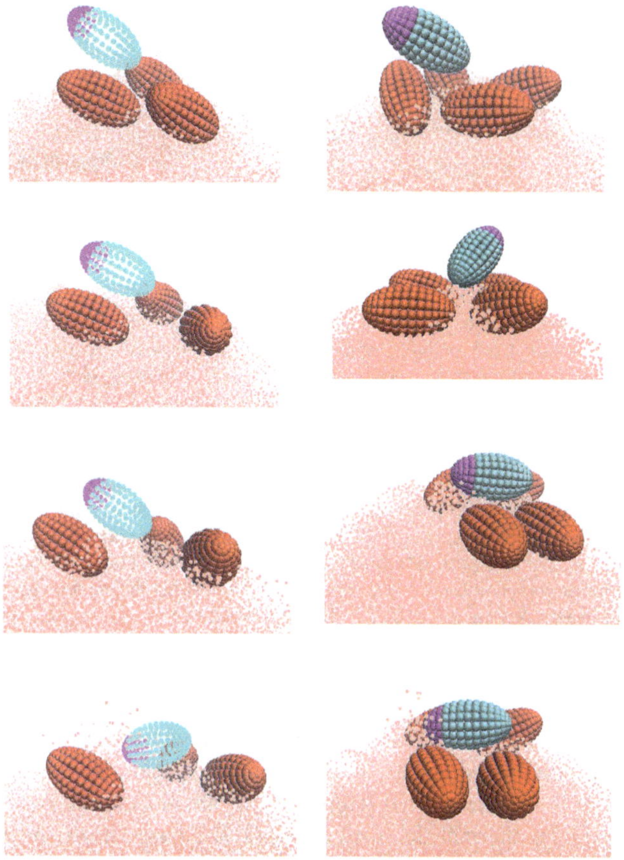

Fig. 6 Detailed sequence of simulation snapshots representing typical coalescence processes of two water droplets stabilized by 160 ellipsoidal nanoparticles of type 70JP during pulling (left panel) and ABMD simulations (right panel). Cyan and purple spheres represent polar and apolar beads, respectively. Considering the pulling simulation, one NP on the incoming droplet interacts first with a NP on the other droplet before it starts interacting with the solvent shell around the water molecules. In ABMD simulation, the incoming NP interacts almost immediately with the solvent shell. The rest of the coalescence process is similar to what is observed for both pulling and ABMD simulations. As described in the text, the longitudinal orientation of the Janus ellipsoidal NPs yields a peculiar coalescence mechanism.

As discussed in the ESI†, we consider NP types and concentrations such that based on our preliminary simulations they are strongly effective at preventing droplet coalescence, and such that the shape of the droplet be spherical before contacting the neighbour droplet. In the simulation discussed here, we consider spherical water droplets in a decane solvent stabilized with 160–165 NPs.

2.3 ABMD algorithm

Adiabatic biased molecular dynamics[17–19] (ABMD) is an algorithm developed to connect any two points in the conformational space of a given system. The method is based on the introduction of a biasing potential, which is a function of

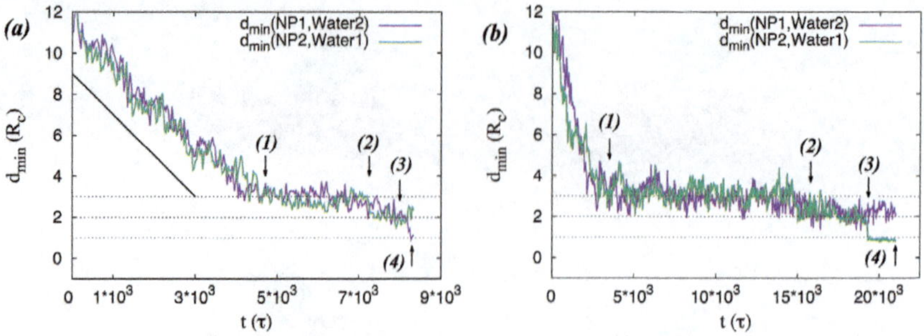

Fig. 7 Temporal evolution of the minimal distance d_{min} between the apolar extremity of the incoming droplet (NP1 or NP2) and the water molecules from the other droplet (Water2 or Water1, respectively). Two water droplets stabilized with 160 ellipsoidal nanoparticles of type 70JP in a decane solvent are considered. During the pulling simulation (panel a) the distance decreases consistently with the simulation pulling rate (black plain line). The distance d_{min} reaches a plateau (1) when the incoming NP interacts with a NP belonging to the other droplet. The incoming NP then interacts with the solvent shell surrounding the droplet (2) and the water molecules (3). The last step differs significantly from the one described for spherical NPs, as described in the text (4). Considering the ABMD simulation (panel b), the distance d_{min} does not show a clear NP–NP interaction plateau. The incoming NP interacts with the solvent shell surrounding the second droplet and the water molecules, *i.e.* $2R_c \leq d_{min} \leq 3R_c$. The last step is similar to the one described for pulling simulations.

chosen coordinates of the system and which is zero when the system is moving toward the desired target point, while disfavoring motions in the opposite direction. This is similar to what happens in a *ratchet-and-pawl* system, which undergoes random thermal fluctuations, while the pawl allows the ratchet to move only in one direction.[18] In ABMD the chosen coordinate plays the role of the ratchet and the biasing potential that of the pawl. In this respect, the biasing potential does not exert any work to direct the system toward the target conformation, as would happen when pulling the system with a force; on the contrary the system makes moves toward the target conformation under the driving effect of the potential of the force-field alone. The biasing potential is implemented as

$$V(\rho(t)) = \begin{cases} \frac{\alpha}{2}(\rho(t) - \rho_m(t))^2 & \rho(t) > \rho_m(t) \\ 0 & \rho(t) \leq \rho_m(t) \end{cases} \tag{5}$$

where

$$\rho(t) = (S(t) - S_{target})^2 \tag{6}$$

is the distance along the coordinate S of the actual configuration of the system with respect to a target value S_{target}, α is a damping constant, and

$$\rho_m(t) = \min_{0 \leq \tau \leq t} \rho(\tau) \tag{7}$$

is the minimum distance reached until time t. The algorithm is defined by the choice of the coordinate S and of the damping constant α. The ABMD algorithm is routinely implemented by the biophysical community to explore free-energy paths associated with protein folding and unfolding.[17,18,35] In such applications, the most probable sequence of events generated by the ABMD simulations is expected to coincide with the minimum free-energy path, independent of the damping constant. The outcome of a set of ABMD simulations is a number of trajectories mapping the free-energy minimum of the system, in which time and the statistical weight of the conformations along the trajectories are unphysically modulated. One of the unique features of this method is that the transformation connecting the two points of interest in the configurational space occurs following natural fluctuations. Additionally, the algorithm minimizes the sum of the potential and kinetic energy of the particles and of the bias potential. This provides means of controlling the quality and *adiabaticity* of the simulation.

In this study, we use the efficiency of the ABMD algorithm within a different perspective, *i.e.*, to understand the complex *dynamic* processes of the early stage of droplet coalescence. The delicate issue is the choice of the damping constant α in eqn (5) that drives the mechanism under study. According to the classical theory of fluid particle coalescence,[36] we aim at tackling, at a mesoscopic scale, the mechanism associated with the drainage of the liquid films of the continuous phase between two approaching droplets. These liquid films should drain to allow the dispersed fluids to make contact and fuse. The coalescence mechanism strongly depends on the velocity of the two approaching droplets, which can be responsible for a large deformation of the droplet shape.[15] In the following, we consider ABMD parameters that allow us to study coalescence while keeping the shape of the droplets unaltered (*i.e.*, the radius of gyration of the droplets). This means we shall use a value of α small enough so as to provide the correct sequence of events associated to coalescence, but large enough to make this happen in a computationally reasonable time. The second ABMD parameter of choice is the target value S_{target}, which is set on the distance between the two water droplets before coalescence occurs. S_{target} is set as equal to the radius of the isolated droplets, and determined by the positions of the NPs. These data are reported in Table 3. Biased simulations were performed using version 2.1 of the plugin for free-energy calculations, named PLUMED.[37,38]

2.4 Pulling algorithm

Pulling dynamics, also named steered molecular dynamics (SMD) is an algorithm which takes inspiration from single-molecule pulling experiments[39] and forces a system to evolve away from its initial equilibrium condition. This is intended to accelerate transitions between different energy minima and to estimate the free-energy difference between two (or more) states *via* the Jarzynski equality.[40] Pulling simulations have become very popular and are extensively applied in studying many biophysical processes[41,42] and phenomena in chemical physics.[15] In SMD, the Hamiltonian of the system, $\mathscr{H}_0$, is modified into a new Hamiltonian, $\mathscr{H}$, which contains a term which depends on time *via* a harmonic spring potential centered on a point which moves linearly with time, *i.e.*,

$$\mathscr{H} = \mathscr{H}_0 + \frac{k_P}{2}(S(t) - S_0 + v_P t)^2, \tag{8}$$

where $S(t)$ is the chosen coordinate of the configuration of the system that is biased, $S_0 = S(t = 0)$, v_P is the pulling velocity, and k_P is the spring constant. For an appropriate set of SMD parameters (k_P, v_P), the system follows closely the center of the moving harmonic spring. In our prior numerical investigation on the stability of Pickering emulsions we employed the pulling algorithm.[15]

3 Results and discussion

In the following, we consider two types of numerical simulations, (1) *pulling simulations* where the two droplets are moved towards each other using one harmonic spring with a force constant of 5000 k_BT/R_c tethered to their center of mass, at a constant relative velocity (pulling rate) $v_P = 0.001R_c$ or $0.0005R_c$ per time step, and (2) *ABMD simulations* where we consider the distance between the center of mass of the two droplets as the biased coordinate S, a damping constant $\alpha = 1000k_BT/R_c^2$ and a target value $S_{target} = 39R_c$.

In Fig. 1 (left panel) we show simulation snapshots obtained during the two coalescence simulations: pulling simulations (left column) and ABMD simulations (right column). In both cases we consider water droplets stabilized by 160 NPs of type 55-JP in decane. The interfacial area per NP equals $27.61R_c^2$. Visual inspection of the simulation snapshots highlights the fundamental difference between the two processes. For the ABMD simulation, the shape of the water droplets remains unaltered, while in pulling simulations a clear alteration of the shape of the two coalescing droplets is observed as the distance between their centres of mass decreases. This behaviour is quantitatively shown in Fig. 1 (right panel) where we show the temporal evolution of the distance between the centers of mass d and the radius of gyration R_G of the two coalescing droplets as a function of simulation time during ABMD and pulling simulations. Before the two droplets come into contact, the radius of gyration remains constant as the distance between the two droplets decreases. When the distance is such that the two droplets start interacting with each other *via* NP–NP interactions ($d = 40R_c$), R_G increases during the pulling simulation, meaning the droplets deform as the distance decreases. R_G keeps increasing until a minimal distance is reached ($d = 32R_c$), when the two droplets coalesce. If one considers the ABMD simulation, the distance between the two droplets reaches a plateau as soon as the two droplets come into contact, at around $d = 39R_c$. This plateau is slightly lower than the value $d = 40R_c$. This difference indeed can be explained by the fact that ABMD simulations drive the system such that the NPs surrounding the droplet interact with the solvent shell that surrounds the water molecules of the second droplet and thus avoid NP–NP interactions that are more energetically costly. The shape of the two droplets remains unaltered until the two droplets coalesce.

It is necessary to further quantify the difference between these two coalescence mechanisms at a mesoscopic scale. In Fig. 2 we show detailed simulation snapshots obtained during both pulling and ABMD simulations. Considering first qualitatively the pulling simulation, the two droplets start interacting when one NP in the incoming droplet (blue sphere in Fig. 2, panel b) comes into contact with one or more NPs from the opposite droplet (red spheres in Fig. 2, panel b). This NP–NP interaction initiates the shape alteration of the interacting droplets. We identify an interaction area as the set of NPs on the neighbour droplet (red spheres in Fig. 2) that defines the first neighbouring shell (*i.e.*, when the distance between blue NP and red NPs is less than $7R_c$) around the incoming NP before merging. The $7R_c$ threshold is set based on observations from prior work.[22] Then the incoming NP shifts towards the center of the interaction area delimited by the red NPs in Fig. 2. This is quantitatively described in Fig. 3 (panel a) where we follow the evolution of the minimal distance $d_{\min}$ between the center of one incoming droplet and the water molecules belonging to the other droplet. The first part of the evolution is representative of the pulling rate, until the distance reaches a plateau around $d_{\min} \approx 5R_c$ corresponding to where the NP–NP interactions start. Then the incoming NP starts interacting with the solvent in the interacting area and keeps compressing the droplet. This evolution goes on until $d_{\min}$ reaches the minimal solvent shell, *i.e.*, $3.5 \leq d_{\min} \leq 4R_c$, and the incoming NP starts coming into contact directly with the water molecules ($d_{\min} = 3R_c$). Merging of the two droplets occurs when two NPs (each belonging to different droplets) interact with water molecules belonging to the other droplet (*cf.* arrows (4) and (5) in Fig. 3 panel a). A water bridge is thus formed between the droplets, which allows the flow of water molecules.

Let us now compare the behaviour just described with the one observed when running the ABMD simulations. Considering the simulation snapshots in Fig. 2 and the quantitative analysis in Fig. 3, we first see that the system does not spend a significant amount of time in the plateau at $d_{\min} \approx 5R_c$. Analysis of the trajectories suggests that the NP in the incoming droplet avoids frontal interactions with NPs belonging to the other droplet. Indeed, as prescribed by the ABMD algorithm, the system follows a minimal free-energy path such that the incoming droplet mainly interacts with the solvent molecules located around the interaction area delimited by the red subset of NPs in Fig. 2. The system progressively reaches the plateau corresponding to the minimal solvent shell, *i.e.*, $3.5 \leq d_{\min} \leq 4R_c$, and the incoming NP eventually comes into contact with the water molecules ($d_{\min} = 3R_c$). The final stages of the process are similar to what is observed during pulling simulations, where a set of two NPs must come into contact with the water molecules of the other droplet to initiate coalescence. Unlike the results observed during the pulling simulations, this evolution happens without an increase of the droplets' radius of gyration.

This difference is due to the pulling parameters chosen, which allow the system to sample two different mechanisms during coalescence. The parameters chosen for the pulling simulations are such that the system remains in the advection regime. This can be highlighted with the use of the Péclet number[43]

$$\mathrm{Pe} = \frac{vL_c}{D}, \tag{9}$$

where v is the effective advection velocity, L_c is a characteristic length, and D is the diffusion coefficient. Considering the pulling velocity in the center of mass reference frame, v_P, the DPD characteristic length, R_c, and the simulated water self-diffusion coefficient D_{sim}, one obtains $\mathrm{Pe} = \frac{v_P R_c}{D_{\text{sim}}} = \frac{0.002R_c/\tau \times R_c}{0.0063{R_c}^2/\tau} \equiv 0.3$. This value for the Péclet number indicates that during the pulling simulations the system is dominated by the pulling velocity. On the contrary, in the ABMD simulations, the evolution follows the natural fluctuations of the system, a regime by design dominated by diffusion.

Our results suggest that when pulling simulations are considered, the coalescence process depends on parameters such as NP contact angles and concentrations. On the contrary, the results obtained from ABMD simulations do not show such dependencies. We illustrate this difference with the example in

Fig. 4 (left panel) where we keep the interfacial area per NP constant (160 NPs) and consider spherical 60JP NPs. In this figure the contact point (C) corresponds to the first NP–NP interaction between the two droplets and the merging point (M) identifies the formation of the water bridge between the two droplets. In Fig. 4 we illustrate that both C and M are different in either pulling or ABMD simulations due to the shape deformation of the droplets. The system seeks to keep the NP contact angles as close as possible to their equilibrium value (data consistent with Table 2). Our simulations suggest that under these conditions, the droplets deform their shape and increase the water–oil interfacial area rather than allowing the NPs to change their contact angle or desorb from the interface. The increased interfacial area, not covered by NPs as shown in Fig. 4, favours coalescence.

The differences between the results obtained from ABMD and pulling simulations also increase when the surface density of the NPs on the droplets increases. We illustrate this difference with the example in Fig. 4 (right panel), where we increase the interfacial area per NP (from 160 NPs to 165 NPs) and consider spherical 60JP NPs. Under these conditions, the NP density becomes sufficiently high that the droplet cannot create a large NP-free area (as we observed for 160 NPs) to facilitate coalescence. When the distance between the two droplets decreases, one NP on the incoming droplet comes in contact with the solvent shell and then the water molecules from the other droplet. We find it interesting that in this scenario coalescence initializes in an area different from the initial interaction area (yellow spheres in Fig. 4). This NP is different from the NP that induces the compression of the neighbour droplet (red spheres in Fig. 4). The contact point (C) and the merging point (M) are different and the merging does not require the contact of two NPs originally on the two separable droplets.

In the ESI,† we report the results obtained when homogeneous NPs were simulated. The results are consistent with pulling simulations reported previously.[15] The key point is the difference in the interaction between the NP on the incoming droplet, and the other droplet. Due to the homogeneous distribution of polar and apolar beads on the NP surface, the homogeneous NP from the incoming droplet can adsorb simultaneously on the two interfaces, yielding the same contact angle on both droplets. As a consequence, in both pulling and ABMD simulations, the last step of the coalescence mechanism differs from the one observed with spherical Janus NPs. When homogeneous NPs are used, the coalescence requires only the contact of one NP to induce the merging of the two water droplets.

We also considered the impact of different NP shapes on the coalescence mechanisms, focusing our attention on ellipsoidal NPs of aspect ratio $c/b = 2$ (242 beads around the ellipsoidal NP). In Fig. 5 we first consider the qualitative evolution of the system. As in the case with spherical NPs, the ABMD simulation does not alter the shape of the droplet, while the pulling simulation induces an increase of the droplets' gyration radius. This is visually investigated in Fig. 6 and more quantitatively shown in Fig. 7. The compression of the droplet observed during pulling simulations can be explained in the first instance by the same mechanism we previously described, *i.e.*, the pronounced NP–NP interaction due to the non-diffusive behaviour of the system. The incoming NP interacts with the solvent shell around the water molecules and eventually comes into contact with the water molecules themselves. However, the last step of the coalescence mechanism differs from the one observed with spherical Janus NPs (*cf.* ESI† for visual inspection). When the droplets are stabilized by ellipsoidal NPs, the coalescence process does not require the contact of two NPs (one on each droplet). Due to the longitudinal orientation of the ellipsoidal NPs with respect to the interface, contact happens between the apolar NP beads (cyan spheres in Fig. 6) on the incoming droplet and the water molecules of the second droplet. When this contact is secured, water molecules from the second droplet can travel a short distance and interact with the polar NP beads (purple spheres in Fig. 6), originally exposed to the decane solvent. Once this interaction is established, a molecular bridge is formed between the two droplets that leads to coalescence. It is worth repeating that this coalescence mechanism is facilitated by the longitudinal orientation of the Janus ellipsoidal NPs on the droplet interface.

4 Conclusions

In this paper, we implemented Adiabatic Biased Molecular Dynamics simulations to investigate the early stage of the coalescence process between two water droplets in a decane solvent at the mesoscopic level. The droplets are stabilized with various NPs (Janus and homogeneous) of varying contact angles and shapes, and different NP densities. In ABMD simulations, the system is forced to coalesce *adiabatically*. This allows us to study how the ability of the NPs to stabilize Pickering emulsions is discriminated by NP mobility and the associated caging effect. We compared how the coalescence process differs when one considers pulling simulations with sufficiently high steering speeds. When ABMD simulations are considered, we showed that the coalescence mechanism does not depend on parameters such as NP contact angles and concentrations. However, it is sensitive to the NP type (Janus NP *vs.* homogeneous NP) and shape (spherical *vs.* ellipsoidal). The coalescence occurs when a water bridge is formed between the two droplets, which drives the subsequent flow of water molecules. The formation of the water bridge depends on the shape of the NPs used as stabilizers. Considering spherical NPs, the water bridge is formed with two NPs (each belonging to different droplets) that interact with water molecules from the other droplet. This differs from ellipsoidal NPs of aspect ratio $c/b = 2$. Due to the longitudinal orientation of the ellipsoidal NPs with respect to the interface, water molecules from the second droplet can interact more easily with the polar NP beads on the incoming droplet originally exposed to the decane solvent. They form a molecular bridge with a single ellipsoidal NP between the two droplets that leads to coalescence.

When pulling simulations are considered, we observed a wider range of coalescence processes. When the system remains in the advection regime, as defined by the Péclet number, the droplet has the ability to undergo large elastic deformations before coalescing. This is due in the first instance to the NP–NP interactions, which are very sensitive to the NP density on the droplets. This is related to the caging effect associated with NP mobility. The collision speed also influences the alteration of the shape of the water droplet, as the coalescence process is sensitive to the time scale associated with the liquid film drainage. These different parameters, associated with the fact that the system tries to keep the NP contact angles as close as possible to their equilibrium value when the shape of the system changes, are responsible for a wide range of coalescence processes.

These results are consistent with the recent microfluidic collision experiment of Zhou *et al.*[44] studying head-to-head moving water droplets in a microfluidic chip. Following preliminary investigations on droplet coalescence,[45,46] the authors analysed the coalescence process using the droplet contact time and the liquid film drainage time (*i.e.*, the time interval from droplet contact to fusion). They defined a coalescence percentage to indicate how coalescence is prone to occur under different flow conditions. However, as the droplets have the opportunity to detach from each other when the droplet velocity is too high, the coalescence percentage can eventually reach zero. The authors highlighted that droplets coalesce effectively at low velocities and suggested that this is related to the importance of the liquid film drainage between two approaching droplets and the droplets' faculty to undergo elastic deformation. Our numerical analysis allows us to investigate the early stage of the coalescence process at a mesoscopic scale and to propose detailed coalescence mechanisms, the observation of which remains at a scale inaccessible through experiments. Although the present work only considers frontal droplet–droplet collisions, our results highlight the importance of droplet speed in modulating the coalescence process.

To go further, it would be useful to estimate the energy barriers encountered by the two droplets as they coalesce following the mechanisms illustrated in this paper. It would also be interesting to modify the chemistry of the NPs to study their ability to enhance or to perturb emulsion stability. Considering both ABMD and pulling simulations, we showed that the two water droplets start coming into contact through NP–NP interactions. Adjusting the apolar–apolar interaction between NPs might have an impact on the coalescence process, perhaps increasing or shortening the NP–NP interaction time and manipulating the effective interaction with the solvent shell. Similarly, it would be interesting to study the effect of particle diameters on the coalescence process. Given a sufficiently high NP density around the droplets, increasing the diameter of the Janus NPs would also increase the NP–NP interaction time without changing the NP chemistry.

Acknowledgements

We acknowledge X-C. Cuong, V. Garbin, and L. Botto for useful discussions. F. S. thanks M. Salvalaglio for fruitful discussion concerning the ABMD algorithm. *Via* our membership of the UK's HEC Materials Chemistry Consortium, which is funded by EPSRC (EP/L000202), this work used the ARCHER UK National Supercomputing Service (http://www.archer.ac.uk).

References

1 S. Pickering, *J. Chem. Soc.*, 1907, **91**, 2001–2021.
2 S. Drexler, J. Faria, M. Ruiz, J. Harwell and D. Resasco, *Energy Fuels*, 2012, **26**, 2231–2241.
3 A. Angelova, B. Angelov, R. Mutafchieva, S. Lesieur and P. Couvreur, *Acc. Chem. Res.*, 2011, **44**, 147–156.
4 E. Shchukina and D. Shchukin, *Curr. Opin. Colloid Interface Sci.*, 2012, **17**, 281–289.
5 C. Puglia and F. Bonina, *Expert Opin. Drug Delivery*, 2012, **9**, 429–441.

6 R. Miller, V. Fainerman, V. Kovalchuk, D. Grigoriev, M. Leser and M. Michel, *Adv. Colloid Interface Sci.*, 2006, 128–130.
7 D. Wang, S. Yordanov, H. Paroor, A. Mukhopadhyav, C. Li, H. Butt and K. Koynov, *Small*, 2011, **7**, 3502–3507.
8 K. Zahn, J. MendezAlcaraz and G. Maret, *Phys. Rev. Lett.*, 1997, **79**, 175–178.
9 D. Cheung, *Chem. Phys. Lett.*, 2010, **495**, 55–59.
10 E. Weeks and D. Weitz, *Chem. Phys.*, 2002, **284**, 361–367.
11 J. Sonnenburg, D. Kremp and R. Sandig, *Mol. Phys.*, 1991, **74**, 649–664.
12 Y. Peng, W. Chen, T. Fischer, D. Weitz and P. Tong, *J. Fluid Mech.*, 2009, **618**, 243–261.
13 S. Cheng and G. Grest, *J. Chem. Phys.*, 2012, **136**, 214702–214709.
14 X.-C. Luu, J. Yu and A. Striolo, *J. Phys. Chem. B*, 2013, **117**, 13922–13929.
15 H. Fan and A. Striolo, *Soft Matter*, 2012, **8**, 9533–9538.
16 H. Fan and A. Striolo, *Phys. Rev. E.*, 2012, **86**, 05610.
17 E. Paci and M. Karplus, *J. Mol. Biol.*, 1999, **288**, 441–459.
18 C. Camilloni, R. Broglia and G. Tiana, *J. Chem. Phys.*, 2011, **134**, 045105.
19 M. Marchi and P. Ballone, *J. Chem. Phys.*, 1999, **110**, 3697–3702.
20 R. Groot and P. Warren, *J. Chem. Phys.*, 1997, **107**, 4423–4435.
21 S. Plimpton, *J. Comput. Phys.*, 1995, **117**, 1–19.
22 X.-C. Luu, J. Yu and A. Striolo, *Langmuir*, 2013, **29**, 7221–7228.
23 R. Groot and K. Rabone, *Biophys. J.*, 2001, **81**, 725.
24 M. Calvaresi, M. Dallavalle and F. Zerbetto, *Small*, 2009, **5**, 2191–2198.
25 J. Partington, R. F. Hudson and K. Bagnall, *Nature*, 1952, **169**, 583.
26 X.-C. Luu and A. Striolo, *J. Phys. Chem. B*, 2014, **118**, 13737–13743.
27 A. Mejia, A. Diaz, S. Pullela, Y.-W. Chang, M. Simonetty, C. Carpenter, J. Batteas, M. Mannan, A. Clearfield and Z. Cheng, *Soft Matter*, 2012, **8**, 10245–10253.
28 D. Dendukuri, T. Hatton and P. Doyle, *Langmuir*, 2006, **23**, 4669–4674.
29 F. Qi, J. Wu, G. Sun, F. Nan, T. Ngai and G. Ma, *J. Mater. Chem. B*, 2014, **2**, 7605–7611.
30 D. French, P. Taylor, J. Fowler and P. Clegg, *J. Colloid Interface Sci.*, 2015, **441**, 30–38.
31 T. Mason and J. Bibette, *Phys. Rev. Lett.*, 1996, **77**, 3481–3484.
32 T. Delmas, H. Piraux, A. Couffin, I. Texier, F. Vinet, P. Poulin, M. E. Cates and J. Bibette, *Langmuir*, 2011, **27**, 1683–1692.
33 J. Vymetal and J. Vondrasek, *J. Phys. Chem. A*, 2011, **115**, 11455–11465.
34 K. Solc, *J. Chem. Phys.*, 1971, **55**, 335.
35 D. Provasi and M. Filizola, *Biophys. J.*, 2010, **98**, 2347–2355.
36 D. Chan, E. Klaseboer and R. Manica, *Soft Matter*, 2011, **7**, 2235–2264.
37 M. Bonomi, D. Branduardi, G. Bussi, C. Camilloni, D. Provasi, P. Raiteri, D. Donadio, F. Marinelli, R. B. F. Pietrucci and M. Parrinello, *Comput. Phys. Commun.*, 2009, **180**, 1961.
38 G. Tribello, M. Bonomi, D. Branduardi, C. Camilloni and G. Bussi, *Comp. Phys. Comm.*, 2014, **185**, 604.
39 H. Grubmuller, B. Heymann and P. Tavan, *Science*, 1996, **271**, 997–999.
40 S. Park, F. Khalili-Araghi, E. Tajkhorshid and K. Schulten, *J. Chem. Phys.*, 2003, **119**, 3559–3566.
41 B. Isralewitz, J. Baudry, J. Gullingsrud, D. Kosztin and K. Schulten, *J. Mol. Graphics Modell.*, 2001, **19**, 13–25.

42 H. Lu, B. Isralewitz, A. Krammer, V. Vogel and K. Schulten, *Biophys. J.*, 1998, **75**, 662–671.
43 M. Huysmans and A. Dassargues, *Hydrogeol. J.*, 2005, **13**, 895–904.
44 Q. Zhou, Y. Sun, S. Yi, K. Wang and G. Luo, *Soft Matter*, 2016, **12**, 1674–1682.
45 Y. Liao and D. Lucas, *Chem. Eng. Sci.*, 2010, **65**, 2851–2864.
46 L. Leal, *Phys. Fluids*, 2004, **16**, 1833–1851.

Faraday Discussions

ROYAL SOCIETY OF CHEMISTRY

PAPER

Wetting and orientation of catalytic Janus colloids at the surface of water

Xiaolu Wang,[a] Martin In,[a] Christophe Blanc,[a] Paolo Malgaretti,[bc] Maurizio Nobili[a] and Antonio Stocco*[a]

Received 25th February 2016, Accepted 11th April 2016

DOI: 10.1039/c6fd00025h

Janus colloidal particles show remarkable properties in terms of surface activity, self-assembly and wetting. Moreover they can perform autonomous motion if they can chemically react with the liquid in which they are immersed. In order to understand the self-propelled motion of catalytic Janus colloids at the air–water interface, wetting and the orientation of the catalytic surface are important properties to be investigated. Wetting plays a central role in active motion since it determines the contact between the fuel and the catalytic surface as well as the efficiency of the transduction of the chemical reaction into motion. Active motion is not expected to occur either when the catalytic face is completely out of the aqueous phase or when the Janus boundaries are parallel to the interfacial plane. The design of a Janus colloid possessing two hydrophilic faces is required to allow the catalytic face to react with the fuel (*e.g.* H_2O_2 for platinum) in water and to permit some rotational freedom of the Janus colloid in order to generate propulsion parallel to the interfacial plane. Here, we discuss some theoretical aspects that should be accounted for when studying Janus colloids at the surface of water. The free energy of ideal Janus colloidal particles at the interface is modeled as a function of the immersion depth and the particle orientation. Analytical expressions of the energy profiles are established. Energetic aspects are then discussed in relation to the particle's ability to rotate at the interface. By introducing contact angle hysteresis we describe how the effects of contact line pinning modifies the scenario described in the ideal case. Experimental observations of the contact angle hysteresis of Janus colloids at the interface reveal the effect of pinning; and orientations of silica particles half covered with a platinum layer at the interface do not comply with the ideal scenarios. Experimental observations suggest that Janus colloids at the fluid interface behave as a kinetically driven system, where the contact line motion over the defects decorating the Janus faces rules the orientation and rotational diffusion of the particle.

[a]*Laboratoire Charles Coulomb (L2C), UMR 5221 CNRS-Univ. Montpellier, Montpellier F-34095, France. E-mail: Antonio.Stocco@umontpellier.fr*

[b]*Max Planck Institüt für Intelligente Systeme, Heisenbergstrasse 3, D-70569 Stuttgart, Germany*

[c]*IV Institüt für Theoretische Physik, Universität Stuttgart, Pfaffenwaldring 57, D-70569 Stuttgart, Germany*

1 Introduction

In ancient Roman religion, the god depicted with two faces called Janus is the god of the beginnings (*e.g.* the first month of the year January is named after Janus), journeys and transitions: future and past, war and peace; and being the god of the doorway and gate can look both inside and outside. A Janus colloidal particle can also undergo (interfacial) transitions and be both inside and outside a fluid, when attached to the fluid interface.

In recent literature, several investigations reported Janus colloidal particles of different sizes, shapes and functionalities.[1–3] Both faces of a Janus particle can be now designed and fabricated with a multiplicity of chemical and physical functions. New self-assemblies, interactions and wetting properties have been realized.[2–6] If one of the faces is hydrophilic and the other is hydrophobic, Janus particles can combine the properties of surface activity and self-assembly of molecular surfactants with the wetting properties of solids. Hence the behavior of Janus particles at fluid interfaces reflects those features, which have an impact in the field of thin films, foams and emulsions.[7–9]

If one of the faces of the Janus particle is chemically reactive and the other is inert, the particle can perform autonomous motion in the presence of a reactant fuel.[10,11] Self-generated phoretic fields are responsible for the particle self-propulsion. For instance, the platinum face of catalytic Janus colloids can transform dissolved hydrogen peroxide into water and oxygen. Hence, a concentration gradient of reactants and products between the reactive face and the inert one will be established and self-propulsion will be generated.[12] Janus colloids become active particles able to transform chemical energy into motion, mimicking the behavior of biological microswimmers. This enhanced particle motion has been exploited to perform tasks such as transporting cargo and purifying water by adsorbing or reducing contaminants.[13,14]

Recently Wang *et al.*,[15] on the experimental side, and Malgaretti *et al.*,[16] on the theoretical side, have investigated the motion of catalytic Janus colloids at the air–water interface in order to elucidate the role of partial wetting confinement in active motion realization. Janus colloids were trapped at the interface and both the persistence length and velocity of the self-propelled motion within the interfacial plane were enhanced with respect to the bulk.[15] The realization of active motion at the interface is governed by wetting and by the orientation of the Janus colloid. The catalytic face reacts with hydrogen peroxide only if it is wetted by the aqueous solution. Directional motion parallel to the interface would also be strongly hindered if the catalytic face is oriented parallel to the interface.

For active colloidal systems, the persistence of the directional motion is ruled by the Brownian rotational diffusion which acts towards randomizing the motion directions.[12,17] For particles attached onto a fluid interface, rotational diffusion coefficients vary significantly with respect to the bulk rotational diffusion coefficient ($D_{r,B} = kT/(8\pi\eta R^3)$, where kT is the thermal agitation energy, η is the dynamic viscosity of the liquid and R is the radius of the particle).[18] The enhanced persistence of the motion of Janus colloids at the air–water interface was explained in terms of a slowing down of the rotational motion at the air–water interface due to partial wetting.[15]

In this article, we focus our attention on the wetting and orientation of Janus colloids at the air–water interface in order to elaborate on the active motion recently observed.[15] The theoretical and experimental results presented here are obtained for catalytic Janus particles but in the absence of fuel. These findings shed light on the fundamental aspects which need to be clarified in order to understand the behavior of Janus particles both in passive and active conditions. Therefore our results will serve as a starting point for more detailed numerical as well as experimental studies aiming at understanding how the activity of the particle breaks the equilibrium picture as well as how the details of fluid rearrangement close to particle surface[19] can modulate the overall transport properties of the particle.

The structure of the text is the following. In the next section, we discuss some theoretical aspects that should be accounted for when studying Janus colloids trapped at the fluid interface. An analytical expression of the free energy profile as a function of the orientation of the Janus colloid is reported; and the role of hydrodynamic rotational friction and surface defects is discussed. Experimental methods and results obtained for spherical silica colloids half covered by platinum are described and discussed in Sections 3 and 4. Section 5 is dedicated to the significance of our investigations for self-propelled Janus colloids trapped at the surface of water.

2 Theoretical aspects

In this section we describe the free energy of an isolated Janus colloid as a function of the immersion depth and the orientation at the air–water interface accounting only for the interfacial energy terms. Interfacial friction terms of the Janus colloids at the interface are also introduced together with the role of surface defects.

2.1 Free energy of a Janus colloid as a function of the immersion depth

Here we consider a spherical particle of radius R possessing two hemispherical faces of equal area. The particle surface is smooth and homogeneous in each hemispherical face. Within these assumptions we neglect for the moment the effect of pinning of the contact line. Moreover, we do not discuss the effect of line tension, since it does not affect significantly the energy of Janus colloids in the micrometer scale $\gamma R^2 \gg k_L R$ (where $\gamma = 10^{-1}$ to 10^{-2} N m^{-1} is a typical interfacial tension), for accepted values of line tension $k_L = 10^{-9}$ to 10^{-12} N.[20–22] The line tension effect starts becoming significant for nanoparticles of a few tens of nanometers or smaller. Using similar scaling arguments, gravity can be neglected here.

The description of the free energy of such a Janus particle at the interface is different from that of a bare colloid, because the two faces of the Janus particle, in general, possess different surface energies. The free energy of a Janus particle will therefore depend not only on the immersion depth but also on the orientation of the two faces and the area ratio of the two faces.[5]

We begin our analysis by considering the case of a Janus colloid with two distinct faces of equal area ($2\pi R^2$) at the air–water (A–W) interface under the assumption of a flat fluid interface and with the boundary of the two faces parallel to the interfacial plane. The faces are denoted as the P and S face respectively, see Fig. 1. Such conditions, which have been firstly considered by Ondarçuhu *et al.*,[5] allow us to both introduce the basic concepts and to use them as a benchmark to compare with the regimes where the constraint on the orientation of the Janus particle is relaxed. Following Ondarçuhu *et al.*,[5] we encode in the particle free energy $E(\alpha)$ all the contributions to the free energy of the system that depend on the particle properties and/or the particle immersion depth z. By introducing the contact angle $\alpha = \arccos(-z/R)$ (see Fig. 1) the particle energy reads as:

If $\alpha < 90°$,

$$E(\alpha) = 2\pi R^2\left[\gamma_{P-W} + \gamma_{S-W}(\cos\alpha) + \gamma_{S-A}(1-\cos\alpha) - \frac{1}{2}\gamma_{A-W}\sin^2\alpha\right] \tag{1}$$

If $\alpha > 90°$,

$$E(\alpha) = 2\pi R^2\left[\gamma_{S-A} + \gamma_{P-A}(-\cos\alpha) + \gamma_{P-W}(1+\cos\alpha) - \frac{1}{2}\gamma_{A-W}\sin^2\alpha\right], \tag{2}$$

where γ_{S-A} is the S face–air interfacial tension, γ_{S-W} is the S face–water interfacial tension, γ_{P-A} is the P face–air interfacial tension, γ_{P-W} is the P face–water interfacial tension, and γ_{A-W} is the air–water interfacial tension; $\cos\alpha_{S,eq} = (\gamma_{S-A} - \gamma_{S-W})/\gamma_{A-W}$ and $\cos\alpha_{P,eq} = (\gamma_{P-A} - \gamma_{P-W})/\gamma_{A-W}$. In Fig. 2 we plot some profiles calculated for typical values of the interfacial tensions.

Configurations of single Janus particles at a fluid interface have been discussed by Casagrande *et al.*[6] When the two faces possess very distinct contact angles: the hydrophilic face (P) $\alpha_{P,eq} < 90°$ and the hydrophobic one (S) $\alpha_{S,eq} > 90°$, the equilibrium contact is 90°, see Fig. 1A and 2A. In this configuration, the hydrophobic face is immersed in the hydrophobic medium and the hydrophilic

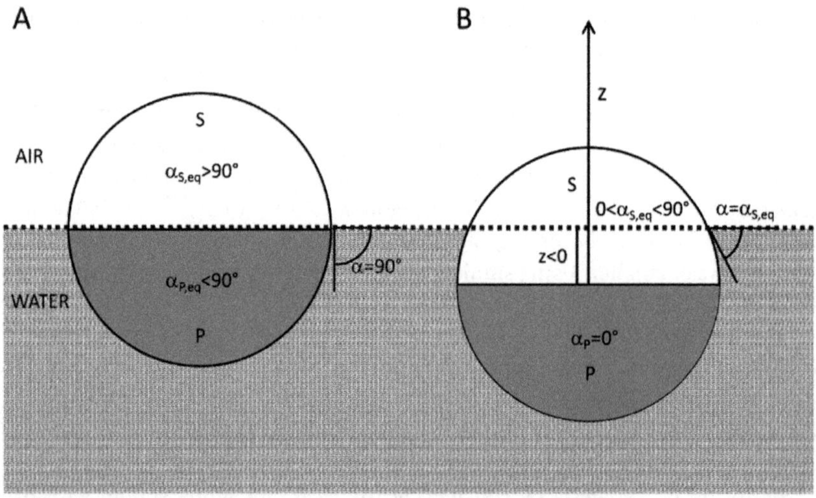

Fig. 1 Sketch of an amphiphilic Janus colloid (A) and a double hydrophilic Janus colloid (*i.e.* possessing two hydrophilic faces) (B) at the air–water interface.

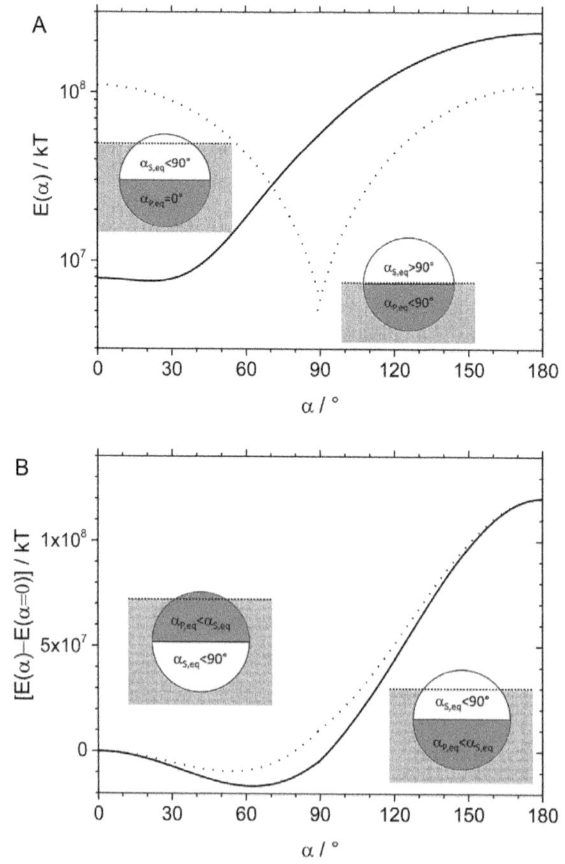

Fig. 2 (A) Energy as a function of the contact angle α for an amphiphilic (dotted line) and a double hydrophilic Janus colloid (solid line). Here the energy profiles were calculated using eqn (1) and (2): $R = 1$ µm, $\gamma_{A-W} = 72.5$ mN m^{-1}, (1) solid curve: $\alpha_{S,eq} = 21°$, $\alpha_{P,eq} = 0°$ ($\gamma_{S-A} = \gamma_{A-W} = 72.5$ mN m^{-1}, $\gamma_{S-W} = 5$ mN m^{-1}, $\gamma_{P-A} = \gamma_{A-W}$, $\gamma_{P-W} = 0$); (2) dashed curve: $\alpha_{S,eq} = 114°$, $\alpha_{P,eq} = 66°$ ($\gamma_{S-A} = \gamma_{P-W} = 20$ mN m^{-1} and $\gamma_{P-A} = \gamma_{S-W} = 50$ mN m^{-1}). (B) Energy profile of a platinum–silica Janus colloid oriented with the platinum face towards the air (dotted line) and with the silica face towards the air (solid line), $E(\alpha = 0) = 2.12 \times 10^8 kT$.

face is immersed in the hydrophilic medium. The immersion depth is just determined by the minimization of the area of unfavorable contact and none of the faces respects its equilibrium contact angle: the contact line wets the Janus boundary with the unique geometry of a quadruple line, where four different media meet. Both faces pull the Janus particle towards their side and the free energy profile shows a singularity at the minimum as shown in Fig. 2.

When both the P and S faces prefer to be wetted by water, $\alpha_{P,eq} < 90°$ and $\alpha_{S,eq} < 90°$, the Janus colloid can be defined as double hydrophilic and the configuration of the Janus particle will be similar to that of a bare colloid, see Fig. 1B. However, also in this case the free energy profile shows a discontinuity in the first derivative when the fluid interface crosses the Janus boundary. The energy profile of a double hydrophilic Janus colloid possessing $\alpha_{S,eq} = 21°$, $\alpha_{P,eq} = 0°$ is also shown in Fig. 2A. It is worth noting that the adsorption energy for a double hydrophilic particle $E(\alpha = 0) - E(\alpha_{eq}) = 0.3 \times 10^6 kT$ is much lower than that for the amphiphilic case showing a value of about $10^8 kT$.

The double hydrophilic system is representative of platinum–silica colloids able to perform autonomous motion when immersed in a hydrogen peroxide solution.[15,23] Considering the latter system and using literature values of the interfacial tensions (silica–air $\gamma_{S\text{-}A} = 58.6$ mN m^{-1}, silica–water $\gamma_{S\text{-}W} = 25$ mN m^{-1};[24] platinum–air $\gamma_{P\text{-}A} = 153$ mN m^{-1} and platinum–water $\gamma_{P\text{-}W} = 110$ mN m^{-1} (ref. 25)) we calculate the free energy profiles in the two possible configurations when the Janus boundaries are parallel to the interface. The first one is the configuration describing the platinum face (P) oriented towards the water and the second one is the one with the platinum face oriented towards the air. The two energy profiles were calculated using eqn (1) and (2) and are shown in Fig. 2B.

In this example, given that the contact angle of platinum is smaller than that of silica,[24–26] the energy profile with the platinum oriented towards the water leads to a smaller free energy as compared to the case with the opposite orientation. The minimum of the energy corresponds to the contact angle of the Janus colloid which is just the silica equilibrium contact angle $\alpha_{S,eq} = 61°$, and the platinum is fully immersed in the water.

Note that if the Janus colloid is unable to change its orientation, stable contact angles may be found whenever equilibrium on the triple line will be satisfied. This occurs not only at $\alpha = 61°$ on the silica face but also at $\alpha_{P,eq} = 52°$ on the platinum face.

Having assumed that the Janus boundary remains parallel to the interface, we did not consider yet any change of orientation of the faces of the Janus particle. The calculations presented just above are actually valid for any orientation that prevents the Janus boundary from intersecting the interface.

2.2 Free energy of a Janus colloid as a function of the orientation

Now we start to investigate the change of the free energy as a function of the orientation of a Janus colloid when the Janus boundary is not parallel to the

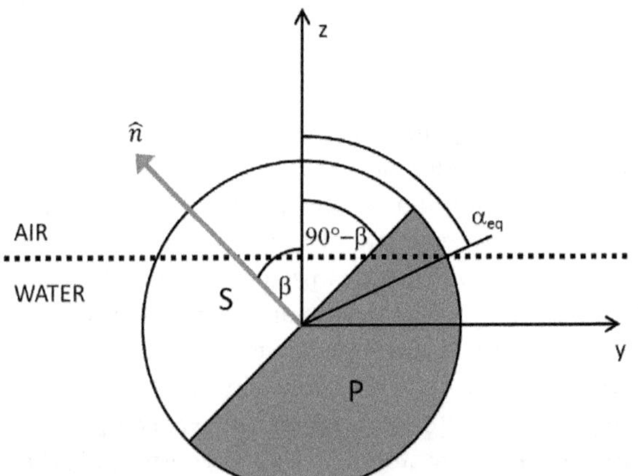

Fig. 3 Schematic representation of a Janus colloid at the air–water interface with the Janus boundary normal forming an angle β with the z-axis.

interface. Here we consider the case of a double hydrophilic colloid and assume that the equilibrium contact angle α_{eq} is the same along the three phase contact line and that the two Janus faces possess the same area ($2\pi R^2$). The orientation is defined by the angle β between the interface normal (*i.e.* z-axis) and the Janus boundary normal $\hat{n}$, see Fig. 3.

Even though the free energy of spheroidal particles has been calculated analytically,[27,28] so far the free energy landscape as a function of the orientation of Janus particles has been calculated only numerically or using computer simulations.[29–31] Here, we present an analytical expression of the free energy assuming a flat interface at a given equilibrium contact angle $\alpha = \alpha_{eq}$. In this view our model can be considered as the extension of the works of Ondarçuhu *et al.* to the case in which the Janus equator is not parallel to the interface.

In the range $0 < 90° - \beta < \alpha_{eq}$, the free energy $E(\beta) = E_T - \gamma_{A\text{-}W}A_i$ can be written as:

$$E(\beta) = \gamma_{S\text{-}A}A_{S\text{-}A} + \gamma_{P\text{-}A}A_{P\text{-}A} + \gamma_{S\text{-}W}A_{S\text{-}W} + \gamma_{P\text{-}W}A_{P\text{-}W} - \pi R^2\gamma_{A\text{-}W}\sin^2\alpha_{eq} \quad (3)$$

where A_i are the areas of the S or P face immersed either in air or water. The conservation of the total area of the particle and the conservation of the area of each Janus face yield:

$$\begin{aligned} A_{S-A} + A_{P-A} &= 2\pi R^2(1 - \cos\alpha_{eq}) \\ A_{S-W} + A_{P-W} &= 2\pi R^2(1 + \cos\alpha_{eq}) \\ A_{S-A} + A_{S-W} &= A_{P-A} + A_{P-W} = 2\pi R^2 \end{aligned} \quad (4)$$

A fourth relation is obtained by expressing, for instance, $A_{P\text{-}A}$ in spherical coordinates (r, θ, φ):

$$A_{P-A} = R^2 \int_{\theta=\frac{\pi}{2}-\beta}^{\alpha_{eq}} \int_{\varphi=\arcsin(1/\tan\beta\tan\theta)}^{\pi-\arcsin(1/\tan\beta\tan\theta)} \sin\theta \mathrm{d}\theta \mathrm{d}\varphi$$

$$A_{P-A} = \pi R^2(1 - \cos\alpha_{eq}) + 2R^2\left[\cos\alpha_{eq}\arcsin(\mathrm{cotan}\,\beta\,\mathrm{cotan}\,\alpha_{eq}) - \arctan\frac{\cos\beta}{\sqrt{\sin^2\beta - \cos^2\alpha_{eq}}}\right] \quad (5)$$

Fig. 4 shows two profiles calculated for two different immersion depths corresponding to the equilibrium contact angles of silica and platinum $\alpha_{eq} = 61°$ and $52°$ (Fig. 2). Three different regimes are distinguished depending on the β dependence of E. The first regime (I) starts at $90° - \beta = \alpha_{eq}$, when the whole P face is wetted by water. This regime describes a minimum energy plateau and it covers the range $\alpha_{eq} < 90° - \beta < 180° - \alpha_{eq}$. It means that the orientation angles corresponding to the energy minimum cover a range of $180° - 2\alpha_{eq}$, *i.e.* the smaller the α_{eq} the larger the range! The following regime (II) $180° - \alpha_{eq} < 90° - \beta < 180° + \alpha_{eq}$ covers a range of $2\alpha_{eq}$. Both P and S faces are partially wetted. The last regime (III) corresponds to the range of maximum energy with the S face completely wetted by water and the P face towards the air.

As sketched in Fig. 4, isolated Janus colloids may rotate only in the range of angles $\alpha_{eq} < 90° - \beta < 180° - \alpha_{eq}$ where the free energy is at the minimum.

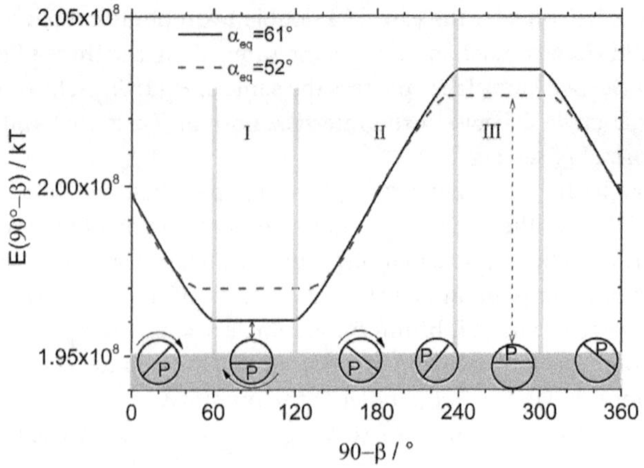

Fig. 4 Two profiles calculated for two different vertical positions corresponding to the equilibrium contact angles of silica ($\alpha_{eq} = 61°$, solid line) and platinum ($\alpha_{eq} = 52°$, dashed line). The scheme at the bottom shows the Janus colloid rotation at the air–water interface. I, II, III correspond to the three regimes described in the text.

A further rotation would correspond to an energy cost of about a million kT for typical values of the interfacial tension.[24,25] It is also worth pointing out that upon rotation, the Janus colloid may change its equilibrium contact angle α_{eq} to minimize its energy. Hence, a Janus colloid oriented with the P face towards the water, which starts rotating with an equilibrium contact angle as the one of the S face (see the solid up-down arrow, Fig. 4) may switch to the equilibrium contact angle of the P face (see the dashed up-down arrow, Fig. 4) when the whole S face is wetted by water.

Note that in all these calculations we assume that the air–water interface remains flat. With this assumption the Janus colloid will be subjected to a torque in regime II since the actual contact angle on one of the faces is not the equilibrium contact angle. If the contact line lies on both the silica and platinum surfaces, a complete analysis would show a dipolar distortion of the water surface to satisfy the two different contact angles.[29–31] This would yield a restoring torque for colloids with non-pinned lines. Deformation of the fluid interfaces due to non-equilibrium orientations have been investigated using computer simulations in the minimum surface energy requirement.[29] At the micrometer scale, accounting for the fluid deformation, one calculates slightly lower energies than when assuming a flat interface. Hence, our model provides the free energy only when the energy associated with the deformation of the interface remains small when compared to the total energy.

2.3 Role of surface defects

When accounting for the roughness and inhomogeneity of the surface, the ideal scenario described above may change significantly. Here we assume that the fluid interface remains macroscopically flat but the three phase contact line can be deformed due to nanometric defects. Weak or strong pinning of the contact line on a single defect may occur depending on the amplitude of the defect force $F_D = F_1(y_m - y_d)$ with respect to the restoring force $F_R = k(y_m - y_L)$ of the liquid tail (F_1 is the defect force function, y_m is the actual position of the line, y_d is the defect position, k is the spring constant of the restoring force and y_L is the position of the contact line in the absence of the defect).[32–34]

For weak defects, F_1/k is small and no hysteresis of the contact angle is expected. Nonetheless the contact line may show some displacement, whose mean squared amplitude is $\langle \Delta y^2 \rangle \sim R\xi$, where ξ is the correlation length of the defect and R is the radius of the particle.[32] If $\xi = 1$ nm and $R = 1$ μm, the amplitude of the contact line displacement is about 30 nm. Note that the latter value is much greater than the amplitude of the thermal capillary waves[35] ($\approx$0.2 nm) or the characteristic displacement in the molecular-kinetic theory of partial wetting (0.1–1 nm).[36]

For strong defects, F_1/k is large and contact angle hysteresis occurs. The contact line is pinned on the defects and only if enough energy is injected into the system, will a line displacement occur towards a new position for which $F_D = F_R$. The limiting values of stable macroscopic contact angles are the advancing α_A and receding α_R contact angles. For a spherical colloid, α_A is the angle reached after a displacement of the contact line resulting in an increase of the colloid wetted area; whereas α_R is the angle reached after a displacement of the contact line resulting in a decrease of the wetted area of the colloid. A useful expression for the contact angle hysteresis for isolated and non-interacting defects reported for structured hydrophobic surfaces made of micropillars is:[37]

$$\Delta\cos\alpha = \cos\alpha_R - \cos\alpha_A \approx \frac{1}{4}a\phi\ \ln\left(\frac{\pi}{\phi}\right), \tag{6}$$

where a is a dimensionless geometrical parameter and ϕ is the density of the micropillars or defects. Eqn (6) may also be interpreted in terms of the influence of surface roughness defects on the contact angle hysteresis. Thus, for non-interacting defects one expects an increase of the contact angle hysteresis if the density of defects increases.

In the following we discuss how the free energy profile $E(\alpha)$ would be affected by surface defects. We consider only the case of Janus colloid faces decorated by topographical defects, *i.e.* roughness. The only consequence of a defect is to increase the surface area. The simplest case is that of cylindrical defects of height h_d and radius R_d. Those defects decorate each face of the Janus colloid with a defect surface fraction $\phi_d = n_d\pi R_d{}^2/(2\pi R^2)$, where n_d is the number of defects. The modified energy profile becomes:

If $\alpha < 90°$,

$$E(\alpha) = 2\pi R^2\left[\left(1 + 2\phi_{d,P}\frac{h_{d,P}}{R_{d,P}}\right)\gamma_{P-W} + \left(1 + 2\phi_{d,S}\frac{h_{d,S}}{R_{d,S}}\right)\gamma_{S-W}(\cos\alpha)\right.$$
$$\left.+\left(1 + 2\phi_{d,S}\frac{h_{d,S}}{R_{d,S}}\right)\gamma_{S-A}(1-\cos\alpha) - \frac{1}{2}\gamma_{A-W}\sin^2\alpha\right] \tag{7}$$

If $\alpha > 90°$,

$$E(\alpha) = 2\pi R^2\left[\left(1 + 2\phi_{d,S}\frac{h_{d,S}}{R_{d,S}}\right)\gamma_{S-A} + \left(1 + 2\phi_{d,P}\frac{h_{d,P}}{R_{d,P}}\right)\gamma_{P-A}(-\cos\alpha)\right.$$
$$\left.+\left(1 + 2\phi_{d,P}\frac{h_{d,P}}{R_{d,P}}\right)\gamma_{P-W}(1+\cos\alpha) - \frac{1}{2}\gamma_{A-W}\sin^2\alpha\right] \tag{8}$$

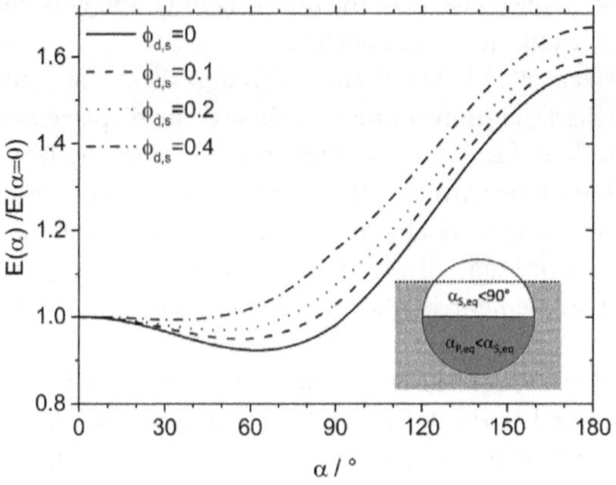

Fig. 5 Free energy profile accounting for topographical cylindrical defects of a given area fraction $\phi_{d,S}$ and $\phi_{d,P} = 0$, $h_d = R_d$. Solid line in the figure corresponds to the solid line in Fig. 2B.

The effect of increasing the defect density is shown in Fig. 5, where we consider a scenario accounting for defects only in the S face and $h_d/R_d = 1$. As expected, the presence of defects increases the energy but more importantly it deeply modifies the α dependence. As the specific area of the colloid increases, the extra energy cost due to the colloid–water contact gets closer and closer to the energy gain from the reduction of the air–water contact and this progressively shifts the $E(\alpha)$ minimum and makes it more shallow. One may compare the angles of the energy minima calculated in eqn (7) and shown in Fig. 5 with the Wenzel equilibrium contact angle α_W for rough surfaces characterized by a ratio r of the actual area and the geometric area.[38] $\alpha_W = \arccos\left(\frac{\gamma_{S-A} - \gamma_{S-W}}{\gamma_{A-W}} r\right) = \arccos\left[\frac{\gamma_{S-A} - \gamma_{S-W}}{\gamma_{A-W}}\left(1 + 2\phi_{d,S}\frac{h_{d,S}}{R_{d,S}}\right)\right]$ agrees well with the values calculated in eqn (7): $\alpha_W = 61°$ for $\phi_{d,S} = 0$, $\alpha_W = 54°$ for $\phi_{d,S} = 0.1$, $\alpha_W = 47°$ for $\phi_{d,S} = 0.2$ and $\alpha_W = 29°$ for $\phi_{d,S} = 0.4$, see Fig. 5.

However it is important to recall that the main consequence of the presence of strong defects is still the pinning of the contact line. Hence, it may be very slow for the Janus colloid to find the equilibrium position. In the next section, kinetic aspects related to the interfacial friction of the Janus colloid rotational motion are discussed.

2.4 Interfacial rotational friction

The rotational friction (about an axis parallel to the interface) of a colloid trapped at the surface of water is related to the hydrodynamic dissipation described in partial wetting when dealing with the motion of a liquid drop approaching its equilibrium contact angle.[39] In both cases, the dissipation is determined by the relative motion of the liquid and the solid at the contact line. A no-slip condition would lead to an infinite dissipation.[39] Hence a colloidal particle does not rotate at all in no-slip conditions! However as in partial wetting dynamics, a slip length or a molecular cut-off can be introduced to account for large but finite dissipation.

Recent numerical studies have characterized the diffusion of homogeneous[40] and Janus[19] nanoparticles adsorbed at a fluid–fluid interface. On the other hand, very few theoretical predictions on the rotational motion of even bare spherical particles partially immersed in a liquid can be found in the literature. O'Neill and coworkers theoretically investigated the rotational and translational friction for the special geometry of a bare spherical particle half-immersed in water (contact angle $\alpha_{eq} = 90°$).[18] Here the rotational friction coefficient $f_{r,\perp}$ describes a rotation of an angle β (see Fig. 3) about an axis perpendicular to the interface normal as sketched in Fig. 4.

A non-dimensional sliding friction coefficient $B = R/b$, where b is the slip length, has been introduced by O'Neill and coworkers to evaluate the rotational friction.[18] The case of $B = \infty$ corresponds to the no-slip condition, whereas $B = 0$ corresponds to the perfect slip condition. The rotational friction coefficient is:[18]

$$f_{r,\perp} = k^{r}_{\perp}\eta R^{3}, \tag{9}$$

where η is the liquid viscosity. The expression of the friction factor $k^{r}_{\perp}$ is rather complex. It changes from zero in the perfect slip condition to infinity in the no-slip condition. The following empirical formula describes the asymptotic behavior for $B > 1$:

$$k^{r}_{\perp} = 4.5 \ln B + 2.5 \tag{10}$$

Note that in the bulk the rotational friction is $8\pi\eta R^3$. A typical value of the slip length $b = 1$ nm (0.1 nm)[41,42] for a colloidal particle of $R = 1$ μm yields $B = 1000$ (10 000) and $k^{r}_{\perp} = 33.6$ (43.9), which is 34% (75%) higher than the bulk value. Hence the rotation of a bare colloid at the interface will be slowed down with a rotational diffusion time $\tau_{r,\perp} = f_{r,\perp}/kT = 8.4$ s (11.0 s) significantly higher (but of the same order of magnitude) than the rotational diffusion time in bulk water $\tau_{r,W} = 6.3$ s.

For Janus colloids the scenario becomes more complicated because the free energy of the system changes significantly with the orientation (see Fig. 4). The rotational diffusion occurring in a harmonic potential can be described by Arrhenius behavior:

$$D_{r,\perp} = \frac{kT}{k^{r}_{\perp}\eta R^{3}}\exp(-|\Delta E_{\beta}|/kT), \tag{11}$$

where ΔE_{β} is the activation energy for the change of orientation. Note that if only thermal agitation kT drives the rotation, the Janus colloid can theoretically rotate only in the region where the free energy remains constant ($\alpha_{eq} < 90° - \beta < 180° - \alpha_{eq}$) as the activation energy is very high for the typical values considered in Fig. 4.

Pinning of the contact line also has a strong impact on the dynamics of Janus colloids at air–water interfaces. If strong defects are present, the contact line pinning leads to additional friction and a severe slowing down of the rotational diffusion. In fact, if the driving energy of rotation is only the thermal agitation (kT), one expects a suppression of rotational motion if the defect energy $\gg kT$.

When the defect energy (*i.e.* the total energy dissipated by a single defect around a hysteresis cycle[43]) is comparable to kT, a slowing down of the rotational diffusion is also expected as observed experimentally for a colloidal particle adsorbing or translating at the water interface and explained in terms of the thermal hopping of the contact line.[44,45]

A Janus colloid may rotate in the region of the minimum free energy orientation but the contact line is expected to be bound to the Janus boundary. It could even be completely stuck to the boundary since topological and chemical defects are expected to accumulate there. Hence, the orientation $\beta = 90° - \alpha_{eq}$ or $-(\alpha_{eq} + 90°)$ may represent the most probable configuration, since the free energy is at a minimum and the contact line would be pinned, preventing any further rotation.

3 Experimental methods

3.1 Particle fabrication

Pt–SiO_2 Janus colloids were fabricated following the procedure by Love *et al.*[46] Silica beads (purchased from Microparticles GmbH, radius $R = 1.06 \pm 0.03$ μm, zeta potential = −13 mV) were cleaned by centrifugation/dilution cycles using Millipore water. This deionized water was produced by a Millipore Milli-Q filtration system with a resistivity of 18 MΩ cm.

First, a monolayer of silica beads was prepared on a silicon wafer (diameter 10 cm) by drop-casting: drops of particle suspension at a concentration of 0.1 mg mL^{-1} were regularly deposited with a syringe onto a silicon wafer to reach an average coverage of about 10%. Observation of the prepared sample shows that the particles form clusters of 4 to 30 particles rather than being homogeneously dispersed (as shown in Fig. 6A). This is possibly due to capillary forces during the evaporation of residual water.

Using plasma bombarded metal sputtering (ALCATEL SCM 400 system), first 10 nm of titanium and then 20 nm of platinum was deposited onto the silica bead monolayer. Using 30 minutes of sonication, Pt coated silica beads were freed into Millipore water. These Pt–SiO_2 Janus colloids were then cleaned and collected by centrifugation/dilution cycles using Millipore water.

Scanning electronic microscopy (SEM, FEI Quanta 200F) was used to observe the as-prepared Janus colloids. The sample was prepared by placing a drop of Janus colloid suspension onto a silica wafer and drying. As shown in Fig. 6B, the SEM images show that the Pt-coated silica particle has two distinct faces. The white faces of the particles correspond to the Pt-coated surfaces. The Janus boundary where the Pt-coated face and the bare silica face meet can be "linear" or "wavy". The wavy Janus boundary probably results from the shadowing effect during the metal deposition procedure.[47] When silica beads are closely packed, as sketched in Fig. 6A, the nearby beads act as shields preventing the deposition of metal atoms.

Looking into the finer details of the Pt-coated face, as shown in Fig. 6C, the Pt layer presents some roughness at a characteristic length scale of about 10 nm. When metal atoms are deposited onto a substrate which is rugged, the atoms do not arrive at the same time uniformly at the surface. This random fluctuation, which is inherent in the process, may create the surface roughness.

3.2 Contact angle of single Janus colloids

The contact angle of a single colloid at the water surface was measured using an optical microscopy technique and a gel trapping method.

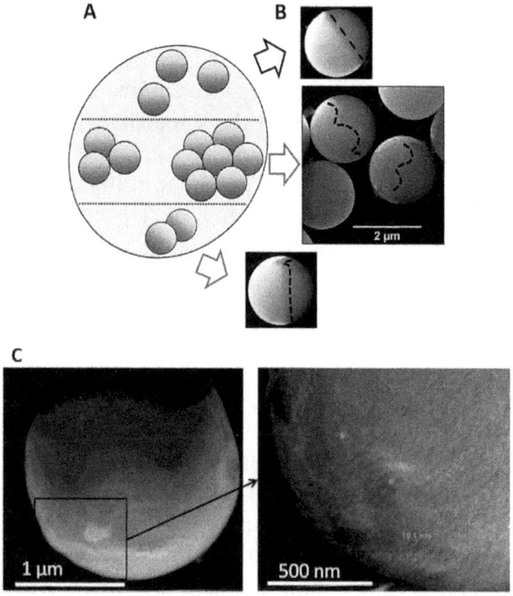

Fig. 6 (A) Sketch of the silica bead monolayer on a silica wafer. (B) SEM images of Pt–SiO_2 colloids. (C) SEM images of the Pt-coated surface.

The immersion of colloids in water was measured using an *in situ* method as described by Hórvölgyi *et al.*[48] A Janus colloid suspension in water was placed in between two parallel optical microscopy slides. The space between the two slides was ~4 µm. Some Janus particles attached onto the air–water interface. For observation, the sample cell was laid down under an inverted microscope equipped with an oil-phase objective (magnification ×100).

Due to the fact that the water layer between the two slides is rather thin, evaporation results in contact line motion of the water suspension on the glass slides. Concomitant motion of the particles demonstrated that particles were trapped at the air–water interface and not stuck onto the glass wall.

Another indirect way to measure the particle–water contact angle is to use a gel trapping method.[49] A 2 wt% Gellen (Phytagel, Sigma-Aldrich) solution in Millipore water was prepared by heating at 90 °C under magnetic stirring. After cooling down to room temperature the gel was formed (gel point 27–32 °C). The bead suspension in water was spread at the gel surface and again heated at 90 °C to let particles be trapped at the interface. The immersion depth of the beads at the gelled water surface is expected to be similar to that at the liquid water surface since the two interfaces have the same surface tension.

To get a replica of the gel interface with the beads, Norland Optical Adhesive 81 (NOA81) was poured over the gelled water surface and then photopolymerized by ultraviolet light for 2 minutes. The solidified NOA81 layer was peeled off taking particles at complementary positions with respect to those at the air-gelled water interface. These samples were finally observed using both optical microscopy and scanning electron microscopy.

3.3 Sessile drop contact angle on planar surfaces

We measured the contact angles of macroscopic water sessile drops on planar silica and Pt-coated surfaces. The planar silica surface is that of a silicon wafer and the platinum coating was made on the same silicon wafer surface. Silicon wafers were first cleaned with ethanol and acetone, and finally with Millipore water. The Pt-coated planar surface was prepared in the same way as the Janus colloids using the metal deposition method.[15] First 10 nm of titanium was deposited onto the wafer and then 20 nm platinum was deposited. The droplet profile was measured using a profile analyzer tensiometer (PAT, Sinterface, Germany).

3.4 Particle tracking and image analysis

Tracking of isolated particles was achieved using a Basler Scout CCD camera equipped Leica inverted microscope mounted on a Melles Griot optical table and a Leica objective of magnification ×32. A video with a field of view of 125 μm × 94 μm was recorded at a rate of 30 frames per second. The tracking was performed using Labview (National Instruments) using an image correlation-based approach ("Stat Tracker St. Andrews") to obtain the particle position over time [time t (s), x (μm), y (μm)]. Using IDL software, we treated the raw image sequences and by inputting a threshold on the grayscale level we detected the Pt-cap. Counting the number of elements in pixels, the area of the Pt-cap detected could be evaluated.

4 Experimental results and discussion

4.1 Janus colloid contact angle and orientation

Representative optical microscopy images displaying the immersion depth and orientation of Janus colloids at the air–water interface are shown in Fig. 7. The black region of the Janus colloid is the platinum face.

We evaluated the contact angle α of the particle from the image using the software ImageJ (and the "Contact Angle" plug-in) as shown in Fig. 7. The accuracy of the method is about 10° in the range of contact angles considered here.

Since the Janus colloid attached onto the interface from the water, where they were originally dispersed, the contact angle measured in Fig. 7 can be regarded as the receding contact angle of the Janus colloids. The average contact angle measured for different Janus colloids is 50° with a standard deviation of 6°.

We observed different possible orientations of the Janus colloids. This result is somewhat surprising considering that the literature equilibrium contact angle of

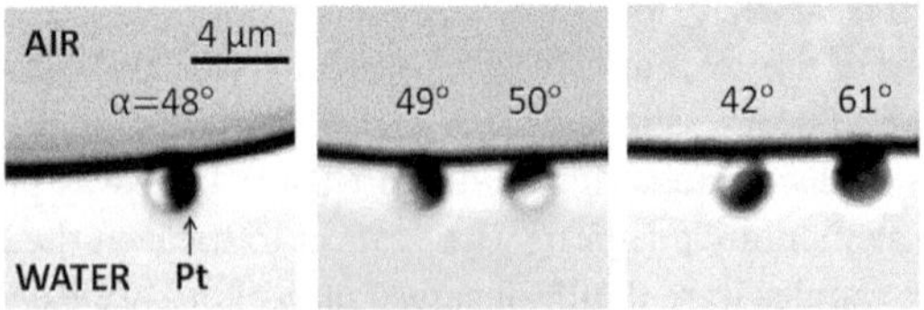

Fig. 7 Microscopic images of Janus colloids at the air–water interface. The black surfaces of the colloids are coated with Pt. Contact angles (α) measured by image analysis are shown.

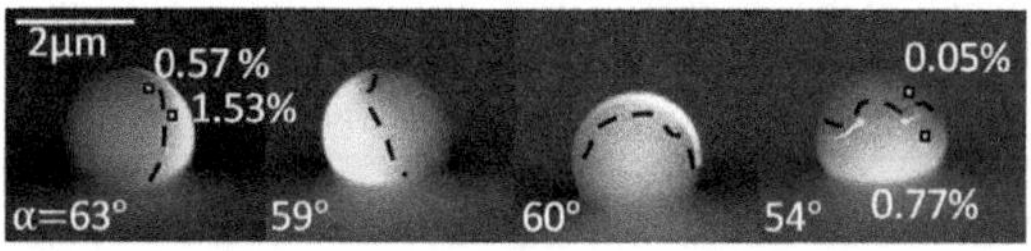

Fig. 8 SEM images of Pt–SiO_2 beads trapped in the NOA81 layer. Pt atomic percentages are shown for two particles with the white parts corresponding to Pt-coated surfaces.

platinum is lower than that of silica.[24–26] As pointed out in Section 2.1 and 2.2 for literature values of the interfacial tensions of silica and platinum one expects the platinum face to be completely wetted by water.

In Fig. 8 SEM images of a Janus colloid at the interface of the solidified NOA81 layer are also shown. Note that the visible part of the colloid is the one previously immersed in gelled water. Since Janus colloids were deposited on top of the gelled water and hence attached onto the interface from the air, the contact angle measured in Fig. 8 can be considered as the advancing contact angle of the Janus colloids, $\alpha_{\text{Janus,A}} = 64° \pm 2°$.[15] Experimental contact angles $\alpha_{\text{Janus,R}} = 50° \pm 6°$ and $\alpha_{\text{Janus,A}} = 64° \pm 2°$ can be compared to the theoretical equilibrium contact angles of silica, $\alpha_{\text{S,eq}} = 61°$, and platinum, $\alpha_{\text{P,eq}} = 52°$, shown in Fig. 2B.

The orientation of the Janus colloids also shows a wide range. Some Janus colloids were straddling the water surface with the air–water interfacial plane across their Janus boundaries. This can be seen clearly from the SEM image shown in Fig. 8. Elemental analysis was used to qualitatively assess that the region of high Pt% corresponded to the brighter face of the Janus colloid. For some other Janus colloids, the interfacial plane can even cross only the Pt-coated surface (see the right SEM image in Fig. 8). Note that we did not observe any significant pinning of the contact line on the Janus boundary, see Fig. 7 and 8. Hence the "wavy" form of the Janus boundary cannot be invoked to explain the orientation of the Janus colloids at the surface.

A random orientation was observed for other Janus systems such as gold–polystyrene particles at the oil–water interface[50] or gold–silica particles at the air–water and oil–water interfaces.[51] It was believed that the difference in equilibrium contact angles could reveal the difference of free energy for the different orientations[50] and the effect of pinning and surface roughness was qualitatively discussed.

Here we measure in Fig. 7 and 8 a hysteresis of the Janus colloid contact angle: $\Delta\cos\alpha_{\text{Janus}} = \cos\alpha_{\text{Janus,R}} - \cos\alpha_{\text{Janus,A}} = 0.2$, which points to a significant pinning of the contact line on the surface defects, see Section 2.3. These defects are clearly observed on the platinum coating as surface nanoroughness and on the Janus boundary. Note also that we did not find any significant change of the contact angle with the orientation of the Janus colloid. Thus, contact line pinning strongly affects the interfacial behavior of Janus colloids and the energy profiles based on ideal surfaces with single equilibrium contact angles cannot fully describe our results. In this sense, future theoretical investigations should take into account the influence of the contact angle hysteresis in the free energy.

4.2 Effect of the platinum coating on wetting

In order to further check the effect of contact line pinning we measured the liquid drop contact angles on solid planar surfaces. We aimed at investigating the effect of the coating layer on a silica substrate.

Table 1 Contact angles of the liquid drop with the silica surface and the Pt-coated surface

Contact angle	Advancing (α_A)	Receding (α_R)
Water–silica surface	$56.9 \pm 1.0°$	$44.3 \pm 0.2°$
Water–Pt-coated surface	$59.1 \pm 1.3°$	$40.1 \pm 0.4°$
Heptane–silica surface/Pt-coated surface	0	0

Results are listed in Table 1. They show that the advancing angle of water with the platinum coated surface is slightly higher than with the silica surface, while the receding contact angles show the opposite trend.

For other high surface tension liquids like glycerol we observed the same increase of contact angle hysteresis $\Delta\cos\alpha$ for the coated layer. Instead, low surface tension liquids like heptane show complete wetting. Note also that the absolute values of the advancing and receding contact angles are not very different between the bare silica and the platinum coated surface. Hence the energy of the two surfaces are somewhat similar. However as we pointed out before, the hysteresis of the contact angle $\Delta\cos\alpha_{\text{Silica}} = 0.16$ and $\Delta\cos\alpha_{\text{Pt-coat}} = 0.25$ reveals the presence of contact line pinning on strong defects. Using eqn (6), and assuming the geometrical factor $a = 1$ and $\Delta\cos\alpha = 0.2$, a defect density $\phi = 0.4$ is calculated. Note that for this defect density (40%) the energy profile may change dramatically, see Fig. 5.

4.3 Rotational friction of Brownian Janus colloids at the air–water interface

The presence of surface defects as discussed in Section 2.4 may also significantly slow down the rotational diffusion of the Janus colloid. Therefore, the orientation of a Janus colloid once attached onto the interface may not change significantly with time if the contact line is strongly pinned.

By analyzing the images taken during the tracking of the Brownian motion of Janus colloids at a planar air–water interface, we were able in only a few cases to evaluate the area of the platinum face observed by optical microscopy from the top. Hence, we calculate the orientation angle β from the area of the platinum face

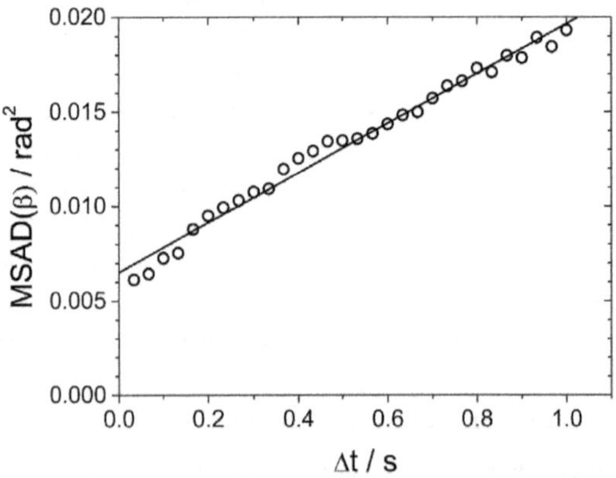

Fig. 9 Mean squared angular displacement as a function of the lag time for a Brownian Janus colloid diffusing at the air–water interface.

obtained by image analysis, and plot the mean squared angular displacement MSAD in Fig. 9 for one specific dataset for which $40° < \beta < 90°$.

We fitted the MSAD results by $\text{MSAD}(\beta) = \text{MSAD}_0 + 2D_{r,\perp}\Delta t$, where MSAD_0 accounts for the error introduced by the image treatment. The rotational diffusion time $\tau_{r,\perp} = 1/D_{r,\perp} = 152$ s measured in Fig. 9 is much larger than the time predicted accounting only for hydrodynamic friction, $\tau_{r,\perp} = 8.4\ldots11.0$ s, see Section 2.4. Note also that an energy ΔE_β of only few kT in eqn (11) could explain the experimental rotational diffusion time $\tau_{r,\perp} = 152$ s. However, for the angular displacement shown in Fig. 9, one expects $\Delta E_\beta \approx 10^5 kT$ in the ideal case shown in Fig. 4.

As discussed in Section 2.3, a slowing down of the rotational diffusion can be attributed to the dynamics of pinning–depinning of the contact line in the presence of topographic defects as the platinum face shows roughness on the nanometric scale (see Fig. 6).[34,44]

5 Self-propelled Janus colloids at the surface of water

Catalytic Janus colloids attached onto the water surface may show autonomous motion if a reactant is present in the aqueous phase as fuel. For platinum coated Janus colloids, the fuel is hydrogen peroxide and concentrations of a few percent are enough to show directional trajectories and a speed of 10 μm s^{-1}.[15] To realize self-propulsion parallel to the interface plane several requirements have to be fulfilled. In this section, theoretical and experimental results shown before will be used to discuss the ability of Janus colloids to perform self-propulsion at the surface of water.

First, the largest contact area between the catalytic surface and the liquid containing the fuel is sought. Furthermore the Janus boundary should not remain parallel to the interface, since no propulsive force can then be generated in a direction parallel to the interfacial plane. The latter requirement excludes amphiphilic Janus colloids from performing self-propulsion parallel to the interface. In fact, as discussed in Section 2.1, the Janus boundary of amphiphilic Janus colloids remains parallel to the interface and the equilibrium contact angle is 90°, see Fig. 1. The double hydrophilic Janus colloid instead is able change its orientation between the limiting cases where the Janus boundary touches the air–water interface without increasing its free energy. In fact, all these configurations result in the same free energy of the system, see Fig. 4. Hence if the colloids' surfaces are defect free, double hydrophilic colloids with low equilibrium contact angles are expected to show a better ability to self-propel at high speeds parallel to the interface with respect to double hydrophilic colloids with high equilibrium contact angles. In this regard, a Janus micrometric colloid with the catalytic P face possessing $\alpha_{P,eq} = 0$ and $\alpha_{S,eq}$ of few degrees shows the optimal conditions for the realization of self-propulsion at the surface of water. Note that when both faces of the Janus colloid possess the same equilibrium contact angle $\alpha_{P,eq} = \alpha_{S,eq}$ the colloid is equivalent to a bare colloid and all orientations correspond to the same energy. In the latter system for relatively high contact angles, one expects an active motion interrupted by a thermal Brownian motion corresponding to the orientations for which the catalytic face is pulled out of the aqueous phase. Finally, the presence of the fuel may affect the interfacial tensions and the free energy landscape and the overall dynamics may depend non trivially on the fuel concentration.[52] Note that for hydrogen peroxide solutions the surface tension change is very small.[53]

Experimental observations for the silica–platinum Janus colloids presented in Section 4 show the importance of surface roughness, which leads to contact line pinning resulting in contact angle hysteresis and random orientations being observed at the interface. Pinning of the contact line also plays a crucial role in the slowing down of the rotational diffusion, see Sections 2.4 and 4.3. Contact line pinning over the defects present on the Janus colloid surface may be invoked to explain the slowing down of the rotational diffusion observed in the absence of catalytic fuel. This result also agrees with the observation that in presence of hydrogen peroxide, the propulsion speed remains constant with time, which corresponds to a small variation of the orientation angle β.[15]

6 Conclusion

We have shown that the behavior of a Janus particle trapped at the air–water interface is controlled by the wettability of its two faces. At equilibrium, stable configurations can be determined by minimizing the free energy of the system upon variation of the relative orientation of the Janus boundary with respect to the interface normal. By assuming a flat interface, we have been able to derive an analytical expression for the free energy as a function of the orientation. It is worth noting that our model provides the exact free energy difference between states in which both halves of the particle are wetted by a single fluid phase, hence providing bounding values for the free energy once deformation of the interface is taken into account.

Comparing our analytical results to our experiments we realized that contact line pinning plays a crucial role in the description of the interfacial behavior of Janus colloids. In particular, contact line pinning due to surface roughness or defects probably prevents the predicted equilibrium position and orientation at the air–water interface to be reached. Surface roughness and defects, which are more dense at the Janus boundary and on the catalytic face, cause random orientations of Janus colloids at the interface. Surface defects also significantly slow down the rotational diffusion of the Janus colloids.

Our results shed light on the behavior of Janus colloids self-propelling at the air–water interface. Janus colloids at the fluid interface may behave as a kinetically driven system, where the contact line motion over the defects decorating the Janus faces rules the contact angle hysteresis, orientation and rotational diffusion of the particles. Such a feature is crucial not only for the dynamics of a single particle, but also for their collective behavior and therefore for the development of self-assembly at fluid–fluid interfaces.

Acknowledgements

We thank Michel Gross for discussions and helping us with some calculations. Financial support from Labex Chemisyst and Labex Numev are also acknowledged.

References

1 A. Walther and A. H. E. Müller, *Soft Matter*, 2008, **4**, 663.
2 S. Jiang, Q. Chen, M. Tripathy, E. Luijten, K. S. Schweizer and S. Granick, *Adv. Mater.*, 2010, **22**, 1060–1071.
3 A. Perro, S. Reculusa, S. Ravaine, E. Bourgeat-Lami and E. Duguet, *J. Mater. Chem.*, 2005, **15**, 3745.
4 Q. Chen, J. K. Whitmer, S. Jiang, S. C. Bae, E. Luijten and S. Granick, *Science*, 2011, **331**, 199–202.
5 T. Ondarçuhu, P. Fabre, E. Raphaël and M. Veyssié, *J. Phys.*, 1990, **51**, 1527–1536.
6 C. Casagrande, P. Fabre, E. Raphaël and M. Veyssié, *Europhys. Lett.*, 2007, **9**, 251–255.
7 S. Fujii, M. Kappl, H.-J. Butt, T. Sugimoto and Y. Nakamura, *Angew. Chem., Int. Ed. Engl.*, 2012, **51**, 9809–9813.
8 R. Aveyard, *Soft Matter*, 2012, **8**, 5233.
9 A. Kumar, B. J. Park, F. Tu and D. Lee, *Soft Matter*, 2013, **9**, 6604–7202.
10 R. Golestanian, T. B. Liverpool and A. Ajdari, *New J. Phys.*, 2007, **9**, 126.
11 W. Wang, W. Duan, S. Ahmed, T. E. Mallouk and A. Sen, *Nano Today*, 2013, **8**, 531–534.
12 J. Howse, R. Jones, A. Ryan, T. Gough, R. Vafabakhsh and R. Golestanian, *Phys. Rev. Lett.*, 2007, **99**, 048102.
13 B. Jurado-Sánchez, S. Sattayasamitsathit, W. Gao, L. Santos, Y. Fedorak, V. V. Singh, J. Orozco, M. Galarnyk and J. Wang, *Small*, 2014, **11**, 499–506.
14 L. Soler, V. Magdanz, V. M. Fomin, S. Sanchez and O. G. Schmidt, *ACS Nano*, 2013, **7**, 9611–9620.
15 X. Wang, M. In, C. Blanc, M. Nobili and A. Stocco, *Soft Matter*, 2015, **11**, 7376–7384.
16 P. Malgaretti, M. N. Popescu and S. Dietrich, *Soft Matter*, 2016, **12**, 4007–4023.
17 R. Golestanian, T. Liverpool and A. Ajdari, *Phys. Rev. Lett.*, 2005, **94**, 1–4.
18 M. E. O'Neill, K. B. Ranger and H. Brenner, *Phys. Fluids*, 1986, **29**, 913.
19 H. Rezvantalab, G. Drazer and S. Shojaei-Zadeh, *J. Chem. Phys.*, 2015, **142**, 014701.
20 J. Drelich, *Colloids Surf., A*, 1996, **116**, 43–54.
21 L. Schimmele, M. Naplórkowski and S. Dietrich, *J. Chem. Phys.*, 2007, **127**, 164715.
22 T. Getta and S. Dietrich, *Phys. Rev. E: Stat. Phys., Plasmas, Fluids, Relat. Interdiscip. Top.*, 1998, **57**, 655–671.
23 J. G. Gibbs and Y.-P. P. Zhao, *Appl. Phys. Lett.*, 2009, **94**, 163104.
24 B. P. Binks and J. H. Clint, *Langmuir*, 2002, **18**, 1270–1273.
25 V. Leo, A. Tusa, Y. C. Araujo and V. Leo, *Colloids and Surfaces A: Physicochem. Eng. Aspects*, 1999, **155**, 131–136.
26 J. R. Gardner and R. Woods, *J. Electroanal. Chem. Interfacial Electrochem.*, 1977, **81**, 285–290.
27 J. Faraudo and F. Bresme, *J. Chem. Phys.*, 2003, **118**, 6518–6528.
28 H. Lehle, E. Noruzifar and M. Oettel, *Eur. Phys. J. E*, 2008, **26**, 151–160.
29 H. Rezvantalab and S. Shojaei-Zadeh, *Soft Matter*, 2013, **9**, 3640.
30 D. L. Cheung and S. a. F. Bon, *Soft Matter*, 2009, **5**, 3969.

31 B. J. Park and D. Lee, *ACS Nano*, 2012, **6**, 782–790.
32 J. F. Joanny and P. G. de Gennes, *J. Chem. Phys.*, 1984, **81**, 552.
33 V. Hisler, L. Vonna, V. Le Houerou, S. Knopf, C. Gauthier, M. Nardin and H. Haidara, *Langmuir*, 2014, **30**, 9378–9383.
34 S. Ramos and A. Tanguy, *Eur. Phys. J. E: Soft Matter Biol. Phys.*, 2006, **19**, 433–440.
35 D. Bonn and G. Wegdam, *J. Phys.*, 1992, **2**, 1755.
36 T. D. Blake, *J. Colloid Interface Sci.*, 2006, **299**, 1–13.
37 M. Reyssat and D. Quéré, *J. Phys. Chem. B*, 2009, **113**, 3906–3909.
38 R. N. Wenzel, *J. Ind. Eng. Chem.*, 1936, **28**, 988–994.
39 D. Bonn, J. Eggers, J. Indekeu, J. Meunier and E. Rolley, *Rev. Mod. Phys.*, 2009, **81**, 739–805.
40 D. L. Cheung, *Chem. Phys. Lett.*, 2010, **495**, 55–59.
41 P. G. de Gennes, *Langmuir*, 2002, **18**, 3413–3414.
42 C. Sendner, D. Horinek, L. Bocquet and R. R. Netz, *Langmuir*, 2009, **25**, 10768–10781.
43 J.-M. Di Meglio and D. Quéré, *Europhys. Lett.*, 2007, **11**, 163–168.
44 G. Boniello, C. Blanc, D. Fedorenko, M. Medfai, N. Ben Mbarek, M. In, M. Gross, A. Stocco and M. Nobili, *Nat. Mater.*, 2015, **14**, 908–911.
45 D. M. Kaz, R. Mcgorty, M. Mani, M. P. Brenner and V. N. Manoharan, *Nat. Mater.*, 2012, **11**, 138–142.
46 J. C. Love, B. D. Gates, D. B. Wolfe, K. E. Paul and G. M. Whitesides, *Nano Lett.*, 2002, **2**, 891–894.
47 T. Karabacak, *J. Nanophotonics*, 2011, **5**, 052501.
48 S. N. Z. Hórvölgyi and J. H. Fendler, *Colloids Surf., A*, 1993, **71**, 327.
49 V. N. Paunov, *Langmuir*, 2003, **19**, 7970–7976.
50 B. J. Park, T. Brugarolas and D. Lee, *Soft Matter*, 2011, **7**, 6413.
51 D. J. Adams, S. Adams, J. Melrose and A. C. Weaver, *Colloids Surf., A*, 2008, **317**, 360–365.
52 A. Domínguez, P. Malgaretti, M. N. Popescu and S. Dietrich, *Phys. Rev. Lett.*, 2016, **116**, 078301.
53 M. Phibbs and P. Giguere, *Can. J. Chem.*, 1951, **29**, 173.

Faraday Discussions

PAPER

From near hard spheres to colloidal surfboards

Ljiljana Palangetic, Kirill Feldman, Raphael Schaller, Romana Kalt, Walter R. Caseri and Jan Vermant*

Received 15th March 2016, Accepted 21st April 2016
DOI: 10.1039/c6fd00052e

This work revisits the synthesis of the colloidal particles most commonly used for making model near hard suspensions or as building blocks of model colloidal gels, *i.e.* sterically stabilised poly(methyl methacrylate) (PMMA) particles. The synthesis of these particles is notoriously hard to control and generally the problems are ascribed to the difficulty in synthesising the graft stabiliser (PMMA-*g*-PHSA). In the present work, it is shown that for improving the reliability of the synthesis as a whole, control over the polycondensation of the 12-polyhydroxystearic acid is the key. By changing the catalyst and performing the polycondensation in the melt, the chain length of the 12-polyhydroxystearic acid is better controlled, as confirmed by ^{1}H-NMR spectroscopy. Control over the graft copolymer now enables us to make small variations of near hard sphere colloids, for example spherical PMMA particles with essentially the same core size and different stabilising layer thicknesses can now be readily produced, imparting controlled particle softness. The PMMA spheres can be further employed to create, in gram scale quantities, colloidal building blocks having geometrical and/or chemical anisotropy by using a range of mechanical deformation methods. The versatility of the latter methods is demonstrated for polystyrene latex particles as well.

1 Introduction

Over the past few decades, the study of model colloidal dispersions has been shown to be essential in furthering our understanding of a wide range of physical phenomena and engineering problems. As early as 1986, Pusey and van Megen demonstrated that suspensions of sterically stabilised poly(methyl methacrylate) (PMMA) colloids exhibited nearly perfect hard sphere equilibrium phase behaviour, with the volume fraction being the single control parameter.[1,2] A key feature for having near hard sphere behaviour was the presence of a short, dense brush of polyhydroxystearic acid on the particles with a sufficiently large diameter (in the micrometer range). Further studies showed how modulating this near hard sphere behaviour leads to changes in properties, for example various aspects of

Department of Materials, ETH Zurich, Vladimir-Prelog-Weg 5, CH-8093 Zurich, Switzerland. E-mail: jan.vermant@mat.ethz.ch

the rheology can be tailored introducing softness or weak attractions.[3] These observations increased the efforts to produce colloids with varying softness, different shapes and directional interactions or with a combination of these,[4–11] to further the array of colloidal building blocks employed to engineer colloid based soft matter.

Poly(methyl methacrylate) (PMMA) particles sterically stabilised with poly(methyl methacrylate)-*graft*-polyhydroxystearic acid (PMMA-*g*-PHSA) have been used as models to mimic atomic systems.[12] The phase transitions, the equation of state and the divergence of the viscosity with increasing volume fraction can be captured by hard sphere behaviour.[13] The colloidal scale enables studies of an array of fundamental phenomena, such as nucleation and crystallization,[14–16] dislocation dynamics[17] glass transitions,[15,16] flow in glasses,[18] and gelation[19,20] by relatively simple techniques such as light microscopy and scattering.

These particles are produced using a dispersion polymerisation that is notoriously difficult to reproduce and control. The synthesis was introduced during the 1970s,[4,21,22] but the one step procedure described by Antl *et al.*[23] is the most commonly used.[24–29] Recently, Elsesser and Hollingsworth reviewed the synthesis of the graft polymer stabiliser, PMMA-*g*-PHSA, pointing out the difficulties in its preparation and the importance of its properties in the dispersion polymerisation.[25] They showed that the graft copolymers manufactured in different labs, according to similar or even identical procedures, are not equally efficient in the production of the particles. One is thus left with an uneasy feeling that the most popular model system in colloid science is rather difficult to obtain with sufficient control.

The stabiliser is generally produced in three steps:[25]

(I) polycondensation of 12-hydroxystearic acid (HSA) to produce the "teeth" of the comb-like graft polymer;

(II) conversion of the 12-polyhydroxystearic acid (PHSA) with glycidyl methacrylate (GM) to yield PHSA-GM and

(III) synthesis of the PMMA backbone and simultaneous incorporation of the PHSA-GM polymer.

The first step, which is the most crucial step of the synthesis of the graft polymer, is done in solution and the water formed during the polycondensation is removed as an azeotrope, typically using a Dean–Stark apparatus. Both catalysed reactions, using methanesulfonic acid or tin(II)chloride, and noncatalysed reactions at a higher temperature have been employed, none giving satisfactory control. Most surprisingly it is typically recommended to carry out the polycondensation with technical grade 12-hydroxystearic acid, as higher purity leads to poorer control of the PHSA.[25]

The second aspect that makes the poly(methyl methacrylate) (PMMA) particles stabilised with PMMA-*g*-PHSA interesting is the possibility to deform them into anisometrically shaped colloids. The method to produce ellipsoids out of spherical PMMA particles was first introduced by Keville *et al.*[5,30] They embedded the colloids into a cross-linked polydimethylsiloxane (PDMS) film that was then uniaxially stretched at a temperature above the glass transition temperature (T_g) of poly(methyl methacrylate). After the cooling and subsequent degradation of the matrix they recovered highly monodisperse prolate ellipsoidal particles. This procedure is robust and ensures a relatively high yield. Moreover, it helped to create another type of colloid which can be considered as a model system for the investigations of particle shape effects.[31,32] Various methods for producing anisotropic particles have been proposed since then, which include making oblate particles[10,33,34] or even chemically patterned anisotropic particles.[32] PDMS cannot be used for hydrophilic particles, as they cannot be dispersed in it. This issue was resolved by Ho *et al.*[35] who produced ellipsoidal particles by uniaxial stretching of poly(vinyl alcohol) (PVA) films loaded with sulfonated polystyrene (PS) particles. The range of obtainable particle shapes was expanded by employing biaxial stretching[10,33] and adding chemical liquefaction of the PS particles to the stretching procedure.[36]

The aim of the present work is to first and foremost optimise the synthesis of the graft stabiliser PMMA-*g*-PHSA in order to facilitate the reproducible and controlled production of spherical PMMA particles, beyond hard spheres. Second we will describe methods for making large quantities of a variety of model anisotropic particles, beyond ellipsoids. The paper is divided into four sections. The first part focuses on the production and characterisation of the comb-like polymer (PMMA-*g*-PHSA) which is being used as the stabiliser of the particles. ^{1}H-NMR spectroscopy is used to monitor the polycondensation of the 12-hydroxystearic acid in detail, and to evaluate how to suppress the side reactions during the polycondensation with the goal of being able to vary the chain length of the PHSA chains on the graft copolymer. Next, the different comb polymers are used to make particles with different stabiliser layer thicknesses. The third section is devoted to the characterisation of the colloidal particles and briefly compares the effect of the length of the PHSA chains. The final part of the paper presents methods for producing gram scale quantities of disk-like (oblate ellipsoids) and surfboard-like (triaxial) particles.

2 Materials and methods

2.1 Stabiliser and particle synthesis

For the synthesis of the 12-polyhydroxystearic acid (PHSA), two grades of 12-hydroxystearic acid (HSA) were employed: 95% purity (Alfa Aesar, Karlsruhe, Germany) and >75% purity (TCI Deutschland GmbH, Eschborn, Germany). The reaction was catalysed with *p*-toluenesulfonic acid (TSA) or methanesulfonic acid (MSA) from Sigma-Aldrich. Glycidyl methacrylate (97%) (GM) was purchased from Acros Organics and used as received. Since the stability of the epoxy ring is crucial in the synthesis of the PMMA-*g*-PHSA, the glycidyl methacrylate was examined using ^{1}H-NMR spectroscopy prior to use. Monomers methyl methacrylate (MM) and methacrylic acid (MA) were also products of Acros Organics. Methyl methacrylate was distilled before it was used, but methacrylic acid was not further purified. The solvents *n*-dodecane, *n*-hexane and *n*-butyl acetate were purchased from Sigma-Aldrich, while toluene came from Acros Organics and ethyl acetate came from Fischer Scientific UK. All the solvents had the highest available purity. A radical scavenger *t*-butyl catechol (*t*BC) and a chain transfer agent 1-octanethiol were also received from Sigma-Aldrich. The reaction between PHSA and glycidyl methacrylate was catalysed by 1-(dimethylamino)dodecane (DDA) purchased from Fluka. The initiator in the polymerisation of methyl methacrylate, azobisisobutyronitrile (AIBN) (Fluka), was recrystallised from acetone before use.

The first step of the stabiliser synthesis (Fig. 1), *i.e.* the polycondensation of 12-hydroxystearic acid, is the most critical. An essential aspect for controlling the polycondensation is efficient water removal from the reaction mixture. When the reaction was performed in solution with toluene as a solvent, as in the reference method described in the literature,[25] water was removed using a Dean–Stark apparatus. For comparison, the reactions were also carried out in the melt state. Additionally different catalysts were used, and we compared the efficiency of methanesulfonic and toluenesulfonic acid. In these cases a vacuum (10^{-3} mbar) or strong nitrogen flow (1.35 l h^{-1}) was used for removing the water formed during the polycondensation. In all three protocols, the reaction temperature and the monomer/catalyst ratios were varied in the range 100–150 °C and 68–811 (wt/wt), respectively, in order to find reliable conditions which favour the main polycondensation reaction. The average number molecular weight of the molecules was based on the analysis of the ^{1}H-NMR spectra.

For the second step, *i.e.* reacting glycidyl methacrylate with the different PHSAs to prepare for the attachment to the back-bone of the comb (Fig. 1), we modified the reference procedure given by Hollingsworth and Elsesser[25] by using a better method for water removal and by using the ^{1}H-NMR results to calculate, for each chain length, the required quantities of PHSA/GM. Briefly, 3 g of PHSA with 5.1 monomer units was weighed into a reaction flask and placed in an oil bath heated to 140 °C. Once the PHSA had melted, a mixture of toluene (1.1012 g), glycidyl methacrylate (0.3324 g), 1-dimethylaminododecane (16 μl) and *t*-butyl catechol (0.0052 g) was added to the PHSA. The flask was connected to a reflux condenser and a nitrogen line which facilitated water removal with a nitrogen flow of 0.084 Nl min^{-1}. The refluxing took 7 hours and the reaction progress was followed by analysing the ^{1}H-NMR spectra of samples taken every hour. During the cooling process 2.2 g of toluene were added to compensate for the solvent that was lost during the reaction and to dilute the product to ~50 wt%. The exact concentration of the solution was determined by drying and weighing a small aliquot and the product was stored in a refrigerator.

In the third step (see Fig. 1), the stabiliser is connected to a PMMA backbone and a comb-copolymer was made using the standard procedure[25] and taking into account the variation in the molecular weight of the PHSA. Methyl methacrylate was first mixed with the initiator (AIBN) and then the product of the second step (toluene solution of PHSA-GM) was introduced. The mixture was put into a glass syringe and within 2–3 h was added to the reaction flask containing a mixture of ethyl acetate and butyl acetate in a ratio of 2 : 1 (wt/wt) and heated to 108 °C. The molar ratio between methyl methacrylate and PHSA-GM was kept between 13 and 15.5. The amount of AIBN was calculated based on the amount of methyl methacrylate; for every gram of methyl methacrylate, 0.017 g of initiator was used. Furthermore, the mass of the acetate mixture was 1.14 times higher than the mass of methyl methacrylate. Once the content of the syringe was completely transferred to the reaction flask, the refluxing continued for 6 additional hours. Within that time the initiator was added two times at intervals of 2 h. The respective aliquots were 2.2 times smaller than the amount added in the initial mixture. When the reaction was finished and the temperature was lowered to about 80 °C, the product was diluted with an ethyl acetate/butyl acetate mixture in order to avoid micelle formation which can occur when the concentration of the graft polymer exceeds 40 wt%.[37] The mass of the added ethyl acetate was equal to the mass of MM and the amount of butyl acetate was half the mass of methyl methacrylate. The exact concentrations of the stabilisers were determined by drying and weighing small aliquots of the solutions.

The PMMA particles were then produced following the procedure of Klein *et al.*[38] Hexane and dodecane were weighed into a 3-neck reaction flask in a ratio of 2 : 1 (wt/wt) and then the stabiliser solution was added to the mixture. The monomers methyl methacrylate and methacrylic acid were weighed in a beaker in a weight ratio of 98 : 2. The initiator and chain transfer agent were mixed with the monomers and added to the reaction flask which was placed above an oil bath heated to 95 °C. Once the AIBN was completely dissolved, the 3-neck flask was lowered into the oil bath and the reaction was allowed to proceed for 2 hours. When the mixture was cooled down to room temperature, the particles were filtered and washed extensively with heptane and dodecane. After five washes, the solvent was replaced with decalin.

2.2 Producing oblate ellipsoids and surfboard-like particles

Poly(methyl methacrylate) (PMMA) particles. To produce the films the particles were embedded into polydimethylsiloxane (PDMS) films by adjusting the procedure introduced by Keville *et al.*[5] and later improved by Mohraz *et al.*[31] and Zhang and collaborators.[32] In our specific case, polydimethylsiloxane ($M_n \sim$ 100 000, viscosity ~ 50 000 cSt, Sigma-Aldrich) was first mixed with *n*-heptane (Acros Organics) in a 1 : 2 (wt/wt) ratio. The particles were added to the PDMS/*n*-heptane mixture as a 6 wt% suspension in *n*-heptane. Poly(dimethylsiloxane-*co*-methylhydrosiloxane) (trimethylsilyl terminated, $M_n \sim 950$, Aldrich-Fine Chemicals), and the catalyst, tin(II)-2-ethylhexanoate (Aldrich-Fine Chemicals) were introduced. Both chemicals were first dissolved in a few ml of *n*-heptane. The amount of cross-linker and catalyst was based on the mass of PDMS and it was 0.3 and 0.36 wt%, respectively.[32] After adding all the components, the mixture was vigorously shaken and poured into circular Teflon mold(s) each with a diameter of 12 cm and left at room temperature for about 20 hours. During this time the solvent evaporated completely and precured films were formed. The films were then cured in a vacuum oven at 120 °C and 4 mbar for 6 hours. This resulted in ~800 μm thick rubbery PDMS film(s). In order to produce 6 of these sheets, 65.5 g of PDMS was needed. To test the production of the particles, two batches with different stabilisers, but with approximately the same size ~2 μm, were compared.

Before deformation, the edges of the films were removed and circular films with a diameter of 10 cm were made. The disk-like particles were obtained by lubricated compression of particle filled PDMS films using a Zwick/Z020 (Zwick Roell Group) as a mechanical platform. For this, a new geometry was designed in-house (Fig. 2a). The top and the bottom surfaces were lubricated with fluoroether based grease (Krytox, DuPont) to ensure uniform deformation of the films. The compression was carried out at 160 °C, well above the glass transition temperature (T_g) of PMMA. The whole procedure consisted of five steps. First, the setup was preheated to 160 °C. Once this temperature is reached, the PDMS is placed on the bottom plate of the lubricated geometry (see Fig. 2). The sample was conditioned for 15 minutes. The next step was applying a predefined force, which was chosen based on the desired final aspect ratio. The films were kept under a constant force (<±5%) for 7 h, with the magnitude of the force depending on the

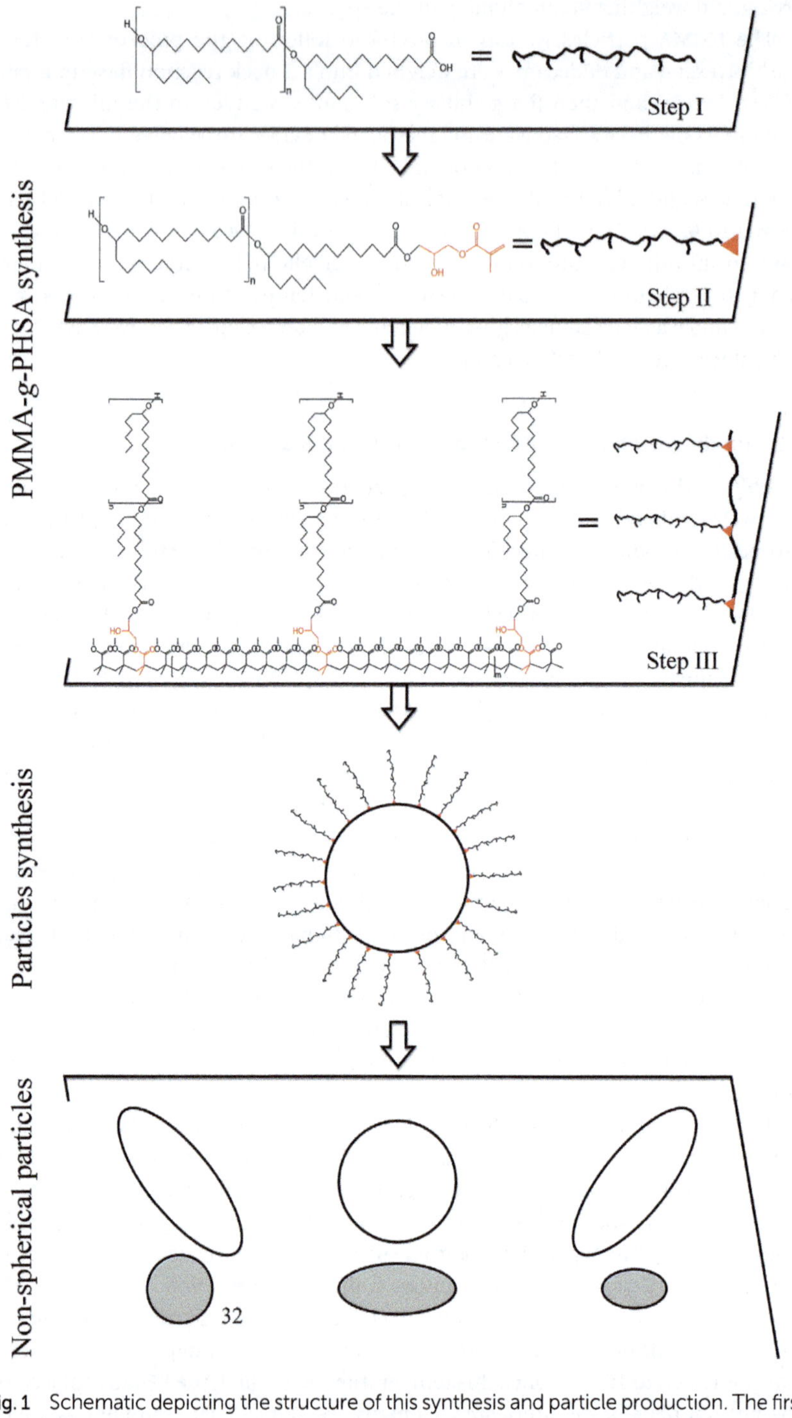

Fig. 1 Schematic depicting the structure of this synthesis and particle production. The first part deals with the synthesis of the graft polymer that is used for PMMA particle production. The spherical particles are further deformed into ellipsoids (see ref. 32),and disk-like and surfboard-like particles (top view and cross sections in grey).

Fig. 2 Geometries for the thermo-mechanical deformation of the latex particles: (a) lubricated compression and (b) plastic drawing of a particle-filled polymer tube over a heated fork.

diameter of the films and the desired aspect ratio. In the first hour the temperature was kept at 160 °C and then the sample was slowly cooled down to room temperature. Since the PDMS exhibits rubber-like behaviour, the films had to be kept under stress while cooling to ensure that the particles kept the new shape. After deformation, the sample was placed between absorbing paper sheets to remove some of the lubricating oil residues. Additionally, the films were soaked in perfluoroheptane for 5–10 minutes which sufficed to remove all traces of Krytox oil. In order to avoid edge effects, 2–3 mm were removed from the edges. Before proceeding to the degradation step, the PDMS sheets were weighed, cut into ~1 cm^2 pieces and swollen in *n*-hexane or *n*-heptane for 24 h.

The procedure to make the surfboards is essentially the same, but the films are now produced in a rectangular trough. After removing the edges the films are cut into 2 cm wide strips and they are compressed in lubricated channel dies of the same width.

A degrading reagent for PDMS is sodium methoxide. Similar to the procedure shown by Zhang, *et al.*[32] a 0.22 wt% stock solution of Na-methoxide in isopropanol and methanol was prepared. This was done by dissolving sodium in methanol to obtain a 0.5 M solution of Na-methoxide in methanol and by further dilution with isopropanol to the final concentration of 0.22 wt%. A given amount of the stock solution was added to the 4.3 wt% mixture of PDMS films and *n*-hexane or *n*-heptane to obtain a final component concentration (wt%) of 3.5% PDMS film, 77.9% *n*-heptane (or *n*-hexane), 17% isopropanol, 2% methanol and 0.04% Na-methoxide. The degradation proceeded for at least 15 h before the recovered particles were washed and re-grafted with a fluorescently labeled (with 7-nitro-benzo-2-oxa-1,3-diazol (NBD)) PMMA-*g*-PHSA (provided by Dr Andrew Schofield, University of Edinburgh) following the procedure from the work of Zhang *et al.*[32] The re-grafted particles were washed extensively with heptane and decalin and stored in decalin.

Polystyrene (PS) particles. Due to the hydrophilic nature of the highly charged sulfate-PS particles, the particles were embedded into polyvinyl alcohol (PVA) films. In order to obtain an 80 μm thick film of 36 × 32 cm^2, 12 g of matrix polymer (Mowiol 40–88, $M_w = 205\ 000$ g mol^{-1}, Sigma Aldrich) was dissolved in 480 g of MilliQ water for 24 h and then 3.45 g of an 8% (wt/v) particle (0.99 μm) suspension (Life Technologies) was added. The mixture was poured into a Teflon coated mold and left to dry at room temperature.

The films were cut into strips which were glued into tubes of different initial diameters, which were then deformed by plastic drawing using an in-house designed geometry[39] (Fig. 2b). The working temperature was fixed to 140 °C, unless otherwise specified. In our experiments, the initial tube length was fixed to 7 cm, while the diameter was varied from 30 to 50 mm. The film thickness was in the range of 0.045 to 0.25 mm before stretching. The tubes were drawn at speed of ~1 m min^{-1} and cooled down to room temperature in the stretched state to prevent relaxation. The final width of the tubes was 7 cm. The films were cut into ~1 cm^2 pieces and degraded in a mixture of water and isopropanol (7/3 v/v). After 12 hours of degradation, the particles were washed three times with the same mixture of water and isopropanol and two times with MilliQ water.

2.3 Characterisation

Nuclear magnetic resonance (NMR) spectroscopy. ^{1}H- and ^{13}C-NMR spectra were recorded using Bruker AV300 MHz and AV500 MHz spectrometers, using deuterated chloroform and perdeuterated dimethyl sulfoxide (DMSO), respectively, as solvents.

Gel permeation chromatography (GPC). The average number (M_n) and weight (M_w) molecular masses were determined using gel permeation chromatography (GPC) (GPCmax VE2001, Viscotek) equipped with two sets of detectors, TDA302

(Viscotek) and UV Detector 2500 (Knauer). The column was calibrated with polystyrene (PS) and chloroform was used as the running solvent.

Dynamic light scattering (DLS). The particles' size and stability was followed using dynamic light scattering. An ALV/CGS-3 compact goniometer system (ALV, Langen) connected to an ALV-7004 multiple tau digital real time correlator was used to perform these measurements. The size of the particles was determined by obtaining the correlation function at different angles from 30° to 150° with steps of 10°. The suspension stability was determined by following the changes in the correlation function with time. For the latter set of experiments, the angle was not changed (90°) and between each measurement the waiting time was fixed to 30 minutes.

Rheology. A rotational rheometer (MCR502, Paar Physica) was used with a double wall Couette geometry, and was thermostated using a Peltier system at 25.0 °C.

Scanning electron microscopy (SEM). The instrument used to acquire micrographs of the particles is a LEO 1530 Gemini (Zeiss, Oberkoch). The samples were deposited on silicon wafers and coated with gold or platinum. The working voltage was set to 1 kV. The images were used to determine the particles' dimensions.

Confocal laser scanning microscopy (CLSM). A multi-beam confocal scanning system VT-Infinity3 (VisiTech international) is used to perform the confocal imaging. The imaging is done using an EM-CCD camera (ImagEM X2 C9100-23B, Hamamatsu) and Nikon Eclipse Ti-E microscope equipped with a 100× oil immersion objective (CFI Plan Apo λ, 1.45 NA, ∞/0.17, 0.14 mm WD, Nikon). The samples were deposited on a glass cover slip and excited at 488 nm with a solid state laser (Laser engine, VisiTech international).

3 Results and discussion

3.1 Synthesis of the poly(methyl methacrylate)-*graft*-polyhydroxystearic acid (PMMA-*g*-PHSA)

Step I. In this stage the soluble part of the stabiliser is produced. The esterification reaction between the hydroxy and carboxylic groups of the 12-hydroxystearic acid (12-HSA) takes place in the presence of an acid catalyst (Fig. 3). The constitutional repeat unit with 12 carbon atoms is linked at the ester bond and

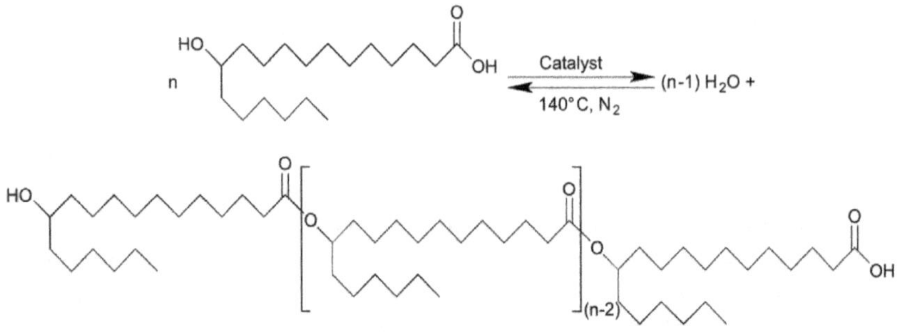

Fig. 3 Polycondensation of 12-hydroxystearic acid.

has a side chain of 6 carbon atoms. Typically the reaction is carried out in solution either without a catalyst, using a higher temperature,[37,40] or employing methanesulfonic acid (MSA)[23,25,29] or tin(II)chloride.[41] However, the control over this reaction is clearly limited. The most important issues are possible side reactions such as the elimination of water in 12-HSA. In an acid environment, the hydroxyl group on the C_{12}-atom can be removed leaving a double bond between the C_{11}- and C_{12}- or C_{12}- and C_{13}-atoms. As the elimination of water is reversible, subsequent addition and elimination of water can lead to a "migration" of the –OH group and of the double bonds, which can change the length of the side chains, and the level of branching. It has been reported that changing the position and/or the length of the branch affects the subsequent particle stability.[42] So it seems that a first key aspect in obtaining precise control over the stabiliser synthesis lies in preventing these elimination reactions. Accordingly, emphasis was put on the choice of catalyst.

Esterification reactions are typically catalysed by sulfuric acid, sulfonic acids or Lewis acids. Sulfonic acids are strong organic acids which are readily soluble in organic and inorganic mixtures. For the polycondensation of 12-hydroxystearic acid two acids from the family of sulfonic acids have been mentioned in the literature – methanesulfonic acid (MSA)[23,25,29] and toluenesulfonic acid (TSA).[43] Even though they are very akin, they affect the reaction in different ways. For example, the condensation between lauraldehyde and glycerol was shown to be more efficient when MSA was used as a catalyst instead of TSA under similar reaction conditions.[44] To the best of our knowledge, there is no analogous study for the case of the polycondensation of 12-HSA.

Another important aspect in the synthesis of the stabiliser is the removal of the reaction product, water, which is one of the key parameters controlling the chain length. To remove the water, some protocols prescribe carrying out the reaction in solution,[23,29] whereas other have performed the reaction in the melt state.[40,43] In solution, the water is typically removed by creating an azeotrope from which the water is separated using a Dean–Stark apparatus.[45] When the reaction is performed in the melt state, water is removed using a vacuum.[43]

Somewhat surprisingly, the majority of groups who have synthesised PHSA for the purpose of making near hard sphere particles seem to agree that technical grade (75–85% pure) 12-hydroxystearic acid leads to better results, when the focus is on making near hard sphere dispersions.[25] The main contaminant in the 12-HSA is stearic acid, but palmitic acid can also be present. These impurities can also participate in the esterification reaction and limit the chain growth (see Elsesser and Hollingsworth[25] for more details), and moreover they affect the branching of the PHSA. Broadening the range of chain lengths that could be produced would facilitate synthesis of PMMA particles of various sizes and different softness levels, but this would be difficult to achieve with the lower grade 12-HSA.

To obtain a deeper insight and better control over the synthesis, we recorded ^{1}H-NMR spectra as a function of reaction time for different reaction protocols. As a reference procedure the most commonly used method of Antl *et al.*[23] was used. The synthesis is performed using toluene as a solvent with methanesulfonic acid as a catalyst and 12-hydroxystearic acid of 95% purity, the ^{1}H-NMR spectrum of which is shown in Fig. 4a. Samples taken at various reaction times for temperatures ranging from 100 °C to 150 °C revealed the presence of considerable

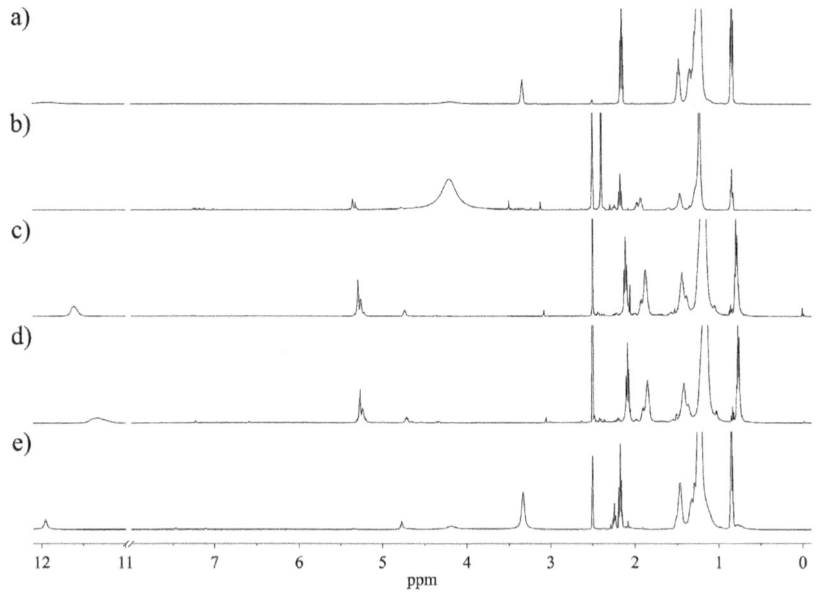

Fig. 4 ^{1}H-NMR spectra of (a) the 95% purity 12-hydroxystearic acid and (b–e) the products of the different polycondensation protocols. The reaction temperature was 140 °C in all cases except in (b) where it was 150 °C. The products shown in (b–d) were obtained after 1 hour using methanesulfonic acid as a catalyst. The ^{1}H-NMR spectrum in (e) represents a product of a prolonged reaction time (6 h) where the catalyst was toluenesulfonic acid. The amount of catalyst was varied only in (b) where it was 1.5 wt% with respect to the monomer; in all other cases it was 0.15 wt%. The water was removed using different methods: (b) as an azeotrope with toluene using Dean–Stark apparatus, (c) employing a vacuum and (d) and (e) with strong nitrogen flow. The solvent was perdeuterated DMSO.

quantities of olefins, products of the elimination reactions of water, as can be seen in Fig. 4b. Varying the temperature at which the catalyst was added also had no significant effect on the suppression of the unwanted side reactions. Following the reports that the reaction in solution is more prone to lactone formation[43] and that the production of light coloured (*i.e.* purer) fatty acid esters of higher alcohols should be done under vacuum to facilitate water removal,[46] the polycondensation of 12-HSA was performed in the melt state under vacuum. However, considerable quantities of elimination products arose also under these conditions (Fig. 4c), and once again varying the temperature and catalyst quantities showed no effect on the final product.

Since further modification of the procedure by employing a strong nitrogen flow to remove the water brought no improvements in the resulting product (Fig. 4d), the focus was turned to the chemical nature of the catalyst.

Toluenesulfonic acid (TSA) has been shown to be an alternate catalyst in the condensation of 12-hydroxystearic acid.[43] In our experiments, catalysis with TSA gave a ^{1}H-NMR spectrum of the product with a very low quantity of side products, even after 6 hours of reaction at 140 °C (Fig. 4e). Note that the only difference in the way the products shown in Fig. 4d and e were obtained was the catalyst used. Hence, from the NMR data it can be concluded that *p*-toluenesulfonic acid shows higher selectivity in the polycondensation of 12-hydroxystearic acid than methanesulfonic acid. Since the conditions in the polycondensation with *p*-toluenesulfonic acid resulted in a product with a very limited amount of olefins, the conditions under which this reaction was carried out were chosen as optimal, *i.e.* 140 °C was set as the reaction temperature and continuous nitrogen flushing as the means to remove the water produced.

In the literature, the success of the polycondensation reaction was typically determined by titration with alcoholic potassium hydroxide (KOH) solution. However, the precision with which transitions in the titration curves could be detected (either by colour or conductometry) was limited. To investigate the evolution of chain length, as well as the occurrence of side reactions, in more detail we used ^{1}H-NMR spectroscopy. Fig. 5 shows the spectra as a function of reaction time from 0 to 48 hours. The peak at 4.86 ppm in the ^{1}H-NMR spectra corresponds to the newly formed ester bond; accordingly this signal increases with time. The peak at 3.60 ppm, which represents the H-atom at the unreacted C_{12}-atom, decreases. Another change can be observed in the range of 2.00–2.50 ppm where the signals from the hydrogen atoms at the C_2-atom are positioned. The signal at 2.35 ppm represents a methylene group right next to the carboxyl group in the 12-HSA molecule, while the signal at 2.27 ppm depicts the methylene group next to the ester group. Based on the analysis of the ^{1}H-NMR spectra, or more precisely, by comparing the integrated intensities of the peaks at 3.60 and 4.85 ppm we see that it took 48 h to obtain PHSA with an average of 5.1 monomer units per chain under the given conditions.

The effect of purity was investigated by comparing the results obtained by employing 12-HSA with 75% purity under the same reaction conditions. The ^{1}H-NMR spectra of the two grades of 12-HSA and the products obtained within 48 h showed that the side reactions are suppressed in both cases. The products

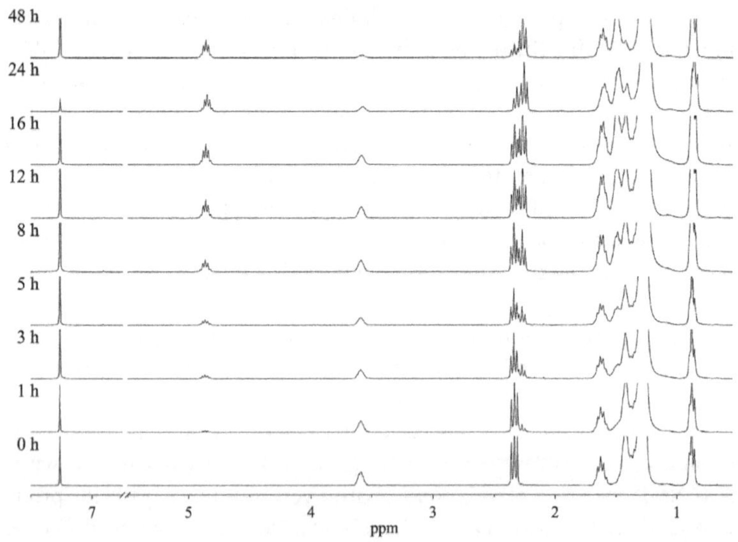

Fig. 5 ^{1}H-NMR spectra of the 95% 12-hydroxystearic acid and the reaction mixtures at different reaction times. The solvent was deuterated chloroform.

from both reactions were further compared by performing ^{13}C-NMR spectroscopy as shown in Fig. 6. In contrast to the literature, where a lower grade 12-HSA is generally considered to yield appreciable results, we found that less pure monomers give ill defined products. Namely, the region between 170 and 180 ppm in the ^{13}C-NMR spectrum reveals a range of carboxylic acids present in the sample prepared from the technical grade 12-HSA (inset in Fig. 6). As the spectra of the monomers do not show so many peaks in this region, some of the groups were obviously generated during the reaction. This implies that the stabiliser will be composed of a mixture of ester units.

To verify that the chain length of the PHSA for the pure 12-HSA can be simply controlled by varying the reaction time, additional experiments were carried out. The reaction proceeded for 24 hours under the selected conditions, then half of the reaction mixture was removed from the flask, the amount of TSA was doubled and the reaction was continued for an additional 24 hours. Based on the calculations from the recorded ^{1}H-NMR spectra, the product of the first 24 h and the PHSA obtained after an additional 24 h with an increased amount of catalyst had 3.6 and 15.7 monomer units, respectively. The ^{1}H-NMR spectra of the two products and the PHSA with 5.1 monomer units as well as the spectrum of the product obtained using 75% pure monomer are compared in Fig. 7.

Using the ^{1}H-NMR spectra of the samples taken at given time intervals, the conversion rate and reproducibility of the reactions were compared. The diagram in Fig. 8 shows how the chain size, as determined from the ratio between the free and reacted hydroxyl groups, changed with time. The repeatability was good. Moreover, the graph depicts that the addition of fresh catalyst has a significant

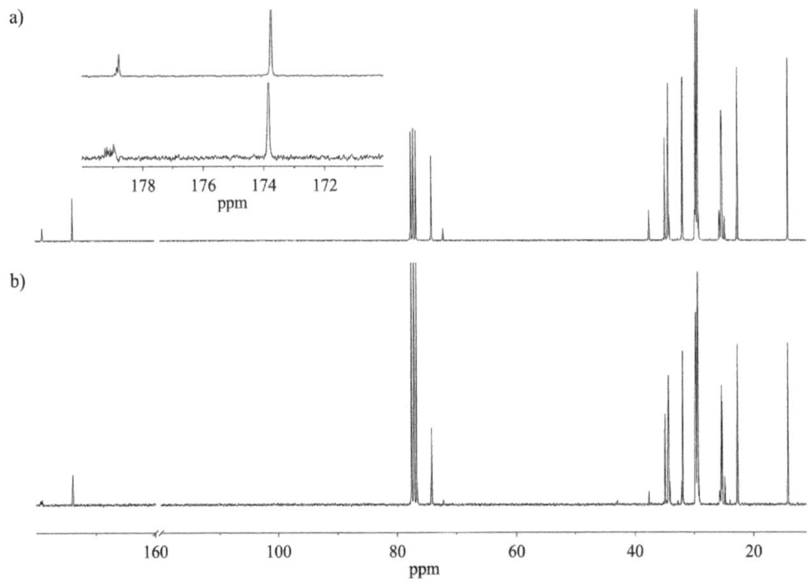

Fig. 6 Comparison of the ^{13}C-NMR spectra of the PHSA produced from the two grades of 12-HSA: (a) 95% and (b) 75% purity. The inset shows the expansion of the region where the signals of the C-atoms of the carboxyl groups are appearing. The solvent was deuterated chloroform.

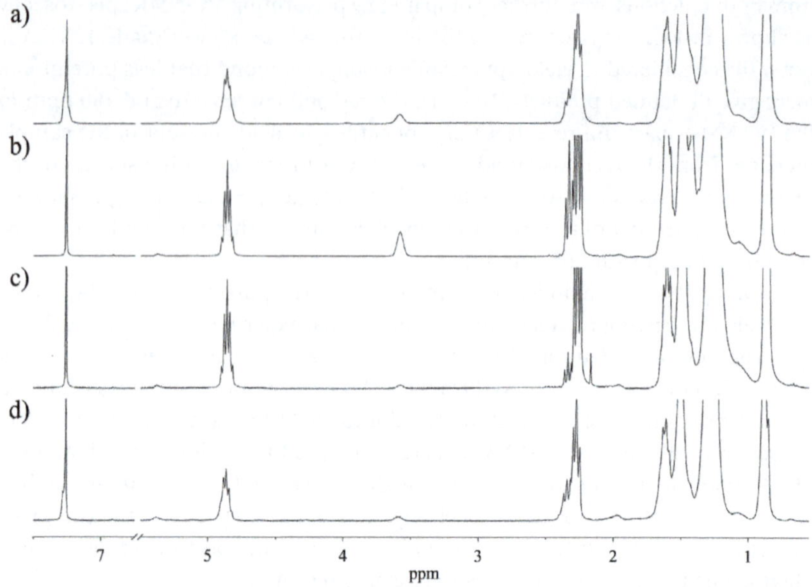

Fig. 7 ^{1}H-NMR spectra of PHSA obtained with different reaction times: (a) 48 h reaction time (5.1 monomer units); (b) 24 h reaction time (3.6 monomer units); (c) 48 h reaction time with the addition of catalyst after the first 24 h (15.7 monomer units); (d) 48 h reaction time with 12-HSA with 75% purity (11.7 monomer units). The number of monomer units was determined from the ratio of the areas under the peaks indicating the free and –OH groups that entered the ester bond. The solvent was deuterated chloroform.

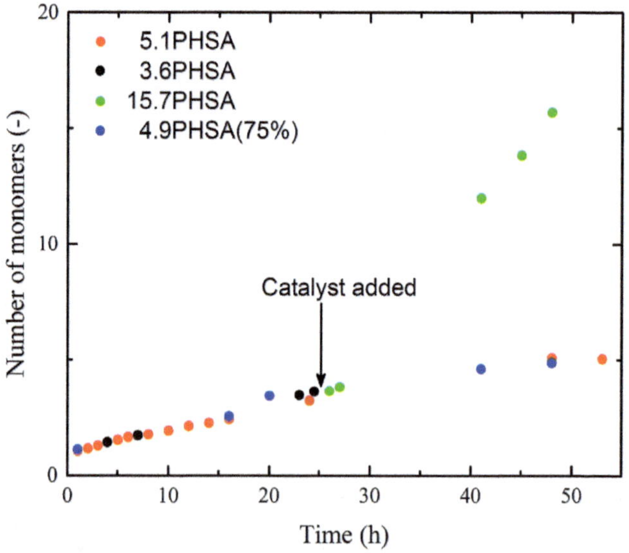

Fig. 8 Progress of the polycondensation reaction of different batches of 12-HSA. Note that the catalyst concentration was doubled after 24 h in the case of 15.7PHSA. 5.1PHSA refers to the sample with a final degree of polymerisation of 5.1, and 3.6PHSA and 15.7PHSA are designated accordingly.

impact on the reaction, in agreement with data from Bawn and Huglin.[43] However, they have observed an increase in the reaction rate after removing the used catalyst and introducing the same amount of fresh catalyst, implying that over time the catalyst loses efficiency.

The ^{1}H-NMR spectra were used to determine the number average molecular weight (M_n) of the different polyhydroxystearic acids. The average number of monomer units per chain of PHSA can be calculated as

$$\overline{x}_n = \frac{1}{\frac{n_{OH}}{rn_{ester}} + \frac{1}{r} - 1} + 1 \tag{1}$$

where n_{ester} and n_{OH} represent the number of ester and OH-groups, respectively, determined by the integration of the corresponding peaks in the ^{1}H-NMR spectra. The purity of the monomer is indicated by r which represents the initial ratio of hydroxyl and carboxylic acid groups. The parameter r of the two grades of 12-hydroxystearic acid used in this work was also determined from the ^{1}H-NMR spectra and it was 98% for the 95% 12-HSA and 87% for the >75% 12-HSA. The values of M_n calculated with $r = 1$ and the determined r are included in Table 1.

All the final products of the first step were also analysed using gel permeation chromatography (GPC) to obtain an alternative way to measure the size of the molecules. The number average molecular weights (M_n) determined by this method showed good agreement with the results acquired by ^{1}H-NMR spectroscopy when the purity of the monomer was taken into account (Table 1). Note that M_n of 15.7PHSA determined by ^{1}H-NMR spectroscopy is subject to a relatively high uncertainty due to the small interpreted intensity of the OH-groups. The polydispersity indices (PDI) obtained using GPC are relatively high which is typical for polycondensation reactions. However, the GPC results should be taken with caution. Namely, the molecular weight curve distribution seems to be composed of several overlapping distributions (Fig. 9a), which makes the determination of M_n less reliable. We performed repeated measurements and checked the effects of small changes in data interpretation on the results obtained from GPC. The collected data showed that the values of the peak average molecular weight (M_p) and weight average molecular weight (M_w) were always close to each other and that they were independent of the analysis protocol. However, the number average molecular weight and consequently the PDI varied somewhat depending on the data interpretation.

Table 1 Molecular weights of different 12-PHSA determined by ^{1}H-NMR spectroscopy and gel permeation chromatography

	NMR		GPC				
Sample	M_n ($r = 1$)/ g mol^{-1}	M_n ($r = r_{real}$)/ g mol^{-1}	M_n/ g mol^{-1}	M_w/ g mol^{-1}	M_z/ g mol^{-1}	M_p/ g mol^{-1}	PDI
5.1PHSA	1559	1459	1167	3794	6863	3749	3.3
3.6PHSA	1080	1035	820	2097	3632	1677	2.6
15.7PHSA	6353	4453	5445	10 345	15 396	10 516	1.9
4.9PHSA (75%)	3317	1401	1323	3589	6726	3083	2.7

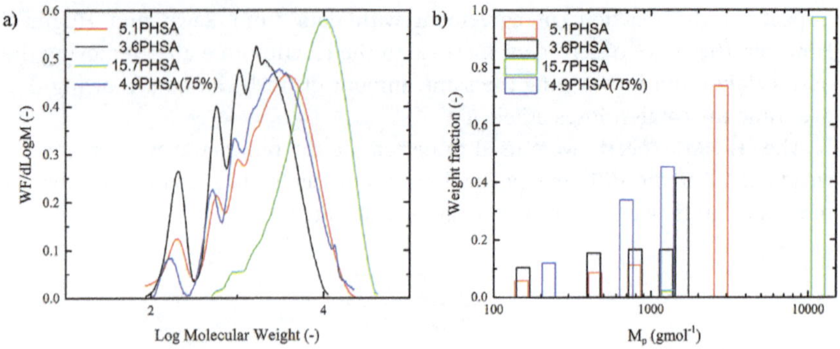

Fig. 9 Gel permeation chromatography results for the PHSA with different chain lengths. (a) The molecular weight distribution. (b) Weight fractions of certain peak molecular weights in the different samples. Note that some samples show a polymodal molecular weight distribution.

Another interesting observation is that the sample prepared from the lower grade monomer had no detectable number of chains with more than 7 monomer units, while the PHSA prepared under the same conditions with purer starting compounds contained a significant quantity of longer molecules (Fig. 9b). In addition, as the reaction time was prolonged and the amount of catalyst increased, the fraction of lower molecular weights became insignificant.

Step II. The next reaction in the synthesis of the PMMA-*g*-PHSA is the addition of glycidyl methacrylate to the PHSA produced in the first stage. The reaction with the –COOH groups in the polyester takes place in a basic environment, by opening the epoxy ring. For this step, the recipe introduced by Elsesser and Hollingsworth was slightly modified (see "Materials and methods").[25] Due to the previously mentioned difficulties observed with the titration method, here, the number of reacted groups was estimated from ^{1}H-NMR spectra, analogous to in the first step. Samples taken at different times were also analysed using NMR spectroscopy to verify that the yield of the side products stays within the limits of <5%. Ideally, after the end of the reaction, all the carboxyl groups should disappear and the double bond in the GM should be intact in order to facilitate the following step. In order to verify whether the –COOH groups have reacted, ^{13}C-NMR spectra of the final products were recorded. The PHSAs obtained from the 95% pure 12-HSA with 3.6, 5.1 and 15.7 monomer units were successfully attached to the glycidyl methacrylate permitting the synthesis to proceed to the next step.

Step III. In the final step of the stabiliser synthesis, the PMMA backbone is formed and the PHSA chains are grafted. The backbone can be produced from methyl methacrylate (MM) only or by using a mixture of MM and glycidyl methacrylate (GM). The latter is employed when the stabiliser needs to be covalently bonded to the particle surface, which was not a requirement for our study, since the particles lose a portion of the stabiliser during the degradation of the matrix after deformation (see "Particle deformation" and Zhang *et al.*[32]). Hence, for simplicity, GM was not included here. Based on the number average molecular weights obtained from the ^{1}H-NMR spectra and the concentrations of the PHSA-GM in the mixture obtained in Step II, we were able to keep the MM/PHSA-GM molar ratio similar for all three PHSAs.

The ^{1}H-NMR spectra confirmed the preservation of the molar ratio of the MM and the PHSA. The molecular weights of the polymers were determined using gel permeation chromatography (Table 2 and Fig. 10). The molecular weights of the comb polymer obtained *via* GPC are only indicative, but it is useful to compare the values measured with those found in the literature; they are in the range M_n = 6000–15 000 and M_w = 20 000–35 000,[37] with exceptions having $M_w \sim$ 60 000–75 000.[25] Our results agree well with these values, with the exception of the stabiliser that contains the shortest PHSA chains. Considering that the molar ratio between the methyl methacrylate and PHSA-GM was kept constant, the number of MM molecules in the reaction mixture was the highest in the case of the shortest PHSA and *vice versa*. This can further result in either longer stabiliser molecules or a larger number of shorter molecules. The molecular weight distribution curves indicate that the stabiliser chain molecules grow longer with an increase in the absolute number of methyl methacrylate units. The efficiency of the produced stabilisers was tested by employing them in particle synthesis.

3.2 Synthesis of spherical particles

The PMMA particles are produced by a dispersion polymerisation.[4] Different groups used various procedures, which mainly differ in the ways in which the components are added to the reaction mixture.[4,23] Antl *et al.* simplified the synthesis by turning it into a one step procedure, followed by a "locking" step in which the stabilisers were covalently bonded to the particles. Here, the particles were produced using the method of Klein *et al.*,[38] omitting the locking step. Particles were synthesised with stabilisers containing 3.6, 5.1 and 15.7 monomer units in the PHSA chains. The first batch was produced with PMMA-*g*-PHSA that has the "standard" length – 5.1 monomers. An SEM image of the particles is shown in (Fig. 11a). Since the results were satisfactory, the amounts of the other stabilisers were calculated in such a way that the reaction mixtures contained the same number of moles of PMMA backbones with 15.7 monomer units as in the case of the "standard" stabiliser. This should result in the same number of PHSA chains per number of MM moles or the same grafting density. The micrographs of the particles are presented in Fig. 11b and c.

In all three cases stable particles were produced. However, in the case of short PHSA chains, the particles exhibited a certain level of roughness. Since the only difference in the three batches is in the length of the soluble moieties, we assumed that the roughness might be caused by a solubility issue. Namely, the

Table 2 Molecular weights of the different stabilisers

Sample	M_n (GPC)/ g mol^{-1}	M_w (GPC)/ g mol^{-1}	M_z (GPC)/ g mol^{-1}	M_p (GPC)/ g mol^{-1}	PDI
PMMA-*g*-5.1PHSA	6151	29 625	59 340	25 395	4.8
PMMA-*g*-3.6PHSA	28 772	79 384	193 063	47 775	2.8
PMMA-*g*-15.7PHSA	8803	23 850	55 914	14 258	2.7

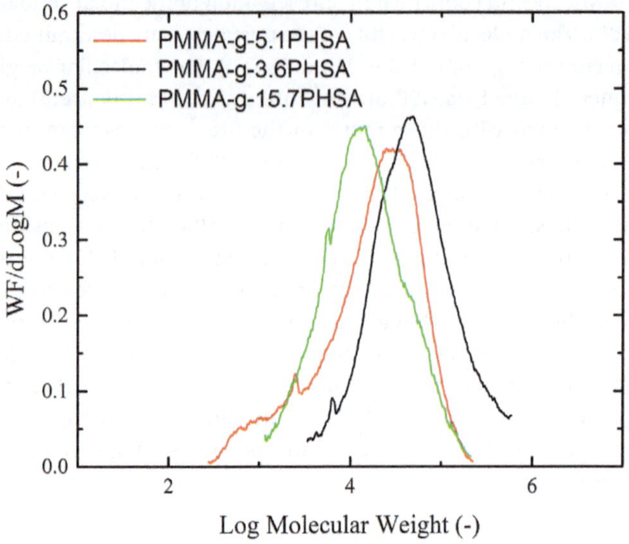

Fig. 10 Molecular weight distributions of the stabilisers containing PHSA with different numbers of monomer units.

PHSA chains are the components that keep the stabiliser dispersed in the hexane/dodecane mixture at the beginning of the dispersion polymerisation because the PMMA backbone is not soluble in these solvents. Since temperature is one of the parameters affecting the solubility of polymers, the particle synthesis in the presence of PMMA-*g*-PHSA was carried out at different temperatures. This indeed confirmed that the roughness on the particles can be reduced as shown in Fig. 12. Moreover, these results indicate that it is possible to produce PMMA particles of the same size, with a controlled level of surface roughness, just by varying the reaction temperature. We conducted further tests in order to test the hypothesis that the roughness comes from the stabiliser. PHSAs containing various numbers of monomer units were dissolved in five solvents with different polarities: hexane ($\varepsilon_r = 1.88$), dodecane ($\varepsilon_r = 2.0$), toluene ($\varepsilon_r = 2.38$), decanol ($\varepsilon_r = 8.1$) and iso-propanol ($\varepsilon_r = 18$). As expected, as the polarity of the solvent increased, the solubility of the PHSA with 3.6 monomer units on average was improved. One batch of particles was produced with half of the hexane being replaced with *tert*-butylmethyl ether ($\varepsilon_r = 4.5$). This led to the formation of smooth particles, but

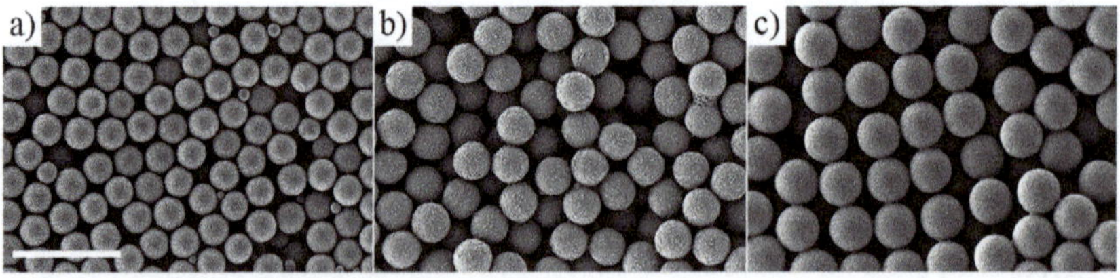

Fig. 11 Particles produced with different stabilisers: (a) PMMA-*g*-5.1PHSA, (b) PMMA-*g*-3.6PHSA and (c) PMMA-*g*-15.7PHSA. Scale bar is 5 μm.

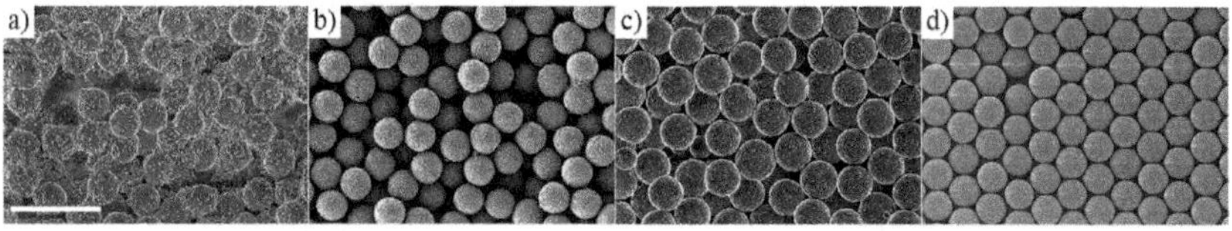

Fig. 12 Effect of the reaction temperature on the particle roughness: (a) 70 °C, (b) 95 °C and (c) 105 °C. All the particles are produced with the stabiliser containing PHSA consisting of 3.6 monomer units. (d) The particles synthesised at 95 °C after 10 minutes of sonication. Scale bar is 5 μm.

a significant polydispersity because the specific solvent mixture was partially dissolving PMMA. Interestingly, when the rough particles were subjected to sonication for >10 minutes, a great deal of the roughness was removed, but the particles remained stable in suspension (Fig. 12d).

The size of the PMMA particles prepared with different stabilisers was determined from SEM micrographs and DLS measurements. Since the former gives an estimate of the core diameter and the latter gives the hydrodynamic dimensions, comparing the diameters determined by the two methods allows estimation of the stabilising layer thickness. The results presented in Table 3 show that the calculated hydrodynamic layer thicknesses agree better with the chain length computed based on the number of monomer units in the soluble moieties than with the estimated radii of gyration, indicating that the grafting density is high. Based on a simple mass balance, the grafting densities are estimated to be on the order of 75 Å^2 on average, which agrees rather well with literature values.[47]

Particle stability was tested using DLS and rheology measurements. DLS measurements as a function of extended periods (of days) showed no tendency to aggregate. This suggests that a chain length of 3.6 suffices to produce hard sphere behaviour, meaning that near hard sphere behaviour can be retained at smaller particle diameters than compared to the PMMA-*g*-5.1PHSA. For the rheology measurements, three different volume fractions ($\Phi = 0.04, 0.09$ and 0.16.) were prepared for every particle type. Samples were subjected to an increasing shear

Table 3 Particle diameters determined by SEM and DLS. The hydrodynamic layer thickness is calculated as the difference in the radii determined by the two methods. For comparison, the computed contour lengths of the PHSA and the radii of gyration of the PHSA in solvents of different quality are shown

Stabiliser	SEM d/μm	DLS d/μm	DLS PDI	Layer thickness/nm	Contour length of PHSA/nm	R_g/nm Good solvent	R_g/nm θ solvent	R_g/nm Poor solvent
PMMA-*g*-5.1PHSA	1.594 ± 0.042	1.621	0.032	14	7.9	2.7	2.3	1.7
PMMA-*g*-3.6PHSA	1.883 ± 0.028	1.897	0.032	7	5.8	2.2	1.9	1.5
PMMA-*g*-15.7PHSA	2.286 ± 0.054	2.566	0.065	105	22.4	5.2	4.0	2.5

rate ($\dot{\gamma}$) until 1000 s^{-1}, when the $\dot{\gamma}$ was decreased until reaching the initial value. All the samples exhibited fully Newtonian behaviour under the chosen shearing conditions, as expected for this range of volume fractions. Moreover, calculated effective volume fractions (from the relative viscosity, using the Krieger Dougherty relation[3]) agreed well with the experimental values for volume fractions, confirming that the particles are exhibiting near hard sphere behaviour.

3.3 Large scale production of non-spherical particles

The production of prolate ellipsoidal particles has been discussed extensively in previous work.[30,32] In the present work we investigate how other shapes can be

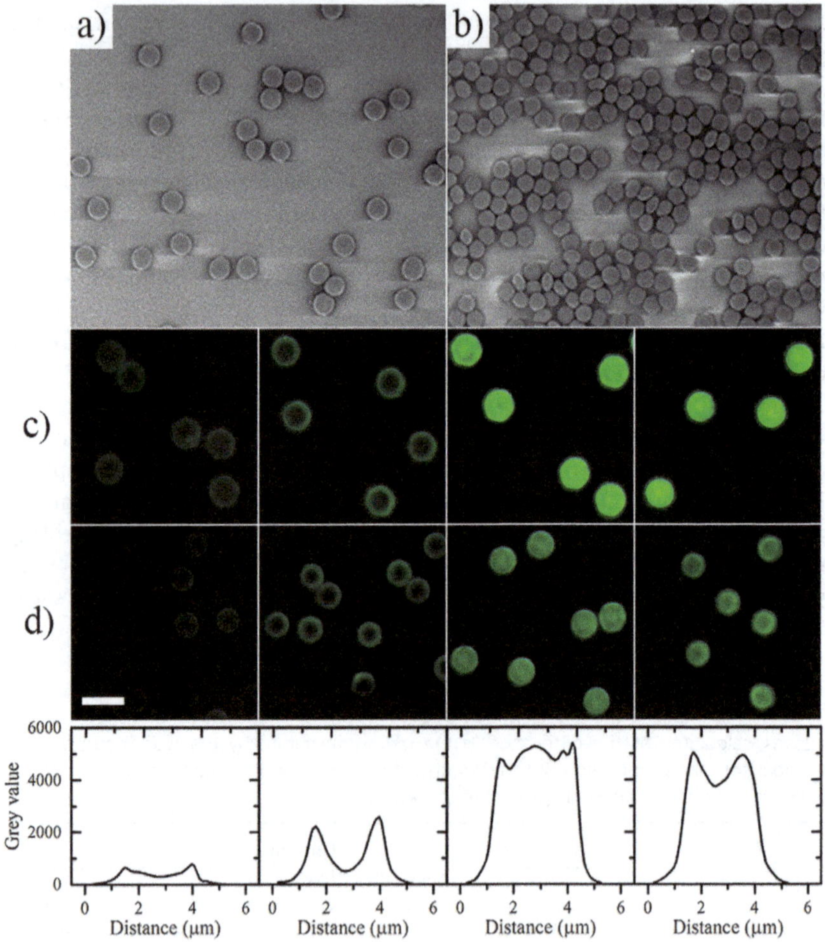

Fig. 13 SEM micrographs of disk-like particles produced by compressing PDMS films laden with PMMA particles stabilised with (a) PMMA-*g*-15.7PHSA and (b) PMMA-*g*-3.6PHSA. Images in (c) and (d) correspond to the particles shown in (a) and (b), respectively, and they show the effects of the degradation time on the stabiliser etching and the subsequent regrafting with a fluorescently labeled stabilizer. The graphs indicate the progression of the fluorescent area of the particles shown in (c). The scale bar represents 5 μm.

made, in large quantities. It has been shown before that biaxial stretching of films laden with particles produces disk-like particles.[10,33,34] The challenge is to produce these in larger quantities. Following Hsiao *et al.*[48] we explored the use of lubricated compression of PMMA laden PDMS films, a method which can easily be scaled up. Preliminary tests varying the PDMS film diameter and the applied pressure revealed a linear dependency between the applied pressure and the final aspect ratio of the particles and no correlation with the film dimensions. Moreover, the tests showed that with a relatively low pressure, 0.36 MPa, it is possible to reach relatively high aspect ratios of 4.3. The PMMA particles synthesised with the stabilisers containing PHSA with short chains (3.6 monomer units) and long chains (15.7 monomer units) subjected to the same amount of pressure (0.1 MPa) using the set-up shown in Fig. 2a were deformed into disk-like particles with aspect ratios of 1.6 and 1.5, respectively (Fig. 13a and b).

Depending on the time of degradation of the PDMS matrix, the PHSA or parts of the PHSA can be cleaved from the particle surface and chemically patterned particles can be produced.[32] The degradation is surface-curvature-dependent, as the flux near regions of higher curvature is higher. Moreover, it can be controlled by the concentration of the degradation agent and the degradation time and facilitates further modification of the anisotropic particles to obtain chemically patterned colloids.[32] The degradation process was quenched at different times and the recovered disk-like particles were re-grafted with a fluorescently labeled stabiliser, to show where the etching had taken place. Initially the degradation occurs at the rim of the particles, and the particles show up as fluorescent circles under the confocal microscope (Fig. 13c and d). With prolonged degradation times, the regions covered with the labeled PMMA-*g*-PHSA grew progressively. The length of the PHSA on the surface of the initial spherical particles had no impact on the subsequent grafting after the degradation (Fig. 13c and d). The same trend

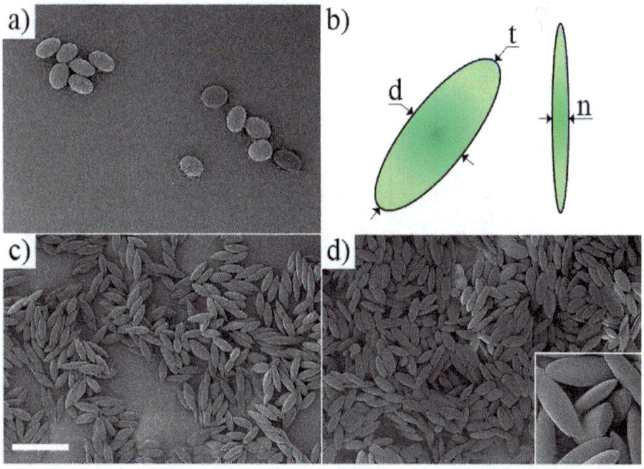

Fig. 14 (a) PMMA particles obtained by compressing PDMS films in a channel die. (b) A schematic representation of surfboard-like particles. (c) and (d) PS particles recovered after the biaxial stretching of PVA films. The draw ratios of the films together with the deformation and final aspect ratio of the particles are shown in Table 4 (experiment (a and b), respectively). The inset in (d) shows a close-up on the circumference of the deformed particles. The scale bar indicates 5 μm and the width of the inset image is 3 μm.

Table 4 The draw ratios of the PVA films laden with PS particles in the transverse, drawing and normal direction to the stretching direction, particle deformation in the three directions and the final aspect ratios of the recovered particles

Experiment	Temperature/°C	Film draw ratio			Particle deformation			Particle final aspect ratio/— : — : —
		Transverse/—	Drawing/—	Normal/—	Transverse/—	Drawing/—	Normal/—	
(a)	120	2.0	1.0	0.4	2.27	0.78	0.58	4.0 : 1 : 1.4
(b)	140	2.3	1.0	0.4	2.40	0.87	0.48	5.0 : 1 : 1.8
(c)	140	2.0	0.8	0.6	2.16	0.73	0.65	3.4 : 1 : 1.2
(d)	140	3.5	0.6	0.4	4.29	0.68	0.34	12.6 : 1 : 2.0
(e)	140	1.3	1.0	0.6	2.32	0.80	0.55	4.3 : 1 : 1.5
(f)	140	1.3	1.0	0.5	1.30	0.90	0.85	1.5 : 1 : 1.1

was observed with the disk particles produced from the fluorescently labeled spheres. However, in this case, the difference between the re-grafted area and the initial stabiliser is very subtle and it is difficult to observe. The spreading of the fluorescent area over time was quantified by evaluating the grey values of the disks produced from PMMA spheres stabilised with the PMMA-*g*-3.6PHSA (Fig. 13).

Uniaxial extension or compression of PDMS films leads to the formation of prolate or oblate ellipsoids, *i.e.* particles with the same dimensions in two directions. However, combining the effects of extension and compression should result in particles with different dimensions in all three directions. In order to achieve this, the set-up used to produce disk-like particles was equipped with channel dies that could be placed between the compressing plates. Since the film width was equivalent to the channel width, the film deformation in the direction of the short axis was constrained, but the compression caused a decrease in sample thickness and an increase in length which resulted in shapes as in Fig. 14a that resemble surfboards (Fig. 14b) due to the triaxial deformation. Such particles may be of interest for studies of interfacial rheology, as there should be no strong lateral capillary forces. The production of the particles by lubricated compression can be scaled up to arbitrary quantities and the deformation is rather uniform.

For PS particles which need to be embedded in PVA, the squeeze flow method is less suitable, due to the different mechanical properties of the PVA. We resorted to plastic drawing of a particle-filled polymer tube over a heated fork (Fig. 2a). The change in shape and diameter between the original tube and the fork sets the deformation of the film and the particles. Some examples of the particles are shown in Fig. 14c and d. The inset in Fig. 14d shows a particle which was broken during processing which nicely shows the cross section. The relation between the draw ratio of the films and the aspect ratio of the particles was investigated for 5 different conditions, and the resulting aspect ratios are given in Table 4. The first two rows correspond to the images in Fig. 14c and d, respectively. Also this method can be scaled up to produce particles in large quantities.

4 Conclusions

The synthesis of model near hard sphere colloids, consisting of PHSA stabilized PMMA colloids, has been revisited. H-NMR spectroscopy was extremely helpful in elucidating the problems which can occur during the polycondensation of the 12-HSA. Performing the polycondensation in the melt state and with a different catalyst compared to the commonly used ones enabled a better control over the synthesis of the model near hard spheres. The properties of the particles can be modulated to deviate from near hard sphere behaviour; for example, the better control over the synthesis enables us to vary the layer thickness at will. Moreover, we can vary the shape of the particles by mechanical deformation, making prolate or oblate ellipsoids or even triaxial, surfboard-like particles with methods which apply to different types of latex particles. Finally, using a localised etching, we can make these anisotropically shaped particles also chemically patterned.

Acknowledgements

The authors thank Dzana Durovic for the practical help with the synthesis and Dr Andrew Schofield (University of Edinburgh) for providing the fluorescently labeled stabiliser. Prof. M. Solomon (University Michigan) is thanked for stimulating discussions. We gratefully acknowledge the Swiss National Science Foundation for funding through Grant SNF 200021-157147.

References

1 P. N. Pusey and W. van Megen, *Nature*, 1986, **320**, 340–342.
2 C. P. Royall, W. C. K. Poon and E. R. Weeks, *Soft Matter*, 2013, **9**, 17–27.
3 J. Mewis and N. J. Wagner, *Colloidal Suspension Rheology*, Cambridge University Press, 2012.
4 K. E. J. Barrett, *Br. Polym. J.*, 1973, **5**, 259–271.
5 K. M. Keville, E. I. Franses and J. M. Caruthers, *J. Colloid Interface Sci.*, 1991, **144**, 103–126.
6 C. C. Ho, A. Keller, J. A. Odell and R. H. Ottewill, *Polym. Int.*, 1993, **30**, 207–211.
7 J. J. Crassous, H. Dietsch, P. Pfleiderer, V. Malik, A. Diaz, L. A. Hirshi, M. Drechsler and P. Schurtenberger, *Soft Matter*, 2012, **8**, 3538–3548.
8 H. Bao, T. Bihr, A.-S. Smith and R. N. Klupp Taylor, *Nanoscale*, 2014, **6**, 3954–3966.
9 V. R. Dugyala, S. V. Daware and M. G. Basavaraj, *Soft Matter*, 2013, **9**, 6711–6725.
10 H. Shin and C. Kim, *Colloid Polym. Sci.*, 2012, **290**, 1309–1315.
11 S. Sacanna, M. Korpics, K. Rodriguez, L. Colón-Meléndez, S.-H. Kim, D. J. Pine and G.-R. Yi, *Nat. Commun.*, 2013, **4**, 1688.
12 W. van Megen and P. N. Pusey, *Phys. Rev. A: At., Mol., Opt. Phys.*, 1991, **43**, 5429–5441.
13 S.-E. Phan, W. B. Russel, Z. Cheng, J. Zhu, P. M. Chaikin, J. H. Dunsmuir and R. H. Ottewill, *Phys. Rev. E*, 1996, **54**, 6633.
14 W. K. Kegel, *Science*, 2000, **287**, 290–293.
15 A. Yethiraj and A. van Blaaderen, *Nature*, 2003, **421**, 513–517.
16 P. Pusey, E. Zaccarelli, C. Valeriani, E. Sanz, W. C. Poon and M. E. Cates, *Philos. Trans. R. Soc., A*, 2009, **367**, 4993–5011.
17 P. Schall, I. Cohen, D. A. Weitz and F. Spaepen, *Nature*, 2006, **440**, 319–323.
18 P. Schall, D. A. Weitz and F. Spaepen, *Science*, 2007, **318**, 1895–1899.
19 C. J. Dibble, M. Kogan and M. J. Solomon, *Phys. Rev. E*, 2006, **74**, 041403.
20 L. C. Hsiao, M. J. Solomon, K. A. Whitaker and E. M. Furst, *J. Rheol.*, 2014, **58**, 1485–1504.
21 A. Doroszkowski and R. Lambourne, *J. Polym. Sci., Part C: Polym. Symp.*, 1971, **34**, 253–264.
22 R. Cainrs, R. H. Ottewill, D. Osmond and I. Wagstaff, *J. Colloid Interface Sci.*, 1976, **54**, 45–51.
23 L. Antl, J. W. Goodwin, R. D. Hill, R. H. Ottewill, S. M. Owens, S. Papworth and J. A. Waters, *Colloids Surf.*, 1986, **17**, 67–78.
24 M. K. Klein, N. Klinkenberg, S. Schuetter, N. Saenger, P. Pfleiderer and A. Zumbusch, *Langmuir*, 2015, **31**, 2655–2661.
25 M. T. Elsesser and A. D. Hollingsworth, *Langmuir*, 2010, **26**, 17989–17996.
26 M. T. Elsesser, A. D. Hollingsworth, K. V. Edmond and D. J. Pine, *Langmuir*, 2011, **27**, 917–927.
27 R. Rice, R. Roth and C. P. Royall, *Soft Matter*, 2012, **8**, 1163–1167.
28 R. P. A. Dullens, *Soft Matter*, 2006, **2**, 805–810.

29 C. Pathmamanoharan, K. Groot and J. Dhont, *Colloid Polym. Sci.*, 1997, **275**, 897–901.
30 K. M. Keville-Polizopoulos, PhD thesis, Purdue University, West Lafayette IN, 1988.
31 A. Mohraz and M. J. Solomon, *Langmuir*, 2005, **21**, 5298–5306.
32 Z. Zhang, P. Pfleiderer, A. B. Schofield, C. Clasen and J. Vermant, *J. Am. Chem. Soc.*, 2011, **133**, 392–395.
33 Y. Hu, J. Ge, T. Zhang and Y. Yin, *Adv. Mater.*, 2008, **20**, 4599–4602.
34 D. Florea and H. M. Wyss, *J. Colloid Interface Sci.*, 2014, **416**, 30–37.
35 C. C. Ho, A. Keller, J. A. Odell and R. H. Ottewill, *Colloid Polym. Sci.*, 1993, **271**, 469–479.
36 J. A. Champion, Y. K. Katare and S. Mitragotri, *Proc. Natl. Acad. Sci. U. S. A.*, 2007, **104**, 11901–11904.
37 K. E. J. Barrett, *Dispersion Polymerization in Organic Media*, John Wiley & Sons, Inc., 1975.
38 M. K. Klein, A. Zumbusch and P. Pfleiderer, *J. Mater. Chem. C*, 2013, **1**, 7228–7229.
39 R. Schaller, PhD thesis, ETH Zurich, Switzerland, 2015.
40 K. Kargupta, P. Rai and A. Kumar, *J. Appl. Polym. Sci.*, 1993, **49**, 1309–1329.
41 Y. Wang, W. Eli, A. Nueraimaiti and Y. Liu, *Ind. Eng. Chem. Res.*, 2009, **48**, 3749–3754.
42 A. Doroszkowski and R. Lambourne, *Faraday Discuss. Chem. Soc.*, 1978, **65**, 252.
43 C. Bawn and M. Huglin, *Polymer*, 1962, **3**, 257–262.
44 T. Zhang, X. Li and J. Dong, *Tenside, Surfactants, Deterg.*, 2014, **51**, 516–520.
45 E. W. Dean and D. D. Stark, *J. Ind. Eng. Chem.*, 1920, **12**, 486–490.
46 A. Bloomfield and W. Christian, Preparation of light colored fatty acid esters, *US Pat.*, 3,071,604, 1963.
47 D. J. Walbridge and J. A. Waters, *Discuss. Faraday Soc.*, 1966, **42**, 294–300.
48 L. C. Hsiao, B. A. Schultz, J. Glaser, M. Engel, M. E. Szakasits, S. C. Glotzer and M. J. Solomon, *Nat. Commun.*, 2015, **6**, 8507.

Faraday Discussions

PAPER

Anisotropic surface functionalization of Au nanorods driven by molecular architecture and curvature effects†

Estefania Gonzalez Solveyra,*[a] Mario Tagliazucchi[b] and Igal Szleifer[a]

Received 22nd February 2016, Accepted 15th March 2016
DOI: 10.1039/c6fd00020g

This work suggests a novel strategy to coat the caps and body of Au-nanorods (Au-NRs) with end-grafted polymer layers of different compositions by taking advantage of the different curvature of these two regions. A molecular theory was used to theoretically investigate the effect of local curvature and molecular architecture (intramolecular connectivity of the monomers) on the adsorption of polymer mixtures on cylindrical (Au-NR body) and spherical (Au-NR caps) surfaces. The adsorption process was systematically studied as a function of the backbone length, number and position of branches, quality of the solvent and total number of monomers of the polymer molecules in the mixture. The balance between repulsive forces and polymer–surface and polymer–polymer attractions governs the amount and composition of the adsorbed layer. This balance is in turn modulated by the architecture of the polymers, the curvature of the surface and the competition between the different polymers in the mixture for the available area. As a result, the equilibrium composition of the polymer layer on spheres and cylinders of the same radius differs, and in turn departs from that of the bulk solution. Curvature plays a major role: the available volume at a given distance from the surface is larger for spherical surfaces than for cylindrical ones, therefore the surface density of the bulkier (more branched) polymer in the mixture is larger on the Au-NR caps than on the Au-NR body. These results suggest that the combination of curvature at the nanoscale and tailored molecular architecture can confer anisotropic nanoparticles with spatially enriched domains and, therefore, lead to nanoconstructs with directional chemical interactions.

[a]*Department of Biomedical Engineering, Department of Chemistry and Chemistry of Life Processes Institute, Northwestern University, Evanston, Illinois 60208, USA. E-mail: estefania.solveyra@northwestern.edu*

[b]*INQUIMAE-CONICET, Facultad de Ciencias Exactas y Naturales, Ciudad Universitaria, Pabellón 2, Ciudad Autónoma de Buenos Aires C1428EHA, Argentina*

† Electronic supplementary information (ESI) available: Molecular details of the polymers used in our calculations, figure showing the density and conformational properties of 20l–(35-2b) mixture on cylindrical surfaces, and additional figures expanding on the effects of the polymers' molecular architecture and the solvent quality on the adsorption process. See DOI: 10.1039/c6fd00020g

1 Introduction

Gold nanorods (Au-NRs) are great candidates for building multifunctional agents with anisotropic surface functionalization. In the past two decades, their unique electronic and optical properties, reproducible synthesis, stability, biosafety and ease of surface modification[1] have multiplied their applications in catalysis,[2] sensing and molecular recognition,[3] and biomedical therapeutics and diagnosis.[4] The different curvatures of the bodies and the caps of Au-NRs offer a great opportunity to engineer anisotropic nanoconstructs. In such spatially heterogeneous NRs, both the location and the composition of the surface coating are important for their use in biomedical applications. The size, local geometry and orientation of the NRs at the contact point with cells play a crucial role in transport, delivery and cellular internalization.[5–7] Thus, controlling the chemistry and location of the grafted ligands on the NR surface can enable chemical directionality, with the ultimate goal of controlling interactions at the nano–bio interface.[8]

Surface modification is also key to the modulation of interactions at the interface of Au-NRs and their surrounding media.[9] Following synthesis, Au-NRs must be functionalized with biocompatible molecules to prevent uncontrolled protein adsorption (biofouling) on the NRs' surface when introduced into biological matrices.[10] The most widely used antifouling ligands include poly(ethylene glycol) (PEG),[11] thiol self-assembled monolayers (SAMs),[12] zwitterionic polymers,[13] peptoids[14] and glycopeptoids.[15] Specific-targeting ligands, such as antibodies, proteins, nucleic acids or small ligands like sugars can also be incorporated into the Au-NRs in order to allow molecular recognition at the desired site.[16]

So far, significant progress has been achieved in the anisotropic functionalization of Au-NRs based on the distinct crystallographic faceting of their surface, since different crystallographic facets exhibit different chemical reactivity.[17,18] Subsequent functionalization with thiol-terminated molecules allows Au-NRs end-terminated with diverse ligands to be obtained[17,19–21] for applications in plasmonic metamaterials[22] and supracolloidal polymers.[23] However, in spite of the directionality incorporated in such strategies and the great control over the interparticle interactions, the as-modified Au-NRs in those works still present a CTAB bilayer along their longitudinal side, which poses serious problems for their utilization in biological and biomedical media.[4]

In this work, we propose to anisotropically modify Au-NRs by taking advantage of the different types of curvatures of their bodies and caps. For the adsorption of a single type of polymer in solution, increasing the surface curvature (*i.e.* decreasing its radius) increases the amount of adsorbed polymer.[24,25] This effect results from the fact that in curved convex surfaces, the available volume increases as the distance from the surface increases, and it scales as r/R for cylinders and as $(r/R)^2$ for spheres (where r is the radial direction and R the radius of the surface). Based on this fact, we hypothesize that in a system combining both cylindrical and spherical surfaces, such as an idealized NR (see Fig. 1b), the surface curvature can be used for spatially selective functionalization. In order to explore this hypothesis, we study the adsorption of mixtures of a linear and a branched polymer (with a varying degree of branching) in order to improve the partitioning of the branched polymers between the caps and body and maximize

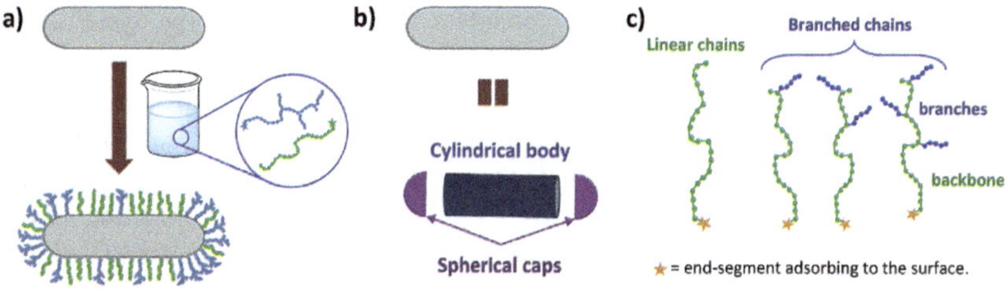

Fig. 1 Modeling strategy. (a) Spatially heterogeneous surface-functionalized nanorod (NR) prepared by adsorbing a mixture of polymers of different molecular architectures from aqueous solution (scheme not to scale). (b) Schematic representation of a NR as a combination of surfaces of different curvatures: cylinders and spheres, representing the NR's body and caps respectively. (c) Schematic representation of the polymer architectures and relevant parameters analyzed in this work.

the effect of the curvature, as schematically shown in Fig. 1a. Keeping in mind the intended purpose of designing Au-NRs for biomedical applications, the polymers explored in this work are neutral and hydrophilic, as numerous publications have reported that hydrophobic or charged coatings tend to adsorb more proteins than neutral or hydrophilic surfaces.[26,27]

In the next section, we present the theory for polymer adsorption from binary solutions on curved surfaces. All polymers studied in this work adsorb on the surface exclusively through a functional end group. Then, we present a discussion of the effect of different variables on the adsorption process: composition of the solution, radius of the NR, the molecular architecture of the polymer molecules and solvent quality. In the last section, we outline the design rules that resulted from our study and discuss the potentials of the methodology employed.

2 Theoretical methods

In order to model the adsorption of polymer mixtures on the surface of the NRs, we independently study the adsorption on the spherical caps and the cylindrical body (Fig. 1b). We consider a curved surface (either cylindrical or spherical) in contact with aqueous solutions containing a mixture of two polymer molecules with different molecular topologies. We model the adsorption of the end group of these polymers on the cylindrical and spherical surfaces with a typical radius, R, from 1 to 20 nm, as depicted in Fig. 1b and c.

2.1 Molecular theory

Our theoretical approach is based on a molecular theory developed to describe a wide range of soft matter phenomena at interfaces, including the structure and thermodynamics of tethered polymer layers,[24,28] the adsorption of polymers and proteins,[14,29–31] ligand–receptor binding,[32–35] acid–base equilibrium of weak polyelectrolytes grafted on surfaces of different curvature[36–38] and interfacial behavior of diverse nanomaterials.[39–41] The predictive power of the molecular theory on the behavior of soft matter in a variety of systems (as compared with experiments[42–47]) gives us confidence about its potential to outline design rules in the synthesis of anisotropically coated NRs.

The employed theory explicitly considers the geometry of the system, the shape, molecular volume and conformations of all molecular species and the repulsive excluded-volume and attractive van der Waals (vdW) interactions. The input of the theory is a set of conformations for each macromolecule in the system and the values of the controlled thermodynamic variables (temperature, composition of the bulk solution, adsorption energy, *etc.*). Minimization of the system's free energy yields structural and thermodynamic properties of the system at equilibrium, including the probability of each conformation, the amount of adsorbed polymer and the spatial distribution of solvent molecules and polymer chains.

The starting point of the theory is an approximate free-energy functional for the system. We are studying a surface in contact with an aqueous solution having a constant composition of linear (l) and branched (b) polymers (we also assume that the composition of the solution is unaffected by adsorption), therefore the calculations have to be performed in the grand canonical ensemble. The grand potential per unit area can be expressed as:

$$\begin{aligned}\frac{\beta W}{A} = & \sum_{i=\mathrm{l,\,b}} \sigma_i \sum_{\{\alpha_i\}} P(\alpha_i) \ln P(\alpha_i) + \sum_{i=\mathrm{l,\,b}} \sigma_i \left[\ln\left(\sigma_i \Lambda_i^{2}\right) - 1\right] \\ & + \sum_{i=\mathrm{l,\,b}} \sigma_i \beta \varepsilon_\mathrm{s} - \sum_{i=\mathrm{l,\,b}} \sigma_i \beta \mu_i^\mathrm{bulk} \\ & + \int \mathrm{d}\mathbf{r} \rho_\mathrm{w}(\mathbf{r}) [\ln(\rho_\mathrm{w}(\mathbf{r}) v_\mathrm{w}) - 1] \\ & + \frac{\beta}{2} \iint \mathrm{d}\mathbf{r}\mathrm{d}\mathbf{r}' \frac{\chi\, a(|\mathbf{r}-\mathbf{r}'|)}{v_\mathrm{p}} \left[\left\langle \phi_\mathrm{p}^\mathrm{l}(\mathbf{r}) \right\rangle + \left\langle \phi_\mathrm{p}^\mathrm{b}(\mathbf{r}) \right\rangle\right] \left[\left\langle \phi_\mathrm{p}^\mathrm{l}(\mathbf{r}') \right\rangle + \left\langle \phi_\mathrm{p}^\mathrm{b}(\mathbf{r}') \right\rangle\right]\end{aligned} \quad (1)$$

with $\beta = 1/k_\mathrm{B}T$ (k_B is the Boltzmann constant, T is the absolute temperature) and $\mathbf{r}$ the position vector, $\mathbf{r} = (x, y, z)$. The amount of adsorbed polymer is quantified through its surface density $\sigma_i = N_i/A(R)$ where N_i is the number of adsorbed molecules of the linear or branched polymer, $A(R)$ corresponds to the area and R to the radius of curvature of the surface, which can be planar ($R \rightarrow \infty$), cylindrical or spherical.

The first term in eqn (1) corresponds to the conformational entropy of the polymer chains, where $P(\alpha_i)$ (with $i = \mathrm{l}$ or b) is the probability of having a linear or branched polymer molecule tethered to the surface in a conformational state α. The second term is the two-dimensional (surface) translational entropy of the polymers, where $\Lambda_i = (h^2/2\pi m_i k_\mathrm{B}T)^{1/2}$ is the polymer de Broglie wavelength, h is Planck's constant and m_i is the polymer molecular mass. The third term corresponds to the adsorption energy between the end-segment of the polymer chain and the surface, with ε_s being the adsorption energy per chain. The fourth term ensures that the chemical potential of each type of polymer, μ_i^bulk, is the same at the surface and in the bulk solution. The fifth term is the translational (mixing) entropy of the water molecules, where $\rho_\mathrm{w}(\mathbf{r})$ is the position-dependent water density and v_w is the molecular volume of a water molecule. The last term is the vdW effective attractive energy between polymer segments.[24] In this term, $\langle \phi_\mathrm{p}{}^i(\mathbf{r}) \rangle$ is the volume fraction of the polymer at $\mathbf{r}$ (with $i = \mathrm{l}$ or b), χ is a prefactor that determines the strength of the polymer–polymer interaction and $a(|\mathbf{r} - \mathbf{r}'|)$ is a distance-dependent vdW attractive function of the form:

$$a(|\mathbf{r}-\mathbf{r}'|) = -\left(\frac{l_{\text{seg}}}{|\mathbf{r}-\mathbf{r}'|}\right)^6 \quad \text{for } l_{\text{seg}}|\mathbf{r}-\mathbf{r}'| < 1.5\delta$$
$$a(|\mathbf{r}-\mathbf{r}'|) = 0 \quad \text{otherwise} \tag{2}$$

with l_{seg} being the segment length and 1.5δ a cut-off parameter (δ is the discretization step, see below).

The repulsion between molecules is accounted for by an incompressibility condition ensuring that the total available volume is occupied either by the polymers or solvent molecules at all distances from the surface:

$$\langle \phi_\text{p}^\text{l}(\mathbf{r})\rangle + \langle \phi_\text{p}^\text{b}(\mathbf{r})\rangle + \phi_\text{w}(\mathbf{r}) = 1 \tag{3}$$

where $\phi_\text{w}(\mathbf{r})$ is the position-dependent volume fraction of water and $\phi_\text{p}^i(\mathbf{r})$ are the position-dependent volume fractions of the polymer (for $i = \text{l, b}$), given by:

$$\left\langle \phi_\text{p}{}^i(\mathbf{r})\right\rangle = \sigma_i \sum_{\{\alpha_i\}} P(\alpha_i) n(\alpha_i;\mathbf{r}) \tag{4}$$

where $n(\alpha_i;\mathbf{r})\text{d}\mathbf{r}$ is the number of segments that the chain in conformation α_i has within the volume between $\mathbf{r}$ and $\mathbf{r} + \text{d}\mathbf{r}$.

The system's free energy is a functional of $\phi_\text{w}(\mathbf{r})$, $P(\alpha_i)$, and σ_i. These quantities are found by the minimization of eqn (1) subject to the packing constraint (eqn (3)), introducing a set of Lagrange multipliers $\beta\pi(\mathbf{r})$. This procedure yields:

$$\phi_\text{w}(\mathbf{r}) = \rho_\text{w}(\mathbf{r})v_\text{w} = \exp[-\beta\pi(\mathbf{r})v_\text{w}] \tag{5}$$

for the solvent density, and:

$$P(\alpha_i) = \frac{1}{q_i}\exp\left\{-\int \beta\pi(\mathbf{r})\ v_\text{p}\ n(\alpha_i;\mathbf{r})\ \text{d}\mathbf{r} - \iint \text{d}\mathbf{r}\ \text{d}\mathbf{r}'\chi a(|\mathbf{r}-\mathbf{r}'|)n(\alpha_i;\mathbf{r})\left[\left\langle \phi_\text{p}^\text{l}(\mathbf{r}')\right\rangle + \left\langle \phi_\text{p}^\text{b}(\mathbf{r}')\right\rangle\right]\right\} \tag{6}$$

for the probability of chain conformations, where $i = \text{l}$ or b, q_i is a normalization constant assuring $\sum_{\{\alpha_i\}} P(\alpha_i) = 1$ and v_p is the volume of a monomer.

Minimization with respect to surface coverage gives:

$$\sigma_i = \frac{q_i}{\Lambda_i^2}\exp\left[\beta\mu_i^{\text{bulk}} - \beta\varepsilon_\text{s}\right] \tag{7}$$

This equation guarantees that the chemical potential of the linear or branched polymer molecules ($i = \text{l}$ or b) on the surface (μ_i^{surf}) is the same as in the bulk solution (μ_i^{bulk}), as required by thermodynamic equilibrium.

The only unknowns in the above equations are the position-dependent repulsive interactions $\pi(\mathbf{r})$, which correspond to the osmotic pressure along the system.[24,36] They are determined by placing the expressions for $P(\alpha_i)$, σ_i, and $\phi_\text{w}(\mathbf{r})$ into the packing constraint (eqn (3)). To implement the theory, we exploit the cylindrical or spherical symmetry of the system by adopting the corresponding coordinate system in our calculations.[35,36] Furthermore, we assume that the system is laterally homogeneous, with the inhomogeneities explicitly considered only in the radial direction. Thus, we change our variable from $\mathbf{r}$ to r, and define r as the distance from the surface (*i.e.* r is the radial coordinate for a cylinder or sphere). For each specific geometry, we discretize the space in layers of dimension δ, transforming the integral equation for $\pi(r)$ into a set of nonlinear equations that we solve using standard numerical methods. For further details on solving the theory we refer the interested reader to ref. 36 and 48.

2.2 Molecular model

In order to solve the theory, we need sets of chain conformations for the linear and branched polymers. While these sets include in principle all allowed conformations that do not collide with the surface for the geometry under study, in practice it is enough to use a very large set of randomly chosen conformations, which we generate in free space using the rotational-isomeric model (RIS).[49] To avoid biases, each random bond sequence is rotated using randomly chosen Euler angles.[50] Only self-avoiding conformations that do not overlap with the surface are considered. We have used a set of 10^6 different configurations for each polymer type and each geometry.

To study the effect of the bulk solution composition, we varied the molar fraction of the branched polymer ($x_\mathrm{b}^\mathrm{bulk}$) in a mixture of branched and linear polymers between zero (solution of linear polymer only) and one (solution of branched polymer only). For all cases, we kept the total volume fraction of polymer constant ($\phi_{\mathrm{p_l}}{}^\mathrm{bulk} + \phi_{\mathrm{p_b}}{}^\mathrm{bulk} = \phi_{\mathrm{p_{tot}}}{}^\mathrm{bulk} = 0.015$). For each composition of the solution, we computed the bulk chemical potential of the polymers (μ_i^bulk) within the same theoretical approach as that of the tethered chains described above, following the expression:[24]

$$\beta\mu_i^\mathrm{bulk} = -\ln\left(N_i^\mathrm{bulk}\right) - \frac{v_\mathrm{p}}{v_\mathrm{w}} n_{\mathrm{seg}_i} \ln\phi_\mathrm{w}^\mathrm{bulk} + \ln\left(\frac{\phi_{\mathrm{p}_i}^\mathrm{bulk}}{n_{\mathrm{seg}_i}}\right) + \beta\chi a n_{\mathrm{seg}_i}\phi_{\mathrm{p}_i}^\mathrm{bulk} \tag{8}$$

where $i = \mathrm{l}$ or b, N_i^bulk is the total number of self-avoiding conformations for each type of polymer and n_{seg_i} is the total number of segments in the chain. All the other quantities have been defined above. Note that since the bulk solution is an homogeneous system, the attractive function a corresponds simply to $a = \iint \mathrm{d}\mathbf{r}\mathrm{d}\mathbf{r}' a(|\mathbf{r} - \mathbf{r}'|)$, and that ϕ_i are the volume fractions of species $i = \mathrm{p_l}, \mathrm{p_b}$, w for the linear polymers, branched polymers and water in the solution respectively, with $\phi_\mathrm{w}^\mathrm{bulk} + \phi_{\mathrm{p_l}}{}^\mathrm{bulk} + \phi_{\mathrm{p_b}}{}^\mathrm{bulk} = 1$.

We chose the parameters for the monomers of the polymer to be those of PEG (segment length $l_\mathrm{seg} = 0.3$ nm and segment volume $v_\mathrm{p} = 0.065$ nm^3) based on a previous implementation of the molecular theory.[28] The solvent molecular volume was set to 0.03 nm^3 (based on water's density at 298 K) and the thickness of the discrete layers to $\delta = 0.3$ nm. The adsorption energy of the functionalized end group (ε_s) for all polymers considered in this work was set to $-13.2k_\mathrm{B}T$ in order to reproduce the experimental surface density achieved for the adsorption of a 20-monomer linear chain on a planar surface, obtained experimentally in a previous work.[14] We conducted calculations considering different values of $\frac{\chi_\mathrm{c}}{\chi}$, where χ_c is the critical parameter defined as the value required to induce microphase segregation in a neutral polymer layer in the limit of vanishing surface coverage[51] and it is close to θ conditions.[52] Values of $\frac{\chi_c}{\chi}$ greater and lower than 1 correspond to good and poor solvent conditions, respectively.

Using the theory and molecular model described above, we performed a systematic analysis on the adsorption process varying the backbone length, number of branches, branching positions, total number of monomers in the polymer molecules (see Fig. 1c) and quality of the solvent. In particular, we are interested in exploring routes that would allow the enrichment of the spherical caps with the branched polymers. The number and position of the branches have been selected given the interest in having a large density of free ends close to the solvent for further chemical functionalization. Details of the chains can be found in Table S1 in the ESI.†

3 Results and discussion

We start our analysis discussing polymer adsorption from mixtures in good solvent conditions. We reserve the discussion of polymer–polymer interactions and the role of solvent quality on the adsorption process for the last section of our results. It is worth mentioning that we allow adsorption only between the end segments of the polymers and the Au-NR. We do not consider mechanisms encompassing backbone- and branch-mediated adsorption. For detailed studies of such systems, including polymer architecture effects, we refer the reader to ref. 53 and the references therein.

The nomenclature used henceforth for the different polymer chains comprise the total number of segments in the molecule, followed by the number of branches. In the case of linear polymers, the code we use is simply the number of segments followed by the letter "l", for "linear". So, "20l" corresponds to a linear polymer of 20 monomers, while "35-2b" corresponds to a branched polymer of 35 total monomers, and 2 branches. Note that for all branched polymers, the number of segments on the branches is the same and equal to 5 (see Table S1 in the ESI†).

3.1 Single-polymer solutions

The thermodynamic behavior of polymers adsorbing on surfaces is intimately related to the molecular organization of the chains at the grafting surface. Carignano and Szleifer have previously addressed the conformational properties of linear and branched polymers on surfaces of planar, cylindrical, and spherical geometries,[24] and have also provided a detailed analysis on the adsorption of linear chains on curved surfaces.[24,48] So, we focus our discussion on the interplay of chain architecture and polymer adsorption in systems containing only one type of polymer as an introduction for the polymer adsorption from binary mixtures in the following sections.

Fig. 2a shows the equilibrium amount of polymer adsorbed as a function of the radius of the surface, for both cylinders and spheres. The amount of adsorbed polymer results from the balance between competing processes, modulated by the curvature of the surface: the attractive forces between the polymer functionalized end segment and the surface, the loss of conformational entropy induced by packing of the chains at the surface and the excluded volume repulsions between the monomers that are confined to the surface vicinity. It is worth recalling that in all the cases studied, the attraction between the end group of the

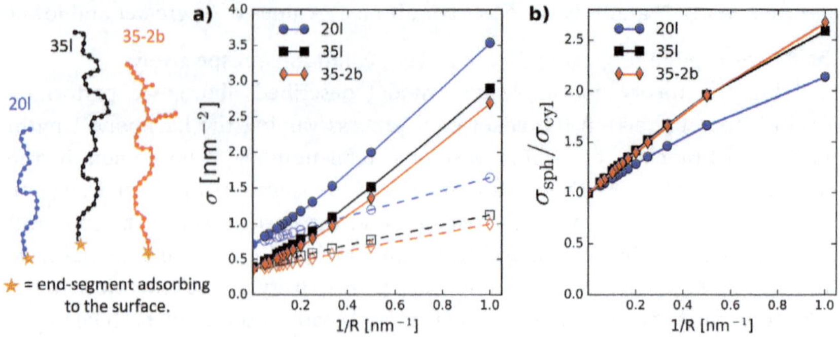

Fig. 2 Curvature effect on polymer adsorption from one-polymer solutions. (a) Equilibrium surface coverage of adsorbed polymer as a function of the curvature for spherical (full lines and solid symbols) and cylindrical (dashed lines and empty symbols) surfaces. Results correspond to the adsorption from aqueous solutions containing one type of polymer of different architectures (blue and circle symbols: linear polymer of 20 monomers, 20l; black and square symbols: linear polymer of 35 monomers, 35l; red and diamond symbols: branched polymer of 35 total monomers, 25 segments in the backbone and 2 branches of 5 segments each, 35-2b). (b) Partition between spheres and cylinders as a function of radius, for the same mixtures as in panel (a).

polymer and the surface is the same for all molecules. Hence, the difference in adsorption derives from the entropic repulsive contributions.

The adsorption isotherms for the different polymers on the same geometry and radius in Fig. 2a, reflect that the repulsions depend very strongly on the shape and size of the molecules. Increasing the molecular weight of the polymers (*i.e.* the total number of segments), leads to an increase in the steric repulsion within the adsorbed layer and, therefore, to a decrease of the adsorbed density (35l *vs.* 20l). The intrinsic effect of the molecular topology is reflected in the adsorption of 35l *vs.* 35-2b, where the only difference is the arrangement of the monomers in the chain. The branched chain is bulkier than the linear one; hence, in the good solvent regime under discussion, the latter experiences stronger intramolecular repulsions and reaches a lower grafting density at equilibrium than the linear polymer.

Regarding the effect of the surface curvature, we observe that the adsorption increases as the radius of the curvature of the surface decreases, irrespective of the molecular topology or the morphology of the surface. This is due to the fact that in curved convex surfaces, the available volume increases as the distance from the surface increases, scaling as r/R for cylinders and as $(r/R)^2$ for spheres (where r is the radial direction and R the radius of the surface). This decreases the steric repulsions within the layer, giving rise to a greater adsorption on spheres than on a cylindrical geometry with the same radius. In a NR, which has spherical caps and a cylindrical body, curvature leads to denser films on the caps than on the body. Therefore, the partition between the caps and the body (which we define as the ratio of the equilibrium surface densities on the sphere and the cylinder, $\sigma_{\text{sph}}/\sigma_{\text{cyl}}$) is always larger than one (Fig. 2b). Long polymer chains adsorb preferentially in regions with a large available volume and, therefore, they have a larger sphere/cylinder partition than short polymers, as observed by comparing 20l and

35l in Fig. 2b. However, we do not observe large partition differences between linear and branched polymers, as seen when comparing 35l and 35-3b in Fig. 2b.

Based on these results, we will focus on the possibility of using mixtures of polymer chains with different molecular topologies in order to enhance the curvature-driven partition. This effect can be useful to engineer surface-modified nanomaterials where different types of molecules enable multifunctional platforms.

3.2 Polymer adsorption from binary mixtures: combining molecular topology and surface curvature

3.2.1 General description of adsorption. Fig. 3a shows the equilibrium amount of linear and branched polymers adsorbed as a function of the bulk

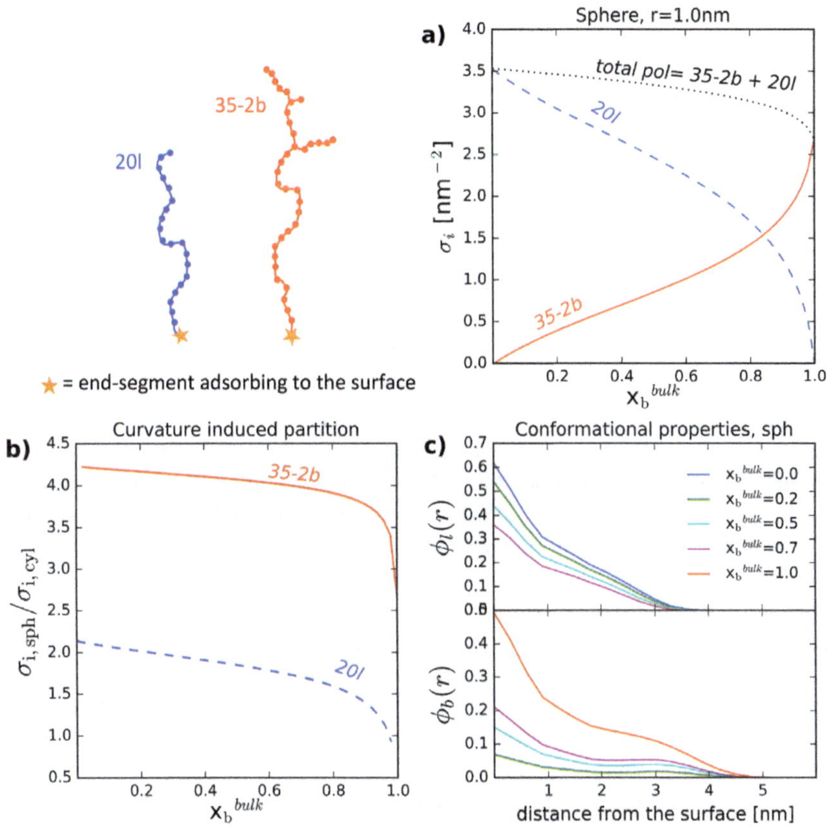

Fig. 3 (a) Density of surface adsorbed polymers as a function of the molar fraction of the branched polymer in the bulk for a mixture comprising a linear polymer of 20 monomers (20l, blue line dashed) and a branched polymer of 35 total monomers, 25 segments in the backbone and 2 branches (35-2b, red solid line) on a sphere of $R = 1$ nm (see scheme on the left). (b) Partition of the linear, the branched and the total amount of polymer between a sphere and a cylinder of 1 nm radius. Colors and lines as in panel (a). (c) Volume fraction of the adsorbed mixture as a function of the distance from the tethering surface. Results correspond to a spherical surface of radius 1 nm. Upper and lower graphs correspond to the volume fraction of the linear and the branched polymer, respectively. Colors correspond to different compositions of the bulk solution, as indicated in the legend.

composition of the solution for spherical surfaces, for a binary mixture of polymers comprising the polymers 20l and 35-2b. We show the results for $R = 1$ nm, but they are representative of the qualitative behavior in the other systems studied.

The adsorption isotherms demonstrate the complexity in the balance of forces that determine adsorption: the attraction between the end-segment of the polymers and the surface, the repulsive interactions between adsorbed molecules and the entropy of mixing. There is a monotonic decrease (increase) in the adsorbed amount of the linear (branched) polymer, as the solution becomes more concentrated in the branched polymer. However, the linear polymer 20l is the major component of the adsorbed film until the solution contains 80% of branched polymers. The preferential adsorption of 20l over 35-2b is due to excluded-volume interactions, since the short linear polymer can be better accommodated than the long branched one in the proximity of the surface. The fraction of the branched polymer in solution that leads to a layer with 50% branched polymer is very sensitive to the molecular details of the polymer chains and to the relative difference in the size and shape between the linear and the branched chains.

The effect of bulk composition on the overall adsorption is also shown in Fig. 3a. We observe that the total amount of adsorbed polymer, $\sigma_{\text{tot}} = \sigma_{\text{l}} + \sigma_{\text{b}}$, is a monotonic decreasing function of the bulk composition for the branched polymer.

The interplay between curvature, molecular architecture and bulk composition is reflected in the curvature-driven partition (Fig. 3b). The sphere/cylinder partition is always larger for the branched polymer than for the linear one due to the bulkier nature of the branched chains. This drives its partition towards the sphere, demonstrating the possibility of enriching the spherical end of a NR with the bulkier polymer in the mixture. The sphere/cylinder partition for both the linear and branched polymers decreases monotonically with the content of branched polymer in solution. As $x_{\text{b}}^{\text{bulk}}$ increases, the competition between the flexible linear chains and the bulky branched ones hinders the adsorption of the linear polymer on regions of larger available volume and, thus, the sphere/cylinder partition for the linear polymer decreases. Regarding the branched polymer, the amount adsorbed on both surfaces increases with increasing bulk concentration (Fig. 3a), but this increase, relative to the maximum surface density at $x_{\text{b}}^{\text{bulk}} = 1$, is higher on the cylinder than on the sphere. Thus, the occupancy of the available surface increases faster with $x_{\text{b}}^{\text{bulk}}$ for the cylinder than for the sphere and the branched polymer partition towards the sphere decreases for increasing $x_{\text{b}}^{\text{bulk}}$. Also, we observe that the partition for the branched polymer does not depend very strongly on the composition of the solution, which could be of interest when engineering nanomaterials for diverse applications as it would allow the surface density to be maximized by adjusting the bulk solution without significantly affecting the spatial partition. It is interesting to note that for the limiting value of $x_{\text{b}}^{\text{bulk}} \sim 1$, the decrease in the curvature-induced partition for the linear polymer falls below 1. While for the adsorption of a single type of polymer in good solvent the partition should be always larger than one, there exists conditions where the partition of one of the polymers in a mixture falls below one (*i.e.* that polymer is enriched at the surface of the cylindrical body in comparison with the spherical tip). The effect results from the fact that the adsorption of mixtures is a competition for the available surface area; therefore, one of the polymers may strongly adsorb, thus depleting the other one from the spherical surface and making its sphere/cylinder partition smaller than one. Changing the quality of the solvent can enhance this effect further, as we discuss below.

The differences in available volume for each geometry dictates the molecular organization of the layer and defines the amount of adsorption for each polymer. The conformational properties of the layer strongly depend on the architecture of the chains. The variation of the polymer volume fraction as a function of the distance from the tethering surface accounts for the interplay between the curvature and the molecular architecture, as shown in Fig. 3c. It is worth noting that the density profiles depend on the grafting density, and that this quantity in turns changes for each geometry (Fig. S1†). Increasing the fraction of branched polymer in solution increases the amount of the branched polymer on the surface, and with that the steric repulsions in the layer. The system lowers these repulsions by the combination of two mechanisms: (i) stretching both the linear and branched chains towards regions of greater available volume (at the cost of conformational entropy) and (ii) decreasing the total amount of adsorbed polymer (at the cost of decreasing the total adsorption energy, Fig. 3a).

The results from our calculations show that the composition of the surface-adsorbed layer departs strongly from the bulk composition for all the geometries studied, as shown in Fig. 4. Moreover, in curved surfaces of increasing radius, the composition of the adsorbed mixture tends to that of the planar surface. The composition of the branched polymer in the bulk required to obtain a 1 : 1 mixture on the surface ($x_b^{surf} = 0.5$) is always larger than 0.5 and shifts to higher values as the radius of the surface increases (Fig. 4b). In the limit of highly curved surfaces, ($R = 1$ nm, Fig. 4a), we observe that for a solution with $x_b^{surf} = 0.9$ the surface fraction of the branched polymer on the sphere is 60% whereas for

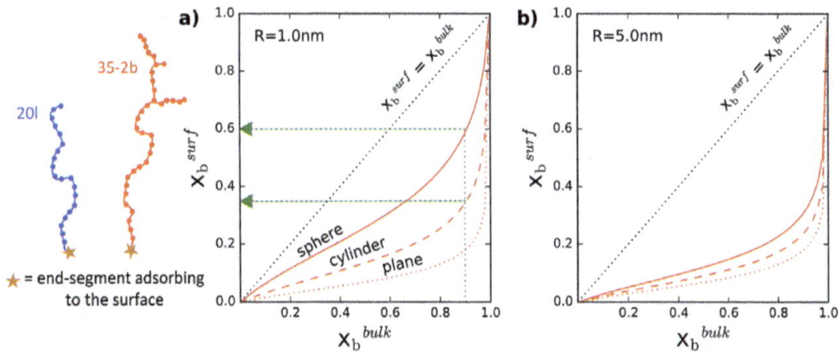

Fig. 4 Composition of the surface-adsorbed mixture as a function of the molar fraction of the branched polymer in the bulk for different geometries and curvatures. Results correspond to a mixture comprising a linear polymer of 20 monomers (20l) and a branched polymer of 35 total monomers, 25 segments in the backbone and 2 branches (35-2b, see scheme on the left). (a) Full, dashed and dotted red lines correspond to a spherical, cylindrical, and planar surface respectively. The black dotted line corresponds to a surface composition equal to that of the bulk solution. Green dotted arrows indicate the partitioning for a solution with $x_b^{bulk} = 0.9$: the surface fraction of the branched polymer on a sphere of $R = 1$ nm is ~60% whereas for a cylinder of the same radius it is ~35%. (b) Same as panel (a), but $R = 5$ nm for the curved surfaces.

a cylinder of the same radius it is 35%. This result strongly suggests the possibility of enriching the adsorbed mixture in the branched polymer in surfaces of high curvature.

3.2.2 Molecular architecture effects. We now provide an analysis of the effect of the number of branches, the branching positions, the backbone length and the topology of the polymer molecules on the competitive adsorption of binary polymer mixtures on surfaces of different geometries. Due to the complexity of the system, the effect of these variables is highly non-additive (see below). Thus, we will not attempt to analyze all possible combinations of these different variables, but rather focus our efforts on systematically analyzing combinations that can improve the partition of the branched polymer toward the Au-NR tips. Given the dimensions of commonly synthesized Au-NRs,[4] we will study systems of 5 nm radius.

Branching effect. Fig. 5 shows the effect of increasing the number of branches at a constant backbone length of the branched polymer. In all cases, the linear polymer corresponds to the polymer 20l. The branched polymer has a backbone of 20 segments, and one, two, or three branches. We observed that the surface density of the branched polymer systematically decreases with an increasing number of branches (Fig. 5a). This is due to the increase in the size and bulkiness of the branched polymer, which results in an increase of the steric repulsions that hinder adsorption in good solvent conditions. This in turn favors the adsorption of the linear polymer (Fig. 5a), yielding an overall adsorption ($\sigma_{tot} = \sigma_b + \sigma_l$) that decreases slightly on increasing the number of branches at a constant backbone length, both on spheres and cylinders (Fig. 5b). One could, therefore, optimize the molecular structure of the chains in the mixtures without compromising the total amount of adsorbed polymer (σ_{tot}), which can be very beneficial when engineering materials for biological applications. The increasing crowding near the surface with an increased number of branches at a constant backbone length also results in a more favorable partition of the branched polymer towards the regions of higher curvature (Fig. 5c). The linear polymer in the mixture is also preferentially partitioned towards the sphere (Fig. S2†) although to a much lesser extent than the branched polymers. For both the branched and the linear polymer, the partition is maximal for $x_b^{bulk} = 0$ (*i.e.* pure linear polymer solution), as we discussed above.

The effect of increasing the number of branches on polymer adsorption depends strongly on the length of the backbone of the branched polymer. Fig. S3† shows the analysis for mixtures comprising the same linear polymer as in Fig. 5, but branched polymers with 35 monomers in the backbone instead of 20, with various numbers of branches. The qualitative effects are the same as discussed above, but much less dramatic, and become even less important for branched polymers of even longer backbones (data not shown). The position of the branches along the polymer backbone is also very important. Fig. 6 shows the surface density of each individual polymer in mixtures comprising the same linear polymer of 20 segments and branched chains containing three branches in different positions with respect to the surface. It can be appreciated that short branches near the surface are very detrimental for the adsorption process of the branched polymer, favoring the adsorption of the linear polymer over almost the whole range of bulk solution compositions. Increasing the spacing between the position of the first branch and the surface strongly favors its adsorption, despite

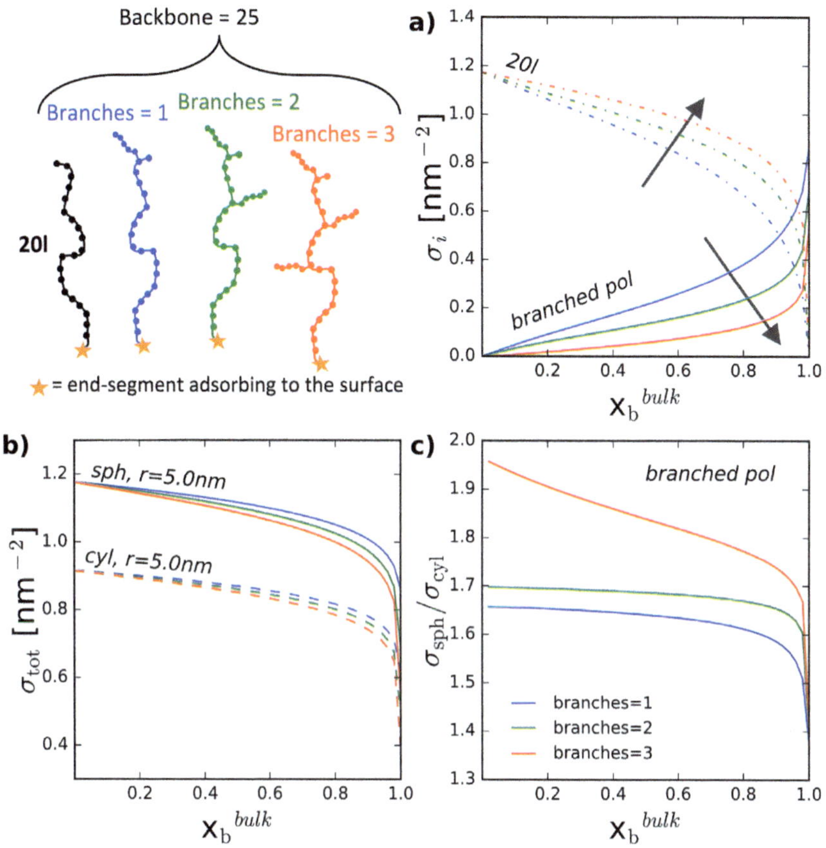

Fig. 5 Effect of polymer branching on the adsorption for polymer mixtures comprising a linear polymer of 20 monomers (20l) and a branched polymer with backbone length = 25 segments and one, two or three branches (colors blue, green, and red respectively, as shown in panel (c), and in scheme on the upper left panel). All results correspond to curved surfaces of $R = 5$ nm. (a) Surface density of the adsorbed linear (dash-dot lines) and branched polymers (full lines) as a function of the bulk molar fraction of the branched polymer for a spherical surface. (b) Total surface density of adsorbed polymers ($\sigma_{tot} = \sigma_l + \sigma_b$) as a function of the bulk molar fraction of the branched polymer for spherical (full lines) and cylindrical (dashed lines) surfaces. (c) Sphere–cylinder partition for the branched polymer in the mixture as a function of the bulk molar fraction of the branched polymer.

the increase in the total number of monomers. This behavior depends on the number of branches of the polymer, as we discuss below.

Backbone length effect. In this section we explore the effect of changing the backbone length of the branched polymers for a constant number of branches (the length of the backbone changes below the position of the first branch), see the scheme in Fig. 7. Increasing the backbone length for the one-branch polymer decreases the surface density of the branched polymer and increases that of the linear chain (Fig. 7a). On the other hand, when the branched polymer has three branches, increasing the backbone length increases the amount of the adsorbed branched polymer and decreases that of the linear chain (Fig. 7b). Mixtures

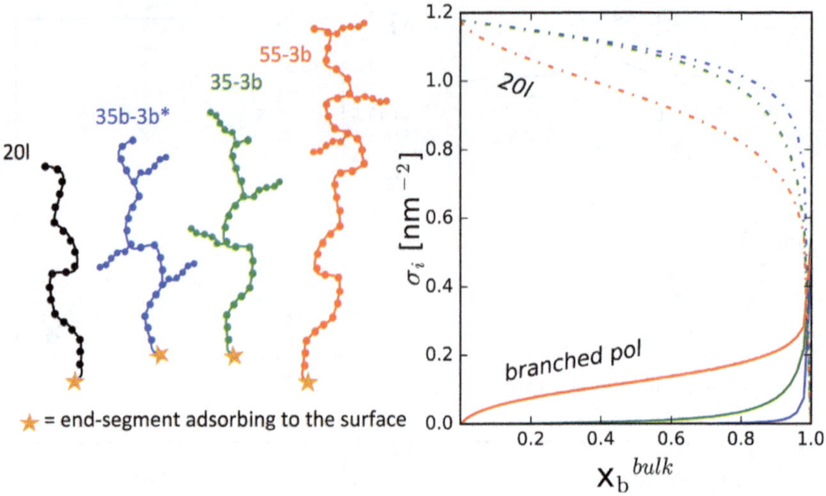

Fig. 6 Effect of the position of the branches with respect to the surface on the adsorption for polymer mixtures comprising a linear polymer of 20 monomers and the branched polymers depicted in the scheme, on a spherical surface of $R = 5$ nm. The surface densities of the adsorbed linear (dash-dot lines) and branched polymers (full lines) are plotted as a function of the bulk molar fraction of the branched polymer. The colors of the lines are the same as those in the scheme.

containing two-branch chains show an intermediate behavior between the one-branch and the three-branch polymers (Fig. S4†).

The molecular organization of the adsorbed layer is the result of the balancing forces governing the adsorption process. In the case of the one-branch polymers, where the branching point is near the end of the chain (Table S1 in the ESI† and scheme in Fig. 7), the repulsive forces with neighboring molecules result in stretched configurations, in which the branch orients towards the solution (Fig. S5†). So, their behavior is similar to that of a linear polymer. In this scenario,

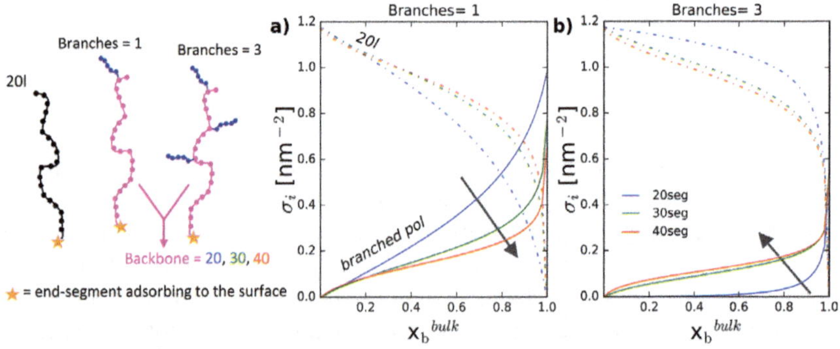

Fig. 7 (a and b) Surface density of adsorbed polymers as a function of the bulk molar fraction of the branched polymer for mixtures comprising a linear (20l, dash-dot lines) and a branched polymer (full lines) with different backbone lengths (20, 30, and 40 monomers shown by blue, green, and red lines respectively) and containing either one or three branches (see scheme on the left). Results correspond to a spherical surface of $R = 5$ nm.

increasing the backbone of the branched chain does not significantly modify the region of the chain that competes for the available volume with the short linear chain near the surface. On the other hand, in the three-branch polymers, the short branches cannot be accommodated in a way that resembles a linear chain (see Fig. 8 and S6†). When the backbones of the linear and the branched polymers have a similar length, there is a competition for the available space between the linear chain and the much bulkier branched chain. However, when the backbone of the branched chain is longer than the linear chain, the competition for the available volume occurs between the linear chain and a region of the linear backbone of the branched chain. This last scenario also improves the accessibility of the free-end segments from the solution. Fig. 8 shows the density of free ends as a function of the distance from the surface. This quantity reflects the most probable position of the free-end segments for the polymers in the mixture. It can be seen that making the backbone of the branched chain longer than the linear polymer allows for the free ends of the branched polymer to be located towards the solution. On the other hand, in branched chains with a short 20-segment backbone, the free ends remain buried within the layer and, thus, are inaccessible from solution (Fig. S6†). The concern about the accessibility of the free ends is of prime importance for conferring further chemical functionalities to these segments for improved biotargeting or tailored chemical reactivity of the nanoconstruct.

The effects of the molecular architecture of the polymers (backbone length and number of branches) on the curvature-induced partition are summarized in Fig. 9, where we show the ratio between the surface density on spheres and cylinders of radius 5 nm (panel a) and 1 nm (panel b) for mixtures of a 20-segment linear polymer (20l) with branched polymers with different backbone lengths and numbers of branches. The bulk composition plotted, $x_b^{bulk} = 0.9$, was chosen to have a significant amount of adsorbed branched polymer. In general, the partition between spheres and cylinders increases when increasing the number of

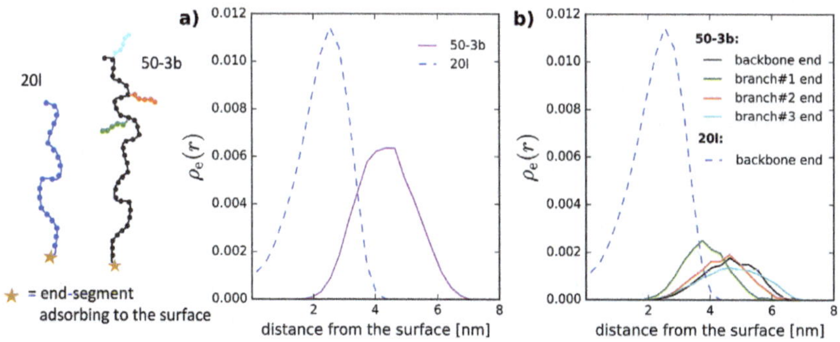

Fig. 8 Density of free ends as a function of the distance from the tethering surface for a spherical surface of $R = 5$ nm. Plots correspond to a mixture of a linear polymer of 20 monomers (20l) and a branched polymer with 3 branches and a backbone length of 35 segments (50-3b, see scheme on the left). All the cases are for a fixed bulk composition with $x_b^{bulk} = 0.9$. Dashed lines show the end-segment density of the backbone of the linear polymer (panels a and b). Full lines show the sum of the end-segment densities of the backbone and each branch of the branched polymer in panel (a) and details of the individual densities of each type of end segment in panel (b).

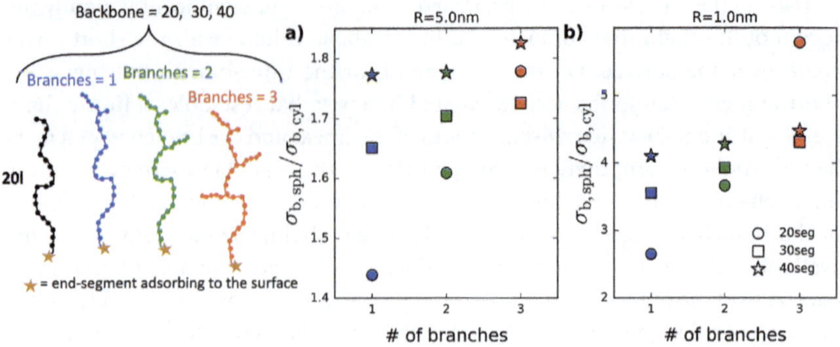

Fig. 9 Partition of the branched polymer between a sphere and a cylinder of radius 5 nm (panel a) and 1 nm (panel b). Polymer mixtures comprise a linear polymer of 20 monomers (20l) and a branched polymer with one, two, or three branches (in blue, green or red respectively) and backbone lengths of 20, 30, and 40 monomers (denoted by circle, square, and star symbols respectively, see scheme on the left). All the cases are for a fixed bulk composition with $x_b^{bulk} = 0.9$.

branches or the backbone length. The outlier behavior of the three-branch polymer with a backbone of 20 segments is derived from the strong repulsions with the linear polymer, as we discussed above. It is worth keeping in mind that in that case, the total amount of polymer adsorbed is significantly diminished, so there is a compromise between achieving a curvature-induced partition and reaching an appropriate surface density. While the partition for $R = 5$ nm, which is in the range of usual sizes of NRs for biomedical applications,[4] is rather large, it is possible to increase it further by increasing the curvature of the surface; see the results for $R = 1$ nm in Fig. 9b.

The strategies we have discussed so far were targeted towards favoring a partition of the branched polymer to spherical surfaces with respect to cylinders of the equivalent radius. It is worth mentioning that different partition scenarios are possible based on the molecular architecture of the chains. For example, in binary mixtures comprising a linear chain longer than the backbone of the branched polymer, we observed that these long linear chains end up partitioning to the regions of greater curvature, *i.e.* away from the surface (Fig. S7†).

Molecular topology. Strategically combining the molecular architecture for each chain could allow the competitive adsorption of the molecules on surfaces of different curvature to be tuned. But, now we ask: what is the gain in resorting to a mixture of linear and branched chains instead of mixtures of two linear polymers of different lengths? To that end, we analyzed the differences between a mixture comprising a linear polymer of 20 monomers (20l) and a three-branch polymer of 50 monomers total (50-3b), and a mixture comprising the same linear polymer (20l) and another linear polymer of 50 monomers (50l). Note that the polymers 50-3b and 50l have exactly the same number of monomers, but different topologies. We observed that the surface density of 50-3b is slightly smaller than that of 50l (Fig. 10a), which results from the fact that the first branching position of 50-3b is at segment 24, while the length of the 20l polymer is 20 segments (see scheme in Fig. 10). Therefore, the 50-3b polymer experiences greater steric repulsions than the 50l, hindering its adsorption. In the case of the 20l + 50l mixture, competition for the available space between the linear chains results in a more favorable adsorption than for the 20l + (50-3b) mixture. However, the curvature-induced partition between spherical and cylindrical surfaces is similar for both mixtures (Fig. 10b). We also observed this behavior for the adsorption of polymers with an equal number of total monomers from solutions containing one single polymer type (Fig. 2b, 35l *vs.* 35-2b). The end-segment accessibility from the solution for the 50-3b polymer (Fig. 10c, upper panel) is more hindered by the linear 20l polymer than that for the 50l chains (Fig. 10c, lower panel), but the overall accessibility is rather good in both cases. Thus, resorting to a branched polymer instead of a linear polymer with the same number of total monomers results in an increase in the density and accessibility of available end-segments,

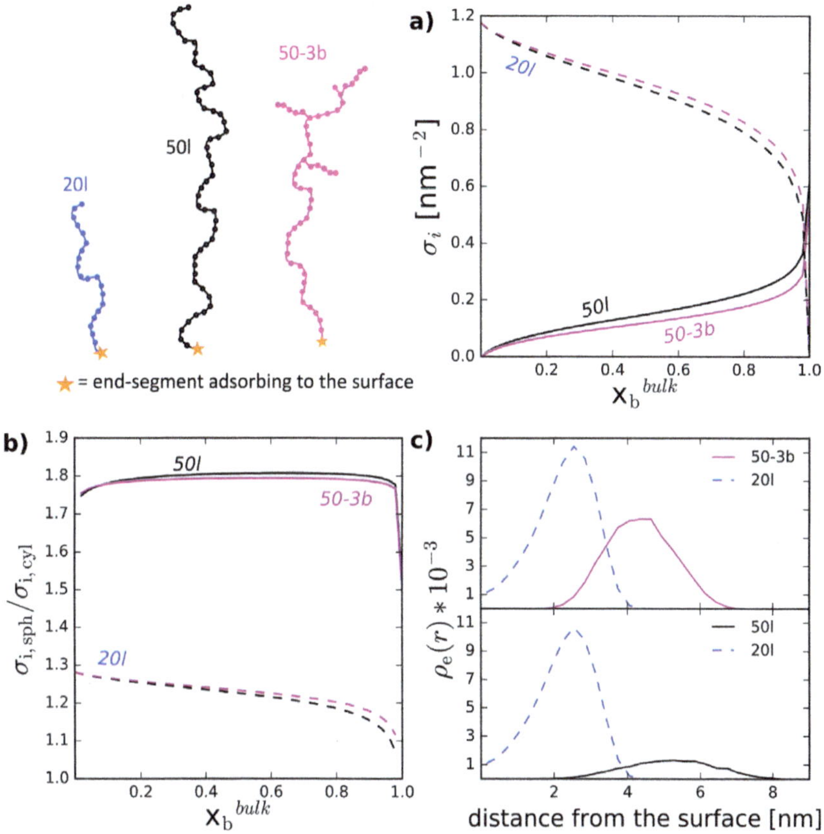

Fig. 10 Effect of the molecular topology on the competitive adsorption of binary mixtures. (a) Surface density of adsorbed polymers as a function of the bulk molar fraction of the branched polymer. The full magenta and full black lines correspond to the 50l and 50-3b polymers respectively. Similarly, the dashed magenta and dashed black lines correspond to the 20l polymer in the 20l–50l and the 20l–(50-3b) mixtures respectively (see scheme in the upper left panel). Results correspond to a spherical surface of $R = 5$ nm. (b) Partition of the linear and branched polymers between a sphere and a cylinder of 5 nm radius for the 20l–50l (magenta) and the 20l–(50-3b) (black) mixtures. (c) Density of free ends as a function of the distance from the tethering surface for a spherical surface of $R = 5$ nm. All the cases are for a fixed bulk composition with $x_b^{bulk} = 0.9$. Lines and colors are indicated in the legend.

but with a small decrease in the amount of adsorbed polymers. Once again, this behavior is strongly dependent on the branching positions of the polymer, for instance a 3-branch chain with two branch points below the end-backbone of the linear chain (Fig. S8†) shows a smaller surface density at equilibrium and fewer accessible free ends than those reported for the branched polymer in Fig. 10. As a conclusion, we propose a general design rule: in order to optimize polymer adsorption and free end accessibility, the length of the backbone between the surface-bound segment and the first branching position in the branched chain should be longer than the length of the linear polymer.

3.2.3 Role of polymer–polymer interactions. The results presented so far corresponded to polymer mixtures in good solvent (*i.e.* $\frac{\chi_c}{\chi} = \infty$, where a higher $\frac{\chi_c}{\chi}$ corresponds to better solvent quality). To address the role of the solvent quality in the competitive polymer adsorption process, we analyzed the adsorption from a binary polymer mixture considering values of $\frac{\chi_c}{\chi}$ equal to ∞, 1.43 and 0.83 (Fig. 11). Note that the parameter $\frac{\chi_c}{\chi}$ is a proxy for the reduced temperature of the system, *i.e.* the polymer is in a good solvent for $\frac{\chi_c}{\chi} > 1$ and in a poor solvent for $\frac{\chi_c}{\chi} < 1$. The parameter $\frac{\chi_c}{\chi}$ results from normalizing the strength of segment–segment attractions, χ, by the critical value χ_c (computed in a separate set of calculations), which is defined as the value of χ required to trigger microphase separation for a neutral, planar polymer brush in the limit of vanishing surface coverage.[52,54] It is worth mentioning that all the polymer mixtures studied here are thermodynamically stable as homogeneous solutions, *i.e.* for the moderate poor solvent conditions that we use, the bulk solution does not phase separate.

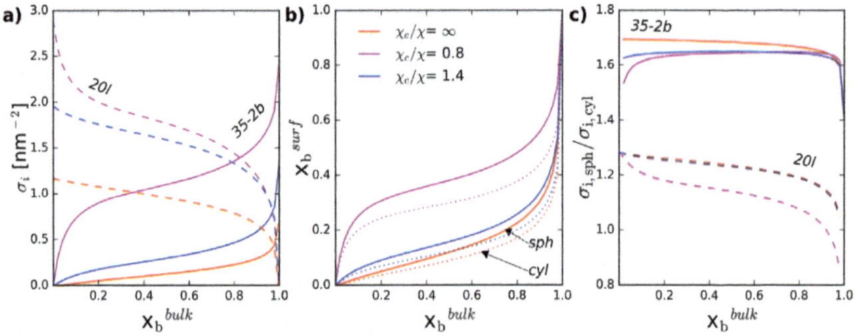

Fig. 11 Effect of solvent quality on the adsorption from polymer mixtures comprising a linear polymer of 20 monomers (20l) and a branched polymer with a backbone length of 25 segments and 2 branches (35-2b). Red, magenta, and blue lines correspond to values of $\chi_c/\chi = \infty$, 1.4, and 0.833, respectively (see legend of panel b). (a) Surface density of adsorbed linear (dashed lines) and branched (full lines) polymers as a function of the bulk molar fraction of the branched polymer for a spherical surface of $R = 5$ nm. (b) Molar fraction of the branched polymer adsorbed on the surface as a function of the bulk molar fraction of the branched polymer for spherical (full lines) and cylindrical (dotted lines) surfaces of $R = 5$ nm. (c) Partition of the linear (dashed line) and the branched (full line) polymer between a sphere and a cylinder of 5 nm radius as a function of the bulk molar fraction of the branched polymer.

As the quality of the solvent decreases, we observe that polymer adsorption increases, due to the higher attractive forces between monomers in the polymers with respect to the solvent. This increase in the amount of adsorbed polymers changes dramatically when $\frac{\chi_c}{\chi} < 1$ (Fig. 11a). At the same time, the fraction of the branched polymer in the adsorbed layer increases because this polymer has a greater number of segments than the linear one (35 *vs.* 20) and, therefore, experiences stronger polymer–polymer attractions that compete with the steric repulsions. Interestingly, while the quality of the solvent affects the amount and composition of the adsorbed polymer layer, it has little effect on the curvature-induced partition between spheres and cylinders for $R = 5$ nm (Fig. 11c). It is also worth noting that the curvature-induced partition for the linear polymer falls below 1 for high values of x_b^{bulk} (Fig. 11c). This is due to the fact that the partition results from the competition between both types of polymers for the available surface area and, therefore, it can become smaller than one for a given polymer if the other polymer in the mixture adsorbs strongly to the spherical surface, blocking it, as discussed above.

Finally, we observed few qualitative differences on the effect of the molecular architecture of the polymer chains in the competitive adsorption process when decreasing the quality of the solvent. As discussed in the above paragraph, there is a change in the adsorbed amount, the surface composition and the curvature-induced partition for each value of $\frac{\chi_c}{\chi}$, but the trends we observed for the good solvent conditions regarding branching and backbone effects do not show significant changes (Fig. S9†).

4 Conclusions

In the present work, we propose a novel strategy to construct gold nanorods with chemically inhomogeneous coatings by using the different curvature of the tips and the body to direct the adsorption of the components of a polymer mixture. To test this strategy, we analyzed the effect of surface curvature and molecular topology on polymer adsorption from binary solutions on Au-NRs modeled as the combination of spherical and cylindrical surfaces, representing the tips and body of the NR, respectively. Our calculations suggest that a spatially resolved surface modification could be attainable by adjusting the surface curvature, the molecular architecture of the polymer chains, and the concentration of the bulk solution. Combining polymers of different architectures introduces a competition between them for the available surface. A complex balance between repulsive forces and polymer–surface and polymer–polymer attractions, modulated by the underlying curvature, governs the overall adsorption process. The composition of the surface mixture departs strongly from that of the bulk solution, and it depends on the curvature, morphology and specific details of the polymers in the mixture. We observed a curvature-induced partition between spherical and cylindrical surfaces that strongly depends on the radius of the curved surface, but can be adjusted by tailoring the molecular topology of the chains in the mixture. For the smallest dimensions of Au-NRs reported in the literature ($R = 5$ nm),[1,2] the enrichment effect for the branched polymer in the mixture is typically a factor of 1.5–2 (*i.e.* we predict the adsorption of approximately twice the amount of branched polymer on the Au-NR tip than on its body). This effect can be further increased to factors of 4–5 for $R = 1$ nm (Fig. 9a), which may be achievable with nanoparticles with very sharp tips such as Au nanostars[55] or with semiconductor NRs, whose diameters can be controlled in the few-nm range.[56] Another possible route to increase the curvature-induced partition is to use Au nano-dumbbells,[47] whose bodies have a concave curvature and, therefore, exhibit a smaller available volume near the surface than the cylindrical body of Au-NRs.

The strength of the theoretical strategy employed in this work lies in the explicit description of the different competing interactions, while taking full advantage of the symmetry of the system in order to reduce the computational cost of the calculations. Our methodology allowed us to perform a systematic analysis of the relevant parameters on polymer adsorption from binary mixtures, which included calculations on over 100 different polymer mixtures in 50 solutions of different compositions and solvent qualities. This type of systematic analysis would be out of reach using molecular dynamics computer simulations, which are currently limited to study a few cases.[57] Our methodology also allowed us to ask questions about structure–property relationships that are very hard to observe experimentally. Particularly, we can outline some useful design rules that would lead to a partition of the branched polymer towards regions of higher curvature:

• The positions of the branching points are crucial: short branches near the surface have a drastic negative effect on the adsorption of the branched polymer, so in order to optimize the adsorption of the branched polymer, the length of the chain segment between the surface and the position of the first branch for a branched polymer should be larger than the length of the linear chain.

• Increasing the number of branches at constant backbone length decreases the adsorbed amount, but at the same time, it favors the partition between spheres and cylinders.

• Using mildly poor solvent conditions allows the amount of adsorbed branched polymer to be increased without drastically decreasing the sphere/cylinder partition.

Future work will encompass modeling systems that simultaneously combine surfaces of different curvatures in two dimensions, which would allow us to explore Au-NRs of different aspect ratios as well as nanoparticle shapes beyond Au-NRs, such as nano-dumbbells.

Acknowledgements

This research was supported by Grant No. EB005772 from the National Institute of Biomedical Imaging and Bio-engineering (NIBIB). MT is a fellow of CONICET (Argentina). The authors greatly appreciate the assistance of Ingrid Goobar-Szleifer during the writing of the manuscript.

References

1 N. Li, P. Zhao and D. Astruc, *Angew. Chem., Int. Ed.*, 2014, **53**, 1756–1789.

2 J. Pérez-Juste, I. Pastoriza-Santos, L. M. Liz-Marzán and P. Mulvaney, *Coord. Chem. Rev.*, 2005, **249**, 1870–1901.

3 X. Huang, I. H. El-Sayed, W. Qian and M. A. El-Sayed, *J. Am. Chem. Soc.*, 2006, **128**, 2115–2120.
4 E. C. Dreaden, A. M. Alkilany, X. Huang, C. J. Murphy and M. A. El-Sayed, *Chem. Soc. Rev.*, 2012, **41**, 2740–2779.
5 J. A. Champion and S. Mitragotri, *Proc. Natl. Acad. Sci. U. S. A.*, 2006, **103**, 4930–4934.
6 X. Shi, A. von dem Bussche, R. H. Hurt, A. B. Kane and H. Gao, *Nat. Nanotechnol.*, 2011, **6**, 714–719.
7 R. Vácha, F. J. Martinez-Veracoechea and D. Frenkel, *Nano Lett.*, 2011, **11**, 5391–5395.
8 C. Lee, W. B. Lee and T. Kang, *Appl. Phys. Lett.*, 2013, **103**, 091602.
9 E. Gonzalez Solveyra and I. Szleifer, *Wiley Interdiscip. Rev.: Nanomed. Nanobiotechnol.*, 2016, **8**(3), 334–354.
10 G. Palui, F. Aldeek, W. Wang and H. Mattoussi, *Chem. Soc. Rev.*, 2015, **44**, 193–227.
11 H. B. Lu, C. T. Campbell and D. G. Castner, *Langmuir*, 2000, **16**, 1711–1718.
12 P. Harder, M. Grunze, R. Dahint, G. M. Whitesides and P. E. Laibinis, *J. Phys. Chem. B*, 1998, **102**, 426–436.
13 K. H. A. Lau, T. S. Sileika, S. H. Park, A. M. L. Sousa, P. Burch, I. Szleifer and P. B. Messersmith, *Adv. Mater. Interfaces*, 2015, **2**, 1400225–1400235.
14 K. H. A. Lau, C. Ren, S. H. Park, I. Szleifer and P. B. Messersmith, *Langmuir*, 2012, **28**, 2288–2298.
15 H. O. Ham, S. H. Park, J. W. Kurutz, I. G. Szleifer and P. B. Messersmith, *J. Am. Chem. Soc.*, 2013, **135**, 13015–13022.
16 K. E. Sapsford, W. R. Algar, L. Berti, K. B. Gemmill, B. J. Casey, E. Oh, M. H. Stewart and I. L. Medintz, *Chem. Rev.*, 2013, **113**, 1904–2074.
17 K. K. Caswell, J. N. Wilson, U. H. F. Bunz and C. J. Murphy, *J. Am. Chem. Soc.*, 2003, **125**, 13914–13915.
18 C. J. Murphy, L. B. Thompson, A. M. Alkilany, P. N. Sisco, S. P. Boulos, S. T. Sivapalan, J. A. Yang, D. J. Chernak and J. Huang, *J. Phys. Chem. Lett.*, 2010, **1**, 2867–2875.
19 Z. Nie, D. Fava, E. Kumacheva, S. Zou, G. C. Walker and M. Rubinstein, *Nat. Mater.*, 2007, **6**, 609–614.
20 Z. Nie, D. Fava, M. Rubinstein and E. Kumacheva, *J. Am. Chem. Soc.*, 2008, **130**, 3683–3689.
21 D. Fava, M. A. Winnik and E. Kumacheva, *Chem. Commun.*, 2009, 2571–2573.
22 S. Mühlig, A. Cunningham, J. Dintinger, T. Scharf, T. Bürgi, F. Lederer and C. Rockstuhl, *Nanophotonics*, 2013, **2**, 211–240.
23 A. Lukach, K. Liu, H. Therien-Aubin and E. Kumacheva, *J. Am. Chem. Soc.*, 2012, **134**, 18853–18859.
24 I. Szleifer and M. A. Carignano, *Adv. Chem. Phys.*, 1996, **94**, 165–260.
25 C. D. Walkey, J. B. Olsen, H. Guo, A. Emili and W. C. W. Chan, *J. Am. Chem. Soc.*, 2012, **134**, 2139–2147.
26 M. Lundqvist, J. Stigler, G. Elia, I. Lynch, T. Cedervall and K. A. Dawson, *Proc. Natl. Acad. Sci. U. S. A.*, 2008, **105**, 14265–14270.
27 C. D. Walkey, J. B. Olsen, F. Song, R. Liu, H. Guo, D. W. H. Olsen, Y. Cohen, A. Emili and W. C. W. Chan, *ACS Nano*, 2014, **8**, 2439–2455.
28 I. Szleifer, *Curr. Opin. Colloid Interface Sci.*, 1996, **1**, 416–423.
29 F. Fang and I. Szleifer, *Langmuir*, 2002, **18**, 5497–5510.

30 I. Szleifer, *Biophys. J.*, 1997, **72**, 595–612.
31 I. Szleifer, *Phys. A*, 1997, **244**, 370–388.
32 G. Longo and I. Szleifer, *Langmuir*, 2005, **21**, 11342–11351.
33 G. S. Longo, D. H. Thompson and I. Szleifer, *Langmuir*, 2008, **24**, 10324–10333.
34 R. J. Nap and I. Szleifer, *Biomater. Sci.*, 2013, **1**, 814–823.
35 M. Tagliazucchi and I. Szleifer, *J. Am. Chem. Soc.*, 2015, **137**, 12539–12551.
36 R. Nap, P. Gong and I. Szleifer, *J. Polym. Sci., Part B: Polym. Phys.*, 2006, **44**, 2638–2662.
37 R. J. Nap, A. L. Božič, I. Szleifer and R. Podgornik, *Biophys. J.*, 2014, **107**, 1970–1979.
38 D. Wang, R. J. Nap, I. n. Lagzi, B. Kowalczyk, S. Han, B. A. Grzybowski and I. Szleifer, *J. Am. Chem. Soc.*, 2011, **133**, 2192–2197.
39 R. J. Nap, S. H. Park and I. Szleifer, *J. Polym. Sci., Part B: Polym. Phys.*, 2014, **52**, 1689–1699.
40 R. Shvartzman-Cohen, E. Nativ-Roth, E. Baskaran, Y. Levi-Kalisman, I. Szleifer and R. Yerushalmi-Rozen, *J. Am. Chem. Soc.*, 2004, **126**, 14850–14857.
41 M. Tagliazucchi and I. Szleifer, *Soft Matter*, 2012, **8**, 7292.
42 A. Andrieu-Brunsen, S. Micoureau, M. Tagliazucchi, I. Szleifer, O. Azzaroni and G. J. D. A. A. Soler-Illia, *Chem. Mater.*, 2015, **27**, 808–821.
43 C.-L. Ren, D. Carvajal, K. R. Shull and I. Szleifer, *Langmuir*, 2009, **25**, 12283–12292.
44 M. Tagliazucchi, O. Azzaroni and I. Szleifer, *J. Am. Chem. Soc.*, 2010, **132**, 12404–12411.
45 M. Tagliazucchi, E. J. Calvo and I. Szleifer, *Langmuir*, 2008, **24**, 2869–2877.
46 M. Tagliazucchi, E. J. Calvo and I. Szleifer, *J. Phys. Chem. C*, 2008, **112**, 458–471.
47 D. A. Walker, E. K. Leitsch, R. J. Nap, I. Szleifer and B. A. Grzybowski, *Nat. Nanotechnol.*, 2013, **8**, 676–681.
48 M. A. Carignano and I. Szleifer, *J. Chem. Phys.*, 1995, **102**, 8662–8669.
49 P. J. Flory, *Statistical mechanics of chain molecules*, Interscience Publishers, New York, 1969.
50 M. A. Carignano and I. Szleifer, *J. Chem. Phys.*, 1993, **98**, 5006–5018.
51 M. Tagliazucchi, M. O. de la Cruz and I. Szleifer, *Proc. Natl. Acad. Sci. U. S. A.*, 2010, **107**, 5300–5305.
52 P. Gong, J. Genzer and I. Szleifer, *Phys. Rev. Lett.*, 2007, **98**, 018302.
53 A. Naderi, J. Iruthayaraj, T. R. Pettersson, R. A. Makuska and P. M. Claesson, *Langmuir*, 2008, **24**, 6676–6682.
54 O. Peleg, M. Tagliazucchi, M. Kröger, Y. Rabin and I. Szleifer, *ACS Nano*, 2011, **5**, 4737–4747.
55 D. H. M. Dam, J. H. Lee, P. N. Sisco, D. T. Co, M. Zhang, M. R. Wasielewski and T. W. Odom, *ACS Nano*, 2012, **6**, 3318–3326.
56 D. B. Tice, D. J. Weinberg, N. Mathew, R. P. H. Chang and E. A. Weiss, *J. Phys. Chem. C*, 2013, **117**, 13289–13296.
57 M. D. Donakowski, J. M. Godbe, R. Sknepnek, K. E. Knowles, M. Olvera de la Cruz and E. A. Weiss, *J. Phys. Chem. C*, 2010, **114**, 22526–22534.

Faraday Discussions

PAPER

Anisotropic optical and conductive properties of oriented 1D-nanoparticle thin films made by spray-assisted self-assembly†

S. Sekar,[ab] V. Lemaire,[ab] H. Hu,[ab] G. Decher*[abcd] and M. Pauly*[ab]

Received 18th February 2016, Accepted 2nd March 2016
DOI: 10.1039/c6fd00017g

We report on the fabrication of oriented anisotropic metal nanoparticle thin films made by Grazing Incidence Spraying (GIS) and on the anisotropic plasmonic properties of the resulting thin films. Gold nanorods of two different aspect ratios and silver nanowires were self-assembled as a uniaxially aligned monolayer with the GIS approach. In particular, we examine the influence of the nanowire/nanorod length and diameter on the degree of ordering determined by electron microscopy pictures. Furthermore, we show that the anisotropy of the optical properties (probed by polarized UV-visible-near infrared spectroscopy) strongly depend on the quality of alignment. The prepared monolayer thin films have an orientation order parameter of up to 0.83 for silver nanowires, which is reflected in an optical anisotropy of 0.57 in the UV-visible and 0.76 in the near infrared through the selective excitation of transverse and longitudinal surface plasmon resonance modes. The electronic transport in oriented silver nanowire monolayers is also shown to be highly directional, with the sheet resistance varying over almost an order of magnitude depending on the transport direction. Such anisotropic conductive plasmonic thin films may find applications in various fields like biochemical sensing, energy transport and harvesting or optoelectronic devices.

Introduction

Anisotropic materials possess physical properties, for example optical, electronic or mechanical properties that differ in different directions of space.[1,2] While

[a]*Institut Charles Sadron, CNRS (UPR22), 23 rue du Loess, F-67034 Strasbourg, France. E-mail: matthias.pauly@ics-cnrs.unistra.fr; decher@unistra.fr*

[b]*Faculté de Chimie, Université de Strasbourg, 1 rue Blaise Pascal, F-67008 Strasbourg, France*

[c]*International Center for Frontier Research in Chemistry, 8 allée Gaspard Monge, F-67083 Strasbourg, France*

[d]*Excellence Cluster "Nanostructures in Interaction with their Environment" (LabEx NIE), 23 rue du Loess, F-67034 Strasbourg, France*

† Electronic supplementary information (ESI) available: QCM analysis of the adsorption of gold nanorods on a substrate before and after functionalization with PVP, and TEM picture of the gold nanorods. See DOI: 10.1039/c6fd00017g

materials science has made enormous progress over the last few decades, especially with respect to composites, and nowadays also nano-composites, the methods available for preparing materials that are anisotropic at the nano- to micro-scale are still quite limited. Another roadblock in materials science lies in the controlled preparation of self-assembled materials composed of more than two components, and this holds especially true for nanoscale multicomposites. This is due to the fact that phase separation phenomena at the micro- and nanoscale often prevent the organization of individual components in a composite with precise distances between them. One strategy for building well-controlled multimaterial thin films is the so-called Layer-by-Layer (LbL) assembly approach,[3,4] in which a sequence of different materials and the distances between them are precisely controlled along the layer normal.[5–7]

Particularly interesting components for the preparation of anisotropic multicomponent materials are one-dimensional (1D-) nanoparticles, such as nanofibrils, nanowires or nanorods whose physical properties can easily be tuned by varying their aspect ratio.[1,2] While the physical properties of such nano-objects are often highly anisotropic at the single particle level, their controlled integration into multi-composites is even more difficult than the integration of their spherical counterparts. This is due to the fact that one does not only have to control the dispersion of the particles in a for example polymeric matrix, but to do so while additionally controlling the anisotropy of the composite.

In the present manuscript we use metallic nanorods and nanowires as anisotropic particles, because the anisotropy of nanocomposites of metallic components can easily be probed optically and electrically. In particular, the geometry dependent optical response of the electron cloud of metallic nanorods and -wires, which gives rise to distinct so-called surface plasmon resonances (SPR) along and perpendicular to their long axis, and the fact that the local plasmons of different rods or wires can couple as a function of their distance makes them valuable probes for studying anisotropic materials properties. In particular, metallic nanowires and nanorods have at least two absorbance bands which can be associated with the resonant oscillation within the nanoparticle of metallic electrons along the long axis (longitudinal SPR mode, LSPR) or along the short axis (transverse SPR mode, TSPR).[8]

Complementally, the direction dependent electrical conductivities of composites depend on macroscopic percolation pathways arising from a network of nanowire/nanowire or nanorod/nanorod contacts across the entire film. Consequently, both the plasmonic properties and the electrical conductivities depend on the in-plane densities and on the degree of alignment of the nanorods and -wires. However, plasmonic and polarized plasmonic phenomena arise predominantly from coupling occurring over some tens of nanometers whereas electrical conduction depends on the homogeneity of a composite at the macroscopic scale. Depending on the percolation geometry, a single break in the millimeter to centimeter long chain of wire/wire contacts between the test electrodes can be sufficient to interrupt percolation completely and will thus greatly reduce electrical conductivities. In this perspective one needs less wire/wire contacts in the direction parallel to the nanowire alignment in order to establish a percolation pathway than in the direction perpendicular to the alignment direction, which can be probed by measuring the directional conductivity of the thin films.

There are numerous applications in which anisotropic properties are needed,[9] and this interesting anisotropy in the properties at macroscopic scales requires the nanoparticles to be deposited as an oriented thin film over macroscopic distances. One strategy in order to tackle this issue is to set up self-assembly approaches that lead to large arrays of highly ordered structures. Indeed, bottom-up fabrication offers great advantages as compared to top-down approaches for the new generation of functional devices that will include nanoparticle building blocks.[9–11] Oriented self-assembly is especially important for plasmonic particles, as it allows fine-tuning of the so-called hotspots,[12–15] which are regions (often located at the gap between nanoparticles) in which the high electromagnetic field created by coupled SPR modes in neighbouring particles enhances the intrinsic properties of an analyte, such as the Raman absorption bands in Surface Enhanced Raman Scattering (SERS).[16] Many other applications also make use of this localized amplified field, for example in optoelectronics, photonics and nanoplasmonics as waveguides, optical filters and sensors.[17,18] These exciting properties indeed have attracted a huge research effort in recent years on the (bottom-up) fabrication of plasmonic nanoparticle arrays.[2,19]

A variety of methods have been proposed in order to build self-assembled oriented arrays of nanomaterials, including Langmuir–Blodgett,[20] electric[21] or magnetic[22] fields, microfluidic-assisted deposition,[23] mechanical stretching,[24] blown-bubble-assisted ordering[25,26] and spray coating.[27] While these techniques have proven to be efficient for the orientation of specific types of nanowires, they all suffer from several drawbacks, such as complex substrate preparation, multistep processes, the use of external fields, deposition on small areas only, *etc.* To keep the alignment method as general as possible we have recently developed grazing incidence spraying as a LbL-compatible technique based on liquid shear for aligning anisotropic nanoparticles with moderate to high aspect ratios in the form of monolayers on surfaces;[28,29] a first example using cellulose nanofibrils[30] was recently reported in more detail. This unique approach overcomes the drawbacks of existing methods and offers the greatest advantages to prepare oriented mono- and multilayer thin films in combination with the well-established Layer-by-Layer technique.[3,31] Additionally, the deposition density and thereby the plasmonic coupling in mono and multilayer can be independently well controlled through easily adjustable spray parameters.

In this article, we present the construction of oriented monolayer thin films of anisotropic metallic nanoparticles of increasing length (gold nanorods (AuNRs) of two different aspect ratios and silver nanowires (AgNWs)) using the GIS approach. As the orientation is due to shear forces exerted by the flowing liquid on the nano-objects, we show that the degree of orientation increases with the length of the anisotropic material. The resulting thin films show strong anisotropic optical and electronic transport properties, and the optical anisotropy reflects the degree of ordering of the thin films.

Results and discussion

Characterization of anisotropic nanoparticles

1D nanomaterials were synthesized following protocols previously reported in the literature[32–34] with slight modifications (for details, see the Experimental section). The TEM and SEM images clearly demonstrate the successful synthesis of AuNRs and AgNWs (Fig. 1A). The length and diameter distributions are given in Fig. 1B and Table 1. The relative size distribution spans from 10–25% for the various nanoparticles. Despite this length distribution, the three different nano-objects used here have a very distinguishable length, ranging from ~130 nm for the Short Au NanoRods (SAuNRs), ~250 nm for the Long Au NanoRods (LAuNRs) and up to ~5.7 μm for the AgNWs.

Only very few spherical impurities are seen in the SAuNRs, whereas other shapes such as nanoplates, triangles and short nanorods are present in the LAuNRs suspension besides long nanorods. The fraction of shape impurities other than nanorods has been estimated to ~10% for SAuNRs and ~15% for LAuNRs, while only nanowires are present in AgNWs. The increasing size and shape distribution for high aspect ratio gold nanorods is a well-known generic problem faced in the synthesis of high aspect ratio rods.[33,35] A closer look at both short and long nanorods TEM images reveal that the rods self-assemble during the TEM-grid preparation as a very dense array triggered by a strong side-by-side packing.[36]

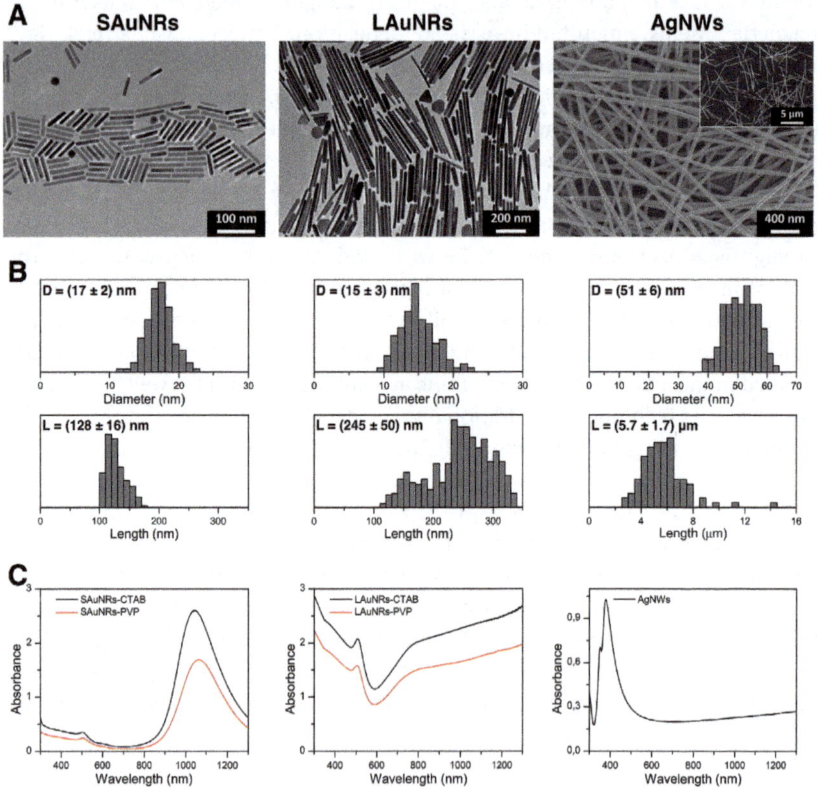

Fig. 1 Electron microscopy (A), diameter and length distribution (B), and UV-vis-NIR spectra (C) of the nanoparticle suspensions for SAuNRs (AR ~ 7, left panels), LAuNRs (AR ~ 16, middle panels) and AgNWs (right panels). The images have been acquired by TEM for AuNRs and by SEM for AgNWs. The size distributions are measured on >100 nanoparticles. The spectra before and after ligand exchange with PVP are measured in water for the AuNRs suspensions and in ethanol for the AgNWs.

Table 1 Length, diameter and aspect ratio of AuNRs and AgNWs

	Length/nm	Diameter/nm	Aspect ratio
SAuNRs	128 ± 16	17 ± 2	~7
LAuNRs	245 ± 50	15 ± 3	~16
AgNWs	$(5.7 \pm 1.7) \times 10^3$	51 ± 6	~110

The optical spectra of both AuNRs and AgNWs are presented in Fig. 1C. While small spherical metal nanoparticles usually exhibit only one single SPR absorption band, more modes can be excited in anisotropic nano-objects. For instance, at least two resonance modes can be excited in 1D nanoparticles such as nanowires and nanorods:[8] the first one, associated with transverse dipolar oscillations along the short axis of the nanorod (TSPR), appears at 503 nm for the SAuNRs and at 507 nm for the LAuNRs. Two TSPR bands are observable for the AgNWs at 349 nm and 378 nm, associated with two different transverse modes for Ag nanowires with a pentagonal cross section.[37] Indeed, nanowires with non-circular cross-sections are known to display multiple resonance modes, and the two bands correspond to the out-of-plane quadrupole resonance and dipole resonance of nanowires respectively.[38–40] The resonance band associated with the Longitudinal Surface Plasmon Resonance (LSPR) mode along the long axis of the nanorod/nanowire appear in the near infrared (NIR). The position of this LSPR band is known to be highly dependent on the shape and size of the nanocrystals, and the absorption band red-shifts and broadens with increasing aspect ratio.[41,42] Indeed, the LSPR of the SAuNRs reflects as a broad absorbance peak centered at 1040 nm. For the higher aspect ratio LAuNRS and for AgNWs, the longitudinal band(s) appears in the NIR range, at a wavelength that is beyond the investigated spectral range. Indeed, the solvent absorption around 1300 nm does not allow for a clear measurement of the nanoparticle absorbance band.

As spherical nanoparticle impurities have their absorbance band in the same range as the TSPR of the nanorods, the high absorbance intensity ratio of 7.6 of LSPR to TSPR of SAuNRs indicates that the contribution of the spherical impurities on the optical absorbance spectrum is low.[32,43] Furthermore, the LSPR peak position depends on the diameter and length, and it has been shown empirically[32,44] that the position of the LSPR band of gold nanorods λ_{LSPR} varies linearly with the aspect ratio (AR):

$$\lambda_{LSPR} = 95AR + 420. \tag{1}$$

This leads to an AR estimated from the optical properties of ~6.5, which is in good agreement with the AR determined from TEM measurements (~7). The same correlation can not be made for the higher AR nano-objects as the λ_{LSPR} can not be measured.

For the LAuNRs, in addition to the TSPR at 507 nm and the broad LSPR extending towards the NIR, one can notice a broad shoulder around 800 nm that can be related to the presence of shape-impurities, such as platelets and triangles.[33] Several strategies have been proposed in order to selectively remove those shape impurities.[45–47] However, the shape selectivity is low and we had to apply several purification cycles in order to increase significantly the nanorod to impurity ratio, which drastically reduced the final synthesis yield. For this reason, we choose to use the LAuNRs as synthesized and without any size/shape selection.

AgNWs are coated *in situ* by polyvinylpyrrolidone (PVP),[48] whereas AuNRs are stabilized by a Cetyl TrimethylAmmonium Bromide (CTAB) bilayer.[49] Exchange of the CTAB bilayer coating by PVP is requisite to facilitate the adsorption of AuNRs. Indeed, Quartz Crystal Microbalance (QCM) experiments (see ESI, Fig. S1†) show that the as-prepared positively charged CTAB capped AuNRs failed to adsorb on the oppositely charged quartz substrates. This might be due to the preferential adsorption of the free CTAB present in the suspension on the substrate which hinders the further deposition of AuNRs and thus leads to lower AuNR coverage. As there is a fast equilibrium between free CTAB in AuNRs dispersions and CTAB adsorbed on the NRs surfaces, a complete removal of free CTAB would require repeated centrifugation cycles, which induces the destabilization of the colloidal dispersion. Therefore, we choose rather to coat the AuNRs with PVP which provides stability through steric interactions. PVP functionalization provides sufficient stability to the AuNRs to maintain their colloidal stability without any aggregation for several months. TEM measurements (ESI, Fig. S2†) shows a 2.5 nm thick PVP coating on the rod surface. The optical spectra of both SAuNRs and LAuNRs after PVP functionalization show a slight red shift in comparison to the native CTAB coating. For instance, the LSPR band of the SAuNRs shifts from 1040 to 1067 nm. This can be ascribed to the local changes in the refractive index due to the PVP coating in the vicinity of the particle surface and to a small extent to the different solvent composition, as ~20 vol% ethanol is present in the AuNRs aqueous suspension after surface modification (see Experimental section).[50,51] QCM experiments demonstrate that the adsorption of AuNRs on the substrate is successful after they have been coated by PVP.

Following the recent reports by Polte and Liz-Marzán,[52,53] the [Au^0] concentration in SAuNRs-PVP and LAuNRs-PVP is deduced from the absorbance intensity at 400 nm and for all our experiments the concentration of [Au^0] in water was set to 0.5 mM. AgNWs were dispersed in ethanol at a concentration [Ag^0] of 0.1 mM.

Oriented monolayer deposition of anisotropic metal nanoparticles

As depicted in Fig. 2A, we used Grazing Incidence Spraying (GIS) in order to build monolayer thin films in which the 1D-nanoparticles are oriented along the spraying direction. As we have shown recently,[29,30] this shear-induced alignment technique is very simple, versatile and applicable to various types of anisotropic materials. In the present case, PVP coated AuNRs suspensions [$Au^0 = 0.5$ mM] and AgNWs suspensions [$Ag^0 = 0.1$ mM] were sprayed on a polyethyleneimine-coated glass substrate at respective air flow rates of 40 L min^{-1} and 30 L min^{-1} and at a liquid flow rate of 1 mL min^{-1} for 250 s with a constant angle of 15–20° and distance of 1 cm between the receiving substrate and spraying nozzle.

The deposited layers were imaged by SEM (Fig. 2B), and the images were analysed using the OrientationJ plugin[54] developed for ImageJ[55] (see Experimental section for details). During this process, each pixel is color-coded according to its orientation angle (Fig. 2C). The angular distribution of the nanorods/nanowires orientation can thus be measured (Fig. 2D). In order to compare the quality of

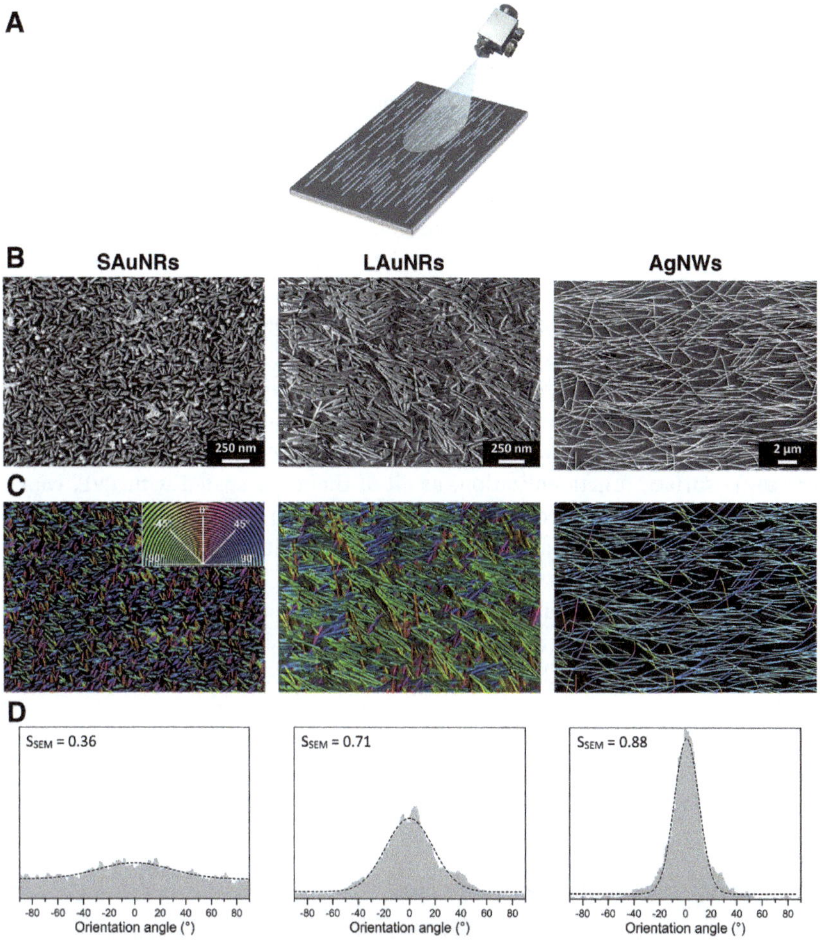

Fig. 2 (A) Schematic representation of the GIS approach. (B) SEM picture of a deposited monolayer, (C) orientation analysis in which the nanoparticles are coloured according to their orientation and (D) angle distribution with respect to the spraying direction (the dashed line represents a Gaussian fit of the distribution, $S_{\text{microscopy}}$ is the nematic order parameter, see eqn (2)) of SAuNRs (left panel), LAuNRs (middle panel) and AgNWs (right panel) deposited on a PEI coated glass substrate.

alignment between the different samples, we calculated a 2D nematic order parameter $S_{\text{microscopy}}$ from microscopy images, defined as:

$$S_{\text{microscopy}} = \left\langle 2\cos^2\theta - 1 \right\rangle \tag{2}$$

where θ is the angle between the anisotropic nanoparticle orientation direction and the main orientation direction as measured on electron microscopy pictures. Indeed, $S_{\text{microscopy}} = 0$ for an isotropic ensemble of 1D-nanoparticles and $S_{\text{microscopy}} = 1$ for a perfectly oriented system.

As the spray jet is directed with a low angle towards the substrate, the nano-object suspension forms a constantly flowing liquid film at the substrate/suspension interface, which induces a shear force on the 1D-nanoparticles and orients the nanorods/nanowires along the spraying direction during the adsorption process. As the shear force exerted by a viscous liquid on an object scales with the interface area, one can expect that the shear force and thus the degree of ordering increases with the length of the deposited object. Indeed, this effect can clearly be observed in Fig. 2: the SAuNRs ($L = 130$ nm) show a very low orientation ($S_{\text{microscopy}} = 0.15$), whereas the longer nanorods LAuNRs ($L = 250$ nm) are much better oriented ($S_{\text{microscopy}} = 0.61$, *i.e.* that 48% of the LAuNRs are oriented within ±15° from the main direction). The presence of shape impurities may partially impair the alignment and probably limit the ordering quality. Furthermore, these shape impurities are less anisotropic than the nanorods, and this results in a broadening of the orientation angle distribution determined from the microscopy pictures. Finally, a further increase of the AgNWs length to the micrometer range ($L = 5.7$ μm) allow to reach a high order parameter of $S_{\text{microscopy}} = 0.83$, which corresponds to 73% of the AgNWs being oriented at ±15° from the main direction. It must also be noted that all the nanoparticles used in this study have the same surface functionalization, as all of them are coated with PVP, which excludes differences that could arise from the different surface chemistry of the nanoparticles and thus a different interaction with the PEI-functionalized substrate.

A recent report by Assad *et al.*[27] on a spray coating route used to orient Si nanowires shows that the quality of alignment highly depends on the compatibility of droplet size with the dimensions of nanomaterials and the temperature of the receiving substrate. However, the process these authors describe is very different from what we present here, as the orientation does not occur upon droplet evaporation as for their study, but we rather keep a liquid film wetting the substrate at all time during the orientation process.

Optical properties of plasmonic thin films

As shown previously, we have been able to assemble the 1D-nanoparticle as a single monolayer in which the nanowire/nanorods are oriented and all point in a single direction. As we have seen in Fig. 1C, the plasmonic 1D-nanoparticles have at least two plasmon modes located in the UV-vis and NIR that are linked to the short and long axis of the nanoparticles, respectively. The anisotropy in the nanoparticle ordering induced by the GIS approach should thus reflect in an anisotropy in the optical properties. This has been probed by measuring the polarized UV-vis-NIR extinction spectra, and by varying the polarizing angle φ between the incident light polarization plane and the nanoparticle orientation direction in the thin film (Fig. 3B). For a perfectly oriented thin film, it can be expected that the LSPR and the TSPR modes can be excited selectively, which means that the extinction spectra should vary drastically with the polarizing angle.[8,24] This is indeed what is observed for all the polarized spectra to different extents (Fig. 3A): when light is polarized along the long axis of the nanorods/nanowires ($\varphi = 0°$, black curves), the LSPR band is excited and gives rise to a broad extinction in the NIR, whereas the TSPR band is excited in the UV-vis when light is polarized along the short axis ($\varphi = 90°$, red curves). All three samples presented here thus show a certain polarizing property, the direction of polarization being orthogonal in the UV-vis and in the NIR range as they are linked to two orthogonal SPR modes.

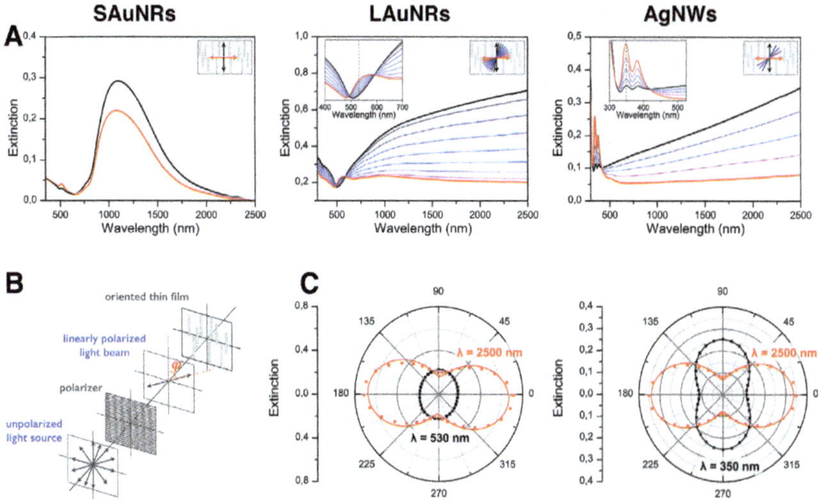

Fig. 3 (A) UV-vis-NIR extinction spectra of SAuNRS-PVP (left panel), LAuNRS-PVP (middle panel) and AgNWs (right panel) oriented thin films for a varying polarizing angle φ, (B) schematic representation of the polarized UV-vis-NIR set-up, in which φ (polarizing angle) is defined as the angle between the orientation direction in the film and the polarization direction of the incident light, and (C) polar plot of the extinction of the LAuNRs-PVP (middle) and the AgNWs (right) thin films as function of φ in the UV-vis (black curves) and in the NIR (red curve). The line is a fit to a sine square function.

The SAuNRs-PVP spray-coated thin film (Fig. 3A) measured at a polarizing angle of $\varphi = 0°$ and $90°$ shows only a small difference between the two spectra. This indicates that the anisotropy of the optical properties is low, as expected from the low degree of ordering of the nanoparticles ($S_{\mathrm{microscopy}} = 0.15$). A slight red shift and a significant broadening of both TSPR and LSPR peaks as compared to the suspension is also observed, which is due to the inter-particle coupling.[50] For instance, the TSPR band shifts from 1067 nm for the suspension to ~1080 nm for the monolayer.

As the degree of ordering increases for the LAuNRs ($S_{\mathrm{microscopy}} = 0.61$) and for the AgNWs ($S_{\mathrm{microscopy}} = 0.83$) thin films, the anisotropy in the optical properties is more and more obvious, as shown by the increasing difference between the polarized spectra at $\varphi = 0°$ and $90°$. The polarized spectra for a varying polarizing angle have been measured (Fig. 3A, middle and right panels), and the extinction at selected wavelength have been represented as a polar plot as function of the polarizing angle φ (Fig. 3C). These polar plots are given at the wavelength close to the TSPR at which the polarization is highest ($\lambda = 530$ nm for the LAuNRs and $\lambda = 350$ nm for the AgNWs, see the dashed line in the insets of Fig. 3A). As the LSPR peak is beyond the investigated spectral range, we have plotted the NIR-extinction at 2500 nm for both LAuNRs and AgNWs. The polar plots evidence clearly that the extinction in the UV-vis is a maximum at $\varphi = 90°$ when the transverse modes are excited, whereas it reaches its minimum at $\varphi = 0°$. On the contrary, the extinction in the NIR is a maximum at $\varphi = 0°$ (*i.e.* that the LSPR is selectively excited) and a minimum at $\varphi = 90°$. The variation of the extinction fits well with a sine square function.

For the LAuNRs thin film in the UV-vis (inset of Fig. 3C), the apparent position of the TSPR peak is red-shifted as compared to the suspension ($\lambda = 507$ nm) and varies with the polarizing angle. For instance, the "peak" appears at ~570 nm for $\varphi = 90°$; this can be explained by a plasmonic coupling along the short axis of neighbouring rods when light is polarized perpendicularly to the alignment direction.

It can also be further observed that the spectra are very sensitive to the polarization direction in the NIR range for both the LAuNRs and the AgNWs, whereas the behaviour is different in the UV-vis range: the TSPR bands are very sensitive to the polarizing angle for the AgNWs but it is much less responsive for the LAuNRS. This can be clearly observed on the polar plots: the extinction as function of φ in the UV-vis and in the NIR ranges are of the same order of magnitude for the AgNWs, whereas the extinction for the LAuNRs samples in the UV-vis is much lower and less responsive to the variation of φ than the extinction in the NIR. This weak amplitude and low polarization is caused by the broadening of the TSPR peak due to its overlap with interband transition of metallic gold at ~400 nm.[8]

The optical anisotropy can be quantified using an optical order parameter S_{opt} defined as:

$$S_{opt} = 1 - \frac{A_{min}}{A_{max}} \quad (3)$$

with $A_{min} = A_{90°}$ and $A_{max} = A_{0°}$ for the LSPR and $A_{min} = A_{0°}$ and $A_{max} = A_{90°}$ for the TSPR. S_{opt} is thus a direct measurement of the polarisation efficiency as $S_{opt} \rightarrow 0$ for an isotropic sample (*i.e.* when $A_{min} \sim A_{max}$), whereas $S_{opt} \rightarrow 1$ for a highly anisotropic sample (*i.e.* when $A_{min} \ll A_{max}$). The values of the optical order parameter S_{opt} are summarized in Table 2 and compared to the 2D nematic order parameter determined from the SEM pictures $S_{microscopy}$. The optical anisotropy increases with the nematic order parameter, reaching values higher than 0.7 for the LAuNRs and the AgNWs thin films in the NIR. This confirms that the optical anisotropy is directly linked to the order parameter. It must be noted that as the peak of the LSPR is beyond the investigated wavelength range, the optical anisotropy further increases for $\lambda > 2500$ nm, and the maximum of optical anisotropy is thus probably much greater than 0.7 further in the NIR. However, the optical anisotropy close to the transverse band for the LAuNRs ($S_{opt} = 0.18$) is lower than for the SAuNRs and for the AgNWs. This means that the extinction is less responsive to the variation of φ, because the overlap of the TSPR with the gold interband transitions that are not dependent on the polarization state of the incoming light.

Table 2 Nematic order parameter $S_{microscopy}$ and optical order parameter S_{opt} measured close to the TSPR and to the LSPR for the oriented SAuNRs, LAuNRs and AgNWs oriented thin films

	$S_{microscopy}$	S_{opt} (TSPR)	S_{opt} (LSPR)
SAuNRs	0.15	0.32@527 nm	0.30@1700 nm
LAuNRs	0.61	0.18@530 nm	0.71@2500 nm
AgNWs	0.83	0.57@350 nm	0.76@2500 nm

Finally, it must be noted that although the same concentration of $[Au^0] = 0.5$ mM has been used for the SAuNRs and LAuNRs monolayer buildup, the optical density of the SAuNRs monolayer is relatively smaller than that of LAuNRs. This is an indirect effect of the lower ordering of the nanorods in the SAuNRs layer, as a disordered monolayer has a lower packing density than an ordered monolayer. The average AuNRs surface density (and consequently the optical density) is thus higher for the ŞAuNRS compared to the LAuNRS.[50]

All these data clearly show that the anisotropy in the optical properties directly arises from the alignment of the nanorods/nanowires along one single direction. As the alignment by GIS is more efficient with high aspect ratio nano-objects, the optical anisotropy is higher for the LAuNRs and AgNWs compared to the SAuNRs.

Electronic transport properties of AgNWs monlayer thin films

Silver nanowires can also be used for their conductive properties, as transparent conductive thin films of AgNWs can have conductivities and transparency values challenging the ones reached by doped oxide thin films, such as the widely used Indium Tin Oxide (ITO).[56] In order to probe if the electronic conduction properties of oriented silver nanowires monolayers are anisotropic, we have measured the sheet resistance in the 4-probe configuration in various directions with respect to the nanowire orientation direction. The $I(V)$ curves are characteristic of an ohmic behaviour (Fig. 4), with a direct proportionality of the measured voltage drop between the two inner pins of the 4-probe head and the current applied between the two outer pins. The sheet resistance can be extracted from the $I(V)$ slopes for each direction, and the values span from 64 Ω sq.$^{-1}$ when the current is measured along the orientation direction (black curve in Fig. 4) and 540 Ω sq.$^{-1}$

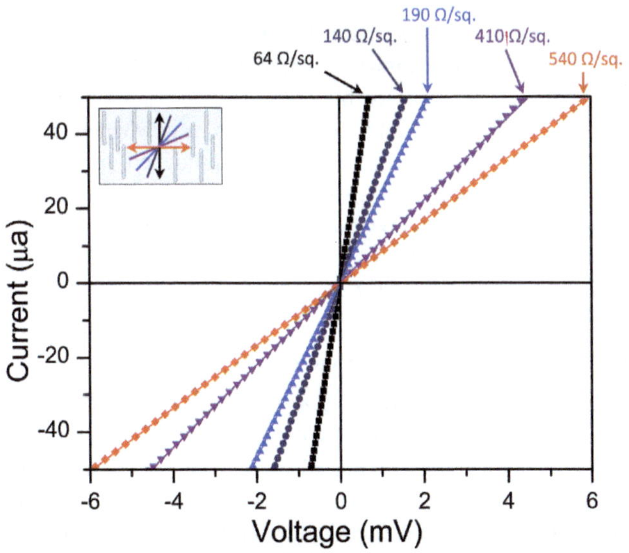

Fig. 4 Current as function of the applied voltage measured in a 4-probe configuration on a AgNW monolayer in different directions with respect to the orientation direction. The sheet resistance is >8 times lower along the nanowire orientation direction compared to the orthogonal direction.

for the perpendicular direction (red curve in Fig. 4), *i.e.* that the sheet resistance is almost one order of magnitude lower in the nanowire orientation direction compared to the perpendicular one. It must be noted that these low sheet resistance values are obtained at a pretty low nanowire surface density, which ensures a high transparency of the samples in the visible range (~90% at 550 nm).

Conclusions

We have shown that oriented thin films of 1D-nanoparticles such as nanorods and nanowires can be efficiently oriented using Grazing Incidence Spraying (GIS). The use of gold nanorods and silver nanowires of increasing aspect ratio but with the same surface chemistry allows us to investigate clearly the effect of the nano-object length. We have thus shown that the quality of alignment increases with the length. Accordingly, highest order parameter ($S_{microscopy} = 0.83$) was obtained for AgNWs of AR > 100 followed by LAuNRs of AR ~ 16. The high anisotropy in the nanoparticle ordering results in highly anisotropic optical properties: the highest optical anisotropy of $S_{opt} = 0.76$ is obtained for AgNWs in the near infrared. Finally, we also show that the electronic transport properties of AgNWs oriented thin films are highly anisotropic, with the sheet resistance being almost one order of magnitude lower in the nanowire orientation direction compared to the perpendicular one. Such oriented plasmonic and/or conductive thin films may find applications in a variety of emerging fields, such as optical filters, on-chip waveguides, (bio)chemical sensors, transparent conductive electrodes or metamaterials.

Experimental

Materials

Chemicals such as hexadecyltrimethylammonium bromide (CTAB H6269, >99.9%), hydrogen tetrachloroaurate trihydrate ($HAuCl_4 \cdot 3H_2O$, >99.9%), sodium borohydride ($NaBH_4$, >99%), L-ascorbic acid (AA, >99%), silver nitrate ($AgNO_3$, 99%), hydroquinone (HQ, 99%) and polymers such as poly(vinylpyrrolidone) (M_w ~ 10 000: PVP10K or M_w ~ 40 000: PVP40K), branched poly(ethyleneimine) solution (50 wt% in H_2O) with average M_n ~ 60 000 (PEI60K), were obtained from Sigma Aldrich and were used as received. Nitric acid (68%) was purchased from Prolabo. Ethylene glycol (>99%) was purchased from JT Baker. MilliQ water (18.2 MΩ cm) was used in all our experiments.

Synthesis of gold nanorods and silver nanowires

Synthesis of short gold nanorods (AR ~ 7, SAuNRs). AuNRs of short AR were synthesized using a protocol reported by Zubarev *et al.*[32] where the conventional ascorbic acid reducing agent was replaced by hydroquinone. In a first stage, the fast addition of alkaline $NaBH_4$ solution (0.01 M $NaBH_4$ in 0.01 M NaOH, 460 μL) to $HAuCl_4$–CTAB complex (10 mL mixture containing 0.5 mM gold salt and 0.1 M CTAB) leads to the formation of gold seeds indicated by a rapid colour change to brown and this was aged for 30 min without any disturbance at 27 °C.

In a second step, 250 μL of $HAuCl_4$ (0.1 M stock solution) was added to 49.75 mL of 0.1 M CTAB with stirring being continued for 15 min at 150 rpm.

Next, 350 μL of 0.1 M $AgNO_3$ and 2500 μL of 0.1 M HQ were added and hand-shaken until the solution became clear. AuNRs formation was initiated by the addition of 800 μL of seed solution with vigorous stirring at 1200 rpm for 30 s and aged overnight with no stirring at 27 °C. The as-prepared rods were centrifuged twice at 10 000 rpm for 30 min.

Synthesis of long gold nanorods (AR ~ 16, LAuNRs). High AR AuNRs were synthesized by a typical three-step seed mediated protocol adopted from Huang *et al.*[33] In brief, gold seeds were prepared by the one-shot injection of freshly prepared 0.2 M $NaBH_4$ aqueous solution (450 μL) into a 10 mL mixture consisting 0.1 M CTAB and 0.25 mM $HAuCl_4$ at 27 °C. Seed formation was indicated by a quick colour change from yellow–orange to brown due to the complete reduction of Au^{3+} by $NaBH_4$. After 5 min of vigorous stirring, the CTAB stabilized gold seeds were aged for 2 h without any disturbance.

Meanwhile, the so-called *growth solution* was prepared by introducing 25 mL of 0.25 mM gold solution into 973 mL of 0.1 M CTAB. This mixture was stirred for 15 min at 27 °C followed by addition of 2 mL of 1 mM HNO_3 in order to modify the pH value. This 1 L growth solution was divided as follows: 83.33 mL were poured into flasks A & B and the remaining 833.33 mL into flask C. Then, 463 μL, 463 μL and 4.63 mL of freshly prepared 0.1 M AA were added to flasks A, B and C, respectively. The solutions became clear upon gentle shaking indicating the reduction of Au^{3+} to Au^{+}. Finally, 7.4 mL of the seed solution was added into flask A and shaken for 10 s. Then, 7.4 mL of the solution from flask A was quickly added to flask B and shaken for 10 s. Then, 74 mL of solution B was introduced into flask C and shaken for 30 s. The appearance of a purple colour denotes the formation of gold nanostructures and this became darker with time and finally turned to reddish brown. Flask C was thermostated for 5 days at 27 °C. Eventually, the upper solution mostly containing spherical gold nanoparticles and by-products were carefully pipetted out. A thin brown layer at the bottom of the flask was collected and dispersed in 50 mL water. The resulting dispersion was purified by centrifugation at 2000 rpm for 30 min twice to remove excess CTAB.

Surface functionalization of both *short and long* gold nanorods. 50 mL of the CTAB stabilized gold nanorods suspensions (either SAuNRs or LAuNRs) were added dropwise to 500 mL of a pre-sonicated 10 wt% PVP10K aqueous solution under ultrasonication followed by gentle overnight stirring (150 rpm) at 27 °C. The PVP coated AuNRs were centrifuged at 3000 rpm for 1 h to remove uncoated PVP. The purple precipitate present on the sidewalls of the tubes were redispersed in ethanol by brief sonication (30 s). The resulting brown solution was centrifuged in the same conditions to remove excess PVP. The precipitate was finally dispersed in a mixture of 10 mL water and 2 mL ethanol. The PVP-coated short and long AuNRs are denoted as *SAuNRs-PVP* and *LAuNRs-PVP* respectively.

Synthesis of silver nanowires. Silver nanowires were synthesized by a polyol synthesis[57] adapted from Wiley *et al.*[34] which consists in the reduction of $AgNO_3$ with Ethylene Glycol (EG). In a typical procedure, stock solutions of the different reagents were first freshly prepared in EG: 0.1285 g of NaCl were dissolved in 10 mL EG (220 mM), 0.0405 g of $Fe(NO_3)_3$ in 5 mL EG (33.5 mM), 0.6720 g of $AgNO_3$ in 16 mL of EG (24.7 mM) and 0.6720 g of PVP in 16 mL of EG. 78.5 mL of EG were introduced in a four-neck round bottom flask and stirred at 250 rpm. 0.1 mL of the NaCl solution and 0.05 mL of the $Fe(NO_3)_3$ solution were added into the flask. The solution was homogenized for 10 min, and the round bottom flask was placed on a heating mantle with the temperature being measured within the solution, and a condenser was placed on the flask. Two 20 mL syringes were then filled with 10 mL of the $AgNO_3$ solution and 10 mL of the PVP solution, respectively. Both solutions were simultaneously injected in the round bottom flask at room temperature under stirring using syringe pumps at a rate of 1.66 mL min^{-1}, *i.e.* the total injection time was 6 min. After complete addition of $AgNO_3$ and PVP, the temperature was increased up to 192 °C at a rate of 9 °C min^{-1}. The solution was then cooled to room temperature with a water bath, and was transferred to an Erlenmeyer flask and left undisturbed for two days, in order to allow the longest AgNWs to precipitate. The supernatant was removed and the precipitate was dissolved in 100 mL of acetone. The AgNWs quickly precipitate, and the supernatant was removed after 5 min. This acetone washing procedure without centrifugation was applied twice. Finally, the AgNWs were suspended in 100 mL of ethanol and were washed three times by centrifugation at 2000 rpm for 30 min. Finally AgNWs were kept in ethanol for further use.

Oriented deposition of anisotropic nanoparticles

Substrate preparation. Oriented 1D-nanoparticle thin films were deposited on glass substrates as these allow polarized UV-vis-NIR spectroscopy measurements. All glass substrates were first cleaned by initial blowing with N_2 and subsequent sonication in water for 10 min followed by copious ethanol rinsing, drying under a N_2 stream and plasma activation for 3 min. The glass substrates were then coated with a PEI60K (2.5 mg mL^{-1} in H_2O) layer by spraying for 10 s and subsequently spray rinsed with water for the same time. Manual air-pump spray cans (Roth) were used for PEI deposition and for rinsing.

Grazing incidence spraying. Oriented monolayers of 1D-metal nanoparticles were deposited on the PEI-coated glass slides by Grazing Incidence Spraying (GIS)[29,30] using a semi-automated home-made setup equipped with a two-fluid spray nozzle (Spraying Systems, internal diameter: 0.3 mm, for AuNRs and Aztek A480 airbrush, internal diameter: 1.03 mm, for AgNWs). The angle between the central axis of the cone of the spray jet and the receiving substrate was fixed at $\alpha =$ 15–20°, and the distance between the nozzle and the substrate was held at a constant distance of 1 cm. The liquid flow rate was fixed to 1 mL min^{-1} and the air flow rate to 40 L min^{-1} for AuNRs and 30 L min^{-1} for AgNWs which produces a cloud of droplets with a diameter in the range of 5–20 μm as determined by Phase Doppler Interferometry (PDI, data not shown here). Following deposition, the thin film was rinsed with Milli-Q water in order to remove any weakly adsorbed nanoparticles and dried using N_2 flow.

Characterization

Electron microscopy. TEM was performed by depositing 10 μL of the colloidal suspension on a carbon coated copper grid which is left to dry in air for a few hours. The nanoparticles were imaged with a Tecnai G2 FEI microscope operated at 120 kV. AgNWs, as well as the oriented monolayer thin films were imaged by SEM using a JEOL 6700F microscope operating at 1 kV and 10 μA. The orientation analysis was done on SEM pictures using OrientationJ,[54] a plugin developed for ImageJ,[55] which is based on the computation of the structure tensor in the local neighbourhood of each pixel. The original SEM picture is first binarized, and the

OrientationJ plugin is applied with a Gaussian gradient on a window of 3 pixels. As a result, an orientation angle is given for each pixel, and a color-code which represents each angle by a color is used to build the images given in Fig. 2C. Finally, the angle distribution with respect to the main orientation direction can be extracted (Fig. 2D). The angle distribution is non-weighted (*i.e.* 1 pixel = 1 count), but the pixels for which the values of the coherency or the energy are lower than 10% have been ignored. Indeed, these pixels are often artifacts, for instance noisy pixels or located at edges and corners.

UV-vis-NIR spectroscopy. UV-visible-near infrared spectroscopy was carried out with a Cary 5000 spectrometer (Agilent) in dual beam configuration. The extinction for the thin film samples was measured through a circular aperture (diameter = 5 mm) and a blank substrate was used as a reference. A Glan–Taylor polarizer has been used for polarized UV-vis-NIR spectroscopy.

Sheet resistance measurements. The oriented AgNWs were annealed at 175 °C for 1 min and 45 s in order to reduce the NW/NW contact resistance.[58] The sheet resistance was measured on monolayer thin films using a 4-probe head (Signatone 302 Resistivity Stand) connected to a Keithley 2400 sourcemeter. $I(V)$ sweeps were obtained by applying current between the two outer pins ($-50\ \mu A < I < 50\ \mu A$) and measuring the voltage drop between the inner pins. The sheet resistance can be calculated from the $I(V)$ slope using:

$$R_s = 4.5324\frac{dV}{dI}. \tag{4}$$

Acknowledgements

The authors thank Marc Schmutz and Christian Blanck (ICS, France) for TEM measurements, Cedric Leuvrey (IPCMS, France) for SEM measurements. The sheet resistance measurements were done in the STnano cleanroom facility in Strasbourg, France. S. S. thanks the IDEX program of the Strasbourg University for funding. V. L. thanks the Labex CSC "Chemistry of Complex Systems" for financial support. M. P. thanks the "Chaire d'Excellence" program of CNRS and the Université de Strasbourg for partial funding. G. D. thanks the Institut Universitaire de France for financial support.

Notes and references

1 P. R. Sajanlal, T. S. Sreeprasad, A. K. Samal and T. Pradeep, *Nano Rev.*, 2011, **2**, 5883.
2 K. Thorkelsson, P. Bai and T. Xu, *Nano Today*, 2015, **10**, 48–66.
3 G. Decher, *Science*, 1997, **277**, 1232–1237.
4 J. Schmitt, G. Decher, W. J. Dressick, S. L. Brandow, R. E. Geer, R. Shashidhar and J. M. Calvert, *Adv. Mater.*, 1997, **9**, 61–65.
5 S. Srivastava and N. A. Kotov, *Acc. Chem. Res.*, 2008, **41**, 1831–1841.
6 G. Schneider, G. Decher, N. Nerambourg, R. Praho, M. H. V. Werts and M. Blanchard-Desce, *Nano Lett.*, 2006, **6**, 530–536.
7 B. P. Pichon, P. Louet, O. Felix, M. Drillon, S. Begin-Colin and G. Decher, *Chem. Mater.*, 2011, **23**, 3668–3675.

8 O. L. Muskens, G. Bachelier, N. D. Fatti, F. Vallée, A. Brioude, X. Jiang and M.-P. Pileni, *J. Phys. Chem. C*, 2008, **112**, 8917–8921.
9 Z. Liu, J. Xu, D. Chen and G. Shen, *Chem. Soc. Rev.*, 2015, **44**, 161–192.
10 Y. Xia, P. Yang, Y. Sun, Y. Wu, B. Mayers, B. Gates, Y. Yin, F. Kim and H. Yan, *Adv. Mater.*, 2003, **15**, 353–389.
11 S. Lal, J. H. Hafner, N. J. Halas, S. Link and P. Nordlander, *Acc. Chem. Res.*, 2012, **45**, 1887–1895.
12 L. Guerrini and D. Graham, *Chem. Soc. Rev.*, 2012, **41**, 7085–7107.
13 K. A. Willets, *Chem. Soc. Rev.*, 2014, **43**, 3854–3864.
14 A. Lee, G. F. S. Andrade, A. Ahmed, M. L. Souza, N. Coombs, E. Tumarkin, K. Liu, R. Gordon, A. G. Brolo and E. Kumacheva, *J. Am. Chem. Soc.*, 2011, **133**, 7563–7570.
15 H. Wei and H. Xu, *Nanoscale*, 2013, **5**, 10794–10805.
16 A. Campion and P. Kambhampati, *Chem. Soc. Rev.*, 1998, **27**, 241–250.
17 H. Chen, L. Shao, Q. Li and J. Wang, *Chem. Soc. Rev.*, 2013, **42**, 2679–2724.
18 L. Duempelmann, A. Luu-Dinh, B. Gallinet and L. Novotny, *ACS Photonics*, 2016, **3**, 190, DOI: 10.1021/acsphotonics.5b00604.
19 A. Klinkova, R. M. Choueiri and E. Kumacheva, *Chem. Soc. Rev.*, 2014, **43**, 3976–3991.
20 L. Mai, Y. Gu, C. Han, B. Hu, W. Chen, P. Zhang, L. Xu, W. Guo and Y. Dai, *Nano Lett.*, 2009, **9**, 826–830.
21 S. Raychaudhuri, S. A. Dayeh, D. Wang and E. T. Yu, *Nano Lett.*, 2009, **9**, 2260–2266.
22 C. M. Hangarter and N. V. Myung, *Chem. Mater.*, 2005, **17**, 1320–1324.
23 Y. Huang, X. Duan, Q. Wei and C. M. Lieber, *Science*, 2001, **291**, 630–633.
24 J. Pérez-Juste, B. Rodríguez-González, P. Mulvaney and L. M. Liz-Marzán, *Adv. Funct. Mater.*, 2005, **15**, 1065–1071.
25 G. Yu, X. Li, C. M. Lieber and A. Cao, *J. Mater. Chem.*, 2008, **18**, 728–734.
26 G. Yu, A. Cao and C. M. Lieber, *Nat. Nanotechnol.*, 2007, **2**, 372–377.
27 O. Assad, A. M. Leshansky, B. Wang, T. Stelzner, S. Christiansen and H. Haick, *ACS Nano*, 2012, **6**, 4702–4712.
28 G. Decher, R. Blell, H. Hu, M. Pauly, O. Felix, D. Martel, X. Lin, S. Sekar, S. Diabang and J. Baer, *US Pat.*, WO2015173432 A1, 2015.
29 H. Hu, M. Pauly, O. Felix and G. Decher, 2016, submitted.
30 R. Blell, X. Lin, T. Lindström, M. Ankerfors, M. Pauly, O. Félix and G. Decher, 2016, submitted.
31 E. Seyrek and G. Decher, in *Polymer Science: A Comprehensive Reference*, ed. K. M. Möller, Elsevier, Amsterdam, 2012, vol. 7, pp. 159–185.
32 L. Vigderman and E. R. Zubarev, *Chem. Mater.*, 2013, **25**, 1450–1457.
33 Y.-N. Wang, W.-T. Wei, C.-W. Yang and M. H. Huang, *Langmuir*, 2013, **29**, 10491–10497.
34 S. M. Bergin, Y.-H. Chen, A. R. Rathmell, P. Charbonneau, Z.-Y. Li and B. J. Wiley, *Nanoscale*, 2012, **4**, 1996–2004.
35 B. D. Busbee, S. O. Obare and C. J. Murphy, *Adv. Mater.*, 2003, **15**, 414–416.
36 B. Nikoobakht, Z. L. Wang and M. A. El-Sayed, *J. Phys. Chem. B*, 2000, **104**, 8635–8640.
37 X. M. Sun and Y. D. Li, *Adv. Mater.*, 2005, **17**, 2626–2630.
38 M. S. Goh, Y. H. Lee, S. Pedireddy, I. Y. Phang, W. W. Tjiu, J. M. R. Tan and X. Y. Ling, *Langmuir*, 2012, **28**, 14441–14449.

39 M. Chen, I. Y. Phang, M. R. Lee, J. K. W. Yang and X. Y. Ling, *Langmuir*, 2013, **29**, 7061–7069.
40 J. P. Kottmann, O. J. F. Martin, D. R. Smith and S. Schultz, *Phys. Rev. B: Condens. Matter*, 2001, **64**, 235402.
41 X. Ye, L. Jin, H. Caglayan, J. Chen, G. Xing, C. Zheng, V. Doan-Nguyen, Y. Kang, N. Engheta, C. R. Kagan and C. B. Murray, *ACS Nano*, 2012, **6**, 2804–2817.
42 B. Nikoobakht and M. A. El-Sayed, *Chem. Mater.*, 2003, **15**, 1957–1962.
43 L. Vigderman, B. P. Khanal and E. R. Zubarev, *Adv. Mater.*, 2012, **24**, 4811–4841.
44 S. Link, M. B. Mohamed and M. A. El-Sayed, *J. Phys. Chem. B*, 1999, **103**, 3073–3077.
45 K. Park, H. Koerner and R. A. Vaia, *Nano Lett.*, 2010, **10**, 1433–1439.
46 B. P. Khanal and E. R. Zubarev, *J. Am. Chem. Soc.*, 2008, **130**, 12634–12635.
47 N. R. Jana, *Chem. Commun.*, 2003, 1950–1951, DOI: 10.1039/b303103a.
48 W. A. Saidi, H. Feng and K. A. Fichthorn, *J. Phys. Chem. C*, 2013, **117**, 1163–1171.
49 S. Gómez-Graña, F. Hubert, F. Testard, A. Guerrero-Martínez, I. Grillo, L. M. Liz-Marzán and O. Spalla, *Langmuir*, 2012, **28**, 1453–1459.
50 S. Vial, I. Pastoriza-Santos, J. Pérez-Juste and L. M. Liz-Marzán, *Langmuir*, 2007, **23**, 4606–4611.
51 I. Pastoriza-Santos, D. Gomez, J. Perez-Juste, L. M. Liz-Marzán and P. Mulvaney, *Phys. Chem. Chem. Phys.*, 2004, **6**, 5056–5060.
52 T. Hendel, M. Wuithschick, F. Kettemann, A. Birnbaum, K. Rademann and J. Polte, *Anal. Chem.*, 2014, **86**, 11115–11124.
53 L. Scarabelli, A. Sánchez-Iglesias, J. Pérez-Juste and L. M. Liz-Marzán, *J. Phys. Chem. Lett.*, 2015, **6**, 4270–4279.
54 R. Rezakhaniha, A. Agianniotis, J. T. C. Schrauwen, A. Griffa, D. Sage, C. V. C. Bouten, F. N. Vosse, M. Unser and N. Stergiopulos, *Biomech. Model. Mechanobiol.*, 2012, **11**, 461–473.
55 C. A. Schneider, W. S. Rasband and K. W. Eliceiri, *Nat. Methods*, 2012, **9**, 671–675.
56 S. Ye, A. R. Rathmell, Z. Chen, I. E. Stewart and B. J. Wiley, *Adv. Mater.*, 2014, **26**, 6670–6687.
57 Y. Sun, Y. Yin, B. T. Mayers, T. Herricks and Y. Xia, *Chem. Mater.*, 2002, **14**, 4736–4745.
58 D. P. Langley, M. Lagrange, G. Giusti, C. Jimenez, Y. Brechet, N. D. Nguyen and D. Bellet, *Nanoscale*, 2014, **6**, 13535–13543.

Faraday Discussions

PAPER

Behaviour of hybrid inside/out Janus nanotubes at an oil/water interface. A route to self-assembled nanofluidics?†

P. Picot,[a] O. Taché,[a] F. Malloggi,[a] T. Coradin[b] and A. Thill*[a]

Received 1st March 2016, Accepted 12th April 2016
DOI: 10.1039/c6fd00034g

Imogolites are natural aluminosilicate nanotubes that have a diameter of a few nanometers and can be several microns long. These nanotubes have different chemical groups on their internal (Si–OH) and external (Al–OH–Al) surfaces, that can be easily functionalised independently on both surfaces. Here we show that taking advantage of the particular shape and chemistry of imogolite, it is possible to prepare inside/out Janus nanotubes. Two kinds of symmetric Janus nanotubes are prepared: one with an external hydrophilic surface and an internal hydrophobic cavity (imo-CH_3) and one with an external hydrophobic surface and a hydrophilic internal cavity (OPA-imo). The behaviour of such inside/out Janus nanotubes at oil/water interfaces is studied. The OPA-imo adsorbs strongly at the oil/water interface and is very efficient in stabilising water-in-oil emulsions through an arrested coalescence mechanism. Imo-CH_3 also adsorbs at the oil/water interface. It stabilises oil-in-water emulsions by inducing slow oil-triggered modifications of the viscosity of the continuous phase. The possible transport of small molecules inside the imo-CH_3 nanotubes is evidenced, opening up routes towards self-assembled nanofluidics.

Introduction

Pickering emulsions are oil-in-water (O/W) or water-in-oil (W/O) dispersions stabilised by solid particles.[1] The stabilisation is due to the preferential adsorption of particles at the interface between two immiscible liquids. Such adsorption occurs for homogeneous particles when they are partially wetted by both fluids.[2] In this case, the contact angle controls the positioning of the particles at the interfacial layer. Homogeneous functionalisations are generally performed in order to control this contact angle and therefore the structure of the interfacial layer.[3] Another strategy enabling tuning the behaviour of particles at an oil/water

[a]LIONS, NIMBE, CEA, CNRS, Université Paris-Saclay, 91191 Gif sur Yvette, France. E-mail: antoine.thill@cea.fr
[b]Sorbonne Universités, UPMC Univ. Paris 6, CNRS, Collège de France, LCMCP, Paris, France
† Electronic supplementary information (ESI) available. See DOI: 10.1039/c6fd00034g

interface is to use inhomogeneous (or asymmetric) surfaces. The association of two surfaces with different wettability on a single particle gives the so-called Janus particles. For example, a spherical particle with hydrophilic and hydrophobic hemispheres will strongly adsorb at an oil/water interface and efficiently stabilises Pickering emulsions.[4] The positioning of the particle at the interface is then defined by the extent of the different surfaces. Many methods have been developed to produce large quantities of Janus nanoparticles with various shapes like sheets, ribbons or rods.[5] However, there are only very few examples of Janus nanotubes used for Pickering emulsions. Stabilisation of emulsions using polymer grafting on multi-wall carbon nanotubes was achieved but the "nanotubular" (*i.e.* the existence of both an internal and an external surface) nature of the particles was not evidenced to be a key parameter in the stabilisation process.[6,7] Using nanotubular clay, Janus nanotubes are also prepared and used to stabilise emulsions but, here again, the importance of the tubular shapes was not clearly identified.[8]

In this work, we wanted to use a system where the nanotubular morphology of the nanoparticles plays a crucial role in the behaviour of the particles at an oil/water interface. Among inorganic nanotubes, aluminosilicates are particularly attractive because of the wide availability and low cost of their constitutive elements. Imogolite is an example of aluminosilicate nanotube whose synthesis is relatively easy. Interestingly, imogolites have been discovered well before carbon nanotubes in 1962 by Yoshinaga and Aomine.[9] Synthetic imogolites exist in a form mimicking natural nanotubes with an average composition of $(OH)_3Al_2O_3Si(OH)$. The diameter of synthetic imogolites is monodisperse and can be tuned depending on the synthesis conditions from 2 to 2.8 nm.[10,11] They are formed thanks to the curving of a gibbsite layer. The bonding of silicon tetrahedron *via* three Al–O–Si bonds is one of the main reasons explaining their well-defined curvature (Scheme 1). In recent years, it has been realised that many structural and chemical modifications of imogolite-like structures can be achieved. It is possible to (i) replace Si by Ge,[12] (ii) vary the morphology from nanospheres to nanotubes,[13] (iii) control the formation of single wall or double wall nanotubes,[14,15] (iv) control the diameter of the nanotubes,[11] (v) modify their external surface through chemical grafting,[16] (vi) modify the wall composition *via* incorporation of iron atoms up to 1%,[17,18] and (vii) modify the internal surface through post functionalisation or directly during the synthesis.[19,20] These latter modifications are possible because, in contrast to CNT, imogolite nanotubes are naturally open at both ends and their growth mechanism does not involve the use of any catalytic processes which could lead to unwanted clogging of the cavity.

All the advantages mentioned above make imogolite-derived nanoparticles a whole family of nanobuilding blocks. Most of them are probably not yet discovered and their potential applications are largely unexplored.[21] Among possible modifications, some lead to the formation of what we will call hereafter inside/out Janus nanotubes. Indeed, the grafting of alkyl-phosphonic acids onto the external surface of imogolite nanotubes gives a hydrophobic external surface while the internal surface remains hydrophilic (Scheme 1).[22] The reverse is possible when using triethoxymethylsilane as a chemical precursor leading to nanotubes having a hydrophobic internal cavity covered by Si–CH_3 groups and a hydrophilic external surface providing colloidal stability in water (Scheme 1).[19]

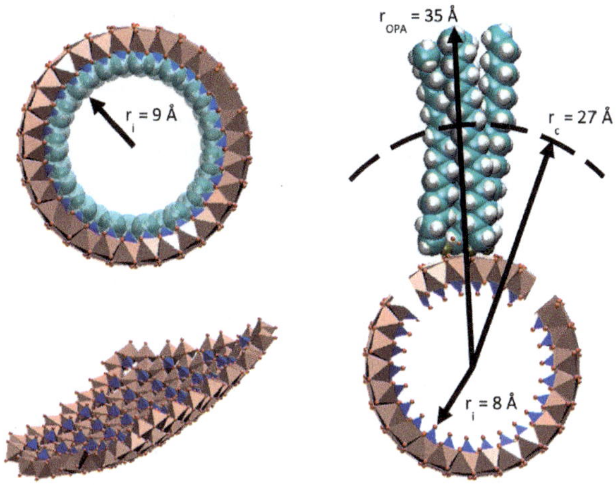

Scheme 1 Side view of imo-CH_3 (top left), proto-imogolite (bottom left), portion of bare imogolite (bottom right) and OPA-imo (top right). Light blue and white spheres are respectively carbon and hydrogen atoms of the methyl groups (hollows for imo-CH_3 and filled for OPA-imo), red and brown spheres are oxygen and phosphorous atoms respectively. Pink octahedra and blue tetrahedra are, respectively, aluminium octahedra and silicon tetrahedra. For clarity, only a few OPA molecules are represented and hydrogen of imogolites hydroxyl groups are omitted. r_i, r_{OPA} and r_c are respectively the radii of the internal cavity, and of OPA-imo without and with inter-penetration of alky chains.

In this work, we hypothesised that such inside/out Janus nanotubes could exhibit very interesting properties and especially that their nanotubular nature can be fully involved in the control of their behaviour at oil/water interfaces. Here, the modification of both the external and internal surfaces of imogolites is described, producing nanotubes with shape and chemical anisotropy. Their adsorption at an oil/water interface and its influence on emulsion stability are studied.

Methods

Synthesis of imogolite nanotubes and hybrid nanotubes

Imogolite nanotubes are obtained through the controlled hydrolysis of a mixture of aluminium and silicon precursors. An acidic solution of aluminium is obtained by dissolving 2 mM $AlCl_3 \cdot 6H_2O$ in water. Tetraethoxysilane is added under vigorous stirring and the obtained solution is stirred for 30 min at room temperature (RT). Then, a 0.1 M sodium hydroxide solution is added at a rate of 0.1 mL min^{-1} until an OH/Al ratio of 2 is obtained. The solution is further stirred until it becomes clear. It is then kept in a close recipient at 90 °C for 5 days. After this heating period, the solution is allowed to cool to RT and concentrated by a custom ultrafiltration system (Polymem, Toulouse, France) using a 10 kDa persulfone membrane. The concentrated solution is dialysed against MilliQ water using an 8 kDa membrane until the conductivity of the permeate reaches 2 µS cm^{-1} or less. The final dispersion of bare imogolite, with both internal and external hydrophilic surfaces, is transparent and birefringent.

The preparation of imogolites with a hydrophobic cavity (imo-CH_3) is performed using a protocol adapted from Bottero *et al.*[19] Briefly, aluminium-*sec*-butoxide is added to a HCl solution (75 mM) in the presence of trimethoxymethylsilane (TMMS) with Al : HCl : Si molar ratios of 2 : 2 : 1. The solution is vigorously stirred at ambient temperature for an hour. After this initial mixing period, the clear dispersion is kept in an oven for 5 days at 90 °C. The solution is then allowed to cool to RT and is concentrated using tangential ultrafiltration. The resulting concentrate is centrifuged at 9000 *g* for 1.5 hours to remove a small proportion of aluminium oxides. The supernatant is a clear, viscous and birefringent dispersion that is further purified by dialysis against milliQ water until the conductivity of the permeate reaches 2 μS cm^{-1} or less.

The external surface of bare imogolite nanotubes is modified by octadecylphosphonic acid (OPA) (Scheme 1). For this, 10 mL of concentrated imogolite dispersion is added dropwise under vigorous stirring into 100 mL of absolute ethanol. Then 100 mL of ethanol with OPA is added until an imogolite : OPA molar ratio of 1 : 5 is reached. The obtained dispersion is sonicated and then stirred for 24 hours at RT. The product is then washed 5 times by centrifugation with ethanol and finally transferred into toluene.

The modified imogolite are characterised by Small Angle X-ray Scattering (SAXS) in toluene (OPA-imo) or water (imo-CH_3). Additionally, the dispersions are dried and the resulting powders are analysed by Fourier Transform Infrared spectroscopy (FTIR) in KBr. The content of organic matter is assessed by Thermogravimetric Analysis (TGA) using a heating rate of 10 °C min^{-1} and a maximum temperature of 1000 °C under N_2.

Preparation of Pickering emulsions stabilised by inside/out Janus nanotubes

OPA-imo in toluene are mixed with water for 2 min at 11 000 rpm using an UltraTurax model IKA T10 basic rod model 10N-8G. Two parameters are systematically varied: the oil/water volume ratio and the concentration of OPA-imo. The emulsions are left to stand overnight before measuring the volume of emulsified phase and the diameter of the droplets. A similar protocol is used for imo-CH_3 nanotubes except that imo-CH_3 dispersed in water are mixed with hexane. As for the previous system, the oil/water volume ratio, the time before stirring as well as the concentration of imo-CH_3 are varied.

Fourier transform infrared spectroscopy

Infrared (IR) spectra are recorded using a Bruker Vertex 70 FTIR spectrometer at RT. About 1.5 mg of dry sample powder is mixed with 130 mg of KBr powder and then pressed into an IR transparent disk. 200 scans with a resolution of 4 cm^{-1} are averaged between 400 cm^{-1} and 4000 cm^{-1}.

Small angle X-ray scattering (SAXS)

SAXS experiments are carried out using a rotating anode generator (Mo Kα = 0.711 Å) collimated by a multilayer optics and hybrid crystal slits, providing a monochromatic beam of 1 × 1 mm^2 at the sample position with a total incident flux of 2 × 10^7 photons per s. A vacuum chamber behind the sample allows minimisation of the X-ray scattering signal from air. A MAR research X-ray sensitive imaging plate detector with a pixel size of 100 μm is placed after the output window of the vacuum chamber at a distance of 725 mm from the sample. Dispersions are filled in Kapton capillaries with a diameter of 0.24 cm. The typical counting time per sample is 3 h. Scattered intensity as a function of the wave-vector modulus $\boldsymbol{Q}$ is obtained by angular integration over the two-dimensional scattering pattern. $\boldsymbol{Q}$ is defined as $\boldsymbol{Q} = |\boldsymbol{k}_d - \boldsymbol{k}_i|$ ($\boldsymbol{k}_d$ and $\boldsymbol{k}_i$ are the wave-vectors of the scattered and incident beams, respectively) and is equal to $4\pi \sin \theta/\lambda$, where λ is the incident wavelength and 2θ is the scattering angle. Data are corrected for water, Kapton scattering and for electronic background.

TGA measurements

TGA spectra are recorded using a TGA Q50 device using a 10 °C min^{-1} temperature increase rate up to 1000 °C under nitrogen.

Viscosity measurements

Viscosity is recorded using an Anton Paar Lovis 2000 M/ME microviscometer. Measurements are made with 1.59 mm capillary, a 1.5 mm diameter steel ball, with the angle varied from 50° to 80°. We prepared three identical dispersions of imo-CH_3 in contact with hexane. A single measurement is performed on each sample to minimise disturbances due to the sample preparation. The dispersion is taken just below the hexane/imo-CH_3 interface.

Atomic force microscopy measurements

The nanotube lengths are measured using Atomic Force Microscopy (AFM, Digital Instruments, Nanoscope V). AFM characterisation is employed in tapping mode to image imogolites adsorbed on mica sheets. To obtain a homogenous deposition, 20 μL of the solutions are deposited on freshly cleaved mica (yielding a smooth negatively charged surface). Positively charged nanotubes irreversibly adsorb on the surface. Mica sheets are dried at RT over 2 days before AFM measurements.

Results

Nanotubes characterisation

The synthesised nanotubes were characterised by IR, SAXS, AFM and TGA. Fig. 1 shows the IR spectra of bare imogolite, imo-CH_3, OPA-imo and OPA. The absorption bands at 417, 552 and 690 cm^{-1} and the doublet at 935 and 977 cm^{-1} are characteristic of imogolite with a tubular structure.[23,24] Concerning imo-CH_3 the two peaks at 1275 and 781 cm^{-1} attributed to the bending of Si–CH_3 and Si–C, respectively, confirm the presence of methyl groups bound to the nanotubes.[25] Furthermore, additional peaks at 2974 and 2921 cm^{-1} corresponding to asymmetric and symmetric stretching of methyl group corroborate the presence of methyl groups for imo-CH_3 nanotubes. In addition, the imo-CH_3 nanotubes are easily dispersed in water but not in toluene, indicating that methyl groups are not located on the outside of the nanotubes but inside the cavity, in agreement with the literature.[17] In parallel, the comparison of the IR spectra of OPA-imo and OPA confirm successful grafting by the presence of CH_2 symmetric and asymmetric stretching at 2850 and 2918 cm^{-1}, the vibration at 1468 cm^{-1} and P–C vibration at 721 cm^{-1} in addition to the imogolite bands.[22,26] Such OPA-imo samples are easily

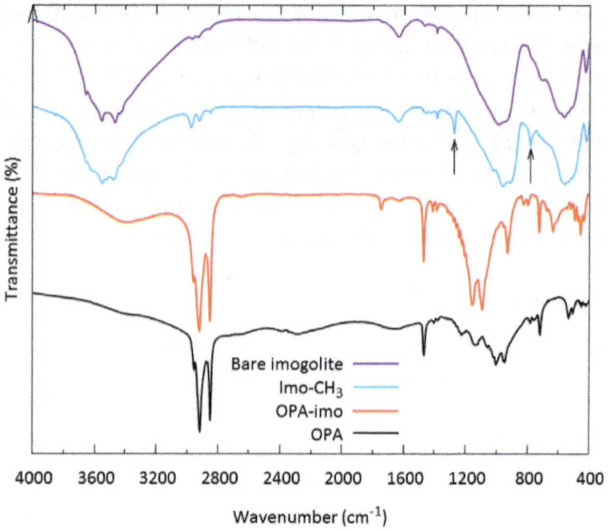

Fig. 1 FTIR spectra of bare imogolite, imo-CH_3, OPA-imo and OPA. The two arrows at 1275 and 781 cm^{-1} are attributed, respectively, to the bending of Si–CH_3 and Si–C for the imo-CH_3 curve.

dispersed in toluene but not in water, indicating that the hydrophobic moieties are present on the external surface of the nanotubes.

The TGA analyses of the bare imogolite, OPA and OPA-imo are shown in Fig. S1 in ESI† and could be used to determine the efficiency of the grafting (details of calculation are given in ESI†). The obtained mass ratio is 1.79 g OPA per g of imogolite and corresponds to 1.05 OPA molecules per imogolite structural unit. The same coverage is obtained, when computed using the same hypothesis, from the measurements of Bac *et al.* on OPA grafted AlGe imogolite.[22] This value has to be compared to the maximum coverage of 2 moles of OPA per mole of imogolite for a complete bi-dentate grafting and 1 mole of OPA per mole of imogolite for a tri-dentate grafting.

SAXS curves of OPA-imo, imo-CH_3 and bare-imo with their respective models are presented in Fig. S2.† These analyses provide information about the radius, the number of structural units per circumference, the concentration and internal electronic density of the nanotubes. The fit parameters, considering hollow cylinders and assuming a homogenous scattering length density, are presented in Table 1. It should be noted that for bare imogolite the internal electronic density is equal to the bulk water electronic density (0.33 e^- $Å^{-3}$), *i.e.* the nanotubes are full of water. For imo-CH_3, it is one third of the water value (0.11 e^- $Å^{-3}$) indicating that the nanotubes are not filled with bulk water. Although the precise physical state of water inside imo-CH_3 is difficult to determine at this time, previous studies have shown that low-density and even vapour-phase water may exist in such confined hydrophobic environments.[27] However, the sole SAXS analysis does not allow to identify the nature of the internal solvent. Noticeably, concentrations obtained from SAXS are confirmed by dry mass measurements (4.35 g L^{-1} for bare imogolite and 14.34 g L^{-1} for imo-CH_3).

Table 1 Fit parameters of bare imogolite and imo-CH_3 SAXS curves. r_i is the radius of the internal cavity, N the number of structural units per circumference, T the period of the repeat distance along the tube axis and C the concentration

Fit parameter	Bare imogolite	Imo-CH_3
r_i/nm	0.8	0.9
N	13	15
Wall thickness/nm	0.6	0.6
T/nm	0.85	0.85
Internal electronic density/e^- $Å^{-3}$	0.33	0.11
C/g L^{-1}	4.35	13.72
Electronic density of the wall/e^- $Å^{-3}$	0.7	0.7

In contrast, OPA-imo nanotubes are not perfectly dispersed and the resulting unstable dispersions do not allow a quantitative concentration measurement by SAXS. According to dry mass determination, the concentration for OPA-imo is 2.56 g L^{-1}. The SAXS curve of OPA-imo (Fig. S2†) reveals that the nanotubes form organised bundles. Considering a bi-dimensional hexagonal phase, the lattice parameter is 5.4 nm. This value is close to the one obtained by Park *et al.* (5.1 nm) and smaller than the diameter of isolated OPA-imo (7.1 nm, Scheme 1), suggesting that the alkyl chains of neighbour nanotubes can interpenetrate.[28]

Since length variation above 100 nm would modify the scattering curves mainly below the low angle limit of our SAXS apparatus, the scattered intensity could not provide reliable information on nanotube length. We have therefore used AFM to image imogolite deposited on freshly cleaved mica sheets. The concentration of the deposited dispersion was tuned to have a sub-monolayer of nanotubes. Analysing longitudinal height sections of the deposited nanotube allowed deduction of their length. Fig. 2 shows the obtained images and histograms of the average length that compile the analysis of 1100 nanotubes. It should be noted that, on these histograms, only nanotubes with a length above 100 nm are reported. The length histogram of bare imogolite (Fig. 2A) has a typical asymmetric shape with a significant number of nanotubes longer than 1.5 μm. As previously mentioned, nanotubes with a length inferior to 100 nm are not reported, although the number concentration of these smaller objects can be estimated to be more than 90% on the mica when 1 μm × 1 μm images are analysed. This is because the AFM resolution does not allow to determine accurately their length distribution. This observation is in agreement with SAXS results (Fig. S2†) showing a damping of the nanotube form factor. It suggests that the sample contains a significant proportion of proto-imogolite (Scheme 1). Proto-imogolites are "tills" and correspond to the first nanoblocks which, by assembling with another or by curving, lead to the formation of imogolite. Therefore the average length of the particles in the bare imogolite dispersion can be estimated to be below 100 nm. The length of the imo-CH_3 nanotubes could be measured with the same method. The length histogram is shown on Fig. 2B indicating an average length of 378 nm. In contrast, it was not possible to obtain a sub-monolayer of OPA-imo on mica due to the poor dispersion and low adherence on the surface. However, whereas the grafting reaction itself should not impact on

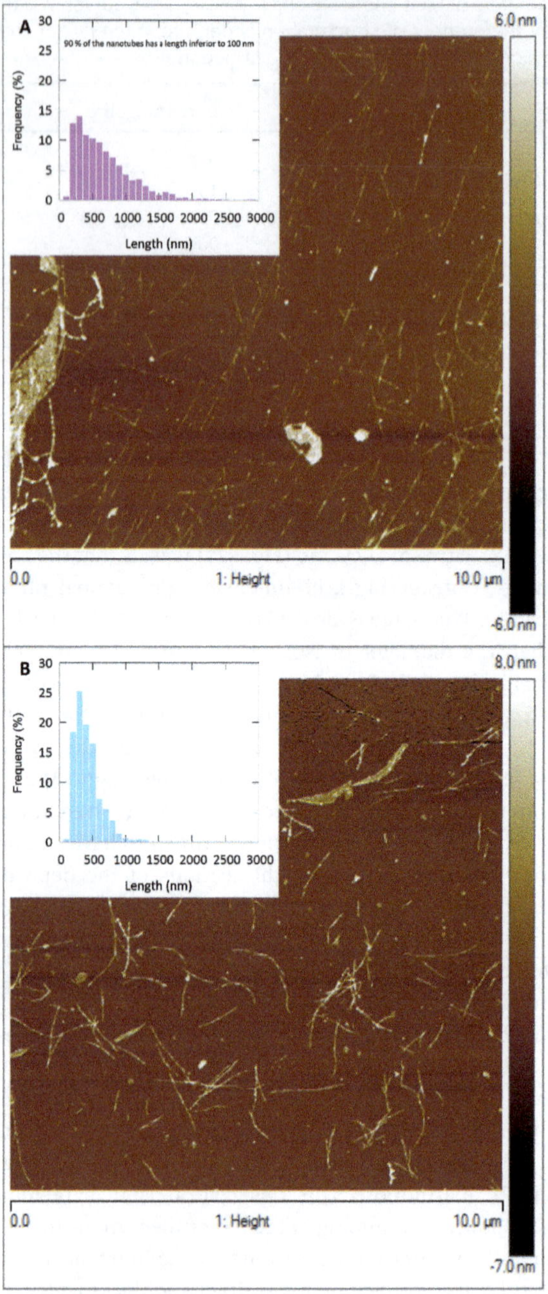

Fig. 2 AFM images and length distribution of bare imogolite (A) and imo-CH_3 (B). Only nanotubes with a length superior to 100 nm are reported on both histograms. For bare imogolite, 90% of the nanotubes has a length inferior to 100 nm.

the nanotube length, the overall procedure that involves many centrifugation and redispersion steps most probably decreases the proportion of smaller objects in the dispersion.

Pickering emulsions with Janus nanotubes

Fig. 3 shows photos of the emulsions prepared with various oil/water ratios and OPA-imo concentrations, taken one day after stirring. Stable water-in-toluene emulsions are obtained for $\Phi_w \leq 0.5$. For $\Phi_w > 0.5$, the stability of the droplets is quickly compromised and no emulsion is observed for the highest water content. The radius of the water droplets decreases when the concentration of OPA-imo increases and/or when Φ_w decreases. When left undisturbed, these emulsions are stable for months. For comparison, emulsions prepared with bare imogolite are presented in Fig. S3.† A marginal stability of the droplets is observed and attributed to the hydrophobicity of the (Al–OH–Al) surface.[29]

In the case of experiments with a dispersion of imo-CH_3 in contact with hexane, no stable emulsion is observed one day after stirring. However, when the imo-CH_3 dispersion has been in contact with hexane for one week before stirring, an emulsion is obtained that remains stable over one week (Fig. 4). Noticeably, during the one-week ageing period, the viscosity of the imo-CH_3 dispersion (and therefore of the emulsion continuous phase) has increased compared to its initial value.

Therefore, OPA-grafted imogolite can stabilise W/O, but not O/W, Pickering emulsions very efficiently. It seems that the droplet size is controlled by an arrested coalescence mechanism. In the case of imo-CH_3, the stability of the emulsions is more complex and apparently linked to the viscosity of the continuous phase. These two mechanisms are discussed in more detail in the following sections.

Discussion

OPA-imo: arrested coalescence by anisotropic particles

From an energetic perspective, the driving force towards particle adsorption at the oil/water interface is due to the gain of interfacial energy between oil and the

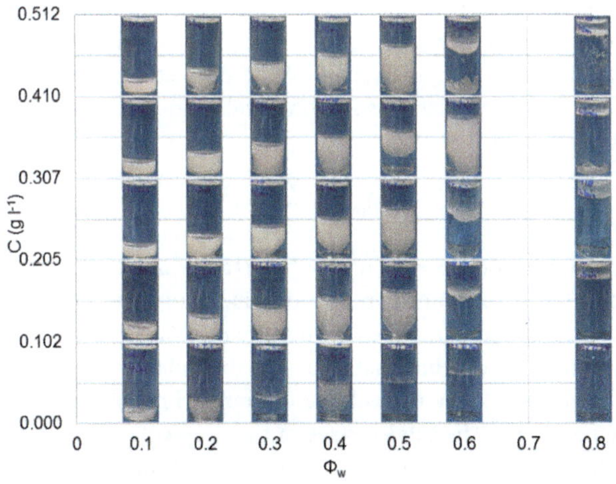

Fig. 3 Emulsion test with OPA-imo, toluene and water. C corresponds to the concentration of OPA-imo in the total volume.

Fig. 4 Emulsion test with hexane and imo-CH_3. Top line: samples is stirred just after preparation; and bottom line: one week after preparation. Photos taken one week after stirring.

water trapped inside OPA-imo tubes. Indeed, the perpendicular position of OPA-imo at the interface with one end of the nanotube in contact with the oil/water interface is the only possibility to minimise this unfavourable contact between the hydrophilic nanotube ends and oil. Considering the diameter of the nanotube ($d = 1.6$ nm) and the toluene/water interfacial tension ($\sigma = 36.1$ mN m^{-1}) the gain in energy upon adsorption is of the order of $E_a = \sigma d^2\pi/2 = 27$ kT.[30] Moreover, in the case of OPA-imo, there is probably also an inter-particle favourable attraction due to the combination of van der Waals attraction and osmotic pressure associated with interpenetration of the grafted OPA chains. The quantification of this attractive energy goes beyond the scope of this work but a rough estimate of a few kT nm^{-1} would be reasonable.[31] Therefore, we can expect a total gain in energy for the OPA-imo nanotube of at least 100 kT making adsorption a probably irreversible process. When particles irreversibly adsorb at the oil/water interface, it is classically observed in Pickering emulsions that the increase of adsorbed particles controls the size of the droplets through an arrested coalescence mechanism.[32,33]

From a structural point of view, there are two extreme options for the adsorption of nanotubes at the oil/water interface: perpendicular or parallel to the interface. When the adsorbed particles are highly anisotropic, these two adsorption modes lead to full coverage of the droplet surface at very different concentrations. Indeed, for an adsorption parallel to the interface, the surface coverage is given by (eqn (1)):

$$\delta_{\text{par}} = \frac{NDL}{S_d} \tag{1}$$

where N is the number of adsorbed nanotubes, D and L are the respective diameter and length of the adsorbed nanotubes, and S_d is the total surface of the droplets. The number of adsorbed nanotubes is given by $N = C_m N_{avo} TV/(ML)$ with C_m the mass concentration of adsorbed nanotubes, M the molar mass per repeat period T along the nanotube axis, V the volume of grafted imogolite added, and L the length of the nanotube. The total surface of the droplets is given by $S_d = 3V_d/R_d$ where V_d is the volume of the dispersed phase and R_d the average radius of the droplets. Combining these expressions, it follows that the droplet diameter obtained through an arrested coalescence mechanism will follow (eqn (2)):

$$R_{\rm d} \sim \frac{V_{\rm d}}{DC_{\rm m}V} \tag{2}$$

If we now assume that the nanotubes adsorb in a perpendicular configuration, then the surface coverage is given by (eqn (3)):

$$\delta_{\rm perp} = \frac{N\pi D^2}{4\ S_{\rm d}} \tag{3}$$

Using the same expressions for N and S_d in this equation, we can show that the arrested coalescence will give a droplet radius with (eqn (4)):

$$R_{\rm d} \sim \frac{V_{\rm d}L}{D^2 C_{\rm m}V} \tag{4}$$

It is easy to understand that in the case of the parallel adsorption the arrested coalescence does not depend on the nanotube length (*i.e.* R_d is independent of L in eqn (2)). For instance, if all adsorbed nanotubes are cut in half, the same surface will be occupied at the interface. On the contrary, in the case of perpendicular adsorption at the interface, length matters. If the nanotubes are cut in their middle, the new nanotubes, not attached to the interface, can stabilise the same interfacial area. Finally, twice the surface can be covered for the same mass of nanotubes when their length is divided by two. The ratio of droplet size at arrested coalescence in the two extreme configurations is given by $4L/(\pi D)$, which is close to 100 in the case of imogolite. Therefore, measuring the average size of the droplets as a function of the OPA-imo concentration should give some clues to conclude on the adsorption geometry. Fig. 5 shows the evolution of the experimentally measured droplet radius (software imageJ analysis) as a function of the

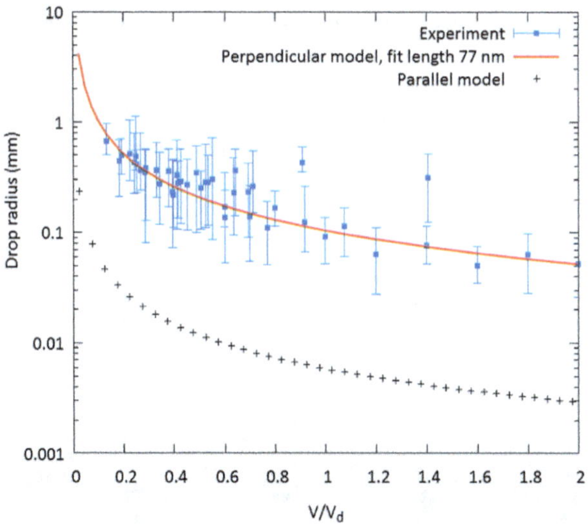

Fig. 5 Evolution of the drop radius *versus* ratio V/V_d for experiment, the perpendicular model with a fit length of 77 nm and the parallel model.

ratio V/V_d, where V_d corresponds to the volume of emulsified water. In the same Figure, the theoretical radii given for the two extreme adsorption scenarios are reported. First of all, we can notice the gap of almost two orders of magnitude in the droplet radius values between the parallel and perpendicular model. Clearly, the experimental values fit very well the perpendicular adsorption model, in agreement with the above-discussed energetic considerations. Second, the fact that the best fit for the perpendicular model is obtained for an average tube length inferior to 100 nm indicates that these smallest OPA-imo contribute to the stabilisation of the emulsion.

Imo-CH_3: gel formation, viscosity effects and transport properties

Imo-CH_3 nanotubes are prepared in water and do not initially contain oil in their cavity. However, it is known that they can capture small hydrophobic molecules.[34] In order to follow this process, Nile Red dye was selected as a probe and added to the oil phase.[35] Fig. 6 shows photos of the Nile-Red-in-hexane solution in contact with water and with an imo-CH_3 dispersion just after preparation for pictures (a) and (b) and after three days of stirring for pictures (c) and (d). The water phase on images (a) and (c) remains colourless. In the presence of imo-CH_3, a pink colour is first observed at the oil/water interface (b) and then the water phase turns pink/violet whereas the oil phase is colourless. This can be explained considering that the hydrophobic inner cavity of imo-CH_3 is able to be filled with the oil phase, also extracting the dye. When the dye is initially confined within the nanotubes located at the oil/water interface, it can then be transferred to the imo-CH_3 particles present in the bulk water phase, resulting in its full coloration after 3 days.

Such filling should provide a driving force toward nanotube adsorption at the oil/water interface. Considering a diameter of 1.8 nm and an interfacial tension between water and hexane of 50 mN m^{-1}, the gain in interfacial energy upon adsorption can be estimated by $\sigma d^2\pi/2 = 49$ kT.[30] However, the imo-CH_3 nanotubes are stable in aqueous solution thanks to interparticle electrostatic repulsion. This repulsion prevents their close packing and therefore the full coverage of the oil/water interface. In this case, we can observe an equilibrium between dispersed and adsorbed nanotubes. The surface coverage can therefore be enhanced by increasing the nanotube concentration or by screening the electrostatic repulsion through an increase of the ionic strength. Notably, it is expected that the same amount of energy is gained when the nanotubes adsorb on the O/W

Fig. 6 Dye test: hexane/Nile Red in contact with water and hexane/Nile Red in contact with imo-CH_3 just after preparation (a) and (b) and three days after stirring (c) and (d).

interface or when the total length of the nanotubes is increased through tip–tip contacts. The presence of oil may therefore trigger an increase of the average length of the unabsorbed nanotubes. It is possible to evidence this phenomenon by following the viscosity of the imo-CH_3 dispersion in contact with oil. Fig. S4† shows that the relative viscosity of the imo-CH_3 dispersion increases from 8.79 to 12.10 after three days of contact with a hexane phase. So far, the rheological behaviour of imogolite dispersions has been sparingly studied.[36–40] Nevertheless, general laws for rod-like particles are well-known and can be tentatively used to understand the observed trends. Using the Einstein equation, for a diluted solution of particles ($\phi_v \ll 1$) the relative viscosity (defined as the ratio between the viscosity of the solution and the viscosity of the solvent, here water) is:

$$\eta_r = 1 + [\eta]\phi_v$$

With $[\eta]$ the intrinsic viscosity defined as:

$$[\eta] = \lim_{\phi \to 0} \frac{\eta_r - 1}{\phi}$$

For uncharged, rigid and non-interacting spheres the intrinsic viscosity equals 2.5. As mentioned previously, imogolites are anisotropic particles. In this situation, the Onsager equation should be applied to calculate the intrinsic viscosity:[40,41]

$$[\eta] = \frac{4f^2}{15\ln(f)}$$

where f is the aspect ratio (ratio between the length and the diameter). Using the diameter of 3 nm given by SAXS analysis and length of 378 nm by AFM measurements, f equals 126, intrinsic viscosity equals 875 and $\eta_r = 8.72$ (for 8.79 measured). Following this equation, the increase of the relative viscosity from 8.79 to 12.1 would correspond to an increase of the aspect ratio from 126 to 154. This slow increase in viscosity is compatible with the slow nanotube growth kinetics through tip–tip collision described by Maillet *et al.*[42] When left undisturbed for very long period (about one month), imo-CH_3 dispersions in contact with oil can even form gels. This gel formation mechanism is very different from the ones induced by pH or salinity increase. In these cases, electrostatic repulsion between nanotubes is reduced up to the destabilisation of the nanotubes dispersion.

Hence, the progressive filling of the internal cavity of imo-CH_3 by the oil phase can contribute to the observed ability of nanotubes to stabilise O/W emulsion after one week of contact with hexane by two factors: favoured nanotube adsorption at the oil/water and increase of the water phase viscosity due to the growth of nanotubes. The relative contributions of the two mechanisms are difficult to establish at this time. Indeed, in this complex situation, it is not possible to assess precisely the number of adsorbed nanotubes and to determine whether perpendicular or parallel adsorption occurs. However, since the imo-CH_3 interactions with an oil-in-water interface should mirror the OPA-imo/water-in-oil situation, the former configuration is the most likely.

Conclusions

In conclusion, we report here the preparation of inside/out Janus nanotubes. Two symmetric systems are obtained: (i) external modification of bare imogolites using a phosphonic acid leads to OPA-imo with an external hydrophobic surface and a hydrophilic cavity; (ii) synthesis of imo-CH_3 leads directly to hydrophilic nanotubes with a hydrophobic cavity. These inside/out Janus nanotubes allow the stabilisation of Pickering emulsions thanks to the presence of an internal solvent. In the case of OPA-imo, the tubes are initially filled with water, their adsorption at the interface is irreversible and an arrested coalescence mechanism is observed. In contrast, the ability of imo-CH_3 to stabilise emulsions requires an aging period allowing to fill the nanotubes' interior with oil. The increase of nanotube length by tip–tip contacts resulting in the enhancement of the viscosity of the continuous phase occurs simultaneously, also favouring emulsion stability. Moreover, these nanotubes can extract a dye from an organic phase.

Structural modelling and energetic considerations suggest that the OPA-imo nanotubes adopt a perpendicular orientation at the oil/water interface. A similar organisation is expected for imo-CH_3, although the complex behaviour of these particles remains to be unravelled. Yet, the here-obtained Pickering emulsions could open the route toward self-assembled nanofluidics if two or more droplets could share connections with the same Janus nanotubes or through a network of connected Janus nanotubes.

Acknowledgements

This research has been funded by an ANR grant HIMO2 (ANR11-BS08-0002) and by the Region Ile-de-France in the framework of C'Nano IdF, the nanoscience competence center of Paris Region (HIPOP grant). The laboratory equipment for the production of imogolites is funded by a Labex NanoSaclay grant for the project PRODIGE. We want to thank Andrea Kolb for precious help and support. We also want to acknowledge the precious discussions with Elodie Bourgeat-Lami, Cécile Cottin-Bizon, Anne-Laure Biance and Christophe Ybert.

Notes and references

1 R. Aveyard, B. P. Binks and J. H. Clint, *Adv. Colloid Interface Sci.*, 2003, **100–102**, 503–546.
2 B. P. Binks and J. H. Clint, *Langmuir*, 2002, **18**, 1270–1273.
3 P. A. Kralchevsky, I. B. Ivanov, K. P. Ananthapadmanabhan and A. Lips, *Langmuir*, 2005, **21**, 50–63.
4 T. Tanaka, M. Okayama, H. Minami and M. Okubo, *Langmuir*, 2010, **26**, 11732–11736.
5 A. Walther and A. H. E. Muller, *Chem. Rev.*, 2013, **113**, 5194–5261.
6 H. Wang and E. K. Hobbie, *Langmuir*, 2003, **19**, 3091–3093.
7 N. M. Briggs, J. S. Weston, B. Li, D. Venkataramani, C. P. Aichele, J. H. Harwell and S. P. Crossley, *Langmuir*, 2015, **31**, 13077–13084.
8 C. Y. Zhou, H. Li, H. Zhou, H. Wang, P. J. Yang and S. A. Zhong, *J. Sep. Sci.*, 2015, **38**, 1365–1371.
9 N. Yoshinaga and S. Aomine, *Soil Sci. Plant Nutr.*, 1962, **8**, 22–29.

10 S. Wada, *Clays Clay Miner.*, 1987, **35**, 379–384.
11 G. I. Yucelen, D. Y. Kang, R. C. Guerrero-Ferreira, E. R. Wright, H. W. Beckham and S. Nair, *Nano Lett.*, 2012, **12**, 827–832.
12 S. Wada and K. Wada, *Clays Clay Miner.*, 1982, **30**, 123–128.
13 B. H. Bac, Y. Song, M. H. Kim, Y. B. Lee and I. M. Kang, *Chem. Commun.*, 2009, 5740–5742.
14 P. Maillet, C. Levard, E. Larquet, C. Mariet, O. Spalla, N. Menguy, A. Masion, E. Doelsch, J. Rose and A. Thill, *J. Am. Chem. Soc.*, 2010, **132**, 1208–1209.
15 A. Thill, P. Maillet, B. Guiose, O. Spalla, L. Belloni, P. Chaurand, M. Auffan, L. Olivi and J. Rose, *J. Am. Chem. Soc.*, 2012, **134**, 3780–3786.
16 W. Ma, J. Kim, H. Otsuka and A. Takahara, *Chem. Lett.*, 2011, **40**, 159–161.
17 A. Avellan, C. Levard, N. Kumar, J. Rose, L. Olivi, A. Thill, P. Chaurand, D. Borschneck and A. Masion, *RSC Adv.*, 2014, **4**, 49827–49830.
18 E. Shafia, S. Esposito, M. Armandi, M. Manzoli, E. Garrone and B. Bonelli, *Microporous Mesoporous Mater.*, 2016, **224**, 229–238.
19 I. Bottero, B. Bonelli, S. E. Ashbrook, P. A. Wright, W. Z. Zhou, M. Tagliabue, M. Armandi and E. Garrone, *Phys. Chem. Chem. Phys.*, 2011, **13**, 744–750.
20 D. Y. Kang, N. A. Brunelli, G. I. Yucelen, A. Venkatasubramanian, J. Zang, J. Leisen, P. J. Hesketh, C. W. Jones and S. Nair, *Nat. Commun.*, 2014, **5**, 9.
21 N. Arancibia-Miranda, S. Lillo and M. Escudey, *J. Chil. Chem. Soc.*, 2013, **58**, 2061–2066.
22 B. H. Bac, Y. Song, M. H. Kim, Y. B. Lee and I. M. Kang, *Inorg. Chem. Commun.*, 2009, **12**, 1045–1048.
23 V. C. Farmer, A. R. Fraser and J. M. Tait, *Geochim. Cosmochim. Acta*, 1979, **43**, 1417–1420.
24 J. L. Bishop, E. B. Rampe, D. L. Bish, Z. Abidin, L. L. Baker, N. Matsue and T. Henmi, *Clays Clay Miner.*, 2013, **61**, 57–74.
25 M. Boyer, E. Paineau, M. Bacia-Verloop and A. Thill, *Appl. Clay Sci.*, 2014, **96**, 45–49.
26 R. Luschtinetz, G. Seifert, E. Jaehne and H. J. P. Adler, *Macromol. Symp.*, 2007, **254**, 248–253.
27 S. Smirnov, I. Vlassiouk, P. Takmakov and F. Rios, *ACS Nano*, 2010, **4**, 5069–5075.
28 S. Park, Y. Lee, B. Kim, J. Lee, Y. Jeong, J. Noh, A. Takahara and D. Sohn, *Chem. Commun.*, 2007, 2917–2919.
29 B. Creton, D. Bougeard, K. S. Smirnov, J. Guilment and O. Poncelet, *Phys. Chem. Chem. Phys.*, 2008, **10**, 4879–4888.
30 J. F. Fu, B. Q. Li and Z. H. Wang, *Chem. Eng. Sci.*, 1986, **41**, 2673–2679.
31 C. L. Kitchens, M. C. McLeod and C. B. Roberts, *J. Phys. Chem. B*, 2003, **107**, 11331–11338.
32 B. P. Binks and C. P. Whitby, *Langmuir*, 2004, **20**, 1130–1137.
33 B. P. Binks, J. Philip and J. A. Rodrigues, *Langmuir*, 2005, **21**, 3296–3302.
34 M. S. Amara, E. Paineau, S. Rouziere, B. Guiose, M. E. M. Krapf, O. Tache, P. Launois and A. Thill, *Chem. Mater.*, 2015, **27**, 1488–1494.
35 P. Greenspan, E. P. Mayer and S. D. Fowler, *J. Cell Biol.*, 1985, **100**, 965–973.
36 K. Egashira, *Clay Sci.*, 1977, **5**, 87–95.
37 K. Egashira, *Clay Sci.*, 1978, **5**, 137–144.
38 K. Egashira and K. Iemura, *Soil Sci. Plant Nutr.*, 1979, **25**, 21–27.

39 Y. Tsujimoto, A. Yoshida, M. Kobayashi and Y. Adachi, *Colloids Surf., A*, 2013, **435**, 109–114.

40 N. Donkai, H. Inagaki, K. Kajiwara, H. Urakawa and M. Schmidt, *Makromol. Chem.*, 1985, **186**, 2623–2638.

41 A. M. Wierenga and A. P. Philipse, *Colloids Surf., A*, 1998, **137**, 355–372.

42 P. Maillet, C. Levard, O. Spalla, A. Masion, J. Rose and A. Thill, *Phys. Chem. Chem. Phys.*, 2011, **13**, 2682–2689.

Faraday Discussions

DISCUSSIONS

Particles at interfaces: general discussion

Alberto Striolo, Jongwook Kim, Catherine Murphy, Luis Liz-Marzán, Joerg Lahann, Javier Reguera, Yu Zhou, Mathias Brust, Antoine Thill, Leonardo Scarabelli, Tobias A. F. König, Martin Buzza, Christian Kuttner, Estefania Gonzalez Solveyra, Heiko Wolf, Jan Vermant, Matthias Pauly, Andrew Harvie, Lucia Pasquato, Antonio Stocco, Hedi Mattoussi, Eugenia Kumacheva, Kelley Heatley, Christoph Hanske, Roland Faller, David French, Andrei Honciuc, Bernard Binks and Francois Sicard

DOI: 10.1039/C6FD90050J

Martin Buzza opened discussion of the paper by Bernard Binks: Why aren't the PECs always neutral? Also, why is there a well defined size for the PECs? Do you know what determines the characteristic size? Is it basically the collapsed size of one anionic and one cationic chain or is it bigger than that?

Bernard Binks responded: It is believed that PEC complexes consist of a charge neutralized core surrounded by a shell of an excess component which stabilizes the particles against further aggregation.[1] This repulsion may be the reason why the obtained PECs have a characteristic size. There is no evidence in the literature that each PEC entity is the collapsed size of one anionic and one cationic chain.

1 H. Dautzenberg and W. Jaeger, *Macromol. Chem. Phys.*, 2002, **203**, 2095–2102.

Andrei Honciuc asked: Do these particles exhibit some anisotropy in terms of geometry and patchiness or could they possess a polarity contrast between the surface regions? Are they amphiphilic?

Bernard Binks replied: The structure of the PEC is considered to be a charge neutralized core surrounded by a shell of the component in excess. Nearly all of the particles reported so far are spherical in shape. It is doubtful that they possess regions of different polarity and hence they are not amphiphilic.

Andrei Honciuc continued: There is evidence that the air/water interface is negatively charged, that could be true also for the oil/water interface. Do you see a difference in the interfacial activity between the positively and negatively charged nanoparticles? Secondly, while you reported an enhanced activity of the neutral charged nanoparticles, could you observe interfacial activity at the air/water interface?

Bernard Binks answered: Concerning the interfacial activity, there is not a remarkable difference between the positively and negatively charged PEC nanoparticles. This is in line with work on anionic or cationic particles of oxides or polymers. Regarding the air–water surface, no reduction in surface tension at any mixing ratio is observed compared to that for pure water. Moreover, no air bubbles were formed upon hand-shaking these aqueous dispersions. Therefore, no interfacial activity at the air–water interface is detected. Clearly, the wettability of the PEC particles is different at the air–water and oil–water interfaces.

Andrei Honciuc remarked: The nanoparticles obtained using a 1 : 1 ratio of the positive and negative polymer chains have the highest interfacial activity in comparison to any other ratio. What is the mechanism of adsorption at the interface that supports the observed behavior?

Bernard Binks answered: As reported in the paper, all PEC dispersions exhibit a lower oil–water tension than the bare interface, and the value is minimal around charge neutralisation. However, the maximum reduction in the interfacial tension is only about 10 mN m^{-1}. In solid-stabilised emulsions, it is accepted that the reduction in the interfacial tension is not the operative mechanism. It has been found both experimentally and theoretically that surfaces become more hydrophobic as the degree of charge decreases, *e.g.* approaching the isoelectric point of an oxide. This enhances the adsorption of such particles to the interface.

Joerg Lahann asked: In terms of the particle diameters for these nanoparticles, they are independent of the mole fraction of the electrolytes, especially after 20 days; how do you explain that? You have deliberately used the same molecular fraction, do you think that if you change this fraction you will get different results?

Bernard Binks responded: The particle diameter of the obtained PEC seems to decrease around 40 nm 20 days after preparation when PDADMAC is in excess. The same tendency was observed for the PEC dispersions prepared from a PEL of a higher concentration (0.5 g L^{-1}, data shown in Fig. S1). For this higher concentration, the measurement was also conducted 2 months after preparation and no further change in the size was noticeable (data not shown). We do not have an explanation as to why the size of the PEC particles is little affected by the mixing ratio. The effect of the molecular weight (M_w) of one of the polymers was briefly evaluated. As discussed in the paper, the M_w can influence PEC formation. In general, bigger PEC structures are achieved with a PEL of higher molecular weight, although there are some exceptions in the literature for the studied polyelectrolyte system. We measured the size of the PEC particles prepared from PELs of M_w differing by a factor of 6.1 (data shown in Fig. S4). No remarkable differences in either the size or PDI were observed in comparison to the previous results with polymers of comparable M_w. However, the value of x_{PSSNa} at neutralisation is considerably lower as expected.

Joerg Lahann continued: In this work you have deliberately used the same packing density, do you think that if you change this fraction, *i.e.* induce a transition from a brush to mushroom-like conformation, you will get different results?

Bernard Binks replied: The PEC formation can be influenced by structural parameters such as the charge density, molecular weight or PEL topology (linear or branched). In terms of the charge density, many examples in the literature show that more compact PEC structures are achieved using a PEL of equal charge density.[1,2] Regarding the molecular weight, in general, bigger PECs are achieved with a PEL of higher molecular weight.[3] The polyelectrolyte topology also influences the properties of the resulting assemblies. Many examples are found in the literature on the interaction of linear/branched PELs[4] or bottle-brush/linear PELs[5,6] amongst others. Each system should be studied separately under defined experimental conditions to understand their formation in the aqueous phase as it seems difficult to state a general pattern of behaviour. All these variables will affect the emulsion stability.

1 H. Dautzenberg and W. Jaeger, *Macromol. Chem. Phys.*, 2002, **203**, 2095–2102.
2 M. Mende, G. Petzold and H. Buchhammer, *Colloid Polym. Sci.*, 2002, **280**, 342–351.
3 Y. Hu, T. Yang and X. Hu, *Polym. Bull.*, 2012, **68**, 1183–1199.
4 M. Müller, B. Keßler, J. Fröhlich, S. Poeschla and B. Torger, *Polymers*, 2011, **3**, 762–778.
5 S. V. Larin, D. V. Pergushov, Y. Xu, A. A. Darinskii, A. B. Zezin, A. H. E. Müller and O. V. Borisov, *Soft Matter*, 2009, **5**, 4938–4943.
6 Y. Xu, O. V. Borisov, M. Ballauff and A. H. E. Müller, *Langmuir*, 2010, **26**, 6919–6926.

David French asked: Do you know why there is a very sudden switch in the zeta potential when the proportion of anionic polymer is increased, in contrast to the smooth transition observed when mixing oppositely charged colloidal silica particles?

Bernard Binks answered: We do not have an explanation for the dramatic change in the sign of the zeta potential value. However, the systems are different; ours involves the mixing of molecules while the one referred to above concerns the mixing of solid particles. To find the PEC with more or less zero charge, we tried to prepare aqueous PEC dispersions around the change point. However, either positive or negative values were achieved.

Javier Reguera addressed Bernard Binks and Francois Sicard: Regarding the deformability of the nanoparticles, what nanoparticles are better stabilizers, soft, hard or hybrid nanoparticles?

Bernard Binks responded: Emulsions stabilised by soft nanoparticles are quite recent and have been extensively studied by, for example, Richtering *et al.* with microgel particles (crossed-linked latex particles that are swollen with a good solvent). In this case, the deformability, swelling and viscoelastic properties at the oil–water interface are the most important features controlling the emulsion stability.[1] It has been found that when microgels are not strongly swollen and therefore less deformable, no stable emulsions are obtained. However, when they swell they become more deformable and stable emulsions can be prepared.[2] Janus-like nanoparticles, whose surfaces show regions of different hydrophobicity, attach more strongly to the interface than homogeneous particles. However, in none of the cases reported do particles deform at the interface. Therefore, the mechanism of emulsion stabilisation is not comparable. Soft, hard and Janus nanoparticles can be considered good emulsion stabilisers.

1 W. Richtering, *Langmuir*, 2012, **28**, 17218–17229.
2 S. Schmidt, T. Liu, S. Rütten, K. Phan, M. Möller and W. Richtering, *Langmuir*, 2011, **27**, 9801–9806.

Francois Sicard replied: Our study focused on the coalescence mechanism within the ABMD (adiabatic biased molecular dynamics) framework. This only allows us to extract the equilibrium free-energy path and to compare them when we consider different types of particle stabilizers. However, it does not give us any information about any free-energy barrier the system must pass. Actually, these paths extracted with ABMD will be considered and used with metadynamics and/or umbrella sampling simulation to reconstruct the free-energy landscape. This will allow us to compare free-energy barriers and discriminate quantitatively what particles are better stabilizers.

Jan Vermant continued discussion of the paper by Bernard Binks: In your paper and in your earlier work on silica particles you suggest that flocculation in the bulk is useful for making stable Pickering emulsions. Why is this? Is this related to the easier adsorption of aggregates relative to individual particles or is this related to the interparticle forces? On a related note, why don't we see voids at the interface but nice densely packed systems when the particles flocculate in the bulk?

Bernard Binks responded: Flocculated dispersions are known to be more efficient at stabilizing emulsions. Emulsions stable to both creaming and coalescence have been achieved when the particles are weakly flocculated, while less stable emulsions result when the particles are highly flocculated. However, flocculation is usually achieved through the addition of salt and so it is difficult to distinguish the effect of increased "particle" size from reduced charge. Larger "particles" are more difficult to remove from an interface once adsorbed. A reduction in the zeta potential of the particles on adding salt increases their hydrophobicity which enhances adsorption. This adsorption, together with aggregation and the formation of a three-dimensional network in the continuous phase, promotes the stability of emulsions as the mobility of the emulsion droplets is reduced. On the other hand, if the size of the floc is too large, the particles sediment due to gravity and the adsorption of particles at the interface becomes difficult.[1,2] Regarding the second question, on the cryo-SEM image densely packed particle layers can be seen. This is due to the fact that the particle concentration is high. Therefore, particles are both densely packed at the interface and form a three-dimensional network in the continuous phase.

1 F. Yang, S. Liu, J. Xu, Q. Lan, F. Wei and D. Sun, *J. Colloid Interface Sci.*, 2006, **302**, 159–169.
2 N. P. Ashby and B. P. Binks, *Phys. Chem. Chem. Phys.*, 2000, **2**, 5640–5646.

Yu Zhou asked: For emulsions, have you evaluated the stability of the emulsion system except for the zeta potential measurement?

Bernard Binks replied: The emulsion stability in this work has been evaluated through monitoring the release of oil above the cream (coalescence), measuring the release of water below the emulsion (creaming) and checking the droplet size with time. However, this methodology can be time consuming and may not be suitable in industry. Therefore, to monitor the emulsion stability, a stability analyser can be used. This instrument allows accurate and rapid stability testing by detecting changes in the light transmission at local positions in the vessel. If the droplets are stable, the change in the transmission is low. However, if creaming or coalescence occurs, more light reaches the detector resulting in a rise in the transmission.

Yu Zhou continued: What are the potential applications of this system?

Bernard Binks responded: In this work we show the possibility of stabilising oil/water emulsions with a new type of particle, the so-called polyelectrolyte complexes. Applied to the cosmetic industry, emulsions stabilised by particles offer several advantages compared to surfactant-stabilized systems. They have long term stability as the particles are irreversibly attached at the oil–water interface and produce less skin irritation.

Estefania Gonzalez Solveyra commented: When analyzing the effect of pH on the aqueous PEC dispersions and emulsions you find similar results to the ones obtained after the addition of salt. This is somewhat surprising, because the polyelectrolytes used in the PEC are both strong polyelectrolytes, so no great dependence on pH should be observed. In the paper you relate this to changes in the ion concentrations due to the addition of HCl or NaOH to prepare the solutions. However, no difference with salt concentration was seen in that same range of concentration. So my question would be, what do you think the reported effects could be due to? Also, have you considered studying the effect of pH in more detail?

Bernard Binks replied: The experiment with HCl and NaOH was performed in order to prove that the pH is not affecting the PEC formation as both polyelectrolytes are strong. However, as both HCl and NaOH are electrolytes, they cause the PEC to aggregate. In the literature, despite the fact that the effect of NaCl is extensively studied, studies on the effect of the nature of the salt are not reported. A study exists on the response of polyelectrolyte complexes to the subsequent addition of salts with different cations.[1] They found that multivalent cations remarkably suppressed the tendency to coagulate compared to monovalent cations. However, to the best of our knowledge, there is no evidence on the effect of electrolytes such as HCl or NaOH on PECs prepared from strong polyelectrolytes. Therefore, in our system we cannot compare directly the effect of the three electrolytes as they differ in both the cation and the anion and that could influence the aggregation process. Hence, this should be studied in more detail.

1 H. Dautzenberg and J. Kriz, *Langmuir*, 2003, **19**, 5204–5211.

Antoine Thill asked: For the most stable emulsions you obtained, what would happen if you introduce afterwards a small amount of PSSNa or PDADMAC? Do you think it will destroy the emulsions or will it only give a charge to the droplets? Behind this question is of course the idea of making layer-by-layer structures on liquid droplets.

Bernard Binks responded: We have not tried this experiment. However, it might not have a remarkable influence. We did a comparable test to evaluate the effect of the addition of salt after emulsification. Salt has shown an important effect on the stability of PEC aqueous dispersions as it increases particle aggregation. The emulsion was completely broken at intermediate salt concentrations as shown in Fig. 12. Therefore, to study the influence of salt after emulsification, an intermediate salt concentration was added to a freshly prepared emulsion. However, in this case, this did not cause the complete destabilisation of the emulsion. We explained this behaviour by the fact that the added electrolyte is not accessible to the particles, which are already placed at the droplet interface. The same explanation could be applied to the addition of individual PELs to a stable emulsion. However, the experiments should be carried out to confirm or disprove this hypothesis.

Alberto Striolo enquired: Is it safe to assume that the nano-structures you obtained for the neutral system with positive and negative polymers assembled in water are composed of 1 chain of each polymer? Is there any evidence for this? Also, the model you used to describe the size of the droplets in the emulsion as a function of the polymer concentration seems to assume that all the nanoparticles assemble at the oil–water interface. How does this model reconcile with the cryo-TEM images you showed in which the polymer-based nanoparticles are visible in the region between the droplets?

Bernard Binks replied: We cannot ensure that each PEC particle is obtained from the interaction of 1 chain of each polymer. To the best of our knowledge, no study exists that proves how many polymer chains interact to form the PEC aggregate. Moreover, this will depend on the PEL type, structure, concentration and other external parameters such as the salt concentration or pH. The limited coalescence model describes the influence of the initial emulsifier concentration on particle-stabilised emulsions. In the emulsifier-poor régime droplets are partially covered with the emulsifier. By increasing the stabiliser concentration the degree of coverage by particles increases and the mean droplet size rapidly decreases. In contrast, at high emulsifier concentration (emulsifier-rich régime) the average droplet size is practically independent of the stabiliser concentration. In our case, the droplet size decreases along all the concentration range. The emulsifier-rich régime seems not to be achieved. However, as seen from the cryo-SEM images at high concentration, the particles apart from being densely-packed at the interface are also present in the aqueous phase. Regarding the limited coalescence model, as all the droplet surface is fully covered by particles, one would expect that the droplet size remains constant. However, this dual location of particles can trap the droplets and influence their size. We have to take into account that the limited coalescence model was designed for solid particles. Therefore, the principles of Pickering emulsions could not be entirely applied to this system. The same was stated for emulsions stabilised by microgels.[1]

1 R. W. Style, L. Isa and E. R. Dufresne, *Soft Matter*, 2015, **11**, 7412–7419.

Roland Faller opened discussion of the paper by Francois Sicard: One of the main themes is comparing equilibrium with non-equilibrium systems. How do you pull in your non-equilibrium system, at the centre of mass of the molecule or somewhere else? What is the path you are pulling along? Do you run the pull with a constant force or constant velocity? Did you test several settings (*e.g.* velocities)?

Francois Sicard responded: Considering the pulling simulations, we pulled at a constant velocity at the centre of mass of each droplet along an arbitrary axis (*z*-axis). We tried different pulling velocities and saw different coalescence mechanisms (*cf.* Fig.1 in the article) considering the shape deformation of the droplets. We explained that one must consider the Peclet number to compare the advection/diffusion properties of the system (which is related to the NP/NP interaction).

Martin Buzza said: If I understand you correctly, the merging of two water droplets is initiated when a nanoparticle from one droplet touches the water phase of the other droplet. If this is the case, Janus particles should resist merging more effectively compared to homogeneous particles since it is not favourable for the hydrophobic patch of the Janus particles to touch the water phase of the other droplet. Have you tried simulating this with homogeneous nanoparticles as well and do you see this difference between homogeneous and Janus particles?

Francois Sicard answered: We considered different types of NP stabilizers: Janus NPs with spherical and ellipsoidal shapes. We also considered homogeneous NPs with a spherical shape (*cf.* supporting information). The main difference is the formation of the bridge that initiates the merging. Unlike Janus NPs, we only need one NP from the first droplet contacting the second droplet to form the "bridge" and initiate the merging.

Antoine Thill commented: Upon coalescence the interfacial energy is reduced and it is transformed into kinetic energy. The fluid and the particles attached to the interface move. Do you think that this kinetic energy could be enough to detach the particles?

Francois Sicard answered: In our simulations we only considered the early stage of the coalescence process where the evolution remains at equilibrium. We do not consider any further evolution in the first instance where the evolution can strongly be out of equilibrium and where particles can detach. In our study, we drive the coalescence process within the ABMD framework where we only consider the thermal fluctuation of the system and where we apply a bias to prevent the system moving backwards.

Antonio Stocco asked: Can you compare the collision speed in your simulations with the velocity encountered under experimental conditions (diffusion, convection and velocity induced by shear forces during emulsification)?

Francois Sicard responded: Actually, we compared the collision speed in our pulling simulations with experimental conditions. As stated in the main text, we compared our pulling rates with representative shear rates that can be reached in a conventional mechanical emulsification or ultrasonification process. We also compared our results with recent microfluidic collision experiments studying head-to-head moving water droplets in a micro chip. Unlike pulling simulations where this comparison is possible, in the ABMD simulations the evolution follows the natural fluctuations of the system, a regime dominated by diffusion by design.

Luis Liz-Marzán remarked: I was trying to find a correlation between your simulations and the experiments presented by Eugenia Kumacheva (DOI: 10.1039/C6FD00057F). If you do not consider droplets but stabilised particles, in case they diffuse and find each other, could you predict what Kumacheva showed? You may imagine a sort of polymer droplet that is surrounding one particle, if two of these droplets come into contact, with the various possible forces, you could even think of a multiparticle problem. Depending on the interparticle forces, could you determine the final configuration?

Alberto Striolo replied: To answer the question of whether we can study the self-assembly of nanoparticles decorated with polymers or surfactants, the answer is yes, provided the systems are close enough. We have for example just reported the coalescence of a nanoparticle made of hydrates and a water droplet, both surrounded by surfactants and immersed in a hydrocarbon phase.[1] This study was conducted at the molecular level. We, and others, have used coarse-grained models for investigating the self-assembly of particles grafted with short polymers.

1 A. Phan, T. Bui, E. Acosta, P. Krishnamurthy and A. Striolo, *Phys. Chem. Chem. Phys.*, 2016, **18**, 24859–24871.

Joerg Lahann commented: You could proceed to change the path size, geometry and affinity, and you could introduce facets; what do you expect to happen, how would it effect the merging of the particles?

Francois Sicard responded: Indeed, the chemical and geometrical properties of nanoparticle stabilizers change the NP/NP interaction (and thus the coalescence process). We tried different types of stabilizers: Janus NPs, Homogeneous NPs and both with different geometries, spherical and ellipsoidal (*cf.* main text and supplemental material). We saw that increasing the NP/NP interaction would be an important step in preventing coalescence. It might also be interesting to play with different NP magnetic properties in order to modulate the NP/NP interaction.

Joerg Lahann opened discussion of the paper by Antonio Stocco: With respect to the two compartments in your system, you could chemically modify them to make the two compartments different? How would that affect your outcomes?

Antonio Stocco replied: Two different scenarios can be discussed. For a Janus particle possessing a hydrophilic face (contact angle lower than 90°) and a hydrophobic face (contact angle higher than 90°), the triple line is expected to be pinned at the Janus boundary and rotation suppressed. Considering self-propelled Janus particles, the propulsion direction in this case is expected to be normal to the interface and no active motion parallel to the interface is expected. Alternatively, one could also imagine that the two faces have the same wetting properties even if chemically different. In this second scenario and in the absence of strong contact line pinning, the Janus particle is expected to rotate as a bare particle. Particles able to self-propel (*e.g.* platinum–silica Janus particles) may show an alternating active motion depending on whether the reactive face (*e.g.* platinum) is wetted by the fluid containing the chemical fuel (*e.g.* hydrogen peroxide) or not.

Martin Buzza asked: What factors control the strength of contact line pinning? Can we control these factors or at least estimate the strength of contact line pinning from first principles for a given particle?

Antonio Stocco answered: Contact line pinning is controlled by the density of defects and by their energy, which depends on the size, shape (roughness) and the chemistry of the defect. As for the wetting of a planar surface, the contact line pinning can be weak or strong depending on the balance between the defect force and the restoring force of the liquid tail. For strong pinning, the advancing and receding contact angles are different and these values define the hysteresis of the contact angle which is by itself a measure of the strength of the contact line pinning.

Joerg Lahann asked: The SEM images you showed indicate differences in the coverage of Pt on the different particle formulations. Do you expect any influence on the pinning of the contact line?

Antonio Stocco replied: The roughness and coverage of platinum on silica strongly affect the contact angle hysteresis as shown for planar surfaces. For Janus particles, we followed just one protocol for the fabrication but we observed a distribution of platinum coverage and shapes of the Janus boundary. At the moment, from the SEM analysis, on a single particle level we can not find a correlation between the roughness or coverage and contact angle hysteresis but we can just discuss the difference between bare and Janus particles, which agrees with the results obtained for planar surfaces.

Mathias Brust commented: Can you comment on the role of gravity please? Beads of the size you are using should precipitate if not pinned to the surface/interface.

Antonio Stocco answered: In bulk water, micron sized silica beads indeed tend to sediment under gravity. However, particles deposited at the water–air interface possess very high wetting energy and gravity effects are negligible with respect to capillary ones when the particles size is typically much smaller than the capillary length (about 3 mm for the water–air interface). The deformation of the interface due to the weight of the trapped micron sized beads is, for instance, negligible.

Andrei Honciuc opened a general discussion of the papers by Bernard Binks and Antonio Stocco: I would like to understand the aspect of size-scale for the particles at interfaces. For example, surfactants are small molecules in comparison with nanoparticles and we never worry about their size or any size-effects for that matter. But for particles this is different. Depending on the particle size one wonders, can a 500 nm particle be just as good in generating and stabilizing

Pickering emulsions as a 10 nm nanoparticle? What can be expected? Up to what size can the particles still emulsify?

Antonio Stocco responded: The size strongly affects the adsorption energy due to wetting. For a spherical particle this energy scales with the square of the radius and it is usually very large *e.g.* 10^6 kT for a micron sized particle. Approaching the nanometer scale, the adsorption energy due to wetting becomes comparable to the thermal agitation energy and 1–10 nm sized particles may not adsorb irreversibly at the interface. Hence Pickering emulsions made of 1–10 nm sized particles may be not stable. However at the nanoscale, long range surface forces such as Van der Waals and electrostatic forces strongly affect the wetting behavior and the interactions at the interface. One could then expect that 1–10 nm sized particles will stabilize emulsions if a stabilization mechanism other than the irreversible adsorption can occur. This could be for instance, increasing the surface potential of the drop interfaces using electrostatically charged nanoparticles.

Bernard Binks answered: It is known that the particle size has a remarkable influence on emulsion stability but obtaining definitive evidence for this is not straightforward as many other properties of the particles must be fixed as the particle size is changed. This dependence can be evaluated by analysing the equation for the energy of detachment E of a spherical solid particle to a fluid–fluid interface (eqn (1)), where R is the particle radius, γ_{ow} is the oil–water interfacial tension and θ is the contact angle the particles make with the interface. By working with a defined system (particle, oil type and water) some assumptions can be made to examine equation 1. If we consider that the particles have the same surface properties, then the contact angle remains unchanged. Moreover, by using defined oil and water phases the interfacial tension is unaltered. Therefore, the particle size plays an important role in the value of E. An increase in particle size increases the detachment energy and potentially enhances the emulsion stability. For a better understanding it is worth checking Fig. 2 in **ref. 1** below. From that review, it is seen that particles with a radius of around 50 nm have a detachment energy of around 100 000 kT (for the conditions chosen). This high value of E means that the particles are considered to be irreversibly adsorbed. On the contrary, particles with a radius < 0.5 nm (size of surfactant molecules) display low energies (several kT) and are not too effective as stabilisers. Evidence exists in this case for the easy desorption of particles from emulsion drop interfaces. The general rule of thumb is that larger particles stabilise larger drops. However, such drops can be as stable as smaller drops, unlike in the case of surfactant-stabilised drops. In addition, cm-sized emulsion drops stabilised by nanoparticles have been reported, prepared by gentle hand-shaking.[2]

$$E = \pi R^2 \gamma_{ow} (1 \pm \cos\theta)^2 \quad (1)$$

1 B. P. Binks, *Curr. Opin. Colloid Interface Sci.*, 2012, **7**, 21–41.
2 S. Arditty, C. P. Whitby, B. P. Binks, V. Schmitt and F. Leal-Calderon, *Eur. Phys. J. E*, 2003, **11**, 273–281.

Kelley Heatley addressed Antonio Stocco: You stated that you are interested in particle propulsion at the interface. Have you explored controlling the particle orientation such that the two different surfaces are both at the interface so the particle could propel due to a difference in surface tension instead of a surface reaction?

Antonio Stocco answered: One could imagine controlling the orientation of a passive Janus particle with an external magnetic or electric field, and imagine exploiting a lateral imbalance of the force on the contact line. However at equilibrium the fluid surface will not remain flat but will deform to cancel this force imbalance. The dynamic of the restoring surface deformation is governed by the contact line motion and the dynamic of partial wetting. In order to produce persistent propulsion (but not necessarily stationary), one could indeed imagine using an alternating external field which varies at a certain high frequency able to keep a force imbalance on the contact line. However, this temporary force imbalance may only lead to rotation and not necessarily to lateral propulsion.

Joerg Lahann addressed Bernard Binks: You have worked with strong polyelectrolytes so far. I would imagine that one would be able to replicate some of these findings with weak electrolytes. What would you expect with respect to the dynamics of the system, can you comment on that?

Bernard Binks replied: In this paper, we report the first results for the PEC system between PSSNa and PDADMAC, both strong polyelectrolytes. We have recently commenced investigations with a new pair of PELs, sodium poly(acrylate) (PAANa) and PDADMAC. In this case, we combine a weak with a strong polyelectrolyte. At pH 10, where both PELs are completely ionised and of opposite charge, one would expect to obtain similar results to the first system. However, under these conditions, instead of PEC particles a coacervate complex phase arises. Coacervation is a liquid–liquid phase separation in which a dense polymer-rich phase (coacervate phase) coexists in equilibrium with a very dilute polymer-deficient phase (aqueous phase). The phenomenon of complex coacervation was first observed by Bungenberg de Jong and Kruyt in 1949.[1] The formation of either a PEC precipitate or coacervate complex depends on many factors such as the pH, PEL concentration or salt content, amongst others. Despite the early discovery and the numerous studies presented by several authors, the conditions that give rise to each of these structures remain unclear. With this new polyelectrolyte combination, no stable emulsions have been achieved for the studied mole fraction variation of PAANa. We need to establish whether the coacervate complex is capable of acting as a stabiliser of emulsions or not.

1 H. G. Bungenberg de Jong and H. R. Kruyt, *Colloid Science*, Elsevier Publishing Co., Amsterdam, 1949, vol. II, pp. 335–432.

Martin Buzza opened discussion of the paper by Jan Vermant: Why surfboards?

Jan Vermant answered: The motivation to make surfboards is – what else could it be – to study the assembly and rheology of the particles at interfaces. Ellipsoidal particles have strong shape induced lateral capillary interactions, as reviewed by Botto *et al.*, leading to self-assembled networks.[1] These lateral capillary interactions will be weaker in the surfboards, which would make for an easier to understand model system where the effects of flow on the surface rheological properties can be evaluated. Recent calculations by Luo *et al.* are among the first to explore non-linear rheological effects at interfaces, but await experimental evaluation.[2]

1 L. Botto, E. P. Lewandowski, M. Cavallaro and K. J. Stebe, *Soft Matter*, 2012, **8**, 9957–9971.
2 A. M. Luo, L. M. C. Sagis, H. C. Öttinger, C. De Michele and P. Ilg, *Soft Matter*, 2015, **11**, 4383–4395.

Heiko Wolf asked: Does this method allow for the fabrication of further shapes (*e.g.* curved disks or twisted surfboards) which could lead to special packing geometries (chains and helices)?

Jan Vermant responded: As long as the deformation is homogeneous, large quantities of a wide array of shapes can be made. This is indeed an interesting route to pursue. An alternate way to put stress on particles in deformed films was pursued by Champion *et al.*, where solvent swelling in the deformed state was used and complex shapes could be generated, but complex mechanical deformation may be better scalable to higher yields.[1]

1 J. A. Champion, Y. K. Katare and S. Mitragotri, *Proc. Natl. Acad. Sci. U. S. A.*, 2007, **104**, 11901–11904.

Luis Liz-Marzán commented: Toward the end of your presentation you showed that the particles could form rings. Could these rings be made sticky? Can you selectively functionalize the rings?

Jan Vermant replied: The rings on the oblate and prolate particles are the result of the curvature dependent etching of the stabilizer during the degradation process. The subsequent regrafting was done with a fluorescent stabilizer, but indeed, different surface chemistries are possible. So far, particles of this nature have been made sticky by not regrafting anything, as was touched upon in our earlier work with Zhang *et al.*[1]

1 Z. Zhang, P. Pfleiderer, A. B. Schofield, C. Clasen and J. Vermant, *J. Am. Chem. Soc.*, 2011, **133**, 392–395.

Joerg Lahann asked: Will changing the molecular weight of the polymer alter the T_g?

Jan Vermant responded: Indeed, although this does not affect the stability of the dispersions at room temperature. Moreover, we should be able to have better control over the stickiness of the particles when we change the solvent quality for the PHSA.

Andrei Honciuc enquired: You have created a nice range of particles with different shapes. Which shapes are more effective for adsorption at the air/water or oil/water interface?

Jan Vermant replied: Particle anisotropy helps particles to be adsorbed into the interfaces. Particles with a sharp tip will be most effective; we refer to the work of de Graaf *et al.* for theoretical work using free energy calculations.[1]

1 J. de Graaf, M. Dijkstra and R. van Roij, *J. Chem. Phys.*, 2010, **132**, 164902.

Joerg Lahann commented: It is intriguing that you have established NMR as a way to estimate the molecular weight of your polymers. For some of the polymers the NMR data correlates well with the GPC, but for some of them, in particular the more branched polymers, the correlations are poor. What is the standard in the GPC? Do you think the reason for this deviation has to do with the limits of GPC for branched polymers? How else do you explain this?

Jan Vermant answered: The NMR method indeed gives an alternate and very precise way to measure the molecular weight of the PHSA chains. As you suggest, GPC is indeed less adequate in the present case due to the fork-branched nature of the PHSA with possible variations in the branching. We include the GPC data to compare with earlier literature work. The NMR method is however more accurate.

Leonardo Scarabelli said: From the electron microscopy images in your manuscript, your objects are still in the micron dimension. Do you think it would be possible to reduce their size down to the nanoscale?

Jan Vermant responded: The mechanical stretching method requires the particles to be deformed homogeneously. There are two aspects that will limit this. First, the mesh size of the polymer film in which the particles are embedded needs to remain a few times smaller than the particle size. Second, as the particles get smaller the capillary pressure inside the particles (when in the molten state) increases. The smallest particles we have stretched were PS particles with a diameter of 75 nm, but we never explored the lower limits.

Jan Vermant opened discussion of the paper by Estefania Gonzalez Solveyra: In your paper you describe an equilibrium calculation that shows curvature; there may also be a kinetic effect as the flux near the regions of higher curvature is expected to be higher compared to the sides. Is this going to affect the distribution as well?

Estefania Gonzalez Solveyra answered: Kinetic effects may alter the distribution of the polymers before equilibrium is reached; however, for very long times, the equilibrium results predicted by our theory should hold. The molecular theory we derived in this work explicitly considers the geometry of the system, the shape, molecular volume and conformations of all molecular species and the repulsive and attractive interactions between them. The theory employed assumes an equilibrium adsorption process dictated by the equality of the chemical potential of the polymers on the surface and in solution. As such, the kinetic details of the process, as the different fluxes as a function of curvature mentioned in the question, are out of reach of the theoretical approach as it is. However, the theory can be modified to study the kinetics of the polymer adsorption process, as has been done previously in our group for analyzing the kinetics of the protein adsorption on planar surfaces.[1,2] Future work encompasses deriving the molecular theory to study the kinetics of the polymer adsorption process.

1 F. Fang and I. Szleifer, *Biophys. J.*, 2001, **80**, 2568–2589.
2 F. Fang and I. Szleifer, *J. Chem. Phys.*, 2003, **119**, 1053–1065.

Luis Liz-Marzán asked: I would like your opinion about how simplified your model is. I looked at your main parameters to predict the adsorption of different polymers and I missed the differences in energies between lateral and tip facets, can you expand on this?

Estefania Gonzalez Solveyra responded: This is a good observation. In order to isolate the interplay between the curvature and molecular architecture, we employed the same adsorption energy for spheres and cylinders. However, the conformations of the polymers change depending on the surface curvature and this impacts the polymer adsorption process even if the adsorption energy is the same regardless of the curvature. In the theoretical approach used in this work, we account for these differences in polymer conformations by providing a very large number of possible conformations as input. Solving the theory then assigns a probability for each possible conformation, which is different depending on how linear and branched chains share the available volume near the surface, modulated by the curvature of the surface. Particularly, we were interested in studying an alternative route to creating anisotropic functionalized nanorods based on the differential curvature between their cylindrical body and spherical end caps, as opposed to previous strategies reported in the literature and also discussed at this conference based on distinct crystallographic faceting on the nanorod surface (DOI: 10.1039/C6FD00057F). We are aware that differences in the adsorption energies between the lateral and tip facets exist and will incorporate them in future systems that simultaneously combine surfaces of different curvatures in two dimensions (as opposed to the 1D calculations performed in this work).

Catherine Murphy commented: There have been quite a few experiments on the surface coverage of ligands on spherical nanoparticles of different sizes. People use the "cone angle" of the ligand as a way to predict, or rationalize, the coverage; branched ligands would have a larger cone angle, so you can fit fewer of them on a surface depending on the curvature *etc.* It seems like your calculations give similar results. How does the cone angle idea map onto your calculations?

Estefania Gonzalez Solveyra answered: Indeed we observe that bulkier branched polymers tend to adsorb less than linear chains of the same molecular weight, as can be seen in Fig. 2a comparing the 35l *vs.* 35-2b polymers for spherical surfaces of different size. The cone angle idea maps well when considering the competitive adsorption from binary polymer solutions. In this case, the competition for the available area to adsorb and the available volume to accommodate the linear and branched chains dictates the adsorption, modulated by the curvature and radius of the surface, as reflected in Fig. 3. The "cone angle" argument qualitatively explains our results. Note, however, that such argument is based on the packing of rigid cones, while the ligands in our calculations are soft due to their conformational degrees of freedom. Therefore the cone model packing argument is only a very approximate guide to explain our calculations.

Joerg Lahann commented: You have both a branched and linear polymer which are different molecular weights, do they have an effect, could you look at other sizes? You could have a linear polymer that may not take on a linear conformation, but may form a coiled structure, which will alter the cone occupied by the polymer.

Estefania Gonzalez Solveyra replied: Regarding the effect of the molecular weight of the polymers in the mixture, our calculations show that this parameter of the polymers plays a crucial role in their adsorption. This can be seen, for example, in Fig. 2 when comparing the adsorption of the linear polymers with 20 and 35 monomers in good solvent conditions (35l *vs.* 20l). The molecular theory we derived in this work explicitly considers the molecular volume, weight and conformations of the polymers, so other combinations can be studied, and we have done so in this work when analyzing the effect of the branching and the backbone length of the chains.

Now, addressing the coiled *vs.* stretched conformations of the chains, we first should mention that we do not fix the conformation of the chain, but rather provide the theory with a very large number of possible conformations (which include extended and coiled ones among many others). The theory then assigns a probability to each possible conformation. In our work, we observed that the resulting molecular organization of the chains is one of the main factors determining the adsorption. It is a matter of how linear and branched chains share the available volume near the surface, modulated by the curvature of the surface. In this regard, given the high values of the total polymer surface density (that is, adding the contribution from the linear and the branched chains), both the linear and branched chains stretch towards regions of greater available volume further away from the surface, resulting in polymer molecules that form brushes in all the mixtures and conditions studied. This can be seen in the profiles of the volume fraction of the polymers as a function of the distance to the surface (see Fig. 3a in the main text or Fig. S1b in the supplementary information). The coiled configurations have a very low probability of occurring in these conditions, for both the linear and the branched chains. We do observe that the linear and shorter chains organize in such a way as to maximize the available space for the branched and more bulky polymers on the surface. Particularly, we found that the optimum conditions for adsorption and spatial partition are obtained when competition for the available volume occurs between the linear chain and a region of the linear backbone of the branched chain, both in a rather stretched configuration.

Matthias Pauly said: How would you deal with polymer mixtures that inherently phase-separate on a flat surface? Would you expect a different composition of the adsorbed layer between the tips and along the cylindrical body of the nanorod?

Estefania Gonzalez Solveyra answered: Regarding phase separation, we are considering 1D systems that are laterally homogeneous, so we do not see that phenomenon with the current model. However, the theoretical approach can be extended to study phase separation, and it has been done in the past for grafted polymers on planar surfaces,[1] cylindrical pores,[2] and nanoparticles.[3] But even for homogeneous surface mixtures, we do observe differences in the surface composition between the tips and the body of the NRs, as reflected in Fig. 4 and Fig. 11b for good and poor solvent conditions, respectively.

1 M. Tagliazucchi, M. Olvera de la Cruz and I. Szleifer, *Proc. Natl. Acad. Sci. U. S. A.*, 2010, **107**, 5300–5305.
2 O. Peleg, M. Tagliazucchi, M. Kröger, Y. Rabin and I. Szleifer, *ACS Nano*, 2011, **5**, 4737–4747.
3 M. Tagliazucchi, M. G. Blaber, G. C. Schatz, E. A. Weiss and I. Szleifer, *ACS Nano*, 2012, **6**, 8397–8406.

Eugenia Kumacheva remarked: How do you know that polymer molecules are forming brushes, instead of having a mushroom configuration?

Estefania Gonzalez Solveyra responded: You pose an interesting question. Indeed we observe that all systems studied in this work corresponded to the brush regime. We can state this based on the volume fraction profiles we obtain for the polymers in the different mixtures. As can be seen in Fig. 3a in the main text or in Fig. S1b in the supplementary information, polymer chains are stretched and expanded towards the solution, even for the systems in poor solvent conditions (although this data is not shown). This is due to the high values of the total surface coverage obtained by summing the contributions for the linear and the branched polymers (Fig. 3a, 5b, S1a, S3b and S9b). In order to lower steric repulsions within the polymer layer, the system stretches both the linear and branched chains towards regions of greater available volume further away from the surface. It is worth noting that changing the composition of the solution or the molecular architecture of the polymers in the mixture results in changes in adsorption that are compensated between the linear and the branched polymer chains, rendering the total polymer surface density rather insensitive to these parameters. This ultimately results in polymer molecules forming brushes in all the mixtures and conditions studied. The high surface coverage of the polymers results from our choice of the attraction energy between the surface and the anchor group in the polymer. We chose this attraction energy in order to reproduce previous experimental results.

Eugenia Kumacheva continued: I would expect linear molecules to adsorb first, then branched polymers would arrive at the surface and replace the linear molecules. What does your model predict will happen?

Estefania Gonzalez Solveyra replied: The theoretical results of this work were obtained through employing a molecular theory based on minimizing the thermodynamic potential of the free energy functional of the system. This means that our results correspond to the limit of thermodynamic equilibrium. In other words, our calculations predict the composition of the system at very long times. The kinetic details of the process, as the different time scales referred to in the question, could not be studied within our framework. We are currently working on expanding our theoretical framework to study the kinetics of the process, as has been done previously in our group for analyzing the kinetics of the protein adsorption of planar surfaces.[1,2]

1 F. Fang and I. Szleifer, *Biophys. J.*, 2001, **80**, 2568–2589.
2 F. Fang and I. Szleifer, *J. Chem. Phys.*, 2003, **119**, 1053–1065.

Yu Zhou questioned: You discuss simulations for the nanospheres with positive curvature, can you have a look at nanowells with negative curvature and compare the polymer density of the two structures?

Estefania Gonzalez Solveyra responded: This is a very good question. Our results show that a complex balance of forces modulated by the surface curvature and polymer architecture dominates the adsorption process. The available surface to adsorb and the available volume to accommodate the chains in play a crucial role. For surfaces with positive curvature (spheres and cylinders in this work), the available volume increases as one moves away from the surface. However, in nanowells or nanopores, the opposite occurs. Based on this, we expect that surfaces with negative curvature such as nanowells and nanopores will have an important effect favoring the adsorption of the linear chains *vs.* the branched more bulkier ones. Previous work from our group extended the theoretical framework employed in this work to study the interplay between the particle shape and local electrostatics in nano-objects with positive and negative curvatures (nanorods and nano-dumbbells, respectively) functionalized with acidic ligands.[1]

1 D. A. Walker, E. K. Leitsch, R. J. Nap, I. Szleiferand B. A. Grzybowski, *Nat. Nanotech.*, 2013, **8**, 676–681.

Yu Zhou remarked: Does the length of the polymer have a role in the coverage?

Estefania Gonzalez Solveyra answered: Our calculations show that the molecular weight of the polymers plays a crucial role in their adsorption. For example, in Fig. 2 when comparing the adsorption of the linear polymers with 20 and 35 monomers in good solvent conditions, we observe that increasing the total number of segments leads to an increase in the steric repulsion within the adsorbed layer and, therefore, to a decrease in the adsorbed density (35l *vs.* 20l). This effect can also be seen in the competitive adsorption from binary polymer solutions shown in Fig. 3a, although in this case the molecular architecture also plays a role. It is worth noting that in poor solvent conditions the analysis is not so simple, because increasing the number of monomers in the chains leads to stronger polymer–polymer attractions that compete with the steric repulsions.

Alberto Striolo said: The results you presented remind me of the work conducted by Francesco Stellacci in collaboration with Sharon Glotzer a few years back for short polymers of different lengths grafted on a spherical particle.[1] There is also a poster outside by Paola Posocco in which short surfactants of different lengths were simulated on a surface. The simulations I am referring to show that a phase segregation can occur, sometimes yielding stripes enriched with the short (or long) compound. I was wondering whether the theory you have implemented for your paper is able to describe these phase separations, which seem to be due to entropy rather than enthalpy. If you are able to observe these phenomena, I would expect stripes of the branched polymers and stripes of the linear polymers, both oriented parallel to the main axis of the cylindrical objects you considered. Can you comment on this?

1 C. Singh, P. K. Ghorai, M. A. Horsch, A. M. Jackson, R. G. Larson, F. Stellacci and S. C. Glotzer, *Phys. Rev. Lett.*, 2007, **99**, 226106.

Estefania Gonzalez Solveyra responded: In the present work, we are not able to observe phase segregation, since we are studying 1D systems, assuming lateral homogeneity of the system and considering inhomogeneities only in the radial direction. However, the theoretical approach can be extended to 2D and 3D systems to study phase separation and it has been done in the past for grafted polymers on planar surfaces[1] and in cylindrical pores.[2]

1 M. Tagliazucchi, M. Olvera de la Cruz and I. Szleifer, *Proc. Natl. Acad. Sci. U. S. A.*, 2010, **107**, 5300–5305.
2 O. Peleg, M. Tagliazucchi, M. Kröger, Y. Rabin and I. Szleifer, *ACS Nano*, 2011, **5**, 4737–4747.

Joerg Lahann remarked: This is a comment about the packing densities on your particles. Once you transition into the mushroom regime, do you expect to observe phase segregation or would you expect more of a mixed monolayer?

Estefania Gonzalez Solveyra replied: Based on the conditions in which we performed our calculations, we do not see a transition into the mushroom regime. Our results show that all systems studied in this work corresponded to the brush regime, as can be confirmed by the volume fraction profiles in Fig. 3a in the main text or in Fig. S1b in the supplementary information. Changing the composition of the solution or the molecular architecture of the polymers results in changes in adsorption that are compensated between the linear and the branched polymer chains, with the total polymer surface density changing slightly. This determines that the polymer molecules form brushes in all the mixtures and conditions studied, without a transition to the mushroom regime.

Regarding the observation of phase separation, such a phenomenon cannot be analyzed with the current implementation of the molecular theory, since we are studying 1D systems, assuming lateral homogeneity of the system and considering inhomogeneities only in the radial direction. Within our theoretical framework, we observe that the monolayer consists of a homogeneous mixture of linear and branched polymers.

Hedi Mattoussi commented: Is the arrangement of the polymer on the surface of the nanoparticle controlled by the solvent quality? When considering polymer mixing do you use the Flory–Huggins theory to help you decide in what regime you need to work?

Estefania Gonzalez Solveyra responded: Indeed, solvent quality affects the surface density and the composition of the surface mixture, as can be seen in Fig. 11a and b, respectively. However, it has little effect on the partition between spherical and cylindrical surfaces (Fig. 11c) and on the branching and backbone effects, as compared to the good solvent conditions (Fig. S9). In this work, we incorporated the solvent quality into our calculations by adding a Van der Waals effective attractive energy between polymer segments, without resorting to the Flory–Huggins theory of polymer mixing. The parameter we changed was the value of χ_c/χ, representing a proxy for the reduced temperature of the system; the polymer is in a good solvent for values greater than 1 and in a poor solvent for values lower than 1. This parameter results from normalizing the strength of segment–segment attractions, χ, by the critical value χ_c, which is defined as the value of χ required to trigger microphase separation for a neutral, planar polymer brush in the limit of vanishing surface coverage.[1] Using this approach, we analyzed the effect of solvent quality on the adsorption from polymer binary mixtures considering different values for χ_c/χ. We found that mildly poor solvent conditions could allow for increasing the amount of adsorbed branched polymer, without drastically decreasing the sphere/cylinder partition.

1 O. Peleg, M. Tagliazucchi, M. Kröger, Y. Rabin and I. Szleifer, *ACS Nano*, 2011, **5**, 4737–4747.

Hedi Mattoussi said: Do you also assume that the polymer grafting on the nanoparticle surfaces is stable or it will occasionally come off? Dynamic desorption of the capping ligands (here the polymer) is an integral and unavoidable property of these systems.

Estefania Gonzalez Solveyra answered: The starting point assumption of the theory employed in this work is that the polymers on the surface are in thermodynamic equilibrium with a polymer bulk solution. This is applied when minimizing the thermodynamic potential of the free energy functional in order to study the thermodynamic behavior of the system considering that the chemical potential of the linear or branched polymer molecules on the surface is the same as in the bulk solution, as required for thermodynamic equilibrium. In thermodynamic equilibrium, the rate of desorption of polymers from the surface is equal to the rate of adsorption of polymers from solution to the surface. As such, the amount of adsorbed polymers is stable and does not vary with time. If the Au-NRs are transferred to a polymer-free solution, ligand desorption will occur. We cannot predict the rate of the process with our current theory, but we plan to modify our theoretical framework to study kinetic effects such as the dynamic desorption of the polymers. Previous calculations of this sort have been performed before in our group when studying the kinetics of the protein adsorption of planar surfaces.[1,2] Developing the molecular theory required to study the kinetics of the adsorption process is one of the future steps we are taking.

1 F. Fang and I. Szleifer, *Biophys. J.*, 2001, **80**, 2568–2589.
2 F. Fang and I. Szleifer, *J. Chem. Phys.*, 2003, **119**, 1053–1065.

Catherine Murphy opened discussion of the paper by Matthias Pauly: I am thinking of Onsager's papers. He has a paper from long ago (1940's) discussing orientational order in concentrated solutions of rods – he was thinking of virus particles at the time. He developed a quantitative relationship between the aspect ratio of the rods and the concentration at which you need to have a solution in order to organize the rods into one dimensional order. I notice in your results that the longer the rods/wires are the more order they have. Does this mean that in the spray solution you have more order of the longer wires because they need a lower concentration to order and that the concentration is below the Onsager prediction for the shorter rods? Or, in the process of drying on the substrate, the concentration increases enough to form ordered structures?

Matthias Pauly responded: In his paper from 1949 (see **ref. 1** below), Onsager investigates the isotropic–nematic phase transition of concentrated suspensions of nanorods. However, the concentration at which this transition happens is much higher than the one we are using: the silver nanowire suspension has a concentration of 0.1 mM, *i.e.* around 0.011 mg mL^{-1}. The resulting average nanowire concentration in a 10 μm droplet is around 10^{-8}, which means that statistically there is only one nanowire per droplet, and most of the droplets do not contain any nanowires. Furthermore, the time scales at play in the spraying process itself mean that the system is far from equilibrium, and it is unlikely that such a phase transition will happen in the nanowire suspension while spraying. The drying of the suspension on the substrate is a different issue: it is indeed possible that due to the fast evaporation of the suspension upon drying, the concentration suddenly increases above the nematic critical concentration predicted by Onsager. It is also possible that the evaporation-induced temperature decrease could shift the transition to a different concentration. However, the substrate is rinsed after spraying (while keeping the substrate wet during transfer), and finally it is dried using N_2 flow. The kinetics of drying using such a procedure makes it very unlikely that the ordering seen on the microscopy pictures is due to such an isotropic–nematic phase transition during evaporation.

1 L. Onsager, *Ann. N. Y. Acad. Sci.*, 1949, **51**, 627–659.

Christian Kuttner asked: You have mentioned that for the spray-assisted self-assembly it is important to have a sticky surface. How do you control the stickiness/adhesion? I presume that the self-assembly at the surface follows a hit-and-stick mechanism. Is this correct and what is the maximum surface coverage (jamming limit) that can be obtained?

Matthias Pauly replied: Indeed, it is important to ensure that there is a strong and fast attractive interaction between the nano-objects and the substrate in order to obtain monolayer thin films with a significant density in a reasonable spraying time. This is actually a general requirement for layer-by-layer assembly, there must always be an attractive interaction between the components in solution/suspension and the surface, this holds up for any layer in a given deposition sequence. Using Poly(ethylene imine) (PEI) as a first (and adhesion promoting) layer is very practical as PEI adheres well on many different surfaces and because most LbL-components also adhere well on a layer of PEI. The PVP-coated nano-objects interact strongly with the PEI layer and allow a monolayer to be built within a few minutes. This is also why we have to functionalize the gold nanorods with PVP: we show in the supplementary material of this paper with Quartz Crystal Microbalance (QCM) measurements that the free CTAB present in the gold nanorod suspensions before ligand exchange adsorbs quickly and irreversibly to the substrate and thus hinders the adsorption of the nanorods. This problem is suppressed after exchanging CTAB with PVP. We have studied the variation of density as a function of spraying time and the density increases quickly for short spraying times before reaching a plateau. We believe that the nanowires adsorb on the substrate at a random position and this creates a kind of "excluded area" around the nanowire on the substrate. While the spraying process is continued there is less and less "free space" for new nanowires to interact with the substrate, thus decreasing the deposition rate until a plateau is reached, which corresponds to the jamming limit. The maximum coverage that can be obtained depends on the dimensions of the nanowires and the quality of the ordering (a well-oriented array will have a higher jamming limit). For silver nanowires sprayed under the conditions described in the paper, the surface density is above 40% after spraying for 1 minute and the maximum density of 60–65% is attained after 5 minutes.

Christian Kuttner commented: Fig. 2 shows that some nanowires exhibit full contact with the surface while others lie over preassembled ones in a bridging fashion. As the latter exhibit a smaller contact area with the surface one could expect that they also possess a weaker attachment to the substrate. Do you expect that to some extent such nanowires are removed during longer processing times and become substituted by more robust attached ones, similar to natural selection?

Matthias Pauly responded: Actually, very few nanowires are bridging ones already there and the large majority are in contact with the substrate. This is related to the previous comment, we functionalize our substrate to ensure a strong binding affinity towards the nano-objects, while the functionalization of the nanorods/nanowires ensures a stable suspension, *i.e.* that the nano-objects are repelling each other. Besides this chemical functionalization, thinking strictly geometrical, one would expect that bridging nanowires have a lower contact area to the substrate and thus a weaker attachment (although silver nanowires have a certain flexibility). The scenario you propose is very likely to happen, and I would even go further and speculate that the "natural selection" can also be applied to the orientation: weakly bound nanowires, but also mis-oriented ones, can be desorbed and replaced with a nanowire with stronger binding and better orientation, thus leading to an increase in the average orientation and in the jamming limit. As you can see, the oriented adsorption process should be governed by the interplay between the attractive forces between the nano-objects and the substrate, which can be tuned by surface functionalization, the colloidal stability in suspension and the mechanical shear exerted by the flowing liquid. Increasing the ordering degree would require that the nanowires are still able to move while adsorbed, or that they can be easily replaced by a better oriented nanowire, but on the other had one wants to ensure a binding probability that is strong enough to obtain thin films that reach a significant density within a few minutes.

Antonio Stocco asked: What are the spatial and temporal distributions of directions in grazing incidence spraying? Does an upper limiting value of the nematic order parameter (0.9?) exist for grazing incident spraying?

Matthias Pauly replied: The spraying process is very stable over time, this is not a problem we are facing. On the contrary, it is perfectly true that there is a spatial distribution of the orientation direction: the spray cone of the nozzle we use here has an opening angle of around 15°, and thus the average direction of orientation, which is parallel to the flow lines on the substrate, has a certain spread at macroscopic distances (the orientation varies over a few degrees when moving 1 cm away from the central axis). However, this distribution is not relevant regarding the calculation of the nematic order parameter that is done from electron microscopy pictures, *i.e.* on a scale of a few tens of micrometers. At such a mesoscale, it is mainly the local inhomogeneities arising from local flow turbulence, defects in the nanowires themselves (some of them are kinked for instance) or the nanowire flexibility that limit the quality of ordering. There is in theory no reason why a higher nematic order parameter could not be reached with a set of "ideal" nano-objects, but in practice all the limitations mentioned above make it apparently difficult to obtain a thin film with a nematic order parameter higher than 0.9.

Tobias A. F. König commented: My question is pointing towards the application of this research for electronic devices. In your paper you have shown an anisotropic observation in sheet resistance. The lower sheet resistance is competitive with indium tin oxide (ITO); however, the resistance perpendicular to the wires is only 5 times larger. How can this anisotropic sheet resistance be improved?

Matthias Pauly responded: Silver nanowires are widely explored in academic research for their ability to form transparent, conductive and flexible substrates, which is important for flexible electronics, as ITO is too brittle to sustain multiple deformations. We are working on those properties, especially by making multilayer structures that allow for a low sheet resistance while maintaining high transparency. In such conductive nanowire arrays, the number and quality of Ag NW/Ag NW junctions dictates the sheet resistance. As shown in the present paper, a monolayer thin film has a conductance more than 8 times higher along the orientation direction compared to the perpendicular one. This anisotropy arises from the differences in the respective percolation pathways, which correspond to the difference in the number of junctions that are necessary to bridge a macroscopic distance in one or the other direction. Increasing the degree of orientation would definitely increase this anisotropy in the sheet resistance. It is also possible to decrease the average density, as the sheet resistance should decrease faster in the direction perpendicular to the orientation compared to the parallel one. But this would come with an overall reduction of the sheet resistance. Finally, the nanowires used here are pretty short (5.7 μm), and working with nanowires of higher aspect ratio would also increase the electronic transport anisotropy (Ag NWs can be synthesized with lengths greater than 20 μm).

Yu Zhou asked: Does the evaporation speed affect the alignment of the nanorods and nanowires?

Matthias Pauly responded: As discussed previously, we believe that the adsorption of nanorods/nanowires on the substrate happens mostly at the liquid/solid interface while spraying and that the evaporation does not play a big role. Our process is in that sense very different from another spray-induced orientation technique presented recently in the literature, in which the authors work with volatile solvents that are sprayed on a heated substrate to quickly evaporate the solvent.[1] In our case, the substrate stays wet during the entire spraying process and drying is only done after rinsing with a gas flow. The solvent has some influence on the colloidal stability of the nanoparticles in suspension, and this also determines the equilibrium between nano-objects in suspension and deposited in the substrate, and thus the deposition rate, but it is rather an effect of solvent polarity than evaporation speed.

1 O. Assad, A. M. Leshansky, B. Wang, T. Stelzner, S. Christiansen and H. Haick, *ACS Nano*, 2012, **6**, 4702–4712.

Christoph Hanske enquired: You mentioned that it is possible to build up multilayers with this spray coating technique. What kind of optical response would you expect from a thick particle stack if the orientation of the substrate was changed gradually from layer to layer?

Matthias Pauly answered: The ability to build multilayer stacks of nano-objects on large areas is a huge advantage of the grazing incidence spraying approach. Furthermore, the composition, density and orientation of the nano-object can be chosen almost independently for each layer. It is thus possible, for instance, to prepare multilayer films of silver nanowires or gold nanorods in which the nanowires are oriented perpendicularly in each new layer. Another interesting structure, as you mention, is a multilayer thin film in which the orientation direction is regularly changed from one layer to the other. This leads to a helical arrangement of nanowires/nanorods, and right-handed or left-handed helices can be made. Such chiral assemblies of plasmonic particles have been shown to exhibit strong circular dichroism due to the coupling of surface plasmons arising from metallic particles arranged in a chiral architecture.

Jan Vermant asked: Where and when is the orientation induced? Is the elongation flow in the nozzle important or is the orientation induced during the shear flow in the thin film? If the latter is the case do you need the spraying mechanism or would a more classical coating technique work as well?

Matthias Pauly responded: The place at which the shear force is highest is within the nozzle during the droplet production. The characteristic time for a nanowire to diffuse and loose its orientation is much longer that the flight time of a droplet between the nozzle and the substrate. It is thus possible that the nanowires keep the orientation they have acquired in the nozzle on the substrate. However, not only diffusion plays a role: the fluid dynamics of a flying droplet are very complex and other mass transport mechanisms within a single droplet must be taken into account. Furthermore, the droplets landing on the surface coalesce on the substrate to form a flowing liquid film and there again the orientation may be lost. This is why we think that the shear force within the flowing liquid on the substrate is the main force directing the orientation. As discussed previously, this is also evidenced from the fact that the local orientation direction (the "nematic director") follows the flow lines and the pattern induced by the cone opening on the substrate. Actually, the use of shear forces in thin liquid films has already been applied in a variety of techniques, not only in spraying. In particular, it is well known that nanowires adopt a preferential radial orientation within spin-coated films, as the solvent is ejected radially from the substrate during the spinning process. For more concentrated suspensions, *i.e.* closer to the gelation point or to the critical concentration that would lead to a spontaneous nematic ordering for example, much weaker shear forces can be used (this is speculation, we do not know the values!), for instance with doctor-blading. Such orientation has also been reported within microfluidic channels, or even within the glass capillary of Pasteur pipettes!

Jongwook Kim said: Your spray droplet size is only several microns and the length of your nanowires is comparable to this. The orientation mechanism might be that, when a droplet is landing, the tip of a nanowire sticks on the substrate surface and the other side lies toward the jet direction due to the mechanical torque, not from the shear. What do you think about this interpretation?

Matthias Pauly replied: We have measured the droplet size using phase Doppler interferometry, and the spray nozzles we use produce droplets with quite a broad size distribution centred around 10–20 μm. It is thus true, as you point out, that the longest silver nanowires and the smallest droplets have the same size (gold nanorods, on the contrary, are much smaller). It is thus possible that the droplet is deformed during its flight due to the presence of the nanowire, and possibly one end of the nanowire might "look outside" the droplet. But remember that these droplets are actually "feeding" a thin liquid film flowing on the substrate, and this liquid film is much thicker (probably a few hundreds of micrometers). The nanowire has then to diffuse or be transported through local turbulence to the solid/liquid interface, where the adsorption process occurs. Thus, I still think that shear forces within the liquid film play an important role. However, the scenario you propose can be envisioned at the solid/liquid interface as well: the nanowires may stick with one end first and then be oriented and brought parallel to the substrate through the mechanical torque.

Andrew Harvie commented: It seems it is not entirely clear which effect dominates the ordering of the nanowires on the substrate; whether it is that the order of the wires as they exit the nozzle is transferred to the substrate, or whether it is the continuous washing of liquid over the substrate that orientates the adsorbed particles. Would it be possible to conduct an experiment where the time of flight of the spray is compared with the order parameter on the surface? Ideally the velocity of the spray at the time of contact with the surface would be kept constant. This experiment may help you to comment on the mechanism of the ordering of the wires, given that after longer flight times the wires will tend to be less ordered in suspension.

Matthias Pauly responded: The experiment you suggest would be very interesting to understand better which forces are at play and over which time scale the different phenomena that can influence the orientation take place. Unfortunately, it is practically impossible to change independently the distance over which the droplets are transported in the gas flow after having left the nozzle and before reaching the liquid film on the substrate while keeping the same droplet velocity and density. The spray droplets exit the nozzle at a speed of more than 50 m s^{-1}, but this speed drastically and quickly falls below 10 m s^{-1}, and thus it is not easily possible to work at a constant droplet speed. One way to keep the droplet speed constant would be to increase the gas flow as the substrate is moved away from the nozzle to ensure that the droplet speed at the point of impact stays the same, but changing the gas and/or liquid flow also changes the speed and size of the droplets, and thus the whole process would be modified. Furthermore, keep in mind that the nozzle is held at 1–2 cm away from the substrate, *i.e.* that the time of flight for a droplet is typically less than 1 μs, whereas the typical time scales for rotational diffusion are much longer.

Joerg Lahann remarked: Can you not directly visualize the spray jet with a high speed camera?

Matthias Pauly replied: We haven't tried that yet, and it is probably possible to visualize the droplets. They are big enough to be seen with a good camera, the challenge is probably that the cloud of droplets is very dense at the exit of the nozzle.

Antoine Thill asked: Would it be possible to use drop-on-demand systems to explore the interactions of the droplets and the rods with the support?

Matthias Pauly responded: Yes, this would also be an interesting experiment to conduct, but the droplets produced by a drop-on-demand system are ejected from the nozzle at low velocity thanks to a piezoelectric or thermal stimulus. The velocity of the droplet would be much lower than for droplets produced using grazing incidence spraying with a two-fluid pneumatic nozzle. Furthermore, the liquid flow rate of drop-on-demand systems is much lower than for spraying, and the droplet quickly evaporates on the surface, thus the entire surface would not be wetted by the nanowire suspension, which is another major difference with the process we introduce in this paper.

Javier Reguera opened discussion of the paper by Antoine Thill: Could you comment on how this kind of Janus configuration, as compared with an external Janus surface, can help to stabilize the emulsions?

Antoine Thill responded: The situation is different for OPA-imo (external hydrophobic/internal hydrophilic) and imo-CH_3 (external hydrophilic and external hydrophobic). OPA-imo attach strongly to the interface and form a solid shell which prevents contact between the droplets. It is thus acting in a similar way compared to a normal Janus configuration. The only puzzling question is that in the simplest configuration, if a nanotube is connected to two different droplets, then there will be continuity of the dispersed phase through the inside of the nanotube. We believe that this configuration is possible with such a system. In the case of imo-CH_3, the stabilization is not so good and comes from, at least in part, a modification of the viscosity of the continuous phase. Therefore, it is very different from standard Janus configurations.

Antonio Stocco commented: What liquids are inside the tubes before adsorption onto the fluid interface? Can you expect a defined immersion depth of the tube at the interface also due to the different contact angles of the external and internal tube surfaces?

Antoine Thill replied: For OPA-imo the nanotube is initially filled with water inside. In the case of imo-CH_3, we don't really know what is inside the nanotube initially. We just know that the electronic density of the internal cavity is 0.1 e^- $Å^{-3}$, which is one third of the bulk water electronic density. I have always though that the partial immersion in water for OPA-imo or in oil for imo-CH_3 would always be unfavourable. I would be curious to see theoretical arguments related to this matter. The situation is rather complex because three surfaces could be defined, external/internal and the borders. The thickness of the nanotube wall is about 6 Å therefore it can hardly be defined a surface. Your question is really interesting and it also deserves more experiments to have more direct ideas about the configuration of the nanotubes at the oil/water interface.

Martin Buzza asked: How surface active are these nanotubes? Given that their diameter is around 1 nm, their desorption energy is only a few kT and naively I would have said that they should easily detach from the interface. Do you know why they attach so strongly to the interface?

Antoine Thill answered: The situation is different for OPA-imo and imo-CH_3. OPA-imo attach strongly at the interface between water and oil (on the oil side). Upon adsorption it forms a solid shell at the interface. Experiments were performed with the injection of a toluene droplet containing OPA-imo in water. After some time, if the toluene is sucked back, the interface shrinks like a balloon showing the presence of a new solid material at the interface. This is not the case for imo-CH_3. It seems that in this case you are right, the attachment is not as strong and the nanotube can detach from the interface. The stabilization of the emulsion in this case comes from the strong increase of the continuous phase viscosity after prolonged contact with the oil phase.

Luis Liz-Marzán said: You mentioned an old paper of mine, related to the synthesis of metal nanoparticles in the presence of imogolite.[1] When I started the experiments that resulted in that paper, the objective was the growth of a silver nanowire inside the imogolite tube. I always thought it could not work, but this is actually getting done in carbon nanotubes. Do you see a possibility to get this to work with your selective functionalization strategy?

1 L. M. Liz-Marzán and A. P. Philipse, *J. Phys. Chem.*, 1995, **99**, 15120–15128.

Antoine Thill responded: You were right concerning your doubts. With normal imogolite the inside of the nanotube is barely accessible. With the imogolite covered on the inside surface with Si–CH_3 groups, we have shown that it is possible to have small molecules enter the nanotube.[1] I think it should be possible to find proper strategies to have metal precursors entering the nanotubes. We will try such things in the near future.

1 M. S. Amara, E. Paineau, S. Rouzière, B. Guiose, M.-E. M. Krapf, O. Taché, P. Launois and A. Thill, *Chem. Mater.*, 2015, **27**, 1488–1494.

Joerg Lahann asked: Since we are talking about the chemistry, I do not understand why TMMS is reacting with the aluminium surface?

Antoine Thill answered: The synthesis is performed in water at low pH. TMMS (trimethoxymethylsilane) is hydrolyzed under this condition. OH groups replace the methoxy ligand. These OH groups can perform nucleophilic attack at the surface of aluminium oxide creating Si–O–Al bonds. In the case of imogolite, three such bonds are formed per silicon atom.

Lucia Pasquato said: If you move towards nanofluidics, is it possible to use a fluorinated silane in order to have a fluorinated (lipophobic) surface inside the cavity to avoid friction?

Antoine Thill responded: This is an excellent idea. I have not yet tried to synthesize such nanotubes. I am only aware of a PhD that has been dedicated partly to this goal. To my knowledge it has not been successful. I will probably look into this direction.

Christoph Hanske asked: The size distributions of the nanotubes shown in your paper are rather broad. Is it possible to control their average length and dispersity during the synthesis?

Antoine Thill replied: The dispersion in length is broad and comes from the growth mechanism of the nanotubes.[1] It is possible to reduce or increase the average length of the nanotubes but still have a very large polydispersity. Reducing this poydispersity would probably require to imagine some sorting methods. For example, some Japanese researchers have proposed that the use of the isotrope/nematic phase separation could be used to select the longest nanotubes.

1 P. Maillet, C. Levard, O. Spalla, A. Masion, J. Rose and A. Thill, *Phys. Chem. Chem. Phys.*, 2011, **13**, 2682–2689.

Christian Kuttner continued discussion of the paper by Matthias Pauly: For the spray-assisted self-assembly a shallow spraying angle of 15 to 20° was used. Is this the optimal angle and what happens if you go even shallower? Fig. 2 shows that some nanowires are slightly bent after spraying. Is this bending related to the angle at which the particles impact the surface?

Matthias Pauly answered: The spraying angle has been systematically varied and the orientation is optimum for a spraying angle of 10–20°. Interestingly, the order parameter is not significantly reduced when the spraying angle is varied between 5 and 60°, which is an advantage as the angle does not need to be finely optimized for each system. For angles higher than 60°, the velocity of the flowing liquid film is reduced, which results in a lower orientation, whereas most of the suspension does not hit the substrate when spraying is applied at a very low angle.

Christian Kuttner asked: Does humidity affect the spray-assisted self-assembly and could pre-wetting of the substrate influence the self-assembly process?

Matthias Pauly replied: We have never investigated such effects specifically, so it is hard to give a definite answer. Humidity will have an influence on the evaporation speed of the droplets during their flight, and thus may slightly change the droplet characteristics upon landing on the substrate, but I don't expect this to have a major influence. Pre-wetting of the substrate may also have an influence, as it is known that some polymeric layers change their conformation depending on humidity or to which liquid phase they are exposed to. It is thus possible that the dry PEI layer somehow changes when exposed to the suspension, and that this may slightly modify the adsorption process.

Luis Liz-Marzán remarked: This is a comment regarding your way to determine the orientation of the nanorods. Looking at the changes in the plasmon with polarized illumination, it seems that you are analysing the collective plasmon that arises from the coupling of the nanorods. For nanowires this is shifted to the IR, so you may not easily see it. Do you think that this is a fair comparison?

Matthias Pauly responded: It is perfectly correct that the longitudinal plasmonic absorption bands are shifted towards the near infrared when the wires are brought closer in the thin film. The peak of the longitudinal SPR can only be measured for the short aspect ratio gold nanorods (it shifts from 1067 nm for the suspension to 1080 nm for the ordered monolayer), it is outside of the measurable range for the longer gold nanorods and for the silver nanowires. What we are measuring is indeed a collective plasmon mode arising from the coupling of the longitudinal SPR of neighbouring nanorods/nanowires. However, even if it is a collective mode, it is still highly directional, as shown by the strong variation of the extinction with the polarizing angle, and the far-field optical properties can still be used as a probe of the macroscopic effect of a local (mesoscopic) orientation.

Faraday Discussions

PAPER

Surface heterogeneity: a friend or foe of protein adsorption – insights from theoretical simulations

Matthew Penna,[ac] Kamron Ley,[a] Shane Maclaughlin[bc] and Irene Yarovsky*[ac]

Received 15th March 2016, Accepted 8th April 2016
DOI: 10.1039/c6fd00050a

A lack in the detailed understanding of mechanisms through which proteins adsorb or are repelled at various solid/liquid interfaces limits the capacity to rationally design and produce more sophisticated surfaces with controlled protein adsorption in both biomedical and industrial settings. To date there are three main approaches to achieve anti biofouling efficacy, namely chemically adjusting the surface hydrophobicity and introducing various degrees of surface roughness, or a combination of both. More recently, surface nanostructuring has been shown to have an effect on protein adsorption. However, the current resolution of experimental techniques makes it difficult to investigate these three phase systems at the molecular level. In this molecular dynamics study we explore in all-atom detail the adsorption process of one of the most surface active proteins, EAS hydrophobin, known for its versatile ability to self-assemble on both hydrophobic and hydrophilic surfaces forming stable monolayers that facilitate further biofilm growth. We model the adsorption of this protein on organic ligand protected silica surfaces with varying degrees of chemical heterogeneity and roughness, including fully homogenous hydrophobic and hydrophilic surfaces for comparison. We present a detailed characterisation of the functionalised surface structure and dynamics for each of these systems, and the effect the ligands have on interfacial water, the adsorption process and conformational rearrangements of the protein. Results suggest that the ligand arrangement that produces the highest hydrophilic chain mobility and the lack of significant hydrophobic patches shows the most promising anti-fouling efficacy toward hydrophobin. However, the presence on the protein surface of a flexible loop with amphipathic character (the Cys3–Cys4 loop) is seen to facilitate EAS adsorption on all surfaces by enabling the protein to match the surface pattern.

[a]*School of Engineering, RMIT University, GPO Box 2476, Melbourne, VIC, 3001, Australia. E-mail: irene.yarovsky@rmit.edu.au*
[b]*BlueScope Steel Research Laboratories, Port Kembla, NSW, Australia*
[c]*ARC Research Hub for Australian Steel Manufacturing, Australia*

Introduction

Protein adsorption at solid/liquid interfaces is a phenomenon observed across nature and technological processes. The ability to design material surfaces which can prevent, facilitate or, generally, control the adsorption of proteins from solution has applications in diverse fields from medicine[1–3] to large scale industrial coatings.[4,5] Despite experimental studies dating back to the 1960s[6] there is not yet a detailed understanding of the mechanisms through which proteins adsorb or are repelled at various solid/liquid interfaces.[7,8] One reason for this is the difficulty in investigating these three phase systems at the molecular level using experimental techniques.[9] This lack of understanding limits the capacity to employ rational design to produce more sophisticated coatings which can control protein adsorption.[4]

Protein adsorption is dependent on both the morphology and chemistry[10,11] of the surface, both of which can be manipulated through functionalization of either extended surfaces or nanoparticles, including technologically relevant materials such as metals,[12,13] polymers[14,15] and oxides, *e.g.* silica.[16,17] In broad terms proteins are believed to typically adhere more strongly to hydrophobic surfaces,[7] however this may not be a reflection of the total amount of adsorbed proteins, as hydrophilic surfaces allow limited surface induced conformational changes, which typically result in much weaker adherence.[18] To prevent the undesired protein adsorption and avoid biofouling in both biomedical and industrial contexts there are several approaches which can be separated into three main themes, micro/nano roughness and morphology, different chemical functionalities, and a combination of both. Current experimental and theoretical approaches can be found in great detail in recent reviews.[19,20] As with a lot of current technologies, the inspiration for these surfaces comes from natural phenomena. A classic example is the lotus leaf,[21] which combines a super hydrophobic surface with micro/nano scale roughness. These properties have been implemented on a variety of surfaces and materials like silica and glass,[22] which use high surface roughness to lower the strength of adherence, combined with a high water contact angle allowing water to bead along the surface, collecting contaminants. Alternative methods have been drawn from the lady's mantle leaf[21] where common organic chain molecules are chemically grafted to surfaces to substantially reduce the amount of adsorbed protein compared to the base surface.[23,24] Typical characteristics of these protein-resistant chains such as poly(ethylene glycol) (PEG) and poly(2-oxazoline) (POX) are their hydrophilic nature, the presence of hydrogen bond acceptors, few hydrogen bond donors, and charge neutrality (Fig. 1).[19,25] The origin of this resistance has been explained on the basis of hydration theory,[26–28] steric repulsion[23,29,30] or as a combination of the two.[31,32]

In recent years with the emergence of nanotechnology, an alternative approach to altering surface resistance to protein adsorption was proposed, including the manipulation of the nanoscale morphology of chemically identical materials.[33,34] Nanopatterned surfaces of increasing morphological sophistication have been employed to study and exploit the interactions between nanostructured materials and biological systems.[35] An example of such protein resistant materials are monolayer protected nanostructured surfaces.[34] Two component ligand mixtures form monolayers of striped domains on the surface of nanoparticles through

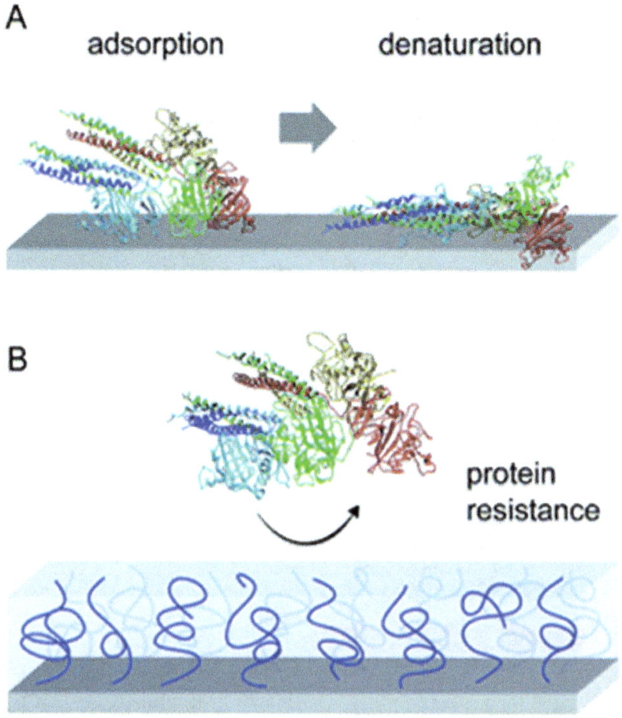

Fig. 1 (A) Dynamic adsorption and denaturation of proteins on a bare surface; (B) protein resistance of polymer-coated surfaces. Reproduced with permission.[19] Copyright 2014, John Wiley and Sons.

entropy driven self-assembly[36,37] with the length scales of the domains (stripes) being 1–2 nm. These nanoparticle surfaces exhibit low protein binding despite having a substantial hydrophobic component[34,38] and counter intuitively show increased protein adsorption with an increased fraction of the hydrophilic functional chains.[39] While the mechanism of this uncorrelated binding affinity remains unclear the adsorption behaviour of proteins has been shown to be significantly influenced by the nano-scale patterning[39,40] and amphipathic amino acid residues on the protein surface have been shown to be able to drive the adsorption on domain separated nanoparticles due to their inherent ability to interact with both the hydrophobic and hydrophilic domains but most importantly, with the domain boundaries.[39] It has been proposed that surfaces can be functionalised to create a functional and morphological heterogeneity (nano-pattern) that will either facilitate or prevent protein adsorption by matching or mismatching the protein surface functionality.[40] Most of these applications involved close-packed self-assembled monolayers (SAMs) on regular (crystal) surfaces or nanoparticles. Here we propose to utilise all-atom Molecular Dynamics (MD) to systematically test the role of surface chemistry and morphology, including nano-patterning that is commensurable with the protein surface features, on protein adsorption to chemically functionalised amorphous silica surfaces using some common functional chains and a model protein

Table 1 Summary of the molecular models representing silica surface functionalization patterns with chemical and morphological heterogeneity

Model nomenclature	Surface pattern schematic[a] with a density of 3.0 chains per nm^2	Morphology	Chemical composition
B[64,72]	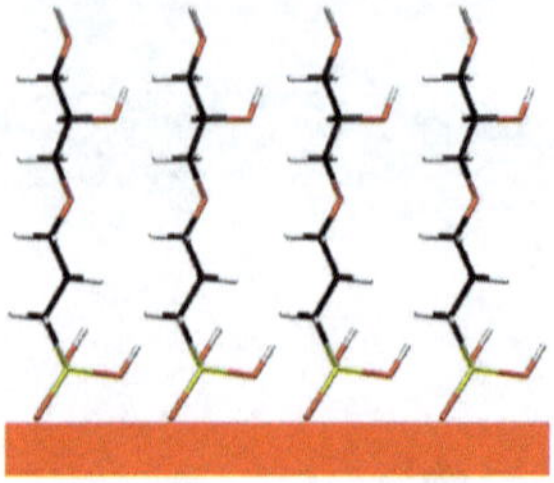	Homogeneous hydrophilic chains	$-Si(OH)_2-(CH_2)_3-O-CH_2-CHOH-CH_2OH$
F	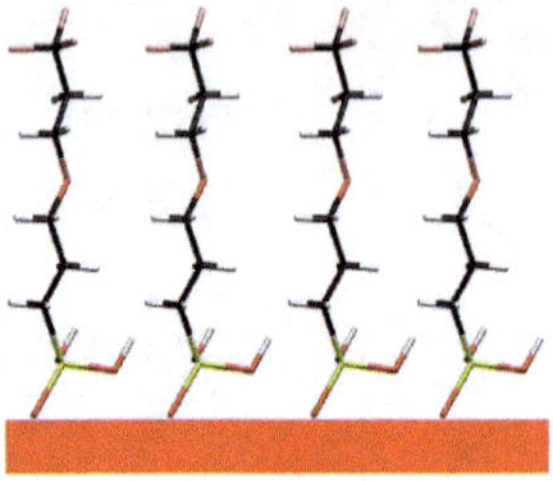	Homogeneous hydrophobic chains	$-Si(OH)_2-(CH_2)_3-O-(CH_2)_2-CF_3$
BF1	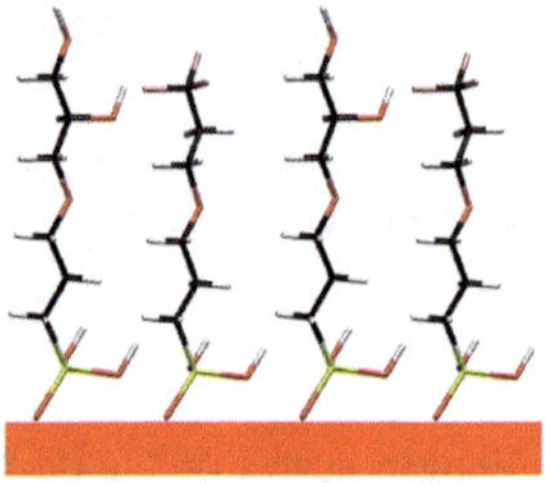	1 × 1 domains of equal length hydrophilic/hydrophobic chains Domain length ~ 6 Å	$-Si(OH)_2-(CH_2)_3-O-CH_2-CHOH-CH_2OH$ $-Si(OH)_2-(CH_2)_3-O-(CH_2)_2-CF_3$
BF2	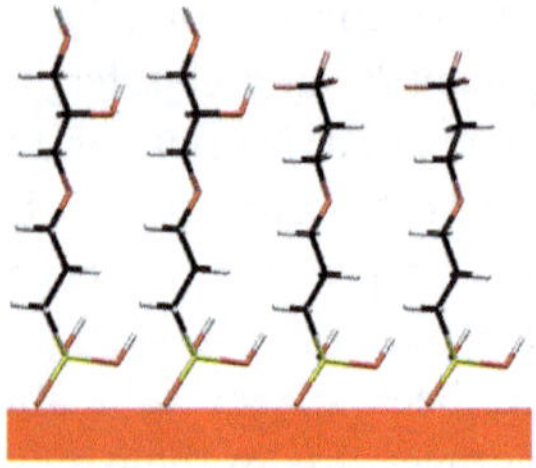	2 × 2 domains of equal length hydrophilic/hydrophobic chains Domain length ~ 12 Å	$-Si(OH)_2-(CH_2)_3-O-CH_2-CHOH-CH_2OH$ $-Si(OH)_2-(CH_2)_3-O-(CH_2)_2-CF_3$
BFs1	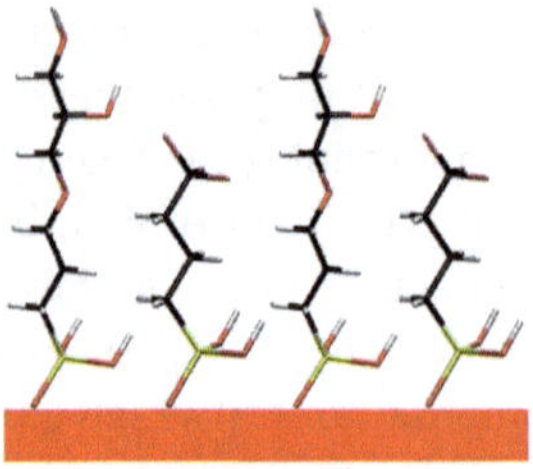	1 × 1 domains of same length hydrophilic & short (s) hydrophobic chains Domain length ~ 6 Å	$-Si(OH)_2-(CH_2)_3-O-CH_2-CHOH-CH_2OH$ $-Si(OH)_2-(CH_2)_3-CF_3$

Table 1 (*Contd.*)

Model nomenclature	Surface pattern schematic[a] with a density of 3.0 chains per nm^2	Morphology	Chemical composition
BFs2	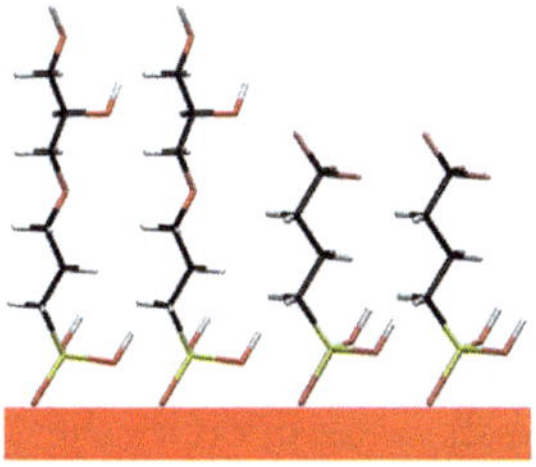	2 × 2 domains of same length hydrophilic & short (s) hydrophobic chains Domain length ~ 12 Å	$-Si(OH)_2-(CH_2)_3-O-CH_2-CHOH-CH_2OH$ $-Si(OH)_2-(CH_2)_3-CF_3$

[a] Colour scheme for schematics: carbon (black), oxygen (red), silicon (yellow), hydrogen (white).

implicated in the ability to spontaneously adsorb onto surfaces of varying nature.[41,42]

MD is a powerful tool able to provide complementary insights to increase the molecular level understanding of protein adsorption at solid/liquid interfaces at the time and length scales not yet accessible to experiments.[43–45] Despite current MD limitations in accessing biologically relevant timescales for protein adsorption and induced conformational changes,[8,43,46,47] statistical sampling over many independent repetitions of standard MD simulations of a single system can provide a detailed picture of the adsorption process of biological molecules at liquid/solid interfaces.[48,49] Alternatively, simulations involving individual trajectories must be sufficiently long to allow for all phases of the adsorption process to be observed, yet this has hardly been achieved.[50] Recent simulation studies modelled the behavior of the short chains tethered to a silica surface in an aqueous environment,[51–57] focusing on the chains' responsive behavior to hydration and their ability to retain aqueous compartments and hydration layers. The interaction of peptides[58,59] and proteins[60,61] with the bare silica surface have also been reported, largely focusing on the initial adsorption stages with limited adsorption induced conformational changes in the protein.

Here we propose to study in atomistic detail the initial protein adsorption process at chemically and morphologically different short chain domains tethered to an amorphous silica substrate. The surfaces to be considered include traditional homogeneous chains known to resist protein adsorption,[62,63] and short hydrophilic[64] and hydrophobic chains chemically grafted onto amorphous silica surfaces forming either uniform or regularly alternating hydrophobic/hydrophilic domains, with the chain spacing determined by the highest stoichiometrically possible density of silanols on the silica surface;[65] see schematics in Table 1. The protein used in this study is the type I hydrophobin EAS.[66] Unique to filamentous fungi, hydrophobins have dual functionality in both protecting and facilitating the adhesion of fungi to surfaces.[67] Hydrophobins are secreted as monomers, and possess the ability to spontaneously adsorb into stable amphipathic monolayers. For class I hydrophobins, these monolayers self-assemble into ordered rodlets with an amyloid-like structure,[41] reducing the surface tension and altering the wettability of both hydrophobic and hydrophilic surfaces alike. Like all class I hydrophobins, EAS has a β-core region, composed of three sets of anti-parallel β-sheets, with the overall structure maintained through four disulfide bonds[66] and four flexible and unstructured loops that make up the protein surface. The largest of the loops, Cys3–Cys4, is largely hydrophobic, however, it shows highly amphipathic regions with alternating hydrophobic and hydrophilic residues. The Cys7–Cys8 loop contains a large amount of polar residues, with a large hydrophobic section between F72–N79. Recent studies have identified several key regions for EAS. The flexible, intrinsically disordered Cys3–Cys4 loop is believed to prevent protein aggregation in bulk solution.[68,69] Although not required for monolayer formation,[70] at the air–water interface the Cys3–Cys4 loop was theoretically shown to stabilize into surfactant-like conformations. Also of interest is the Cys7–Cys8 loop, specifically residues F72–I75, which is believed to be the amyloidogenic region responsible for rodlet formation.[71]

Although experimental studies of hydrophobin at the surface–water interface are limited, we have recently theoretically studied the adsorption of EAS on silica surfaces[61] where we identified two possible adsorption mechanisms. This preliminary work on the initial steps of hydrophobin adsorption on bare silica surfaces in solution formed a necessary benchmark for our current study where we investigate the effects of the chemically and morphologically heterogeneous patterns introduced by functional chain monolayers on silica surfaces on the protein adsorption. This modelling study aims to test the hypothesis that protein adsorption can be facilitated through the matching surface–protein hydrophobic/hydrophilic nanopatterns, or alternatively inhibited by the surface–protein nanopattern mismatch by exploring the adsorption of a versatile surface active protein like EAS on surfaces with varying heterogeneity. We also aim to further explore the role of surface hydration and functional chain dynamics in the ability of organic ligand functionalised surfaces to resist protein adsorption.

Methodology

Model details

Surface patterns of different chemical and morphological characteristics have been created on an amorphous silica substrate (Table 1). Hydrophobic/hydrophilic functionalities have been introduced to create domain sizes potentially commensurate with the size of the hydrophobic and hydrophilic domains on the protein surface[61] as well as the size of individual amphipathic amino acids, such as Lys and Arg[39,40] to test the hypothesis on the role of surface–protein hydrophobicity pattern matching or mismatching in protein adsorption. The rationale behind the choice of the functional chains is presented below.

Silica substrate. To allow for both experimental comparison and future surface modification we used a previously modelled silica surface[73,74] originating from Feuston *et al.*[75] This represents a realistic, highly hydrated amorphous silica surface with a surface silanol density of 4.7 OH groups per nm^2.[65] The amorphous silica substrate displays a density of 2.6 g cm^{-3} (comparable to experiment), an average film thickness of 17 Å (in the z direction) and lateral dimensions of 81 × 81 Å^2 or 81 × 128 Å^2 depending on the required grafting as described below. After adding a vacuum spacer in the z direction, the film was packed in a periodically replicated three-dimensional cell with periodic boundary conditions. During the subsequent simulations, the surface OH groups remained free to move, while the underlying SiO_2 atoms were kept fixed at their initial x, y and z coordinates.

Hydrolysed glycidoxypropylsilane/fluoro functionalized surface models. The summary of all surface models and selected patterns is presented in Table 1. Hydrolysed glycidoxypropylsilane (B) functional chains were chosen for their highly hydrophilic characteristics[64,72] and potential ability to retain a dense hydration layer on the surface.[28,51] The fluorinated chains (F or Fs depending on length) represent the opposite extreme hydrophobic counterparts. A total of six separate interfaces were formed through a siloxane link of each chain to the silica surface. The grafted chain density was kept consistent across these systems at 3.0 chains per nm^2 which is approximately half the grafting density of typical SAMs on quartz and gold surfaces.[39,40,51] Relatively short chains (C4–C6) were used here due to their relevance to industrial anti-fouling systems.[51,57,64,72] Chains were added to the bare silica surface replacing surface hydroxyl groups as described previously.[53–56,76–78] The systems were subsequently hydrated and equilibrated in the NVT ensemble with the z direction adjusted such that the density of the water 15 Å from the interface was equal to that of bulk water at 298 K and 1 atm.

Initial protein structure. The NMR solution structure of the class I hydrophobin EAS, determined by Kwan *et al.*, was obtained from the PDB structure 2FMC.[66] The protein was protonated in zwitterionic form and initially equilibrated in bulk solvent[61] with two counter-ions added to maintain system neutrality.

Adsorption simulation settings. An all-atom representation of the protein and the chains was used with the intra- and inter-molecular interactions being modelled by the CHARMM36 potential[79,80] with additional parameters taken from existing literature for the siloxane bond.[81] Water molecules were treated explicitly using the TIP3P water model.[82]

Molecular dynamics (MD) simulations as implemented in NAMD[83] were used for all the work reported here. In all simulations, a cut-off distance of 12 Å was applied for non-bonded interactions with switching applied between 10 and 12 Å. Long range electrostatic interactions were treated using the particle mesh Ewald method.[84] A timestep size of 2 fs was similarly used in all simulations; this was made possible by fixing the O–H bond lengths of the water molecules using the SHAKE algorithm.[85] NPT simulations were undertaken using a Langevin thermostat[86] and piston to control the temperature and pressure, respectively. The NVT simulations were undertaken using a Langevin thermostat with a damping coefficient of 1 ps^{-1} to control the temperature. In the adsorption simulations, the water volume above the surface was first prepared at 1 bar and 298 K in the absence of the protein *via* a 1 ns NPT simulation. This formed the basis for all the adsorption simulations. The equilibrated NMR structure of the protein was taken from the NPT simulation in the bulk phase and inserted above the centre of the interface in 4 orientations such that the atom of the protein closest to the surface was outside the cut-off length. Any water molecules overlapping the protein following its insertion were eliminated and the height of the simulation cell was adjusted marginally to ensure the density of the water 15 Å or more from the surface was equal to that of bulk water at 298 K and 1 atm. The system was then subject to local relaxation at 0 K before heating up to 298 K in an NVT MD simulation of 300 000 timesteps. The protein was constrained within the vicinity of its initial insertion point during this phase of the simulation by a harmonic constraint attached to its Centre of Mass (CoM). A single NVT simulation was then run with the harmonic constraint still attached to generate the 20 starting configurations for each adsorption system simulation, each separated by 0.5 ns. The adsorption simulations, which were started by releasing the harmonic constraint, were all undertaken in the NVT ensemble at 298 K for between 75 and 150 million timesteps (*i.e.* 150–300 ns) depending on the time taken to move from the starting point to the final adsorbed state. These series of MD simulations generated 3–6 microseconds of real time adsorption statistics for each system (Table 1).

Results

Table 2 shows the states of the 20 MD simulations (runs) for each of the 6 systems at the termination of the runs. The "non-interacting runs" (NIRs) did not evolve to the point where the protein diffused toward the interface such that the minimum distance between the protein and the functional chains was less than 1 nm meaning there was essentially no non-negligible interaction between the protein and the surface. Although possible reasons for non-adsorption behaviour are discussed below, the adsorption analysis is based only on the "interacting runs" (IRs) where the protein did engage with the interface. Further, we define the "adsorbed runs" (ARs) as the number of runs where the protein was found to be in direct contact with the surface at the termination of the run. We define the adsorbed runs with an average contact area greater than 750 Å^2 over the last 50 ns of run time as "AA" and only those were used for adsorbed state analysis. We define the "association events" (AEs) as an event where the minimum of the protein falls within 0.6 nm of the functional chains and "contact events" (CEs) where the protein falls within 0.3 nm of the functional chains. These definitions were introduced for ease of analysis and discussion presented below.

The analysis performed here is split into three sections. First, a characterisation of the patterned solid/liquid interfaces, including hydration layer behaviour and chain dynamics, is provided as this relates to the foundation for two theories of protein resistance. Second, the process of protein adsorption at the solid/liquid interfaces is described. A comparison between the solid/liquid interfaces is made to identify features of the functional chain patterns which may

Table 2 Simulated system data summary: R = total number of runs performed; IR = interacting runs; AR = adsorbed runs; AA = high contact area ARs used for adsorbed state analysis; AEs/IR = average number of association events per IR; CEs/IR = average number of contact events per IR; AE/CE = ratio of association events to contact events

System	R	IR	AR (%)	AA (%)	AEs/IR	CEs/IR	AE/CE
B	20	16	75	56	2.6	1.1	2.3
F	20	17	94	82	2.2	0.4	5.4
BF1	20	19	84	32	2.2	0.7	3.2
BF2	20	19	84	42	3.1	0.4	7.4
BFs1	20	18	72	39	4.5	1.7	2.7
BFs2	20	19	89	47	3.7	0.6	6.5

be influential in limiting protein adsorption. The final section of the results focuses on the adsorbed states of the protein at each of the solid/liquid interfaces compared to the previously observed behaviour of EAS hydrophobin at air/water[61,68,70] and silica/water[61] interfaces. The analysis will focus on outlining features which may inhibit the conformational rearrangement of the protein and subsequently restrict the formation of the undesirable self-assembled protein monolayer implicated in biofouling.

Interfacial characterisation

Solvent exposed surface. Fig. 2(a)–(d) shows the top view of the equilibrium MD snapshots of the solvent exposed surfaces of the four patchy surfaces investigated here with the red and blue surfaces representing the B and F/Fs chains respectively. These snapshots bear a qualitative resemblance to the equilibrium surface structures obtained from mesoscale simulations of the formation of striped SAM nanoparticles reported previously.[87,88] The average solvent accessible

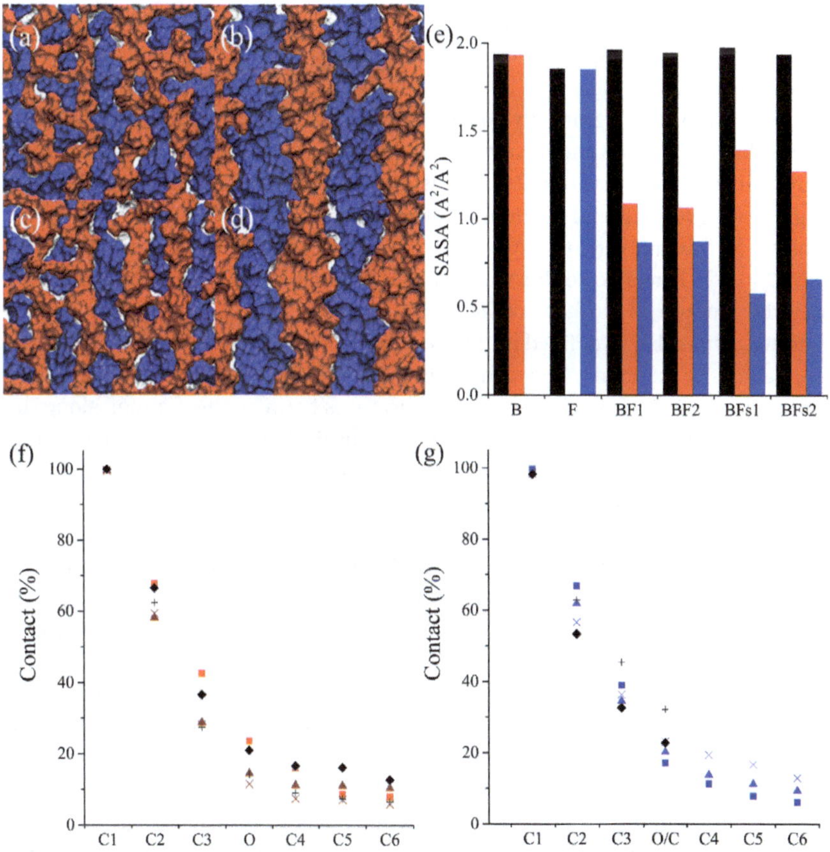

Fig. 2 (a–d) Surface representations of BF1, BF2, BFs1 and BFs2 respectively: B (red); F/Fs (blue); (e) solvent accessible surface area of the surfaces (normalised against the *xy* plane area): total (black), F (blue); B (red); (f and g) contact percentages for atoms along the chain length, contact is defined as being within 5 Å of the silica substrate: (f) B chains; B (■), BF1 (▲), BF2 (×), BFs1 (◆), BFs2 (+); (g) F chains F (■), BF1 (▲), BF2 (×), BFs1 (◆), BFs2 (+).

surface areas (SASAs) for these systems and the pure B and F systems are presented in Fig. 2(e). The total area (black) is shown along with that associated with the B (red) and F/Fs (blue) chains separately. For consistency the values are normalised by the area of the *xy* plane in which the surface is placed.

The SASAs for the pure B and F systems are statistically indistinguishable when taking into account the fluctuations about the mean value. The result of this is that when the protein is exposed to these surfaces the area available for protein surface contact should not be a variable that influences the adsorption behaviour meaning that the surface chemistry should be the dominating factor. Significant differences are observed for the four patchy interfaces. While the average SASA distributions for BF1 and BF2 are statistically indistinguishable from one another, visual inspection of the snapshots shows that the domain ordering is far more distinct at the BF2 interface. In fact the solvent exposed surface of BF1 appears to lack the desired ordered heterogeneity. The large increase in both the total and B exposure for the BFs1 and BFs2 systems occurs as the tops of these chains are more exposed to the solvent due to the shorter length of the fluorinated chains. The increase in fluorinated contact chain exposure in BFs2 compared to BFs1 arises as the increased domain width restricts the amount of overlap the longer B chains can achieve. From Fig. 2(c) it can be seen that the longer B chains are capable of crossing over the shorter fluorinated chains limiting their exposure to the solvent.

The ligand packing density considered here is the highest stoichiometrically possible on the amorphous silica surfaces. The grafting densities are however approximately half that of SAMs. Whilst the reduced coverage density allows for a greater conformational freedom, as discussed below, it also creates potential for chains to interact with the substrate. The lack of substrate exposure, seen in Fig. 2(a)–(d), is related to this chain-substrate interaction. In Fig. 2(f) and (g) we see the amount of surface contact for the carbon (and in-chain oxygen), with surface contact being defined as the atom being within 5 Å of the silica substrate, for each of the systems. This contact between the functional chains, the decays along the chain length and the average distances between these atoms along the length of the chain and the substrate monotonically increase for all systems (data not shown). Less than 15% of chains have a substantial interaction with the silica substrate along their entire length. This enables us to conclude that the vast majority of chains still maintain the "brush-like" character, with the functional heads pointing away from the surface and able to interact with solvent phase.

Chain flexibility. Fig. 3 shows the RMSD of the functional chains in the B, F, BF1, BFs1 and BFs2 systems. The data is split with individual plots presented for B and F chains.

RMSD profiles provide a semi-quantitative description of the chain dynamics. The hydrophobic chains form a less dynamic interface than their hydrophilic counterpart which more readily engages with the interfacial water molecules. From the plots of B and F compared to BF1:B and BF1:F it can be seen that there is little variation in the dynamics of the individual chains if the patterning consists of functional chains of the same length. However, there is a substantial increase in motion of the B chains in the BFs1 systems compared to that in the B or BF1 systems. Below we show that this increased motion plays a substantial role in the protein adsorption process.

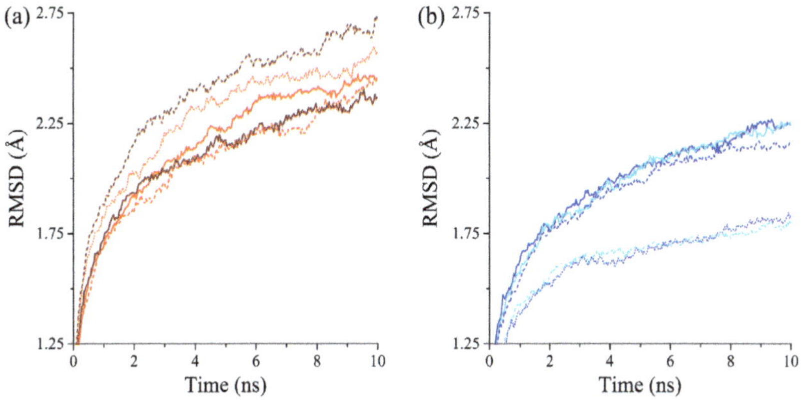

Fig. 3 Functional chain RMSD profiles: (a) B; (b) F: single chains (red/blue solid), BF1 (maroon/light blue solid); BF2 (red/blue dash), BFs1 (maroon/light blue dash) and BFs2 (red/blue dots).

Interfacial water structure. Fig. 4(a) shows the normalised average water density profile as a function of distance from the functional chains tethered to the amorphous silica for the B, F and BFs1 systems. The data for BF2, BF1 and BFs2 (not displayed) showed the same trends as presented in Fig. 4(a) for the BFs1 system.

The two peaks observed in the B and BFs1 systems are indicative of water molecules arranged such that they have the capacity to act as either hydrogen bond donors or acceptors with the hydroxyl groups at the terminus of the B chains. This behaviour is consistent with the hydration behaviour of similar functional chains tethered at a packing density of 4.2 residues per nm^2 to a α-quartz substrate.[51] The small first peak observed at 1.9 Å for the F system results from water molecules embedded within the chains interacting with the hydroxyl groups of the siloxane group at the base of the functional chain. The second peak for the F system is slightly further from the chains than that of the B system indicative of the presence of the large fluorine atoms at the chain terminus rather than the hydroxyl group. Despite displaying a hydrophobic character on the macroscopic scale (in preparation for publication elsewhere), the interfacial water density profile above the fluorinated chains shows an increase in water density compared to the bulk, indicative of a hydration layer. This observation is consistent with interfacial water structuring at hydrophobic SAM/water interfaces.[89] However, in all systems, unlike the interfacial water structuring observed at SAMs,[89] the interfaces examined here exhibit limited water structuring propagating into the solvent phase. This is a reflection of the lower chain density on the amorphous silica surface (<3 residues per nm^2) compared to the density of SAMs (typically >5 residues per nm^2) which results in increased interface roughness and increased chain dynamics (see discussion below). These two factors produce a more physically heterogeneous interface for water to interact with compared to the tightly packed SAM systems resulting in a less cohesive interfacial water structure on functionalised amorphous silica substrates.

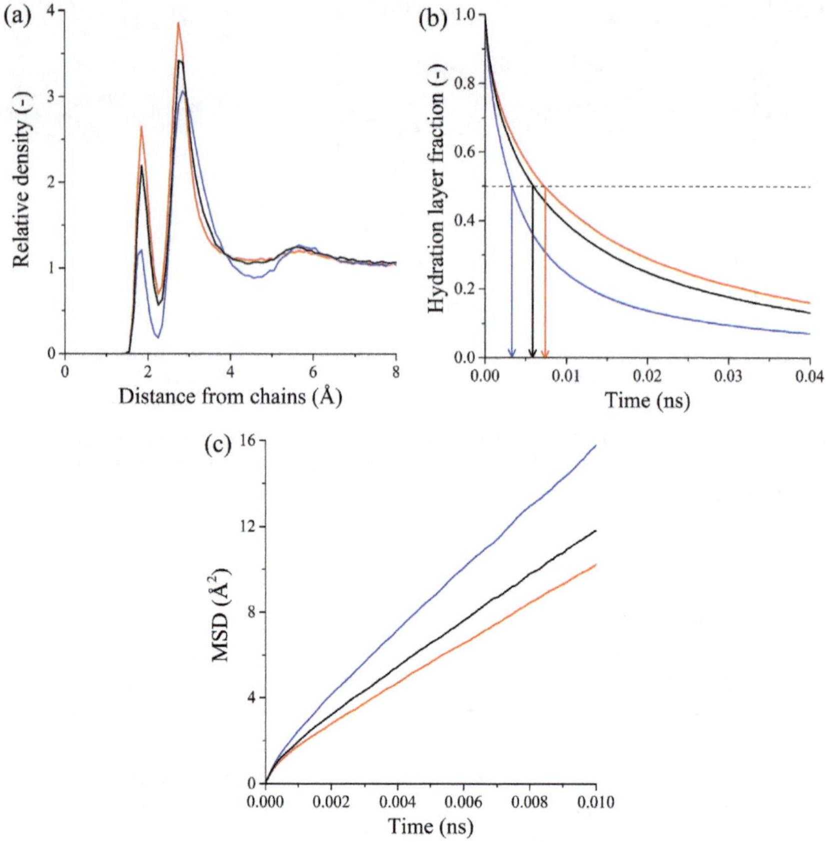

Fig. 4 Water structuring at interface: (a) relative density profile of water radiating away from functional chains: B (red), BFs1 (black), F (blue); (b) hydration layer decay profile: B (red), BFs1 (black), F (blue); arrows indicate the half-life of the hydration layers which are 3.3 ps, 5.7 ps and 7.4 ps for F, BFs1 and B respectively; (c) MSD of water in the first hydration layer at B (red), F (blue) and BFs1 (black) surface.

Fig. 4(b) shows the decay rate of water molecules within the first water layer for the B, F and BFs1 systems. BF2, BF1 and BFs2 systems (not shown) were indistinguishable from that of BFs1. The decay plots are consistent with the concept of individual water molecules interacting with either hydrophilic or hydrophobic chains. The hydrophobic chains, with the half-life of the hydration layer being 3.3 ps, exhibit a decay rate twice that of the purely hydrophilic B chains (half-life = 7.0 ps) indicating that the interface is a less hospitable environment for the solvent despite the appearance of a hydration layer. The half-life of the hydration layer in the four patterned systems ranges between 5.7 and 6.1 ps. This is a further indication that there is little cooperation between the interfacial water molecules and that their dynamics are influenced only by species within an atomic radius.

Fig. 4(c) shows three-dimensional mean squared displacements (MSDs) for water molecules at the 3 interfaces investigated. Curves are generated over a short time period (10 ps) with data taken every 40 fs to capture the dynamics of the chain bound water molecules (*i.e.* an "instantaneous" hydration layer).[53,61] The data concurs with that presented above showing that water molecules in the hydration layer of B chains are more tightly held than in the hydration layer above the fluorinated chains. Again the data for the four patchy surfaces falls between the two single chain systems and no distinction can be made between these four systems using this analysis technique.

The analyses of the average nano structures and dynamics of the interfacial water layers for the four patterned systems which include both B (hydrophilic) and F or short F (hydrophobic) chains indicated that these properties are largely independent on the nature of the nanoscale patterning of the surface. This can be rationalised when taking into account the small size and fast dynamics of individual water molecules, the behaviour of which is largely influenced by short range forces acting on length scales of less than 1 nm. We suggest that the relatively large spacing between the chains contributes to the lack of differentiation between the behaviour of the water layer at the 4 patterned interfaces containing both hydrophilic and hydrophobic chains suggesting a lack of cooperative interactions between the water molecules within this layer. This view is consistent with water behaviour at the hydrophobic/hydrophilic interface (albeit of a different kind to those considered here) proposed by Granick *et al.*[90–93] These works suggested that the interplay of water at a hydrophobic/hydrophilic interface results in the formation of a flickering vapour phase where water molecules are unable to structure in accordance to either surface. These water molecules are believed to diffuse through several favourable metastable energy states, with various stability times and frequencies of contact with the surface. Due to the size and instability of these water layers, there is little to no ability for water molecules to form ordered hydration shells, as expected around a homogeneous surface.

Adsorption process

To the best of our knowledge the mechanisms of the resistance of proteins to adsorb onto dynamic solid/liquid interfaces have not been elucidated on a molecular level. Developing a detailed picture of the adsorption process can help identify the key events along the adsorption pathway which are important for instilling the repulsive character of various interfaces, especially at the initial stages of the adsorption process. Given that both hydration theory and steric repulsion considerations (DLVO theory[94–96]) have been utilised to explain this phenomenon, analysis here will be directed towards identifying molecular mechanisms by which either of these two factors can play a role in the adsorption process. It has been discussed in the previous section that due to the moderate packing density and flexibility of the chains there is only one structured persistent hydration layer present at the interface which the protein must penetrate in order to form direct contact with the functional chains before it can possibly target the underlying substrate surface. Therefore, the analysis is focused on protein interactions with this hydration layer or the functional chains themselves. Potential energy barriers beyond the first hydration layer are not discussed below as well as the influence of the hydration shell around the protein which was not observed to have a significant impact on the adsorption process.

Fig. 5 shows an exemplar adsorption trajectory taken from a run of EAS at the B/water interface. The general molecular level features of protein adsorption have

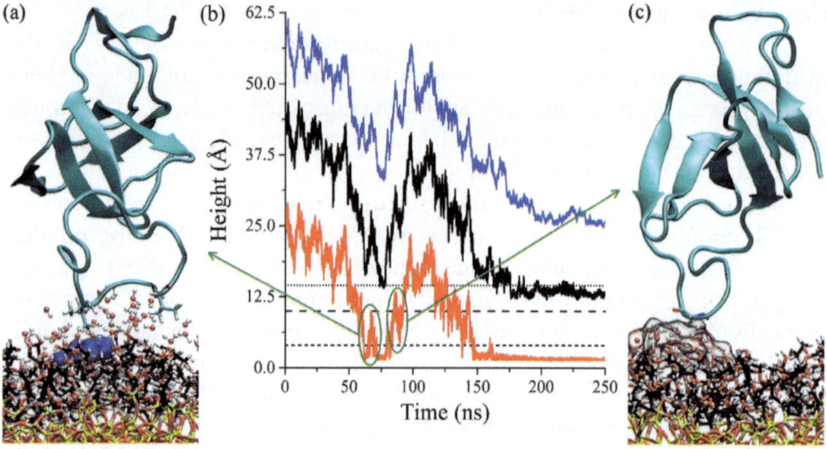

Fig. 5 (a) Snapshot of the system for the contact event (CE) between 60 and 65 ns, protein shown as cyan new cartoon representation, functional chains shown within same colour scheme as Table 1; functional chains within 5 Å of protein represented as blue surface, and water molecules within 5 Å of both protein and functional chains in CPK representation. (b) Exemplar temporal behaviour of the height of the centre of mass of the protein above the functional chains B (blue); height of minimum of the protein (black); and the minimum distance between the protein and the chains (red). The first contact (CE) and association (AE) events are circled for clarity. Also shown at 4 Å is the outer distance of the first hydration layer at the interface between the functional chains and water (black short dash); the mean height of the terminal (6th) carbon atom in the B chains (black dots); and the 10 Å cut-off for contact and association events is also shown (black long dash). (c) Snapshot of the system for the association event (AE) between 85 and 90 ns, water molecules within 5 Å of both protein and functional chains are represented as a surface indicating the barrier between protein and the functional chains.

been previously described[48–50,97] and are observed here. At the beginning of the run EAS is well outside the cut-off distance from the surface and freely diffuses in the bulk solvent until it undergoes diffusion towards the interface at circa 45 ns from the start of the simulation (the exact timing here and throughout the following discussion is not relevant/significant and is only used for the process description purpose). There is a period of reversible engagement with the interface from ~45 ns to ~150 ns. In this phase regions of the protein directly interact with the hydration layer above the interface, exemplified at 55 ns, 85 ns and between 125 and 145 ns, or come into direct contact with the functional chains, as seen between 60 and 75 ns. After 150 ns the protein stabilises at the interface, the height of the centre of mass (CoM) of the protein decreases and the contact area between the protein and the chains increases.

In this adsorbed phase there is still fluctuation in both the protein CoM and minimum highlighting the dynamic character of the adsorption. This point is discussed below in the adsorbed state section. A final point is that, at 175 ns there is a step drop in the trajectory of the protein minimum (black dotted line). This corresponds to a region of the protein penetrating into the body of the functional chains and becoming embedded below the centre of mass of the chains. This is a result of the lower grafting density of the functional chains on the amorphous silica substrate and the associated reduced rigidity compared to the behaviour observed at typical SAMs.[32,98,99] Substrate rigidity has been previously presented as one of three critical features for limiting adsorption.[54,76–78,100,101] Embedding the protein into the body of the chains introduces an additional energy barrier which must be overcome in order for the protein to disengage with the interface.

It should be stressed that the number of runs where the protein does not engage with the interface does not provide an indication of the ability of that interface to resist protein adsorption. Neither is the time taken for a protein to engage with the interface in a given run a useful measure to determine the affinity between the protein and the surface due to the dependence on the random diffusion of the protein at the start of the run when the protein is placed such that there is negligible interaction between the protein and the surface.

However, the percentage of the runs where the interface is able to disrupt the adsorption process such that the protein diffuses away from the interface, exemplified multiple times in Fig. 5 between 50 and 150 ns, does provide an indication of the adsorption resistance capacity of the functional chains. Herein, two types of interaction events are used to characterise the adsorption behaviour of EAS at the various interfaces (Table 1). An association event (AE) is defined as an event where the minimum of the protein falls within 0.6 nm of the functional chains and can thus directly interact with the first hydration layer but then subsequently exceeds 1 nm (*i.e.* the protein disassociates from the surface). An example of an AE can be seen at 85 ns in Fig. 5. A contact event (CE) is defined as an event where the minimum of the protein falls within 0.3 nm of the functional chains and forms direct contact with the functional chains but then subsequently the region of the protein shifts away from the interface such that this minimum exceeds 1 nm. An example of a CE can be seen at 60 ns in Fig. 5.

The average number of AEs per interacting run for each system is shown in Table 2. The number of AEs for interfaces consisting of functional chains of the same length (B, F, BF1 and BF2) ranges between 2.2 and 3.1. The similarity in the hydration structure and dynamics of the ligands makes this unsurprising. There is, however, a noticeable relationship between the number of association events per run and the increased mobility of the systems consisting of chains of varying length. Specifically, for the patterned surfaces BFs1 and BFs2 there are 4.5 and 3.7 association events, respectively, per interacting run. The additional conformational freedom afforded to the longer B chains in these systems, as evidenced by the larger RMSD discussed in the interfacial characterisation section above, appears to enable these chains to act in a brush-like manner and "sweep" the associating protein residue away from the interface. This suggests that *one design strategy* for protein resistant surfaces is to imbue the interface with increased flexibility which can be achieved by creating patterns of hydrophobic/hydrophilic ligands of varying length which imparts increased conformational flexibility of the terminal region of the (longer) hydrophilic chains. The role of the slightly shorter hydrophobic chains would be to reduce the likelihood of the hydrophilic chain systems forming a 'mushroom' like structure at the interface due to excluded volume effects.

Table 2 also displays the average number of CEs per interacting run and the ratio of AEs to CEs for each system. In the case of CEs the systems can be grouped in terms of domains rather than chain length. At interfaces where substantial hydrophobic domains may exist (fluorinated, BF2 and BFs2), the average number of contact events ranges between 0.4 and 0.6. This is approximately half that of the

B and single chain width domains (BF1 and BFs1) which range from 0.7 to 1.7 events per interacting run. The trend is made clearer when considering the ratio of AEs to CEs where systems B, BF1 and BFs1 have AE/CE ratios ranging from 2.3 to 3.2 while the ratios for systems F, BF2 and BFs2 are substantially higher ranging from 5.4 to 7.4. This indicates that there is a substantial decrease in the ability to resist protein adsorption once direct contact has been made by the protein to surfaces with significant hydrophobic domains. This likely arises due to additional stabilisation of the formed contact caused by the depletion of water from the hydrophobic domain whereas at hydrophilic domains this does not arise. The CE data suggests the *second design strategy* for disruption of the early phase of protein adsorption, namely ensuring the length scale of the surface domains is sufficiently small, on the order of a single chain width (<0.5 nm). This prevents formation of large hydrophobic patches on the surface where hydrophobic protein–surface association can be enhanced due to water depletion, whilst also enabling the increased mobility for chains to exhibit the "brush" like behaviour, especially for the short chain hydrophobic – longer chain hydrophilic patterns if using *both design principles together*.

Adsorbed state

The limited number of adsorbed runs for each system is insufficient to completely describe the adsorbed state of a protein at a solid/liquid interface. Furthermore, due to the significant energy barriers associated with the conformational rearrangement of the protein at the interface, some runs were observed to be kinetically trapped in adsorbed conformations with limited protein–surface interaction. Therefore, the analysis of the adsorbed state is focused on systems where the contact area between the protein and the surface was, on average, greater than 750 Å^2 over the last 50 ns of the run (defined as AA in Table 1). Unless otherwise stated, results are averaged over the final 50 ns of the runs. The adsorbed protein behaviour across the systems will be compared to previously reported work describing the adsorption of EAS hydrophobin at air/water[61,68,70] and silica/water[61] interfaces.

Foundational hypothesis and other considerations for EAS biofouling resistant surface design. The foundational hypothesis of this work was to determine whether it was possible to design an interface consisting of hydrophobic and hydrophilic domains which would disrupt the initial adsorption of EAS. The versatile ability of this protein to adsorb on essentially any surface is likely to be related to the fact that the exposed surface of EAS consists of hydrophobic/hydrophilic domains on a variety of length scales. For example, the Cys3–Cys4 loop is hypothesised, due to its inherently disordered conformation and high flexibility, to prevent aggregation in the bulk phase.[68] At the same time it has been theoretically shown that at the air/water interface, the loop displays surfactant like behaviour[68] and potentially stabilises adsorption[61] due to a mixture of hydrophobic and hydrophilic residues capable of interacting with non-homogeneous interfaces. Amphiphilic domains on the amino acid length scale have previously been shown to aid adsorption at close packed hydrophobic/hydrophilic domain separated surfaces.[39,40] In addition, the overall surface of EAS has been shown to have large hydrophobic and hydrophilic domains.[102] Thus, the goal here was to test a range of chemically patterned surfaces with domains not commensurable to either length scale presented by the protein surface.

It must be stated that from the evidence available through our relatively extensive simulations a definitive statement on this hypothesis still cannot be made. However, upon comparing the adsorption behaviour at the purely hydrophobic and hydrophilic chains some clear trends emerge. Specifically, while the fraction of contact provided by the neutral sidechains or protein backbone differs by less than 2.5% for these two non-patterned systems there is a clear preference for hydrophobic association on the fluorinated chains with 54% of contact between the protein and surface being provided by the hydrophobic sidechains compared to only 44% in the hydrophilic B system. On the patterned surfaces the distinction between the adsorption behaviour in terms of hydrophobic–hydrophobic or hydrophilic–hydrophilic association could not be made due to the loose mobile domain boundaries on both the protein and the functional ligand protected surfaces. Interestingly, one notable distinction between the adsorption behavior of EAS at BFs2 compared to other patchy surfaces was that at this surface the hydrophobic domain was observed to begin to influence adsorption behavior in a directional way. In 4 of 9 adsorbed runs at BFs2 the protein was observed to be directed to the hydrophobic domain. This behavior is exemplified by Fig. 6 where it can be seen from the protein footprint on the surface that the protein adsorbed lengthwise along the hydrophobic domain. Similar behavior has been reported on the micrometer length scale for the adsorption of spores on layered hydrophobic/hydrophilic surfaces.[103] This behavior raises some alternative considerations for EAS adsorption. Specifically, the mechanism for monolayer formation implicated in biofouling has been shown to require the beta barrel to be positioned on opposite sides of the adsorbed protein to enable the association of the Cys7–Cys8 loop and fibril propagation.[71] It can be reasoned that trapping the protein in a monolayer incompatible conformation on the surface (*e.g.* as shown in Fig. 6) can serve as a surface design strategy where the subsequent self-assembly is

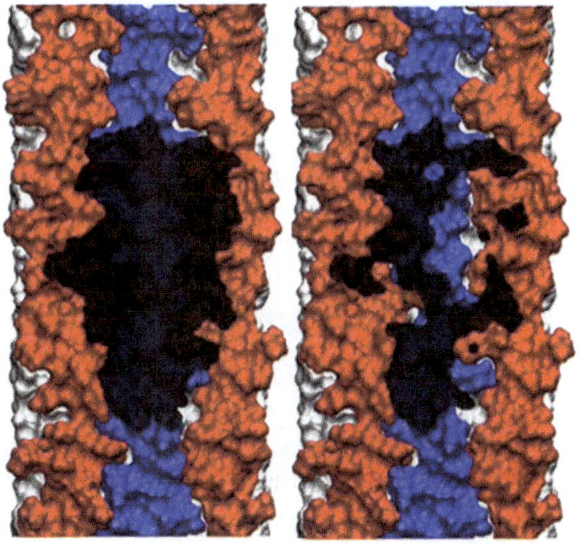

Fig. 6 (a) EAS surface above BFs2; (b) contact region (within 5 Å of the surface) of EAS above BFs2.

unlikely. On the other hand, it can also be reasoned that hydrophobic patches able to accommodate the protein footprint in the observed way should be avoided to prevent the non-specific adsorption, in principle.

Protein–surface contact area. To determine the extent of adsorption at the various solid/liquid interfaces the contact area between EAS and the functional chains was calculated. Fig. 7 shows the value of the average contact area and maximum contact area for each system with the fluctuation across runs shown by error bars.

Across all systems the difference between the average and maximum contact areas is relatively consistent with an average of 285 $\pm$ 21 Å^2. The consistency of this difference can be taken as an indication that EAS is not interacting with the different surfaces in vastly different manners. The average contact areas are higher for both the fluorinated (1255 Å^2) and BFs2 (1223 Å^2) systems suggesting that the protein is more likely to spread across a hydrophobic surface, in line with the general understanding of protein adsorption. This also provides further evidence that the domains on the BFs2 surface are sufficiently large and distinct that the protein can differentiate between them and potentially target the hydrophobic regions which are more vulnerable to adsorption. There is a reduction in the contact areas of around 100 Å^2 for the B, BF1 and BF2 surfaces which are 1121 Å^2, 1154 Å^2, and 1115 Å^2 respectively. A further decrease to 1025 Å^2 is seen for the BFs1 system. The implication of the contact area data is that there are essentially three different classes of surface with F and BFs2 being the most hospitable to the protein and the BFs1 surface being the least with the remaining surfaces falling between these extremes. Further discussion of this finding and possible implications to the biofouling resistance of surfaces with the BFs1 type design are presented below.

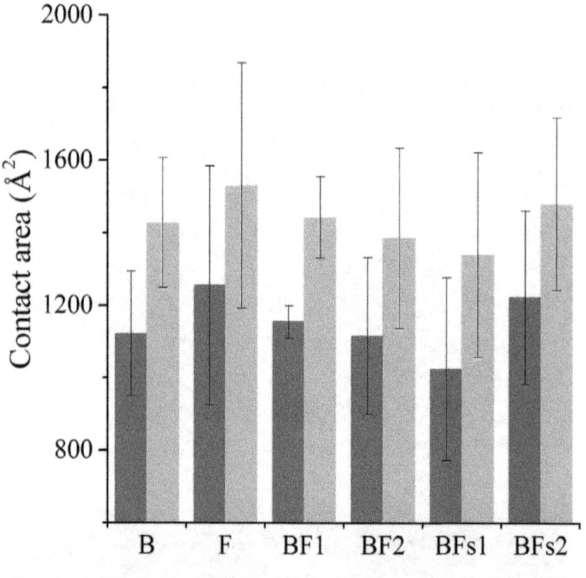

Fig. 7 Contact area of EAS at the solid/liquid interfaces found by taking the average (dark) and maximum (light) of the contact area of the adsorbed runs. The errors associated with the averages are the standard deviations across all such runs and do not consider fluctuations within individual runs.

Protein conformational transitions. The preceding discussion has focused on the average adsorbed behaviour of EAS at the various interfaces and as such does not take into account conformational variation within a single run. Peptide[48,49,97] and protein[104,105] adsorption occurs in a stepwise manner at solid/liquid interfaces and it is expected that desorption will occur through a similar mechanism. Therefore, the ability of a surface to enable conformational transitions which reduce the interaction between the protein and the surface could serve as an indication of the surface's ability to reject a protein. Analysis of the fluctuations in the adsorbed behaviour during individual runs provides a significant insight into the conformational behaviour and imparted (in)stability of EAS at the interfaces. Fig. 8(a) shows the contact area as a function of time for EAS at the B/water interface of an exemplar trajectory.

The positive and negative step changes in the contact area indicate the dynamic character of adsorption and emphasise that water molecules are capable of competing with the protein for surface contact and as such with sufficient time might be able to completely replace the protein at the surface. Fig. 8(b) shows the frequency of transition events plotted against the size of the event (*i.e.* the change in contact area between the transitions) for B/F, BFs1 and BFs2 systems. A transition event is considered to occur if the difference between the running average of the contact area and a previously reached minimum or maximum exceeds a defined value. For example, in Fig. 8(a) if the change in contact area for a transition event is required to be 200 Å^2 then the first three transition events will occur between 250 and 252 ns, 255–260 ns and 265–270 ns as indicated. The running average is used to smooth sharp but brief fluctuations and focus on true transitions. There is a clear distinction in the EAS–surface contact area variations on B compared to the F chains, indicating more stable protein conformations when adsorbed at the hydrophobic interface. The increased fluctuations for the B system indicate that, although the underlying driving force for the conformational transitions cannot be elucidated here, they arise through more than

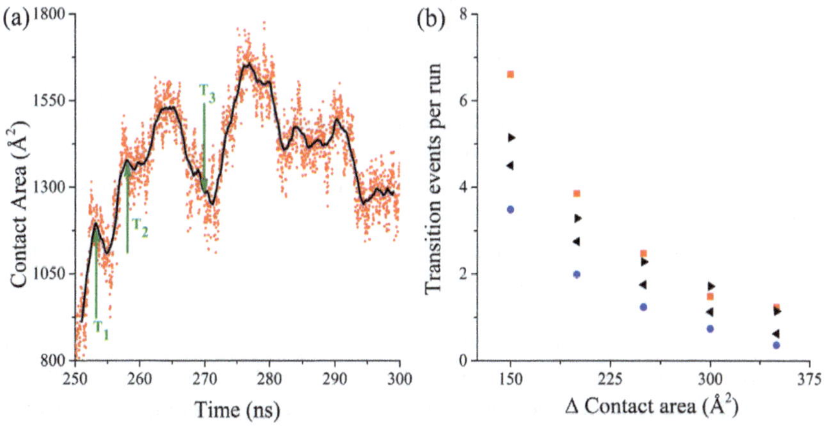

Fig. 8 (a) Exemplar trajectory of the contact area (red) and a running average over 1 ns of the contact area (black) for EAS at the B/water interface; the first three conformation transitions with required contact area change of 200 Å^2 are shown; (b) average number of conformational transitions per run as a function of the contact area variation in the running average of the contact area: B (■), F (●), BFs1 (▶), BFs2 (◀).

thermal fluctuations meaning that the surface is not passive. The two fold decrease in fluctuations on F surfaces corresponds with the reduction in the half-life of the water molecules in the hydration layer for these chains, as discussed above, and is indicative of the decreased competition for surface contact between the protein and water molecules at this hydrophobic interface. The data for the patchy surface again indicates that there is little variation in the observable behaviour at BFs1, BF1 and BF2 (the latter two systems are not shown in Fig. 8(b) for clarity). While there is a slight increase in the number of transition events per run at smaller contact area changes (<200 Å^2) the transition rate above this value is largely indistinguishable for these three systems and B meaning each surface has the potential to reject an adsorbed protein. The BFs2 system shows differentiation suggesting that the domains for this system are beginning to present some degree of recognisable heterogeneity to the protein and have an observable effect on the protein dynamics.

Cys3–Cys4 and Cys7–Cys8 loops. As discussed above there are two key surface regions present in EAS associated with its adsorption versatility. The Cys3–Cys4 loop, between Cys19 and Cys45, hypothesised to stabilise adsorption at non-homogeneous interfaces[68,61] and the second Cys7–Cys8 loop, between Cys60 and Cys80, for which it is hypothesised that its region F72–I75 has an amyloidogenic character and is responsible for rodlet formation[71] when EAS self-assembles into monolayers.

Fig. 9 shows average contact fractions for these two regions of EAS with the surface. Due to the dynamic character of the adsorbed behaviour a residue is considered to be in contact with the surface when a minimum of 3 atoms in a residue are within 5 Å of the surface for more than 75% for the last 50 ns of the run. This definition enables average contact fractions for different regions of the protein to be considered without the need to describe individual runs. The Cys3–Cys4 loop is conspicuous as a contact region regardless of the underlying chemical character of the surface. At all surfaces this loop has a substantially higher contact fraction than the Cys7–Cys8 loop.

This point is emphasised by the data presented in Table 3 which distinguishes between those runs where the fraction of residues within each loop in contact out of the total number of residues in the loop was less than 0.2 and those where the fraction was greater than or equal to 0.2. For the Cys3–Cys4 loop only 13% of all runs (7 of the 53) had less than 20% of the residues in contact with the surface compared to 43% (23 of 53) of runs for the Cys7–Cys8 loop. It should be recognised that the interaction of the Cys3–Cys4 loop is more limited on the surfaces made of 1 × 1 domains. This might be an indication that the lack of well-defined domains caused by the dynamics and inability of self-association between the functional chains are disrupting the surface recognition capacity of the Cys3–Cys4 loop. However, due to the limited adsorbed phase data, further investigation on this point is required before a definitive conclusion can be drawn. Finally, the behaviour of these two loops is in good agreement with the understanding of their function discussed above and is consistent with the limited existing experimental and theoretical data relating to EAS adsorption.[61,68,70]

A critical feature of the Cys3–Cys4 loop is the flexibility created by the collection of small amino acid residues at both ends of the loop, residues 20 to 24 and 40 to 44, which appear to be critical in enabling the larger middle residues to form direct contact with the surface regardless of the behaviour of the β-core of EAS. We

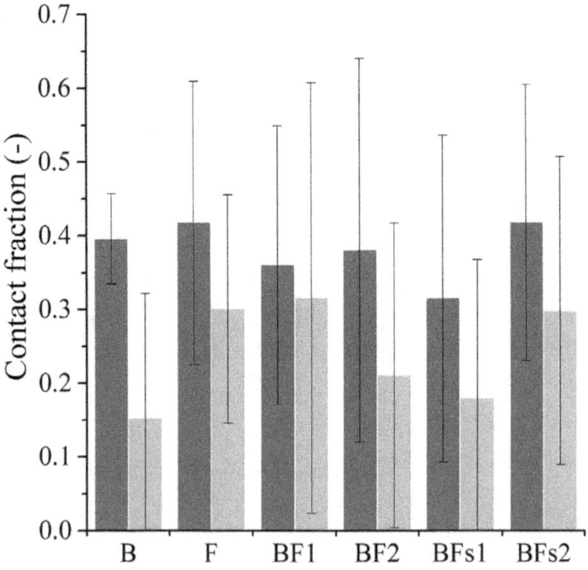

Fig. 9 Fraction of residues within protein regions that are in direct contact with the functionalised surface, defined as having a minimum of 3 atoms in a residue within 5 Å of the surface for more than 75% of the last 50 ns of a run: Cys3–Cys4 loop (dark grey), Cys7–Cys8 loop (light grey). Error bars show the variation between runs for the various systems.

Table 3 Data relating to Cys3–Cys4 and Cys7–8 loop surface connectivity

		Contact fraction			
		Cys3–Cys4		Cys7–Cys8	
System	Runs	<0.2	≥ 0.2	≤ 0.2	>0.2
B	9	0.00	1.00	0.56	0.44
F	14	0.14	0.86	0.36	0.64
BF1	6	0.33	0.67	0.50	0.50
BF2	8	0.13	0.87	0.50	0.50
BFs1	7	0.29	0.71	0.43	0.57
BFs2	9	0.00	1.00	0.33	0.67

term these regions of the protein "flexible hinges" for the ability they impart to enable the bulk of the Cys3–Cys4 loop to swing into contact with the surface. This ability of the Cys3–Cys4 loop to arrange itself to find favourable interactions with the underlying surface is reflected in its hydrogen bonding with the functional chains. Specifically, we calculated at the B surface that on average there are 2.25 hydrogen bonds between the chains and the Cys3–Cys4 loop. It might be expected that the number of hydrogen bonds would halve at the patchy surfaces given the presence of hydrophobic chains not capable of forming hydrogen bonds. Indeed, there is a notable decrease; however, it is not proportional to the reduction in potential hydrogen bonding sites on the surface. The lowest average number of

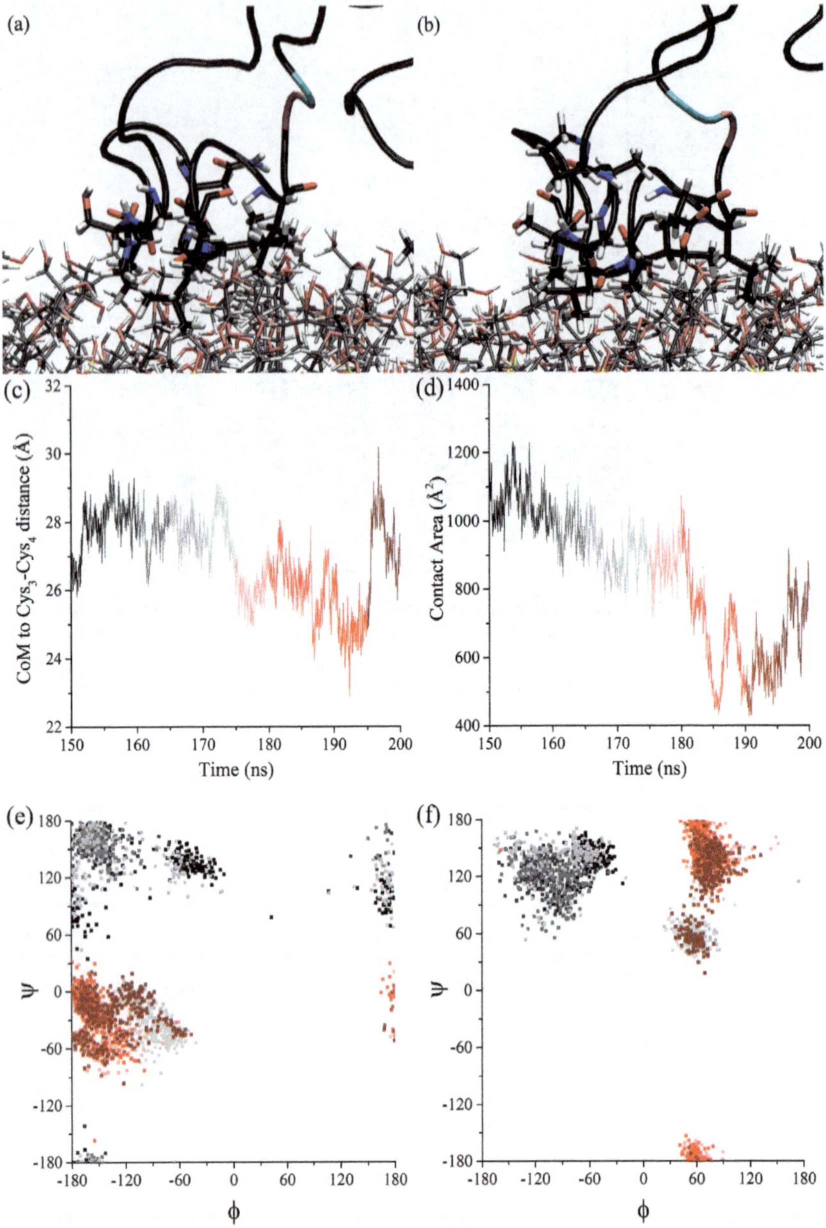

Fig. 10 Transition event with the Cys3–Cys4 loop isolated from the bulk motion of the rest of the protein through a shift in the phase space at one end of the loop. Snapshots of the interaction between the Cys3–Cys4 loop and the functional chains at (a) 155 ns and (b) 192 ns: protein residues in direct contact with the surface shown in licorice representation with carbon, nitrogen, oxygen and hydrogen atoms coloured black, blue, red and white respectively; the remaining protein is shown as a tube with flexible hinge residues Ala_{41} (pink) and Ser_{42} (cyan) highlighted; B chains shown in thin licorice representation with carbon, oxygen and hydrogen atoms coloured grey, pink and white. (c) Distance between the centre of mass of the protein (not including residues 20 to 44) and the region of the Cys3–Cys4 loop between the flexible hinges (residues 25 to 39); (d) contact area between the protein and the surface; (e) Ramachandran plot for Ala_{41} between 150 and 200 ns; and

hydrogen bonds amongst the patterned surfaces is 1.54 at the BF2 surface while the highest is 2.12 at BFs2 (BF1 = 1.73; BFs1 = 1.74).

A second feature of the flexible hinges appears to be the ability to isolate the Cys3–Cys4 loop from the motion of the core of the protein, thus allowing the loop to maintain direct contact with the surface. Fig. 10 exemplifies this behaviour. The snapshots in (a) and (b) show the system at 155 and 192 ns respectively. This transition event can be described in terms of a varying distance between the bulk protein (all residues of the protein except residues 20 to 44) and the region of the Cys3–Cys4 loop between the flexible hinges (residues 25 to 39). This motion results in the core of the protein detaching from the surface as reflected by the decrease in the contact area (Fig. 10(d)). The dihedral maps for Ala_{41} and Ser_{42} between 150 and 200 ns are shown in Fig. 10(e) and (f). We believe this clear shift in dihedral space at the time of transition would have isolated the remainder of the loop from the bulk motion of the protein reducing the magnitude of the internal stress due to protein–surface interactions ensuring the loop region of the protein could maintain direct contact with the surface while maintaining the structural integrity of the core.

Chain dynamics. From the discussion above the BFs1 systems appears to resist protein adsorption more than any of the other systems investigated here. As previously discussed, despite the competition between the protein and water for surface contact, the hydration layer at the interface at this surface is not sufficiently different to be the cause of this behavior. However, an explanation might arise in terms of the entropic cost associated with the protein adsorption on the more mobile hydrophilic chains on the BFs1 surface. To confirm this, Fig. 11 shows the RMSD of the B and short fluorinated chains with and without the protein present at the surface for BFs1.

The RMSD in the presence of the protein includes only those chains which come into direct contact with the protein. The overall chain mobility reduction is consistent across all ligand systems, with chain dynamics being restricted due the presence of the protein. A reduction in chain mobility can be qualitatively described as an entropic penalty taken by the chains associated with protein adsorption. Therefore, comparison of the loss of motion across the systems can provide a measure of the degree of this penalty. The relative reduction in chain dynamics can be measured as the percentage drop off in the RMSD over the last 2.5 ns of the RMSDs for the chains due to protein adsorption (Fig. 11). We found that in the BFs1 system when the protein is adsorbed the B chains lose ~17% of the mobility compared to the surface without the protein. In contrast the four other surfaces lose ~12–13% of B chain mobility due to protein adsorption compared to the free chains. The fluorinated chains, on average, lose 10% of the free motion when the protein is present. This more significant drop in ligand dynamics for the BFs1 pattern may indicate an additional entropic barrier for the adsorbing protein which makes it more difficult to overcome and may be indicative of an increased resistance of the BFs1 surface pattern to EAS adsorption.

(f) Ramachandran plot for Ser_{42} between 150 and 200 ns. For Fig. (c–f), the black region represents the accessed dihedral space pre-transition while the red is that accessed after the shift in the bulk protein.

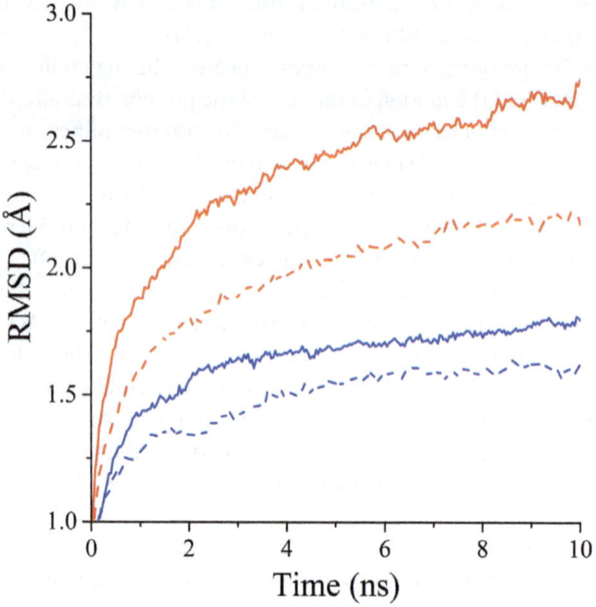

Fig. 11 RMSD of the functional chains for the BFs1 system without the protein (solid) and in the presence of the protein (dashed); B (red), F (blue).

While we lack a direct experimental counterpart to the modelling work presented here there is a trend in the experimental literature which supports the hypothesis that unfavourable entropic effects arising from a loss of chain dynamics may play a role in protein resistance. For example, it has been reported that well-ordered OEG SAMs formed on silver show a reduced ability to resist protein adsorption compared to those formed on gold substrates or under conditions which introduced disorder/defects in the SAMs.[106] The authors inferred that the improved adsorption resistance was predicated on a relaxed packing density or disorder/defects in the SAM both of which would result in increased chain flexibility consistent with the conclusion we draw here. Similarly increased disorder of OEG SAMs of various lengths on gold substrates was reported to, in conjunction with increased water retention within these disordered SAMs, improve the SAMs' protein resistant character.[32]

Conclusions and discussion remarks

In this work we have employed all-atom MD simulations to examine the adsorption process and early phase of the adsorbed behaviour for EAS hydrophobin at the interface between short functional chains tethered to an amorphous silica surface and water. Six systems were considered consisting of either hydrophilic or hydrophobic chains or mixtures thereof with varying degrees of nanoscale patterning. We have explored the impact of surface hydration, dynamics, chemistry and roughness on the protein adsorption to elucidate design principles for producing anti biofouling surfaces by grafting short functional chains with chemical and morphological heterogeneity.

It was found that a combination of both nanoscale chemical patterning and surface roughness, in the form of functional chain length variation, provided the most noticeable effect on protein adsorption. This result is predominately attributed to an increased entropic penalty arising from a more pronounced loss of conformational freedom by exposed (longer) hydrophilic chains upon protein adsorption compared to surfaces that do not possess this type of roughness. Furthermore, it was noticed that surfaces with larger domains displayed a reduced ability to deter protein adsorption once direct contact was made. We infer that key design strategies for silica based antifouling surfaces utilising short functional chains would involve a combination of short hydrophobic chains with longer hydrophilic chains, whilst ensuring a domain size on the order of a single chain width (noting the lower inherent chain density for the amorphous silica substrate employed here compared to typical SAMs on crystalline surfaces).

We found that for all studied surfaces no significant structuring of interfacial water was observed. This is attributed to the relatively high mobility of the chains, the majority of which had brush-like characteristics with significant conformational freedom, creating a dynamic interface not suitable for the propagation of interfacial water structuring through cooperative interactions between individual water molecules. However, the degree of conformational transitions for EAS in the adsorbed phase correlated with the dynamics of the hydration layer at the interface, indicating that competition for surface contact also influences the capacity of the ligands to reduce adsorption. Therefore, enhancing interfacial hydration by modifying the functional groups on the hydrophilic chains may enhance the capacity of the chains to reject the protein from the surface.

Our analysis of the adsorbed state of EAS hydrophobin revealed that due to the mixture of hydrophobic/hydrophilic domains on the surface of EAS, there was reasonably consistent adsorption behaviour across all systems, thus confirming the versatile ability of this protein to adsorb on most surfaces. We noticed a common adsorption motif whereby the highly flexible amphipathic Cys3–Cys4 loop stabilises the protein adsorption. Flexibility near the disulphide bond isolates this region from fluctuations in global protein motion, allowing independent movement of the loop to that of the β-core. It was also found that across all systems the Cys7–Cys8 loop had much less affinity towards the surface compared to the Cys3–Cys4 loop. This is consistent with the view that this loop is integral to monolayer formation through association with other EAS proteins, rather than being involved in protein–surface interactions. While there is limited experimental detail on the behaviour of EAS at the surface water interface to compare to, it is believed that further conformational rearrangement of the EAS would be undertaken when contact with another protein occurs. However, it should be noted that determining the comprehensive mechanism for EAS monolayer formation directly from the simulation is currently computationally prohibitive.

It is also important to highlight that it is currently computationally prohibitive to perform comprehensive conformational sampling to completely describe the individual protein's adsorbed state, and we noted that several runs were kinetically trapped in conformations with limited protein–surface interactions. Whilst enhanced sampling techniques are available, a combination of large system sizes, and the correlation of protein adsorption with the chain chemistry and morphology prevent adequate application of the standard enhanced sampling techniques for such study. Future simulations must be extended into microsecond time scales in order for slow and rare events to be observed such that the mechanism of surface expulsion can be elucidated in real time. Currently this can only be achieved through a coarse-graining approach which will lead to the undesirable loss of the atomistic detail of the systems.

From this work there are several directions for future study. We have identified the Cys3–Cys4 loop's ability to "neutralize" surface heterogeneity, so it would be useful to study the mutated EASΔ15 protein, which has 15 residues from the Cys3–Cys4 loop removed. Furthermore, it would be useful to investigate multiple proteins on bare and modified surfaces to understand the monolayer formation process of EAS and, in particular, if these surfaces are able to disrupt or prevent rodlet formation. It would also be useful to explore variations in interfacial rigidity through cross-linking or increased chain density to further study effects on interfacial water and the ability for EAS monomers to penetrate into more dense protective layers. Finally, more complex nanodomain patterning should also be explored, investigating patches/checkers or other regular or irregular arrangements.

Acknowledgements

The authors thank the Australian Research Council Industrial Transformation Research Hub for Australian Steel Manufacturing (IH130100017) for financial support of this research. KL acknowledges his postgraduate scholarship from the Commonwealth Scientific and Industrial Research Organisation (CSIRO). We also acknowledge the generous allocation of high performance computational resources from the Australian National Computational Infrastructure (NCI), the Western Australian computational facility (iVEC), the Victorian Partnership for Advanced Computing (VPAC) and the Victorian Life Sciences Computational Initiative (VLSCI).

References

1 P. Thevenot, W. Hu and L. Tang, *Curr. Top. Med. Chem.*, 2008, **8**, 270–280.
2 I. Lynch, A. Salvati and K. A. Dawson, *Nat. Nanotechnol.*, 2009, **4**, 546–547.
3 A. A. Shemetov, I. Nabiev and A. Sukhanova, *ACS Nano*, 2012, **6**, 4585–4602.
4 J. A. Callow and M. E. Callow, *Nat. Commun.*, 2011, **2**, 244.
5 I. Banerjee, R. C. Pangule and R. S. Kane, *Adv. Mater.*, 2011, **23**, 690–718.
6 L. Vroman, *Nature*, 1962, **196**, 476–477.
7 N. P. Boks, W. Norde, H. C. van der Mei and H. J. Busscher, *Microbiology*, 2008, **154**, 3122–3133.
8 J. J. Gray, *Curr. Opin. Struct. Biol.*, 2004, **14**, 110–115.
9 B. Kasemo, *Surf. Sci.*, 2002, **500**, 656–677.
10 F. A. Denis, P. Hanarp, D. S. Sutherland, J. Gold, C. Mustin, P. G. Rouxhet and Y. F. Dufrêne, *Langmuir*, 2002, **18**, 819–828.
11 N. Mitik-Dineva, P. R. Stoddart, R. Crawford and E. P. Ivanova, *Wiley Encyclopedia of Biomedical Engineering*, 2006, DOI: 10.1002/9780471740360.ebs0027.
12 A. Ulman, *Chem. Rev.*, 1996, **96**, 1533–1554.

13 J. C. Love, L. A. Estroff, J. K. Kriebel, R. G. Nuzzo and G. M. Whitesides, *Chem. Rev.*, 2005, **105**, 1103–1169.
14 S. Krishnan, C. J. Weinman and C. K. Ober, *J. Mater. Chem.*, 2008, **18**, 3405–3413.
15 S. A. Umoren and M. M. Solomon, *Open Mater. Sci. J.*, 2014, **8**, 39–54.
16 R. P. Bagwe, L. R. Hilliard and W. Tan, *Langmuir*, 2006, **22**, 4357–4362.
17 P. M. Price, J. H. Clark and D. J. Macquarrie, *J. Chem. Soc., Dalton Trans.*, 2000, **2**, 101–110.
18 P. Roach, D. Farrar and C. C. Perry, *J. Am. Chem. Soc.*, 2005, **127**, 8168–8173.
19 Q. Wei, T. Becherer, S. Angioletti-Uberti, J. Dzubiella, C. Wischke, A. T. Neffe, A. Lendlein, M. Ballauff and R. Haag, *Angew. Chem., Int. Ed. Engl.*, 2014, **53**, 8004–8031.
20 L. Shen and J. Zhu, *Adv. Colloid Interface Sci.*, 2016, **228**, 40–54.
21 A. Otten and S. Herminghaus, *Langmuir*, 2004, **20**, 2405–2408.
22 K. L. Cho, I. I. Liaw, A. H. F. Wu and R. N. Lamb, *J. Phys. Chem. C*, 2010, **114**, 11228–11233.
23 S. I. Jeon, J. H. Lee, J. D. Andrade and P. G. De Gennes, *J. Colloid Interface Sci.*, 1991, **142**, 149–158.
24 J. H. Lee, J. Kopecek and J. D. Andrade, *J. Biomed. Mater. Res.*, 1989, **23**, 351–368.
25 E. Ostuni, R. G. Chapman, M. N. Liang, G. Meluleni, G. Pier, D. E. Ingber and G. M. Whitesides, *Langmuir*, 2001, **17**, 6336–6343.
26 R. Wang, K. Hashimoto, A. Fujishima, M. Chikuni, E. Kojima, A. Kitamura, M. Shimohigoshi and T. Watanabe, *Nature*, 1997, **388**, 431–432.
27 P. Harder, M. Grunze, R. Dahint, G. M. Whitesides and P. E. Laibinis, *J. Phys. Chem. B*, 1998, **102**, 426–436.
28 N. M. Pawlowska, H. Fritzsche, C. Blaszykowski, S. Sheikh, M. Vezvaie and M. Thompson, *Langmuir*, 2014, **30**, 1199–1203.
29 I. Szleifer, *Phys. A*, 1997, **244**, 370–388.
30 I. Szleifer, *Biophys. J.*, 1997, **72**, 595–612.
31 S. Pasche, M. Textor, L. Meagher, N. D. Spencer and H. J. Griesser, *Langmuir*, 2005, **21**, 6508–6520.
32 L. Li, S. Chen, J. Zheng, B. D. Ratner and S. Jiang, *J. Phys. Chem. B*, 2005, **109**, 2934–2941.
33 S. Mitragotri and J. Lahann, *Nat. Mater.*, 2009, **8**, 15–23.
34 A. M. Jackson, J. W. Myerson and F. Stellacci, *Nat. Mater.*, 2004, **3**, 330–336.
35 M. D. Mager, V. LaPointe and M. M. Stevens, *Nat. Chem.*, 2011, **3**, 582–589.
36 S. C. Glotzer and M. J. Solomon, *Nat. Mater.*, 2007, **6**, 557–562.
37 J. J. Kuna, K. Voitchovsky, C. Singh, H. Jiang, S. Mwenifumbo, P. K. Ghorai, M. M. Stevens, S. C. Glotzer and F. Stellacci, *Nat. Mater.*, 2009, **8**, 837–842.
38 A. M. Jackson, Y. Hu, P. J. Silva and F. Stellacci, *J. Am. Chem. Soc.*, 2006, **128**, 11135–11149.
39 A. Hung, S. Mwenifumbo, M. Mager, J. J. Kuna, F. Stellacci, I. Yarovsky and M. M. Stevens, *J. Am. Chem. Soc.*, 2011, **133**, 1438–1450.
40 A. Hung, M. Mager, M. Hembury, F. Stellacci, M. M. Stevens and I. Yarovsky, *Chem. Sci.*, 2013, **4**, 928–937.
41 H. A. Wösten, *Annu. Rev. Microbiol.*, 2001, **55**, 625–646.
42 M. B. Linder, *Curr. Opin. Colloid Interface Sci.*, 2009, **14**, 356–363.
43 R. A. Latour, *Colloids Surf., B*, 2014, **124**, 25–37.

44 O. Cohavi, S. Corni, F. De Rienzo, R. Di Felice, K. E. Gottschalk, M. Hoefling, D. Kokh, E. Molinari, G. Schreiber, A. Vaskevich and R. C. Wade, *J. Mol. Recognit.*, 2010, **23**, 259–262.

45 P. Charchar, A. Christofferson, N. Todorova and I. Yarovsky, *Small*, 2016, **12**, 2395–2418, DOI: 10.1002/smll.201503585.

46 R. A. Latour, *Biointerphases*, 2008, **3**, FC2–FC12.

47 L. M. Szott and T. A. Horbett, *Curr. Opin. Chem. Biol.*, 2011, **15**, 683–689.

48 M. J. Penna, M. Mijajlovic, C. Tamerler and M. J. Biggs, *Soft Matter*, 2015, **11**, 5192–5203.

49 M. J. Penna, M. Mijajlovic and M. J. Biggs, *J. Am. Chem. Soc.*, 2014, **136**, 5323–5331.

50 T. Wei, M. A. Carignano and I. Szleifer, *Langmuir*, 2011, **27**, 12074–12081.

51 S. Sheikh, C. Blaszykowski, R. Nolan, D. Thompson and M. Thompson, *J. Colloid Interface Sci.*, 2015, **437**, 197–204.

52 Z. Benková, B. Szefczyk and M. N. l. D. S. Cordeiro, *Macromolecules*, 2011, **44**, 3639–3648.

53 G. Yiapanis, A. J. Christofferson, M. Plazzer, M. P. Weir, E. L. Prime, G. G. Qiao, D. H. Solomon and I. Yarovsky, *Langmuir*, 2013, **29**, 14451–14459.

54 G. Yiapanis, D. J. Henry, E. Evans and I. Yarovsky, *J. Phys. Chem. C*, 2010, **114**, 478–486.

55 G. Yiapanis, D. J. Henry, S. Maclaughlin, E. Evans and I. Yarovsky, *Langmuir*, 2012, **28**, 17263–17272.

56 G. Yiapanis, S. Maclaughlin, E. J. Evans and I. Yarovsky, *Langmuir*, 2014, **30**, 10617–10625.

57 K. J. Ley, L. A. Shaw, G. Yiapanis, S. MacLaughlin and I. Yarovsky, *Mol. Simul.*, 2015, **42**, 563–572.

58 A. A. Skelton, T. Liang and T. R. Walsh, *ACS Appl. Mater. Interfaces*, 2009, **1**, 1482–1491.

59 I. Yarovsky, M. T. W. Hearn and M. I. Aguilar, *J. Phys. Chem. B*, 1997, **101**, 10962–10970.

60 N. Hildebrand, S. Köppen, L. Derr, K. Li, M. Koleini, K. Rezwan and L. Colombi Ciacchi, *J. Phys. Chem. C*, 2015, **119**, 7295–7307.

61 K. Ley, A. Christofferson, M. Penna, D. Winkler, S. Maclaughlin and I. Yarovsky, *Front. Mol. Biosci.*, 2015, **2**, 64.

62 R. Konradi, C. Acikgoz and M. Textor, *Macromol. Rapid Commun.*, 2012, **33**, 1663–1676.

63 B. Pidhatika, M. Rodenstein, Y. Chen, E. Rakhmatullina, A. Muhlebach, C. Acikgoz, M. Textor and R. Konradi, *Biointerphases*, 2012, **7**, 1–15.

64 P. Greenwood and B. Gevert, *Pigm. Resin Technol.*, 2011, **40**, 275–284.

65 R. K. Iler, *The chemistry of silica: solubility, polymerization, colloid and surface properties, and biochemistry*, Wiley, New York, 1979.

66 A. H. Y. Kwan, R. D. Winefield, M. Sunde, J. M. Matthews, R. G. Haverkamp, M. D. Templeton and J. P. Mackay, *Proc. Natl. Acad. Sci. U. S. A.*, 2006, **103**, 3621–3626.

67 M. B. Linder, G. R. Szilvay, T. Nakari-Setälä and M. E. Penttilä, *FEMS Microbiol. Rev.*, 2005, **29**, 877–896.

68 A. De Simone, C. Kitchen, A. H. Kwan, M. Sunde, C. M. Dobson and D. Frenkel, *Proc. Natl. Acad. Sci. U. S. A.*, 2012, **109**, 6951–6956.

69 J. P. Mackay, J. M. Matthews, R. D. Winefield, L. G. Mackay, R. G. Haverkamp and M. D. Templeton, *Structure*, 2001, **9**, 83–91.
70 A. H. Kwan, I. Macindoe, P. V. Vukasin, V. K. Morris, I. Kass, R. Gupte, A. E. Mark, M. D. Templeton, J. P. Mackay and M. Sunde, *J. Mol. Biol.*, 2008, **382**, 708–720.
71 I. Macindoe, A. H. Kwan, Q. Ren, V. K. Morris, W. Yang, J. P. Mackay and M. Sunde, *Proc. Natl. Acad. Sci. U. S. A.*, 2012, **109**, E804–E811.
72 W. Yoshida, R. P. Castro, J.-D. Jou and Y. Cohen, *Langmuir*, 2001, **17**, 5882–5888.
73 D. J. Henry, E. Evans and I. Yarovsky, *J. Phys. Chem. B*, 2006, **110**, 15963–15972.
74 I. Yarovsky, M. L. Aguilar and M. T. W. Hearn, *Anal. Chem.*, 1995, **67**, 2145–2153.
75 B. P. Feuston and S. H. Garofalini, *J. Chem. Phys.*, 1989, **91**, 564–570.
76 G. Yiapanis, D. J. Henry, E. Evans and I. Yarovsky, *J. Phys. Chem. C*, 2007, **111**, 6465–6472.
77 G. Yiapanis, D. J. Henry, E. Evans and I. Yarovsky, *J. Phys. Chem. C*, 2007, **111**, 3000–3009.
78 G. Yiapanis, D. J. Henry, E. Evans and I. Yarovsky, *J. Phys. Chem. C*, 2008, **112**, 18141–18149.
79 A. D. MacKerell, D. Bashford, M. Bellott, R. Dunbrack, J. D. Evanseck, M. J. Field, S. Fischer, J. Gao, H. Guo and S. Ha, *J. Phys. Chem. B*, 1998, **102**, 3586–3616.
80 A. D. Mackerell Jr, M. Feig and C. L. Brooks 3rd, *J. Comput. Chem.*, 2004, **25**, 1400–1415.
81 I. Bahar, I. Zuniga, R. Dodge and W. L. Mattice, *Macromolecules*, 1991, **24**, 2986–2992.
82 W. L. Jorgensen, J. Chandrasekhar, J. D. Madura, R. W. Impey and M. L. Klein, *J. Chem. Phys.*, 1983, **79**, 926–935.
83 J. C. Phillips, R. Braun, W. Wang, J. Gumbart, E. Tajkhorshid, E. Villa, C. Chipot, R. D. Skeel, L. Kale and K. Schulten, *J. Comput. Chem.*, 2005, **26**, 1781–1802.
84 T. Darden, D. York and L. Pedersen, *J. Chem. Phys.*, 1993, **98**, 10089–10092.
85 J.-P. Ryckaert, G. Ciccotti and H. J. Berendsen, *J. Comput. Phys.*, 1977, **23**, 327–341.
86 D. A. McQuarrie, *Statistical Mechanics*, New York, 1976.
87 C. Singh, P. K. Ghorai, M. A. Horsch, A. M. Jackson, R. G. Larson, F. Stellacci and S. C. Glotzer, *Phys. Rev. Lett.*, 2007, **99**, 226106.
88 C. Singh, Y. Hu, B. P. Khanal, E. R. Zubarev, F. Stellacci and S. C. Glotzer, *Nanoscale*, 2011, **3**, 3244–3250.
89 R. Godawat, S. N. Jamadagni and S. Garde, *Proc. Natl. Acad. Sci. U. S. A.*, 2009, **106**, 15119–15124.
90 S. Granick and S. C. Bae, *Science*, 2008, **322**, 1477–1478.
91 A. Poynor, L. Hong, I. K. Robinson, S. Granick, Z. Zhang and P. A. Fenter, *Phys. Rev. Lett.*, 2006, **97**, 266101.
92 X. Zhang, Y. Zhu and S. Granick, *Science*, 2002, **295**, 663–666.
93 Y. Zhu and S. Granick, *Phys. Rev. Lett.*, 2001, **87**, 096104.
94 J. T. G. Overbeek, *J. Colloid Interface Sci.*, 1977, **58**, 408–422.
95 F. Wang, S. J. Stuart and R. A. Latour, *Biointerphases*, 2008, **3**, 9–18.

96 J. N. Israelachvili, *Surf. Sci. Rep.*, 1992, **14**, 109–159.
97 J. Yu, M. L. Becker and G. A. Carri, *Langmuir*, 2012, **28**, 1408–1417.
98 S. Chen, L. Li, C. Zhao and J. Zheng, *Polymer*, 2010, **51**, 5283–5293.
99 J. Zheng, L. Li, H. K. Tsao, Y. J. Sheng, S. Chen and S. Jiang, *Biophys. J.*, 2005, **89**, 158–166.
100 D. J. Henry, C. A. Lukey, E. Evans and I. Yarovsky, *Mol. Simul.*, 2005, **31**, 449–455.
101 L. A. Shaw, G. Yiapanis, D. J. Henry, S. MacLaughlin, E. Evans and I. Yarovsky, *Soft Matter*, 2013, **9**, 1798–1806.
102 M. Sunde, A. H. Kwan, M. D. Templeton, R. E. Beever and J. P. Mackay, *Micron*, 2008, **39**, 773–784.
103 J. A. Finlay, S. Krishnan, M. E. Callow, J. A. Callow, R. Dong, N. Asgill, K. Wong, E. J. Kramer and C. K. Ober, *Langmuir*, 2008, **24**, 503–510.
104 T. Wei, M. A. Carignano and I. Szleifer, *J. Phys. Chem. B*, 2012, **116**, 10189–10194.
105 M. Hoefling, S. Monti, S. Corni and K. E. Gottschalk, *PLoS One*, 2011, **6**, e20925.
106 S. Herrwerth, W. Eck, S. Reinhardt and M. Grunze, *J. Am. Chem. Soc.*, 2003, **125**, 9359–9366.

Faraday Discussions

PAPER

Anisotropic lanthanide-based nano-clusters for imaging applications†

Xiaoping Yang,[a] Shiqing Wang,[a] Tyler L. King,[b] Christopher J. Kerr,[c] Clement Blanchet,[c] Dmitri Svergun,[c] Robert Pal,[d] Andrew Beeby,[d] Jamuna Vadivelu,[e] Katherine A. Brown,*[cf] Richard A. Jones,*[c] Lijie Zhang[a] and Shaoming Huang*[a]

Received 18th February 2016, Accepted 31st March 2016
DOI: 10.1039/c6fd00018e

We have developed a new class of lanthanide nano-clusters that self-assemble using flexible Schiff base ligands. Cd–Ln and Ni–Ln clusters, $[Ln_8Cd_{24}(L^1)_{12}(OAc)_{39}Cl_7(OH)_2]$ (Ln = Nd, Eu), $[Eu_8Cd_{24}(L^1)_{12}(OAc)_{44}]$, $[Ln_8Cd_{24}(L^2)_{12}(OAc)_{44}]$ (Ln = Nd, Yb, Sm) and $[Nd_2Ni_4(L^3)_2(acac)_6(NO_3)_2(OH)_2]$, were constructed using different types of flexible Schiff base ligands. These molecular nano-clusters exhibit anisotropic architectures that differ considerably depending upon the presence of Cd (nano-drum) or Ni (square-like nano-cluster). Structural characterization of the self-assembled particles has been undertaken using crystallography, transmission electron microscopy and small-angle X-ray scattering. Comparison of the metric dimensions of the nano-drums shows a consistency of size using these techniques, suggesting that these molecules may share similar structural features in both solid and solution states. Photophysical properties were studied by excitation of the ligand-centered absorption bands in the solid state and in solution, and using confocal microscopy of microspheres loaded with the compounds. The emissive properties of these compounds vary depending upon the combination of lanthanide and Cd or Ni present in these clusters. The results provide new insights into the construction of novel high-nuclearity nano-clusters and offer a promising foundation for the development of new functional nanomaterials.

[a]*College of Chemistry and Materials Engineering, Wenzhou University, Zhejiang Key Laboratory of Carbon Materials, Wenzhou 325035, China. E-mail: smhuang@wzu.edu.cn; Tel: +86 577 88373064*

[b]*The University of Texas at Austin, Department of Chemistry, Austin, Texas, 78712, USA. E-mail: rajones@cm.utexas.edu; Tel: +1 512 4711706*

[c]*European Molecular Biology Laboratory, Hamburg Unit, EMBL c/o DESY, Hamburg, 22607, Germany*

[d]*Department of Chemistry, University of Durham, South Road, Durham, DH1 3LE, UK*

[e]*Department of Medical Microbiology, University of Malaya, Kuala Lumpur 50603, Malaysia*

[f]*Cavendish Laboratory, Department of Physics, University of Cambridge, Cambridge CB3 0HE, UK. E-mail: kb518@cam.ac.uk*

† Electronic supplementary information (ESI) available: Characterization details for clusters **1–3**. CCDC 1450317–1450319. For ESI and crystallographic data in CIF or other electronic format see DOI: 10.1039/c6fd00018e

Introduction

Lanthanide nanomaterials are being increasingly investigated for applications in bio-imaging and molecular recognition,[1] due in part to their advantageous photophysical properties, which include a very large pseudo-Stokes shift between the excitation and emission wavelengths, the absence of photo-bleaching, long lived excited states and narrow emission bands.[1,2] Since emissions *via* 4f–4f transitions are parity forbidden,[3] Ln^{3+} ions usually have a relatively long photoluminescence lifetime of the excited state. This can allow for gated and time-resolved photoluminescence detection and thus can eliminate interference from short-lived autofluorescence.

High-nuclearity heterometallic nano-clusters composed of d-block transition metals and lanthanide ions (Ln) have received extensive attention due not only to their aesthetically stunning molecular structures but also their versatile applications in optoelectronics, magnetism, and as probes in biological systems.[4] Most trivalent lanthanide ions exhibit long-lived (in the microsecond-to-millisecond range) and line-like emission bands at characteristic wavelengths, *e.g.* Tb^{3+} (green), Eu^{3+} (red) and Yb^{3+}, Nd^{3+}, and Er^{3+} (near infra-red, NIR). These characteristics allow time-gated detection. More specifically, the NIR emitters have some potential applications in bioassays and luminescent probes because of the transparency of biological tissues to electromagnetic radiation in the range of 0.8–1.3 μm.[5] In high-nuclearity d–f clusters, light-absorbing d-block metal chromophores (*i.e.* Pt^{II},[6a,b] Ru^{II},[6c,d] Zn^{II},[6e,f] and Cd^{II} [6g,h]) can act as sensitizers for luminescence from Ln(III) centers following d → f energy-transfer. Many reports have so far been focused on high-nuclearity 3d–4f clusters, such as Cu/Ln,[7a] Mn/Ln[7b,c] and Ni/Ln[7d,e] cluster systems, in order to study their magnetic properties as single-molecule magnets. In contrast, high-nuclearity 4d–f systems with luminescence properties have received much less attention.

Compartmental Schiff bases with two dissimilar metal-binding sites, one being specific for the d metal ion and another for the f metal ion, have been employed to synthesize d–f heteronuclear clusters.[8] Recent studies in our laboratories have focused on the construction of luminescent d–f clusters with flexible Schiff base ligands.[9] Flexible ligands may provide more possibilities for the construction of unique frameworks because of their freedom of conformation. In our previous studies, the use of Schiff base ligands which have flexible carbon–carbon backbones containing from 2 to 4 methylene (CH_2) units, resulted in a variety of binuclear,[10a] trinuclear,[10b] tetranuclear,[10c] and hexanuclear[10d,e] 3d–4f complexes (3d = Zn, Cu and Ni). Recently, we discovered a series of 24-metal Cd–Ln complexes $[Ln_6Cd_{18}(L^1)_9Cl_8(OAc)_{28}]$ from reactions of a flexible Schiff base ligand, *N,N'*-bis(5-bromo-3-methoxysalicylidene)hexane-1,6-diamine (H_2L^1), which has a 6 carbon backbone (Scheme 1), with $Cd(OAc)_2\cdot4H_2O$ and $LnCl_3\cdot6H_2O$.[9b] The structures of these metal clusters are often influenced by a variety of factors such as the ligand structure, the nature of the counter anions, and the pH value of the surrounding environment. As part of our continuing studies focused on the construction of luminescent polynuclear lanthanide-based frameworks, we report here seven d–f clusters with H_2L^1, H_2L^2 and long-chain Schiff base ligand *N,N'*-bis(5-bromo-3-methoxysalicylidene) (*N''*-(3-amion-propyl)-propane)-1,6-diamine (H_2L^3), which has a flexible $(CH_2)_3NH(CH_2)_3$ backbone

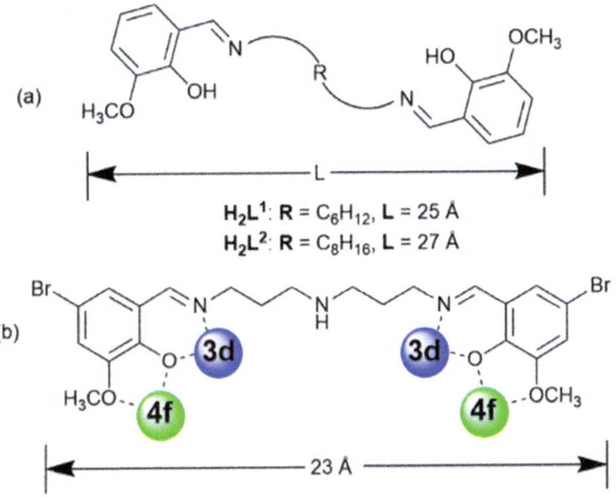

Scheme 1 (a) Flexible Schiff base ligands; (b) the "stretching" coordination mode of H_2L^3.

(Scheme 1b). They are $[Ln_8Cd_{24}(L^1)_{12}(OAc)_{39}Cl_7(OH)_2]$ (Ln = Nd (**1**) and Eu (**2**)), $[Nd_2Ni_4(L^3)_2(acac)_6(NO_3)_2(OH)_2]$ (**3**), $[Ln_8Cd_{24}(L^2)_{12}(OAc)_{48}]$ (Ln = Eu (**4**) and Sm (**5**)) and $[Ln_8Cd_{24}(L^2)_{12}(OAc)_{44}Cl_4]$ (Ln = Nd (**6**) and Yb (**7**)). **1** and **2** were synthesized from the reactions of H_2L^1 with $CdCl_2 \cdot 2H_2O$ and $Ln(OAc)_3 \cdot 4H_2O$ in the presence of NaOH. Interestingly, differing from our previously reported Cd–Ln complexes $[Ln_6Cd_{18}(L^1)_9Cl_8(OAc)_{28}]$, **1** and **2** exhibit 32-metal anisotropic nano-drum-like architectures. Hydroxide (OH^-) anions are found in the structures of **1** and **2**, indicating that the basic environment favors the formation of these complexes. The structures appear to be ligand and anion dependent. Compared with H_2L^1 and H_2L^2, H_2L^3 features a backbone NH group. Thus, **3** shows a 6-metal square-like structure. The flexible Schiff-base ligands H_2L^1, H_2L^2 and H_2L^3 exhibit "stretched" coordination modes with metal ions (Scheme 1b), resulting in the large sizes of **1–7**. For example, the sizes of **1** and **2** are approximately 19 × 20 × 20 Å, which are much larger than most other d–f Schiff base polynuclear complexes thus far reported. To further explore the self-assembly properties and imaging of these molecules, we also report preliminary small angle X-ray scattering measurements and confocal imaging of Cd–Ln complexes containing the H_2L^2 Schiff base ligand.

Experimental section

General information

All reactions were performed under dry oxygen-free dinitrogen atmospheres using standard Schlenk techniques. The Schiff-base ligands $H_2L^{1,2,3}$ were prepared according to well-established procedures.[11] Physical measurements: NMR: VARIAN UNITY-plus 600 spectrometer (^1H, 600 MHz) at 298 K; IR: Nicolet IR 200 FTIR spectrometer. Elemental analyses (C, H, N) were carried out on a EA1112 elemental analyser. Melting points were obtained in sealed glass capillaries under dinitrogen and are uncorrected. Conductivity measurements were carried out with a DDS-11 conductivity bridge for 10^{-3} M solutions in CH_3CN. Transmission electron microscopy (TEM) images were recorded on a JEOL JEM-1200EX transmission electron microscope. Scanning electron microscopy (SEM) images were recorded on a Nova NanoSEM 200 scanning electron microscope. Absorption spectra were obtained on a UV-3600 spectrophotometer, and excitation and emission spectra on a QuantaMaster PTI fluorimeter.

Syntheses

$[Nd_8Cd_{24}(L^1)_{12}(OAc)_{39}Cl_7(OH)_2]$ (1). $CdCl_2 \cdot 2H_2O$ (0.0439 g, 0.20 mmol), $Nd(OAc)_3 \cdot 4H_2O$ (0.0393 g, 0.10 mmol) and H_2L^1 (0.0770 g, 0.20 mmol) were dissolved in MeOH (20 mL) at room temperature, and a solution of NaOH in EtOH (0.04 mol L^{-1}, 10 mL) was then added. The resulting solution was stirred and heated under reflux (30 min). The mixture was allowed to cool and was then filtered. Diethyl ether was allowed to diffuse slowly into the filtrate at room temperature and pale yellow crystals were obtained after one week. The crystals were filtered off, washed with EtOH (5 mL) and dried in the air. Yield (based on $CdCl_2 \cdot 2H_2O$): 0.0623 g (63%). Mp > 217 °C (dec.). Found: C, 36.55; H, 4.71; N, 2.39%. Calc. for $C_{342}H_{428}Cd_{24}Cl_7Nd_8N_{24}O_{128}(EtOH)_6(EtOEt)_2(MeOH)_8(H_2O)_{15}$: C, 37.08; H, 4.56; N, 2.81%. IR (CH_3OH, cm^{-1}): 3091 (w), 2935 (w), 1630 (s), 1575 (m), 1472 (m), 1350 (w), 1309 (m), 1239 (m), 1213 (s), 1082 (m), 1045 (m), 962 (w), 847 (m), 795 (s), 672 (w), 616 (w).

$[Eu_8Cd_{24}(L^1)_{12}(OAc)_{39}Cl_7(OH)_2]$ (2). The procedure was the same as that for **1** using $Eu(OAc)_3 \cdot 4H_2O$ (0.0418 g, 0.10 mmol). Pale yellow single crystals of **2** were formed after one week. Yield (based on $CdCl_2 \cdot 2H_2O$): 0.0519 g (52%). Mp > 220 °C (dec.). Found: C, 36.72; H, 4.57; N, 2.51%. Calc. for $C_{342}H_{428}Cd_{24}Cl_7Eu_8N_{24}O_{128}(EtOH)_5(EtOEt)_2(MeOH)_6(H_2O)_{12}$: C, 37.00; H, 4.43; N, 2.83%. IR (CH_3OH, cm^{-1}): 3099 (w), 1637 (m), 1578 (s), 1470 (s), 1410 (m), 1341 (m), 1318 (w), 1210 (s), 1088 (m), 1039 (m), 962 (m), 855 (m), 798 (s), 673 (m), 638 (m), 615 (w).

$[Nd_2Ni_4(L^3)_2(acac)_6(NO_3)_2(OH)_2]$ (3). The procedure was the same as that for **1** using $Ni(acac)_2 \cdot 2H_2O$ (0.1534 g, 0.52 mmol), $Nd(NO_3)_3 \cdot 6H_2O$ (0.0526 g, 0.12 mmol) and H_2L^3 (0.26 mmol, 0.1448 g). Pale green single crystals of **3** were formed after two weeks. Yield (based on $Nd(NO_3)_3 \cdot 6H_2O$): 0.0756 g (52%). Mp > 178 °C (dec.). Found: C, 26.39; H, 2.13; N, 4.16%. Calc. for $C_{56}H_{50}Br_4Nd_2N_8Ni_4O_{52}$: C, 26.86; H, 2.00; N, 4.48%. IR (CH_3OH, cm^{-1}): 2933 (m), 1631 (m), 1575 (s), 1432 (s), 1400 (s), 1301 (m), 1205 (s), 1072 (m), 1022 (w), 970 (w), 845 (w), 730 (s), 666 (m).

$[Eu_8Cd_{24}(L^2)_{12}(OAc)_{48}]$ (4). $Cd(OAc)_2 \cdot 2H_2O$ (0.1382 g, 0.52 mmol), $Eu(OAc)_3 \cdot 4H_2O$ (0.0502 g, 0.12 mmol) and H_2L^2 (0.1073 g, 0.26 mmol) were dissolved in MeOH (30 mL) at room temperature and 4 drops of Et_3N were added. The resulting solution was stirred and heated under reflux for 1 hour. The mixture was then allowed to cool to room temperature and was gravity filtered. Diethyl ether was allowed to diffuse slowly into the filtrate at room temperature and yellow crystals were obtained after one week. The crystals were washed with EtOH (5 mL) and Et_2O (5 mL) and collected and dried in the air. Yield (based on $Eu(OAc)_3 \cdot 4H_2O$): 0.0478 g (53%). Mp > 216 °C (dec.). IR (neat, cm^{-1}) 2926 (w), 2825 (w), 1630 (m), 1552 (s), 1403 (s), 1345 (m), 1305 (m), 1210 (s), 1167 (m), 1097 (m), 1078 (m), 1017 (w), 964 (w), 852 (w), 735 (s), 666 (m).

$[Sm_8Cd_{24}(L^2)_{12}(OAc)_{48}]$ (5). The synthesis of **5** was the same as that for **4** using $Sm(OAc)_3 \cdot 4H_2O$ (0.0482 g, 0.12 mmol). Yellow crystals of **5** were formed after one week. Yield (based on $Sm(OAc)_3 \cdot 4H_2O$): 0.0419 g (47%). Mp > 207 °C (dec.). IR

(CH_3OH, cm^{-1}): 2926 (w), 1631 (m), 1575 (s), 1558 (s), 1471 (m), 1410 (s), 1306 (m), 1212 (s), 1074 (m), 1016 (w), 966 (w), 854 (w), 734 (s), 665 (w).

[$Nd_8Cd_{24}(L^2)_{12}(OAc)_{44}Cl_4$] (6). The synthesis of **6** was the same as that for **4** using $NdCl_3 \cdot 6H_2O$ (0.0431 g, 0.12 mmol). Yellow crystals were obtained after one week. Yield (based on $NdCl_3 \cdot 6H_2O$): 0.0712 g (45%). Mp > 209 °C (dec.). IR (CH_3OH, cm^{-1}): 3374 (w), 2930 (w), 1631 (m), 1572 (s), 1470 (m), 1410 (s), 1347 (w), 1307 (m), 1238 (m), 1212 (s), 1080 (m), 1048 (m), 963 (w), 849 (m), 738 (s), 673 (s), 640 (m).

[$Yb_8Cd_{24}(L^2)_{12}(OAc)_{44}Cl_4$] (7). The synthesis of **7** was the same as that for **4** using $YbCl_3 \cdot 6H_2O$ (0.0466 g, 0.12 mmol). Pale yellow single crystals of **7** were formed after one week. Yield (based on $YbCl_3 \cdot 6H_2O$): 0.0612 g (37%). Mp > 209 °C (dec.). IR (CH_3OH, cm^{-1}): 2930 (w), 1635 (m), 1576 (s), 1468 (m), 1407 (s), 1340 (w), 1309 (m), 1238 (m), 1212 (s), 1078 (m), 1020 (m), 962 (m), 854 (m), 735 (s), 676 (m), 638 (m).

X-ray analysis

Data were collected using a Smart APEX CCD diffractometer with graphite monochromated Mo-Kα radiation ($\lambda = 0.71073$ Å) at 223 K. The data set was corrected for absorption based on multiple scans and reduced using standard methods. Data reduction was performed using DENZO-SMN.[12] The structures were solved by direct methods and refined anisotropically using full-matrix least-squares methods with the SHELX 97 program package.[12] Coordinates of the non-hydrogen atoms were refined anisotropically, while hydrogen atoms were included in the calculation isotropically but not refined. Neutral atom scattering factors were taken from Cromer and Waber.[14]

For the crystal structures of **1** and **2**, some uncoordinated solvent molecules such as CH_3OH, C_2H_5OH, $C_2H_5OC_2H_5$ and H_2O molecules were found to be badly disordered. Attempts to model the disorder were unsatisfactory. The contributions to the scattering factors due to these solvent molecules were removed by use of the utility SQUEEZE[15a] in PLATON98.[15b,c] PLATON98 was used as incorporated in WinGX.[15d] Crystallographic data for **1–3** (CCDC reference numbers 1450317–1450319) are presented in Table S1 and selected bond lengths are given in Tables S2–S4.†

1: $C_{342}H_{428}Cd_{24}Cl_7N_{24}Nd_8O_{128}$, orthorhombic, space group *Pna*2(1), a = 54.734(11), b = 27.619(12), c = 35.243(18) Å, $\alpha = 90°$, $\beta = 90°$, $\gamma = 90°$, V = 53 277(18) Å^3, $Z = 4$, Dc = 1.374 g cm^{-3}, μ(Mo-Kα) = 1.798 mm^{-1}, $F(000)$ = 21 692, T = 223 K. R_1 = 0.1018, wR_2 = 0.2608 for 90 178 independent reflections with a goodness-of-fit of 1.038.

2: $C_{342}H_{428}Cd_{24}Cl_7N_{24}Eu_8O_{128}$, orthorhombic, space group *Pna*2(1), a = 54.106(11), b = 27.127(5), c = 34.876(7) Å, $\alpha = 90°$, $\beta = 90°$, $\gamma = 90°$, V = 51 189(18) Å^3, $Z = 4$, Dc = 1.438 g cm^{-3}, μ(Mo-Kα) = 2.040 mm^{-1}, $F(000)$ = 21 788, T = 223 K. R_1 = 0.1035, wR_2 = 0.2090 for 86 518 independent reflections with a goodness-of-fit of 0.922.

3: $C_{56}H_{50}Br_4N_8Nd_2Ni_4O_{52}$, monoclinic, space group $P2_1/c$, a = 15.462(3), b = 19.156(4), c = 19.871(4) Å, $\alpha = 90°$, $\beta = 107.63°(3)$, $\gamma = 90°$, V = 5609.2(19) Å^3, Z = 2, Dc = 1.481 g cm^{-3}, μ(Mo-Kα) = 3.073 mm^{-1}, $F(000)$ = 2452, T = 223 K. R_1 = 0.0781, wR_2 = 0.2092 for 9806 independent reflections with a goodness-of-fit of 1.061.

Small-angle X-ray scattering (SAXS)

SAXS data for $[Eu_8Cd_{24}(L^2)_{12}(OAc)_{48}]$ (**4**) were collected on the SAXS beamline P12 at the PETRA III storage ring (Deutsches Elektronen-Synchrotron, Hamburg).[13] Using a PILATUS 2M pixel detector at a sample-detector distance of 3.0 m and at an energy of 9.7 keV ($\lambda = 1.28$ Å), the range of momentum transfer $0.01 < s < 0.45$ Å^{-1} was covered ($s = 4\pi \sin \theta/\lambda$, where 2θ is the scattering angle). Solute concentrations of 1.8 and 3.7 mg mL^{-1} were measured in MeOH/MeCN (50 : 50) at 20 °C. Samples were loaded manually into the observation capillary. Primary data processing steps were performed using the automated data pipeline SASFLOW.[16]

Polystyrene bead loading

Luminex MicroPlex microspheres (6 μm cross-linked polystyrene beads with surface carboxyl groups) were prepared for loading by drying 12.5×10^6 beads under high vacuum for 24 hours. The dried beads were resuspended in a chloroform/MeOH solution (1 mL, 50 : 50) containing 5 mg of sample. This suspension was rotated slowly for 2 days. The suspension was then centrifuged, the supernatant discarded and the beads washed with MeOH (3 × 1 mL) followed by drying under high vacuum (24 h).

Confocal microscopy of polystyrene beads

Images were recorded using a Leica SP5 II confocal microscope equipped with a HCX PL APO 63x/1.40 NA LambdaBlue Oil immersion objective. Data were collected using 5× digital magnification at a 400 Hz/line scan speed (4 line average, bidirectional scanning) at 355 nm (3rd harmonic NdYAG laser) with 3 mW laser power. In order to achieve excitation with maximal probe emission, the microscope was equipped with a triple channel imaging detector, comprising two conventional PMT systems and a HyD hybrid avalanche photodiode detector. The latter part of the detection system, when operated in the BrightRed mode, is capable of improving imaging sensitivity by 25%, reducing signal to noise by a factor of 5. Frame size was determined at 2048 × 2048 pixels, with a 0.6 airy disc unit determining the applied pinhole diameter rendering one voxel to correspond to 24.02 × 24.02 nm (frame size 49.16 × 49.16 μm) with a section thickness of 380 nm. This particular LSCM is equipped with a novel structural illumination module called PhMoNa (achievable resolution 62 × 62 × 280 nm).[17] To detect sensitized Ln emission, a corresponding detection window of 400–800 nm was used to record images using only the above detailed 355 nm laser line.

Results and discussion

In the presence of NaOH (0.01 mol L^{-1}), reactions of H_2L^1 with $CdCl_2 \cdot 2H_2O$ and $Ln(OAc)_3 \cdot 4H_2O$ in refluxing methanol/ethanol produced yellow solutions from which **1** and **2** were isolated as pale yellow crystalline solids. **1** and **2** are isomorphous and have 32-metal drum-like structures. Two views of the crystal structure of **1** are shown in Fig. 1a. The top view is essentially a side-on view while the lower one is looking down into the top of the drum. The ends of the drum are created by two rings of 16 metals (4 Nd(III) and 12 Cd(II)) coordinated to half of the N, O binding groups of the 12 Schiff base ligands. The sides of the drum are

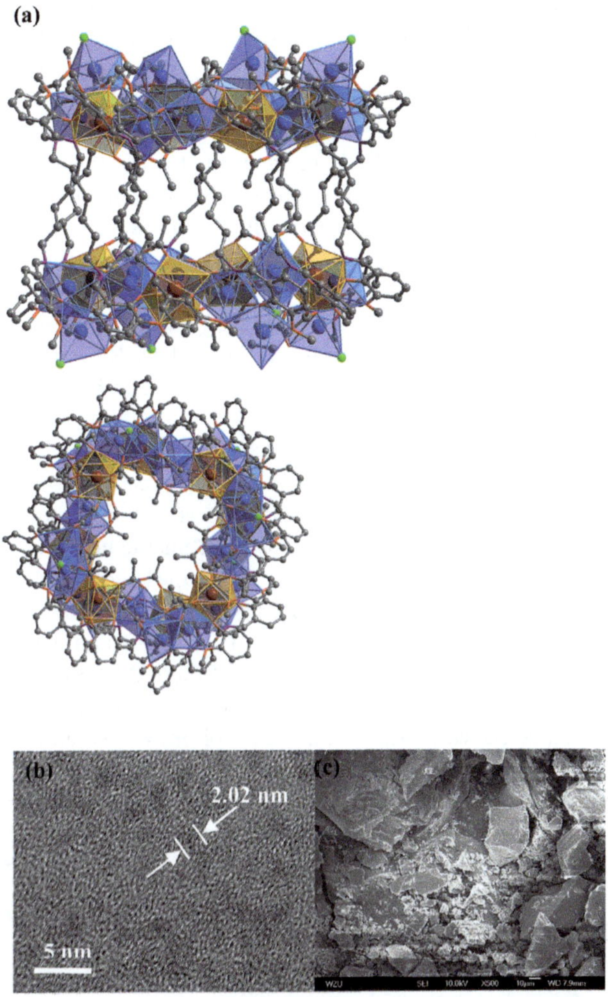

Fig. 1 (a) Two views of the nano-drum-like structure of **1**: viewed along the *b*-axis (top) and the *ac*-axis (lower) (Nd^{3+}: brown; Cd^{2+}: blue; Cl^-: green); (b and c) TEM and SEM images of **1**.

formed by the $-(CH_2)_6-$ linkers of the Schiff base ligands. Regarding the two Nd_4Cd_{12} rings, one includes four Cl^-, one OH^- and nineteen OAc^- anions and the other has three Cl^-, one OH^- and twenty OAc^- anions to balance the charge of the cluster. In **1**, the Cd–O and Cd–N bond lengths range from 2.163 Å to 2.646 Å and 2.210 Å to 2.406 Å, respectively. The Nd–O and Nd–N bond lengths range from 2.208 Å to 2.652 Å and 2.527 Å to 2.650 Å, respectively. In **1** and **2**, each Ln^{3+} ion and its closest two Cd^{2+} ions are linked by phenolic oxygen atoms of the L^1 ligand and OAc^- anions.

The Cd(II) ion tends to have relatively high coordination numbers (6–8) and exhibits variable coordination geometries, somewhat similar to Ln(III). This geometry may favor the formation of high-nuclearity Cd–Ln clusters **1** and **2**. Compared with Cd(II), the Ni(II) ion tends to have lower coordination numbers

(4–6) and exhibits some common coordination geometries such as square-planar, tetragonal pyramid and octahedral. Thus, the reaction of H_2L^3 with Ni(acac)$_2\cdot 2H_2O$ and $Nd(NO_3)_3\cdot 6H_2O$ under the same experimental conditions as above produced green solutions from which **3** was isolated as a green crystalline solid. Two views of the square-like structure of **3** are shown in Fig. 2. The top view is looking down into the top of the rectangle while the lower view is essentially a side-on view. The X-ray structure of **3** reveals a centrosymmetric core with two equivalent $NdNi_2(acac)_3(NO_3)(OH)$ moieties linked by two L^2 ligands. In each $NdNi_2(acac)_3(NO_3)(OH)$ moiety, the Nd^{3+} ion is 8-coordinate, surrounded by eight oxygen atoms from two L^2 ligands, one acac$^-$, one NO_3^- and one OH^- ion. Both Ni^{2+} ions have bi-pyramidal geometries. They are bridged by one acac$^-$ and one OH^- ion with a separation of 3.237 Å. The Nd^{3+} ion and two Ni^{2+} ions are linked by phenolic oxygen atoms of the L^2 ligand and OH^- anions. The average distance between the Nd^{3+} and Ni^{2+} ions is 3.413 Å.

It is noticeable that the Schiff base ligands exhibit a "stretched" coordination mode with metal ions in **1–3** (Scheme 1b). This may be due to the fact that, compared with those Schiff base ligands with shorter carbon-carbon backbones, there is less of a chelating or templating effect.[1,2] Thus the heights of the drum-like and square-like structures are mainly decided by the lengths of the Schiff base ligands, resulting in the formation of nanoscale clusters. The molecular dimensions of **1** and **2** are bigger than those of **3** (*i.e.* 19 × 20 × 20 Å for **1** *vs.* 8 × 12 × 18 Å for **3**). It is also possible to obtain images of these Cd–Ln molecular nanoparticles using transmission electron microscopy (TEM). Dilute solutions of **1** in CH_3CN were contacted with a Cu grid and the solvent was carefully evaporated under vacuum. The TEM images obtained (Fig. 1b) show uniform nanoparticles with diameters measuring approximately 2.02 nm which corresponds well with the diameter of the 16-metal ring end of the drum found in the crystal structure, indicating that the nano-cluster retains its unique molecular structure in

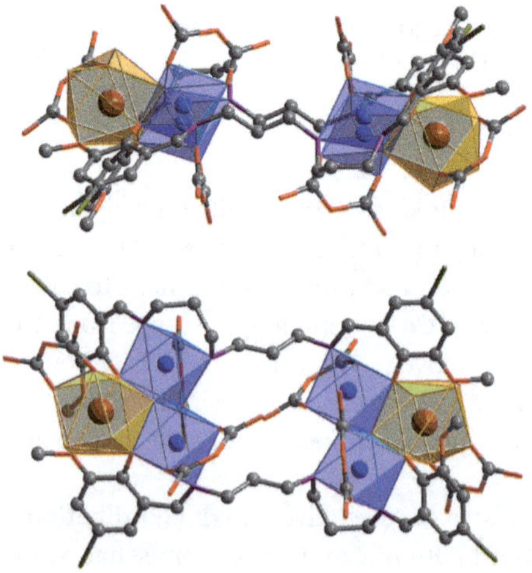

Fig. 2 Two views of the square-like structure of **3**: viewed along the *ac*-axis (top) and the *b*-axis (lower) (Nd^{3+}: brown; Ni^{2+}: blue).

solution. In Fig. 1c a panoramic scanning electron microscopy (SEM) image shows the crystalline nature of **1**.

The nature of **1** in solution has been probed using molar conductivity and ^{1}H NMR studies. Molar conductivity studies of **1** in CH_3CN also confirm that this molecular nanoparticle is neutral in solution, in accordance with the solid state structure. The ^{1}H NMR spectrum of **1** in $CDCl_3$ contains multiple broad peaks ranging from −17 to +18 ppm, due to the paramagnetic Nd^{3+} ion. The ^{1}H NMR spectrum of **1** remains unchanged for several weeks, indicating that it is stable in solution (Fig. S1†).

The nature of **4** in solution has been probed using small-angle X-ray scattering (SAXS). SAXS data were obtained at 1.8 and 3.7 mg mL^{-1}, as described above, and merged using the ALMERGE[18] software as shown in Fig. 3a. Bragg peaks in the scattering data appear to be arising from undissolved material still in a crystalline state. Evaluation of the particle distance distribution function *P*(*r*) was undertaken using GNOM[19] and shows a bi-modal distribution, with a maximum particle dimension of approximately 2.7 nm, consistent with the crystal structure of this nano-drum.[20] However, further studies are required to establish conditions that

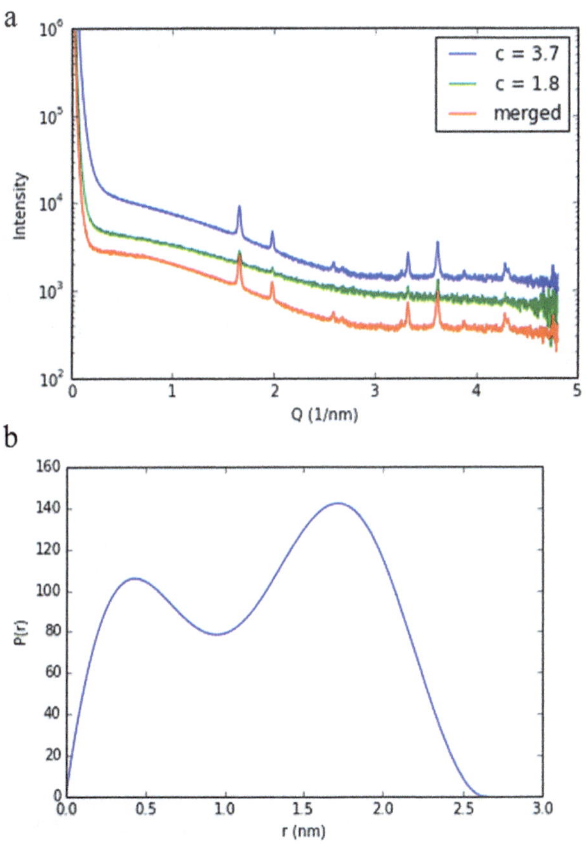

Fig. 3 SAXS analysis of compound **4**. (a) SAXS curves of data collected at 1.8 and 3.7 mg mL^{-1} are shown with the merged curve that was used as an input to derive (b) the particle distance function, *P*(*r*), plot.

are purely monodisperse, without the presence of undissolved crystalline material and/or nano-cluster substructures.

The photophysical properties of **1–3** were studied in both solution and the solid state. The d–f clusters' absorption bands in the UV-vis region are all red-shifted compared to those of the ligands alone ($H_2L^{1,2}$) (Fig. 4). For the free ligands, $H_2L^{1,2}$, excitations produce broad emission bands at $\lambda_{max} = 515$ nm and 509 nm, respectively (Fig. S2†). The emission and excitation spectra of **1–3** in CH_3CN are shown in Fig. 5, 6 and S3.† Upon excitation of the ligand-centered absorption bands, **1** and **3** show the NIR luminescence of Nd^{3+} ($^4F_{3/2} \rightarrow {}^4I_{j/2}$ transitions, $j = 9$, 11 and 13). As shown in Fig. 6, the emissions at 873 nm can be assigned to the $^4F_{3/2} \rightarrow {}^4I_{9/2}$ transition, 1063 nm to the $^4F_{3/2} \rightarrow {}^4I_{11/2}$ transition and 1350 nm to the $^4F_{3/2} \rightarrow {}^4I_{13/2}$ transition of Nd^{3+}. The free Schiff base ligands $H_2L^{1,2}$ and $Nd(OAc)_3 \cdot 4H_2O$ do not exhibit NIR luminescence in CH_3CN under similar conditions. Upon excitation of the ligand centered absorption bands, **2** shows typical visible emission bands of the Eu^{3+} ion ($^5D_0 \rightarrow {}^7F_j$ transitions, $j = 0$, 1, 2, 3 and 4). As shown in Fig. 5, the appearance of the symmetry-forbidden emission $^5D_0 \rightarrow {}^7F_0$ at 579 nm indicates that the Eu^{3+} ions in **2** occupy sites with low symmetry and have no inversion center.[21]

This is further confirmed by the intensity ratio of 3.9 for $I(^5D_0 \rightarrow {}^7F_2)/I(^5D_0 \rightarrow {}^7F_1)$, which is a good measure of the nature and symmetry of the first coordination sphere of the Eu^{3+} ion.[22] The overall quantum yield of **2** in CH_3CN was determined as 0.083 relative to $[Ru(bipy)_3]Cl_2$ in water (bipy = 2,2′-bipyridine; $\Phi_{em} = 0.028$)[23] and corrected for the refractive index of the solvent. **1** and **2** show two excitation bands at approximately 280 nm and 350 nm, while **3** has one at approximately 360 nm, in agreement with their absorption spectra, confirming that the energy transfers from the ligand centers to Ln^{3+} ions occur (Scheme 2). For each Cd–Ln or Ni–Nd nano-cluster the luminescence spectrum in the solid state is similar to that in solution.

For the Nd(III) nano-clusters **1** and **3**, we were naturally interested in the influence of their structures and d metals on their photophysical properties. The relative emission intensities of **1** and **3** were determined under the same

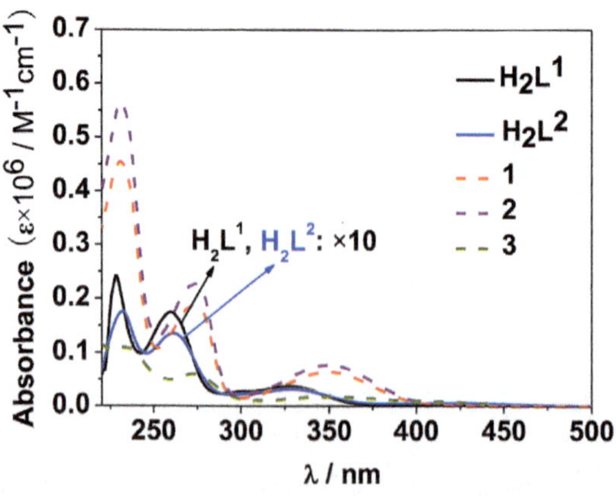

Fig. 4 UV-vis spectra of the free ligands $H_2L^{1,2}$ and clusters **1–3** in CH_3CN ($C = 10^{-8}$ to 10^{-7} M).

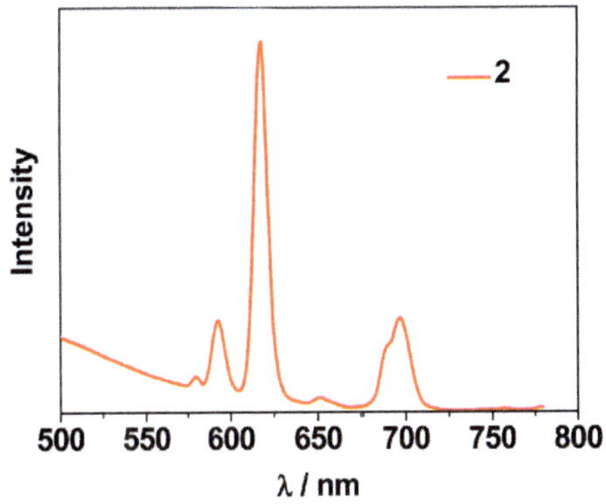

Fig. 5 The visible emission spectrum of **2** in CH_3CN.

experimental conditions. With the same absorbance value at 355 nm, the relative emission intensity at 1068 nm was estimated to be 5.4 for **1** : **3** in CH_3CN (Fig. 6), indicating that with the central metal ions encapsulated by chromophoric Cd/L^1 components (energy transfer donors), **1** has superior luminescence properties compared to **3**. It is also noticeable that, compared to the "enclosed" nano-drum-like structure in **1** the "open" square-like structure of **3** results in metal centers which are less well shielded from outside solvents which may quench lanthanide luminescence.[24] In addition, the d-block metal ions introduced into the lanthanide-based clusters may play different roles in the luminescence properties of the Ln^{3+} ions. For the Cd^{2+} ion, its saturated d^{10} electronic configuration prevents the quenching of lanthanide luminescence through a d–d transition (f → d energy transfer), while the Ni^{2+} ion may quench the luminescence *via* f → d energy transfer.[25] The emission lifetimes (τ) of **1** and **3** are 1.9 μs and 0.5 μs, respectively. The intrinsic quantum yields (Φ_{Ln}) of Nd^{3+} emission in **1** and **3** are calculated as 0.76% and 0.20%, respectively, using $\Phi_{Ln} = \tau/\tau_0$ ($\tau_0 = 250$ μs, the natural lifetime of Nd^{3+}).[26]

To explore the imaging utility of Cd–Ln nano-clusters on an imaging platform used for biological materials, cross-linked polystyrene beads (6–7 μm) containing

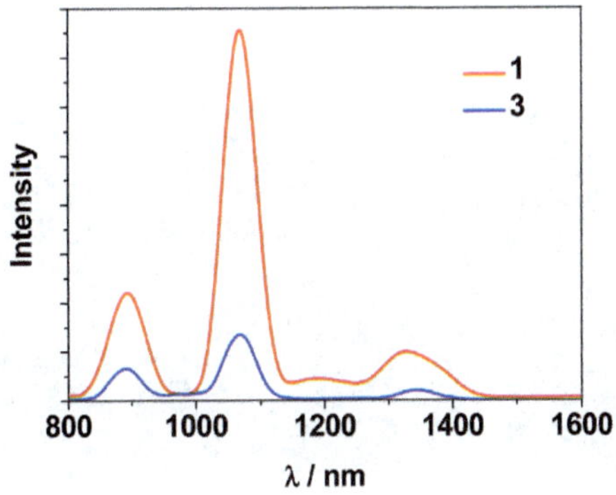

Fig. 6 NIR emission spectra of clusters **1** and **3** with the same absorbance value at λ_{ex} = 355 nm in CH_3CN.

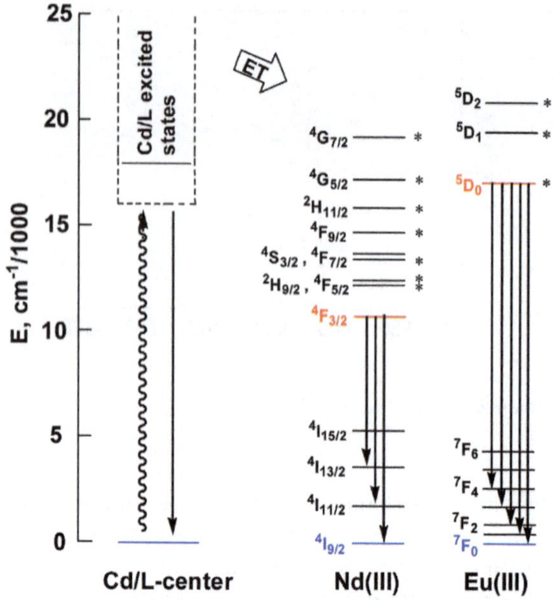

Scheme 2 Relevant energy levels in **1–3**. Those marked with * can act as energy acceptors by either Förster or Dexter mechanisms (the former requiring $|\Delta J| = 2, 4,$ or 6 at the lanthanide, and the latter requiring $|\Delta J| = 0$ or 1 with the exception of $|\Delta J| = |\Delta J'| = 0$, which is forbidden).[27]

complexes **5**, **6** and **7** were visualized using confocal microscopy (Fig. 7). Images were collected to show total emission (400–800 nm). All experimental parameters are identical and bead emissions shown are a true representation of brightness on a color scale of black to red to white to blue, where blue indicates saturation. As shown in Fig. 7, beads loaded with **5** are considerably brighter, suggesting that the Nd–Cd nano-cluster may have enhanced emission properties with the H_2L^2 Schiff base ligand, compared to the Sm–Cd and Yb–Cd nano-clusters. Further analysis of the relative emission intensities between the bead sets can be undertaken to improve the confidence of this observation.

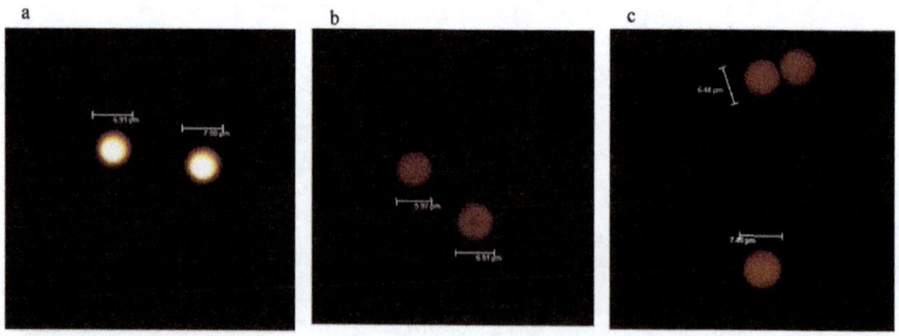

Fig. 7 Confocal images of polystyrene beads loaded with Ln–Cd nano-clusters (a) **5**, Nd; (b) **6**, Sm, and (c) **7**, Yb.

Conclusions

Nanostructured materials are increasingly being developed as imaging agents for both fundamental research imaging and diagnostic imaging. Here, we have presented data characterizing a new class of self-assembling lanthanide nanoclusters. These compounds have structures and imaging properties making them suitable candidates for cell and deep tissue imaging agents. Unlike classical nanoparticles, these materials are self-assembling homogeneous particles with asymmetric structures. The Cd–Ln clusters adopt nano-drum-like structures. The Ni–Ln cluster **3** has a square-like architecture. TEM and SAXS data are consistent with these molecular compounds having solution structures similar to those found in the solid state. The different architectures give rise to different imaging properties. With the Ln(III) centers enclosed within its nano-drum-like structure, Cd–Nd cluster **1** exhibits better NIR luminescence properties than Ni–Nd cluster **3**. The photophysical properties of the nano-drum-like clusters demonstrate their potential as lanthanide emitters, and their utility as bio-imaging agents in confocal microscopy platforms. Future studies will focus on understanding how their molecular structures influence their stability, photophysical properties and ability to be delivered to cells and tissues in nano-formulations.

Acknowledgements

The work was supported by the Welch Foundation (F-816) (RAJ), NSFC (51025207) (SMH), the Ministry of High Education (MOHE), Malaysia under High Impact Research (HIR) – MOHE project (UM.C/625/1/HIR/MoE/CHAN/13/6 H-50001-00-A000034) (KAB), NIH/NIAID 1U01AI078008-3, the Centre for Blast Injury Study at Imperial College London (KAB), the Royal Society (University Research Fellowship to RP), and the EU infrastructure grant BioStruct-X (contract number: 283570) (DIS).

Notes and references

1 (*a*) Y. Sun, W. Feng, P.-Y. Yang, C.-H. Huang and F.-Y. Li, *Chem. Soc. Rev.*, 2015, **44**, 1509–1525; (*b*) L. Prodi, E. Rampazzo, F. Rastrelli, A. Speghini and N. Zaccheroni, *Chem. Soc. Rev.*, 2015, **44**, 4922–4952; (*c*) S. Ranjan, M. K. Jayakumar and Y. Zhang, *Nanomedicine*, 2015, **10**, 1477–1491; (*d*) Q. Liu, W. Feng and F.-Y. Li, *Coord. Chem. Rev.*, 2014, **273/274**, 100–110; (*e*) A. Foucault-Collet, C. M. Shade, I. Nazarenko, S. Petoud and S. V. Eliseeva, *Angew. Chem., Int. Ed.*, 2014, **53**, 2927–2930.

2 (*a*) E. G. Moore, A. P. S. Samuel and K. N. Raymond, *Acc. Chem. Res.*, 2009, **42**, 542–552; (*b*) J.-C. G. Bünzli, *Chem. Rev.*, 2010, **110**, 2729–2755; (*c*) C. P. Montgomery, B. S. Murray, E. J. New, R. Pal and D. Parker, *Acc. Chem. Res.*, 2009, **42**, 925–937; (*d*) A. Thibon and V. C. Pierre, *Anal. Bioanal. Chem.*, 2009, **394**, 107–120.

3 T. Nishioka, K. Fukui and K. Matsumoto, *Handbook on the Physics and Chemistry of Rare Earth*, ed. K. A. Gschneidner, J.-C. G. Bünzli and V. K. Pecharsky, Elsevier Science B. V., Amsterdam, 2007, p. 37.

4 (*a*) J.-B. Peng, Q.-C. Zhang, X.-J. Kong, Y.-Z. Zheng, Y.-P. Ren, L.-S. Long, R.-B. Huang, L.-S. Zheng and Z. Zheng, *J. Am. Chem. Soc.*, 2012, **134**, 3314–3317; (*b*) B. Wang, Z. Zang, H. Wang, W. Dou, X. Tang, W. Liu, Y. Shao,

J. Ma, Y. Li and J. Zhou, *Angew. Chem., Int. Ed.*, 2013, **52**, 3756–3759; (*c*) A. Müller, *Nature*, 2007, **447**, 1034–1035; (*d*) A. Müller, P. Kögerler and A. W. M. Dress, *Coord. Chem. Rev.*, 2001, **222**, 193–218; (*e*) E. G. Mednikov, M. C. Jewell and L. F. Dahl, *J. Am. Chem. Soc.*, 2007, **129**, 11619–11630.

5 (*a*) J.-C. G. Bünzli, *Metal Ions in Biological Systems*, ed. M. Dekker, New-York, 2004, p. 42; (*b*) I. Hemmila and S. Webb, *Drug Discovery Today*, 1997, **2**, 373–380; (*c*) J. W. Stouwdam, G. A. Hebbink, J. Huskens and F. C. J. M. Van Veggel, *Chem. Mater.*, 2003, **15**, 4604–4616.

6 (*a*) X.-L. Li, L.-X. Shi, L.-Y. Zhang, H. M. Wen and Z. N. Chen, *Inorg. Chem.*, 2007, **46**, 10892–10900; (*b*) H.-B. Xu, L.-X. Shi, E. Ma, L.-Y. Zhang, Q.-H. Wei and Z.-N. Chen, *Chem. Commun.*, 2006, 1601–1603; (*c*) H.-M. Wen, Y. Yang, X.-S. Zhou, J.-Y. Liu, D.-B. Zhang, Z.-B. Chen, J.-Y. Wang, Z.-N. Chen and Z.-Q. Tian, *Chem. Sci.*, 2013, **4**, 2471–2477; (*d*) S. G. Baca, H. Adams, C. S. Grange, A. P. Smith, I. Sazanovich and M. D. Ward, *Inorg. Chem.*, 2007, **46**, 9779–9789; (*e*) M. Yin, X. Lei, M. Li, L. Yuan and J. Sun, *J. Phys. Chem. Solids*, 2006, **67**, 1372–1378; (*f*) H.-B. Xu, H.-M. Wen, Z.-H. Chen, J. Li, L.-X. Shi and Z.-N. Chen, *J. Chem. Soc., Dalton Trans.*, 2010, **39**, 1948–1953; (*g*) Y.-X. Chi, S.-Y. Niu, J. Jin, R. Wang and Y. Li, *J. Chem. Soc., Dalton Trans.*, 2009, **47**, 7653–7659; (*h*) Y.-X. Chi, S.-Y. Niu, Z.-L. Wang and J. Jin, *Eur. J. Inorg. Chem.*, 2008, 2336–2343.

7 (*a*) S.-C. Xiang, S.-M. Hu, T.-L. Sheng, R.-B. Fu, X.-T. Wu and X.-D. Zhang, *J. Am. Chem. Soc.*, 2007, **129**, 15144–15146; (*b*) C. M. Zaleski, E. C. Depperman, J. W. Kampf, M. L. Kirk and V. L. Pecoraro, *Angew. Chem., Int. Ed.*, 2004, **43**, 3912–3914; (*c*) V. M. Mereacre, A. M. Ako, R. Clerac, W. Wernsdorfer, G. Filoti, J. Bartolome, C. E. Anson and A. K. Powell, *J. Am. Chem. Soc.*, 2007, **129**, 9248–9249; (*d*) X.-J. Kong, Y.-P. Ren, L.-S. Long, Z. Zheng, R.-B. Huang and L.-S. Zheng, *J. Am. Chem. Soc.*, 2007, **129**, 7016–7017; (*e*) J.-B. Peng, Q.-C. Zhang, X.-J. Kong, Y.-Z. Zheng, Y.-P. Ren, L.-S. Long, R.-B. Huang, L.-S. Zheng and Z.-P. Zheng, *J. Am. Chem. Soc.*, 2012, **134**, 3314–3317.

8 (*a*) M. Sakamoto, K. Manseki and H. Okawa, *Coord. Chem. Rev.*, 2001, **219**, 379–414; (*b*) R. E. P. Winpenny, *Chem. Soc. Rev.*, 1998, **27**, 447–452; (*c*) C. Camp, V. Guidal, B. Biswas, J. Pecaut, L. Dubois and M. Mazzanti, *Chem. Sci.*, 2012, **3**, 2433–2448; (*d*) M. Andruh, *Chem. Commun.*, 2011, **47**, 3025–3042; (*e*) X. Feng, W. Zhou, Y. Li, H. Ke, J. Tang, R. Clerac, Y. Wang, Z. Su and E. Wang, *Inorg. Chem.*, 2012, **51**, 2722–2724; (*f*) T. Yamaguchi, J.-P. Costes, Y. Kishima, M. Kojima, Y. Sunatsuki, N. Brefuel, J.-P. Tuchagues, L. Vendier and W. Wernsdorfer, *Inorg. Chem.*, 2010, **49**, 9125–9135.

9 (*a*) X.-P. Yang, R. A. Jones and S.-M. Huang, *Coord. Chem. Rev.*, 2014, **273**, 63–75; (*b*) X.-P. Yang, D. Schipper, R. A. Jones, L. A. Lytwak, B. J. Holliday and S.-M. Huang, *J. Am. Chem. Soc.*, 2013, **135**, 8468–8471.

10 (*a*) W.-K. Lo, W.-K. Wong, W.-Y. Wong, J. Guo, K.-T. Yeung, Y.-K. Cheng, X.-P. Yang and R. A. Jones, *Inorg. Chem.*, 2006, 9315–9325; (*b*) X.-P. Yang, S. Desmond, A. Liao, J. M. Stanley, R. A. Jones and B. J. Holliday, *Polyhedron*, 2013, **52**, 165–168; (*c*) X.-P. Yang, R. A. Jones, V. Lynch, M. M. Oye and A. L. Holmes, *J. Chem. Soc., Dalton Trans.*, 2005, **5**, 849–851; (*d*) X.-P. Yang, R. A. Jones, W.-K. Wong, V. Lynch, M. M. Oye and A. L. Holmes, *Chem. Commun.*, 2006, 1836–1838; (*e*) X.-P. Yang, C. Chan, D. Lam, D. Schipper, J. M. Stanley, X. Chen, R. A. Jones, B. J. Holliday, W.-K. Wong, S.-C. Chen and Q. Chen, *J. Chem. Soc., Dalton Trans.*, 2012, **41**, 11449–11453.

11 F. Lam, J.-X. Xu and K.-S. Chan, *J. Org. Chem.*, 1996, **61**, 8414–8418.

12 Z. Otwinowski and W. Minor, *Methods in Enzymology, 276: Macromolecular Crystallography, Part a*, ed. C. W. J. Carter, M. I. Simon and R. M. Sweet, Academic Press, 1997, pp. 307–326, DENZO-SMN.

13 G. H. Sheldrick, *SHELX 97, a Software Package for the Solution and Refinement of X-ray Data*, University of Göttingen, Göttingen, Germany, 1997.

14 D. T. Cromer and J. T. Waber, *International Tables for X-Ray Crystallography*, Kynoch Press, Birmingham, 1974, vol. 4, Table 2.2A.

15 (*a*) P. van der Sluis and A. L. Spec, *Acta Crystallogr., Sect. A: Found. Crystallogr.*, 1990, **46**, 194–201; (*b*) A. L. Spec, *Platon, A Multipurpose Crystallographic Tool*, Utrecht, the Netherlands, 1998; (*c*) A. L. Spec, *J. Appl. Crystallogr.*, 2003, **36**, 7–13; (*d*) L. J. Farrugia, *J. Appl. Crystallogr.*, 1999, **32**, 837–838.

16 D. Franke, A. G. Kikhney and D. I. Svergun, *Nucl. Instrum. Methods Phys. Res., Sect. A*, 2012, **689**, 52–59.

17 R. Pal, *Faraday Discuss.*, 2015, **177**, 507–515.

18 M. V. Petoukhov, D. Franke, A. V. Shkumatov, G. Tria, A. G. Kikhney, M. Gajda, C. Gorba, H. D. Mertens, P. V. Konarev and D. I. Svergun, *J. Appl. Crystallogr.*, 2012, **45**(2), 342–350.

19 D. I. Svergun, *J. Appl. Crystallogr.*, 1992, **25**, 495–503.

20 K. A. Brown, X. Yang, D. Schipper, J. W. Hall, L. J. DePue, A. J. Gnanam, J. F. Arambula, J. N. Jones, J. Swaminathan, Y. Dieye, J. Vadivelu, D. J. Chandler, E. M. Marcotte, J. L. Sessler, L. I. R. Ehrlich and R. A. Jones, *Dalton Trans.*, 2015, **44**, 2667–2675.

21 (*a*) D. H. Metcalf, R. G. Ghirardelli and R. A. Palmer, *Inorg. Chem.*, 1985, **24**, 634–636; (*b*) M. Albin, R. R. Whittle and W. D. Horrocks Jr, *Inorg. Chem.*, 1985, **24**, 4591–4594.

22 (*a*) Q. H. Xu, L. S. Li, X. S. Liu and R. R. Xu, *Chem. Mater.*, 2002, **14**, 549–555; (*b*) J.-C. G. Bünzli and C. Piguet, *Chem. Soc. Rev.*, 2005, **34**, 1048–1077; (*c*) Y. B. Katsimirski and N. K. Davidenko, *Coord. Chem. Rev.*, 1979, **27**, 223–273.

23 K. Nakamaru, *Bull. Chem. Soc. Jpn.*, 1982, **5**, 2697–2705.

24 (*a*) F. S. Richardson, *Chem. Rev.*, 1982, **82**, 541–552; (*b*) S. Yanagida, Y. Hasegawa, K. Murakoshi, Y. Wada, N. Nakashima and T. Yamanaka, *Coord. Chem. Rev.*, 1998, **171**, 461–480.

25 (*a*) V. Wen, Y. Zhao, L. Wang, M. Zhang and D. Gao, *J. Rare Earths*, 2007, **25**, 679–683; (*b*) E. Wolcan, L. Villata, A. L. Capparelli and M. R. Feliz, *Photochem. Photobiol. Sci.*, 2004, **3**, 322–328; (*c*) B. C. Barja, A. Remorino, M. J. Roberti and P. F. Aramendia, *J. Argent. Chem. Soc.*, 2005, **93**, 81–88; (*d*) Q.-Y. Chen, C.-J. Li and Q.-H. Luo, *Synth. React. Inorg. Met.-Org. Chem.*, 2003, **33**(9), 1701–1715.

26 (*a*) S. B. Meshkova, Z. M. Topilova, D. V. Bolshoy, S. V. Beltyukova, M. P. Tsvirko, V. Y. Venchikov and V. Ya, *Acta Phys. Pol., A*, 1999, **95**, 983–990; (*b*) S. I. Klink, L. Grave, D. N. Reinhoudt and F. C. J. M. v. Veggel, *J. Phys. Chem. A*, 2000, **104**, 5457–5468.

27 (*a*) D. L. Dexter, *J. Chem. Phys.*, 1953, **21**, 836–850; (*b*) G. F. de Sá, O. L. Malta, C. de Mello Donegá, A. M. Simas, R. L. Longo, P. A. Santa-Cruz and E. F. da Silva Jr, *Coord. Chem. Rev.*, 2000, **196**, 165–195; (*c*) J.-C. G. Bünzli and C. Piguet, *Chem. Soc. Rev.*, 2005, **34**, 1048–1077; (*d*) N. M. Shavaleev, G. Accorsi, D. Virgili, D. R. Bell, T. Lazarides, G. Calogero, N. Armaroli and M. D. Ward, *Inorg. Chem.*, 2005, **44**, 61–72.

Faraday Discussions

PAPER

A multi-coordinating polymer ligand optimized for the functionalization of metallic nanocrystals and nanorods

Wentao Wang, Xin Ji, Hunter Burns and Hedi Mattoussi*

Received 17th March 2016, Accepted 22nd March 2016
DOI: 10.1039/c6fd00056h

We report the design and use of a multi-coordinating polymer ligand that is ideally suited for functionalizing gold nanoparticles and nanorods, and promoting their steric stabilization in buffer media. The ligand is prepared *via* a one-step nucleophilic addition reaction between poly(isobutylene-*alt*-maleic anhydride) and amine-modified anchoring groups and hydrophilic moieties. Surface functionalization of gold nanoparticles and nanorods with this polymer yields nanocrystals that exhibit excellent long-term colloidal stability over a broad range of conditions, including pH changes and in growth media, as verified using dynamic light scattering measurements combined with agarose gel electrophoresis. This polymer coating can also prevent the formation of protein corona. These features bode well for use in biological applications where small size, reduced nonspecific interactions and colloidal stability are highly desired. Furthermore, this design can be easily expanded to functionalize a variety of other inorganic nanocrystals.

Introduction

Inorganic nanomaterials made of metals and metal oxides possess unique photophysical properties that can be tuned *via* size, shape and composition.[1–5] For example, gold nanorods exhibit pronounced longitudinal surface plasmon resonance bands that can be tuned by changing the aspect ratio.[2,6] Similarly, magnetic nanoparticles show size- and composition-tunable coercivity.[7,8] These materials have generated great interest in a variety of applications. Some of those applications center on one's ability to integrate them into biology, for use as drug delivery vehicles and as diagnostic tools.[2,4,5,9–14] This provides platforms with well-defined photo and physical properties that can advance our understanding of various challenging problems in biology. One important prerequisite to achieve some of those goals is to surface-modify the nanomaterials with capping ligands that promote colloidal stability in biological media while endowing them with

Department of Chemistry and Biochemistry, Florida State University, Tallahassee, FL 32306, USA. E-mail: mattoussi@chem.fsu.edu

targeting capabilities.[10,15,16] However, integration of these materials in biology is still constrained by problems such as limited colloidal stability, low targeting efficiency and surface coating that yields hydrophilic nanocrystals with rather large hydrodynamic size.[17–20] The large size could negatively affect their transport properties, biodistribution and blood circulation.[17,19,21] These properties are primarily controlled by the nature of the capping ligands used to functionalize the nanocrystals.

Surface functionalization of metallic nanocrystals has been achieved *via* three strategies: silica coating, encapsulation, or ligand exchange.[10,13,16,22] Encapsulation within amphiphilic polymers and/or micelles and silica coating can preserve the native nanocrystal cap.[23–26] However, these strategies tend to significantly increase the hydrodynamic size of the hydrophilic nanocrystals.[27] In comparison, cap exchange with coordinating bifunctional ligands can substantially reduce the hydrodynamic radius of the nanocrystals.[13,28] Two inherent properties of the ligand greatly determine the ultimate stability of hydrophilic nanoparticles: (1) strength of the anchor coordination onto the metal surface; and (2) affinity of the hydrophilic motif to buffer media.[29]

In this study, we introduce a metal-coordinating polymer ligand ideally adapted for the surface-functionalization of gold nanoparticles and gold nanorods. The ligand synthesis relies on the one-step nucleophilic addition reaction of distinct amine-containing functionalities with poly(isobutylene-*alt*-maleic anhydride). This chemical route provides modular ligands that present multiple lipoic acid anchors for enhanced metal-coordination and several polyethylene glycol (PEG) moieties for water compatibility. The multiple-coordination of the ligand to the nanocrystal surface shifts the binding equilibrium to lower concentrations, which reduces the ligand desorption rate from the nanocrystals. Because heavy atom techniques such as transmission electron microscopy are unable to probe the properties of the organic surface coating,[30] we applied dynamic light scattering and agarose gel electrophoresis to characterize the polymer-coated nanocrystals. We found that ligation with this polymer produced nanocrystal dispersions with long-term stability over a broad range of biologically-relevant conditions. In addition, the introduction of the PEG moieties in the ligand drastically reduced non-specific interactions of the nanocrystals in biological media.

Experimental

Materials

Poly(isobutylene-*alt*-maleic anhydride), PIMA (average M_w: ~6000 Da, 85%), poly(ethylene glycol)methyl ether (HO-PEG-OMe, average M_w: ~750 Da), lipoic acid (LA, >98.0%), hexadecyltrimethylammonium bromide (CTAB, >98.0%), gold(III)chloride trihydrate ($HAuCl_4 \cdot 3H_2O$, >99.9%), sodium citrate tribasic dihydrate ($C_6H_5Na_3O_7 \cdot 2H_2O$), silver nitrate ($AgNO_3$, >99%), sodium borohydride ($NaBH_4$, 99%), sodium salicylate (NaSa, >99.5%), hydrochloric acid (HCl, 37 wt% in water), dimethylformamide, along with most of the chemicals used were purchased from Sigma Aldrich (St Louis, MO). Sodium oleate (NaOL, >97.0%) was purchased from Tokyo Chemical Industry Co, Ltd. Solvents were purchased from Sigma Aldrich (St Louis, MO). Deuterated solvents used for H NMR experiments were purchased from Cambridge Isotope Laboratories (Andover, MA). Ultrapure water (18 MΩ) obtained from a Milli-Q Integral 5 system was used in all experiments.

The precursors H_2N-PEG$_{750}$-OCH_3 was synthesized in our laboratory starting from HO-PEG$_{750}$-OCH_3, following the procedures detailed in previous publications.[31] The amine-terminated lipoic acid (LA-NH_2) was also prepared following literature protocols.[32]

Instrumentation

^{1}H NMR spectra were recorded using a 600 MHz spectrometer (Bruker SpectroSpin, Billerica, MA). Optical spectra of the various AuNP and AuNR dispersions were collected using either a UV-vis absorption spectrophotometer (UV 2450 model, Shimadzu, Columbia, MD) or a UV-vis-NIR spectrophotometer (Perkin Elmer Lambda 950, Shelton, CT). Fourier Transform Infrared (FT-IR) spectra were collected using an FT-IR spectrometer (Perkin Elmer Spectrum 100, Shelton, CT). Transmission electron microscopy (TEM) images were acquired using either a 200 kV JEOL-2010 instrument (JEOL, Peabody, MA) or a Philips FEI CM-120 microscope (Philips, Hillsboro, OR). Samples for TEM were prepared by drop casting the nanocrystal dispersions onto holey carbon film on a fine-mesh Cu grid (400 mesh).

Ligand synthesis

Synthesis of the lipoic acid- and PEG-modified PIMA polymer was carried out following the rationales detailed in our previous work.[29] In a typical reaction, 0.385 g of PIMA (2.5 mmol of monomer units) dissolved in 5 mL of DMF was transferred into a 50 mL three-neck round-bottom flask equipped with a magnetic stirring bar. The solution was purged with nitrogen and then heated up to 60 °C. In a separate vial, lipoic acid-amine (0.31 g, 1.25 mmol) was dissolved in 1 mL of DMF and added dropwise to the PIMA solution (using a syringe), followed by addition of H_2N-PEG-OMe (0.94 g, 1.25 mmol) in 2 mL of DMF. After the addition was complete, the solution was left stirring at 60 °C for 8–10 hours (*i.e.*, overnight). The solvent was removed under vacuum, and 3 mL of chloroform was added. The final compound was purified on a silica column with chloroform as the eluent.

Growth of gold nanoparticles

The hydrophobic oleylamine-capped AuNPs (~9.8 nm in diameter) were synthesized by rapidly injecting $HAuCl_4$ precursor into a pre-heated surfactant solution, following the strategy reported by Swihart and coworkers.[33]

Growth of gold nanorods

Two sets of gold nanorods (109 nm × 20 nm and 36 nm × 14 nm) were prepared following the schemes initially developed by Murray and co-workers using seeded growth in binary surfactant mixtures of CTAB/NaOL and CTAB/NaSa, respectively.[34,35] We have previously utilized this route to grow AuNRs which were tested with the imidazole-coordinating polymer described in ref. 36.

Ligand exchange of oleylamine-capped AuNPs

The stock solution of oleylamine-AuNPs (~0.9 μM, 400 μL) was precipitated using ethanol and redispersed in 400 μL of chloroform. Then, 15 mg of ligands dissolved in 400 μL of chloroform were mixed with the AuNP dispersion. The mixture was stirred at 50 °C overnight. The AuNPs were then precipitated by adding 200 μL of methanol and excess hexane. After sonication, the turbid solution was centrifuged for 5 min at 3700 rpm to provide a dark pellet. The clear supernatant was discarded, and the pellet redispersed in a 50 : 50 mixture of chloroform and methanol (400 μL total), followed by precipitation using excess hexane. After centrifugation, the residual precipitate was dried under vacuum to yield a pellet, which was easily dispersed by adding DI water with sonication. The aqueous dispersion was passed through a 0.45 μm disposable syringe filter, then excess free ligands were removed by applying 3 to 4 rounds of concentration/dilution with DI water using a centrifugal filtration device (Millipore, M_w cutoff = 50 kDa).

Ligand exchange of CTAB-capped AuNRs

The stock solution (2.82 nM, 400 μL) was centrifuged at 7000 rpm for 8 min to remove excess solubilized CTAB, yielding a pellet, which was redispersed in 200 μL of phosphate buffer (pH = 8, 20 mM). 20 mg of the ligands dissolved in 400 μL of buffer were added to the NR dispersion. The mixture was stirred at 50 °C overnight, further diluted by adding 2 mL of DI water with sonication for 2 min. The aqueous dispersion was passed through a disposable syringe filter (0.45 μm), and excess ligands were removed by applying 2 rounds of concentration/dilution with DI water, as described above.

Dynamic light scattering (DLS) study

These measurements were applied to AuNPs only. Dynamic light scattering profiles were collected from dispersions of oleylamine-capped AuNPs in hexane, and from LA-PIMA-PEG-capped AuNPs dispersed in buffer media (pH = 7.5). Data were collected using an ALV/CGS-3 Compact Goniometer System equipped with a He–Ne laser source ($\lambda = 633$ nm) and a variable delay time (multi-tau) ALV-7004 correlator (ALV-GmbH, Langen, Germany). The scattered signal consisting of an average over 3 acquisition periods (10 seconds each) was collected at several angles between 40 and 140 degrees. The intensity count rates were maintained at ~150–300 Hz, achieved through adjustments of the NP concentrations. The resulting auto-correlation function, generated from the above scattered intensity profiles using the ALV correlator, was fitted to a cumulant series using ALV software. For every sample we verified that the measured hydrodynamic size (extracted from the Laplace transform of the auto-correlation function) is independent of the scattering angle, as anticipated.[30] To track possible changes in the colloidal stability of the nanocrystal dispersions with time, the autocorrelation function was collected and analyzed after different storage time periods.

Protein corona tests

The nanocrystal dispersions were run on 0.6% agarose gel using tris borate EDTA buffer (TBE, 89 mM tris, 89 mM boric acid, 1 mM EDTA). The stock dispersion of nanocrystals was mixed with BSA solutions at different molar ratio in Eppendorf tubes, and then diluted with phosphate buffer (pH 7, 20 mM). The final concentration of AuNPs was 35 nM and AuNRs was 1.2 nM. The molar ratio of BSA with respect to nanoparticles was varied from 0 to 1000. The mixtures were incubated for 1 hour at room temperature and then mixed with 2 μL of loading buffer made of Ficoll 400 immediately prior to use. 20 μL aliquots of the mixtures were loaded into the wells, then a voltage of 5.5 V cm^{-1} was applied for 30 min to facilitate the electrophoretic migration of the NPs and NRs in the gel.

Results and discussion

Ligand design and nanocrystal growth

Synthesis of the polymer ligands relies on the one-step nucleophilic addition reaction between poly(isobutylene-*alt*-maleic anhydride), PIMA, and amine-modified LA and PEG moieties. It builds on the chemical rationales described in our recent work.[20,29,37,38] This reagent-free coupling reaction is highly efficient, and yields easy to purify products. Combined with the presence of several maleic anhydrides (~39) along the backbone, this allows for the insertion of controllable numbers of distinct but complementary functionalities into the same polymer chain *via* stable amide bonds. The resulting ligand presents multiple-thiol groups (resulting from the LA groups) for strong coordination on the nanocrystal surface, along with several PEG segments to promote hydrophilicity, as shown in Fig. 1. It also frees several carboxyl groups along the backbone, which provide potential reactive groups for bioconjugation. The architecture of the final ligand can be optimized by controlling the molar ratio of each amine-modified moiety with respect to the maleic anhydride groups in the PIMA chain. In this study, the molar ratio of lipoic acid and PEG was maintained at 50 : 50, which is expected to introduce ~20 LA groups and ~20 PEG moieties per ligand.

The gold nanoparticles with an average diameter of 9.8 nm (measured by TEM) were grown by pyrolysis of gold precursors in oleylamine following literature protocols.[33,36] Gold nanorods with two different aspect ratios (109 nm × 20 nm and 45 nm × 19 nm) were grown using seeded growth in binary surfactant mixtures following the schemes outlined by Murray and co-workers.[34,35] The TEM images in Fig. 2 show that the prepared nanocrystals have uniform shape and reduced size dispersity. The AuNPs exhibit the classical SPR (surface plasmon resonance) peak at 520 nm. Conversely, the AuNRs exhibit a strong longitudinal SPR band that shifts depending on the aspect ratio, in addition to a weaker band

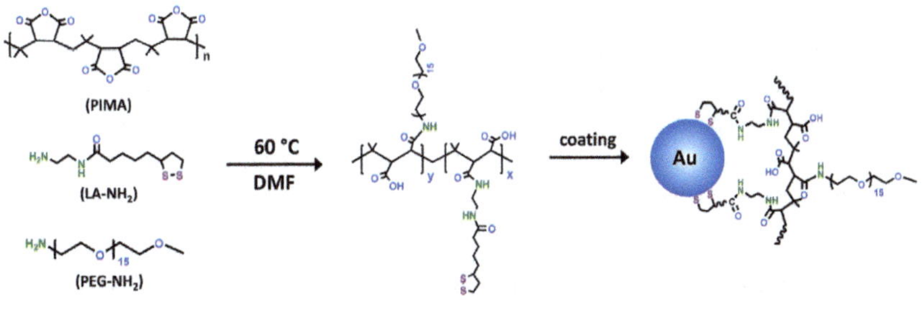

Fig. 1 Schematic representations of the ligand synthesis using the one-step nucleophilic addition reaction along with a polymer-capped AuNP.

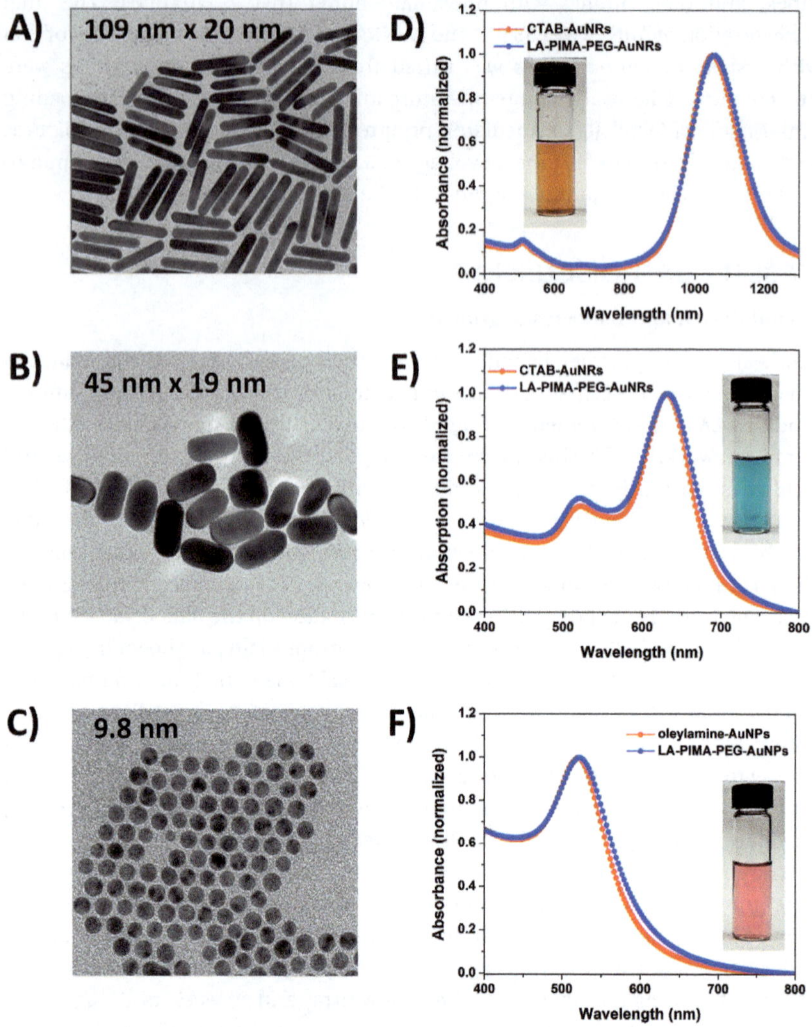

Fig. 2 (A–C) TEM images of CTAB-capped gold nanorods ($L \times D$: 109 nm × 20 nm and 45 nm × 19 nm) and oleylamine-capped gold nanoparticles (D = 9.8 nm). (D–F) UV-vis absorption spectra of gold nanocrystals before and after ligand exchange with LA-PIMA-PEG polymer. Inset shows white light images for three sets of LA-PIMA-PEG-capped nanostructures dispersed in buffer media.

at 520 nm ascribed to the transverse SPR mode; this imparts the AuNR dispersions with distinct colors. For instance, dispersions of gold nanorods with high aspect ratios appear red because they have an SPR band in the near infrared, while shorter nanorods with lower aspect ratios have an absorption band in the red, yielding blue color dispersions.

Characterization of the polymer-coated nanocrystals

The hydrophilic AuNR and AuNP dispersions have been characterized using three complementary analytical techniques: UV-vis absorption spectroscopy, ^{1}H NMR spectroscopy, and FT-IR spectroscopy. In addition, the spherical AuNPs were further characterized by dynamic light scattering to extract information about the hydrodynamic size and colloidal stability before and after phase transfer.

1. **Optical characterization.** Fig. 2A to C show the absorption spectra of the AuNRs and AuNPs before and after ligand exchange with LA-PIMA-PEG. The absorption spectra collected from the hydrophilic dispersions are essentially identical to those measured for the initial CTAB-AuNRs and oleylamine-AuNPs, with clearly defined transverse and longitudinal plasmon bands for the NRs and SPR band for the NPs. These results indicate that the modification of the gold nanocrystals with multi-coordinating polymer preserves the integrity of the Au nanostructures regardless of the initial growth route used.

2. **NMR and FT-IR characterization.** The effectiveness of the ligand exchange with LA-PIMA-PEG was confirmed using pulsed-field gradient-based water suppression ^{1}H NMR spectroscopy (in D_2O). Fig. 3A shows a representative ^{1}H NMR spectrum collected from a dispersion of LA-PIMA-PEG-AuNRs. This spectrum is overall similar to the one collected from the ligands alone.[29] The pronounced resonances shown at 3.25 ppm and 3.58 ppm are respectively attributed to the protons of the terminal methoxy and ethylene glycol monomers in the PEG-OMe moieties. The small peaks observed over the range from 1 to 3 ppm are the characteristic protons of the lipoic acid groups, while the broad peak at 0.9 ppm is ascribed to the methyl protons in the PIMA backbone. Conversely, no proton signals from the CTAB were measured in the NMR spectrum. The FT-IR spectra shown in Fig. 3B further confirm the efficient cap exchange with the polymer. The bands at ~1470 cm^{-1} (ascribed to the trimethylammonium groups of CTAB) and ~1720 cm^{-1} (attributed to the carbonyl groups of NaOL) have disappeared, while a new strong peak at ~1100 cm^{-1} (due to the ether bonds of PEG chains) is measured in the spectrum of the polymer-coated AuNRs. These results combined clearly indicate that the strong thiol affinity to the metal surface of the nanostructures drives the complete substitution of the native capping ligands with the polymer on the nanocrystals. Similar ^{1}H NMR and FT-IR spectra were collected from the LA-PIMA-PEG-AuNPs, confirming the complete removal of the native oleylamine ligands from the NP surfaces (data not shown).

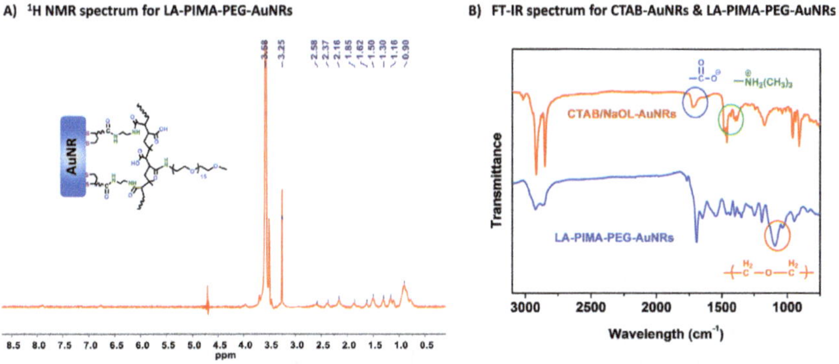

Fig. 3 (A) ^{1}H NMR spectrum of the LA-PIMA-PEG-capped AuNRs in D_2O. (B) FT-IR spectra of CTAB/NaOL- and LA-PIMA-PEG-capped AuNRs. The circled peaks represent the typical signatures ascribed to specific functionalities in the capping ligands (CTAB/NaOL and LA-PIMA-PEG).

3. Characterization using dynamic light scattering. DLS relies on the analysis of the normalized auto-correlation function of the scattered electric field, $g^{(1)}(\tau,q)$, as a function of the decay time, τ, and scattering wavevector, $q = \frac{4\pi n}{\lambda}\sin\left(\frac{\theta}{2}\right)$.[30] For dispersions of colloidal nanoparticles in the dilute regime $g^{(1)}(\tau,q)$ is commonly fit to a cumulants series:

$$g^{(1)}(\tau, q) = a\ \exp\left(-\Gamma\tau + \frac{\mu_2}{2}\tau^2\cdots\right) \tag{1}$$

where a is an experimental constant proportional to the amplitude of the scattered signal, while Γ and μ_2 designate the first and second cumulants. Γ corresponds to the decay rate expressed in terms of q and the apparent diffusion coefficient, D_{app} ($\Gamma = D_{app}q^2$), and the ratio μ_2/Γ^2 is the polydispersity index (PDI). A further refined analysis uses the Laplace transform, $G(\Gamma)$, which entails treating the scattered intensity as a superposition of contributions from all the individual nanoparticles, each with a characteristic size in the sample population, using the relationship:[39]

$$G(\Gamma) = \int g^{(1)}(\tau, q)\exp(\Gamma\tau)d\tau \tag{2}$$

$G(\Gamma)$ can be presented as a histogram of the intensity *vs.* Γ, with the peak location corresponding to $D_{app}q^2$; this yields a measure for the apparent hydrodynamic size, $R_{H,app}$, for the NP dispersion, extracted using the Stokes–Einstein relation: $R_{H,app} = kT/6\pi\eta D_{app}$; here k is the Boltzmann constant, T the temperature in Kelvin, and η is the solvent viscosity. Thus $G(\Gamma)$ can also be represented as intensity *vs.* $R_{H,app}$ profile.

Fig. 4A and B show side-by-side two autocorrelation functions collected from oleylamine-capped AuNPs in hexane (red symbols) and LA-PIMA-PEG-AuNPs in water (blue symbols) at $\theta = 90°$. Data show that cap exchange with the polymer ligands slightly increases the hydrodynamic size of the NPs, as indicated by the smaller slope of the decay curve in a semi logarithmic plot (Fig. 4B) of the correlation function measured for LA-PIMA-PEG-AuNPs, compared with oleylamine-capped AuNPs. Plots of intensity *vs.* $R_{H,app}$ profiles for oleylamine- and LA-PIMA-PEG-AuNP dispersions are shown in Fig. 4C. The narrow profiles (with PDI < 0.1) indicate that after ligand exchange the nanocrystals stay homogeneous and

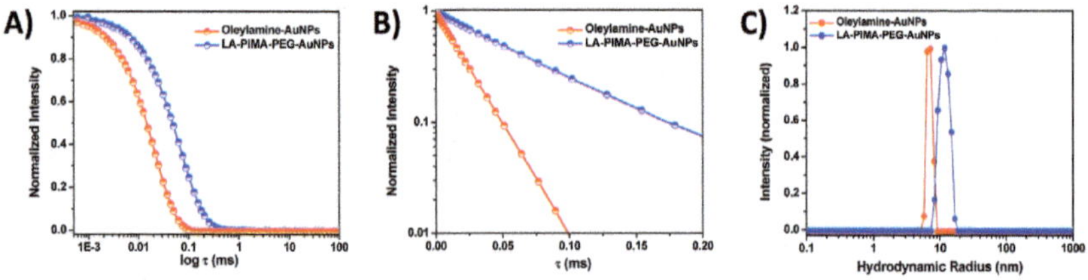

Fig. 4 Representative plots of the autocorrelation function, $g^{(1)}$, collected from dispersions of AuNPs (9.8 nm in diameter) before and after cap exchange with the LA-PIMA-PEG: (A) $g^{(1)}$ *vs.* log(τ) and (B) log($g^{(1)}$) *vs.* τ, (C) histograms showing the distribution of the normalized intensity *vs.* hydrodynamic radius, extracted from the Laplace transform of the auto-correlation functions collected from the dispersions of oleylamine- and LA-PIMA-PEG-capped AuNPs.

aggregate-free. The value of $R_{H,app}$ measured for LA-PIMA-PEG-AuNPs in water is ~12 nm, indicating ~3–4 nm increase compared to the size measured for the hydrophobic dispersion (~8 nm). This compact size reflects the nature of the multi-coordinating ligands, yielding homogeneous nanoparticles with a thin polymer coating.

To gain better understanding of the colloidal stability of the LA-PIMA-PEG-AuNPs in buffer media, we collected the autocorrelation function at six different scattering angles from 40° to 140°. Fig. 5A shows several plots of $\log(g^{(1)}(\tau,q))$ *vs.* τ collected from nanoparticle dispersion with a finite concentration at different scattering angles, along with the corresponding cumulant fits. Clearly, the second-order cumulant (eqn (1)) provides a good fit to the data with low PDI (PDI < 0.1). A plot of decay rate Γ *versus* q^2 shown in Fig. 5B yields a linear fit, as predicted. The linear behavior indicates that the apparent diffusion coefficient (slope in Fig. 5B) is independent of the scattering angle as expected for dispersions in the dilute regime and at small q. Additionally, Fig. 5C shows the dependence of D_{app} on the concentration over the range from 1 nM to 5 nM. A linear plot of the form, $D_{app} = \mathrm{D_z}(1 + k_d c)$, with a positive slope ($k_d > 0$) is assembled, implying that the interparticle interactions are repulsive and the dispersions of the AuNPs are sterically stabilized.

Stability tests

Colloidal stability of the aqueous nanocrystals was evaluated under several biologically-relevant conditions. Fig. 6A–C show representative white-light images of LA-PIMA-PEG-AuNRs and LA-PIMA-PEG-AuNPs dispersed in buffers over the pH range 2–13, in the presence of 1 M NaCl buffer, and when mixed with growth media (100% RPMI-1640). All dispersions stayed stable for at least 8 months storage, with no sign of degradation or aggregate formation.

These colloidal stability results were further corroborated with agarose gel electrophoresis measurements applied to AuNPs and AuNRs dispersed in varying pH buffers. Fig. 6D–F shows that narrow and homogeneous migration bands (no smearing) were measured for the hydrophilic AuNRs and AuNPs either freshly prepared or after 8 months of storage, indicating the uniform distribution in size and excellent colloidal stability. The mobility shift of the NPs and NRs towards the anode reflects the presence of homogeneous negative surface charge, which can

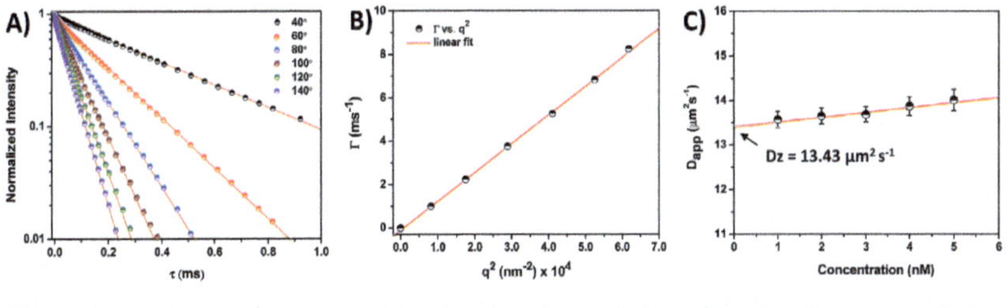

Fig. 5 (A) Representative semi-logarithmic plots of the normalized autocorrelation function, $\log(g^{(1)})$ *vs.* τ, collected from LA-PIMA-PEG-AuNPs at different scattering angles (40–140°), along with fits using eqn (1). (B) A plot of decay rate Γ *versus* q^2, together with a linear fit. (C) The apparent diffusion coefficient (D_{app}) *versus* concentration (c), along with a linear fit.

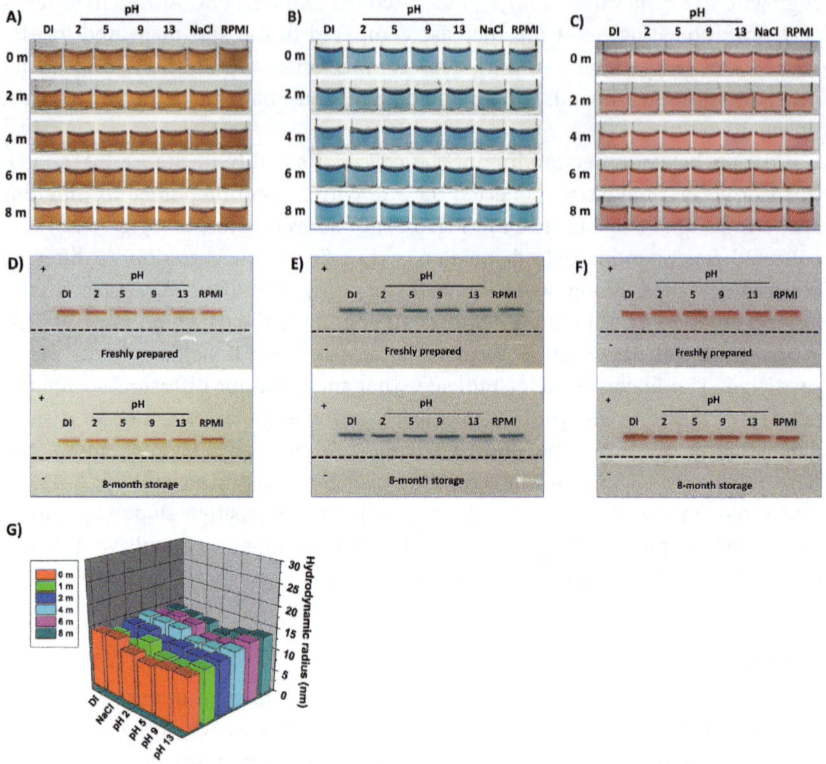

Fig. 6 Colloidal stability tests. (A–C) White light images of dispersions of AuNRs (A and B) and AuNPs (C) coated with LA-PIMA-PEG dispersed in phosphate buffer (20 mM) over the pH range from 2 to 13, in the presence of 1 M NaCl and when mixed with growth media (RPMI-1640). (D–F) Agarose gel electrophoresis images of AuNRs (D and E) and AuNPs (F) ligated with LA-PIMA-PEG and dispersed in pH 2 to pH 13 buffers. The top panels represent freshly-prepared samples while the bottom panels show the images collected from the samples after 8 months storage. The dashed line indicates the location of the loading wells. (G) Time-progression of the apparent hydrodynamic radius measured from LA-PIMA-PEG-AuNPs in different pH buffers.

be attributed to the carboxylic groups on the polymer freed during the addition reaction. DLS measurements complemented the above data by allowing us to monitor the changes in the hydrodynamic size of the AuNPs with storage time under the various conditions. Fig. 6G shows that the apparent hydrodynamic radius measured for the LA-PIMA-PEG-AuNPs stayed constant (~12 nm) across the full pH range and in the presence of 1 M NaCl after storage for 8 months.

Protein adsorption tests

Use of AuNPs and AuNRs along with an array of other inorganic nanocrystals as *in vivo* and *in vitro* probes has attracted active interest in the past decade, due to the promise they offer in a variety of biomedical applications, including drug delivery tracking and photothermal therapy.[2,9,40,41] However, these applications are still hampered by low targeting yields and unfavorable biodistribution, a result primarily attributed to the buildup of protein (and other biomolecules) adsorption on the nanocrystal surfaces (protein corona formation) when these materials are dispersed in biological media.[42–44] Protein adsorption can severely alter the nanoparticle's properties such as cellular internalization, intracellular distribution, and circulation lifetime in the blood vasculature. Additionally, buildup of protein corona compromises the NP targeting efficiency by shielding or preventing interactions between the nanoparticles and target receptors.[45] Thus, developing "non-sticky" coating with minimized protein adsorption is crucially important for the integration of surface-engineered nanocrystals into biology.

Having established that the polymer-coated NRs and NPs exhibit great colloidal stability, we then proceeded to evaluate the ability of our coating design to prevent the adsorption of bovine serum albumin (BSA) using gel electrophoresis measurements. BSA is a thiol-rich soluble protein that is also a major constituent of the circulatory system. It has been found to strongly interact with a wide range of nanoparticles.[44] Conversely, agarose gel electrophoresis is an effective technique for assessing protein adsorption. It was used by a few groups to detect the presence of a few proteins non-specifically adsorbed on nanoparticle surfaces, where corona protein formation can alter the mobility shift and potentially its band shape.[46] Dispersions of LA-PIMA-PEG-capped AuNRs and AuNPs were incubated with increasing concentrations of BSA, *i.e.*, increasing BSA-to-NP/NR molar ratios. Following mixing and incubation for one hour, the samples were loaded onto agarose gels and run under the conditions described in the Experimental section (30 min with 5.5 V cm^{-1}). The white light images in Fig. 7 shows that narrow bands with essentially identical mobility shift were measured for both sets of nanocrystals at all BSA-to-nanocrystal molar ratios used. This indicates that uniform size and surface charge distribution of the nanocrystals persisted at all BSA concentrations, with no signs of corona protein formation or aggregate build up.

Overall the above data collected from NMR, FT-IR, DLS measurements, pH stability tests and gel electrophoresis prove that capping the nanostructures with the polymer ligands yields materials with great colloidal stability over a wide range of conditions. This result can be attributed to the enhanced coordination of the multi-thiol ligands combined with the high affinity of PEG to water.

We should note that phase transfer and functionalization of gold nanoparticles and nanorods using rationales based on polymer coating have been utilized by several groups.[23–26] In those approaches the authors have focused on the design of amphiphilic block-copolymers to encapsulate the nanocrystals within micelle-like structures. The strategy preserves the native cap within the micelle-contained nanocrystals, which can be beneficial if the photophysical properties of the nanomaterial are preserved. Polymer-encapsulated nanocrystals

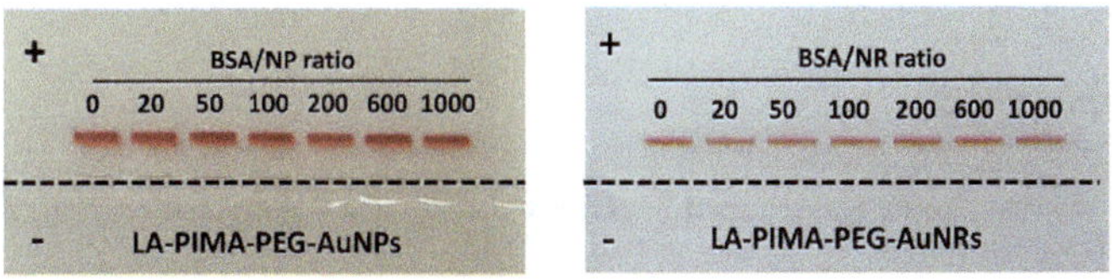

Fig. 7 Agarose gel electrophoresis images of LA-PIMA-PEG-capped AuNPs (left) and AuNRs (right) after incubation with different molar ratios of BSA protein for 1 hour.

have been employed by several groups in a variety of *in vitro* and *in vivo* studies. The present ligand design and surface coating strategy is different from those approaches, as coating of the nanocrystal relies on the competitive substitution of the native cap (such as CTAB and oleylamine for AuNRs and AuNPs). It involves the direct coordination of the polymer ligands on the nanocrystal surfaces. This yields hydrophilic nanocrystals with more compact coating compared to the encapsulation strategy. Additionally, the ability to promote multi-coordination interactions of thiols with the nanoparticle surfaces shifts the binding equilibrium and drastically reduces the desorption rate of the polymer from nanoparticle surfaces, thereby enhancing the colloidal stability of the nanocrystals under several conditions.

Conclusion

We have designed a metal-coordinating polymer and applied it as a ligand to surface-functionalize Au nanoparticles and nanorods and promote their dispersion in buffer media. The polymer ligand combines several lipoic acid anchoring groups that provide strong affinity to metal surfaces, along with hydrophilic PEG segments. The polymer is prepared *via* a nucleophilic addition reaction starting with a poly(isobutylene-*alt*-maleic anhydride) precursor. The surface functionalization and the colloidal properties of the resulting hydrophilic nanocrystals were thoroughly characterized using ^{1}H NMR, FT-IR, gel electrophoresis and dynamic light scattering. Data show that functionalization of the nanocrystals with this polymer ligand provides hydrophilic nanocrystals with excellent colloidal stability over a broad range of biological conditions, a result that can be directly attributed to the enhanced coordination of the ligands combined with the high affinity of PEG to water. The PEG coating also drastically reduces non-specific protein adsorption, yielding nanocrystals that can resist protein corona formation.

These results are greatly promising for the application of metal nanostructures in biology and nanomedicine, where platforms/probes with small size and good colloidal stability in a biological environment are required. We anticipate that the surface coating strategy would be applicable to prepare additional polymers adapted to a variety of inorganic nanocrystals. This can be achieved by the careful selection of a coordinating group that is ideally suited for a particular metal surface, along with hydrophilic moieties and reactive functions for conjugation to biomolecules and targeting.

Acknowledgements

The authors thank FSU and the National Science Foundation for financial support (NSF-CHE #1508501 and #1058957). We also thank Goutam Palui, Naiqian Zhan, Anshika Kapur and Dinesh Mishra for the fruitful discussions.

References

1 C. B. Murray, C. R. Kagan and M. G. Bawendi, *Annu. Rev. Mater. Sci.*, 2000, **30**, 545.

2 P. K. Jain, X. H. Huang, I. H. El-Sayed and M. A. El-Sayed, *Acc. Chem. Res.*, 2008, **41**, 1578.

3 M. Grzelczak, J. Perez-Juste, P. Mulvaney and L. M. Liz-Marzan, *Chem. Soc. Rev.*, 2008, **37**, 1783.

4 N. Lee, D. Yoo, D. Ling, M. H. Cho, T. Hyeon and J. Cheon, *Chem. Rev.*, 2015, **115**, 10637.

5 X. Yang, M. X. Yang, B. Pang, M. Vara and Y. N. Xia, *Chem. Rev.*, 2015, **115**, 10410.

6 R. Sardar, A. M. Funston, P. Mulvaney and R. W. Murray, *Langmuir*, 2009, **25**, 13840.

7 J. Park, K. An, Y. Hwang, J.-G. Park, H.-J. Noh, J.-Y. Kim, J.-H. Park, N.-M. Hwang and T. Hyeon, *Nat. Mater.*, 2004, **3**, 891.

8 J. H. Lee, Y. M. Huh, Y. W. Jun, J. W. Seo, J. T. Jang, H. T. Song, S. Kim, E. J. Cho, H. G. Yoon, J. S. Suh and J. Cheon, *Nat. Med.*, 2007, **13**, 95.

9 D. A. Giljohann, D. S. Seferos, W. L. Daniel, M. D. Massich, P. C. Patel and C. A. Mirkin, *Angew. Chem., Int. Ed.*, 2010, **49**, 3280.

10 J. Nam, N. Won, J. Bang, H. Jin, J. Park, S. Jung, S. Jung, Y. Park and S. Kim, *Adv. Drug Delivery Rev.*, 2013, **65**, 622.

11 P. D. Howes, R. Chandrawati and M. M. Stevens, *Science*, 2014, **346**, 1247390.

12 C. M. Lemon, P. N. Curtin, R. C. Somers, A. B. Greytak, R. M. Lanning, R. K. Jain, M. G. Bawendi and D. G. Nocera, *Inorg. Chem.*, 2014, **53**, 1900.

13 G. Palui, F. Aldeek, W. T. Wang and H. Mattoussi, *Chem. Soc. Rev.*, 2015, **44**, 193.

14 K. Saha, S. S. Agasti, C. Kim, X. N. Li and V. M. Rotello, *Chem. Rev.*, 2012, **112**, 2739.

15 H. Mattoussi, G. Palui and H. B. Na, *Adv. Drug Delivery Rev.*, 2012, **64**, 138.

16 R. Mout, D. F. Moyano, S. Rana and V. M. Rotello, *Chem. Soc. Rev.*, 2012, **41**, 2539.

17 H. S. Choi, W. Liu, P. Misra, E. Tanaka, J. P. Zimmer, B. I. Ipe, M. G. Bawendi and J. V. Frangioni, *Nat. Biotechnol.*, 2007, **25**, 1165.

18 M. Baker, *Nat. Methods*, 2010, **7**, 957.

19 E. Cai, P. H. Ge, S. H. Lee, O. Jeyifous, Y. Wang, Y. X. Liu, K. M. Wilson, S. J. Lim, M. A. Baird, J. E. Stone, K. Y. Lee, M. W. Davidson, H. J. Chung, K. Schulten, A. M. Smith, W. N. Green and P. R. Selvin, *Angew. Chem., Int. Ed.*, 2014, **53**, 12484.

20 W. Wang, X. Ji, A. Kapur, C. Zhang and H. Mattoussi, *J. Am. Chem. Soc.*, 2015, **137**, 14158.

21 H. S. Han, J. D. Martin, J. Lee, D. K. Harris, D. Fukumura, R. K. Jain and M. Bawendi, *Angew. Chem., Int. Ed.*, 2013, **52**, 1414.

22 R. A. Sperling and W. J. Parak, *Philos. Trans. R. Soc. London, Ser. A*, 2010, **368**, 1333.

23 M. Grzelczak, A. Sanchez-Iglesias and L. M. Liz-Marzan, *CrystEngComm*, 2014, **16**, 9425.

24 F. Zhang, E. Lees, F. Amin, P. R. Gil, F. Yang, P. Mulvaney and W. J. Parak, *Small*, 2011, **7**, 3113.

25 D. Huhn, K. Kantner, C. Geidel, S. Brandholt, I. De Cock, S. J. H. Soenen, P. R. Gil, J. M. Montenegro, K. Braeckmans, K. Mullen, G. U. Nienhaus, M. Klapper and W. J. Parak, *ACS Nano*, 2013, **7**, 3253.

26 M. G. Soliman, B. Pelaz, W. J. Parak and P. del Pino, *Chem. Mater.*, 2015, **27**, 990.

27 H. Mattoussi and J. Cheon, *Inorganic Nanoprobes for Biological Sensing and Imaging*, Artech House, Boston, 2009.

28 I. Garcia, A. Sanchez-Iglesias, M. Henriksen-Lacey, M. Grzelczak, S. Penades and L. M. Liz-Marzan, *J. Am. Chem. Soc.*, 2015, **137**, 3686.
29 W. Wang, A. Kapur, X. Ji, M. Safi, G. Palui, V. Palomo, P. E. Dawson and H. Mattoussi, *J. Am. Chem. Soc.*, 2015, **137**, 5438.
30 T. Pons, H. T. Uyeda, I. L. Medintz and H. Mattoussi, *J. Phys. Chem. B*, 2006, **110**, 20308.
31 B. C. Mei, K. Susumu, I. L. Medintz and H. Mattoussi, *Nat. Protoc.*, 2009, **4**, 412.
32 K. Susumu, E. Oh, J. B. Delehanty, J. B. Blanco-Canosa, B. J. Johnson, V. Jain, W. J. Hervey, W. R. Algar, K. Boeneman, P. E. Dawson and I. L. Medintz, *J. Am. Chem. Soc.*, 2011, **133**, 9480.
33 S. Liu, G. Y. Chen, P. N. Prasad and M. T. Swihart, *Chem. Mater.*, 2011, **23**, 4098.
34 X. C. Ye, C. Zheng, J. Chen, Y. Z. Gao and C. B. Murray, *Nano Lett.*, 2013, **13**, 765.
35 X. C. Ye, L. H. Jin, H. Caglayan, J. Chen, G. Z. Xing, C. Zheng, V. Doan-Nguyen, Y. J. Kang, N. Engheta, C. R. Kagan and C. B. Murray, *ACS Nano*, 2012, **6**, 2804.
36 W. Wang, F. Aldeek, X. Ji, B. Zeng and H. Mattoussi, *Faraday Discuss.*, 2014, **175**, 137.
37 W. Wang, X. Ji, H. B. Na, M. Safi, A. Smith, G. Palui, J. M. Perez and H. Mattoussi, *Langmuir*, 2014, **30**, 6197.
38 W. Wang, X. Ji, A. Kapur and H. Mattoussi, *SPIE BiOS*, 2016, **9722**, 972203.
39 B. J. Berne and R. Pecora *Dynamic Light Scattering: With Applications to Chemistry, Biology, and Physics*, ed. Dover, Dover Publications, Mineola, N.Y., 2000.
40 S. Lal, S. E. Clare and N. J. Halas, *Acc. Chem. Res.*, 2008, **41**, 1842.
41 X. H. Huang, P. K. Jain, I. H. El-Sayed and M. A. El-Sayed, *Lasers Med. Sci.*, 2008, **23**, 217.
42 M. Mahmoudi, I. Lynch, M. R. Ejtehadi, M. P. Monopoli, F. B. Bombelli and S. Laurent, *Chem. Rev.*, 2011, **111**, 5610.
43 M. P. Monopoli, C. Aberg, A. Salvati and K. A. Dawson, *Nat. Nanotechnol.*, 2012, **7**, 779.
44 I. Lynch and K. A. Dawson, *Nano Today*, 2008, **3**, 40.
45 E. Mahon, A. Salvati, F. B. Bombelli, I. Lynch and K. A. Dawson, *J. Controlled Release*, 2012, **161**, 164.
46 K. Welsher, S. A. McManus, C.-H. Hsia, S. Yin and H. Yang, *J. Am. Chem. Soc.*, 2015, **137**, 580.

Faraday Discussions

PAPER

Design of artificial membrane transporters from gold nanoparticles with controllable hydrophobicity†

Marcin P. Grzelczak, Alexander P. Hill, Domagoj Belic, Dan F. Bradley, Casper Kunstmann-Olsen and Mathias Brust*

Received 2nd March 2016, Accepted 23rd March 2016
DOI: 10.1039/c6fd00037a

Gold nanoparticles with variable hydrophobicity have been prepared in three different size regimes following established methods. The control of hydrophobicity was achieved by complexation of the 18-crown-6-CH_2-thiolate ligand shell with potassium ions. Potassium dependent phase transfer of these particles from dispersion in water to chloroform was demonstrated, and the equilibrium partitioning of the particles in water–chloroform liquid/liquid systems was quantified by optical spectroscopy. The gradual complexation of the ligand shell with potassium ions was further monitored by zeta potential measurements. Potassium dependent insertion of nanoparticles into the phospholipid bilayer membrane of vesicles in aqueous dispersion has been demonstrated by cryogenic transmission electron microscopy (cryo-TEM). Nanoparticle-dependent potassium ion transport across the vesicle membrane has been established by monitoring the membrane potential with fluorescence spectroscopy using a potential sensitive dye.

Introduction

At a certain level of structural or functional complexity, asymmetry becomes a vital property of the constituents of self-assembled systems.[1] Without delving into theory, this statement is intuitively corroborated by looking at the elemental building blocks of existing assemblies at any scale, not least Lego® bricks, the asymmetry of which forms the basis for their directional assembly. The growth of microtubules, which are part of the cytoskeleton and, among other functions, responsible for directional intracellular transport, represents one of countless biological examples of directional assembly, in this case from bipolar protein subunits.[2] Unlike proteins and other biomolecules, most artificial nano-objects are crystalline groups of atoms, ions or molecules, sometimes protected by

University of Liverpool, Department of Chemistry, Crown Street, Liverpool, UK, L69 7ZD. E-mail: mbrust@liverpool.ac.uk

† Electronic supplementary information (ESI) available. See DOI: 10.1039/c6fd00037a

a molecular ligand shell and usually highly symmetric. In order to break this symmetry, for example by creating a top and a bottom or a left and a right side, these objects have to interact with an asymmetric environment, which could be an external electric[3,4] or magnetic field,[5,6] directional flow[7] or freezing,[8] any type of gradient, or an interface.[9–12] In this contribution we demonstrate the rational design of water dispersible gold nanoparticles that can be programmed to reside at the interfaces formed by phospholipid bilayer membranes in an aqueous dispersion of vesicles. This is achieved by increasing the hydrophobicity of the particles in a controlled fashion *via* complexation of potassium ions in their ligand shell. The complexing ligand chosen here is a thiolated crown ether, which is known to become more hydrophobic upon complexation of potassium ions.[13] This interaction has been used before to control the hydrophobicity of nanoparticles[14,15] but not to target phospholipid membranes with them. However, crown ethers modified with self-assembling moieties[16,17] and with peptides[18] have been reported to reside in bio-membranes and to act as artificial ion channels. Once our particles were bound to the membrane, we could have tried to take advantage of this environment to modify the particles only on one side creating asymmetry in the ligand shell. There are indeed many reports on similar approaches to this end.[10,11,19] Instead of exploring this route, we were interested in probing the functionality of membrane bound particles and found that they act as specific potassium transporters and, in the presence of a potassium concentration gradient, rapidly polarise the membrane to the corresponding potential. In this discussion article we present the design and preparation of crown ether modified gold nanoparticles of three different sizes, detailed phase transfer studies and zeta potential measurements to demonstrate the control of hydrophobicity by complexation of potassium ions to the crown moieties, and finally vesicle membrane polarisation by membrane bound nanoparticles in the presence of a potassium ion concentration gradient across the membrane.

Experimental

Chemicals

Trisodium citrate $Na_3C_6H_5O_7$, hydrogen tetrachloroaurate(III) hydrate $HAuCl_4 \cdot 3H_2O$, sodium borohydride $NaBH_4$ were supplied by Aldrich. *O*-[2-(3-Mercaptopropionylamino)ethyl]-*O*′-methylpolyethylene glycol (mPEG-SH, M_w 5000) was purchased from Sigma-Aldrich. Liposome kit and LiCl, NaCl, KCl salts were purchased from Sigma. 18-Crown-6-CH_2 was supplied by ProChimia. Pure-grade methanol, ethanol and Milli-Q-grade water were used in all preparations.

Synthesis of 2 nm Au/18-C-6-CH_2-SH NPs

Small, 2 nm Au/18-C-6-CH_2-SH NPs were synthesized by a modified literature method[20,21] where a 1 : 1 Au^{III} : capping agent ratio was used. Briefly, 3 mL of a 9 mM solution of hydrogen tetrachloroaurate(III) hydrate in methanol was mixed with 3 mL of a 9 mM solution of 18-C-6-CH_2-SH in methanol resulting in a clear yellow solution. The mixture was stirred for 10 minutes at room temperature and turned almost colorless and slightly cloudy. Then, 3 mL of a freshly prepared 0.056 M solution of $NaBH_4$ in methanol was rapidly added under vigorous stirring. Upon addition, the solution quickly became deep brown in color and was left for 15 minutes to allow the reaction to proceed to completion. After one hour, the resulting Au/18-C-6-CH_2-SH NPs dispersion was rotary evaporated at 30 °C to yield a black residue of NPs which was washed three times with diethyl ether. In the final step, the crude product was redispersed in iso-propanol to remove excess hydrophilic reaction products that were filtered off, and after careful rotary evaporation (30 °C) the purified product was obtained as a black solid, which was dispersed in MQ water for further use.

Synthesis of citrate stabilised particles

Citrate Au NPs were synthesized by a modified Turkevich method reported by Bastús[22] *et al.* Briefly, an aqueous solution of trisodium citrate (2.2 mM, 75 mL) was refluxed for 15 minutes, and aqueous $HAuCl_4$ (0.5 mL, 25 mM) was rapidly added to the solution under vigorous stirring. Over a period of 10 minutes, the colour of the mixture gradually became deeper red. The reaction mixture was left stirring for a further 20 minutes to ensure the reaction had gone to completion and then left to cool to room temperature.

To grow larger particles, the above procedure was repeated but instead of allowing the mixture to cool down, the temperature was reduced to 90 °C, and 1 mL of the dispersion was replaced by 0.5 mL of a trisodium citrate solution (60 mM) and 0.5 mL of an $HAuCl_4$ solution (25 mM). This process was repeated every 4 hours to a total of 4 times. It promoted the gradual growth of particles to a final size of 38 nm.

Two step functionalization of citrate particles with 18-C-6-CH_2-SH

Lack of stability control in direct functionalization of citrate Au NPs with 18-C-6-CH_2-SH led us to introduce an intermediate step which consists of replacing citrate ligand with *O*-[2-(3-Mercaptopropionylamino)ethyl]-*O*′-methylpolyethylene glycol (mPEG-SH, M_w 5000) molecule. This was then followed by ligand replacement in ethanol with a large excess of 18-C-6-CH_2-SH. In order to functionalize Au/citrate NPs with m-PEG-SH 5000 an aqueous dispersion of 15 nm citrate particles (30 mL) was mixed with m-PEG-SH 5000 (4 molecules per nm^2 of total gold NPs surface) under vigorous stirring. The solution was left to stir for 40 minutes at room temperature to ensure complete functionalization occurred. The solution was then centrifuged at 14 500 rpm for 30 minutes, followed by the removal of water and re-dispersion in ethanol (5 mL); this purification process was repeated 3 times.[23]

In the same way, larger, 38 nm citrate Au particles were functionalized with the same number of m-PEG 5000 molecules per nm^2 based on the total gold NPs surface.

In order to replace the intermediate m-PEG 5000 ligand with the desired one, a solution of 18-C-6-CH_2-SH (400 molecules per nm^2 of total gold NPs surface) dissolved in 1 mL of ethanol was slowly added drop wise under vigorous stirring to the ethanolic dispersion of 15 nm Au/m-PEG 5000 NPs (25 mL). The mixture was then sonicated for 20 minutes and left at room temperature for 12 hours. The product was isolated by repeated centrifugation at 12 000 rpm (1 hour) and re-dispersion in MQ water. Samples were then analysed by TEM and UV-vis.

The above process was repeated for the larger 38 nm citrate particles with the same number of 18-C-6-CH_2-SH ligand molecules 400 per nm^2.

Phase transfer of Au/18-C-6-CH_2-SH NPs

Dispersions containing 2 nm Au/18-C-6-CH_2-SH ($(8 \pm 1) \times 10^{14}$ NPs/ml) were mixed with KCl to adjust the following final concentrations of salt in aqueous phases: 0, 1.25, 12.5, 62.5, 200, 400 mM, followed by shaking for 15 seconds. Then each sample was carefully added to chloroform (1 mL) to form a two-phase water/chloroform system in a glass vial (4 mL). Pictures were taken at regular time intervals to monitor the gradual phase transfer of the particle.

Phase transfer experiments with larger particles (15 and 38 nm) were all carried out at a constant KCl concentration of 4 mM. The concentrations of particles were adjusted to give comparable optical densities.

The phase transfer ability of different cations (LiCl, NaCl and KCl) was tested with 2 nm particles ($(8 \pm 1) \times 10^{14}$ Au/18-C-6-CH_2-SH NPs per mL); experiments were performed as described above.

Quantitative particle partitioning experiments were carried out in the same way except for the rapid equilibration, which was achieved by vortexing the two-phase systems for 10 s. After phase separation, samples were analysed by UV-vis.

Zeta potential measurements

All measurements were carried out with a Malvern zetasizer nano ZSP. For this purpose, KCl was added to aqueous dispersions containing 2 nm Au/18-C-6-CH_2-SH ($(3 \pm 0.5) \times 10^{11}$ NPs per mL) to the following final salt concentrations: 0, 0.25, 2.5, 8, 12.5, 20, 30, 40 μM.

TEM and HRTEM

TEM samples were prepared by drop casting 3 μL of Au nanoparticle colloidal solution onto holey carbon TEM grids (Agar Scientific or TAAB). After drying in air, the samples were imaged on an FEI Tecnai Spirit G2 BioTWIN TEM at an accelerating voltage of 120 kV, using an Olympus-SIS MegaView III digital camera. High-resolution TEM was performed on an aberration-corrected JEOL JEM 2100FCs microscope operating in STEM mode at an accelerating voltage of 200 kV.

Vesicles preparation

Vesicles were prepared by a modified variant of the method described by Newcomb[24] *et al.* Briefly, a 10 mM stock solution of phospholipids was made by dilution of liposome kit lyophilized powder (cholesterol 9 μmol, L-α-phosphatidylcholine (egg yolk) 63 μmol, stearylamine 18 μmol) in chloroform. 1 mL aliquots of the stock solution were rotary evaporated for 2 hours in order to remove all residues of organic solvent. MQ water or KCl and NaCl buffer solutions were used to hydrate the phospholipids depending on the requirements of each experiment. In order to make the white suspension of phospholipids more homogeneous, it was heated to 60 °C followed by 4–5 freeze–thaw cycles using liquid nitrogen. Next, the hydrated and homogeneous phospholipid solution was extruded 20 times through a mini-extruder (Avanti) with a 100 nm-pore-size polycarbonate filter (Whatman) to form vesicles of about 100 nm diameter. The vesicle dispersion was purified by dialysis using MWCO 12 000–14 000 dialysis tubing (Serva).

Cryo-TEM

Samples for cryo-TEM were prepared by dropcasting 4 μL of the solution onto freshly glow-discharged holey carbon TEM grids (Agar Scientific or TAAB). After 60 s of incubation, the grids were double-side blotted for 2 × 2 s and plunge-frozen using an FEI Vitrobot Mk2 system working at >96% relative humidity. Vitrified samples were transferred onto a 626 Gatan cryo-holder and imaged on the FEI Tecnai Spirit G2 BioTWIN TEM in low-dose mode at an accelerating voltage of 120 kV.

Membrane potential changes detected by fluorescence spectroscopy

The suspension of vesicles with a final concentration of phospholipids of 1 mM was placed in a fluorimetric cuvette followed by addition of the potential probe dye Safranin O (180 nM final concentration). Then the fluorescence intensity was allowed to equilibrate for 2 minutes. The kinetic mode was used to detect continuously fluorescent intensity changes at 589 nm with an excitation wavelength of 521 nm. Next, 20 nM of valinomycin or different concentrations of 2 nm Au/18-C-6-CH_2-SH NPs was added, and the fluorescence intensity was monitored for 1700 s.

Results and discussion

A modified capping agent, 18-C-6-CH_2-SH, which is a thiol-derivative of a crown ether (18-crown-6) was immobilized on the gold particle surface by binding *via* its thiol functionality as illustrated in Fig. 1a. Gold nanoparticles of three different size regimes, around 2, 15 and 38 nm were prepared and modified in this way. The corresponding TEM or HRTEM images, size distributions and UV-vis spectra are shown in Fig. 1. The differences in size are clearly manifested in the evolution of the UV-vis spectra (Fig. 1b). The smallest NPs show virtually no plasmon peak (black line) whereas the largest exhibit a pronounced and slightly red shifted plasmon band (red line).

The idea of using crown ether 18-crown-6 as capping agent is based on its well established complexation-dependent hydrophobicity. The hydrophilic ethylene oxide ring becomes much more hydrophobic upon complexation of cations, in particular potassium, which binds very strongly to the cavity. The strategy of controlling the hydrophobicity of supramolecular objects in this way has been reported several times.[25,26] Recently M. I. Bodnarchuk[15] *et al.* have shown that the presence of macrocyclic ethers enables the solubilization of inorganic-capped NPs in solvents of any polarity and improves an ability to form NP superlattices. Since our gold NPs are functionalized with 18-crown-6 molecules which are capable of selectively coordinating hard alkali metal ions such as potassium[27,28] the solubility of those NPs in different solvents can be tuned by changing the type and salt concentration in the surrounding medium. This is confirmed by the simple experiment shown in Fig. 2, where the same concentration of the 2 nm Au/18-C-6-CH_2-SH NPs dispersed in water was brought into contact with a chloroform phase. In this way, two immiscible phases were created with a well-defined interface. The only variable component in each vial presented in Fig. 2 is the KCl salt concentration. In Fig. 2a the initial and the final state of phase transfer after 18 hours are shown indicating that even after a long time equilibrium has not yet

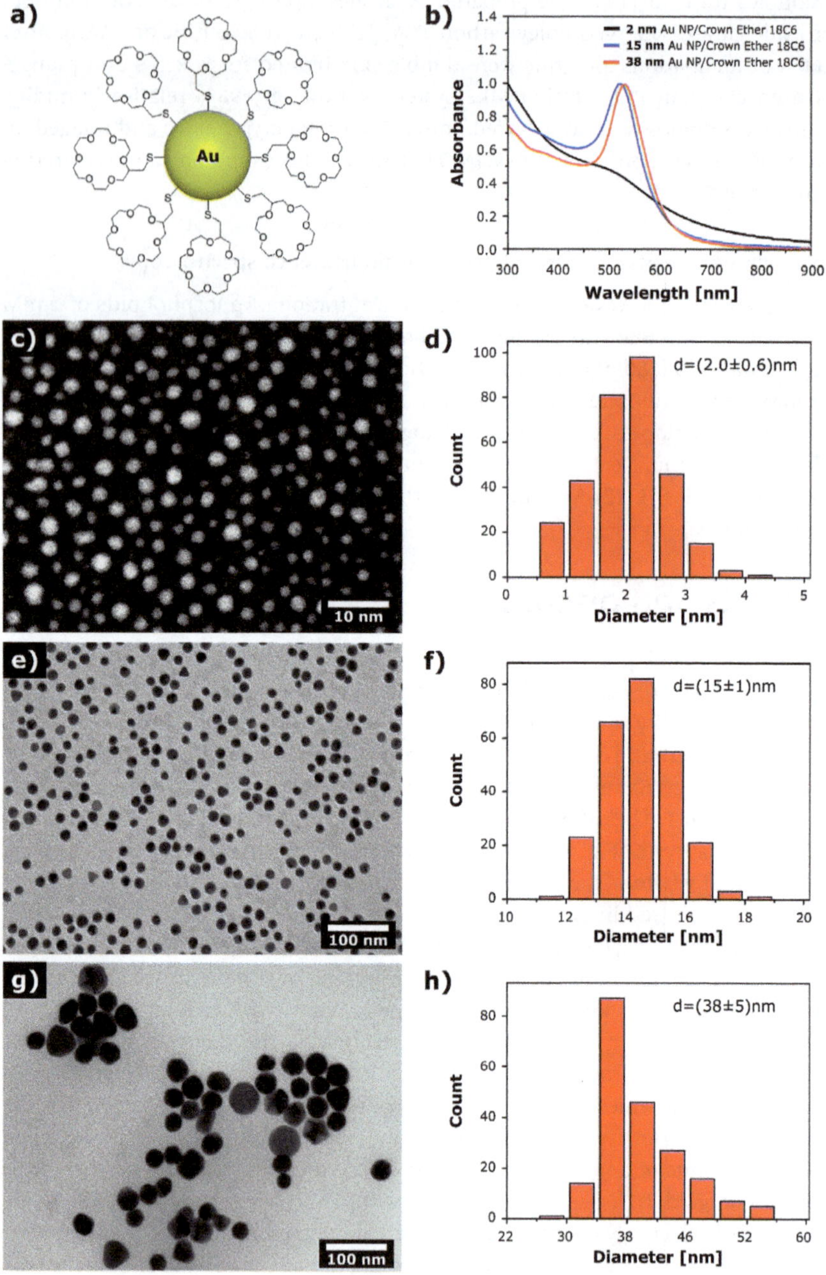

Fig. 1 (a) Model of Au NP functionalized with crown ether 18-C-6-CH_2-SH. (b) UV-vis spectra of three different sizes of Au/18-C-6-CH_2-SH NPs in water. (c) High angle annular dark field (HAADF) STEM image of 2 nm Au/18-C-6-CH_2-SH NPs. (e and g) Bright field TEM images of 15 nm and 38 nm Au/18-C-6-CH_2-SH. (d, f and h) The corresponding size distributions.

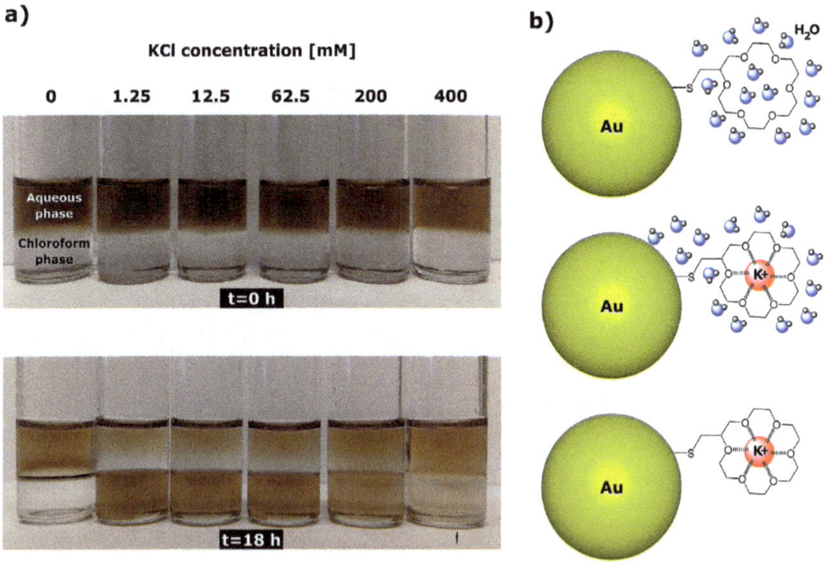

Fig. 2 (a) Photos of phase transfer experiments from water to chloroform of 2 nm Au/18-C-6-CH_2-SH NPs at different KCl salt concentrations. (b) Model of hydrophobicity change upon complexation of potassium ions.

been reached. A more detailed time evolution of this process is presented in Fig. S1.† Both the absence and much higher concentrations of KCl salt either entirely prevent, or dramatically slow down the phase transfer process.

Before complexation, the crown ether 18-crown-6 and hence the Au/18-C-6-CH_2-SH NPs are stable in aqueous media. This is attributable to the oxygen atoms present within the ring (Fig. 2a). These are capable of hydrogen bonding with water molecules resulting in an energetically favorable process for the water to form a solvent shell around the 18-crown-6-CH_2-SH and hence render the particles dispersible in water.

On the other hand, upon complexation with a suitable cation, the electrons of the oxygen atoms are no longer interacting with the water present in the system but with the complexed cation.[29] This results in the aqueous solvent shell being no longer energetically favorable, and thus in the transfer of the particles to the chloroform phase. The data shown so far are qualitative kinetic data of phase transfer, which may not necessarily relate to the thermodynamic properties of the system. For this reason, we have also looked at the equilibrium partition of NPs between the water and the chloroform phase. NPs were first dispersed in water and in aqueous KCl solutions in the same concentration range used in Fig. 2a and then thoroughly vortexed together with a chloroform phase. After this, when phase separation had completed (Fig. 3a and b), UV-vis spectra were taken of both phases to determine the amount of NPs present in both solutions (see Fig. S2†).

Fig. 3a–c shows the equilibrium partitioning of 2 nm nanoparticles between water and chloroform at different potassium chloride concentrations in the aqueous phase, after intense mixing (vortex) and subsequent phase separation. With increasing salt concentration the particles initially become more hydrophobic, reaching maximal hydrophobicity at around 60 mM salt concentration. At

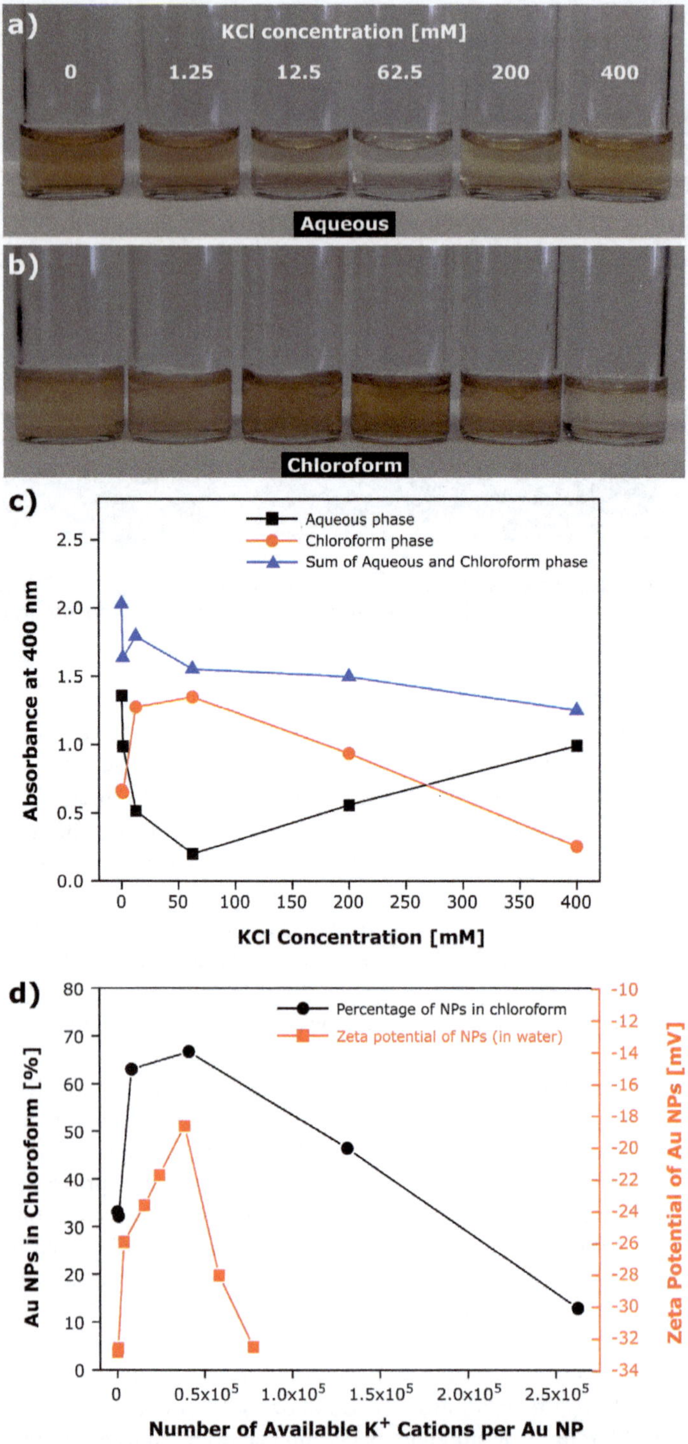

Fig. 3 (a) and (b) Equilibrium partitioning of 2 nm Au/18-C-6-CH_2-SH NPs in water and chloroform phases after vortexing and solvent separation. (c) Partitioning as a function of salt concentration. (d) Partitioning (only chloroform phase shown) and zeta potential of the particles as a function of the potassium per nanoparticle ratio.

further increasing salt concentration this trend is reversed so that at the highest salt concentration used (400 mM) most particles remain in the aqueous phase, comparable to the situation without salt. Thus the empirical optimum of salt concentration for creating hydrophobic particles is at around 60 mM. As we assume that the complexation of potassium ions with the crown ether controls the hydrophobicity of the particles, we have also measured their zeta-potential, which should become more positive with increasing complexation. Unfortunately, the range of salt concentrations of the partitioning experiments (Fig. 3a–c) extends by three orders of magnitude the range in which zeta potential measurements are reliably possible (see Fig. S3†) due to the adverse effect of increased solution conductivity. It is however possible to conduct zeta potential measurements at relatively low particle concentrations compared to the concentrations typically needed for the partitioning experiments. Therefore we can plot both the zeta potential and the partitioning data on the same graph if we use as the *X*-axis the ratio between the number of potassium ions and the number of particles present in solution, instead of the salt concentration. This is shown in Fig. 3d, suggesting indeed a good correlation between the zeta potential and the hydrophobicity of the particles. This notion however has to be regarded with some caution since it implicitly contains the hypothesis that not the absolute salt concentration but the relative amount of salt compared to the number of crown ether moieties matters in determining the hydrophobicity of the particles. We have not yet been able to show unequivocally that this is the case. For example, a change in particle concentration should give a rather different appearance to the graph in Fig. 3c since at lower/higher particle concentration, saturation of the crown ether with potassium ions will occur at lower/higher salt concentrations. So far, we found that this is not really the case over the entire range of salt concentrations shown in Fig. 3c. Note that the increase in hydrophobicity is initially very steep (between 0 and 10 mM KCl). It appears from preliminary studies that this becomes even steeper at lower particle concentrations confirming the model of crown ether saturation with potassium ions. For the nanoparticles used here we would expect no more 300 ligands per particle which means that full saturation with potassium ions may be expected already at very low potassium concentrations depending to what extend their binding constant is influenced by the confinement of ligands to the nanoparticle surface. Above 10 mM KCl concentration the appearance of the graph is quite independent of particle concentration suggesting an important role of other effects that are independent of particle concentration, for example, the ionic strength of the solution, which can lead to a well-known salting out effect. The most curious behavior is that at higher salt concentrations the hydrophilicity of the particles returns. This may be for the same reason we observe a return of the zeta potential to more negative values bearing in mind the above mentioned limitation that zeta potentials could only be measured at very low absolute salt concentrations. We tentatively attribute the reverse trends in both the hydrophobicity, and the zeta potential to the build-up of the electrochemical double layer around the nanoparticles which leads increasingly to a screening of the positive charge associated with the complexed potassium ions. From our current understanding we thus consider that three different phenomena cooperate in determining the partition of nanoparticles between water and chloroform. (i) The complexation of potassium ions causing hydrophobicity and lowering the absolute value of the zeta potential, and (ii) the build-up of the electrochemical double-layer causing hydrophilicity and increasing the absolute value of the zeta potential, depend on the potassium/nanoparticle ratio. (iii) The salting out phenomenon, effectively increasing the hydrophobicity of the particles, only depends on the ionic strength of the solution and does not scale with nanoparticle concentration. This is most likely responsible for the significant broadening of the partitioning graph in Fig. 3d in comparison with the rather sharply decreasing and increasing curve of the zeta potential. Meanwhile, above 60 mM salt concentration the particles become rapidly more hydrophilic with further increasing salt concentration; the salting out effect works in the opposite direction but is overcompensated eventually.

We have also qualitatively compared phase transfer of crown ether modified gold nanoparticles of three different sizes, 2, 15 and 38 nm. The results are shown in Fig. 4a. Particles of all sizes transfer to the organic phase. In all cases, phase transfer is fast compared to diffusion, which leads to depletion of particles in the aqueous phase just above the interface. Even after 23 hours equilibrium is not reached due to the very slow diffusion rate. The UV-vis spectra taken from the both phases show that the particles remain stable and that there is virtually no loss of particles. Plasmon peaks of 15 and 38 nm Au/18-C-6-CH_2-SH NPs are red shifted in comparison to the absorbance of Au NPs in the aqueous phase, which is attributed to changes in the refractive index of the medium.[30]

The small inset in each absorbance spectrum shows the organic phase, which was separated from the bottom of each vial to indicate that NPs are really moving through the interface and are not adhering to the glass wall.

Having established the ability to control their hydrophobicity, the Au/18-C-6-CH_2-SH NPs were now tested for membrane targeting. For this purpose, unilamellar, positively charged vesicles (zeta potential = 36.5 ± 1.0 mV) were prepared in the absence of salt buffers. Nanoparticles dispersed in a 4 mM aqueous KCl solution and added to the same volume of an aqueous dispersion of vesicles expecting that under these conditions the particles are fully complexed with potassium and hence sufficiently hydrophobic to associate preferentially with the vesicle membrane. After 60 min of incubation samples were prepared for inspection by cryo-TEM. As expected, good adhesion of particles to the vesicles was found (Fig. 5c and d). For comparison, control experiments without KCl shown in Fig. 5a and b gave no evidence for interaction of particles with vesicles.

Finally, the high selectivity of 18-crown-6 molecules for potassium allowed us to design experiments where NPs could be used as a new type of artificial ionophore polarizing the phospholipid bilayer membrane of the vesicles. For this purpose, we have adjusted the potassium concentrations inside and outside the vesicles so that in the presence of an ionophore specific for potassium a membrane potential, ψ, is established according to the Nernst–Donnan equation.[31]

$$\Delta\psi = \frac{RT}{zF}\ln\left(\frac{[K^+]_{out}}{[K^+]_{in}}\right) \tag{1}$$

The membrane potential was interrogated by fluorescence spectroscopy using Safranin O as an indicator dye. The fluorescence intensity of this dye depends

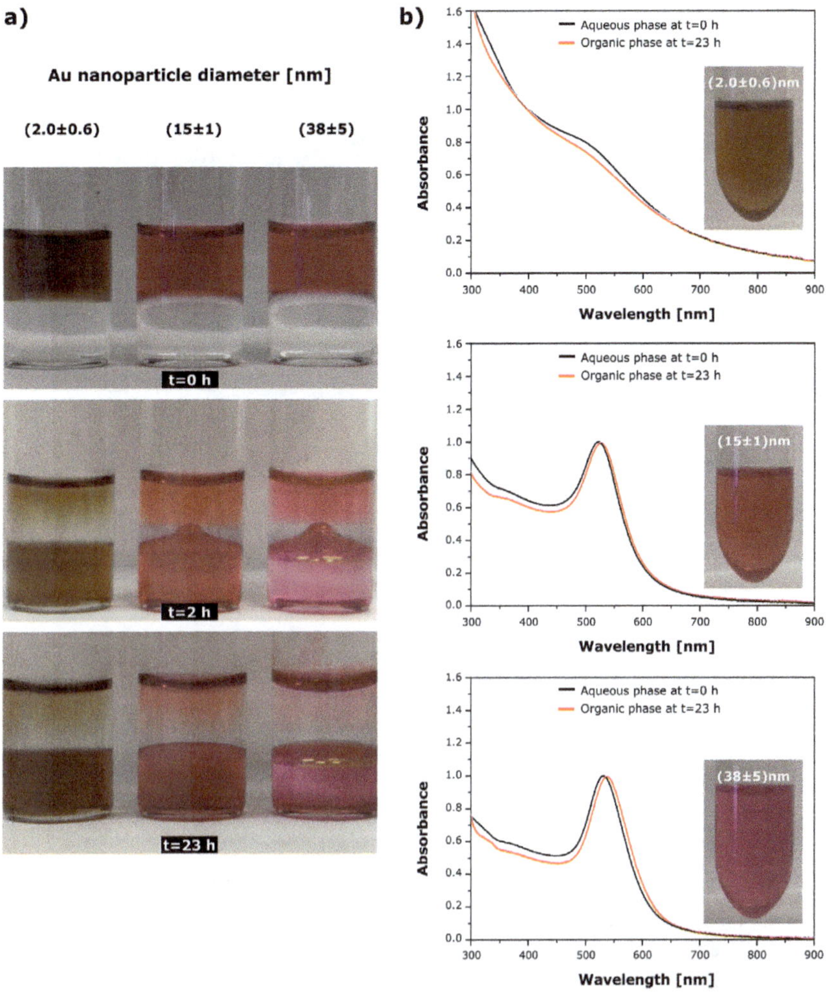

Fig. 4 (a) Photos of phase transfer experiments from water to chloroform of three different sizes of Au/18-C-6-CH_2-SH NPs at the same salt concentrations. (b) UV-vis spectra of the initial aqueous phase and final chloroform phase, for each NP size. The insets show the respective organic phase evidencing fully dispersed and stable nanoparticles.

nearly linearly on the membrane potential (see ESI Fig. S4†). The dye is only sensitive if the polarity of the vesicle membrane inside is negative. It is generally assumed that the positively charged dye is driven into the membrane from outside the vesicle by the electric field across the membrane. The results are shown in Fig. 6. If the potassium concentration is high inside and low outside (Fig. 6a) the polarity of the membrane is inside negative and hence membrane potential can be quantified by the fluorescence intensity increase. While the final membrane potential is given by the Nernst–Donnan equation and should not depend on the concentration of ionophore, the speed of membrane polarization is expected to depend on the amount of ionophore present. This can indeed be clearly seen in Fig. 6c where three different concentrations of nanoparticles are

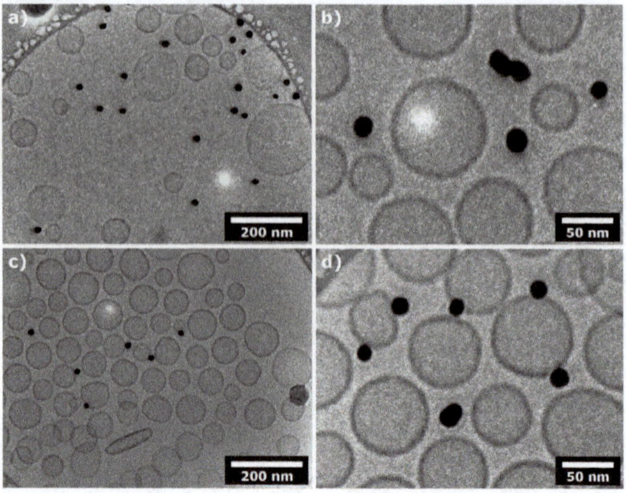

Fig. 5 Cryo-TEM images of phospholipid vesicles incubated with 15 nm Au/18-C-6-CH_2-SH NPs for one hour. (a and b) In the absence of K^+ cations the NPs do not become embedded within the vesicle membrane. (c and d) In the presence of 2 mM K^+ cations the NPs exhibit a tendency to associate with the vesicle membrane.

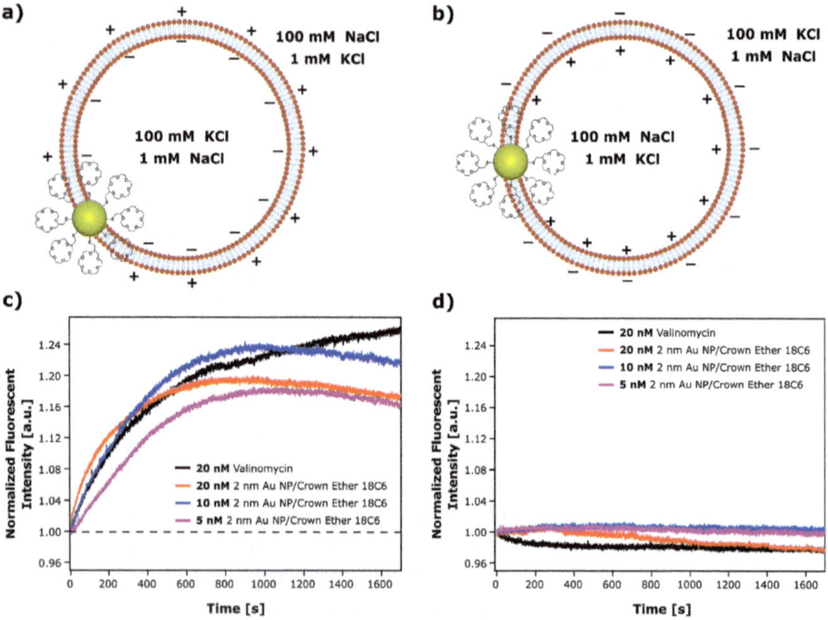

Fig. 6 (a and b) Models of polarised phospholipid bilayer membrane. Note that the membrane polarity is determined only by external and internal potassium ion concentrations. (c) Normalized fluorescent intensity changes caused by membrane polarization after addition of different concentrations of 2 nm Au/18-C-6-CH_2-SH NPs and, for comparison, valinomycin. (d) Opposite polarization of the membrane does not lead to a fluorescence increase due to the nature of Safranin O.

compared. For comparison the membrane was also polarised using the well-established ionophore valinomycin. In the case of the addition of gold nanoparticles, a small and immediate irreversible drop of fluorescent intensity was observed (for clarity, not shown in Fig. 6) that we attribute to quenching by the metal. Otherwise the behavior of valinomycin and gold nanoparticles was near-identical. As expected, reversing the polarity[32] of the membrane potential (Fig. 6b and d) by adjusting the high potassium concentration outside of vesicles does not lead to measurable polarization due to the above-mentioned limitations of Safranin O. This experiment establishes high selectivity for potassium compared to sodium. The same selectivity can be found in phase transfer experiments, where neither sodium nor lithium is capable of increasing the hydrophobicity of nanoparticles (see Fig. S5†).

It is somewhat surprising that our particles appear to act as perfect and selective ionophores comparable to the standard compound valinomycin. The mechanism of ion transfer has not been studied and remains open to speculation. It is most likely that the complex potassium is in relatively fast equilibrium with free potassium and that the transfer of potassium across the membrane is facilitated by a tumbling motion of the particles that randomly exposes crown ether moieties to either side of the membrane. An alternative, perhaps even more appealing mechanism could be that the crowns self-organize on the surface of the particle in such a way that a continuous channel is formed. A comparatively high level of ligand shell organization has been reported before in contributions by Rotello *et al.*[33] and by Stellacci *et al.*[34–36] but the latter's have later been disputed by others.[37,38]

Conclusions

We have demonstrated a rational approach to the design of gold nanoparticles with controllable hydrophobicity. These particles readily transferred from aqueous to organic dispersion and were successfully employed to target vesicle membranes in an aqueous dispersion as evidenced by cryo-TEM. The particles do not only associate, as expected, with the hydrophobic vesicle membrane, but also act as highly efficient and selective ionophores for potassium ions. Membrane polarization by potassium transfer has been achieved. This report establishes functionalized metal nanoparticles as a new class of membrane transporters with potential applications in sensing and biomedical research.

Acknowledgements

We would like to thank A. Beckett and I. Prior from Biomedical Electron Microscopy Unit, University of Liverpool, and T. Heil and S. Romani from the Nanoinvestigation Centre, University of Liverpool. The research was founded through ERC Advanced Grant "PANDORA" under grant number 108269.

References

1 G. M. Whitesides and M. Boncheva, Beyond Molecules: Self-assembly of mesoscopic and macroscopic components, *Proc. Natl. Acad. Sci. U. S. A.*, 2002, **99**(8), 4769–4774.

2 C. Conde and A. Cáceres, Microtubule assembly, organization and dynamics in axons and dendrites, *Nat. Rev. Neurosci.*, 2009, **10**(5), 319–332.

3 J. Kleinert, S. Kim and O. D. Velev, Electric-field-assisted convective assembly of colloidal crystal coatings, *Langmuir*, 2010, **26**(12), 10380–10385.

4 K. M. Ryan, A. Mastroianni, K. A. Stancil, H. Liu and A. P. Alivisatos, Electric-field-assisted assembly of perpendiculary oriented nanorod superlattice, *Nano Lett.*, 2006, **6**, 1479–1482.

5 A. T. Ngo and M. P. Pileni, Assemblies of ferrite nanocrystals: Partial orientation of the easy magnetic axes, *J. Phys. Chem. B*, 2001, **105**(1), 53–58.

6 M. A. Correa-Duarte, M. Grzelczak, V. Salgueiriño-Maceira, M. Giersig, L. M. Liz-Marzán, M. Farle, K. Sierazdki and R. Diaz, Alignment of carbon nanotubes under low magnetic fields through attachment of magnetic nanoparticles, *J. Phys. Chem. B*, 2005, **109**, 19060–19063.

7 Y. Huang, X. Duan, Q. Wei and C. M. Lieber, Directed assembly of one-dimensional nanostructures into functional networks, *Science*, 2001, **291**(5504), 630–633.

8 H. Zhang, I. Hussain, M. Brust, M. F. Butler, S. P. Rannard and A. I. Cooper, Aligned two- and three-dimensional structures by directional freezing of polymers and nanoparticles, *Nat. Mater.*, 2005, **4**, 787–793.

9 K. Nørgaard, M. J. Weygand, K. Kjaer, M. Brust and T. Bjørnholm, Adaptive chemistry of bifunctional gold nanoparticles at the air/water interface. A synchrotron X-ray study of giant amphiphiles, *Faraday Discuss.*, 2004, **125**, 221–233, discussion 293–309.

10 J. Du and R. K. O'Reilly, Anisotropic particles with patchy, multicompartment and Janus architectures: Preparation and application, *Chem. Soc. Rev.*, 2011, **40**(5), 2402–2416.

11 D. Rodríguez-Fernández and L. M. Liz-Marzán, Metallic Janus and patchy particles, *Part. Part. Syst. Charact.*, 2013, **30**(1), 46–60.

12 A. Perro, S. Reculusa, S. Ravaine, E. Bourgeat-Lami and E. Duguet, Design and synthesis of Janus micro- and nanoparticles, *J. Mater. Chem.*, 2005, **15**(35–36), 3745–3760.

13 D. J. Sam and H. E. Simmons, Crown polyether chemistry. Potassium permanganate oxidations in benzene, *J. Am. Chem. Soc.*, 1972, **94**(11), 4024–4025.

14 K. Y. Lee, Y. Bae, M. Kim, G.-W. Cheong, J. Kim, S. S. Lee and S. W. Han, Crown ether derivatives-mediated self-assembly of nanoparticles at the liquid/liquid interface, *Thin Solid Films*, 2006, **515**(4), 2049–2054.

15 M. I. Bodnarchuk, S. Yakunin, L. Piveteau and M. V. Kovalenko, Host–guest chemistry for tuning colloidal solubility, self-organization and photoconductivity of inorganic-capped nanocrystals, *Nat. Commun.*, 2015, **6**, 10142.

16 A. Gilles and M. Barboiu, Highly selective artificial K(+) channels: An example of selectivity-induced transmembrane potential, *J. Am. Chem. Soc.*, 2016, **138**, 426–432.

17 Z. Sun, M. Barboiu, Y.-M. Legrand, E. Petit and A. Rotaru, Highly selective artificial cholesteryl crown ether K^+-channels, *Angew. Chem.*, 2015, **127**(48), 14681–14685.

18 N. Voyer, M. Arseneault and F. Otis, Synthesis and characterization of peptide nanostructures designed for sensing applications, *Photon. Appl. Biosens. Imaging*, 2005, **5969**, 59690M1.

19 M. R. Rasch, E. Rossinyol, J. L. Hueso, B. W. Goodfellow, J. Arbiol and B. A. Korgel, Hydrophobic gold nanoparticle self-assembly with phosphatidylcholine lipid: Membrane-loaded and Janus vesicles, *Nano Lett.*, 2010, **10**(9), 3733–3739.

20 M. Brust, J. Fink, D. Bethell, D. J. Schiffrin and C. Kiely, Synthesis and reactions of functionalised gold nanoparticles, *J. Chem. Soc., Chem. Commun.*, 1995, 1655–1656.

21 A. M. Cioran, A. D. Musteti, F. Teixidor, Ž. Krpetić, I. A. Prior, Q. He, C. J. Kiely, M. Brust and C. Viñas, Mercaptocarborane-capped gold nanoparticles: electron pools and ion traps with switchable hydrophilicity, *J. Am. Chem. Soc.*, 2012, **134**(1), 212–221.

22 N. G. Bastús, J. Comenge and V. Puntes, Kinetically controlled seeded growth synthesis of citrate-stabilized gold nanoparticles of up to 200 nm: Size focusing *versus* Ostwald ripening, *Langmuir*, 2011, **27**(17), 11098–11105.

23 C. Fernández-López, C. Mateo-Mateo, R. A. Álvarez-Puebla, J. Pérez-Juste, I. Pastoriza-Santos and L. M. Liz-Marzán, Highly controlled silica coating of PEG-capped metal nanoparticles and preparation of SERS-encoded particles, *Langmuir*, 2009, **25**(24), 13894–13899.

24 C. J. Newcomb, S. Sur, J. H. Ortony, O.-S. Lee, J. B. Matson, J. Boekhoven, J. M. Yu, G. C. Schatz and S. I. Stupp, Cell death *versus* cell survival instructed by supramolecular cohesion of nanostructures, *Nat. Commun.*, 2014, **5**, 3321.

25 C. J. Pedersen and H. K. Frensdorff, Macrocyclic polyethers and their complexes, *Angew. Chem., Int. Ed. Engl.*, 1972, **11**(1), 16–25.

26 C. J. Pedersen, Cyclic polyethers and their complexes with metal salts, *J. Am. Chem. Soc.*, 1967, **89**, 2495–2496.

27 Y. Inokuchi, T. Mizuuchi, T. Ebata, T. Ikeda, T. Haino, T. Kimura, H. Guo and Y. Furutani, Formation of host–guest complexes on gold surface investigated by surface-enhanced IR absorption spectroscopy, *Chem. Phys. Lett.*, 2014, **592**, 90–95.

28 Y. Inokuchi, T. Ebata, T. Ikeda, T. Haino, T. Kimura, H. Guo and Y. Furutani, New insights into metal ion–crown ether complexes revealed by SEIRA spectroscopy, *New J. Chem.*, 2015, **39**(11), 8673–8680.

29 B. Pullman and N. Goldblum, Metal–ligand interactions in organic chemistry and biochemistry, Part 1, in *Jerusalem Symposia*, ed. A. Pullman and N. Goldblum, 1976, pp. 345–349.

30 S. Underwood and P. Mulvaney, Effect of the solution refractive index on the color of gold colloids, *Langmuir*, 1994, **10**, 3427–3430.

31 G. A. Woolley, M. K. Kapral and C. M. Deber, Potential-sensitive membrane association of a fluorescent dye, *FEBS Lett.*, 1987, **224**(2), 337–342.

32 W. Si, Z. T. Li and J. L. Hou, Voltage-driven reversible insertion into and leaving from a lipid bilayer: Tuning transmembrane transport of artificial channels, *Angew. Chem., Int. Ed.*, 2014, **53**(18), 4578–4581.

33 A. K. Boal and V. M. Rotello, Fabrication and self-optimization of multivalent receptors on nanoparticle scaffolds, *J. Am. Chem. Soc.*, 2000, **122**, 734–735.

34 A. M. Jackson, J. W. Myerson and F. Stellacci, Spontaneous assembly of subnanometre-ordered domains in the ligand shell of monolayer-protected nanoparticles, *Nat. Mater.*, 2004, **3**(5), 330–336.

35 M. Yu and F. Stellacci, Response to "Stripy nanoparticles revisited", *Small*, 2012, **8**(24), 3720–3726.

36 Q. K. Ong, J. Reguera, P. J. Silva, M. Moglianetti, K. Harkness, M. Longobardi, K. S. Mali, C. Renner, S. De Feyter and F. Stellacci, High-resolution scanning tunneling microscopy characterization of mixed monolayer coated gold nanoparticles, *ACS Nano*, 2013, 7(10), 8529–8539.

37 Y. Cesbron, C. P. Shaw, J. P. Birchall, P. Free and R. Levy, Stripy nanoparticles revisited, *Small*, 2012, **8**(24), 3714–3719.

38 J. Stirling, I. Lekkas, A. Sweetman, P. Djuranovic, Q. Guo, B. Pauw, J. Granwehr, R. Levy and P. Moriarty, Critical assessment of the evidence for striped nanoparticles, *PLoS One*, 2014, **9**(11), e108482.

Faraday Discussions

ROYAL SOCIETY OF CHEMISTRY

PAPER

New route toward nanosized crystalline metal borides with tuneable stoichiometry and variable morphologies†

Guillaume Gouget,[a] Patricia Beaunier,[b] David Portehault*[a] and Clément Sanchez*[a]

Received 15th March 2016, Accepted 18th April 2016
DOI: 10.1039/c6fd00053c

Herein we highlight for the first time the ability to tune the stoichiometry of metal boride nanocrystals through nanoparticle synthesis in thermally stable inorganic molten salts. Two metal–boron systems are chosen as case studies: boron-poor nickel borides and boron-rich yttrium borides. We show that NiB, Ni_4B_3, Ni_2B, Ni_3B, and YB_6 particles can be obtained as crystalline phases with good selectivity. Anisotropic crystallization is observed in two cases: the first boron-rich YB_4 nanorods are reported, while boron-poor NiB nanoparticles show a peculiar crystal habit, as they are obtained as spheres with uniaxial defects related to the crystal structure. Crystallization mechanisms are proposed to account for the appearance of these two kinds of anisotropy at the nanoscale.

Introduction

The family of metal–boron alloys spans a wide range of compounds, including various stoichiometries. This diversity results from the high density of covalent bonds in metal boride structures. By increasing the boron content (Fig. 1), the structure evolves from linear B–B chains, to two- and three-dimensional boron frameworks including interconnected boron clusters[1] in which B–B lengths are shorter than in one-dimensional chains. This structural observation is commonly related to the mechanical properties of metal borides, such as super-hardness in WB_4 [2] or high temperature refractory behaviour in metallic ceramics as HfB_2.[3] Covalence also results in high melting temperatures of borides (2600 °C for YB_6)

[a]Sorbonne Universités-UPMC Univ. Paris 06, CNRS, Collège de France, Laboratoire de Chimie de la Matière Condensée de Paris, 11 place Marcelin Berthelot, 75005 Paris, France. E-mail: david.portehault@upmc.fr; clement.sanchez@upmc.fr

[b]Sorbonne Universités-UPMC Univ. Paris 06, CNRS, Laboratoire de Réactivité de Surface (LRS), 4 place Jussieu, F-75252 Paris Cedex 05, France

† Electronic supplementary information (ESI) available: Details of reaction conditions for yttrium boride syntheses; low magnification TEM picture of NiB samples; HRTEM picture of a Ni_4B_3 particle; TEM and HRTEM images of Ni_4B_3, Ni_2B, and Ni_3B particles; EDX spectra of Ni–B samples. See DOI: 10.1039/c6fd00053c

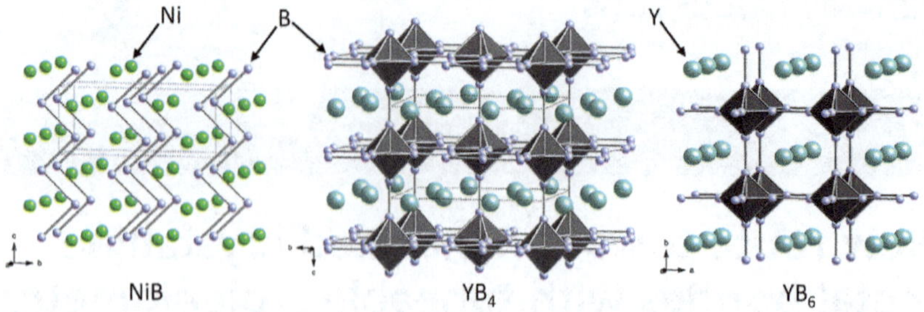

Fig. 1 Crystal structures of NiB (space group *Amam*), YB_4 (*P4/mbm*) and YB_6 ($Pm\bar{3}m$).

and pure boron allotropes (2075 °C for β-B).[4] Notably, the metal–boron interaction also brings specific behaviours. For instance, in OsB, metal atoms bring high electron concentration, so that electrostatic repulsions prevent compression, thus yielding ultra-incompressibility.[2] More generally, boron-poor (elemental ratio $M : B \geq 1 : 1$) borides of transition metals draw attention in a range of fields from magnetism[5] to catalysis.[6–9] In this latter case, B-poor phases such as NiB and CoB are potential low-price catalysts for dehydrogenation and Fisher–Tropsch reactions.[10] In these phases, boron may prevent coking and deactivation of the active sites by sequestering carbon atoms.[11,12] On the other side, boron-rich borides are actively studied and used, such as hexaborides for field emission[13] and thermoelectric energy conversion.[14] Richer borides comprising icosahedral B_{12} boron clusters, for instance YB_{66}, are also strongly investigated for thermoelectricity.[15] Thus, original properties of metal borides are deeply related to both elements and the way they interact. This is particularly exemplified by MgB_2 superconductivity:[16] its remarkably high critical temperature of 39 K for classical conductors was not expected, given the apparent simplicity of the composition[17] compared to usual cuprate superconductors. The superconductivity of MgB_2 is related to both

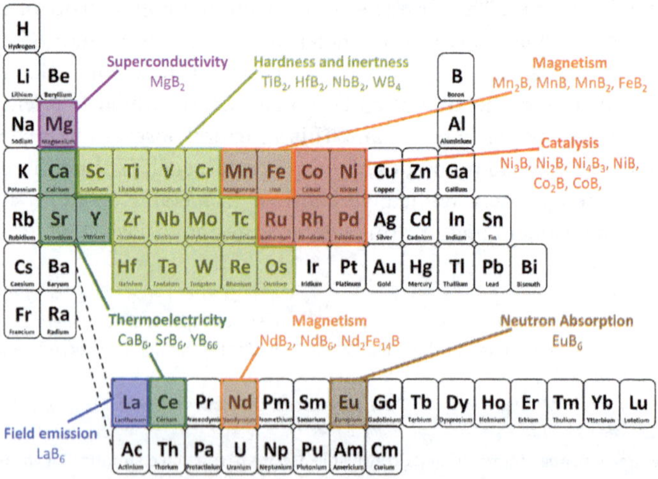

Fig. 2 Attractive properties of metal borides, known for bulk phases. The periodic table of elements is restricted to hydrogen, boron and metals.

elements being light, as suggested in the BCS theory,[18] together with anisotropic boron organisation in metallic honeycomb layers[19] stabilized by electron transfer from magnesium to boron. All in all, metal borides exhibit a large diversity of technologically relevant mechanical and transport behaviours (Fig. 2) that originate from the complex interplay between the covalent boron framework and the metal atoms.

Interestingly, all the above mentioned properties are related to the bulk phase. The example of other materials families, including gold[20–22] and chalcogenides,[23] shows how properties, *e.g.* in catalysis and optics, can be impacted when nanoparticles are scrutinized. However, such studies are still extremely scant for crystalline metal borides.[5,24] Such scarcity originates from the arduous synthesis of crystalline metal borides, which has been discussed recently.[24] Only few phases are known at the nanoscale, especially rare earth and alkaline earth hexaborides, produced by chemical vapor deposition (CVD)[25,26] or solid state reactions, often under autogeneous pressure.[27–29] Some transition metal borides (VB_2,[30] FeB,[5] Co_2B[31]) have been reported at the nanoscale from solid state chemistry, while MgB_2 has been obtained *via* a range of methods involving the reactivity of gaseous magnesium.[32–34] In most cases the synthetic processes cannot be generalized to a wide range of metal borides, do not enable size control (except CVD), and particle sizes are rarely below 30 nm, often above one micrometer. Besides, the control of the final stoichiometry for nanoscaled crystalline metal borides is currently uncharted.

Recently, a route was proposed toward a wide range of metal boride nanocrystals,[35] based on the reaction between metal salts and an alkali borohydride below 900 °C, at atmospheric pressure in an inorganic molten salt mixture composed of LiCl/KCl at the eutectic composition ($T_f = 353$ °C). This one-pot process enables production of nanocrystalline metal borides at temperatures

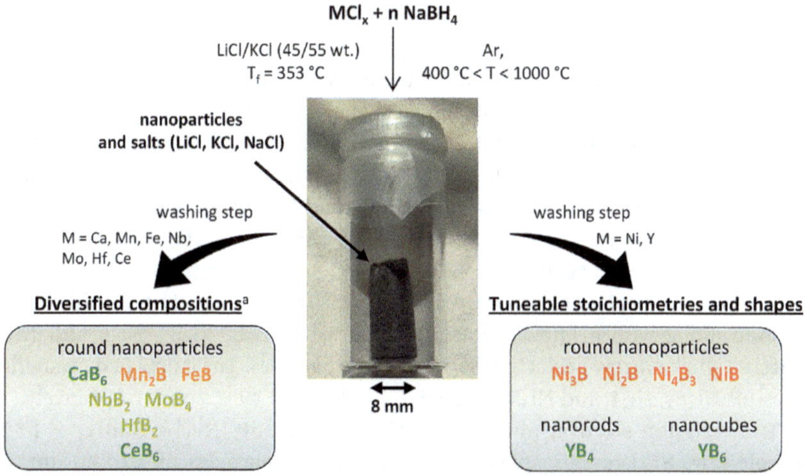

Fig. 3 Synthesis of metal borides in inorganic molten salts. LiCl/KCl (45/55 wt.) melts at $T_f = 353$ °C. After cooling down the reactant medium and washing out the salts, various metal boride nanoparticles are obtained in term of composition, stoichiometry and shape. [a]Results from Portehault *et al.*[35]

between 400 °C and 1000 °C, with a readily available, thermally stable and non-oxygen-containing solvent, easily eliminated after reaction by using water. In a typical synthesis, schematized in Fig. 3, the solid precursors and salts are heated together over the melting temperature of the reactant medium in argon atmosphere. After cooling down to room temperature, a black block is obtained. It contains salts that are dissolved afterwards with a polar solvent such as deionized water. Washing and drying the remaining black suspension led to a black powder consisting of metal boride nanoparticles. The particle size could be controlled and decreased below 20 nm. However, each metal studied was associated with a precise M : B ratio, so that stoichiometry control for a given metal was not provided. Besides, all reported nanoparticles were round in shape.

In this article we present the first evidence of a general synthesis of nanocrystalline metal borides with controlled stoichiometry using syntheses in inorganic molten salts, by focusing on the B-poor Ni–B and B-rich Y–B systems. Depending on the ratio between metal and boron precursors, the temperature and reaction time, NiB, Ni_4B_3, Ni_2B, Ni_3B, YB_6 and YB_4 can be selectively fabricated (Fig. 3). Furthermore, we highlight the occurrence of a peculiar anisotropic uniaxial defective crystal structure within spherical particles that originate from crystallization in a confined space. Finally, we report for the first time the ability to reach anisotropic nanoparticles, as a first step toward morphological control of crystalline metal boride nanostructures.

Results

Briefly, all syntheses were performed under an inert argon atmosphere. Prior to heating, metal chlorides and sodium borohydride were mixed under argon with the powder mixture of LiCl and KCl at the eutectic composition (45/55 wt.). After reaction and cooling, the frozen salt matrix was washed with deionized water, the remaining black powders were dried under vacuum and stored in air.

Nickel borides

Starting from $NiCl_2$ as metal precursor, the reaction medium was heated at 750 °C for 1.5 h. Different $NiCl_2 : NaBH_4$ (Ni : B) molar ratios were investigated. The corresponding X-ray diffraction (XRD) patterns (Fig. 4) show that for Ni : B = 1 : 2, NiB is the major product, with broad reflections suggesting nanosized crystallites. The apparent crystallite size calculated according to the Scherrer formula is 9 ± 2 nm. Minor peaks of Ni_4B_3 are also observed. For Ni : B = 1 : 1.2; 1 : 1 and 0.65, Ni_4B_3, Ni_2B and Ni_3B are isolated as single crystalline phases, respectively. Minor parasite peaks are observed in the two last patterns. They could not be indexed along with a known phase. In contrast to NiB, Ni_4B_3, Ni_2B and Ni_3B patterns show narrow peaks that suggest larger particles. The calculated crystallite sizes for Ni_4B_3, Ni_2B and Ni_3B are 29 ± 6, 33 ± 3 and 40 ± 2 nm, respectively.

Transmission electron microscopy (TEM) of the NiB ($NiCl_2 : NaBH_4$ = 1 : 2) sample (Fig. S1†) reveals spherical crystalline nanoparticles of 2 to 40 nm, in agreement with the crystallite size, dispersed within an amorphous matrix of 1 to 2 nm thickness. The smaller the particles, the more they tend to perfect spheres. Particles bigger than 20 nm are slightly faceted. High resolution TEM (HRTEM) (Fig. 5) highlights lattice fringes in a spherical NiB nanoparticle. Many defects are

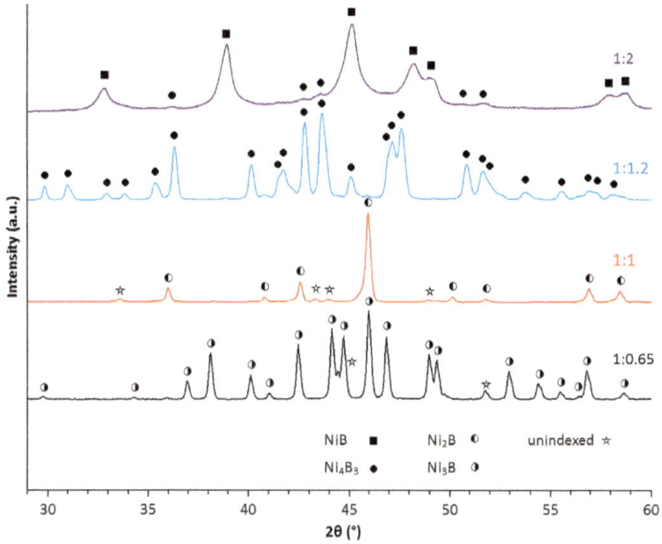

Fig. 4 Powder XRD patterns for $NiCl_2$: $NaBH_4$ molar ratios 1 : 2, 1 : 1.2, 1 : 1 and 1 : 0.65.

observed and demonstrate the polycrystalline nature of each particle. Notably, single crystalline domains appear as parallel stripes within the single isotropic particles. When comparing two domains aside, the orientation of B–B bonds

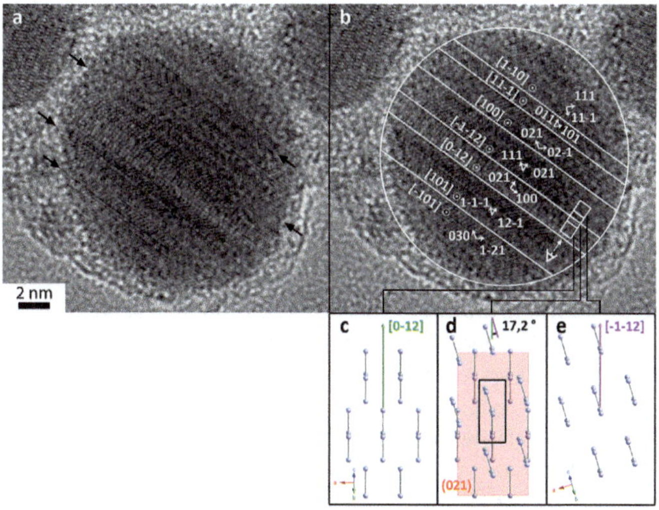

Fig. 5 HRTEM picture of a NiB nanoparticle from the initial ratio $NiCl_2$: $NaBH_4$ = 1 : 2 and orientations of B–B bonds for two consecutive single crystal domains. (a) Black arrows highlight the direction of uniaxial defects within the crystalline NiB particle. (b) For each single domain stripe, the zone axes and lattice planes are indexed according to the NiB structure. (c–e) Example of a misfit at the interface between two domains, viewed orthogonally to the (021) interface: (c) ahead of the interface; (d) at the interface; and (e) behind the interface. Nickel atoms are omitted for clarity.

changes. For instance (Fig. 5c and e) the two domains with [0−12] and [−1−12] zone axes differ from each other by rotating the structure 17.2° along the [021]* direction. Note that such a defect does not yield coherent domains, as the zig-zag chains are discontinued at the interface (Fig. 5d). Each single domain can be indexed according to the NiB structure for all particles with a diameter below 20 nm (Fig. S2†). A minority of bigger faceted particles are indexed as Ni_4B_3 (Fig. S3†). Brunauer–Emmett–Teller analysis of the N_2 sorption isotherms provides a specific surface area value of 57 $m^2\ g^{-1}$ for the NiB sample ($NiCl_2 : NaBH_4 = 1 : 2$), almost three times higher than values reported for analog amorphous[6] or crystalline[36] NiB nanoparticles (20 $m^2\ g^{-1}$).

Fig. S4† displays TEM and HRTEM pictures of the samples obtained for higher proportions of sodium borohydride. In agreement with XRD results, 50–200 nm crystals of Ni_4B_3 and Ni_2B embedded in an amorphous layer (1 to 2 nm thick) are observed for initial $NiCl_2 : NaBH_4$ ratios 1 : 1.2 and 1 : 1. The sample with a $NiCl_2 : NaBH_4$ ratio of 1 : 0.65 is made of particles bigger than 100 nm (Fig. S5†).

SEM-EDX analyses show that all samples are composed of boron, nickel, and oxygen. For $NiCl_2 : NaBH_4 = 1 : 2$, quantitative analysis provides an elemental composition of $B_{58}Ni_{31}O_{11}$. We showed previously that the amorphous matrix embedding the particles is partially oxidized boron.[35] Consequently, by considering that the crystalline phase is in the vast majority NiB, then the material composition is equivalent to $(NiB)_{31}(B_{27}O_{11})$, reflecting the biphasic crystalline-NiB/amorphous-B (B_{am}) character of the sample. 54% of the B atoms are involved in the crystalline phase, in accordance with the initial excess of boron introduced. The O/B content within the amorphous phase is $O/B_{am} = 0.4$, which suggests that the amorphous layer is composed of partially oxidized amorphous boron, in agreement with previous reports. Unfortunately, the boron content could not be quantitatively evaluated by SEM-EDX in the samples from initial $NiCl_2 : NaBH_4$ ratios 1 : 1.2, 1 : 1 and 1 : 0.65 as the B content was too low to deconvolute unambiguously the EDX peaks of B and C Kα edges (Fig. S6†).

Yttrium borides

By using YCl_3 as metal precursor, yttrium borides can be targeted through synthesis in molten salts. In order to reach different boron-rich compounds, extensive screening of experimental parameters was performed, by focusing on the yttrium-to-boron precursor ratio $YCl_3 : NaBH_4$, as in the Ni–B system, but also on the temperature and the dwell time. Below, the different samples are identified as "$YCl_3 : NaBH_4$-T-t", where T and t are respectively the reaction temperature (°C) and duration (h). By screening these conditions, speciation diagrams can be mapped (Fig. 6), where relative amounts of crystalline phases are qualitatively evaluated from the relative intensities of the corresponding phases in the XRD patterns. Overall and as expected, YB_6 is favored at high proportions of $NaBH_4$, but also at high T and t.

The XRD diagrams of three typical powders are displayed Fig. 7. The sample 1 : 12-1000-4 exhibits the cubic structure of YB_6 as the only crystalline phase. For 1 : 8-900-4, the reflections are attributed to YB_4 and YB_6. For 1 : 5-800-1, two crystalline phases are observed: YB_4 is the only metal boride phase, while the yttrium borate $Y(BO_3)$ is also observed. More generally, at low temperature and short time corresponding to a low energy input, oxidized phases (yttrium borate

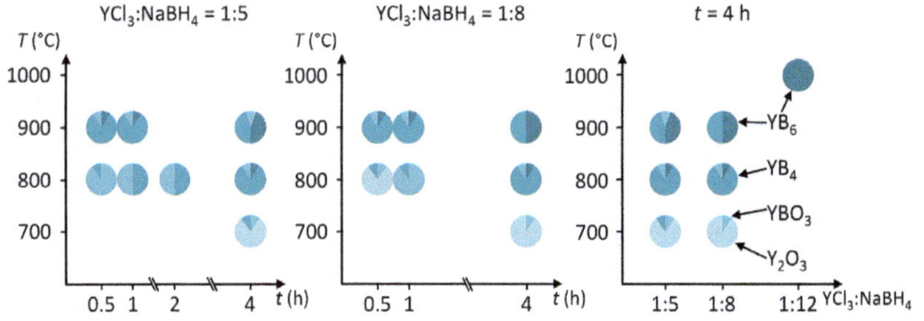

Fig. 6 Speciation diagrams of yttrium borides and other yttrium phases obtained by synthesis in salt melts in respect to the molar ratio of precursors ($YCl_3 : NaBH_4$), the temperature (T) and the time (t) of heat treatment. Left to right: $YCl_3 : NaBH_4$ is fixed at 1 : 5, T and t vary; $YCl_3 : NaBH_4 = 1 : 8$, T and t vary; and $t = 4$ h, $YCl_3 : NaBH_4$ and T vary. For each experiment, the phase proportion is qualitatively represented in a histogram pie.

or oxide Y_2O_3) are obtained. As the reaction media do not contain any source of oxygen, then these byproducts originate from poorly stable yttrium species that react with water during the washing step. Efforts to selectively dissolve this oxidized phase, using acidic or reductive conditions, were inefficient.

Apparent crystallite sizes ε_{hkl} were calculated for the different crystalline phases (Table 1). Smaller YB_6 particles are synthesized at 1000 °C ($\varepsilon_{hkl} = 49$ nm for $YCl_3 : NaBH_4 = 1 : 12$) than 900 °C ($\varepsilon_{hkl} = 59$ nm for $YCl_3 : NaBH_4 = 1 : 8$). In other common systems, an increase in the temperature usually yields extensive growth and bigger particles, thus suggesting in the specific case of this boride synthesis in molten salts that the initial reagent ratio (*e.g.* $YCl_3 : NaBH_4$) is the key parameter for size control. For YB_4 (1 : 8-900-4 sample), anisotropic crystallites are detected, with $\varepsilon_{00l} = 81$ nm and $\varepsilon_{hk0} = 26$ nm, thus indicating faster growth along the [001] axis.

TEM observation of the 1 : 12-1000-4 sample (Fig. 8a–c) highlights cubic particles with diameters ranging from 30 to 200 nm. The selected area electron

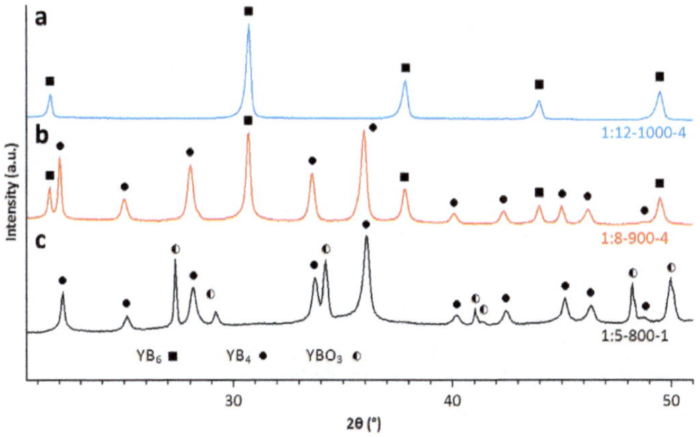

Fig. 7 XRD patterns of yttrium boride samples (a) 1 : 12-1000-4 (YB_6), (b) 1 : 8-900-4 (YB_4 and YB_6) and (c) 1 : 5-800-1 (YB_4 and YBO_3).

Table 1 Apparent crystallite sizes of yttrium borides calculated from the XRD diagrams

Sample	YB_6	YB_4	
	ε_{hkl}/nm	ε_{00l}/nm	ε_{hk0}/nm
1 : 12-1000-4	49	—	—
1 : 8-900-4	59	81	26
1 : 5-800-1	—	51	18

diffraction (SAED) pattern (Fig. 8a inset) on an ensemble of particles shows that the particles are YB_6 cubes, in agreement with XRD. HRTEM (Fig. 8d) of one nanocube highlights 0.41 nm (100) lattice fringes showing that the nanocubes expose (100) faces, which is consistent with the cubic structure of metal hexaborides (Fig. 1) and with previous reports on bigger hexaboride particles.[27] A thin amorphous layer is distinguished at the surface of the crystals, with thickness below 1 nm. Note that although the cubes are well faceted, they exhibit rough surfaces and round angles (Fig. 8b and c). Slight amounts of amorphous matter are also observed next to the cubes (Fig. 8b and c).

TEM of the 1 : 8-900-4 sample (Fig. 9a and c) displays two populations of crystalline particles: approximately 100 nm long and 30 nm large nanorods on one side, and 80 nm nanocubes on the other side. All crystals are embedded in a nanotextured amorphous solid, already encountered for Ni–B systems, that is made of partially oxidized boron. Again, the nanocubes are YB_6 (Fig. 9e) particles.

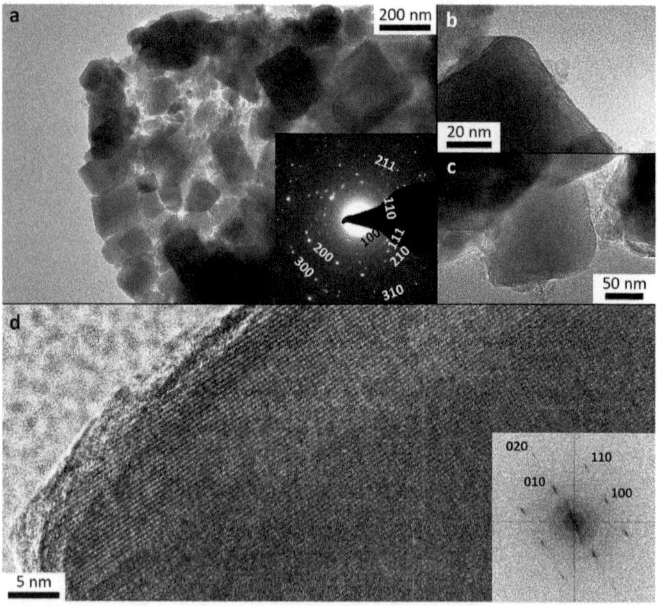

Fig. 8 TEM analysis of YB_6 particles (1 : 12-1000-4 sample). (a–c) Low magnification images of a region exhibiting nanocubes, the SAED pattern in the inset is indexed along the YB_6 structure. (d) HRTEM picture of a nanocube and the corresponding fast Fourier transform in inset.

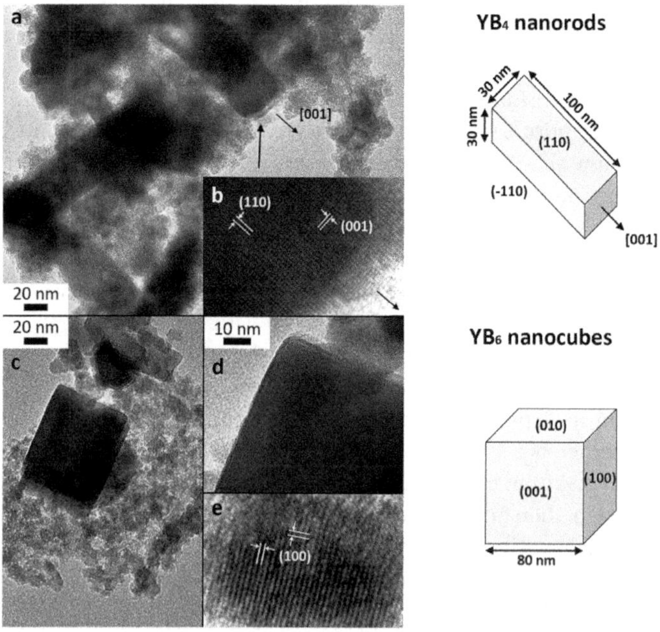

Fig. 9 Left: TEM analysis of the 1 : 8-900-4 sample. (a) Low magnification picture of a region exhibiting nanorods dispersed in an amorphous oxidized boron matrix, (b) HRTEM picture of a nanorod with fringes corresponding to YB_4, (c) and (d) TEM pictures of a region exhibiting a nanocube partially embedded in an amorphous oxidized boron matrix and (e) HRTEM picture of a nanocube indexed along the YB_6 structure. Right: typical morphologies of yttrium borides observed in the 1 : 8-900-4 sample.

On the contrary, HRTEM (Fig. 9b) performed on the tip of one nanorod shows lattice distances of 0.50 nm and 0.40 nm, which correspond to (110) and (001) planes of YB_4, respectively. Consequently, these YB_4 nanorods exhibit preferential growth along the [001] axis, in accordance with XRD results. To our knowledge, we report here the first occurrence of a metal tetraboride at the nanoscale with well-defined morphology.

Discussion

Composition selectivity in the crystalline phase

For a given metal–boron binary system, the results presented herein indicate the feasibility to control the stoichiometry of crystalline nanoparticles by using a set of synthesis parameters, such as the ratio of metal and boron precursors, the temperature and the heating duration. This composition control is first shown for the B-poor nickel borides, with NiB nanospheres of 2 to 20 nm diameter ($NiCl_2 : NaBH_4 = 1 : 2$). These are, to our knowledge, the smallest crystalline nickel boride nanoparticles synthesized so far.[36] Ni_4B_3, Ni_2B and Ni_3B can also be selectively fabricated by only varying the precursor $NiCl_2 : NaBH_4$ ratio from 1 : 1.2, 1 : 1 and 1 : 0.65 ratios, respectively. In these latter cases, particles are bigger than for NiB but are still below 100 nm. To our knowledge, we show here the first occurrences of Ni_4B_3 and Ni_2B nanocrystals with such a small size, while crystalline Ni_3B particles were previously reported.[36] On the side of B-rich phases, we explored the yttrium–boron system. YB_6 nanocubes of *ca.* 80 nm are reachable, while YB_4 nanorods are synthesized for the first time. In this latter case, oxidized phases, especially yttrium borate YBO_3, are observed as side products and suggest that the powder prior to washing exhibits reactive yttrium species, presumably poorly crystalline solid phases that oxidize spontaneously in contact with air and water. Although more efforts are required to isolate single crystalline phases in the case of yttrium, the Ni–B and Y–B systems clearly demonstrate that increasing the relative amount of borohydride precursor favors the synthesis of higher borides, as expected intuitively. In addition, a temperature increase drives the reaction towards B-rich phases, in agreement with the high energy barrier to crystallization exhibited by such compounds that usually require synthesis at high temperature, above 1000 °C in solid state reactions.[37]

Notably, an amorphous phase made of partially oxidized boron[35] is always observed, whether as a 1 to 2 nm-thick shell (Fig. 5 and 8d) or as particles separated (Fig. 8 and 9) from the crystalline phases. As oxidation can only occur during the washing step, then amorphous boron is obtained at the end of the synthesis, prior to purification. The occurrence of this amorphous component is consistent with the fact that an excess of boron precursor is always required to reach a given metal-to-boron ratio in the crystalline phase (*e.g.* $NiCl_2 : NaBH_4 = 1 : 2$ for NiB, $YCl_3 : NaBH_4 = 1 : 12$ for YB_6). Hence, part of the boron atoms initially introduced are segregated into the amorphous shell in the final product. Consequently, we hypothesize that this amorphous phase is reminiscent of the growth mechanism, which will be discussed elsewhere. Nevertheless, this amorphous shell has been shown[35] to be sufficiently porous to provide access for ionic species to the surface of the nanocrystals, so that it enables future investigation of catalytic properties.

Crystal habit: faceting and occurrence of anisotropy in metal boride nanocrystals

Particle size control. As exemplified previously on NbB_2 nanocrystals,[35] an increase of the metal-to-boron ratio in the precursor ratio yields a decrease in the size of crystalline YB_6 nanoparticles. Two explanations can be proposed to account for this size reduction. First, as the concentration of metal precursor is constant, then an increase in the relative amount of borohydride results in an overall increase in the reactant concentrations, thus favoring nucleation over growth and yielding smaller particles. A second explanation may originate from the amorphous boron shell formed in the course of the reaction. Indeed, an increase of the borohydride content should yield an increase in the volume fraction of the amorphous boron phase, which may act as a template restricting nanocrystal growth.

Anisotropic shapes: the case of yttrium boride YB_4

Isotropic YB_6 nanocubes could be obtained from the synthesis in molten salts at ambient pressure. These particles are similar to other hexaboride particles as previously reported from solid-state methods under autogeneous pressure.[27] They exhibit {100} facets and a morphology related to the cubic symmetry of the crystal structure of hexaborides. More interestingly, YB_4 nanocrystals could be observed for the first time. Moreover they exhibit a nanorod shape with an anisotropy factor of *ca.* 3. Notably, the only processes reported so far to produce anisotropic one-dimensional (1D) nanostructures of metal borides employed reactive gaseous species, for instance through chemical or physical vapor deposition.[24] They were mostly focused on metal hexaborides. Herein, we report the first example of 1D crystalline boron-rich metal boride nanoparticles from a liquid synthesis. Contrary to reactions based on gaseous species and that often yield 1D nano-objects through the mediation of a substrate, the YB_4 nanorods are likely to grow through a specific crystal habit related to the crystal structure of YB_4. Indeed, its tetragonal structure differentiates the (001) plane on one side and the ($hk0$) planes on the other side. More in-depth observation of the reported crystal structure (Fig. 1 and 10) shows that chains of B_6 octahedra clusters run along the c axis (orthogonal to the (001) plane), while the (a, b) plane can be considered as sheets of B_6 clusters connected by B_2 dumbbells. The Y atoms are located between the B_6 sheets, on top of the holes left within the sheets between B_6 clusters and B_2 dumbbells (Fig. 10). The nanorods show {001} tip facets and {110} lateral faces. Interestingly, the B-terminated {110} planes show the highest densities of B_6 clusters and B atoms (0.05 B_6 clusters per nm^2 and 0.35 B atoms per nm^2) in YB_4, while {001} and {100} planes show the second (0.04 B_6 clusters per nm^2 and 0.32 B atoms per nm^2) and third (0.03 B_6 clusters per nm^2 and 0.28 B atoms per nm^2) highest B_6 and B densities, respectively. Consequently, one can draw a parallel with the crystal habit of more common compounds such as oxides, for which growth directions in classical crystallization are often driven by surface energies (the Wulff law),[38] in turn related (in the absence of surface directing ligand) to the atom density in each plane. Indeed, the most stable faces are usually those showing the highest atom densities. The correlation of the B_6 and B density with the relative amount of the different facets in YB_4 nanorods ({110} in majority, {001} (tips) in minority, while no {100} face is observed) indicates that the surface

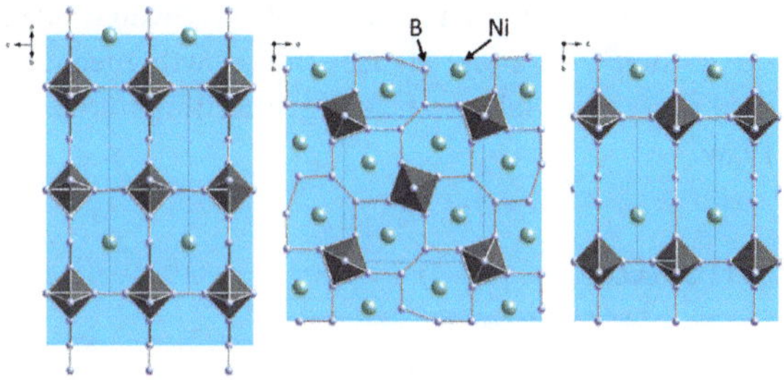

Facet	{110}		{001}		{100}
B_6 density (B_6 clusters.nm^{-2})	0.05	>	0.04	>	0.03
B density (B atoms.nm^{-2})	0.35	>	0.32	>	0.28
Facet proportion (%)	87	>	13	>	negligible

Fig. 10 Facets {110}, {001} and {100} of yttrium tetraboride and density of boron octahedra and boron atoms for the three facets. The facet proportion in YB_4 nanorods is evaluated according to the dimensions observed in Fig. 9. It increases with the boron surface density. In particular, no {100} face is observed by TEM.

energy is driven by the density of B_6 clusters or B atoms, as building-blocks of the YB_4 structure. B-terminated facets are indeed expected from the excess of boron source introduced. Although the consideration of relative surface energies requires confirmation by first principle calculations, it is consistent with the high strength of the B–B bonds in such compounds and with the high stability of B_6 clusters in borides,[39] which suggest that the boron covalent framework should drive the relative surface energies, so that the role of the metal atoms would be secondary. Notably, the largest facets in bulk tetraboride grown by a metal flux method are also {110},[40] confirming their low surface energy.

Anisotropic structures within isotropic nanoparticles: the case of nickel boride NiB

In contrast to YB_4 and YB_6, the spherical morphology of the NiB nanoparticles is not related to the orthorhombic crystal structure (*Amam*). This conclusion is supported by the large density of crystalline defects (Fig. 5) that precludes any structure-driven faceting. The well-shaped spherical particles are unusual in the case of an anisotropic crystal structure, such as NiB. Hence, this morphology suggests solidification from molten metal or alloy droplets.[41] The melting point of bulk NiB is tabulated at 1035 °C,[4] close to the value for bulk gold. As known for gold[42] and covalent solids,[43] strong melting point depression can occur in this size range. Although physical data are missing in the case of NiB to evaluate the expected melting point of 10 nm NiB nanoparticles, one can point out that the synthesis temperature of 750 °C lies in the range where melting can be expected for 10 nm nanoparticles of metallic-covalent cohesion.[42,43] Following the assumption of NiB crystallization from molten droplets and considering the very high melting point of boron allotropes,[4] then the droplets should be confined within the solid amorphous boron matrix. Solidification and crystallization within the confined droplets could explain the high density of crystal defects observed within the NiB nanoparticles after cooling. On the contrary, YB_4 and YB_6 exhibit melting points higher than 2300 °C, which precludes any process involving a molten Y–B alloy. Hence, the growth mechanism follows a more conventional route that yields to faceting according to relative facet energies and the Wulff law.

Experimental

Syntheses

The precursors $NaBH_4$ (min 98%, Alfa Aesar), anhydrous YCl_3 (99.99%, Alfa Aesar) and anhydrous $NiCl_2$ (99.99%, Alfa Aesar) were stored in an argon filled glovebox (M-Braun) and used as received. A LiCl/KCl mixture at the eutectic composition (wt 45/55) was prepared from the commercially available salts (Aldrich). The mixture was evacuated under vacuum at 200 °C for 1 day, and then stored in a glovebox. Powder mixtures were manipulated in an inert atmosphere. They were finely ground in a ball-mill (Retsch MM400) for 2 min at 20 Hz, and then transferred into a glassy carbon crucible. The crucible was maintained in a quartz tube, which was connected to an argon ramp. During heating in a vertical oven (carbolite), the medium was stirred with a glassy carbon rod rotating at 100 rpm. After the apparatus cooled below 100 °C, the frozen salt medium was washed with deionized water, the suspension was centrifuged and the powder was separated from the supernatant. The obtained powder was then washed, centrifuged and recovered from the supernatant three times. The powder was evacuated at 40 °C overnight.

Nickel boride syntheses. Following the general procedure, 2.5 g of eutectic mixture were mixed to 2.0 mmol of nickel(II) chloride and 4.0, 2.4, 2.0 and 1.3 mmol of sodium borohydride for experiments 1 : 2, 1 : 1.2, 1 : 1 and 1 : 0.65 respectively. The mixture was heated from ambient temperature to 750 °C (20 °C.min^{-1}), then maintained at 750 °C for 1.5 h before cooling.

Yttrium boride syntheses. Each synthesis is named after the molar ratio of inserted precursors, the temperature and the dwell time. The parameters of the yttrium borides syntheses are detailed in Table S1.†

Techniques

Powder X-ray diffraction (XRD). XRD measurements were performed on a D8 Bruker apparatus operating at the Cu-Kα1 radiation. Patterns were indexed according to ICDD files: NiB (01-074-1207), Ni_4B_3 (01-073-2551), Ni_2B (04-001-0967), Ni_3B (04-014-0853), YB_6 (04-003-0596), YB_4 (00-007-0057), YBO_3 (01-089-3501), Y_2O_3 (01-071-0049). Apparent crystallite sizes ε were calculated using the Scherrer formula and pseudo-Voigt functions to fit peaks.

Transmission electron microscopy (TEM). A 120 kV Tecnai Spirit microscope was used for TEM and Selected Area Electron Diffraction (SAED) pictures. High Resolution-TEM (HRTEM) was also operated on a 200 kV JEOL JEM 2011 apparatus, at the Institut des Matériaux de Paris Centre. Samples were prepared by evaporating a drop of suspension in ethanol on a carbon-coated copper grid.

Energy dispersive X-ray spectrometry (EDX). Elemental EDX analyses were obtained on a Scanning Electron Microscope (SEM) S-3400-N (Hitachi), equipped with an EDX detector (Oxford) calibrated prior measurements to allow for retrieving quantitative data. 100 × 50 μm^2 areas were probed. Acceleration tension of the source was 10 kV. A titanium grid was used as quantitative reference. Samples were fixed to a conductive carbon scotch.

Nitrogen sorption. Samples were degassed at 150 °C overnight before measuring on an ASAP 2000 apparatus (Micromeritics).

Conclusions

To conclude, this study shows for the first time that the stoichiometry of metal boride nanocrystals can be tuned. To achieve this goal, syntheses are performed in inorganic molten salts that are stable in temperature, with negligible vapor pressure, thus ensuring a convenient, low cost and scalable method. Single crystalline phases can be isolated in most Ni–B systems and in YB_6, although NiB is obtained with slight amounts of Ni_4B_3 as side-product, while YB_4 could not be isolated properly. Besides, the crystalline metal borides are always obtained along with amorphous boron, which in some cases (*e.g.* NiB) play an important role in the particle shape and size. This points out different crystallization mechanisms, which might depend on the melting point of the nanoscaled alloys. When the synthesis temperature is lower than the melting point (*e.g.* the Y–B system), then classical particle growth may occur, then leading to faceting and anisotropic shapes expected from the Wulff law, without any surface-directing ligand. On the opposite, if the synthesis is performed at a temperature exceeding the melting point, then amorphous boron may act as a template confining alloy droplets, which would in turn crystallize upon cooling in the space defined by the droplets, thus yielding metal boride spheres. In the case of an anisotropic crystal structure (*e.g.* NiB), the competition between the structure-driven tendency to anisotropic growth and the spherical shape imposed by the template results in strains, released in the form of uniaxial crystal defects. To confirm these mechanistic conclusions, *in situ* characterizations are under way, using spectroscopic, scattering and electron microscopy techniques.

References

1 N. N. Greenwood, R. V. Parish and P. Thornton, *Q. Rev., Chem. Soc.*, 1965, 441–464.

2 Q. Gu, G. Krauss and W. Steurer, *Adv. Mater.*, 2008, **20**, 3620–3626.

3 W. G. Fahrenholtz, G. E. Hilmas, I. G. Talmy and J. a. Zaykoski, *J. Am. Ceram. Soc.*, 2007, **90**, 1347–1364.

4 *CRC Handbook of Chemistry and Physics*, ed. D. R. Lide, CRC Press/Taylor & Francis Group, Boca Raton, FL, 88th edn, 2007.

5 S. Rades, S. Kraemer, R. Seshadri and B. Albert, *Chem. Mater.*, 2014, **26**, 1549–1552.

6 Y.-C. Liu and Y.-W. Chen, *Ind. Eng. Chem. Res.*, 2006, **45**, 2973–2980.

7 C. Wu, F. Wu, Y. Bai, B. Yi and H. Zhang, *Mater. Lett.*, 2005, **59**, 1748–1751.

8 N. Patel, R. Fernandes, G. Guella, A. Kale, A. Miotello, B. Patton and C. Zanchetta, *J. Phys. Chem. C*, 2008, **112**, 6968–6976.

9 U. B. Demirci and P. Miele, *Phys. Chem. Chem. Phys.*, 2010, **12**, 14651–14665.

10 L. Chen, Y. Lu, Q. Hong, J. Lin and F. M. Dautzenberg, *Appl. Catal., A*, 2005, **292**, 295–304.

11 J. Xu and M. Saeys, *J. Catal.*, 2006, **242**, 217–226.

12 M. Saeys, K. F. Tan, J. Chang and A. Borgna, *Ind. Eng. Chem. Res.*, 2010, **49**, 11098–11100.

13 X. H. Ji, Q. Y. Zhang, J. Q. Xu and Y. M. Zhao, *Prog. Solid State Chem.*, 2011, **39**, 51–69.

14 M. Takeda, M. Terui, N. Takahashi and N. Ueda, *J. Solid State Chem.*, 2006, **179**, 2823–2826.

15 T. Mori and T. Tanaka, *J. Solid State Chem.*, 2006, **179**, 2889–2894.

16 J. Nagamatsu, N. Nakagawa, T. Muranaka, Y. Zenitani and J. Akimitsu, *Nature*, 2001, **410**, 63–64.

17 C. Buzea and T. Yamashita, *Supercond. Sci. Technol.*, 2001, **14**, R115–R146.

18 J. Bardeen, L. N. Cooper and J. R. Schrieffer, *Phys. Rev.*, 1957, **108**, 1175–1204.

19 J. Kortus, I. Mazin, K. Belashchenko, V. Antropov and L. Boyer, *Phys. Rev. Lett.*, 2001, **86**, 4656–4659.

20 L. M. Liz-Marzán, C. J. Murphy and J. Wang, *Chem. Soc. Rev.*, 2014, **43**, 3820–3822.

21 G. L. Nealon, B. Donnio, R. Greget, J.-P. Kappler, E. Terazzi and J.-L. Gallani, *Nanoscale*, 2012, **4**, 5244–5258.

22 C. Goldmann, R. Lazzari, X. Paquez, C. Boissière, F. Ribot, C. Sanchez, C. Chanéac and D. Portehault, *ACS Nano*, 2015, **9**, 7572–7582.

23 D. V. Talapin, J. S. Lee, M. V. Kovalenko and E. V. Shevchenko, *Chem. Rev.*, 2010, **110**, 389–458.

24 S. Carenco, D. Portehault, C. Boissière, N. Mézailles and C. Sanchez, *Chem. Rev.*, 2013, **113**, 7981–8065.

25 J. R. Brewer, R. M. Jacobberger, D. R. Diercks and C. L. Cheung, *Chem. Mater.*, 2011, **23**, 2606–2610.

26 S. S. Amin, S. Li, J. R. Roth and T. T. Xu, *Chem. Mater.*, 2009, **21**, 763–770.

27 R. K. Selvan, I. Genish, I. Perelshtein, J. M. Calderon Moreno and A. Gedanken, *J. Phys. Chem. C*, 2008, **112**, 1795–1802.

28 Y. Liu, W. Lu, J. Qin and D. Zhang, *J. Alloys Compd.*, 2007, **431**, 337–341.

29 M. Zhang, Y. Jia, G. Xu, P. Wang, X. Wang, S. Xiong, X. Wang and Y. Qian, *Eur. J. Inorg. Chem.*, 2010, 1289–1294.

30 L. Shi, Y. Gu, L. Chen, Z. Yang, J. Ma and Y. Qian, *Mater. Lett.*, 2004, **58**, 2890–2892.

31 P. Cai, H. Wang, L. Liu and L. Zhang, *J. Ceram. Soc. Jpn.*, 2010, **118**, 1102–1104.

32 W. N. Kang, E.-M. Choi, H.-J. Kim, H.-J. Kim and S.-I. Lee, *Phys. C*, 2003, **385**, 24–30.

33 S. Lai, S. Liu and M. Lan, *J. Cryst. Growth*, 2007, **304**, 460–463.

34 S.-M. Zhou, P. Wang, S. Li, B. Zhang, H.-C. Gong and X.-T. Zhang, *Mater. Lett.*, 2009, **63**, 1680–1682.

35 D. Portehault, S. Devi, P. Beaunier, C. Gervais, C. Giordano, C. Sanchez and M. Antonietti, *Angew. Chem., Int. Ed.*, 2011, **50**, 3262–3265.

36 Z. L. Schaefer, X. Ke, P. Schiffer and R. E. Schaak, *J. Phys. Chem. C*, 2008, **112**, 19846–19851.

37 B. Albert and H. Hillebrecht, *Angew. Chem., Int. Ed.*, 2009, **48**, 8640–8668.

38 H. Cölfen and M. Antonietti, *Mesocrystals and Non-classical Crystallization*, Wiley, San Francisco, CA, 2008.

39 B. Albert, *Eur. J. Inorg. Chem.*, 2000, 1679–1685.

40 S. Otani, Y. Xuan, Y. Yajima and T. Mori, *J. Alloys Compd.*, 2003, **361**, L1–L3.

41 T.-H.-C. Chan-Chang, G. Gouget, C. Sanchez, S. Carenco and D. Portehault, manuscript in preparation.

42 P. Buffat and J. P. Borel, *Phys. Rev. A: At., Mol., Opt. Phys.*, 1976, **13**, 2287–2298.

43 H. H. Farrell and C. D. Van Siclen, *J. Vac. Sci. Technol., B: Microelectron. Nanometer Struct.–Process., Meas., Phenom.*, 2007, **25**, 1441.

Faraday Discussions

PAPER

Routes to the preparation of mixed monolayers of fluorinated and hydrogenated alkanethiolates grafted on the surface of gold nanoparticles†

Maria Şologan,[a] Cristina Cantarutti,[a] Silvia Bidoggia,[a] Stefano Polizzi,[b] Paolo Pengo[a] and Lucia Pasquato*[a]

Received 16th February 2016, Accepted 8th March 2016
DOI: 10.1039/c6fd00016a

The use of binary blends of hydrogenated and fluorinated alkanethiolates represents an interesting approach to the construction of anisotropic hybrid organic–inorganic nanoparticles since the fluorinated and hydrogenated components are expected to self-sort on the nanoparticle surface because of their reciprocal phobicity. These mixed monolayers are therefore strongly non-ideal binary systems. The synthetic routes we explored to achieve mixed monolayer gold nanoparticles displaying hydrogenated and fluorinated ligands clearly show that the final monolayer composition is a non-linear function of the initial reaction mixture. Our data suggest that, under certain geometrical constraints, nucleation and growth of fluorinated domains could be the initial event in the formation of these mixed monolayers. The onset of domain formation depends on the structure of the fluorinated and hydrogenated species. The solubility of the mixed monolayer nanoparticles displayed a marked discontinuity as a function of the monolayer composition. When the fluorinated component content is small, the nanoparticle systems are fully soluble in chloroform, at intermediate content the nanoparticles become soluble in hexane and eventually they become soluble in fluorinated solvents only. The ranges of monolayer compositions in which the solubility transitions are observed depend on the nature of the thiols composing the monolayer.

1 Introduction

In recent years, a significant interest in the development of gold nanoparticles protected by monolayers comprising fluorinated ligands has started to emerge[1] and the now available strategies to improve their solubility,[2,3] in addition to the

[a]Department of Chemical and Pharmaceutical Sciences and INSTM Trieste Unit, University of Trieste, via L. Giorgieri 1, 34127 Trieste, Italy. E-mail: lpasquato@units.it

[b]Department of Molecular Sciences and Nanosystems, University Ca' Foscari of Venezia, Via Torino 155/b, I-30172 Venezia-Mestre, Italy

† Electronic supplementary information (ESI) available. See DOI: 10.1039/c6fd00016a

remarkable examples of their properties,[4–7] will likely stimulate future developments. Our previous studies, carried out on mixed monolayer gold nanoparticles (MMNPs) protected by blends of fluorinated (F-) and hydrogenated (H-) amphiphilic thiolates, demonstrated that phase segregation triggered by, *inter alia*, the reciprocal phobicity of the immiscible thiolates is operative and can be observed even when using less than 5% F-ligands.[8,9] Self-sorting of hydrogenated and fluorinated compounds is known to occur in the bulk phases,[10,11] in supramolecular assemblies both at the micro-[12,13] and nano-scale[14–19] regime, or even at the molecular level.[20,21] We believe that fluorophilic/fluorophobic interactions will also represent a powerful tool in the controlled formation of domains of well-defined morphology in the monolayer protected gold nanoparticles. The morphological features of mixed monolayers are central in determining the properties of nanoparticles such as their solubility behaviour and wettability;[22] their interaction with biological membranes;[23–25] their assembly properties[26] and their catalytic activity.[27] A large body of experimental evidence[28–35] and theoretical investigations[36–40] provides a valuable framework for the understanding of the factors governing the formation of domains in the monolayer protected gold nanoparticles and other nanosized structures.[41] However, despite these outstanding achievements, there is a lack of comprehensive experimental data concerning the preparation of MMNPs. Even for well characterised MMNPs, the information on how differences in the molecular structures of the ligands used in the synthesis impact on the final composition of the monolayer is sparse. It is often assumed that the stoichiometric ratio of the ligands used in the synthesis is equal to the ratio on the ligand shell,[29] and in some cases evidence does exist.[35] In other cases, only the composition of the MMNPs is reported but not the ratio of the thiols used in the synthesis.[42] In a different route to MMNPs that exploits the assembly of mixtures of thiols on the surface of dioctylamine-capped gold nanoparticles,[43,44] it was also reported that the final composition of the monolayer reflects the initial composition of the ligand mixture. On the contrary, the place exchange reaction is known to be sensitive to the structure of the incoming thiols, with the equilibrium position depending on their length and steric bulk.[45]

Exploiting the fluorophobic/fluorophilic interactions between F- and H-thiolates may represent an advantageous additional tool to control the morphology of mixed monolayers. This approach, if successful, will allow us to bring the "chemical mismatch" between ligands to a maximum; hence maximizing the phase segregation behaviour. However, the preparation of F-/H-MMNPs is, at present, only partly explored. No comprehensive data on the synthetic conditions needed to introduce a given amount of fluorinated components into a monolayer of nanoparticles and their dependence on the thiol molecular structure are presently available. Moreover, the different bulkiness of F- *versus* H-alkyl chains adds complexity to such an investigation. Based on these considerations we realised the urgency of fundamental analysis addressing this gap.

Understanding how the final monolayer composition depends on the initial composition of the reaction mixture in either the direct synthesis or in the place exchange reaction is thus instrumental to any approaches based on the use of F- and H-thiols for the preparation of MMNPs. We reasoned that the only way to investigate this point in detail was to reduce the complexity of the systems to a minimum using simple fluorinated and hydrogenated model thiols of different lengths and steric bulks and exploring a wide range of monolayer compositions. Based on the experimental data, we discuss the correlation between the initial ratio of the thiols and the final composition of the monolayer in relation to the structure of the ligands. We also correlate the core size of the MMNPs to the nature of the thiols. Finally, we analyse the effect of the monolayer composition on the solubility properties of the MMNPs.

2 Results and discussion

2.1 Design of the approach for the preparation of MMNPs

In order to explore a broad range of conditions for the preparation of gold nanoparticles displaying mixed monolayers of H- and F-ligands and analyse how the monolayer composition depends on the ligand structure, we devised the library of simple thiols displayed in Fig. 1A. The fluorinated thiols we used are 1*H*,1*H*,2*H*,2*H*-perfluorooctanethiol (**HF6**)‡ and 1*H*,1*H*,2*H*,2*H*-perfluorododecanethiol (**HF10**) while the hydrogenated thiols are dodecanethiol (**HC12**), octanethiol (**HC8**), hexadecanethiol (**HC16**) and 3-methyldodecane-1-thiol (**HbrC12**). The thiols **HF6**, **HC8**, **HC16** and **HC12** are commercially available; thiol **HbrC12** was synthesised according to Scheme S1 of the ESI† while thiol **HF10** was prepared according to literature procedures.[46] The structural diversity of the thiols reported in Fig. 1A was conceived in order to analyse how structural parameters such as the thiol length, length mismatch and steric bulk influence the outcome of the syntheses.

The combination of thiols **HC12** and **HF10** in nanoparticles **NP-C12/F10**§ or thiols **HC8** and **HF6** in nanoparticles **NP-C8/F6** will allow an analysis of the conditions required for the preparation of MMNPs displaying H- and F-ligands of the same length. The combination of thiols **HC12** and **HF6** in nanoparticles **NP-C12/F6** or thiols **HC16** and **HF6** in nanoparticles **NP-C16/F6**, allows instead an analysis of the role played by the length mismatch (4 atoms and 8 atoms respectively) in the outcome of the syntheses. By combining thiols **HbrC12** and **HF6** in nanoparticles **NP-brC12/F6** we will explore the effect of using a branched thiol, preventing the formation of a monolayer stabilised by van der Waals interactions,[42] in association with a fluorinated thiol that is four atoms shorter. A cartoon representation of the MMNPs discussed is presented in Fig. 1B.

2.2 Nanoparticle synthesis

The preparation of the MMNPs **NP-C16/F6**, **NP-C8/F6**, and **NP-brC12/F6** was achieved using direct synthesis following the Brust–Schiffrin procedure employing a blend of the H- and F-thiols. **NP-C12/F6** and **NP-C12/F10** were prepared using place exchange on nanoparticles bearing monolayers comprising hydrogenated thiolates only. For the preparation of the MMNPs by place exchange, narrowly dispersed **NP-C12** were prepared using the method of Miyake.[47] The use

‡ **HF6** stands for 1*H*,1*H*,2*H*,2*H*-perfluorooctanethiol, with explicit reference to the sulfhydryl proton. The thiolate derived from 1*H*,1*H*,2*H*,2*H*-perfluorooctanethiol is reported as **F6**. All the thiols and thiolates in the text are named accordingly.

§ **NP-C12/F10** stands for mixed monolayer nanoparticles displaying the thiolates derived from thiols **HC12** and **HF10** in the monolayer. All of the nanoparticle systems described in the text are named accordingly.

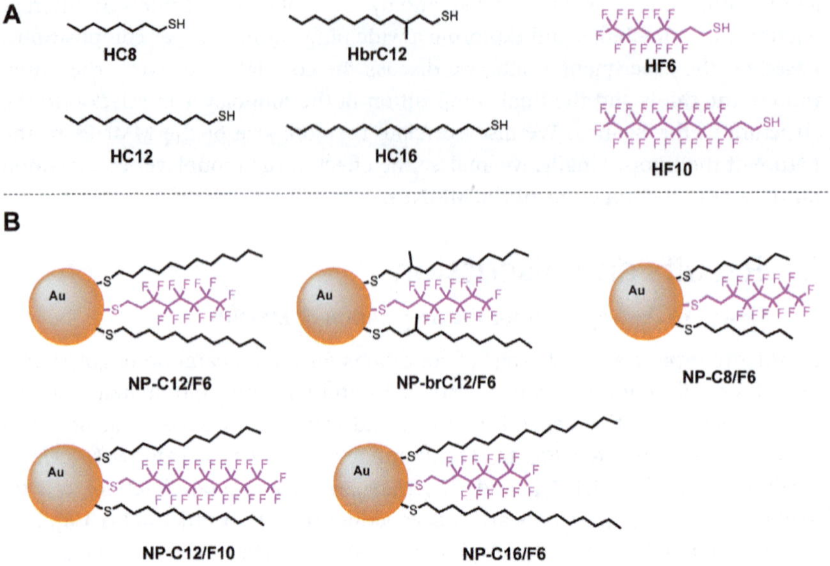

Fig. 1 (Panel A) Library of the hydrogenated and fluorinated thiols used in this study. (Panel B) Structures of the mixed monolayer nanoparticles prepared by combining the thiols reported in panel A.

of two different methodologies for the preparation of the MMNPs is particularly relevant because mass analyses show that mixed monolayers obtained by direct synthesis display little phase segregation, while ligand self-sorting seems to be more pronounced for the systems obtained by place exchange.[48] The ratio between H- and F-thiolates in the monolayer was assessed by decomposing a small amount of the nanoparticles in the presence of excess iodine. The mixture of disulfides thus obtained was analysed using ^{1}H NMR and the ratio between the two thiolates present in the monolayer was determined by integration of the signals due to the methylene group in the alpha position to the sulfur atom.

The diameters of the nanoparticle cores were determined using TEM and the amount of organic material was assessed using thermogravimetric analyses. In some cases in both the direct synthesis and the place exchange reactions, two or three minor fractions of nanoparticles (all fractions from the same synthesis are indicated in the tables with the same lowercase letter) could be separated by exploiting their different solubilities in chloroform, hexane and/or hexafluorobenzene. All of these fractions have been completely characterised. The solubility properties of these nanoparticles are given by adding (C), (H) or (F) to their designation, to indicate solubility in chloroform, hexane or hexafluorobenzene, respectively, as reported in Tables S1, S2, S6 and S8 of the ESI†.

2.3 Analysis of the monolayer composition and composition-related properties

Prior to presenting an analysis of the composition of mixed monolayers as a function of the relative amount of thiols in the initial reaction mixture,[49] we find it useful to define the limiting behaviours expected under place exchange conditions or in direct synthesis. Let us consider the following cases: (i) the preparation of MMNPs by place exchange of thiol B from homoligand nanoparticles comprising thiolates A only. (ii) The preparation of MMNPs obtained by direct synthesis using a blend of thiols A and B. For the sake of simplicity, we shall consider the formation of a single population of nanoparticles.

Mixed monolayer obtained by place exchange. The relationship between the molar fraction of ligand B in the monolayer and its initial molar fraction in the reaction mixture can conveniently be represented by the plot in Fig. 2. In this plot, the x-axis represents the initial molar fraction of thiol B in the reaction mixture, while the y-axis represents the final composition of the monolayer expressed as a molar fraction of the thiolate B.

Three different limiting cases are theoretically possible depending on the relative affinity of thiols A and B and the thermodynamics of the monolayer formation. If we assume that the ligand B has a very high tendency to assemble in the monolayer and that this is much higher than that of thiol A, in the place exchange reaction ligand B will displace ligand A completely. The amount of ligand B introduced into the monolayer is only limited by its initial molar fraction. This process will lead, eventually, to the complete conversion of the homoligand A-monolayer into a homoligand B-monolayer when the initial molar fraction of B is just 0.5 (red curve in Fig. 2); any further increase of the initial molar fraction of thiol B cannot produce further changes. A second limiting behaviour is obtained when no preference between ligand A and ligand B exists in forming the monolayer. The final compositions of the monolayers as a function of

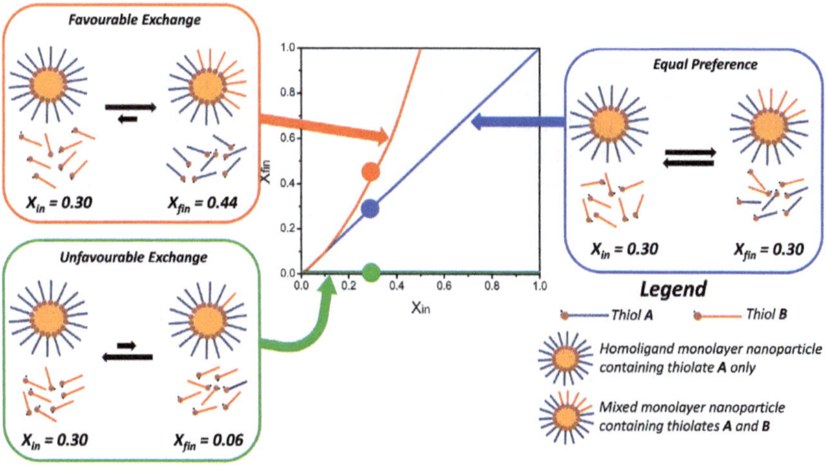

Fig. 2 Limiting cases theoretically possible in the place exchange between thiol B and homoligand nanoparticles featuring thiolate A only. If the grafting of thiol B is favoured (top left of the figure), complete consumption of thiol B can take place with displacement of an equal amount of thiolate A. In this case, the experimental data points will cluster along the red curve. If the grafting of thiol B is disfavoured (bottom left of the figure), only a small amount of exchange will be observed and the experimental data will be found close to the green line of the plot. If there is an equal preference for the grafting of ligands A or B (right hand side of the figure), the composition of the mixed monolayer will reflect the initial composition of the reaction mixture and the experimental data points will cluster along the diagonal of the plot.

the initial compositions will thus cluster along the diagonal of the plot (blue line in Fig. 2). A third limiting case is the one pertaining to a situation in which the grafting of ligand B is strongly disfavoured (green line in Fig. 2). In this case, no B ligands will be found in the monolayer at any initial molar fraction.

Mixed monolayer obtained by direct synthesis. In the direct synthesis of MMNPs, the thiols are in excess with respect to the available grafting sites on the nanoparticle surface.¶ This is a remarkable difference with respect to place exchange. If ligand B has a very high affinity for the monolayer and this is much higher than the affinity of thiol A, it is theoretically possible to end up with homoligand B-monolayers even if the initial molar fraction of thiol B is relatively small. In this case the experimental data points could be found either above the red line of Fig. 2 or close to it. On the other hand, if there is no preference for the grafting of the A or B thiolates, the experimental data points will cluster along the diagonal of the plot. If the grafting of thiol B is disfavoured the experimental points will be found beneath the diagonal or approaching the green line of the plot in Fig. 2.

Analysis of the experimental data for the synthesis of mixed monolayer nanoparticles. For nanoparticles **NP-C12/F6** (Fig. 3A) the experimental data cluster along the diagonal of the plot, with some deviations only at very high initial loading of the fluorinated component. Hence, from a synthetic point of view, the preparation of nanoparticles **NP-C12/F6** containing up to 50% of the fluorinated ligand is straightforward, since the initial molar fraction of the reaction mixture is retained in the final product. This indicates that there is essentially no energetic penalty or gain in introducing thiol **HF6** into the monolayer of the preformed **NP-C12**. Instead, some energetic penalty, resulting in a less facile introduction of the F-ligand into the monolayer, occurs only when the initial molar fraction is very high. In some cases, the synthesis of nanoparticles **NP-C12/F6** gives rise to a sub-population of nanoparticles with different solubility properties that are richer in the fluorinated component and that could be isolated (nanoparticles **NP-C12/F6-i(H)** and **NP-C12/F6-k(H)**, Table S1).† The compositions of these systems are reported as open symbols in the plot of Fig. 3A.

Similarly, for nanoparticles **NP-C12/F10** (Fig. 3B) the experimental data points tend to cluster close to or slightly above the diagonal. As in the previous case, in the preparation of nanoparticles **NP-C12/F10**, a small fraction of nanoparticles with monolayers very rich in the fluorinated component could be isolated (nanoparticles **NP-C12/F10-a(H)**, **NP-C12/F10-b(H)** and **NP-C12/F10-d(H)**, Table S2).† The composition of these nanoparticles is reported with open symbols in the plot of Fig. 3B. These data indicate a facile introduction of the fluorinated ligands into the monolayer of the nanoparticles even at a low molar fraction of the fluorinated component in the reaction mixture.

Upon increasing the length difference between the ligands, as in nanoparticles **NP-C16/F6**, a completely different behaviour was observed (Fig. 3C). In fact, the introduction of a few F-ligands into the monolayer of these nanoparticles proved

¶ The number of alkanethiolate chains on the surface of gold nanoparticles as a function of the total number of gold atoms is fairly well described by the equation: No. chains $\cong 2.5(N\text{Au})^{-0.37}$. For nanoparticles with core size of 3.0 nm (976 gold atoms), the minimum theoretical Au/thiol ratio necessary to achieve passivation leaving no excess thiols is Au/thiols = 1/5. For these calculations, we used the data reported in ref. 50.

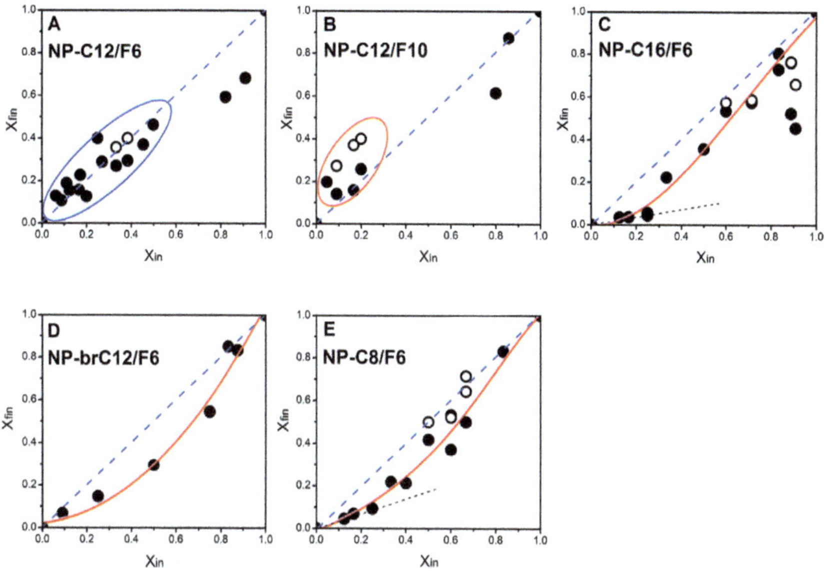

Fig. 3 (Panel A and B) Experimental data of the monolayer compositions for nanoparticles **NP-C12/F6** and **NP-C12/F10**, respectively, as a function of the composition of the initial reaction mixture. (Panel C–E) Experimental data of the monolayer compositions for nanoparticles **NP-C16/F6**, **NP-brC12/F6**, and **NP-C8/F6**, respectively, as a function of the composition of the initial reaction mixture.

to be extremely unfavourable with a strong negative deviation from the diagonal of the plot in the region of a small initial fraction of the F-component. In these syntheses more than 20% of fluorinated ligand in the reaction mixture was necessary to achieve a mixed monolayer containing a mere 5% of fluorinated thiolates. After this threshold, however, the introduction of fluorinated ligands becomes more facile, with the experimental data points approaching the diagonal of the plot. Also in this case we observed the formation of sub-populations of nanoparticles richer in the fluorinated component, but, in contrast with **NP-C12/F6** and **NP-C12/F10**, this was observed only at initial molar fractions of the F-component higher than 60%; the compositions of these systems are reported with open symbols in the plot of Fig. 3C.

For nanoparticles **NP-brC12/F6** (Fig. 3D) yet another behaviour appears. In this case the introduction of F-ligands remains unfavourable in all of the conditions explored.[23] This is not unexpected because, by design and in analogy with literature evidence, these nanoparticles are believed to display a poorly organised monolayer. Notably for nanoparticles **NP-brC12/F6**, we could not identify sub-populations of nanoparticles with different fluorinated ligand content obtained in the same synthesis.

A somewhat intermediate behaviour was observed for nanoparticles **NP-C8/F6** (Fig. 3E) where more than 20% of F-ligands in the reaction mixture was needed to achieve 10% of the F-thiolate in the final monolayer composition. After this threshold the introduction of F-ligands becomes more favourable, with the experimental data points slowly approaching the diagonal of the plot at higher molar fractions of **HF6**. In some cases, the preparation of nanoparticles **NP-C8/F6** yielded a small fraction richer in the fluorinated component when the initial composition of the reaction mixture contained more than 50% of the F-ligand. The compositions of these systems are reported with open symbols in the plot of Fig. 3E.

From these data it is clear that the formation of mixed monolayers comprising fluorinated thiolates may be favoured or disfavoured depending on the structure of the ligands and the degree of substitution that is achieved; relatively subtle structural changes impact considerably on the outcome of the syntheses. Most importantly, these data show that the final composition of the MMNPs cannot be *a priori* predicted on the basis of the composition of the reaction mixture, neither in the direct synthesis, nor in the place exchange reaction.

Effect of the fluorinated ligand loading on the nanoparticle size. It is well established by a large number of experimental results that in the Brust–Schiffrin synthesis, the size of the gold nanoparticles can be tuned by varying the initial gold/thiol ratio; the larger the ratio, the larger the resulting nanoparticles.[50] There is also evidence that bulky thiols tend to favour the formation of smaller gold nanoparticles.[51] Fluorocarbons have a cross-sectional area of 28.3 Å^2 while for hydrocarbons the molecular cross section is only 18.9 Å^2.[52] The fluorinated thiols used in this study are therefore much bulkier (1.5 times larger cross-sectional area) than the hydrogenated ligands, with the possible exception of the branched **HbrC12**. It is expected that the introduction of F-ligands into the monolayer of gold nanoparticles may result in systems of a small size. In the Brust–Schiffrin synthesis of MMNPs we indeed observed a monotonous decrease of the nanoparticle size by increasing the molar fraction of the F-component in the initial reaction mixture while maintaining a constant total gold/thiol ratio. This behaviour was found to be general, regardless of the difference in length between the F- and H-ligands and the steric bulk of the hydrogenated thiols. The experimental data for the nanoparticles **NP-C8/F6**, **NP-C16/F6** and **NP-brC12/F6** are reported in Fig. 4A, B and C, respectively. The size contraction observed when a large molar fraction of the F-ligands is used clearly has a significant impact on the choice of the reaction conditions. In addition, it is also likely to have an impact on the organization of the monolayer since the morphology of mixed

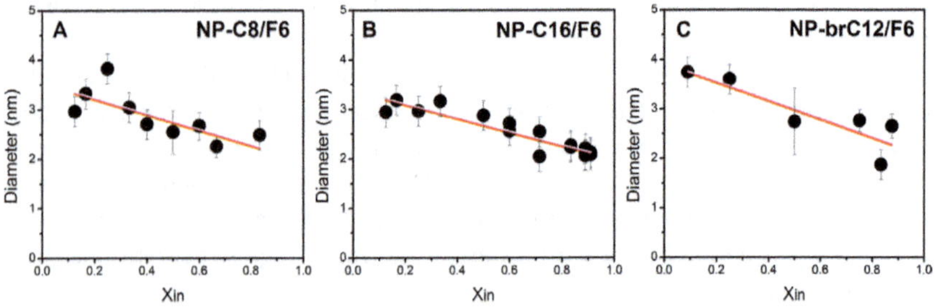

Fig. 4 Dependence of the nanoparticle core diameter on the initial molar fraction of the fluorinated ligand. Error bars represent the standard deviation of the average diameter measured using TEM analyses. In the case of multiple preparations with the same initial loading of the fluorinated component, the experimental points represent the average of the diameter and error bars represent their standard deviation. Panel A–C: experimental data of the diameters for nanoparticles **NP-C8/F6**, **NP-C16/F6** and **NP-brC12/F6** respectively.

monolayers also depends on the nanoparticle curvature radius and in turn on the free volume available per chain.

Solubility behaviour of MMNP. The solubility properties of these nanoparticle systems are very informative and were found to vary according to the amount of fluorinated ligand in the monolayer and the structure of the fluorinated and hydrogenated thiolates. At low molar fractions of the fluorinated component, the nanoparticles were freely soluble in chloroform and methylene chloride. At an intermediate content of the fluorinated component, the nanoparticles were soluble in hexane, while at the higher molar fractions, they were soluble in fluorinated solvents only. In all of the cases, the solubility limit in the different solvents was higher than about 10 mg mL^{-1}. To qualitatively analyse the solubility behaviour in relation to the monolayer composition, we found it convenient to use the solubility in the different solvents as a categorical variable and to plot this variable against the monolayer composition expressed as a molar fraction of the F-ligand (Fig. 5).

To the systems fully soluble in chloroform we arbitrarily assigned a score of 1, the nanoparticle systems soluble in hexane were ranked with a score of 0.5 and those soluble in fluorinated solvents only (hexafluorobenzene was used throughout the study) were ranked with a score of zero. This approach allows the simple comparison of the solubility properties of a different set of nanoparticles. Nanoparticles **NP-C12/F6** display a solubility transition, corresponding to the onset of solubility in hexane, when about 40% of the fluorinated thiolate is present in the monolayer; the same percentage was found for nanoparticles **NP-C12/F10**. A

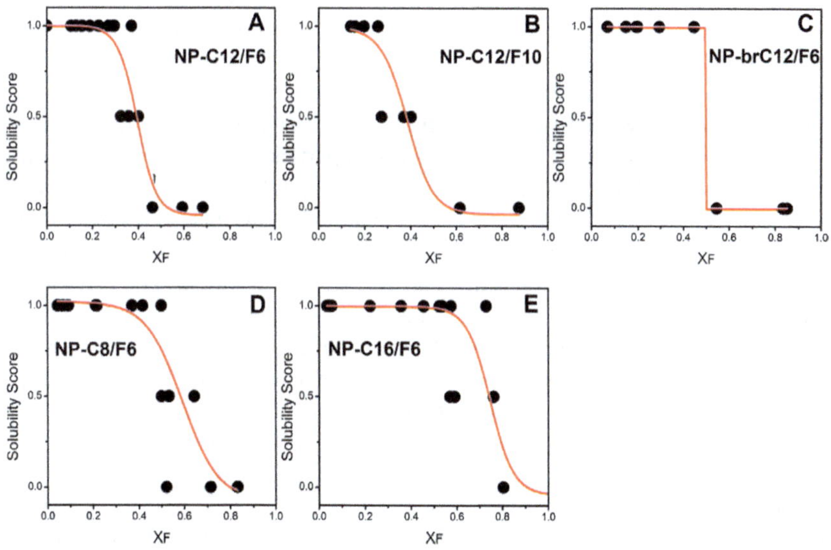

Fig. 5 Comparison of the solubility transitions for the MMNPs as a function of the molar fraction of the fluorinated component in the monolayer. The solubility is expressed according to the following score: score = 1 is assigned to the nanoparticles that are soluble in chloroform. Score = 0.5 is assigned to the nanoparticles that are soluble in hexane, score = 0 is assigned to the nanoparticles that are soluble in hexafluorobenzene. Panel A–C: solubility scores for nanoparticles **NP-C12/F6**, **NP-C12/F10** and **NP-brC12/F6** respectively. Panel D–E: solubility scores for nanoparticles **NP-C8/F6** and **NP-C16/F6**.

significant difference was instead observed for nanoparticles **NP-C16/F6** that remained fully soluble up to a molar fraction of fluorinated ligand of 0.8. For nanoparticles **NP-brC12/F6**, the transition occurred when the molar fraction of the F-ligand was 0.5; the same behaviour was found for nanoparticles **NP-C8/F6**.

2.4 Discussion

A rationalization of the experimental evidence presented above can be put forth by considering that the immiscibility of H- and F-ligands may lead to self-sorting of the two species and that the difference in their length and/or steric bulk may substantially contribute to the driving force of the self-sorting process.

Direct evidence of these phenomena have been already reported for mixtures of H- and F-ligands;[9] further detailed studies addressing the influence of the ligand structure on the determination of the extent of clustering and the monolayer morphologies of the MMNPs presented here are ongoing in our lab. From a general point of view, if short fluorinated thiolates tend to cluster in a monolayer of longer hydrogenated ligands, the introduction of a very small number of F-ligands in the monolayer will be unfavourable. This is because introducing a few fluorinated ligands will decrease the number of van der Waals contacts between hydrogenated chains without offering a significant enthalpic gain derived from the establishment of fluorophilic interactions. In addition an unfavourable H/F interface will be formed. At this stage the entropic gain due to the increased conformational mobility of the hydrogenated thiolates will also be minimal. Only when the amount of F-ligands exceeds a certain threshold should the introduction of more fluorinated ligands become favourable because of the increased number of fluorophilic interactions and the increased entropic gain associated with the conformational mobility of the longer hydrogenated thiolates. This is reminiscent of a cooperative process, where the (unfavourable) introduction of the first few fluorinated ligands generates the conditions for a more favourable assembly. We can clearly trace this phenomenon to the synthetic conditions explored for the preparation of nanoparticles **NP-C16/F6** (Fig. 3C). In this case the introduction of up to 5% fluorinated ligands is strongly disfavoured, while after this threshold it becomes more favourable. The change in slope in the plot of Fig. 3C is consistent with the cooperative mechanism outlined above; in the first phase a few nucleation centres are formed that eventually evolve towards the growth of fluorinated domains. Nucleation and growth of alkanethiolate monolayers by displacement of weakly bound ligands on the surface of gold nanoparticles has indeed been reported.[53] Our data may be taken as an indication that in the case of nanoparticles **NP-C16/F6**, 5% F-ligand is sufficient to trigger the formation of fluorinated domains. This percentage is very close to the results of our previous studies showing that already at 5% loading, domains are formed in mixed monolayers comprising amphiphilic H- and F-ligands.[9] Another remarkable property of **NP-C16/F6** is that up to 80% fluorinated ligand can be introduced without significantly affecting the solubility of the system. This implies that the F-ligands cannot form large solvent exposed domains that would trigger particle aggregation. On the other hand, the data of Fig. 3C do suggest that domains indeed exist implying that these should be relatively small and/or shielded from the solvent and from the fluorinated domains of other nanoparticles by the longer H-ligands.

For nanoparticles **NP-C12/F10**, the experimental data points cluster slightly above the diagonal of the plot of Fig. 3B indicating that the introduction of the F-ligands into the monolayer is favourable even at low loadings of the fluorinated component. It should also be noted that for these nanoparticles we observed the formation of a second population of nanoparticles with a high fluorinated ligand content and different solubility properties even when only 10% of thiol **HF10** was used in the place exchange reaction. Taken together, these data suggest a strong tendency of **F10** to be assembled in the monolayer of nanoparticles **NP-C12/F10.** When, as in this case, there is no length mismatch between the thiolates, the enthalpic balance due to the loss of interactions between hydrogenated ligands and the establishment of fluorophilic interactions will be a significant contribution to the overall ΔG of reaction; a further (unfavourable) contribution to the ΔG of reaction will be the formation of an H/F interface. Based on our experimental evidence, it is reasonable to think that for **NP-C12/F10**, clustering of ligands is likely to occur initially with the formation of small patches, eventually evolving towards larger compact domains, minimising in all cases the unfavourable formation of H/F interfaces.

The TEM image of **NP-C12/F10-b** drop cast from a 10 ng mL^{-1} chloroform solution (Fig. 6) shows the formation of nanoparticle dimers and trimers accounting for 50% of the total population. As a comparison, the TEM image of **NP-C12/F6-k(H)**, in the same conditions, shows 15% of dimers and trimers in the total nanoparticle population. This specific self-assembly pattern supports the existence of a well-defined anisotropy in the nanoparticle monolayer, consistent with the presence of large fluorinated and hydrogenated domains. The aggregation is likely due to the interaction of the fluorinated domains pertaining to different particles.

Given the absence of length mismatch between the two thiols **HC12** and **HF10,** the observed behaviour is consistent with Glotzer's theoretical prediction of Janus nanoparticles.

As in the preceding case, the experimental data for nanoparticles **NP-C12/F6** (Fig. 3A) indicate that there is essentially no energetic penalty or gain in

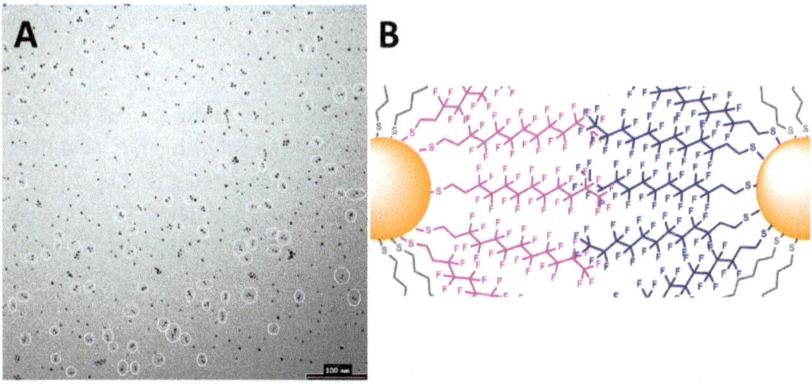

Fig. 6 (a) TEM image (magnification of 140k×) of nanoparticles **NP-C12/F10-b** drop cast from a 10 ng mL^{-1} chloroform solution. Dimers, trimers and some higher oligomers are circled in white. (b) Schematic representation of the interdigitation between fluorinated ligands pertaining to different Janus nanoparticles.

introducing thiol **HF6** into the monolayer of preformed **NP-C12** up to a final composition of 50%. Since, experimentally, the introduction of the fluorinated ligand was found to be easy even at very low loading, we expect the initial formation of small domains as in the case of **NP-C12/F10**. However, in contrast with the previous case, for nanoparticles **NP-C12/F6** a dodecanethiolate unit is replaced by the **F6** unit that is four atoms shorter. If clustering of the fluorinated ligands takes place, this geometrical mismatch should produce an increased conformational freedom for the hydrogenated ligands at the boundaries of the F-ligand clusters. In this case domains with a large interfacial area to surface ratio could be expected.

The solubility properties of both nanoparticles **NP-C12/F6** and **NP-C12/F10** are in keeping with the formation of clusters of ligands, leading to the onset of low solubility already when only about 40% of the F-ligands are introduced into the monolayer (Fig. 5A and B respectively).

For nanoparticles **NP-brC12/F6**, the experimental data points cluster beneath the diagonal of the plot of Fig. 3D, suggesting an unfavourable assembly of the ligand **F6** in the nanoparticle monolayer. Since in this case the branched nature of the **brC12** thiolate hinders the formation of a compact monolayer, the formation of domains will be unlikely and a random distribution of the thiolates on the monolayer will result, in analogy to the observation of Stellacci and co-workers.[42] The absence of fluorinated domains is consistent with the solubility properties of these systems that remain soluble in chloroform up to the introduction of 50% fluorinated component. Experimental evidence that is consistent with a random distribution of the thiolates is also the absence of sub-populations of nanoparticles, obtained in the same synthesis but displaying a different average content of F-ligands. Indeed these sub-populations are likely to be formed only if the introduction of F-ligands results in the formation of clusters in the monolayer. A somewhat intermediate behaviour is displayed by nanoparticles **NP-C8/F6**; indeed, the graph of Fig. 3E shares some features with that obtained for nanoparticles **NP-brC12/F6** and nanoparticles **NP-C16/F6**. This is peculiar since given the absence of length mismatch between the two thiols, the composition of the monolayer as a function of the initial molar fraction of the F-ligand should display a trend similar to that obtained for nanoparticles **NP-C12/F10**.

3 Conclusions

In this study we reported an analysis of the synthetic conditions for the preparation of MMNPs displaying fluorinated and hydrogenated ligands of different lengths and steric bulks. The nanoparticles were synthesised either by exploiting direct synthesis or by place exchange on preformed hydrogenated NPs. In most of the cases, the composition of the monolayer of MMNPs comprising fluorinated and hydrogenated ligands was found to be a non-linear function of the initial composition of the reaction mixture. The amount of fluorinated component introduced into the monolayer depends on the structures of the fluorinated and hydrogenated thiolates. The observed behaviours, including the deviations from linearity, are consistent with the tendency of the fluorinated and hydrogenated species to self-sort and with the role of the length mismatch in triggering the formation of domains by contributing to the overall thermodynamics of the monolayer assembly. The solubility properties of the nanoparticles are also consistent with the formation of fluorinated and hydrogenated domains in the monolayer of these systems. Taken together, the data presented here suggest that for nanoparticles **NP-C16/F6**, that display the largest difference in thiolate length among the systems explored, the formation of the mixed monolayer is dominated by the initial formation of small nucleation centres and the further growth of the domains is cooperative. In the case of nanoparticles **NP-C12/F6** and **NP-C12/F10**, the formation of small domains is likely to occur, and to be favoured, even at very small loadings of the fluorinated component. For nanoparticles **NP-brC12/F6** the experimental evidence fully supports the absence of any organization of the monolayer, in agreement with the observation of Stellacci and co-workers[42] regarding the effect exerted by branched thiolates on the organization of mixed monolayer nanoparticles. Further specific experimental and theoretical investigations are in progress to assess the morphology of the monolayer of these MMNPs. Overall, the study presented here represents the first systematic approach aimed at defining a useful guideline for the design of synthetic strategies for the preparation of MMNPs. This work also shows that for MMNPs, the analysis of the monolayer composition as a function of the molar fraction of the reacting thiols used in the synthesis may provide information on the tendency of the thiolates to cluster on the surface of the nanoparticles.

4 Experimental

4.1 Synthesis

Synthesis of MMNPs. The synthesis of the MMNPs **NP-C12/F6** and **NP-C12/F10** was performed by place exchange from **NP-C12** prepared according to a literature procedure.[47] The synthesis of **NP-C8/F6**, **NP-brC12/F6**, and **NP-C16/F6** was performed using the Brust–Schiffrin[54] method with a mixture of hydrogenated and fluorinated thiols. The general procedure for the syntheses are outlined below; synthetic details are reported in Table S1 for **NP-C12/F6**, Table S2 for **NP-C12/F10**, Table S3 for **NP-brC12/F6**, Table S5 for **NP-C8/F6** and Table S7 for **NP-C16/F6**.† Characterization data are presented in Table S1 for **NP-C12/F6**, Table S2 for **NP-C12/F10**, Table S4 for **NP-brC12/F6**, Table S6 for **NP-C8/F6** and Table S8 for **NP-C16/F6**.†

4.2 General procedure for the place exchange reaction

Nanoparticles NP-C12/F6 and NP-C12/F10. A solution of **NP-C12** dissolved in DCM at a concentration of 2 mg mL^{-1} was deoxygenated and used for the synthesis. To the nanoparticles, a solution of fluorinated thiols in deoxygenated DCM (the proper amount is reported in Table S1 for **NP-C12/F6** and Table S2† for **NP-C12/F10**) was added. The reaction mixture stirred at 40 °C in a pressure-tight screw-capped reaction vessel for three days. After this time the solution was concentrated to a small volume (about 5 mL) and the nanoparticles were precipitated by addition of methanol. The supernatant was discarded and the precipitated nanoparticles were dissolved in 1 mL of $CHCl_3$ and precipitated a second time by addition of methanol. The supernatant was discarded and the solid residue was dissolved in a small amount of $CHCl_3$ and transferred to a centrifuge tube. The solvent was removed with the aid of a gentle argon stream and then the residue was washed with methanol (4 × 20 mL) and acetone (4 ×

20 mL). To improve the purification process, the nanoparticles were dissolved in $CHCl_3$, the solvent was removed under an argon flux, and the residue was washed with methanol (4 × 20 mL) and acetone (4 × 20 mL). The purified nanoparticles were subjected to selective extractions first with $CHCl_3$ and afterwards with hexane. The insoluble material eventually present was tested for solubility in hexafluorobenzene. All of the fractions were characterised using ^{1}H NMR, UV-VIS, TGA, TEM. Characterization data are reported in Table S1 for **NP-C12/F6** and Table S2 for **NP-C12/F10.**†

4.3 General procedure for the direct synthesis of nanoparticles NP-brC12/F6, NP-C8/F6 and NP-C16/F6

A solution of tetraoctylammonium bromide (2.5 equivalents) in DCM was added to an aqueous solution of $HAuCl_4 \cdot 3H_2O$ (1 equivalent) – see Table S3 for **NP-brC12/F6**, Table S5 for **NP-C8/F6** and Table S7 for **NP-C16/F6.**† The mixture was vigorously stirred and the fading of the aqueous phase was observed while the organic phase turned orange. After the phase transfer was complete, a freshly prepared solution of the hydrogenated and fluorinated thiols in DCM was added to the reaction mixture. The concentration of the thiol solution and the volume used varied in the different syntheses. The total amount of thiol and their molar ratios are reported in Table S3 for **NP-brC12/F6**, Table S5 for **NP-C8/F6** and Table S7 for **NP-C16/F6.**† The reaction mixture was left stirring at room temperature for 10 minutes and afterwards, a freshly prepared aqueous solution of $NaBH_4$ was added under vigorous stirring; the time required for adding the $NaBH_4$ solution is reported in Table S3 for **NP-brC12/F6**, Table S5 for **NP-C8/F6** and Table S7 for **NP-C16/F6.**† The reaction mixture was left stirring for 18 hours at room temperature.

4.4 General procedures for work-up

Nanoparticles NP-brC12/F6 and NP-C8/F6. The organic and the aqueous layers were separated, the organic layer was washed with brine (1 × 20 mL) and the nanoparticles were precipitated by addition of methanol to the organic phase. The turbid suspension was transferred into two centrifuge tubes and centrifuged for 30 minutes at 4500 rpm at 15 °C. The supernatant was discarded and the solid residue was dissolved in 1 mL of $CHCl_3$; the nanoparticles were precipitated a second time by addition of methanol and recovered by centrifugation. After removal of the supernatant, the solid was washed with MeOH (3 × 15 mL) and acetone (3 × 15 mL). The purified nanoparticles were subjected to selective extractions with $CHCl_3$ and afterwards with hexane. The insoluble material eventually present was tested for solubility in hexafluorobenzene. All of the fractions were characterised using ^{1}H NMR, UV-VIS, TGA, TEM. Characterization data are reported in Table S4 for **NP-brC12/F6** and Table S6 for **NP-C8/F6.**†

Nanoparticles NP-C16/F6. The organic and the aqueous layers were separated and the organic layer was washed with brine (1 × 20 mL). The nanoparticles were precipitated by addition of methanol. The turbid suspension was transferred into two centrifuge tubes and centrifuged for 3 minutes at 4500 rpm at 15 °C. The supernatant was discarded and the precipitate was washed with methanol. The crude nanoparticle preparation was recovered by centrifugation. The nanoparticles were dissolved in 1.0 mL of $CHCl_3$, the solvent was removed in an argon stream and the residue was washed with methanol (7 × 15 mL). The purified nanoparticles were subjected to selective extractions with $CHCl_3$ and afterwards with hexane. The insoluble material eventually present was tested for solubility in hexafluorobenzene. All of the fractions were characterised using ^{1}H NMR, UV-VIS, TGA, TEM. Characterization data are reported in Table S8.†

Acknowledgements

This work was partly supported by the University of Trieste (FRA 2014, FRA 2015), FIRB 2011 Prot. RBAP11ETKA and MIUR: project MULTINANOITA. TEM images were recorded at the Microscopy Facility of the Department of Molecular Sciences and Nanosystems of the University Ca' Foscari of Venezia or at the Electron Microscopy Facility of the University of Trieste (with the assistance of Mr Claudio Gamboz and Dr Paolo Bertoncin).

References and notes

1 P. Pengo and L. Pasquato, *J. Fluorine Chem.*, 2015, **177**, 2–10.
2 C. Gentilini, F. Evangelista, P. Rudolf, P. Franchi, M. Lucarini and L. Pasquato, *J. Am. Chem. Soc.*, 2008, **130**, 15678–15682.
3 T. Nishio, K. Niikura, Y. Matsuo and K. Ijiro, *Chem. Commun.*, 2010, **46**, 8977–8979.
4 K. Niikura, N. Iyo, Y. Matsuo, H. Mitomo and K. Ijiro, *ACS Appl. Mater. Interfaces*, 2013, **5**, 3900–3907.
5 K. Niikura, N. Iyo, T. Higuchi, T. Nishio, H. Jinnai, N. Fujitani and K. Ijiro, *J. Am. Chem. Soc.*, 2012, **134**, 7632–7635.
6 M. Boccalon, S. Bidoggia, F. Romano, L. Gualandi, P. Franchi, M. Lucarini, P. Pengo and L. Pasquato, *J. Mater. Chem. B*, 2014, **3**, 432–439.
7 M. Boccalon, P. Franchi, M. Lucarini, J. J. Delgado, F. Sousa, F. Stellacci, I. Zucca, A. Scotti, R. Spreafico, P. Pengo and L. Pasquato, *Chem. Commun.*, 2013, **49**, 8794–8796.
8 C. Gentilini, P. Franchi, E. Mileo, S. Polizzi, M. Lucarini and L. Pasquato, *Angew. Chem., Int. Ed.*, 2009, **48**, 3060–3064.
9 P. Posocco, C. Gentilini, S. Bidoggia, A. Pace, P. Franchi, M. Lucarini, M. Fermeglia, S. Pricl and L. Pasquato, *ACS Nano*, 2012, **6**, 7243–7253.
10 M. P. Krafft and J. G. Riess, *Chem. Rev.*, 2009, **109**, 1714–1792.
11 J. D. Dunitz, *ChemBioChem*, 2004, **5**, 614–621.
12 L. D. Zarzar, V. Sresht, E. M. Sletten, J. A. Kalow, D. Blankschtein and T. M. Swager, *Nature*, 2015, **518**, 520–524.
13 N. C. Yoder, V. Kalsani, S. Schuy, R. Vogel, A. Janshoff and K. Kumar, *J. Am. Chem. Soc.*, 2007, **129**, 9037–9043.
14 P. Long and J. Hao, *Adv. Colloid Interface Sci.*, 2012, **171–172**, 66–76.
15 S. De Feyter and F. C. De Schryver, *J. Phys. Chem. B*, 2005, **109**, 4290–4302.
16 M. Kadi, P. Hansson, M. Almgren and I. Furò, *Langmuir*, 2002, **18**, 9243–9249.
17 V. Percec, M. R. Imam, M. Peterca and P. Leowanawat, *J. Am. Chem. Soc.*, 2012, **134**, 4408–4420.
18 V. Percec, M. Glodde, T. K. Bera, Y. Miura, I. Shiyanovskaya, K. D. Singer, V. S. K. Balagurusamy, P. A. Heiney, I. Schnell, A. Rapp, H.-W. Spiess, S. D. Hudsonk and H. Duank, *Nature*, 2002, **419**, 384–862.

19 V. Percec, G. Johansson, G. Ungar and J. Zhou, *J. Am. Chem. Soc.*, 1996, **118**, 9855–9866.
20 B. Bilgicüer, X. Xing and K. Kumar, *J. Am. Chem. Soc.*, 2001, **123**, 11815–11816.
21 B. C. Buer and E. N. G. Marsh, *Protein Sci.*, 2012, **21**, 453–462.
22 A. Centrone, E. Penzo, M. Sharma, J. W. Myerson, A. M. Jackson, N. Marzari and F. Stellacci, *Proc. Natl. Acad. Sci. U. S. A.*, 2008, **105**, 9886–9891.
23 A. Verma, O. Uzun, Y. H. Hu, Y. Hu, H. S. Han, N. Watson, S. L. Chen, D. J. Irvine and F. Stellacci, *Nat. Mater.*, 2008, **7**, 588–595.
24 A. Verma and F. Stellacci, *Small*, 2010, **6**, 12–21.
25 R. C. Van Lehn, P. U. Atukorale, R. P. Carney, Y.-S. Yang, F. Stellacci, D. J. Irvine and A. Alexander-Katz, *Nano Lett.*, 2013, **13**, 4060–4067.
26 S. Jiang, Q. Chen, M. Tripathy, E. Luijten, K. S. Schweizer and S. Granick, *Adv. Mater.*, 2010, **22**, 1060–1071.
27 A. Ghosh, S. Basak, B. H. Wunsch, R. Kumar and F. Stellacci, *Angew. Chem., Int. Ed.*, 2011, **50**, 7900–7905.
28 A. M. Jackson, J. W. Myerson and F. Stellacci, *Nat. Mater.*, 2004, **3**, 330–336.
29 A. M. Jackson, Y. Hu, P. J. Silva and F. Stellacci, *J. Am. Chem. Soc.*, 2006, **128**, 11135–11149.
30 A. Centrone, Y. Hu, A. M. Jackson, G. Zerbi and F. Stellacci, *Small*, 2007, **3**, 814–817.
31 Q. K. Ong, J. Reguera, P. J. Silva, M. Moglianetti, K. Harkness, M. Longobardi, K. S. Mali, C. Renner, S. De Feyter and F. Stellacci, *ACS Nano*, 2013, **7**, 8529–8539.
32 G. A. DeVries, M. Brunnbauer, Y. Hu, A. M. Jackson, B. Long, B. T. Neltner, O. Uzun, B. H. Wunsch and F. Stellacci, *Science*, 2007, **315**, 358–361.
33 H. Kim, R. P. Carney, J. Reguera, Q. K. Ong, X. Liu and F. Stellacci, *Adv. Mater.*, 2012, **24**, 3857–3863.
34 Y. Hu, O. Uzun, C. Dubois and F. Stellacci, *J. Phys. Chem. C*, 2008, **112**, 6279–6284.
35 O. Uzun, Y. Hu, A. Verma, S. Chen, A. Centrone ahd F. Stellacci, *Chem. Commun.*, 2008, 196–198.
36 C. Singh, P. K. Ghorai, M. A. Horsch, A. M. Jackson, R. G. Larson, F. Stellacci and S. C. Glotzer, *Phys. Rev. Lett.*, 2007, **99**, 226106.
37 I. C. Pons-Siepermann and S. C. Glotzer, *ACS Nano*, 2012, **6**, 3919–3924.
38 C. Singh, A. M. Jackson, F. Stellacci and S. C. Glotzer, *J. Am. Chem. Soc.*, 2009, **131**, 16377–16379.
39 A. Santos, C. Singh and S. C. Glotzer, *Phys. Rev. E*, 2010, **81**, 011113.
40 R. P. Carney, G. A. DeVries, C. Dubois, H. Kim, J. Y. Kim, C. Singh, P. K. Ghorai, J. B. Tracy, R. L. Stiles, R. W. Murray, S. C. Glotzer and F. Stellacci, *J. Am. Chem. Soc.*, 2008, **130**, 798–799.
41 H. Singh, Y. Hu, B. P. Khanal, E. R. Zubarev, F. Stellacci and S. C. Glotzer, *Nanoscale*, 2011, **3**, 3244–3250.
42 X. Liu, M. Yu, H. Kim, M. Mameli and F. Stellacci, *Nat. Commun.*, 2012, **3**, 1182.
43 F. Manea, C. Bindoli, S. Polizzi, L. Lay and P. Scrimin, *Langmuir*, 2008, **24**, 4120–4124.
44 G. Guarino, F. Rastrelli, P. Scrimin and F. Mancin, *J. Am. Chem. Soc.*, 2012, **134**, 7200–7203.
45 R. S. Ingram, M. J. Hostetler and R. W. Murray, *J. Am. Chem. Soc.*, 1997, **119**, 9175–9178.

46 R. Da Costa, M. Jurisch and J. Gladysz, *Inorg. Chim. Acta*, 2008, **361**, 3205–3214.
47 T. Shimizu, T. Teranishi, S. Hasegawa and M. Miyake, *J. Phys. Chem. B*, 2003, **107**, 2719–2724.
48 K. M. Harkness, A. Balinski, J. A. McLean and D. E. Cliffel, *Angew. Chem., Int. Ed.*, 2011, **50**, 10554–10559.
49 A. Stewart, S. Zheng, M. R. McCourt and S. E. J. Bell, *ACS Nano*, 2012, **6**, 3718–3726.
50 M. J. Hostetler, J. E. Wingate, C.-J. Zhong, J. E. Harris, R. W. Vachet, M. R. Clark, J. D. Londono, S. J. Green, J. J. Stokes, G. D. Wignall, G. L. Glish, M. D. Porter, N. D. Evans and R. W. Murray, *Langmuir*, 1998, **14**, 17–30.
51 P. J. Krommenhoek, J. Wang, N. Hentz, A. C. Johnston-Peck, K. A. Kozek, G. Kalyuzhny and J. B. Tracy, *ACS Nano*, 2012, **6**, 4903–4911.
52 V. H. Dalvia and P. J. Rosskya, *Proc. Natl. Acad. Sci. U. S. A.*, 2010, **107**, 13603–13607.
53 Y. Wang, O. Zeiri, A. Neyman, F. Stellacci and I. A. Weinstock, *ACS Nano*, 2012, **6**, 629–640.
54 M. Brust, M. Walker, D. Bethell, D. J. Schiffrin and R. Whyman, *J. Chem. Soc., Chem. Commun.*, 1994, 801–802.

Faraday Discussions

PAPER

Amphiphilic brush polymers produced using the RAFT polymerisation method stabilise and reduce the cell cytotoxicity of lipid lyotropic liquid crystalline nanoparticles†

Jiali Zhai,[ab] Randy Suryadinata,[b] Bao Luan,[cde] Nhiem Tran,[acf] Tracey M. Hinton,[g] Julian Ratcliffe,[c] Xiaojuan Hao[c] and Calum J. Drummond*[a]

Received 7th March 2016, Accepted 22nd March 2016

DOI: 10.1039/c6fd00039h

Self-assembled lipid lyotropic liquid crystalline nanoparticles such as hexosomes and cubosomes contain internal anisotropic and isotropic nanostructures, respectively. Despite the remarkable potential of such nanoparticles in various biomedical applications, the stabilisers used in formulating the nanoparticles are often limited to commercially available polymers such as the Pluronic block copolymers. This study explored the potential of using Reversible Addition-Fragmentation chain Transfer (RAFT) technology to design amphiphilic brush-type polymers for the purpose of stabilising phytantriol and monoolein-based lipid dispersions. The synthesised brush-type polymers consisted of a hydrophobic C12 short chain and a hydrophilic poly(ethylene glycol)methyl ether acrylate (PEGA) long chain with multiple 9-unit poly(ethylene oxide) (PEO) brushes with various molecular weights. It was observed that increasing the PEO brush density and thus the length of the hydrophilic component improved the stabilisation effectiveness for phytantriol and monoolein-based cubosomes. Synchrotron small-angle X-ray scattering (SAXS) experiments confirmed that the RAFT polymer-stabilised cubosomes had an internal double-diamond cubic phase with tunable water channel sizes. These properties were dependent on the molecular weight of the polymers, which were considered in some cases to be anisotropically distributed

[a]*School of Science, College of Science, Engineering and Health, RMIT University, PO Box 2476, Melbourne, Victoria, 3001 Australia. E-mail: calum.drummond@rmit.edu.au*

[b]*CSIRO Manufacturing, 343 Royal Parade, Parkville, Victoria, 3052, Australia*

[c]*CSIRO Manufacturing, Private Bag 10, Clayton, Victoria, 3169, Australia*

[d]*Chengdu Institute of Organic Chemistry, Chinese Academy of Sciences, Chengdu 610041, China*

[e]*Graduate University of Chinese Academy of Sciences, Beijing 100049, China*

[f]*Australian Synchrotron, 800 Blackburn Road, Clayton, Victoria, 3168, Australia*

[g]*CSIRO Biosecurity, Australian Animal Health Laboratory, 5 Portarlington Road, East Geelong, Victoria, 3129, Australia*

† Electronic supplementary information (ESI) available. See DOI: 10.1039/c6fd00039h

within the cubosomes. The *in vitro* toxicity of the cubosomes was assessed by cell viability of two human adenocarcinoma cell lines and haemolytic activities to mouse erythrocytes. The results showed that phytantriol cubosomes stabilised by the RAFT polymers were less toxic compared to their Pluronic F127-stabilised analogues. This study provides valuable insight into designing non-linear amphiphilic polymers for the effective stabilisation and cellular toxicity improvement of self-assembled lipid lyotropic liquid crystalline nanoparticles.

1. Introduction

Lipid lyotropic liquid crystalline nanoparticles (LCNPs) have emerged as a unique class of functional nanomaterials in multiple biological applications,[1–6] with the inverse bicontinuous cubic phase (cubosomes) and the inverse hexagonal phase (hexosomes) LCNPs being of particular research interest for the development of next-generation drug delivery systems. Phytantriol and monoolein are common lipids that form the building blocks for LCNPs made from dispersing the bulk lyotropic liquid crystalline phase. These self-assembled isotropic or anisotropic phases contain two- and three-dimensional nanostructures with high surface areas and extensive porosity, thus offering the advantage of enhanced encapsulation and controlled release of both hydrophilic and hydrophobic bioactives/therapeutics.[7–14] Another advantage of the LCNPs as a drug delivery system is that the internal nanostructure and external particle surface can be custom-designed using compositional variations and bioconjugation chemistry, as well as manipulation of temperature, pH, pressure, and ionic strength.[15–19]

While the formulation of LCNPs is largely dependent on multiple compositional and environmental factors, the choice of stabiliser is considered to be an important factor because it controls key properties including colloidal stability, internal nanostructure in terms of phase behaviour and porosity, and external targeting functionality.[20] Previous studies have successfully explored the use of various types of steric stabilisers for formulating LCNPs, including Pluronic polymers,[21–23] polysorbate 80,[24] Myrj 59,[25] modified cellulose,[26] phospholipids with poly(ethylene oxide) (PEO) chains,[27–29] and β-casein.[30] The Pluronic F127 amphiphilic copolymer, which consists of the PEO-poly(propylene oxide)-PEO (PEO-PPO-PEO) triblocks, is the most commonly used stabiliser because it is commercially available and has a hydrophobic block for partitioning into the "parent" lipid layer and two long hydrophilic regions to form a steric corona around the nanoparticle surface (commonly known as PEGylation). In addition to its stabilising effect, PEGylation has also been found to improve the pharmacokinetics and biocompatibility of nanoparticles *in vivo* by interfering with plasma protein adsorption and macrophage-mediated foreign body response.[31,32]

Polymer stabilisers that provide a better PEGylation coverage around the nanoparticle surface also possess a demonstrably reduced cytotoxicity effect.[28,33] Apart from the choice of polymer stabilisers, it is currently considered that LCNP-induced cytotoxicity may also be dependent on several other factors, including cell line, incubation time, the lyotropic liquid crystalline phase, and nanoparticle composition and concentration; and the research into the cellular tolerance towards both cubosomes and hexosomes has recently been gaining momentum.[28,33–37] For example, analyses of a panel of different cell lines revealed that the phytantriol-based LCNPs demonstrate increased cytotoxic and haemolytic activities compared to monoolein-based LCNPs at the same lipid concentration.[35] Additionally, cubosomes have been reported to be slightly more toxic than hexosomes that consist of the same "parent" lipid and stabilisers.[36] These observations of complex factors contributing to the cytotoxicity have been attributed to mechanisms of membrane fusion and/or disruption due to lipid transfer between the lipid in the particles and in the plasma membranes.[28,38] Indeed, monoolein-based cubosomes have been shown to alter the intracellular lipid components and increase the presence of intracellular lipid droplets.[39,40] The link between *in vitro* cytotoxicity tests and *in vivo* behaviour is also unclear.

While a number of studies have certainly provided some insights into the differential effects of LCNP addition towards cellular tolerance, in-depth investigations into the impact of various LCNP formulations, including the use of custom-made brush polymers as stabilisers, towards inducing cell cytotoxicity remains very limited. The application of custom-made, non-linear polymers as alternative stabilisers to generate LCNPs is attractive; and the recent development of the Reversible Addition-Fragmentation chain Transfer (RAFT)-synthesised poly(octadecyl acrylate)-*block*-poly(polyethylene glycol methyl ether acrylate) (P(ODA)-*b*-P(PEGA)) brush copolymer has revealed comparable and/or increased stabilisation effectiveness in generating phytantriol and monoolein-based LCNPs.[41] The RAFT synthesis technique allows control over the molecular weight and polydispersity of the polymers. Additionally, functionalisation of bioactive molecules through the end group is possible, leading to custom-designed non-linear polymers as stabilisers for generating multifunctional nanoparticles as drug delivery systems.[42]

In the current study, we aim to understand how the use of RAFT polymers with a brush structure in the hydrophilic region as stabilisers differ, in regard to their colloidal stability, lyotropic mesophases, and cytotoxicity, from the linear copolymers when formulating LCNPs. The study focused on two aspects: (a) using the RAFT technique to synthesise brush amphiphilic polymers consisting of a C12 short chain and a poly(ethylene glycol)methyl ether acrylate (PEGA, M_n 480) long chain, with molecular weights (MWs) of 5 kDa, 25 kDa and 50 kDa, which were then used as stabilisers to formulate phytantriol or monoolein-based LCNPs and; (b) investigating the internal lyotropic liquid crystalline phase and the cellular tolerance of the formulated nanoparticles *in vitro*. Synchrotron small-angle X-ray scattering (SAXS) and cryo-transmission electron microscopy (Cryo-TEM) were used to study internal liquid crystalline nanostructures. LCNP-induced cytotoxicity was analysed in both human prostate cell adenocarcinoma (LNCaP) and human renal cell adenocarcinoma (ACHN) cell lines, while mouse erythrocytes were used to measure the nanoparticles' haemolytic activities *in vitro*. Analogous studies have been conducted on Pluronic F127-stabilised nanoparticles for comparison, which showed consistency with previously published results.

2. Materials and methods

2.1. Materials

Monoolein was obtained from Nu-check-Prep, Inc with >99% purity as determined by gas liquid chromatography. Phytantriol (98%) was a gift from DSM Nutritional Products. Pluronic F127 and ethanol were purchased from Sigma-

Aldrich, along with poly(ethylene glycol)methyl ether acrylate (PEGA, M_n 480), inhibitor removers, 4,4′-azobis(4-cyanovaleric acid) (V501; 75%), and 4-cyano-4-(dodecylsulfanylthiocarbonylsulfanyl)pentanoic acid (CPADB; 97%). PEGA was passed through a plug of inhibitor removers prior to use. V501 was recrystallised from ethanol twice prior to use. Cyclohexane and dioxane were purchased from Merck. Milli-Q H_2O (18.2 MΩ) was used for all aqueous preparations. All compounds were used without further purification.

2.2. Synthesis and characterisation of P(PEGA) polymers

The RAFT technique was used to synthesise poly(PEGA) (P(PEGA)) polymers (ESI Fig. S1†). P(PEGA) with a 5 kDa molecular weight (MW) was prepared with a feed molar ratio of [PEGA]/[CDP]/[V501] as 50/5/1. PEGA (2.00 g, 4.17 mmol), CDP (116.4 mg, 0.417 mmol), V501 (23.4 mg, 0.083 mmol), and ethanol (3 mL) were mixed and transferred to a microwave reactor vial, equipped with a spin bar and sealed. The reaction mixture was purged with ultra-high purity argon for 30 min. The vial was then placed in an oil bath set at 70 °C for 120 min with stirring. The polymerisation was terminated by placing the vial in an ice bath for 10 min. The crude polymer was purified by precipitation in cyclohexane (200 mL) three times, and the supernatant was removed. The collected polymer was then dried in a vacuum oven at room temperature to yield a light yellow oil. A similar procedure was followed for the synthesis of P(PEGA) with 25 kDa and 50 kDa MWs, but with feed molar ratios of [PEGA]/[CDP]/[V501] as 260/5/1 and 750/5/1, respectively. The reaction times of P(PEGA) with 25 kDa and 50 kDa MWs were 120 min and 65 min, respectively. In this study, P(PEGA) with target MWs of 5 kDa, 25 kDa, and 50 kDa will be abbreviated to $P(PEGA)_{5k}$, $P(PEGA)_{25k}$ and $P(PEGA)_{50k}$ respectively. Synthesised polymers were characterised using proton nuclear magnetic resonance (^1H NMR) spectroscopy and gel permeation chromatography (GPC). NMR and GPC methods are provided in the ESI† section.

2.3. Preparation of lipid lyotropic liquid crystalline nanoparticles

Phytantriol and monoolein were solubilised in ethanol in 96-well plates. Ethanol was then evaporated using a centrifugal evaporator (GeneVac). RAFT polymers and Pluronic F127 were first dissolved in Milli-Q H_2O, and aliquots of the stock solution were then added onto the top of the dried lipid to achieve the desired molar percentage. The total volume of each sample was adjusted to 0.5 mL using Milli-Q H_2O to achieve a final lipid concentration of 60 mM. Mixtures were then sonicated using a 24-tip sonicator (Qsonica, Q700, 200 watts) with a processing sequence of 2 min at 50% amplitude intensity followed by 3 min at 30% amplitude intensity, with a 10 s pulse-on and a 10 s pulse-off regime.

2.4. Visual assessment of the performance of the P(PEGA) polymers as stabilisers

A standard procedure involving visual assessment was performed as an initial screen to rapidly exclude poor dispersions.[41] The well-dispersed lipid lyotropic liquid crystalline nanoparticles exhibit a milky white appearance with no visible lipid aggregates. Therefore, the sonicated samples were examined visually and graded for their dispersion quality using the following scale: "–" indicates a transparent aqueous phase coexisting with a lipid phase with minimal lipid dispersed in the aqueous phase; "+" indicates a translucent solution with a blue tinge where some lipids were dispersed and some lipids had aggregated; "++" indicates a milky white dispersion with some small lipid aggregates apparent; "+++" indicates a homogenous milky white solution with no lipid aggregates apparent; "+++b" indicates a homogenous dispersion with no lipid aggregates apparent that appears translucent with a blue tinge rather than milky white.

2.5. Synchrotron SAXS characterisation

Synchrotron small-angle X-ray scattering (SAXS) measurements were performed at the small-angle/wide-angle X-ray scattering (SAXS/WAXS) beamline at the Australian Synchrotron. The X-ray had a beam with a wavelength of 1.033 Å (12.0 keV) with a typical flux of approximately 10^{13} photons per s. The sample to detector distance was chosen as 1.6 m which provided a q-range of 0.01–0.5 $Å^{-1}$ (scattering vector $q = 4\pi \sin(\theta)/\lambda$ where θ is the scattering angle and λ is the wavelength). Two dimensional X-ray diffraction images were recorded on a Decris-Pilatus 1-M detector using in-house IDL-based ScatterBrain software. The scattering images were integrated into one dimensional plots of intensity *versus* q for phase identification. A silver behenate standard (d = 58.38 Å) was used for calibration. The exposure time for each sample was 1 s. Prepared samples were loaded into a 96-well, half-area UV-clear plate (Greiner Bio-One) and positioned in a custom-designed plate holder with temperatures controlled *via* a recirculating water bath as previously described.[43] SAXS data were analysed using the IDL-based AXcess software package to determine the identity and the lattice parameter of the internal lyotropic liquid crystalline mesophase.[44]

2.6. Dynamic light scattering characterisation

The particle size distribution and polydispersity index were measured using a Malvern Zetasizer Nano ZS (Malvern Instruments, UK). Samples (100 µL) were diluted with Milli-Q H_2O (3 mL) and transferred to disposable low-volume plastic cuvettes. Samples were analysed at 25 °C using the refractive index of water (n = 1.333) to calculate the hydrodynamic size. The temperature was controlled inside the measuring chamber with an equilibration time of 15 min. The hydrodynamic diameter and size distribution (polydispersity index: PDI) of the nanoparticles were averaged from triplicate measurements and the results were reported as a mean ± range.

2.7. Cryo-TEM experiment

300-mesh copper grids coated with lacy formvar-carbon film (ProSciTech, Thuringowa, Queensland) were glow discharged to provide a hydrophilic substrate and these were placed in a laboratory-built vitrification system to prepare the samples for cryo-TEM. A 4 µL aliquot of the sample was pipetted onto the grids. After 10 seconds adsorption time the grid was blotted manually using Whatman 541 filter paper, for between 2 and 5 seconds. Blotting time was optimised for each sample. The grid was then plunged into liquid ethane cooled by liquid nitrogen. Frozen grids were stored in liquid nitrogen until required.

The samples were examined using a Gatan 626 cryoholder (Gatan, Pleasanton, CA, USA) and a Tecnai 12 Transmission Electron Microscope equipped with an FEI Eagle 4k × 4k CCD detector (FEI, Eindhoven, The Netherlands) at an operating voltage of 120 kV. At all times low dose procedures were followed, using an electron dose of 8–10 electrons per Å^2 for all imaging.

2.8. Cell culture and *in vitro* cytotoxicity assay

Human prostate adenocarcinoma (LNCaP) cells were a gift from Dr Elizabeth Williams (APCRC-Q, Queensland, Australia), and were maintained in Roswell Park Memorial Institute-1640 (RPMI-1640) (Thermofisher) supplemented with 5% Foetal Bovine Serum (FBS) (Thermofisher) and 2 mM GlutaMAX (Thermofisher). The human renal adenocarcinoma (ACHN) cells were maintained in Dulbecco's Modified Eagle's Medium (DMEM) (Thermofisher) supplemented in 10% FBS and 2 mM GlutaMAX. Both cell lines were cultured at 37 °C with 5% CO_2.

Exponentially growing LNCaP and ACHN cells were cultured in triplicate at a density of 10×10^3 and 5×10^3, respectively, in 96-well tissue culture plates (Nunc). The following day, cells were treated with increasing concentrations of phytantriol-based (3.125, 6.25, 12.5, 25, 50, 100 μg mL^{-1}) or monoolein-based (6.25, 12.5, 25, 50, 100, 200, 400, 800 μg mL^{-1}) cubosomes, and were cultured further for 48 h. Stock solution (20 μL) of Alamar Blue (Thermofisher) was then added to the cells, and incubated according to the manufacturer's instructions. The data analyses were presented as a percentage of viable, metabolically active cells, which is in direct proportion to the Alamar Blue fluorescence intensity measured by fluorescent emission at 580 nm using a fluorescent excitation at 550 nm ($550_{Ex}/580_{Em}$) on a FluoSTAR Optima plate reader (BMS).

2.9. Haemolysis assay

Mouse blood in EDTA was obtained from C57/BLK6 mice from the small animal facility of the Australian Animal Health Laboratory (AAHL) with Animal Ethics Committee approval. Erythrocytes (red blood cells) were isolated from the blood suspension after centrifugation (1000*g*, 5 min) and were washed in PBS three times, before being resuspended at a concentration of 7.5×10^7 cells per mL. Erythrocyte aliquots of 100 μL were then added to 100 μL of PBS containing the serial dilution of cubosomes in a 96-well microtiter plate and incubated for 1 h at 37 °C with constant shaking. After removal of the unlysed erythrocytes by centrifugation (1000*g*, 5 min), 100 μL of the supernatant were transferred to a new microtiter plate and haemoglobin absorption was determined at 450 nm (background correction at 750 nm) on an EL808 Absorbance microplate reader (BIO-TEK). 100% lysis was determined by adding 5 μL of a 0.1% Triton X-100 solution prior to centrifugation. Results are presented as a percentage of haemoglobin release compared to the PBS control.

3. Results

3.1. P(PEGA) polymer synthesis and characterisation

The RAFT polymerisation technique was used to synthesise three brush-type amphiphilic polymers, $P(PEGA)_{5k}$, $P(PEGA)_{25k}$, and $P(PEGA)_{50k}$, with molecular weights as shown in Table 1. There was a good agreement between the molecular weights determined based on monomer conversion using ^{1}H NMR and GPC for the 5 kDa polymer. As expected, the GPC values for the higher molecular weight polymers were much smaller than the NMR values, due to the difference in the

Table 1 The hydrophilic–lipophilic balances (HLBs), molecular weights and polydispersities for the P(PEGA) polymers

Polymer	HLB[a]	M_n (NMR)[b]	M_n (GPC)[c]	M_w/M_n (GPC)[c]
$P(PEGA)_{5k}$	19.0	5200	5900	1.15
$P(PEGA)_{25k}$	19.8	25 100	10 800	1.30
$P(PEGA)_{50k}$	19.9	50 800	16 500	1.40

[a] HLB – calculated using Griffin's method.[45] [b] Number-average molecular weight determined based on monomer conversion using ^{1}H NMR. [c] Number-average molecular weight and polydispersity determined using GPC with poly(methyl methacrylate) as the calibration standard.

hydrodynamic volumes of the brush-type polymers from those of the linear polymer standard used for GPC calibration. The GPC method was used to measure the polydispersity of the synthesised polymers as an indicator of the efficiency of the RAFT technique to control polymerisation, and all synthesised polymers were found to have a narrow polydispersity ($M_w/M_n \leq 1.40$). Therefore, the molar calculation of the P(PEGA) polymers in the following experiments were based on the molecular weights determined by NMR.

All polymers were synthesised to include a short hydrophobic region (the C12 chain) and a brush-type hydrophilic region that contains multiple 9-unit PEO side chains (brushes) with various target molecular weights (Fig. 1). The hydrophilic–lipophilic balance (HLB) values of the synthesised polymers are in the range of 19.0–19.9 when calculated using Griffin's method.[45] These values are similar to that for the commercial Pluronic F127 polymer (HLB 18-23). HLB values measure the asymmetrical amphiphilic structure of a polymer and previous studies found that a HLB value greater than 17 was a favourable characteristic for a polymer in order to serve as a stabiliser in the generation of LCNPs.[41]

3.2. Performance of P(PEGA) polymers in stabilising LCNPs

The effectiveness of the synthesised P(PEGA) polymers in stabilising LCNPs was first assessed by visual examination of the dispersion quality as described in previous studies.[41] Well-dispersed lipid nanoparticle samples such as cubosomes

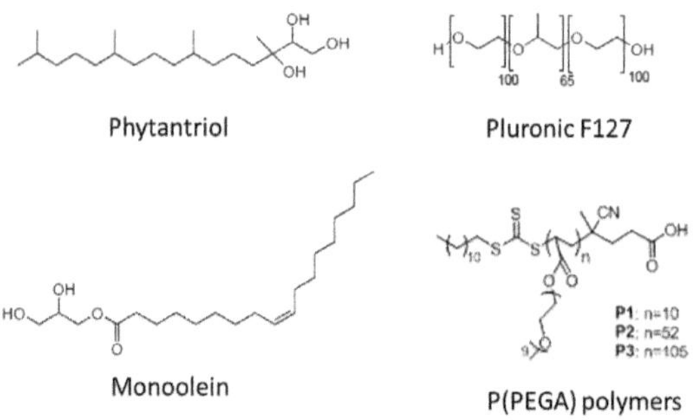

Fig. 1 Chemical structures of phytantriol, monoolein, Pluronic F127, and poly(ethylene glycol)methyl ether acrylate (P(PEGA)) polymers. P1 = $P(PEGA)_{5k}$, P2 = $P(PEGA)_{25k}$, P3 = $P(PEGA)_{50k}$.

and hexosomes have a milky white appearance when they contain the corresponding lyotropic cubic phase and hexagonal phase. The visual assessment of phytantriol or monoolein-based dispersions stabilised using P(PEGA) polymers in a 0.05–3 mol% range is presented in Table 2. The result of a commonly used commercial triblock copolymer for LCNP synthesis, Pluronic F127 with a 12.6 kDa MW, was also included for comparison. Overall, it was found that: (1) the $P(PEGA)_{5k}$ and $P(PEGA)_{25k}$ polymers served as better stabilisers to disperse phytantriol than monoolein at 0.05–3 mol% concentrations; (2) the stabilising function of the P(PEGA) polymers increased as a function of the number of PEO units and thus the hydrophilic brush chain length ($P(PEGA)_{50k}$ > $P(PEGA)_{25k}$ > $P(PEGA)_{5k}$); (3) the $P(PEGA)_{50k}$ polymer had an equivalent stabilising efficiency to that of the triblock copolymer Pluronic F127 and milky white, aggregate-free dispersions were observed at a similar concentration.

Well-dispersed phytantriol LCNPs with a score of "+++" could only be obtained by using the $P(PEGA)_{5k}$ polymer at a 3 mol% concentration. Other polymers with longer PEG chains demonstrated more effective dispersions at lower concentrations (1 mol% for $P(PEGA)_{25k}$; 0.5 mol% for $P(PEGA)_{50k}$ and Pluronic F127). To include the low molecular weight polymer $P(PEGA)_{5k}$ in subsequent physicochemical and biological analyses, phytantriol dispersions containing 3 mol% polymers were selected. One interesting observation is that despite displaying aggregation-free and homogeneous characteristics, phytantriol dispersions that were stabilised by Pluronic F127 at concentrations of 2 mol% or more became translucent with a blue tinge ("+++b" score), indicating changes in the nanoparticle mesophase (Section 3.3). As such, a sample containing phytantriol-dispersed cubosomes that were stabilised with 0.5 mol% Pluronic F127, was also included in the analyses to study the effect of polymer concentration on the phase behaviour and the subsequent biological implications of the phytantriol nanoparticles.

Monoolein was found to be more difficult to disperse than phytantriol when $P(PEGA)_{5k}$ or $P(PEGA)_{25k}$ was used, as evident in the presence of small to medium lipid aggregates even at the higher 3 mol% concentration. In contrast, $P(PEGA)_{50k}$

Table 2 Visual assessment of the performance of the P(PEGA) polymers as steric stabilisers for lipid lyotropic liquid crystalline dispersions[a]

Stabiliser concentration (mol%)	Phytantriol				Monoolein			
	$P(PEGA)_{5k}$	$P(PEGA)_{25k}$	$P(PEGA)_{50k}$	F127	$P(PEGA)_{5k}$	$P(PEGA)_{25k}$	$P(PEGA)_{50k}$	F127
0.05	+	+	+	++	–	–	+	+
0.1	+	+	++	++	–	–	+	++
0.125	+	++	++	++	–	–	++	++
0.25	+	++	++	++	–	–	++	+++
0.5	+	++	+++	+++	–	+	+++	+++
1	+	+++	+++	+++	–	+	+++	+++
2	++	+++	+++	+++b	+	++	+++	+++b
3	+++	+++	+++	+++b	+	++	+++	+++b

[a] Key: "–" transparent samples with separate aqueous and lipid phases; "+" translucent/cloudy samples with some lipid aggregates; "++" milky samples with small visible lipid aggregates; "+++" homogeneous milky samples with no lipid aggregates; "+++b" homogeneous translucent samples with no lipid aggregates. Highlighted samples were further assessed for lyotropic phase behaviour and *in vitro* cytotoxicity.

demonstrated a comparable stabilising function to that of Pluronic F127. Since the synchrotron SAXS study demonstrated that high polymer concentrations adversely affected the internal phase (Section 3.3), monoolein-based LCNPs that were stabilised with $P(PEGA)_{50k}$ and Pluronic F127 at a 0.5 mol% concentration were chosen for further studies.

3.3. Characterisation of nanoparticles: internal lyotropic liquid crystalline structures, lattice parameters and particle size distribution

High-throughput synchrotron SAXS screening was used to assess the internal lyotropic liquid crystalline structures of the phytantriol or monoolein-based nanoparticles stabilised by P(PEGA) polymers. It was found that 3 mol% of P(PEGA) polymers were able to disperse phytantriol into nanoparticles containing scattering peaks with spacing ratios of $\sqrt{2}:\sqrt{3}:\sqrt{4}$, attributed to the double-diamond cubic phase structure (Q_{II}^{D}) (Fig. 2A). The lattice parameters of the Q_{II}^{D} phase had an order of $P(PEGA)_{25k}$ > $P(PEGA)_{5k}$ > $P(PEGA)_{50k}$ (Table 3). The particle size of the $P(PEGA)_{5k}$–phytantriol nanoparticles had a bimodal distribution, indicating poorer dispersing ability, which was not previously recognised by visual examination (Table 3). Cryo-TEM was used to visualise individual cubosomes and the results confirmed the presence of nanostructured particles displaying periodic internal structures (Fig. 3A–C).

Phytantriol nanoparticles stabilised by 3 mol% Pluronic F127 displayed reduced peak intensities corresponding to the Q_{II}^{D} phase as compared to the P(PEGA) nanoparticles. It was reported previously that excess Pluronic F127 could result in a large amount of liposomes rather than cubosomes.[28] In the current study, changes in the visual appearance (Table 2) and reduced particle size of particles stabilised by high concentrations of F127 (Table 3) supported this view. Furthermore, cryo-TEM experiments confirmed the presence of more liposomes in the phytantriol dispersions stabilised by 3 mol% Pluronic F127 (Fig. 3D) than in their 0.5 mol% analogues (Fig. 3E). Therefore, phytantriol dispersed by 0.5 mol % Pluronic F127 was included for further studies, because this polymer concentration was more commonly used in previous studies and the sample displayed a distinct Q_{II}^{D} phase and monomodal particle size distribution.[22,28]

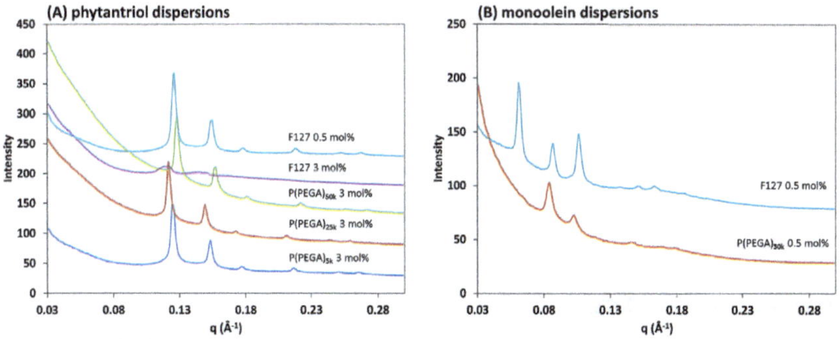

Fig. 2 Synchrotron SAXS profiles of (A) phytantriol-based dispersions stabilised by P(PEGA) polymers (3 mol%) and Pluronic F127 (0.5 mol% and 3 mol%); (B) monoolein-based dispersions stabilised by 0.5 mol% $P(PEGA)_{50k}$ polymer and Pluronic F127.

Table 3 Particle size distributions and lattice parameters of the selected nanoparticles[a]

Sample	Phase[b]	a^c (Å)	Size (nm)	PDI[d]
P(PEGA)$_{5k}$–phyt (3 mol%)	Q_{II}^{D}	71.5	416 ± 10[e]	0.29 ± 0.01
P(PEGA)$_{25k}$–phyt (3 mol%)	Q_{II}^{D}	73.4	251 ± 3	0.16 ± 0.02
P(PEGA)$_{50k}$–phyt (3 mol%)	Q_{II}^{D}	69.8	145 ± 2	0.15 ± 0.01
F127–phyt (3 mol%)	Q_{II}^{D}/L_{α}	—	137 ± 1[e]	0.41 ± 0.03
F127–phyt (0.5 mol%)	Q_{II}^{D}	71.1	199 ± 2	0.11 ± 0.02
P(PEGA)$_{50k}$–MO (0.5 mol%)	Q_{II}^{D}	106.3	362 ± 3	0.22 ± 0.01
F127–MO (0.5 mol%)	Q_{II}^{P}	145.2	185 ± 3	0.17 ± 0.01

[a] Abbreviations: Phyt: phytantriol, MO: monoolein. [b] Phase identity of the internal lyotropic liquid crystalline mesophase: Q_{II}^{D} (the double-diamond cubic phase), Q_{II}^{P} (the primitive cubic phase), L_{α} (the lamellar phase). [c] a refers to the lattice parameter of the non-lamellar bicontinuous cubic phase. [d] PDI refers to the polydispersity index. [e] Indicates a bimodal, rather than monomodal, distribution of the particle size.

In monoolein dispersions stabilised by Pluronic F127, a large population of liposomes was detected to coexist with cubosomes as indicated by the SAXS experiment (ESI Fig. S2†). Similar to the case of phytantriol, the higher the F127

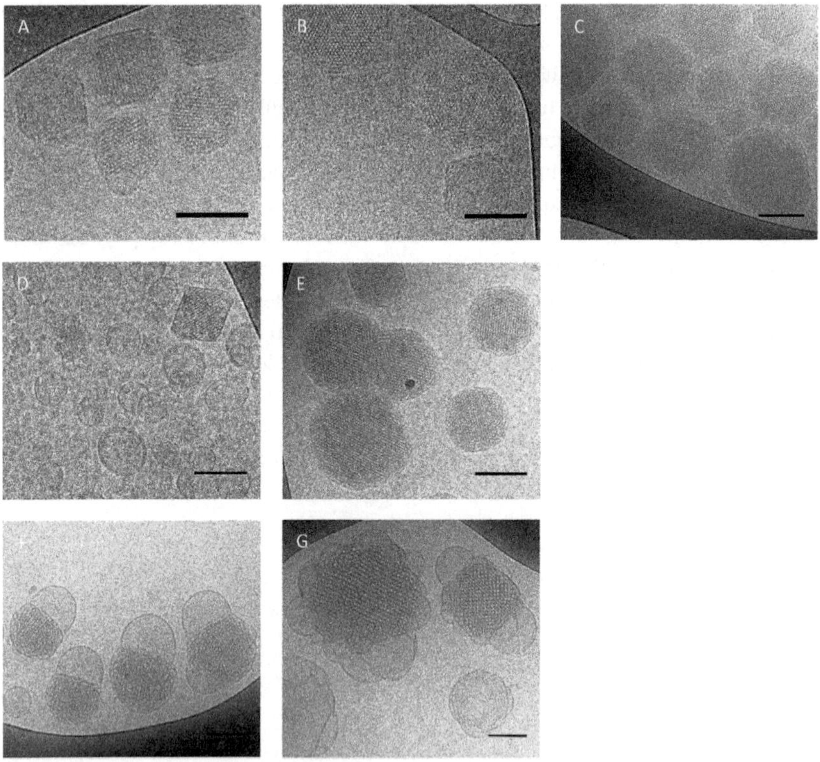

Fig. 3 Cryo-TEM images of phytantriol dispersions stabilised by 3 mol% P(PEGA)$_{5k}$ (A), 3 mol% P(PEGA)$_{25k}$ (B), 3 mol% P(PEGA)$_{50k}$ (C), 3 mol% Pluronic F127 (D), 0.5 mol% Pluronic F127 (E); and monoolein dispersions stabilised by 0.5% mol P(PEGA)$_{50k}$ (F) and 0.5 mol% Pluronic F127 (G). Scale bars = 100 nm.

concentration in the monoolein dispersions, the higher the liposomes to cubosomes ratio. It is clear that monoolein dispersions stabilised by 0.5 mol% $P(PEGA)_{50k}$ and Pluronic F127 are more relevant in order to study comparable lyotropic liquid crystalline nanoparticles, from the internal nanostructure perspective. The $P(PEGA)_{5k}$ and $P(PEGA)_{25k}$ polymers were omitted in the monoolein study due to the poor dispersion quality. Interestingly, at 0.5 mol%, the Pluronic F127 and $P(PEGA)_{50k}$ polymers formed monoolein nanoparticles with the primitive cubic phase (Q_{II}^{P}) and the double diamond cubic phase (Q_{II}^{D}) respectively (Fig. 2B). The SAXS scattering pattern of $\sqrt{2} : \sqrt{4} : \sqrt{6}$ peaks indicated a Q_{II}^{P} phase with a 145 Å lattice parameter for the Pluronic F127-monoolein particles, which was consistent with previous studies.[36] This finding suggested that $P(PEGA)_{50k}$ was able to stabilise monoolein dispersions while preserving the monoolein bulk lipid Q_{II}^{D} phase. Cryo-TEM images showed that the monoolein dispersions stabilised by 0.5 mol% $P(PEGA)_{50k}$ or Pluronic F127 consisted of faceted cubosomes with vesicles attached to their surfaces. The DLS result showed that $P(PEGA)_{50k}$-monoolein nanoparticles were nearly twice the size of the Pluronic F127-stabilised ones (Table 3).

3.4. Cellular cytotoxicity profiles of P(PEGA)-stabilised nanoparticles

A previous study indicated that the Pluronic F127-stabilised phytantriol-based cubosomes became highly toxic to the various cultured cell lines at a relatively low concentration of nanoparticles around 20 μg mL^{-1}.[35] To analyse the impact of P(PEGA)-stabilised cubosomes on cellular tolerance, *in vitro* cytotoxicity assays were performed on LNCaP and ACHN cell lines using well established Alamar Blue staining (Fig. 4). LNCaP cells displayed a similar level of tolerance towards Pluronic F127-stabilised phytantriol-based cubosomes as previously reported,[35] where nanoparticle concentrations of more than 12.5 μg mL^{-1} became toxic to the cells (Fig. 4A). Treatment with P(PEGA)-stabilised phytantriol cubosomes improved the tolerance of the LNCaP cells, with nanoparticles that were stabilised with 3 mol% $P(PEGA)_{25k}$ exhibiting the lowest level of cytotoxicity, demonstrated by a calculated concentration required to induce 50% cell death of 50.8 μg mL^{-1}, which is the IC_{50} value. By comparison, phytantriol cubosomes stabilised by 3 mol% $P(PEGA)_{5k}$, $P(PEGA)_{50k}$ and Pluronic F127 control required concentrations of 26.8 μg mL^{-1}, 14.0 μg mL^{-1} and 9.4 μg mL^{-1}, respectively, to induce the same level of cytotoxicity. Phytantriol dispersions that were stabilised by 0.5 mol% Pluronic F127 were also tested, with an IC_{50} value of 11.5 μg mL^{-1}.

ACHN cells treated with phytantriol cubosomes stabilised by 3 mol% $P(PEGA)_{5k}$ and $P(PEGA)_{25k}$ polymers also demonstrated increased cellular tolerance, compared to control cells that were treated with Pluronic F127-stabilised nanoparticles (Fig. 4C). In contrast, at 3 mol%, $P(PEGA)_{50k}$-stabilised phytantriol dispersions displayed a similar level of cytotoxicity to the control, with an IC_{50} value of 27.4 μg mL^{-1}, compared to 26.3 μg mL^{-1} for the Pluronic F127-stabilised cubosome control. Comparatively, phytantriol cubosomes stabilised by 3 mol% $P(PEGA)_{5k}$ and $P(PEGA)_{25k}$ required double the concentration (55.7 μg mL^{-1} and 54.9 μg mL^{-1}, respectively) to induce the same level of cytotoxicity, while cubosomes stabilised by 0.5 mol% Pluronic F127 required 36.7 μg mL^{-1}.

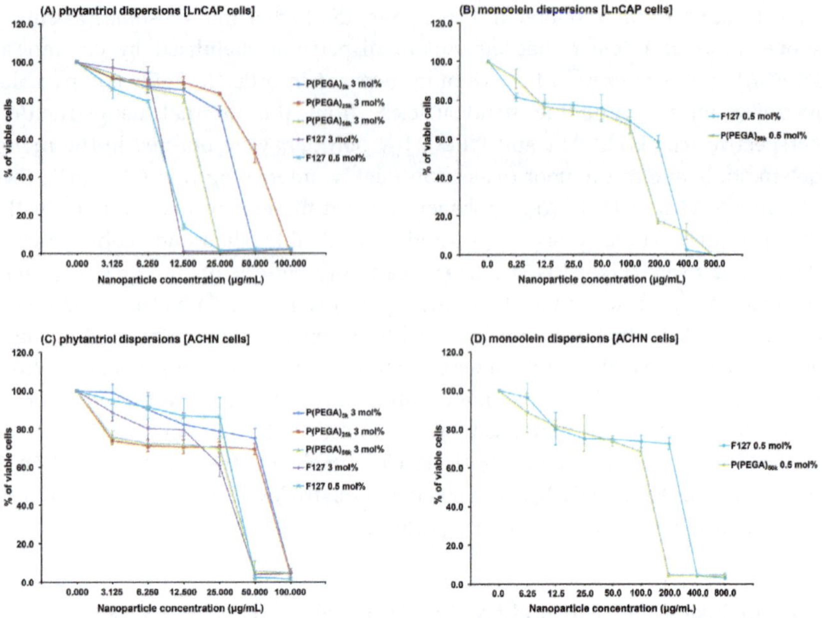

Fig. 4 P(PEGA)- and Pluronic F127-stabilised cubosomes induce cell cytotoxicity in LNCaP and ACHN cells in a concentration-dependent manner. *In vitro* cell cytotoxicity profiles of LNCaP cells following treatment with increasing concentrations of phytantriol- (A) and monoolein-based (B) cubosomes, as well as ACHN cells following treatment with phytantriol- (C) and monoolein-based (D) cubosomes, as measured by the Alamar Blue assay. Alamar Blue fluorescence intensity was measured at $550_{Ex}/580_{Em}$ and is in direct proportion to the metabolically active, viable cells. Data is represented as an average of triplicate experiments ± SD.

Monoolein cubosomes stabilised by 0.5 mol% $P(PEGA)_{50k}$, when added to the LNCaP cells, were more toxic compared to the Pluronic F127-stabilised nanoparticle control (Fig. 4B), as demonstrated by an IC_{50} value of 150 µg mL^{-1} (*versus* 241.2 µg mL^{-1} required for the control). Similarly, $P(PEGA)_{50k}$-stabilised monoolein cubosomes were also more toxic towards ACHN cells, requiring 118 µg mL^{-1} to induce 50% cytotoxicity, compared to a 248 µg mL^{-1} nanoparticle concentration needed by the control (Fig. 4D).

The haemolytic activities of P(PEGA)-stabilised, phytantriol- and monoolein-based cubosomes were also tested in mouse erythrocytes (Fig. 5). At nanoparticle concentrations of less than 2 µg mL^{-1}, phytantriol cubosomes that were stabilised with 3 mol% $P(PEGA)_{5k}$ and $P(PEGA)_{25k}$ demonstrated a lower level of toxicity towards the mouse erythrocytes compared to the nanoparticles that were stabilised with $P(PEGA)_{50k}$ or the Pluronic F127, which were similar in their haemolytic activities towards the erythrocytes (Fig. 5A). Similarly, addition of <2 µg mL^{-1} of monoolein cubosomes that were stabilised by 0.5 mol% $P(PEGA)_{50k}$, resulted in lower haemolytic activities compared to those of the Pluronic F127-stabilised control (Fig. 5B). At higher concentrations, all nanoparticle samples yielded near complete haemolysis of the erythrocytes.

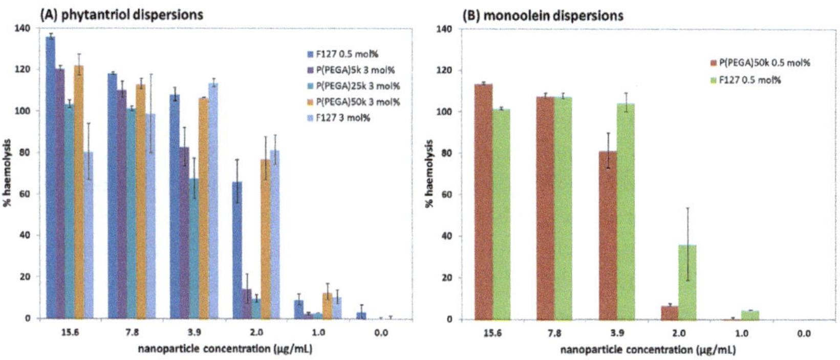

Fig. 5 Haemoglobin release due to the interaction of (A) phytantriol-based and (B) monoolein-based lipid nanoparticles with the mouse erythrocytes. The data were normalised to the haemoglobin release level of the Triton X-100 control. Data = mean ± SD. $N = 3$.

4. Discussion

With the burgeoning of multifunctional LCNPs as drug delivery systems, there is a need to custom design functionalisable stabilisers that provide not only effective colloidal stabilisation but also the flexibility to engineer the internal nanostructures and the external surfaces of the nanoparticles, as these structural features eventually determine the therapeutic encapsulation and active targeting functionalities. However, the majority of LCNP research has been heavily reliant on the use of commercially available linear copolymers such as the Pluronic series.[21–23] This study demonstrated the use of the RAFT polymerisation technique for the synthesis of custom designed amphiphilic brush polymers, which offered a two-fold advantage over the traditional Pluronic F127 triblock copolymers: (1) effective stabilisation of cubosomes that contain an internal Q_{II}^{D} phase and an external functionalisable RAFT end-group for potential conjugation with targeting agents, and in particular, the $P(PEGA)_{50k}$ could disperse monoolein into cubosomes without disrupting the bulk Q_{II}^{D} phase as Pluronic F127 would do; (2) reduced cytotoxicity profiles, making these non-linear polymers attractive for further developing cubosome-based drug delivery systems.

4.1. Performance of P(PEGA) polymers as stabilisers

The synthesised P(PEGA) polymers were designed to have an amphiphilic nature, including a hydrophobic part that anchors into the phytantriol or monoolein lipid bilayer, and a hydrophilic brush region to impart steric stabilisation, with a HLB value that is equivalent to that of the Pluronic F127 copolymer. Given the amphiphilic nature of the synthesised P(PEGA) polymers, it is important to confirm that the cubic phase observed in subsequent lipid dispersions is due to lipid self-assembly behaviour rather than polymer self-assembly. SAXS experiments demonstrated that all polymers formed micellar aggregates in excess water at the highest investigated concentration of 3 mol% (ESI Fig. S3†). Therefore, it can be concluded that the cubic phase observed in lipid dispersions stabilised by P(PEGA) polymers was not interfered with by the polymer self-assembly objects.

It was found that despite the lipid composition (phytantriol or monoolein), the stabilising function of the P(PEGA) polymers was improved with increasing MW and thus the hydrophilic chain length ($P(PEGA)_{50k} > P(PEGA)_{25k} > P(PEGA)_{5k}$). Although the PEO units in the P(PEGA) polymers remained at 9 units on average per brush, the length of the hydrophilic brush region was effectively increased by increasing the number of PEO block brushes along the polymer backbone (Fig. 1), thus providing better steric repulsion coverage for the nanoparticles. This trend is in agreement with previous results describing the Pluronic,[25] PEGylated phospholipids,[27,28] and PEGylated-phytanyl stabiliser series,[46] whereby the increase in linear PEO units led to better colloidal stabilisation. For example, phospholipids conjugated with 70 or 113 PEO units were effective in stabilising phytantriol dispersions, whereas analogues containing 45 PEO units failed to form stable dispersions.[28] A recent study examining an amphiphilic brush copolymer (P(ODA)-*b*-P(PEGA)) that comprises a hydrophobic P(ODA) block and a hydrophilic P(PEGA) block also found that increasing the number of 9-unit PEO brush arms in the P(PEGA) block from 23 to 34 increased the effectiveness of the polymer in stabilising both phytantriol and monoolein dispersions.[41]

For phytantriol dispersions stabilised by the P(PEGA) polymers, it was found that the length of the hydrophilic brush region affected the lattice parameter of the Q_{II}^{D} phase with an order of $P(PEGA)_{25k} > P(PEGA)_{5k} > P(PEGA)_{50k}$. It is known that the internal mesophase of the cubosomes could tolerate internalisation of the stabilisers up to a certain degree depending on the length of the hydrophilic region, which results in a swollen state of the cubic phase. For example, phospholipids conjugated with 70 PEO units were found to be incorporated into the internal phytantriol layers, which resulted in the swelling of the internal phytantriol mesophase into a Q_{II}^{P} phase (lattice parameter a = 99 Å).[28] However, increasing the number of PEO units from 70 to 113 returned the internal nanostructures to the "parent" Q_{II}^{D} phase (a = 70 Å), indicating that the hydrophilic region of the PEO tail and hence the polymer molecule had exceeded the internalisation limit.[28] In this study, the $P(PEGA)_{25k}$-stabilised cubosomes had a larger lattice parameter (73.4 Å) than the $P(PEGA)_{5k}$-stablised cubosomes indicating that the insertion of the hydrophilic brush arms (52 brushes of 9-unit PEO chains) into the internal water channels had occurred, leading to a swollen cubic lattice. On the other hand, $P(PEGA)_{50k}$, which has twice the length of the $P(PEGA)_{25k}$ polymers, behaved similarly to Pluronic F127 in dispersing phytantriol and resulted in a lattice parameter around 70 Å consistent with previous studies.[22] This result suggested that the addition of $P(PEGA)_{50k}$ to cubosomes created a larger anisotropy in polymer distribution, with most $P(PEGA)_{50k}$ immobilised on the nanoparticle external surfaces with limited internalisation.

For monoolein-based nanoparticles, only the $P(PEGA)_{50k}$ polymer was found to have a stabilisation effectiveness comparable to that of the control Pluronic F127. Clearly the "kink" present in the unsaturated oleyl chain of monoolein affects the affinity for binding with the hydrophobic part of the P(PEGA) polymers of lower MW. The phytanyl chain and the C12 chain of the P(PEGA) polymers are both saturated, thus presumably providing a higher affinity. Linear Myrj polymers and P(ODA)-*b*-P(PEGA) brush copolymers have also been reported to have less stabilisation effectiveness for monoolein than for phytantriol.[25,47] The Pluronic F127 copolymer has a good stabilisation effectiveness for monoolein cubosomes because of its unique triblock structure. It has been proposed that the hydrophobic PPO part of the Pluronic F127 polymer resides on the lipid–water interface rather than in the middle of the lipid layer, while the two hydrophilic PEO tails are inserted into the internal water channel.[48,49] The change from the bulk Q_{II}^{D} phase of monoolein to the Q_{II}^{P} phase of the monoolein dispersions stabilised by Pluronic F127 confirms the theory that the 100-unit PEO region of Pluronic F127 inserts into the internal nanochannels. The current study found that, at a 0.5 mol% concentration, the Pluronic F127 cubosomes had a Q_{II}^{P} phase (a = 145.2 Å) that is consistent with previous studies.[36] On the other hand, P(PEGA)$_{50k}$-stabilised monoolein cubosomes had a Q_{II}^{D} phase (a = 106.3 Å) indicating very limited partitioning of its hydrophilic brush region into the internal water channel, as the monoolein bulk Q_{II}^{D} phase reported had a lattice parameter of around 100 Å.[19] This observation demonstrates that the conformation and the length of the hydrophilic region in the stabiliser macromolecule appear to be critical determinants in the ultimate nanoparticle structure.

It is worth noting that the length of the hydrophobic part of a polymer influences its stabilising function for LCNPs. For example, at a similar hydrophilic chain length, increasing the hydrophobic P(ODA) block length increased the stabilisation effectiveness of the P(ODA)-*b*-P(PEGA) brush copolymers.[41] This effect could be attributed to the stronger affinity between the aliphatic chain lengths. Nonetheless, we have demonstrated that phytantriol or monoolein-based cubosomes could be effectively stabilised using a relatively short C12 chain, which could be attributed to the HLB values greater than 19 being a favourable characteristic. Optimising the balance between the hydrophobic and the hydrophilic length of next-generation custom-made brush polymers to create stable and functionalisable LCNPs will be one direction for future studies.

4.2. Cell cytotoxicity and biological implications

Understanding the inherent cytotoxicity of LCNPs as vehicles for drug delivery systems has become a pivotal research focus in recent years.[20] Nevertheless, only a handful of studies have reported cytotoxicity profiles of these nanomaterials towards cells.[28,33–37] It is well-accepted that cellular tolerance towards LCNPs is dependent on a number of factors, including the type of polymer stabilisers used. The current study observed a marked improvement in the cellular tolerance of phytantriol-based cubosomes that were stabilised by P(PEGA) brush polymers, prepared through the RAFT technique, compared with nanomaterials that were stabilised with Pluronic F127, highlighting the importance of polymer stabilisers in providing a sufficient barrier to allow reduced cytotoxicity.[28,33] In-depth analyses further revealed that cubosome-induced cytotoxicity to the LNCaP cells varied depending on the length of the P(PEGA) brush polymers, with increased cytotoxicity obbserved in the order of P(PEGA)$_{25k}$- $<$ P(PEGA)$_{5k}$- $<$ P(PEGA)$_{50k}$- $<$ F127-stabilised phytantriol-derivatives. Interestingly, this observed order of the phytantriol cubosome cytotoxicity effect on LNCaP cells corresponds to the order of the reduced lattice parameter of the Q_{II}^{D} phase in the phytantriol dispersions; that is, the result indicates that the P(PEGA)$_{25k}$-stabilised cubosomes containing the Q_{II}^{D} phase with a larger lattice parameter due to a certain amount of polymer internalisation exhibited the least toxic effect compared to the other analogues and the control sample. Interestingly, monoolein-based cubosomes stabilised by P(PEGA)$_{50k}$ were more toxic to both LNCaP and ACHN cells compared to the nanoparticles that were dispersed by F127, which also corresponds to the F127-stabilised monoolein cubosomes having a larger lattice parameter due to polymer internalisation, which caused the transition of the Q_{II}^{D} phase in the bulk monoolein to the Q_{II}^{P} phase in the dispersed form. This interesting observation indicates that the internalisation of the polymer to the internal lipid cubic phase, while maintaining effective external particle coverage to impart colloidal stability, may affect cubosome cytotoxicity to some extent. It may be attributed to the fact that an external surface coverage and an internal incorporation of the stabilisers further limits the membrane fusion and lipid transfer between the cubosomes and the cell membranes, thus achieving reduced toxicity. Currently, there is a lack of understanding on the mechanism of how internalisation of stabilisers in the internal lyotropic liquid crystalline phases of cubosomes may potentially reduce their cytotoxicity and further studies are necessary for elucidation.

Previous studies have also demonstrated that bulk lipids and their corresponding cubic phases are determinants of cubosome cytotoxicity.[35] The current study revealed consistent results in that the phytantriol-based cubosomes were relatively more toxic than the monoolein derivative of the same concentration regardless of the cell type. Imaging experiments showed that the impermeable fluorescent dye lucifer yellow was uptaken into the cytoplasm of the phytantriol cubosome-treated cells more readily than the monoolein-based analogues. The authors attributed this finding to the fact that Pluronic F127 had a stronger interaction with monoolein than phytantriol, providing better coverage and limiting the exposure of the cubosome lipids to the cells.[3] This could also be viewed from the perspective that Pluronic F127 could insert into the monoolein internal phase but not the phytantriol internal phase, supporting our previous discussion.

Comparative studies on two cell lines also reveal that the cell variant is a determinant in the cell–nanoparticle interaction. Interestingly, $P(PEGA)_{25k}$- and $P(PEGA)_{5k}$-stabilised phytantriol cubosomes were equally the least toxic variants when used in ACHN cells, but not in LNCaP cells. Moreover, ACHN cells showed a better tolerance than LNCaP cells when treated with phytantriol cubosomes. Certainly, the contribution of complex cell membrane variations towards the cubosomes' cytotoxicity levels were also observed in the haemolytic assays, where P(PEGA)- and F127-stabilised phytantriol dispersions were far more toxic to erythrocytes than to the epithelial LNCaP and ACHN cells. These results indicate that the differences in the complexity of the cellular membrane compositions displayed by the different cell types affect how nanoparticles interact with the cell membrane, and therefore the inherent cytotoxicity.[50]

5. Conclusion

In the current study, RAFT-synthesised, P(PEGA) amphiphilic brush polymers were characterised as alternative stabilisers for cubosomes compared to the commonly used Pluronic block copolymer series. Not only did P(PEGA) brush polymers provide effective colloidal stabilisation, these polymers demonstrated a marked improvement in cellular tolerance towards phytantriol-based cubosomes. Certainly, this study highlights the potential attractiveness of the use of these non-commercial P(PEGA) polymers as stabilisers for generating new LCNP drug delivery systems, and warrants additional studies with a focus on further improving the architecture of custom-designed amphiphilic polymers that could provide even more effective stabilisation. In addition, functionalisation of the custom-designed brush polymer end groups with active targeting agents to promote selective targeting and further improve cellular tolerance is also attractive and awaits future investigations.

Acknowledgements

This research was undertaken, in part, on the SAXS/WAXS beamline at the Australian Synchrotron, Victoria, Australia. Jiali Zhai and Randy Suryadinata were supported by a CSIRO OCE Postdoctoral Fellowship. Bao Luan is supported by a CSC scholarship and CAS-CSIRO Collaborative Grant. Nhiem Tran was supported by a SIEF John Stocker Postdoctoral Fellowship. Jiali Zhai is currently a research fellow at RMIT University. Nhiem Tran is currently a Vice-Chancellor research fellow at RMIT University. The authors thank Lynne. J. Waddington for her support on the cryo-TEM experiment, Charlotte E. Conn and Nigel M. Kirby for their support on the SAXS/WAXS beamline, and Xavier Mulet for his useful feedback on this research.

References

1 A. Angelova, B. Angelov, M. Drechsler, V. M. Garamus and S. Lesieur, *Int. J. Pharm.*, 2013, **454**, 625–632.
2 N. Bye, O. E. Hutt, T. M. Hinton, D. P. Acharya, L. J. Waddington, B. A. Moffat, D. K. Wright, H. X. Wang, X. Mulet and B. W. Muir, *Langmuir*, 2014, **30**, 8898–8906.
3 Y.-D. Dong, I. Larson, T. J. Barnes, C. A. Prestidge and B. J. Boyd, *ACS Appl. Mater. Interfaces*, 2011, **3**, 1771–1780.
4 X. Mulet, B. J. Boyd and C. J. Drummond, *J. Colloid Interface Sci.*, 2013, **393**, 1–20.
5 Y. Namiki, T. Fuchigami, N. Tada, R. Kawamura, S. Matsunuma, Y. Kitamoto and M. Nakagawa, *Acc. Chem. Res.*, 2011, **44**, 1080–1093.
6 J. Zhai, J. A. Scoble, N. Li, G. Lovrecz, L. J. Waddington, N. Tran, B. W. Muir, G. Coia, N. Kirby, C. J. Drummond and X. Mulet, *Nanoscale*, 2015, **7**, 2905–2913.
7 B. Angelov, A. Angelova, B. Papahadjopoulos-Sternberg, S. V. Hoffmann, V. Nicolas and S. Lesieur, *J. Phys. Chem. B*, 2012, **116**, 7676–7686.
8 B. W. Muir, D. P. Acharya, D. F. Kennedy, X. Mulet, R. A. Evans, S. M. Pereira, K. L. Wark, B. J. Boyd, T.-H. Nguyen, T. M. Hinton, L. J. Waddington, N. Kirby, D. K. Wright, H. X. Wang, G. F. Egan and B. A. Moffat, *Biomaterials*, 2012, **33**, 2723–2733.
9 S. B. Rizwan, W. T. McBurney, K. Young, T. Hanley, B. J. Boyd, T. Rades and S. Hook, *J. Controlled Release*, 2013, **165**, 16–21.
10 N. K. Swarnakar, K. Thanki and S. Jain, *Pharm. Res.*, 2013, **31**, 1219–1238.
11 Z. W. Yang, M. W. Chen, M. H. Yang, J. Chen, W. J. Fang and P. Xu, *Int. J. Nanomed.*, 2014, **9**, 327–336.
12 X. Mulet, D. F. Kennedy, C. E. Conn, A. Hawley and C. J. Drummond, *Int. J. Pharm.*, 2010, **395**, 290–297.

13 D. Demurtas, P. Guichard, I. Martiel, R. Mezzenga, C. Hebert and L. Sagalowicz, *Nat. Commun.*, 2015, **6**, 8915, DOI: 10.1038/ncomms9915.
14 A. Yaghmur, B. Sartori and M. Rappolt, *Langmuir*, 2012, **28**, 10105–10119.
15 J. Barauskas, M. Johnsson, F. Joabsson and F. Tiberg, *Langmuir*, 2005, **21**, 2569–2577.
16 C. Fong, T. Le and C. J. Drummond, *Chem. Soc. Rev.*, 2012, **41**, 1297–1322.
17 S. Murgia, A. M. Falchi, V. Meli, K. Schillén, V. Lippolis, M. Monduzzi, A. Rosa, J. Schmidt, Y. Talmon, R. Bizzarri and C. Caltagirone, *Colloids Surf., B*, 2015, **129**, 87–94.
18 N. Tran, X. Mulet, A. M. Hawley, C. E. Conn, J. Zhai, L. J. Waddington and C. J. Drummond, *Nano Lett.*, 2015, **15**, 4229–4233.
19 A. I. I. Tyler, H. M. G. Barriga, E. S. Parsons, N. L. C. McCarthy, O. Ces, R. V. Law, J. M. Seddon and N. J. Brooks, *Soft Matter*, 2015, **11**, 3279–3286.
20 J. Y. T. Chong, X. Mulet, B. J. Boyd and C. J. Drummond, "Chapter 5 -Steric Stabilisers for Cubic Phase Lyotropic Liquid Crystal Nanodispersions (Cubosomes)", in *Advances in Planar Lipid Bilayers and Liposomes*, ed. C. V. K. Aleš Iglič and M. Rappolt, Academic Press, 2015, vol. 21, pp. 131–187.
21 J. Y. T. Chong, X. Mulet, L. J. Waddington, B. J. Boyd and C. J. Drummond, *Soft Matter*, 2011, **7**, 4768–4777.
22 Y.-D. Dong, I. Larson, T. Hanley and B. J. Boyd, *Langmuir*, 2006, **22**, 9512–9518.
23 J. Gustafsson, H. Ljusberg-Wahren, M. Almgren and K. Larsson, *Langmuir*, 1997, **13**, 6964–6971.
24 J. Barauskas, A. Misiunas, T. Gunnarsson, F. Tiberg and M. Johnsson, *Langmuir*, 2006, **22**, 6328–6334.
25 J. Y. T. Chong, X. Mulet, L. J. Waddington, B. J. Boyd and C. J. Drummond, *Langmuir*, 2012, **28**, 9223–9232.
26 M. Uyama, M. Nakano, J. Yamashita and T. Handa, *Langmuir*, 2009, **25**, 4336–4338.
27 C. Nilsson, J. Østergaard, S. W. Larsen, C. Larsen, A. Urtti and A. Yaghmur, *Langmuir*, 2014, **30**, 6398–6407.
28 J. Zhai, T. M. Hinton, L. J. Waddington, C. Fong, N. Tran, X. Mulet, C. J. Drummond and B. W. Muir, *Langmuir*, 2015, **31**, 10871–10880.
29 A. Angelova, M. Ollivon, A. Campitelli and C. Bourgaux, *Langmuir*, 2003, **19**, 6928–6935.
30 J. Zhai, L. Waddington, T. J. Wooster, M. I. Aguilar and B. J. Boyd, *Langmuir*, 2011, **27**, 14757–14766.
31 D. Vllasaliu, R. Fowler and S. Stolnik, *Expert Opin. Drug Delivery*, 2014, **11**, 139–154.
32 A. Rauscher, M. Frindel, C. Maurel, M. Maillasson, P. Le Saëc, H. Rajerison, J. F. Gestin, J. Barbet, A. Faivre-Chauvet and M. Mougin-Degraef, *Nucl. Med. Biol.*, 2014, **41**, e66–e74.
33 S. Murgia, A. M. Falchi, M. Mano, S. Lampis, R. Angius, A. M. Carnerup, J. Schmidt, G. Diaz, M. Giacca, Y. Talmon and M. Monduzzi, *J. Phys. Chem. B*, 2010, **114**, 3518–3525.
34 J. Barauskas, C. Cervin, M. Jankunec, M. Špandyreva, K. Ribokaitė, F. Tiberg and M. Johnsson, *Int. J. Pharm.*, 2010, **391**, 284–291.
35 T. M. Hinton, F. Grusche, D. Acharya, R. Shukla, V. Bansal, L. J. Waddington, P. Monaghan and B. W. Muir, *Toxicol. Res.*, 2014, **3**, 11–22.

36 N. Tran, X. Mulet, A. M. Hawley, T. M. Hinton, S. T. Mudie, B. W. Muir, E. C. Giakoumatos, L. J. Waddington, N. M. Kirby and C. J. Drummond, *RSC Adv.*, 2015, **5**, 26785–26795.
37 J. C. Bode, J. Kuntsche, S. S. Funari and H. Bunjes, *Int. J. Pharm.*, 2013, **448**, 87–95.
38 A. Tilley, Y. D. Dong, H. Amenitsch, M. Rappolt and B. J. Boyd, *Phys. Chem. Chem. Phys.*, 2011, **13**, 3026–3032.
39 A. M. Falchi, A. Rosa, A. Atzeri, A. Incani, S. Lampis, V. Meli, C. Caltagirone and S. Murgia, *Toxicol. Res.*, 2015, **4**, 1025–1036.
40 A. Rosa, S. Murgia, D. Putzu, V. Meli and A. M. Falchi, *Chem. Phys. Lipids*, 2015, **191**, 96–105.
41 J. Y. Chong, X. Mulet, A. Postma, D. J. Keddie, L. J. Waddington, B. J. Boyd and C. J. Drummond, *Soft Matter*, 2014, **10**, 6666–6676.
42 G. Moad, E. Rizzardo and S. H. Thang, *Polymer*, 2008, **49**, 1079–1131.
43 X. Mulet, C. E. Conn, C. Fong, D. F. Kennedy, M. J. Moghaddam and C. J. Drummond, *Acc. Chem. Res.*, 2013, **46**, 1497–1505.
44 J. M. Seddon, A. M. Squires, C. E. Conn, O. Ces, A. J. Heron, X. Mulet, G. C. Shearman and R. H. Templer, *Philos. Trans. R. Soc., A*, 2006, **364**, 2635–2655.
45 W. C. Griffin, *J. Soc. Cosmet. Chem.*, 1954, **5**, 249.
46 J. Y. T. Chong, X. Mulet, D. J. Keddie, L. Waddington, S. T. Mudie, B. J. Boyd and C. J. Drummond, *Langmuir*, 2015, **31**, 2615–2629.
47 J. Y. T. Chong, D. J. Keddie, A. Postma, X. Mulet, B. J. Boyd and C. J. Drummond, *Colloids Surf., A*, 2015, **470**, 60–69.
48 T. Landh, *J. Phys. Chem.*, 1994, **98**, 8453–8467.
49 L. Sagalowicz, R. Mezzenga and M. E. Leser, *Curr. Opin. Colloid Interface Sci.*, 2006, **11**, 224–229.
50 Q. Mu, G. Jiang, L. Chen, H. Zhou, D. Fourches, A. Tropsha and B. Yan, *Chem. Rev.*, 2014, **114**, 7740–7781.

Faraday Discussions

DISCUSSIONS

Applications: general discussion

Alberto Striolo, Francois Sicard, Luis Liz-Marzán, Catherine Murphy, Anna Roig, Axel Mueller, Javier Reguera, Yu Zhou, Mathias Brust, Leonardo Scarabelli, Luciano Tadiello, Antoine Thill, Irene Yarovsky, Martin Mayer, M. Arturo López-Quintela, Christian Kuttner, Estefania Gonzalez Solveyra, Heiko Wolf, Euan Kay, Lucia Pasquato, David Buceta, David Portehault, Hedi Mattoussi, Guillermo González, Roland Faller, David French, Benjamin Abécassis, Molly Stevens, Younan Xia, Richard Jones, Marcin Grzelczak, Matthew Penna and Calum Drummond

DOI: 10.1039/c6fd90051h

Molly Stevens opened a general discussion of the paper by Irene Yarovsky: You mention in this work that you use an all atom simulation instead of coarse grained whilst also mentioning that your timescales are too short to see how the hydrophobin comes off again. Can you comment on whether you might benefit from using a combination of coarse grained models too?

Irene Yarovsky responded: Coarse graining would give access to both longer timescales and length scales and may enable investigation into multiple protein systems where the interactions between multiple proteins could also be examined.[1] This would be extremely beneficial. The challenge when implementing coarse grained methods is identifying those degrees of freedom within the system which can be sacrificed without leading to results which are quantitatively/qualitatively different from the atomistic counterpart (making the common assumption that the atomistic simulation more closely approximates the physical reality). With respect to the work presented here, it is questionable whether a coarse grained description of the short functional chains would be sufficient to observe the loss of chain freedom upon adsorption, which aids in elucidating the design principles. A more general example might be the resolution with which water is described within the model. The MARTINI model, a widely utilized coarse grained model for lipids and proteins, utilizes a model which beads 4 water molecules together.[2] Hydration theory suggests a structuring of water molecules at the interface presenting a physical barrier for adsorption,[3] while the kosmotrope hypothesis suggests thermodynamically favourable water/chain complexes preventing adsorption.[4] It is unlikely that a CG model of water would capture these molecular level details. One avenue which may avoid these difficulties is combined atomistic and coarse grained simulations, see for example ref. 5. On the fly, resolution adaptation enabling transitioning between defined regions in the systems depending on their importance for the adsorption process may provide an avenue to access longer timescales by sacrificing the resolution required for antifouling mechanisms or theories to be tested.

1 M. Baaden and S. Marrink, *Curr. Opin. Struct. Biol.*, 2013, **23**, 878–886.
2 S. Marrink and D. Tieleman, *Chem. Soc. Rev.*, 2013, **42**, 6801–6822.
3 J. Zheng, L. Li, H-K. Tsao, Y-J. Sheng, S. Chen and S. Jiang, *Biophys J.*, 2005, **89**, 158–166.
4 R. Kane and G Whitesides, *Langmuir*, 2003, **19**, 2388–2391.
5 M. Praprotnik, L. Delle Site and K. Kremer, *Annu. Rev. Phys. Chem.*, 2008, **59**, 545–571.

Roland Faller asked: You focus on hydrophobin, can you speak about other proteins that might be relevant and how your results might change?

Irene Yarovsky answered: The choice of protein for this work was dictated by the potential future industrial application of the design conceptualized and presented in the paper. A wider field relevant to this is the biofouling of medical devices, implants or drug delivery systems. For example, there are many different proteins present in the biomolecular coronas formed on nanoparticles[1,2] which could be of interest to investigate in this context. The diversity of proteins makes prediction of how the results would vary in a single case impossible. The design considerations presented in the work are not framed with reference to the protein and we are specifically trying to avoid such references due to the aforementioned diversity. What we could say is that depending on the protein, and specifically the exposed surfaces of the protein in the course of adsorption/unfolding, the relative importance of hydration (competition for the surface protective ligand contact between the protein and water molecules) and the entropic sweeping action of the ligands may vary. Therefore, our only 'prediction' is that the goal of any anti-fouling surface functionalisation design should be to maximize both these effects to decrease the likelihood of protein adsorption.

1 M. Lundqvist, J. Stigler, G. Elia, I. Lynch, T Cedervall and K. Dawson, *Proc. Natl. Acad. Sci. U. S. A.*, 2008, **105**, 14265–14270.
2 S. Schlotter, G. Becker, S. Winzen, T. Steinbeck, K. Mohr, K. Landfester, V. Mailander, F. R. Wurn, *Nat. Nanotechnol.*, 2016, **11**, 372–377.

M. Arturo López-Quintela asked: Can you comment on the influence of roughness on your system?

Irene Yarovsky answered: The roughness in the systems considered here is two-fold. First, the underlying amorphous silica is atomically rough. Second, as the lengths of the functional chains are varied in some systems they introduce a nanoscale roughness to the surface. We observe that the disparity in height of the chains, which could be deemed a nanoscale roughness, is critical for the enhanced sweeping action which we observed to improve the protein adsorption resistance. Roughness on larger length scales that has been reported to have antibacterial properties (*e.g.*, ref. 1), is beyond our model scale and is, therefore, not considered here. However, we believe that incorporating roughness on multiple length scales, as is observed in nature,[2] may be critical for the design of antifouling systems. Multiscale models, including the all-atom representation (as employed here) and coarse-grained representation (where groups of atoms are modeled as interaction sites), are required to incorporate the multiple length scale roughness and these approaches are considered in our current and future work.

1 E. Ivanova, J. Hasan, H. Webb, V. Truong, G. Watson, J. Watson, V. Baulin, S. Pogodin, J. Wang, M. Tobin, C. Lobbe and R. Crawford, *Small*, 2012, **8**, 2489–2494.
2 T. Sun, L. Feng, X. Gao and L. Jiang, *Acc. Chem. Res.*, 2005, **38**, 644–652.

Molly Stevens asked: What about the underlying substrate roughness, does this have an influence?

Irene Yarovsky responded: The underlying roughness and rigidity of the silica substrate has been shown to enhance the antifouling capacity of PEG chains.[1] The nanoscale roughness is likely to increase chain dynamics due to height fluctuations in the grafting point which will increase the entropic penalty associated with adsorption.

1 G. Yiapanis, S. Maclaughlin, E. Evans and I. Yarovsky, *Langmuir*, 2014, **30**, 10617–10625.

Molly Stevens commented: You focus on an industrial rather than biomedical applications. It is worth noting that in terms of biomedical applications, for example if you think of hip implants, then being able to replicate the roughness would be very useful for such applications.

Irene Yarovsky answered: Thank you for this comment, we agree entirely that the conceptual design developed here would be useful for biomedical applications where spontaneous and uncontrolled adsorption of proteins on surfaces and nanoparticles is undesirable, including in tissue engineering, biosensing and drug delivery (where protein coronas can interfere with the intended function). The challenge is, however, for synthetic chemists and materials engineers to implement the multiscale roughness and the interface dynamics.

Molly Stevens asked: I don't think there are models for this biomedical application, it could be interesting to see if this could work?

Irene Yarovsky responded: One of the overarching goals of this and subsequent work is to elucidate general design principles which can be translated and implemented in a variety circumstances. Biomedical applications are an area of immense interest and we are currently in the process of testing our design hypothesis against a range of plasma proteins.

Alberto Striolo said: We have been interested in protein adsorption for some time, but only recently we have started to use atomistic simulations for these purposes. In the case of interest to us we are actually focused on conformational changes on the protein surface that is not adsorbed on to a surface. In the case of the system you discuss in your paper, have you checked how much the conformation of the protein changes upon adsorption, but in the region far from the surface? Do you think 200–300 ns simulations are sufficiently long for assessing reliably the conformational changes I am referring to?

Irene Yarovsky responded: While the complete characterization of changes in protein conformation upon adsorption was not undertaken in this work we certainly were able to observe some conformational changes away from the interface at the solvent exposed protein surface. In our detailed analyses we focused on the behavior of two flexible loops on the surface of the protein, which are known to be inherently disordered and implicated in the adsorption process. At the same time, changes in the beta core were minimal due to the high degree of stability of this region owing partly to four disulfide bonds. However, in a general sense, it would certainly be possible to use molecular simulations to identify conformational changes within the protein surface away from the solid substrate it adsorbs to, but the efficiency of such characterization will depend on several factors, including the size of the protein and the degree of unfolding it experiences upon adsorption. For example, Wei *et al.*[1] have highlighted the need for long simulations (greater than those used here) to observe such conformational changes. We would suggest that a large number of independent spontaneous 'brute force' MD simulations or multiple runs employing advanced sampling techniques, such as replica exchange methods (REMD, REST), would be required to obtain an adequate picture of conformational changes in the protein due to adsorption. Sampling is critical as there is unlikely to be a single adsorbed conformation and the distribution of adsorbed states can be separated by large free energy barriers, for discussion on this point see Latour *et al.*,[2] for example. Furthermore, whether or not this sampling would reflect the true distribution of adsorbed states of the protein at the interface would be difficult to verify due to the inability of current experimental techniques to provide detailed characterization of protein structures at interfaces. Another factor might be the accuracy of force fields describing the interfacial behavior of proteins. Typically, most current biomolecular force fields are parameterized to reproduce the native structures of proteins and it is yet to be determined whether or not they are capable of adequately describing the behavior at interfaces.

1 T. Wei, A. Carignano and I. Szleifer, *Langmuir*, 2011, **27**, 12074–12081.
2 R. A. Latour, *Colloids Surf. B: Biointerfaces*, 2014, **124**, 25–37.

Roland Faller commented: What is the distribution of silanols in your system? I am not so much interested in the density but actually the distribution? Did you generate a number of different distributions?

Irene Yarovsky responded: The SiO_2 substrate used here was a slab of depth greater than 2 nm with two faces, such that two surface silanol distributions on the surface were generated with the density of both being approximately 4.7 silanols nm^{-2}. A detailed description of the substrate generation can be found in the work of Garofalini and coworkers[1-4] and Yarovsky and coworkers.[5–10] Other models of amorphous silica surfaces have also been reported, for example Cruz-Chu,[11] Hassanali,[12–13] Emami,[14] to list a few, and Rimola[15] published a detailed review of silica surfaces from both an experimental and computational perspective.

While a detailed comparison of the hydroxyl distribution was beyond the scope of what we were interested in here, in response to this question, we have generated the nearest neighbour distribution (Fig. 1) which shows that on length scales relevant to individual protein adsorption the distribution of silanols on the underlying silica substrate is homogeneous while it is certainly not uniform.

1 S. Garofalini, *J. Chem. Phys.*, 1983, **78**, 2069–2072.
2 S. Levine and S. Garofalini, *Surf. Sci.*, 1985, **163**, 59–66.
3 S. Garofalini, *J. Non-Cryst. Solids*, 1990, **120**, 1–12.
4 D. Athanasopoulos and S. Garofalini, *J. Chem. Phys.*, 1992, 3775–3780.
5 I. Yarovsky, M. Aguilar and M. Hearn, *Anal. Chem.*, 1995, **67**, 2145–2153.
6 I. Yarovsky, M. Aguilar and M. Hearn, *J. Phys. Chem.*, 1997, **101**, 10962–10970.
7 G. Yiapanis, D. J. Henry, S. MacLaughlin, E. J. Evans and I. Yarovsky, *Langmuir*, 2012, **28**, 17263–17272.
8 G. Yiapanis, S. MacLaughlin, E. Evans and I. Yarovsky, *Langmuir*, 2014, **30**, 10617–10625.
9 K. J. Ley, A. Christofferson, M. Penna, S. MacLaughlin, D. Winkler, and I. Yarovsky, *Frontiers Mol. Biosci.*, 2015, **2**, 64.
10 K. Ley, L. Shaw, G. Yiapanis, S. MacLaughlin and I. Yarovsky, *Mol. Simul.*, 2016, **42**, 563–572.
11 E. Cruz-Chu, A. Aksimentiev and K. Schulten, *J. Phys. Chem. B*, 2006, **110**, 21497–21508.
12 A. Hassanali and S. Singer, *J. Phys. Chem.*, 2007, **111**, 11181-11193.
13 A. Hassanali, H. Zhang, C. Knight, Y-K. Shin and S. Singer, *J. Chem. Theory Comput.*, 2010, **6**, 3456–3471.
14 F. Emami, V. Puddu, R. Berry, V. Varshney, S. Patwardhan, C. Perry and H. Heinz, *Chem. Mater.*, 2014, **26**, 2647–2658.
15 A. Rimola, D. Costa, M. Sodupe, J-F. Lambert, P. Ugliengo, *Chem. Rev.*, **113**, 4216–4313.

Roland Faller commented: I actually do not agree with the statement that the distribution is not relevant and also based on your visualizations it appears that there is actually clustering and not a homogeneous distribution.

Irene Yarovsky responded: We agree that the distribution of hydroxyls on the surface is extremely important for (at least) two reasons. First, in our own previous work it has been shown that hydrophobic protein residues can interact with the hydrophobic patches on the silica surface (Ley, 2015). Second, specifically pertinent to design of the SiO2 surface functionalization patterns, the distribution of silanols on the surface would determine the distribution of ligands grafted to the surface. However, in the current study ligands were specifically grafted in layers (1x1 or 2x2 patterns) to test the hypothesis that mismatching hydrophobic/hydrophilic domains on the surfaces of substrate and protein could inhibit protein adsorption. It was concluded that the more randomly packed 1x1 layering was more effective at limiting adsorption due to improved sweeping action and

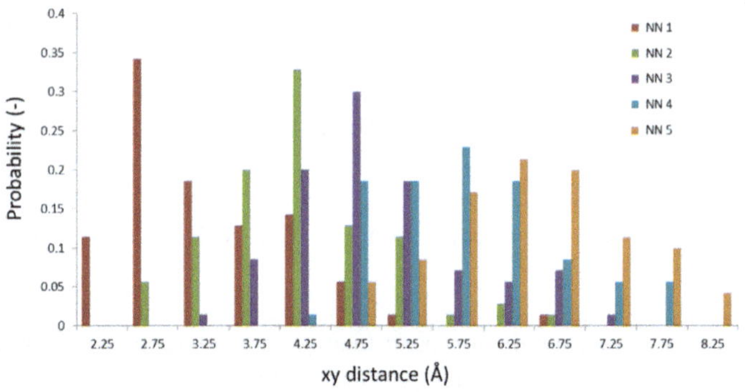

Fig. 1 Nearest neighbour distribution.

lower ligand aggregation propensity (which is indeed identifiable in the 2x2 surface pattern to which you are likely referring to in this question). This suggests that so long as the ligands are grafted in an approximate one for one distribution with a medium density ligand spacing across the surface (a challenge for experimentalists to consider) this should be sufficient to achieve the requisite design goal. One limit to the success of the functionalization will be if there exist hydrophobic patches on the surface large enough (approximately >1 nm2) for chains on the edges of these regions to collapse onto the surface thus losing their responsive ability. In fact, we have recently demonstrated such effects in a separate study (Ley, 2016) and, therefore, we believe, that the goal of designing anti-fouling SiO2 surfaces/nanoparticles would be to avoid such situations.

1 K. J. Ley, A. Christofferson, M. Penna, S. MacLaughlin, D. Winkler, and I. Yarovsky, *Frontiers Mol. Biosci.*, 2015, **2**, 64.
2 K. Ley, L. Shaw, G. Yiapanis, S. MacLaughlin and I. Yarovsky, *Mol. Simul.*, 2016, **42**, 563–572.

Francois Sicard enquired: I am curious to know if you are planning to run some biased numerical simulations to reconstruct and explore the free energy landscape of your system (umbrella sampling, metadynamics, …). This might allow you to improve new or favorable paths related to a specific residue in your protein when it interacts with the surface. Considering metadynamics simulation, I would suggest the use of the number of hydrogen bonds (or any pertinent coordination number) in addition to the distance from the surface as collective variables. You might also try to run some ABMD simulations with these 2 variables to figure out if they are good candidates to define a minimal free-energy path in a first instance.

Irene Yarovsky responded: Yes, we are currently looking into metadynamics as a means of more thoroughly exploring the free energy landscape of these systems. Indeed, we are considering the distance from the surface as the first collective variable. Identification of additional collective variables is more complicated. The number of hydrogen bonds may provide an estimate of hydrophilic/electrostatic attraction but we might be concerned that we are neglecting any hydrophobic driving forces. Perhaps looking at the combined hydration shell of both the protein and the surface or the hydrophobic/hydrophilic contact surface area, accounting for both hydrogen bonding and any hydrophobic association. In reality we would prefer variables which are largely independent of the nature of both the surface and the protein, but beyond the separation distance we are yet undecided on what these might be. A consideration for us is also the ability to identify variables which can be compared with data produced by our experimental collaborators.

Molly Stevens said: Have you combined this work with any experimental measurements?

Irene Yarovsky answered: Our experimental collaborators are in a process of manufacturing and testing prototype systems based on the proposed design considerations. We believe there exists a system with similar interfacial characteristics to those proposed here which outperforms the majority of commonly employed antifouling surfaces.

Heiko Wolf opened a general discussion of the paper by Richard Jones: Looking at the drum-like structures, a thing that comes to mind are connected 'drums' forming a tubular structure. Did you observe any tubes when preparing the drums? In the drum-like structures (Fig. 1 in your paper) all alkyl chains of the ligands point in the same direction linking the two rings of the drums. Is there a way to prepare modified ligands that point their alkyl chain preferably in the other direction to connect to a further ring? A mixture of ligands might then induce the formation of tubes. Or can the drums host any guest molecules in their ring which then might be able to link two or more drums?

Richard Jones answered: Thank you for your interesting questions and the associated suggestions. We have not observed tubular structures yet when preparing the nano-drums; however, by modification of functional groups on some of our Schiff-base ligands and using different counter ions we have been able to generate double-drum structures.[1] The possibility of using mixed ligands and/or counter ions in reactions to generate alternative architectures such as tubular structures is intriguing and we may be able to observe such structures using EM techniques similar to what we show in Fig. 1b and c in our paper. We are exploring the use of small guest molecules in our reaction conditions as a template for self-assembly of drum-architectures. If successful we could, for example, consider linking guest molecules to favor assembly of larger structures such as stacked drums or even tubes.

1 X. Yang, D. Schipper, L. Zhang, K. Yang, S. Huang, J. Jiang, C. Su and R. A. Jones, Anion dependent self-assembly of 56-metal Cd–Ln nanoclusters with enhanced near-infrared luminescence properties, *Nanoscale*, 2014, **6**(18), 10569–10573.

David Portehault remarked: These clusters exhibit an empty site. What is the precise diameter of this inner space, and can you store gas into such pores, potentially under pressure?

Richard Jones responded: The diameters of the inner spaces of the nano-drum will vary from compound to compound since they contain different lanthanides and may have Schiff-base ligands of different dimensions. The closest H–H intramolecular distances across the inner spaces of these molecules vary from ~3.5 to 4 Å. As found in many of our crystal structures, the inner space is large enough to accommodate small disordered solvent molecules and may therefore be of sufficient size to accommodate small molecular gases. The suggestion to study gas storage under pressure is interesting and our crystals should be amenable to diffraction studies using pressure cells at synchrotron sources. The ability to store specific gases may also be dictated by the chemical environment of the empty site and could vary depending upon the nature of the lanthanides and other components present in our nano-drums.

Mathias Brust asked: Some of your structures look like twisted barrels. I was wondering if they were chiral?

Richard Jones answered: You are correct in that the ligands in the nano-drums do assemble with a twist but in fact, all the nanoclusters discussed here crystallize in enantiomorphic space groups and are not chiral.

M. Arturo López-Quintela asked: Can you comment on the luminescence properties of your materials in comparison to other materials, like quantum dots or rare earth-doped silica nanoparticles?

Richard Jones answered: A key difference between our nanoclusters and quantum dots is that our compounds exhibit no 'blinking' – an intermittent fluorescence leading to a dark state that limits the utility of quantum dots for studying time-resolved processes. Although rare-earth doped nanoparticles do not blink, like quantum dots their luminescent properties are highly dependent upon the size, size distribution and particle homogeneity that can affect energy transfer processes and quenching. In comparison, our nanoclusters are perfectly homogeneous and obtained as pure molecular species by crystallization from solvent. The luminescence properties are dependent upon the nature of the lanthanide and associated ligands, but not complicated by inhomogeneity issues or concentration distributions within a host matrix which occur with quantum dots or rare-earth doped nanoparticles.

Catherine Murphy asked: What are the coordination numbers around your lanthanides? Is solvent bound to them?

Richard Jones answered: The X-ray structures of **1–3** reveal 8 coordinate lanthanides in all cases. Although non-coordinating disordered solvent is found in the structures, the Ln(III) ions are not directly coordinated to solvent.

Catherine Murphy asked: I know your lanthanides are paramagnetic, so this might be difficult: but have you done NMR experiments on your particles? Even with lanthanide shifts you might be able to see ligand resonances, and perhaps even tell what part of the ligand is closer to the lanthanide. Also, if you do DOSY (diffusion-oriented spectroscopy) NMR measurements, you could pull out diffusion coefficients and hence the size of object in solution.

Richard Jones replied: Thank you for these questions. You are correct in that interpretation of NMR spectra when lanthanides are present can be extremely challenging. We have, however, done some comparative studies in D_2O of ligands in solution and with the corresponding lanthanide nanocluster. The ligand resonances do appear to differ but because of the complexity of the spectra it is very difficult to interpret which part of ligand is interacting with the metal. Although we rely upon crystal structures to provide us with information about lanthanide–ligand interactions in our nanoclusters, these interactions may change in solution. In fact, the Bragg peaks observed in our SAXS data do not correspond exactly with the crystal structure and suggest that the structural arrangements of our nanoclusters may be at least subtly altered in solution (depending upon the solvent). Your suggestion for using DOSY NMR is good, as it has been used with other polynuclear lanthanide complexes to study the size, shape and stoichiometries in solution. We will look into the feasibility of using

DOSY NMR with our nanoclusters as this technique would be complementary to our SAXS, DLS and TEM studies.

Benjamin Abécassis commented: About your SAXS data, there are some peaks in the diagram which seem unusual for well dispersed objects in solution. Could you comment on that point?

Richard Jones answered: The peaks present in the SAXS data are Bragg peaks that are likely due to the presence of undissolved material in a crystalline state. We do intend to collect SAXS data again soon of lanthanide clusters in better optimized solvents to minimize or eliminate this artefact.

Luis Liz-Marzán asked: When you think about these compounds as nanoparticles, what happens if you increase the size? Do you expect a shift in the luminescence? Any other size effects?

Richard Jones replied: Answer: We have generated a number of nanoclusters with different structural arrangements including double nano-drums (see for example, ref. 1). The excitation of these larger particles differs from the single nano-drums, not because of the size but because of the nature of the ligands that act as antennas. The emissions are essentially unaffected if the same lanthanide is used. Also, other properties such as solubility and melting point change, again dependent primarily upon the chemical nature of the coordinating ligand groups.

1 X. Yang, D. Schipper, L. Zhang, K. Yang, S. Huang, J. Jiang, C. Su and R. A. Jones, Anion dependent self-assembly of 56-metal Cd–Ln nanoclusters with enhanced near-infrared luminescence properties, *Nanoscale*, 2014, **6**(18), 10569–10573.

Luis Liz-Marzán remarked: Lanthanide compounds are used as nanothermometers, on the basis of the effect of temperature on the emission properties. Can you also use your material for this application?

Richard Jones replied: Thank you for this interesting question. Our understanding is that the lanthanide nanothermometers use an upconversion energy transfer between specific lanthanide pairs (*e.g.*, Tb^{3+} to Eu^{3+}, Nd^{3+} to Yb^{3+}). While our nanoclusters are currently synthesized with a single lanthanide ion, it may be possible to produce nanoclusters containing pairs of lanthanides that may be capable of upconverting. If that is the case then we could certainly explore their potential as nano-thermometers.

Guillermo González said: Lanthanide ions exhibit quite interesting optical properties. Have you tried to introduce rationally two different rare earth ions in your molecules to obtain up-conversion nanoclusters?

Richard Jones replied: Thank you for this suggestion. We understand that effective upconversion depends upon a number of factors including the geometric separation between the sensitizer and the activator. We have a number of nanocluster architectures (*e.g.*, rings, drums, double drums) available to us, each with different intermolecular distances between lanthanide ions. Based upon our crystal structures we can attempt to make rational decisions about which lanthanide pairs would be the best choices for creating upconverting complexes. The geometries of our clusters may also prove particularly beneficial for limiting quenching effects by maintaining Ln^{3+} ions at fixed distances and concentrations. However, identifying the reaction conditions to favor generation of upconverting nanoclusters is more empirical. We know little about the mechanisms involved in the self-assembly of our clusters and therefore anticipate that considerable optimization of metal stoichiometries and other reaction parameters will be required.

Molly Stevens remarked: You have done electron microscopy of your materials, but only regular TEM; did you try and do tomography? I don't understand what is happening in your SEM image, please can you expand on this?

Richard Jones replied: Thank you for these questions. We have not yet undertaken any tomographic studies of our materials, though this could be a very interesting technique to use to complement data obtained using electron microscopy and SAXS. Crystals should be suitable for undertaking synchrotron X-ray tomography studies, particularly as the lanthanides display absorption edges that are within the range of tuneable beamlines used for phase contrast methods. As for the SEM images, they were included simply for the sake of completeness; to answer the question "What do SEM images look like?" As you can see they are not particularly informative and show irregular polycrystalline materials.

Leonardo Scarabelli asked: At the end of your presentation you showed a ring-like structure obtained in the presence of cobalt. In this case the cavity looks slightly larger compared to your nano-drums; do you think it would be possible to fit a small nanoparticle inside the ring? A second question regards the possibility of breaking the perfect symmetry of your structure; this would allow you to differentiate different sides, making them interact with different species or maybe even interlocking two or more structures together?

Richard Jones replied: Thank you for these interesting questions. Just to clarify, the ring-like structure was actually obtained in the presence of dysprosium, not cobalt. You are correct in that the cavity in the ring-like structure is slightly larger than the nano-drums, but still only able to accommodate small solvent molecules. We agree with your suggestion that trapping larger molecules, even nanoparticles, would be very interesting. We have been able to interlock the nano-drums to produce double-drum structures,[1] though these are still essentially symmetrical. We are trying to understand more about the self-assembly process involved in generating our nanoclusters. It may be possible using mixed ligand systems, or by synthetically modifying these clusters after self-assembly, to increase the diversity and functional properties of our nanoclusters.

1 X. Yang, D. Schipper, L. Zhang, K. Yang, S. Huang, J. Jiang, C. Su and R. A. Jones, Anion dependent self-assembly of 56-metal Cd–Ln nanoclusters with enhanced near-infrared luminescence properties, *Nanoscale*, 2014, **6**(18), 10569–10573.

Molly Stevens commented: Thinking about the bio-applications of these materials, can you expand on this? For the bio-applications can you functionalise these with targeting moieties or do you think it would upset the assembly?

Richard Jones responded: Great questions! We are currently exploring the potential of these nanoclusters as both imaging and theranostic agents. We are currently pursuing studies to functionalize different nanoclusters for conjugation to bio-molecules, particularly for time-resolved and single molecule studies. Functionalisation may well disrupt the assemblies and so this is something which we hope to explore in future studies.

Molly Stevens commented: Have you examined the uptake mechanism yet? This could be really interesting!

Richard Jones responded: We agree that understanding how these nanoclusters are taken up by cells is very interesting. We suspect that endocytosis mechanisms may be important but we have no definitive data at this point in time.

David Buceta asked: Do the properties change with the length of the chains and if so, do you know how? Can you can fine-tune the size of your drum in order to explore new applications?

Richard Jones answered: A general observation is that increasing the chain length lowers the overall bulk density of the material. In addition, as shown in Fig. 4 in our paper, the UV-vis spectra of the longer chain C_8 ligand is slightly red-shifted compared to the C_6 ligand. Red-shifting of excitation wavelengths is important in some applications involving biological materials where exposure to UV wavelengths should be minimized if possible. Exploiting new image-based applications will benefit from fine-tuning of the size, photophysical charactertics and self-assembly properties of our lanthanide nanoclusters. Our current data suggest that we may be able to optimize the utility of our nanoclusters in a rational manner by judicious choice of ligand chain length and/or functional groups in our reaction conditions, but clearly more research will be required to do this effectively.

Irene Yarovsky asked: Does the chain length effect the mechanical properties of these crystals? Do they become more 'spongey' with longer chains? If so, can this property be exploited?

Richard Jones replied: This is a very interesting question. However, we have not yet undertaken specific studies of the mechanical properties of the molecular nanoclusters. Comparative mechanical properties of these crystals could be evaluated, for example, using nanonindentation techniques to understand hardness and fracture resistance, or resilience to high pressures using diamond anvil cells.

Molly Stevens said: You could potentially do single molecule force spectroscopy; it isn't easy, it might take a long time but it should be possible.

Richard Jones replied: Sounds like an exciting possibility. Single molecule AFM might be possible with the larger forms of our nanoclusters. We will consider this suggestion in the future.

Molly Stevens commented: Once you have these drum structures, can you cage other things inside, so that you can use them as nanodelivery structures?

Richard Jones answered: That's a great observation/question. At present the internal dimensions of these drums are small and at the moment we may only be able to cage very small molecules such as solvents or gases. We are currently trying to understand more about the self-assembly mechanisms of these clusters and how optimization of those mechanisms may allow assembly around a larger target molecule *in situ*.

Yu Zhou opened a general discussion of the paper by Hedi Mattoussi: What are the advantages of the multipolymer ligand modified nanoparticles compared with only PEG coated nanoparticles?

Hedi Mattoussi answered: A polymer coating provides the flexibility for combining multi-functionality and multi-coordination where more than one particular group can be introduced along the same chain. For a coordinating polymer ligand, where one can introduce several coordinating groups in the same chain, this produces a substantially increased affinity between the ligand and nanoparticle surface, yielding a sterically-stabilized dispersion. In comparison, a mono-dentate ligand (say a thiol-terminated single chain PEG) would produce a higher dynamic desorption rate, yielding weaker affinity and limited colloidal stability of the hydrophilic NPs. The effects of coordination number on the dispersion stability has been documented in the literature.

Yu Zhou asked: Can you comment on the toxicity of your nanoparticle system?

Hedi Mattoussi answered: We saw baseline toxicity (*i.e.*, negligible effects) on cell cultures when PEG-modified polymer coatings are used with QDs and AuNPs.

Yu Zhou enquired: What is the coverage density of ligands on the nanoparticle surface?

Hedi Mattoussi replied: The density of polymer ligands may slightly vary from one type of nanoparticle to another. We have estimated the density of those ligands on the surface of green-emitting QDs, and found that nanocrystals have on average ~14–15 ligands.[1]

1 W. Wang, A. Kapur, X. Ji, M. Safi, G. Palui, V. Palomo, P. E. Dawson and H. J. Mattoussi, Photoligation of an amphiphilic polymer with mixed coordination provides compact and reactive quantum dots, *J. Am. Chem. Soc.*, 2015, **137**, 5438.

Yu Zhou asked: If there is a next step such as functionalisation, how difficult is it to add functionalisation to the surface? You can see the launching of the

–COOH on the side chain between PEG chains; how difficult is it to add large functional groups to this site, for example, antibodies?

Hedi Mattoussi answered: Adding other functionalities to the polymer coating and thus the resulting NPs is quite easy using this synthetic route. It is actually one of the benefits of using reactions starting with maleic anhydride rings. The introduction of bio-orthogonal reactive groups (*e.g.*, amine, azide, alkyne) in the same polymer ligand can be achieved by introducing amine-modified moieties containing those groups during the one step synthesis. Introducing large functionalities such as antibodies and other proteins can be carried out on the NP following phase transfer. The ease of this depends on the nature of the coupling reaction. It should be straightforward to introduce proteins, peptides, dyes or redox complexes on the NP surfaces, *via* sequential or simultaneous coupling. We have shown this on luminescent quantum dots.[1]

1 W. Wang, A. Kapur, X. Ji, M. Safi, G. Palui, V. Palomo, P. E. Dawson and H. J. Mattoussi, Photoligation of an amphiphilic polymer with mixed coordination provides compact and reactive quantum dots, *J. Am. Chem. Soc.*, 2015, **137**, 5438.

Mathias Brust asked: You have an impressive track record in developing ligands for quantum dots and for metal nanoparticles. A safe guiding principle appears to be that more binding groups per ligand molecule give better stability of the final product. Can you comment on the limits of this rationale, and, in particular when using polymeric ligands, on potential side reactions that lead to the undesired cross-linking of particles?

Hedi Mattoussi answered: The benefits of using ligands with higher coordination numbers are numerous. However, one should be mindful of getting 'too greedy' by using larger and larger macromolecules in the quest of continuously increasing the coordination number. This will be detrimental as it will produce excessively large hydrophilic NPs, due to the fact that entropy will make the polymer chain coil and prevent several of the coordinating groups from interacting with the metal surface. This increases the overall hydrodynamic size of the NPs and defeats the original purpose.

Cross-linking is usually avoided during the ligand exchange step, because the process relies on mass action effects and, routinely, ligand concentrations are used in excess (*i.e.*, 100-100 time molar excess). If much smaller ligand concentrations are used, cross coordination can in principle occur, resulting in aggregate formation. We never see it within our experimental conditions.[1,2]

1 W. Wang, A. Kapur, X. Ji, M. Safi, G. Palui, V. Palomo, P. E. Dawson and H. J. Mattoussi, Photoligation of an amphiphilic polymer with mixed coordination provides compact and reactive quantum dots, *J. Am. Chem. Soc.*, 2015, **137**, 5438.
2 W. Wang, F. Aldeek, X. Ji, B. Zeng and H. Mattoussi, A multifunctional amphiphilic polymer as a platform for surface-functionalizing metallic and other inorganic nanostructures, *Faraday Discuss.*, 2014, **175**, 137.

Euan Kay asked: The remarkable stability of these polymer-coated nanoparticles is (understandably) ascribed to the multivalent nature of the ligand–nanoparticle interaction, achieved through combining several thiol side-chains on the one polymer backbone. With this design, is there any benefit in using lipoic acid (a bidentate thiol) as the ligating side chain, *versus* a monodentate thiol side chain? Does the fact that each lipoic acid unit is pre-organized for oxidative formation of a 5-membered dithiolane ring harm stability in your system?

Hedi Mattoussi responded: The chemical rationale used for introducing several lipoic acid coordinating groups can be applied to using mono-thiol coordinating groups. We have focused on lipoic acid anchors because the oxidized version is stable and less susceptible to interfere with the coupling schemes. So, issues of sulfur–sulfur bridging during storage are avoided. The nature of the pre-organized dithiolane five member ring, due presumably to ring closing, does not harm the stability. That problem can occur if the ligand fully desorbs and wanders away from the metal surface for extended periods of time. When introduced into a short polymer such events are reduced due to the multiple point of attachment of the same polymer ligand onto the NP surface.

Molly Stevens asked: Have you looked at more complex protein mixtures?

Richard Jones responded: Not yet, however we are very interested in using our nanoclusters as imaging agents in complex cell environments, as well as using proteins and other macromolecules in molecular interaction studies. A key aim of our work is to functionalize the nanoclusters to enable multi-wavelength imaging, taking advantage of the sharp, characteristic emissions of different lanthanides. Examples of studies could include time-resolved studies of protein localization in a cell, or single molecule studies of protein complex formation.

Catherine Murphy enquired: When you say you have 'no' protein adsorption, what do you really mean? What is the detection limit in your technique?

Hedi Mattoussi responded: We mean that within the experimental conditions that we used (namely using a nanoparticle/nanorod concentration of ~1–100 nm while increasing the BSA molar concentration up to 1000 times excess) we do not measure any changes in the NP or NR electrophoretic mobility shift. That would correspond to a protein concentration reaching ~0.01–1 mM. With other less effective coatings one can easily measure changes in the mobility at such equivalent molar concentrations. So, in terms of detection limit for such problem, it is rather how high of a protein concentration one can explore without corona build up; corona formation must be correlated to what is expected in biological media where mM protein concentrations are common. We believe that our coating is able to prevent corona formation at high concentrations, *e.g.*, 1–100 mM.

Molly Stevens commented: In your system, did you look at just the 9 nm nanoparticles, or did you look at other smaller particles? How much does surface curvature affect the results?

Hedi Mattoussi responded: For spherical nanoparticles we only tested the 9-nm size. However, we have tested AuNRs having different aspect ratios coated with these polymers. We did not measure any protein corona build up on either nanostructured platform. In fact we do not anticipate any effects associated with surface curvature on the protein adsorption in the presence of this coating. The polymer ligand should provide effective surface functionalization to any nanostructure, as long as the metal-promoted coordination on the polymer is strong enough.

Molly Stevens added: If curvature is not that key, you could look at planar surfaces and use SPR to probe binding to the surface.

Irene Yarovsky asked: Following on from the discussion of a flat surface, how transferrable is the technology to flat surfaces, in principle, could it be applied at industrial scale processing?

Hedi Mattoussi answered: The ligands would also work for flat surfaces. Application to industrial-scale processing is only limited by the ability to synthesize the ligands in large amounts. In principle, the procedures developed here can be adapted to larger industrial scales.

David Buceta opened a general discussion of the paper by Mathias Brust: Can you explain the mechanism of potassium transport with the gold functionalised nanoparticle; is there is an interchannel or maybe the ligands are transporting the potassium?

Mathias Brust answered: I would tend towards an explanation by a random tumbling or rotational motion that shuttles ions across. Channel formation would require quite a high level of self organisation no matter how many particles are involved. In the absence of any experimental evidence I find it hard to speculate further about this. We should carry out experiments with different crown ligand loadings and perhaps support this by modeling. This may yield some hints on the transport mechanism.

Javier Reguera asked: Has the density of ligands been characterized? This could be indeed an important parameter as it could change the packing of the crown ethers on the surface of the nanoparticle.

The short linker used with this ligand has to force somehow the crown ethers to have a certain organization (for example parallel packing) and would be very interesting to study. From this point of view the density should affect the packing and also the capture of potassium. Then, the change in hydrophobicity, upon interacting with the membrane, could be also due to this rearrangement of the ligands.

Mathias Brust answered: I agree that this is an important point that we have not been able to address yet. The particles are also moderately polydisperse, and it may be the case that only those of certain sizes are active as potassium ion membrane transporters. I personally have difficulty imagining that regular channels can form on the curved surface by the stacking of crown ether moieties, but this may indeed be the case and would be quite an important finding if confirmed.

Lucia Pasquato asked: This question concerns the mechanism of ion transport through the lipid bilayer. Is it possible to argue that a channel is formed by the cooperation between the crown ethers which belong to two different NPs? Of course this may be more easy for NPs of 2 nm protected by a homoligand monolayer of 18-C-6-CH_2-SH and will depend on the concentration of NPs used.

On the other hand for NPs of 15 and 38 nm it will be important to determine the amount of 18-C-6-CH2-SH present in the mixed-monolayer. In any case is the crown ether, in the presence of PEG5000 exposed to the solvent in order to act with the same mechanism as for NPs of 2 nm?

Mathias Brust replied: These are all very valid points. Unfortunately, I cannot give any answers based on experimental observations. The transport mechanism is not clear, and a number of hypotheses can be formulated immediately. My feeling tells me that a shuttle motion may be more likely, since it would not require such an amount of self-organisation of ligands or, as you suggest, even particles with respect to each other. A way forward would be to vary carefully the number of crown moieties on each particle and see how this effects ion transport. If self-organisation of channels is needed, I would expect that below a certain loading with crown ether, transport would no longer be possible.

Lucia Pasquato asked: Considering NPs of 15 or 38 nm in which both PEG5000 and 18-C-6-CH_2-SH are present – do you know the relative amount of the two ligands? Is there any possibility that PEG chains are also taking part in the potassium ion transfer through the lipid bilayer? Are NPs protected by PEG5000 only active in the transport of K ions?

Mathias Brust replied: We do not know the relative amounts of the two PEG species on the particle surface. Particles stabilized with PEG alone, *i.e.* without crown ethers, do not act as potassium transporters across the membrane. We have checked this many times.

Euan Kay commented: In the phase transfer experiments, have you attempted to reverse the process by adding a species that would sequester the K^+ from the nanoparticle monolayer – *e.g.* adding excess unfunctionalized 18-C-6?

Mathias Brust responded: We did some preliminary experiments along this line, I think with Ba ions instead of K and adding sulphate. So far, we could not reverse the phase transfer step in this way.

Euan Kay remarked: Have you examined the membrane transport of any cations other than K^+ and Na^+? For example, Cs^+ should bind to 18-C-6 in a 1 : 2 'sandwich' format; so if Cs^+ is transported then it might support the hypothesis of self-organization of crown ethers on the nanoparticle surface.

Mathias Brust responded: That is an interesting point to consider. We have not tried Cs but, as mentioned before, Ba, which also leads to phase transfer of particles. We have not studied the ability of the particles to transport ions across the membrane though, except of course for K, as reported in our paper.

Yu Zhou said: Can you talk more about the factors affecting the transporting efficiency, the size of the nanoaprticles, pH, temperature *etc.*?

Mathias Brust responded: Surely there will be many factors that play a role here. We have not yet investigated any of these systematically. I believe that particle size is very important but we have no data yet to confirm this.

Yu Zhou said: If you increase the temperature, how stable are the nanoparticles inside the membrane?

Mathias Brust replied: This is a really interesting point. If the hydrophobic effect is responsible for the insertion of particles in the membrane, this should be favoured at elevated temperatures and so should the mobility of the particles inside the membrane. Unfortunately, we have not yet measured membrane polarisation at different temperatures, which is certainly something we should do.

Molly Stevens remarked: You could look at changing the membrane fluidity by changing the composition, this would also help reveal the mechanism.

Mathias Brust replied: Currently, we are using a standard protocol for cationic vesicles, and we have not yet tried to change the membrane fluidity. I agree that this would be an excellent way to distinguish between the channel formation and shuttling action. The former should not depend much on the membrane fluidity whereas the latter would crucially change transport rate with different fluidities. Thanks for this suggestion.

Christian Kuttner addressed Mathias Brust and Marcin Grzelczak: The paper states that an intermediate ligand-exchange step with PEG was required to go from citrate to the crown ether ligation. Was this necessary owing to issues with colloidal stability during the direct exchange? Can it be expected that these exchanges proceed quantitatively and did you try to quantify residual traces of ligands in the final surface coating, *e.g.* by SERS?[1]

1 M. Tebbe, C. Kuttner, M. Männel, A. Fery and M. Chanana, Colloidally stable and surfactant-free protein-coated gold nanorods in biological media, *ACS Appl. Mater. Interfaces*, 2015, 7, 5984–5991.

Mathias Brust answered: Yes, this is due to an issue of colloidal stability. We had problems with aggregation when avoiding the intermediate step, and, no, we cannot assume that these processes are necessarily quantitative. A better characterization of the ligand shells is still much-needed. We have used FTIR but this is limited to functional groups with a clear signature. PEGs and crown ethers cannot be distinguished in this way.

Marcin Grzelczak then replied: The composition of a thiolated ligand on the metal surface of bigger NPs, 15 and 38 nm, was not characterised quantitatively using SERS.

Christian Kuttner addressed Marcin Grzelczak: The applied gold nanoparticles are rather small and possess a limited SERS enhancement, which makes it difficult to detect molecular information at the particle surface. As a suggestion, combining your small nanoparticles with a second type of nanoparticles of larger size could help. In combination core/satellite nanostructures could form, with SERS-active hopspots between the small and large particles; see also R. P. M. Höller *et al.*[1] In this configuration, the larger particles can serve as the reporter component while the small particles carry the molecular component of interest. Depending on the core-to-satellite separation, this could allow for high sensitive sensing of the present ligand corona and possibly even reveal residual ligands.

1 R. P. M. Höller, M. Dulle, S. Thomä, M. Mayer, A. M. Steiner, S. Förster, A. Fery, C. Kuttner and M. Chanana, Protein-assisted assembly of modular 3D plasmonic raspberry-like core/satellite nanoclusters: Correlation of structure and optical properties, *ACS Nano*, 2016, **10**, 5740–5750.

Marcin Grzelczak replied: Synthesis of core/satellite nanostructures using smaller and bigger nanoparticles and further characterisation of the exact ligand composition on the metal particles surface is beyond the scope of this work.

Antoine Thill asked: When exploring the size effects of the gold NPs, the meniscus in your Fig. 4a for the medium and large size has a very strange shape at intermediate times (2 h). Do you have an idea why?

Mathias Brust answered: It is an air bubble. This is not typical, and I do not know the origin of this bubble.

Irene Yarovsky asked: To continue on the topic of the size of the Au clusters, within the 2–3 nm size range they are known to be fluorescent, have you looked at this? It may be exploited to characterise their location in the membrane?

Mathias Brust replied: You are sticking a knife in and old wound! I have never understood the fluorescence of these particles, and I never even believed it until I saw it myself when we prepared very carefully size-selected batches in the one to four nanometre range.[1,2] Orange–red fluorescence under UV illumination was observed for a very small number of these batches. Except for these carefully size-selected samples, in our hands, gold nanoparticles in this size regime do not show any significant fluorescence, certainly not to an extent that it could interfere with the dye we are using to monitor the membrane potential. I believe it is still pretty much an open question which electronic transitions are responsible for the emission that has been observed and reported so far.

1 I. Hussain, S. Graham, Z. Wang, B. Tan, D. C. Sherrington, S. P. Rannard, A. I. Cooper and M. Brust, Size-controlled synthesis of near-monodisperse gold nanoparticles in the 1–4 nm range using polymeric stabilizers, *J. Am. Chem. Soc.*, 2005, **127**, 16398–16399.
2 Z. Wang, B. Tan, I. Hussain, N. Schaeffer, M. F. Wyatt, M. Brust, and A. I. Cooper, Design of polymeric stabilizers for size-controlled synthesis of monodisperse gold nanoparticles in water, *Langmuir*, 2007, **23**, 885–895.

Irene Yarovsky asked: Have you considered models of small ~2 nm nano-clusters? Within this size range Au clusters behave as molecular species with only certain sizes of the clusters being stable ('magic numbers'). Structures of these stable clusters have been characterised experimentally and theoretically. Therefore it is possible to model different locations and attachment densities of the crown ethers on these clusters. From such simulations you can predict the maximum loading capacity, packing and the resultant functionalised cluster sizes. it might help rationalise your observations. This is more of a suggestion than a question.

Mathias Brust responded: This is a very good suggestion and we would certainly be happy to participate in such an effort. To get a realistic model of the particles we are currently using, I think one would have to consider a distribution of sizes, even if each individual cluster is exactly defined as a 'molecule' with a distinct structure and a fixed number of gold atoms and ligands. Clearly, we do not have only one species of clusters in our experimental system.

Irene Yarovsky asked: For your 2 nm nanoparticles functionalised with the crown ether, the size distribution is quite broad, do you think it is the gold cluster size effecting this or instead it is the density/packing of molecules on the surface? Also, some NPs do not appear as near-spherical particles, to what can you attribute these effects?

Mathias Brust answered: Such polydispersity is common for this preparative protocol regardless of the type of thiol ligand used. Also, the appearance of a few non-spherical particles is typical. What is interesting is to consider if all particles are capable to act as ion transporters, or only those that are in the appropriate size range to span across the membrane. We would like to address this question in future work.

Irene Yarovsky remarked: Have you attempted hydrodynamic diameter measurements for the crown ether functionalised NPs? Could it help clarify the crown ether density/packing arrangements on the Au NP surfaces?

Mathias Brust answered: We have not, but we are planning to address this question with differential centrifugal sedimentation (DCS), at least for the larger particles. We have had very good experience with this technique in characterizing ligand shell thickness using a variety of thiols and more recently also protein adsorption.

Molly Stevens addressed Mathias Brust and Marcin Grzelczak: In terms of your EM (Fig 5), you don't have any quantification in the paper, have you looked at doing larger scale analysis and quantification, how many particles are interacting with the membrane, *etc.*?

Marcin Grzelczak responded: We did not perform large-scale analysis and quantification of interacting particles with vesicle membranes. However, the most curious observation was achieved when comparing samples differing in their absence/presence of KCl. The particles showed no interaction at all with the vesicular surface in its absence, yet when presented with an environment imbued with KCl (2 mM) the affinity between the vesicular surface and the particles was increased greatly.

Mathias Brust replied: This is an important point that highlights that our contribution is still a bit preliminary. There are a number of problems associated with this quantification. Firstly, by cryo-TEM we can only really 'see' the large gold particles. Here it would be possible by careful image analysis to estimate the proportion of particles associated with membranes under various different conditions, say depending on potassium concentration in analogy with the phase transfer experiments reported in our paper. We have not had the time to do this since we only recently have taught ourselves how to do cryo-TEM well enough to image the vesicles and the particles. We will certainly do more imaging and image analysis in the near future. As far as the smaller particles are concerned, other techniques would have to be used, probably quantisation of gold by AES or ICP-MS after careful separation of the vesicles from the solvent. In general, we find gold trace analysis quite tricky and prone to large fluctuations due to the experimental difficulties to ensure all gold is present in a form that it will be captured by the analysis. Perhaps the simplest way of tackling this general question is by looking at the optical absorption of the separated vesicles but again this may be much easier for the large particles.

David French enquired: Do your nanoparticles aggregate following the addition of salt? Under what conditions do they adsorb to an oil–water interface?

Mathias Brust answered: They are quite robust against aggregation but we have not tried to add salt until they aggregate, which would most certainly happen at some concentrations of salt. We do believe that there is a salting-out effect that makes particles appear more hydrophobic at higher salt concentrations well above the saturation of the crowns with potassium ions. This is discussed in the paper. We have not tried to direct the particles to an interface by controlling the salt–salt concentration, but I believe that this should be possible.

Guillermo González remarked: In the case of the study of 38 nm gold nanospheres, the nanoparticles were co-functionalized with thiolated PEG to obtain colloidal stability and the thiolated crown ether to obtain variable hydrophobicity. However, the control experiments were realized only with thiolated PEG. Thus, it would be more suitable to use another thiolated hydrophobic molecule without tunable hydrophobicity to co-functionalized the nanoparticles for the control experiment.

Mathias Brust replied: I see your point: how can we be sure that the inability of the 'PEG-alone clusters' to transport potassium across the membrane is not simply due to the lack of interaction with the vesicle membrane. Ultimately, we cannot be absolutely sure and better controls as you suggest may be needed, but we have a host of observations that suggest that these particles do interact with membranes; for example, the observation of some fluorescence quenching of membrane-bound dyes upon addition of the particles (unpublished) and the interpretation of other experiments with very similar particles that adsorb to phospholipid monolayers on mercury surfaces.[1] PEG particles tend to stick to all different kinds of surfaces and interfaces, which also suggests that they are quite likely to associate with the membrane.

1 G. J. Gordillo, Ž. Krpetić and Mathias Brust, Interactions of gold nanoparticles with a phospholipid monolayer membrane on mercury, *ACS Nano*, 2014, **8**, 6074–6080.

Guillermo González commented: Gold nanoparticles are quite attractive nanomaterials not only due to their surface properties but also due to their optical properties. Would you think it is possible to take advantage of the interaction of light with the surface plasmon resonance of gold nanoparticles to modulate the potassium transport, as well as the hydrophobicity of the system?

Mathias Brust replied: I am not sure if I would expect plasmons to interfere with the binding of potassium ions to the crown. On the other hand, I am convinced that your suggestion can lead to very exciting opportunities of optically triggering events by coupling the surface chemistry of gold and silver nanoparticles to plasmons. This is a very new field and little about it is published, mostly in the photocatalysis literature and with only preliminary mechanistic understanding.

Luis Liz-Marzán asked: A number of capping ligands have been presented at this meeting, including crown ethers, polymerizable molecules, proteins and PEG, as well as different distributions as a function of curvature. What would be your rule-of-thumb to select a ligand, for example in biomedical applications? What can we tell students who are starting now to concentrate on?

Mathias Brust responded: Unfortunately, there is no rule-of-thumb other than the fact that larger particles tend to require longer ligands for stability. Even when talking about gold particles only, it is often quite difficult to find ligands and a suitable procedure that lead to stable particles for any specific purpose. PEG-thiol ligands tend to work very well for aqueous systems and so do combinations of PEGs with other ligands that may be needed for a desired function. PEGs for biological applications may have the additional disadvantage that they often prevent efficient cellular uptake of the particles. Another group of ligands that is great for particle stabilisation in aqueous systems are peptides. I would avoid giving a new student the impression that these are trivial tasks. Students need to develop a feeling for this and do many experiments themselves to see how it looks and feels when the final product is either stable or unstable. Intuition is still the best guide, and that usually comes with experience.

Lucia Pasquato communicated: In consideration of a paper by Royce W. Murray,[1] in which it is reported that gold NPs can be protected by HS-PEG5000: "PEG-S-MPCs do not exhibit place-exchange reactions (with dodecanethiol) even at elevated temperatures (35 °C), high alkanethiol concentrations, or long reaction times (4 days)". Can you be more specific about the conditions used for the place-exchange reaction and evaluation of the occurred exchange? Is the 18-C-6-CH_2-SH really linked to the gold surface or rather entrapped in the monolayer of PEG?

1 W. P. Wuelfing, Stephen M. Gross, D. T. Miles and R. W. Murray, Nanometer gold clusters protected by surface-bound monolayers of thiolated poly(ethylene glycol) polymer electrolyte, *J. Am. Chem. Soc.*, 1998, **120**, 12696.

Mathias Brust responded: Murray's paper refers to relatively small particles, below about 4 nm, and they are usually prepared with an excess of PEG ligand. Our particles are much bigger and will be faceted, which should make it easier for the incoming ligand to reach the metal surface. Also, our PEG5000 is slightly different in that it is terminated by methyl groups. The subsequent ligand exchange then takes place in ethanolic solution, not in water. This protocol was developed by others and appears to work well, as cited in our paper.

Molly Stevens opened a general discussion of the paper by Clément Sanchez:† When you make your particles in the amorphous boron matrix, how easy or difficult is it to clean it up?

David Portehault answered: The nanocrystals are embedded into a boron-rich ~2 nm-thick amorphous matrix that is strongly chemically inert. Attempts to dissolve this matrix by various methods resulted in the dissolution of the nanocrystals while the matrix remained intact. Therefore, this matrix cannot be removed at the moment. However, the amorphous matrix seems sufficiently porous to provide access of molecular-scale species to the surface of the nanocrystals. Hence, the systems are suitable for catalysis for instance.

Catherine Murphy asked: What is the size range and polydispersity of your materials?

David Portehault answered: In the article, we describe for nickel monoboride NiB a mean particle size of about 10 nm, from TEM and XRD data, with size ranging between 2 and 40 nm. Hence, the polydispersity is still quite large compared to the exquisite control that can be reached for other systems, such as gold nanoparticles or quantum dots. We are currently developing ways to improve the size polydispersity. Besides, we have already shown in a previous report[1] that it is possible in some cases to tune the particle size by adjusting the reactant ratio. For instance, the mean particle size could be changed from 5 or 10 nm for NbB_2 nanocrystals, as evaluated from TEM data.

1 D. Portehault, S. Devi, P. Beaunier, C. Gervais, C. Giordano, C. Sanchez, and M. Antonietti, A general solution route toward metal boride nanocrystals, *Angew. Chem., Int. Ed.*, 2011, **50**, 3262–3265.

Catherine Murphy asked: Many of these materials have been made as 2D nanoscale thin films. What new properties do you expect when you are able to make 3D colloidal versions of them?

David Portehault answered: Indeed, some of these metal borides have been described as thin films obtained by planar chemical or physical fabrication

† Clément Sanchez's paper was presented by David Portehault, Laboratoire Chimie de la Matière Condensée de Paris, France.

methods, such as PVD, CVD or PLD as described in our extensive review of the topic.[1] For instance, HfB_2 films have been grown because of the high refractory properties of this metallic compound. In this case, the colloidal version should exhibit a strongly decreased electrical conductivity, and perhaps modified transport properties, as suggested by our own results on metal diborides.[2] Likewise, Mattox *et al.*[3] showed recently that localized surface plasmon resonance arises in metallic LaB_6 nanocrystals. Albert *et al.*[4] also showed the impact of nanoscaling on the properties of a-FeB particles. Another field that may be strongly impact is catalysis. Amorphous transition metal borides, especially nickel and cobalt compounds, are well-recognized as precious metal-free catalysts for a range of reactions, especially hydrogenation reactions[1] and hydrolysis of borohydrides for hydrogen production.[1,5] One can expect that crystalline nanoparticles should exhibit enhanced activity and modified reactivity. We are currently investigating such effects.

1 S. Carenco, D. Portehault, C. Boissière, N. Mézailles, and C. Sanchez, Nanoscaled metal borides and phosphides: Recent developments and perspectives, *Chem. Rev.*, 2013, **113**, 7981–8065.
2 D. Portehault, S. Devi, P. Beaunier, C. Gervais, C. Giordano, C. Sanchez, and M. Antonietti, A general solution route toward metal boride nanocrystals, *Angew. Chem., Int. Ed.*, 2011, **50**, 3262–3265.
3 T. M. Mattox, A. Agrawal, and D. J. Milliron, Low temperature synthesis and surface plasmon resonance of colloidal lanthanum hexaboride (LaB_6) nanocrystals, *Chem. Mater.*, 2015, **27**, 6620–6624.
4 S. Rades, S. Kraemer, R. Seshadri, and B. Albert, Size and crystallinity dependence of magnetism in nanoscale iron boride, α-FeB, *Chem. Mater.*, 2014, **26**, 1549–1552.
5 U. B. Demirci and P. Miele, Cobalt in $NaBH_4$ hydrolysis, *Phys. Chem. Chem. Phys.*, 2010, **12**, 14651–14665.

Anna Roig asked: How does the heating rate influences particle size distribution? Have you considered using microwave as a heating source? With microwave radiation, a volumetric heating can be achieved and this can impact on narrowing the particle size distribution. Moreover, microwave heating also allows a fast cooling rate and thus nanocrystal growth quenching.

At present there are microwave ovens, commercialized by Milstone, that can achieve very high temperatures up 900 °C, even allowing sintering processes.

David Portehault responded: Answer to your first question: We did not assess this parameter, but we can expect a burst of nucleation and smaller nanoparticles for increased heating rates.

Answer to your second question: Yes we considered this process. Note that in order to trigger crystallization of these nano-alloys in inorganic molten salts, temperatures above 450 °C are usually required. Commercial microwave ovens dedicated to liquid phase syntheses are often restricted to temperatures below 300 °C. A few less-common models can stand higher temperatures, but those are usually multimode ovens for which sample irradiation is not optimized. Besides, some of these high temperature ovens use susceptors, such as SiC plates that absorb microwave radiation and release heat. Then, heating is no longer volumetric and the advantage of microwave heating becomes less-and-less obvious. All-in-all, to maintain volumetric heating for molten salt synthesis of metal boride nanocrystals, one requires a microwave oven able to increase the temperature of the inorganic salt to above 450 °C without any susceptor. Such a device should be a monomode oven able to maximize irradiation on the sample. Unfortunately, the situation gets even more complicated, as the inorganic salts are solid at syntheses starting point below the melting point of the eutectic mixture. In this physical state, inorganic salts have a very low ability (loss tangent) to convert microwaves into heat. Hence, a high microwave power would be needed to bring the reaction medium to the molten state. All-in-all, our attempts to run such syntheses under microwave heating using a range of commercial ovens were unsuccessful, but we are currently working on a new device that could meet the points discussed above and provide an innovative solution to this problem.

Martin Mayer said: You describe your particles after washing as a "black powder consisting of metal boride nanoparticles". Is it possible to stabilize those nanoparticles in any solvent, *e.g.* with ligands, to form a colloidally stable solution? In that way you could separate the different sizes and morphologies.

David Portehault replied: The nanocrystals are embedded into an ~2 nm-thick amorphous matrix mostly made of boron. Reaching a colloidal suspension of the nanocrystal requires removal of this component. However, amorphous boron is strongly chemical inert. Attempts to dissolve this matrix in various ways resulted in the dissolution of the alloy nanocrystals while the matrix remained intact. Although it suggests that dispersion of the aggregates is not reachable at the moment, this observation also indicates that the amorphous matrix is sufficiently porous to provide access for molecular-scale species to the surface of the nanocrystals. Hence, the systems are suitable for catalysis, for instance.

Leonardo Scarabelli said: How general do you think your approach is? Which parameters do you think you should change to apply a similar protocol to target different elements?

David Portehault replied: In our previous report,[1] we have shown that metal boride synthesis in inorganic molten salts could be applied to a wide range of metals, from the s-block to rare earths through to transition metals (the first, second and third series). In the present article, we demonstrate that for each metal, in case different M : B stoichiometries are known in the bulk scale, they can be selected through molten salt synthesis by adjusting the reaction time, temperature and the reactant ratio. The main limitation up to now lies in the washing process, performed by dissolving the frozen salt in water. Hence, products sensitive to water cannot be obtained readily by applying this process, for instance some alkali boride nanoparticles. Other polar solvents enabling high solubility of the salts may however be used.

1 S. Carenco, D. Portehault, C. Boissière, N. Mézailles, and C. Sanchez, Nanoscaled metal borides and phosphides: Recent developments and perspectives, *Chem. Rev.*, 2013, **113**, 7981–8065.

Luciano Tadiello asked: I am not familiar with shape controlled synthesis in molten salts. Working with solvents it is common to use organic molecules, typically surfactants or amphiphilic block copolymers, as structure-directing agents. Is there a similar concept of a templating system also in the molten salt technique?

David Portehault answered: Organic molecules and polymers are not stable in the synthesis temperature range. It should however be possible to use strongly complexing inorganic anions, such as sulfates or phosphates, as surface complexant and structure directing agents. Other polarizable anions may as well have an impact on the nucleation-growth process.

Antoine Thill said: To understand the formation mechanism of the nanoparticles in hot molten salts, looking at intermediate times could be very useful. Do you have any idea on techniques that could enable such mechanistic exploration *in situ*?

David Portehault responded: Two main procedures may be applied: (1) Using fast cooling rates to reach a temperature below the eutectic melting point. The salts freeze and the system is quenched. Then, regular characterization techniques can be used to investigate the frozen reaction media. These studies are under way. (2) *In situ* studies using synchrotron radiation may also well be facilitated by the low vapor pressure of inorganic molten salts, which do not require pressurized vessels. We are currently developing these approaches.

Molly Stevens opened a general discussion of the paper by Lucia Pasquato: You present really nice data! Can you comment on what's happening during the morphological characterisation of these molecules?

Lucia Pasquato responded: The monolayer morphology of these systems was assessed experimentally by ^{19}F NMR spectroscopy by monitoring the chemical shift changes triggered by changing the monolayer composition. This experimental approach was associated with a theoretical investigation based on multiscale molecular simulations in explicit solvent. The combined results show the formation of stripe-like domains when the H- and F-ligands have a length mismatch of eight carbon atoms. Interestingly, a length difference of only four carbon atoms favors the formation of patches or stripe-like domains. Janus nanoparticles are instead observed when both types of ligands have a chain length of twelve carbon atoms; however, if the ligands are shorter, although of the same length, the morphology is ill-defined. Lastly, in the case of the branched thiolate, the monolayer displays a random organization.

David Portehault asked: What is the evidence that an equilibrium is reached and that the system is not in a kinetically trapped state? To assess this point, one could wonder: (1) if the exchange of ligand A at the surface by ligand B would yield the same surface composition as the exchange of B by A; or (2) if ligand exchange would give the same surface composition as direct bi-ligand synthesis.

Lucia Pasquato responded: Starting from the second point. We addressed this point a few years ago in the context of our analysis of mixed monolayer nanoparticles protected by hydrogenated and fluorinated amphiphilic thiolates; the answer was in the positive. We could not observe differences in the monolayer properties when the preparation was performed by direct synthesis using a blend of thiols or by place exchange. Regarding the first point. What you mention is a place exchange reaction, and its equilibrium nature is well established. Of course the position of the equilibrium will depend on the initial conditions as well as the nanoparticles' composition at equilibrium.

David Portehault asked: In your Fig. 3, where the compositions of the surface and of the whole reaction medium are correlated, red curves are added as guides for the eye. If you assume that the equilibrium is reached, it should be possible to go further in the analysis by fitting the data using a mass law, in order to extract thermodynamic constants of ligand exchange, and eventually to be able to predict the surface composition at a given composition of the medium?

Lucia Pasquato responded: I think this is certainly possible if no domains are formed, because the KB-to-A equilibrium constant will be always the same. Things might be different if domains are formed. Let me assume that the B ligand tends to form domains in a monolayer of A ligands. Let me also assume that the formation of the initial nucleation spot made of a few B ligands is disfavored because of unfavorable B/A interactions. This will appear as a small equilibrium constant KB-to-A. When the nucleation spot is large enough to provide favorable B/B interactions, the KB-to-A equilibrium constant will increase. This will complicate things.

Mathias Brust asked: Two quick questions: (i) do you see any unusual solubility or aggregation behaviour that may be due to ligand segregation on the particle surface; and (ii) are any of your particles with perfluorinated ligands soluble in supercritical carbon dioxide?

Lucia Pasquato answered: Concerning the first question. We indeed observed unusual solubility of these systems when compared one to another and we believe this is correlated to the self-sorting of the ligands and to the size of the domains. The most enlightening example is that of NP-C16/F6, and NP-C12/F10, Fig. 5 of our paper. The former remain soluble in chloroform (the solvent of highest polarity among the ones we used) up to a molar fraction of ligand F6 of roughly 75%, while for the latter already at 30% undergo a solubility transition. For NP-C12/F10 it is reasonable to think at the formation of extended fluorinated domains leading to aggregation in solution with deterioration of the solubility properties. For NP-C16/F6 the fluorinated domains have to be smaller and/or shielded from the fluorinated domains of other nanoparticles. Concerning the second question. We did not test our systems for solubility in supercritical CO_2 mainly because we wanted to achieve solubility in conventional organic solvent. However, supercritical carbon dioxide was used as solvent for nanoparticles protected by fluorinated ligands, and some theoretical analyses were also performed in order to rationalize how the solubility depends on the nature of the passivating ligand.

Javier Reguera remarked: The nanoparticles used in this work are relatively small (below 4 nm); could you comment on the effect of the curvature on the ligand arrangement? What would happen for bigger nanoparticles? Have you observed Janus morphology, as predicted by S. Glotzer, for smaller ones?

Lucia Pasquato replied: In this work we have not explored the effect of the size of the gold core on the monolayer morphology. For NPs with a length mismatch of 4 and 8 carbon atoms – in which formation of domains like patches/stripes may be foreseen – the range of size explored is between 2 and 4 nm, and for an F-/H-ligand ratio between 2 : 1 and 1 : 2 we do not have evidence of Janus organization. I believe that smaller and larger NPs have to be studied: we will do this in the future.

Javier Reguera asked: This question regards reaching of an equilibrium state of the ligand domains. Could you use your NMR methods to characterize nanoparticles that start with a non-equilibrium arrangement of ligands (for example, using masking) and then measure the lateral mobility of ligands on the surface of the nanoparticle? Many researchers are synthesizing nanoparticles stabilized by more than one ligand molecule without previous knowledge of their arrangement or the change of arrangement with time. Could you comment on how the determination of the arrangement and the time-scale for the mobility of ligands could help in the design of better systems?

Lucia Pasquato answered: Concerning the first question: In my opinion this is theoretically possible: Evidence gained so far during our experimental work suggest that different arrangements of the ligands do produce different NMR responses. Hence, if any evolution in the NMR response occurs in time, one could probe the evolution of the domains due to the lateral mobility of the ligands. However I see a critical point: In order to enable such an approach, ligand diffusions should be slow on the NMR timescale; this is likely to be the case for gold nanoparticles, but may not be true in general, especially if weakly interacting ligands are used as stabilisers.

Concerning the second question: Certainly, to have information on the final 'equilibrium' arrangement of the ligands on the surface of a metal nanoparticle, how this arrangement depends on the structure of the ligands and what is the timescale of ligands' mobility is paramount in developing better systems. This is because one may have a design tool: by using a certain pair of ligands one gets an arrangement, while a different arrangement may be obtained by using a different set of ligands. Theoretical work predicts this; several experimental data already provide some practical approaches to anisotropic monolayers, however more has to be done in order to enrich the diversity of the tools available to materials scientists.

Estefania Gonzalez Solveyra spoke: This is a comment in the context of the discussion on the monolayer composition *vs.* composition of the initial reaction mixture (Fig. 3 in your paper). There were questions on whether the systems studied had reached the thermodynamic equilibrium before determining the composition of the surface mixture. My contribution to this discussion is that resorting to equilibrium calculations, you can get results as the one presented in your Fig. 3, where the composition on the surface and in the solution are not the same. This is due to the balance of different forces competing at the nanoscale (loss of conformational entropy, gain in absorption energy, gain in entropy of mixing, polymer–polymer attractions, *etc.*), modulated in turn by the curvature of the NP. This competition would give the kind of results you presented.

Lucia Pasquato added: Thank you for your comment.

Axel Mueller opened a general discussion of the paper by Calum Drummond: You synthesized cylindrical brushes with a hydrophobic acrylate backbone in the centre and hydrophilic PEG sidechains around. I calculated that the sidechains on average consist of about 10 EG units and the three polymers have between 10 and 100 of those sidechains. If those chains now attach with their backbone. I think that rearrangement of the sidechains so that the hydrophobic backbone can attach to the cubosome would suffer from quite a large entropic penalty. Now my question is whether you have any evidence about how the brush is arranged in a molecular way at the interface between the cubosomes and water?

Jiali Zhai communicated in reply on behalf of Calum Drummond: We do not have direct evidence about how the brush polymer is arranged at the cubosome–water interface. However, the hydrophobic effect drives the fundamental self-assembly behaviour and molecular interactions. The interfacial adsorption of amphiphilic brush polymers is expected to result in a molecular orientation where the C11 hydrophobic tail of the polymer resides in the monoolein or phytantriol lipid layer, due to the hydrophobic interaction, and the hydrophilic PEG sidechain is solubilised in the water phase. This molecular orientation is crucial for providing the effective stabilisation for cubosomes because the hydrophilic PEG sidechain in the external water phase will impart steric repulsion to prevent nanoparticle aggregation.

Molly Stevens said: I would like to get your opinion on how one would proceed from cell testing to clinical approval for existing systems without using the RAFT; how are these going?

Calum Drummond answered: Proceeding from cell testing, I suggest one should establish a disease model of interest in animals to test the biological response and the therapeutic efficacy of these nanosystems for drug delivery. In

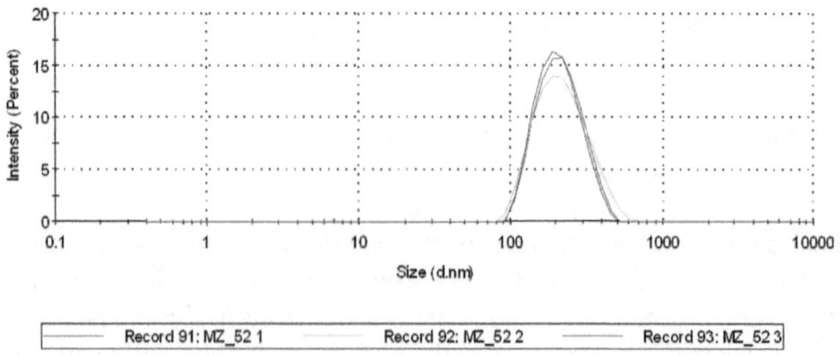

Fig. 2 Size distribution by intensity.

general the animal testing will provide significant data to assess whether the new drug delivery system can be considered safe and can deliver improved efficacy and intervention. Key parameters to investigate can include the toxicity, pharmacokinetics, biodistribution, bioavailability, immune response and the therapeutic effect of the loaded drugs. Based on pre-clinical cell and animal testings, one can proceed to apply for human research ethics and phase 1 clinical trials.

Lucia Pasquato remarked: You mention that there is a difference in toxicitiy in your polymers, is this related to the chemical nature of the monomer or the polymer, or is it related to the roughness of the nanosystem?

Calum Drummond answered: The results in the current study show that the toxicity depends on various factors, including the type and the concentration of the polymers used to stabilise the cubosomes as well as cell lines. Previous studies (cited in our paper) by us also demonstrate these phenomena. The surface coverage of the polymers surrounding the cubosomes plays a significant role in the interaction with cells and hence the toxicity. We hope to differentiate the toxicity effect of the polymers themselves from the polymer-stabilised cubosomes in future studies. At this stage, we cannot evaluate any possible link between toxicity and roughness on the nanoscale.

Yu Zhou remarked: What is the minimum size of the particles that you make?

Calum Drummond answered: Our dynamic light scattering (DLS) experiment measured the particle size distribution and polydispersity of the particles we made. The data suggests that the minimum size of the particles varies with the type and the concentration of the stabilizer in the system. For example, the phytantriol cubosomes stabilized by 0.5 mol% Pluronic F127 had an average particle size of around 199 nm in diameter and a polydispersity of 0.11 (Table 3 in our paper). The raw DLS data (Fig. 2) suggest that the particles followed a monomodal distribution and had a minimum size around 100 nm.

Yu Zhou said: Can you comment on the stability of the materials under different conditions, like temperature?

Calum Drummond answered: We hope to investigate the effect of various temperatures and pHs on the stability of the cubosomes formed by the brush-type P(PEGA) polymers in future studies. The stability of the particles will be assessed by the retention of the particle size distribution (*i.e.* the aggregation behaviour) and the changes of the internal lyotropic liquid crystalline mesophase (*i.e.* the phase transition behaviour). We hypothesise that the stability of the particles under different conditions will largely depend on the interaction of various polymers with the host lipid.

Yu Zhou asked: You say that you will use these materials for drug delivery, what strategy will you use to organise that?

Calum Drummond answered: We plan to use a translational research strategy to investigate the potential of these materials for drug delivery. This would involve loading these cubosomes with model chemotherapeutic drugs such as paclitaxel and investigate the encapsulation efficiency and controlled release of the drugs *in vitro*. We would then use cancer cell lines to investigate the cytotoxicity and the therapeutic efficacy of the drug-loaded materials *in vitro*. Finally we would use tumour xenograft models in nude mice to study the pharmacokinetics, biodistribution and turmour suppression effect of these materials.

David Portehault asked: How robust is the liquid crystalline organisation *versus* the incorporation of drug molecules, and what is the maximum load reachable? How does the latter compare to other cargos, such as mesoporous silica?

Calum Drummond answered: The maximum load capacity for the cubosomes depends on the cargo's size and hydrophobicity. For example, one of our previous studies[1] shows nine small molecule drugs were loaded into cubosomes at very high concentrations (10–15 mol%) and achieved up to 95% encapsulation efficiency. The study also shows that the lyotropic liquid crystalline organisation within the cubosomes can change from the cubic phase to an inverse hexagonal phase of larger surface curvature with the incorporation of drug molecules. Large biomolecules such as peptides, proteins and DNAs are also demonstrated to be effectively loaded into the lyotropic liquid crystalline phases and the maximum load capacity is under investigation. Compared to other nanosystems, for example liposomes, the internal liquid crystalline phase in cubosomes offers high surface-to-volume ratios and hence potentials in loading a high amount of both hydrophobic and hydrophilic drug molecules.

1 X. Mulet, D. F. Kennedy, C. E. Conn, A. Hawley and C. J. Drummond, High throughput preparation and characterisation of amphiphilic nanostructured nanoparticulate drug delivery vehicles, *Int. J. Pharm.*, 2010, **395**(1–2), 290–297.

Antoine Thill opened a discussion of the Concluding Remarks by Younan Xia: In your excellent presentation, the role of the nanoparticle preexisting crystalline facets is important. It is also possible to break the symmetry of amorphous material (like silica, see for example ref. 1). Can you comment on what is known on the role of the capping agents? Is there interaction with specific crystalline faces, or anything else?

1 A. Kuijk, A. van Blaaderen and A. Imhof, Synthesis of monodisperse, rodlike silica colloids with tunable aspect ratio, *J. Am. Chem. Soc.*, 2011, **133**(8), 2346–2349.

Younan Xia replied: The role of a capping agent is to selectively passivate a specific type of facet on a seed and thus slow down its growth. It is most useful for controlling the shape of a nanocrystal, and thus the type of facet expressed on the surface of a product. However, the capping agent is not directly responsible for symmetry breaking during the growth of a seed. It is the limited supply of atoms, relative to the equivalent sites on a seed, that ultimately leads to symmetry breaking or reduction.

Hedi Mattoussi asked: Your ability/capacity to promote the growth on specific facets strongly depends on the quality of the seeds you start with. How do you achieve that for some of the syntheses you discussed? In other words, during the initial growth your perfect cubes are usually mixed with less perfect and undesired nanostructures. Do you have to use post-growth processing techniques to isolate the structure of interest, *e.g.*, the perfect cube?

Younan Xia responded: You can use some tricks to obtain the seeds you want. For example, you can introduce oxidative etching to selectively remove seeds with twin defects, leaving behind single-crystal seeds (like cubes) as the products. Alternatively, you can adjust the reduction rate to selectively obtain single-crystal seeds, decahedral seeds, or icosahedral seeds. Please refer to our recent perspective article.[1]

1 Y. N. Xia, X. H. Xia and H. C. Peng, Shape-controlled synthesis of colloidal metal nanocrystals: Thermodynamic *versus* kinetic products, *J. Am. Chem. Soc.*, 2015, **137**, 7947.

Luis Liz-Marzán said: One of the parameters that we constantly use to explain symmetry breaking is the preferential adsorption of surfactants or polymers to specific facets; you also mentioned this today. Do we have clear evidence to justify these statements?

Younan Xia answered: My personal experience is against such an argument. It is easy to argue that a capping agent can selectively passivate a specific type of facet, but it will be much harder to argue that the capping agent can be unequally distributed among the equivalent facets, for example, the six side faces of a cubic seed. To me, a limited supply of atoms (or precursor) relative to the number of equivalent sites due to kinetic control is a more reasonable mechanism for symmetry breaking. After all, for the comparison experiments I discussed in my presentation, all seeds were essentially the same; the only difference was the rate at which the precursor was introduced, and thus the number of atoms relative to the number of seeds.

Luis Liz-Marzán concluded: You asked who designed the logo for the conference, please let me explain. This is an electron tomography image combined with EDX mapping obtained by the group of Professor Sara Bals (University of Antwerp) on a Janus nanoparticle prepared in my group and reported in ref. 1. The particle is made of gold (yellow part) and silica (in blue). The preparation of this system involves many concepts that were discussed at this conference: it starts by making a gold nanoparticle that is covered by two polymers on two sides (Janus configuration by polymer segregation), then silica grows only on the side that is covered by the polymer with the highest affinity, and finally spikes are grown on the other side.

1 D. Rodríguez-Fernández, T. Altantzis, H. Heidari, S. Bals and L. M. Liz-Marzán, A protecting group approach toward synthesis of Au–silica Janus nanostars, *Chem. Commun.*, 2014, **50**, 79.

Faraday Discussions

PAPER

Concluding Remarks: Anisotropy: the good, the "bad" and ...

Younan Xia

Received 8th August 2016, Accepted 10th August 2016
DOI: 10.1039/c6fd00177g

The 191st Faraday Discussion meeting on "Nanoparticles with Morphological and Functional Anisotropy" offers a broad range of new insights into an interdisciplinary research field converged at the forefronts of chemistry, colloid science, and materials science. It was held on the campus of University of Strathclyde, Glasgow, UK. Across the street of the meeting place, one could see a large portrait of Thomas Graham, the father of physical chemistry who coined the term of "colloid" in the 19th century, painted on the outside wall of a tall building. Without any doubt, he would also have enjoyed listening to the fascinating progress in a research field pioneered by him. The discussion was organized into four sessions, with each session covering a specific theme: (i) Janus and patchy nanoparticles; (ii) anisotropic nanoparticles; (iii) particles at interfaces; and (iv) applications. The meeting was further enriched by the presentation of 79 posters. Instead of giving a summary of what was presented at the meeting, I would like to take this opportunity to provide some personal perspectives, in conjunction with the insights offered by various authors.

Anisotropy *vs.* isotropy

As opposed to isotropy, which implies "identical properties in all directions", anisotropy means directional dependence. It can be simply defined as the difference, when measured along different axes, in a material's properties, including physical dimensions, refractive index, electrical/thermal conductivity, and tensile strength, among others. When applied to the topic of this Discussion, it refers to nanoparticles with a non-spherical shape (*i.e.*, morphological anisotropy) or nanoparticles with a non-uniform distribution of surface properties (*i.e.*, functional anisotropy). In principle, true isotropy can only be observed in an amorphous or non-crystalline material, like a liquid or a glassy solid. As driven by thermodynamics, particles made of such a material (*e.g.*, silica or amorphous polymer) tend to take a spherical shape as a result of minimization of surface area and thereby surface free energy. For a crystalline solid, there always exists anisotropy owing to

The Wallace H. Coulter Department of Biomedical Engineering, Georgia Institute of Technology and Emory University, Atlanta, GA 30332, USA. E-mail: younan.xia@bme.gatech.edu; Fax: +1-404-386-7493; Tel: +1-404-385-3209

the inherent disparity in interaction strength among the constituent atoms/molecules. As a result of the intrinsic difference in free energy among different types of facets, nanoparticles derived from a crystalline material naturally take a non-spherical shape, leading to morphological anisotropy. A typical example can be found in the cuboctahedral morphology exhibited by nanoparticles made of a face-centered cubic (fcc) metal such as gold. Strictly speaking, gold nanoparticles should not take an isotropic or spherical shape unless they are held in a glassy or amorphous state. In literature, however, people tend to approximate cuboctahedral nanoparticles as nanospheres because of the insignificance in anisotropy. In this case, the anisotropy could be greatly augmented using one of the following strategies:[1] (i) inclusion of defects and/or twin boundaries into the crystal structure; (ii) use of a capping agent to increase the proportion of one specific type of facet on the surface; (iii) conduction of the synthesis under a kinetic rather than thermodynamic control; (iv) introduction of an interface to break the symmetric environment surrounding the particle; and (v) modification of the surface or structure using lithography or other fabrication techniques.

Anisotropy means new properties and opportunities

As illustrated by Murphy in her introductory lecture (DOI:10.1039/c6fd00145a), switching from spherical to non-spherical morphology can immediately bring in new properties to the nanoparticles. A classical example can be found in gold nanorods, a popular nanomaterial that was also discussed by many other speakers at the meeting. Different from the case of a gold nanosphere, the free electrons in a gold nanorod can collectively oscillate along two distinctive directions parallel to the short and long axes of the rod, respectively.[2] As a result, the localized surface plasmon resonance (LSPR) peak splits into two modes: a transverse mode with a spectral position similar to that of a gold nanosphere, and a longitudinal mode whose position is highly tunable depending on the aspect ratio (*i.e.*, the ratio of length to diameter) of the nanorod. Significantly, the longitudinal peak can be continuously red-shifted from the visible (around 525 nm) to the infrared region (2 μm and even beyond) by merely increasing the aspect ratio.[3] Although not discussed at this meeting, it is worth pointing out that silver and copper nanowires have received ever increasing interest as conductive elements for the fabrication of transparent electrodes needed for touchscreen panels, flexible displays, and solar cells.[4] As a serious alternative to the indium-tin-oxide (ITO) technology, thin coatings derived from silver or copper nanowires have been demonstrated with remarkable performance under high bend strains. Again, such applications critically depend on our ability to improve the transmittance by increasing the anisotropy (*i.e.*, the aspect ratio) of the nanowires as this parameter is directly proportional to the open area formed among the nanowires, through which the light will be able to pass without experiencing any attenuation.

Anisotropy means diversity and complexity

One of the major motivations for studying anisotropic nanoparticles is the diversity and complexity that can be brought into a colloidal system, including both the individual components and their assembled structures. To this end, Janus nanoparticles with two distinct regions (or patches) on the surface, have emerged as a subject of extensive research in recent years.[5] At the meeting, Granick (DOI:10.1039/c6fd00077k) discussed an intriguing system involving the non-equilibrium assembly of Janus particles (silica beads with half of the surface coated by a metal) into rotating pinwheels when mixed with a trace amount of pristine silica beads in the presence of an AC electric field. The Janus particles were observed to form spiral arms, with the pristine bead being pinned to serve as a central rotation point. To achieve a steady state, the spiral arms had to undergo an evolutionary process to have their motions coordinated. Synytska (DOI:10.1039/c6fd00008h) discussed the fabrication of complex structures with unique properties by exploiting colloidal building blocks with morphological and functional anisotropy. An easy and scalable bottom-up approach was demonstrated for the programmed assembly of Janus particles and hairy, oppositely charged, and homogeneously decorated particles by tuning the electrostatic interactions between the polyelectrolytes grafted onto their surfaces. Two different structures were observed: raspberry-like or half-raspberry-like and dumbbell-like clusters. Most significantly, stable symmetric and asymmetric microstructures could be obtained in a well-controlled manner.

Kotov (DOI:10.1039/c6fd00064a) discussed the adoption of layer-by-layer assembly (LBL) for the fabrication of chiral films from positively charged polyelectrolytes and negatively charged chiral CdS nanoparticles whose surfaces were functionalized with D- and L-cysteine. This work demonstrates the potential to translate the optical activity of chiral semiconductor (or metallic) nanoparticles into thin films needed for the fabrication of optoelectronic devices. Kumacheva (DOI:10.1039/c6fd00057f) discussed the rational design and assembly of polymer-functionalized metal nanoparticles (*e.g.*, gold nanorods with their ends derivatized with polymer ligands) into one-dimensional structures in a fashion similar to the conventional step-growth polymerization. The assembly process could be triggered by inducing solvophobic attraction between the polymer ligands by altering the solvent quality. It was shown that both anisotropic (*i.e.*, patchy) nanoparticles and nanoparticles uniformly covered by polymer ligands could self-assemble into chains with controlled length, morphology, and composition. The systems presented by these authors clearly demonstrate many interesting new features that would be otherwise missing in the self-assembly of isotropic or uniform building blocks.

If there is anything "bad" that one should complain about anisotropy, it is the complexity! For isotropic nanoparticles, it is quite straightforward to characterize and model them using conventional tools. For example, one can easily measure the size of a spherical particle from its projected area using transmission electron microscopy (TEM). The identity and coverage density of surface functional groups can also be determined by analyzing a bulk quantity of the particles using X-ray photoelectron spectroscopy (XPS) or other spectroscopic techniques. For a particle with anisotropic morphology, however, it is a non-trivial task to even just measure its physical dimensions under TEM as the results critically depend on its orientation relative to the electron beam. In some cases, one has to rely on tomography to resolve the three-dimensional morphology of anisotropic nanoparticles. At the meeting, Peddis (DOI:10.1039/c6fd00059b) discussed a new approach based upon Aspect Maps for resolving the shape or morphology of anisotropic nanoparticles through TEM analysis. In a case study, the authors examined the magnetite nanoparticles (45 nm in mean size) extracted from *Magnetospirillum gryphiswaldense* magnetostatic bacteria. As noted by several speakers at the meeting, it still remains a daunting task to resolve the actual distribution of functional groups on the surface of a nanoparticle (either isotropic or anisotropic) due to the limitations in terms of spatial resolution and/or detection sensitivity. In fact, there is very little quantitative information about the coverage density of ligands conjugated to the surface of a nanoparticle.[6]

Anisotropy can also mean simplicity in terms of synthesis

In many cases, the morphological anisotropy exhibited by nanoparticles is encoded into the anisotropic interaction among the atomic/molecular building blocks of the crystalline solid. For these systems, anisotropic morphology would be naturally developed during a synthesis.[7] One of the most prominent examples can be found in inorganic minerals such as asbestos and chrysolite, which are well-documented to exhibit a fibrous growth habit owing to their anisotropic crystal structure. In addition, many polymeric and biological systems are known to preferentially exist in the fibrous form because of anisotropy in bonding, with notable examples including cellulose and collagen. Chalcogens (specifically, selenium and tellurium) represent another interesting system that can naturally grow into highly anisotropic nanostructures with one-dimensional morphology. While oxygen exists primarily as diatomic molecules, selenium and tellurium tend to form polymeric, helical chains through covalent bonding. In the trigonal phase, the helical chains are further packed into a hexagonal lattice through van der Waals interactions. As a result, these two solid materials have a strong tendency to grow into nanorods or nanowires even when they are crystallized in an isotropic medium, favoring the stronger covalent bonds within the spiral chains over the relatively weaker van der Waals forces among the chains.

For metals, although metallic bonding is isotropic, the steric effect only allows each atom to be in contact with a limited number (*e.g.*, 12 for an fcc metal) of nearest neighbors. Such restriction also leads to anisotropy naturally, including the variation in atomic packing density along different directions and thereby disparity in free energy among different types of crystallographic facets. However, such anisotropy is often inadequate for the formation of nanoparticles with noticeably anisotropic morphology. Under thermodynamic control, nanoparticles comprised of an fcc metal are supposed to take a cuboctahedral profile, the so-called Wulff shape, with the surface enclosed by a mix of {111} and {100} facets to minimize the total surface free energy. The high order of symmetry makes a cuboctahedron look like an isotropic rather than an anisotropic structure. In this case, one has to break the thermodynamic confinement using various tricks in order to generate highly anisotropic structure such as nanorods.[1] At the meeting, López-Quintela (DOI:10.1039/c6fd00015k) presented a new seed-mediated route to the facile synthesis of gold nanorods without involving the commonly used silver ions. Their trick was to employ gold clusters, prepared with a very short reduction time (<5 min), as a source of atoms for seed-mediated growth. It was proposed that the gold clusters could adsorb onto particular crystal planes of the seeds, promoting one-dimensional growth and thus the formation of gold nanorods. While this growth mechanism is not unreasonable, another plausible explanation could be ascribed to the slower kinetics associated with the transfer of gold atoms from the clusters to the seeds and thereby symmetry reduction during the growth process. Roig (DOI:10.1039/c6fd00028b) discussed a polyol method based on microwave-assisted heating for the synthesis of triangular gold nanoplates decorated with iron oxide nanoparticles. This simple method is attractive for the fabrication of multi-component, multi-functional nanostructures by enabling selective heating at the desired sites. Overall, it is still a grand challenge to prepare anisotropic nanoparticles from crystalline solids characterized by highly isotropic interactions among the building blocks.

Symmetry breaking or reduction

Symmetry breaking or reduction is a powerful approach to the formation of nanoparticles with morphological anisotropy because the products derived from asymmetric growth naturally display an anisotropic, thermodynamically less favorable shape. This approach is particularly effective in the production of metal nanoparticles *via* seed-mediated growth. By shutting down some of the equivalent sites on the surface of a seed, one automatically produces nanoparticles with anisotropic morphology.[8] Experimentally, symmetry breaking or reduction can be induced using a number of different strategies, with kinetic control being most straightforward and effective. To this end, my group recently reported a quantitative analysis of the symmetry reduction phenomenon involved in the seed-mediated growth of metal nanocrystals under dropwise addition of a precursor solution. After the introduction of several drops, the system will quickly enters into a steady state, in which the number of precursor species in the growth solution oscillates within a narrow range defined by the low and up limits. By quantitatively controlling experimental parameters such as the reduction kinetics, the initial concentration of the precursor solution, and the injection rate, one can manipulate the steady state and thus deterministically control the growth pattern (symmetric *vs.* asymmetric) of a seed. Remarkably, there exists a clear correlation between the growth pattern and the low limit of precursor concentration in the steady state. If the low limit of the steady state is below the number of growing sites on the surface of a seed, it will no longer be possible to have atoms deposited on all the equivalent sites. When the reaction temperature is sufficiently low, the atoms will be confined to the initially deposited sites, causing symmetry reduction. Taken together, it is clear that symmetry breaking or reduction can only be initiated and retained by keeping the atom deposition at a rate slow enough to limit the number of initial nucleation sites on a seed but fast enough to beat the surface diffusion process. This hypothesis has been validated using a number of systems, including the deposition of silver on palladium cubic seeds for the formation of bars and non-concentric cubes; the deposition of silver on silver cubic seeds for the formation of silver bars and semi-truncated octahedra; as well as the deposition of palladium on palladium cuboctahedral seeds for the formation of tetrahedra.

At the meeting, Reguera (DOI:10.1039/c6fd00012f) discussed the synthesis of a new class of Janus particles consisting of gold nanostars and iron oxide nanospheres through two steps of consecutive, seed-mediated growth. A combination of these two materials in a single particle results in both plasmonic and magnetic properties sought for applications in bio-sensing and therapeutics. The authors also removed the iron oxide nanoparticles from the gold nanostars to gain some insights into the nucleation and growth mechanisms. Symmetry reduction was involved in the first step of this synthesis. When $Fe(CO)_5$ was decomposed in the presence of gold seeds, the iron atoms would start to nucleate on the gold seed and it was proposed that the free electrons from gold had to compensate for the charge induced by the polarized plane at the interface. Since the gold seed only has a limited number of free electrons, such compensation would make all other facets of the gold seed kind of electron deficient and thus unsuitable for multi-nucleation, resulting in the formation of a dumbbell particle with a Janus structure. Alternatively, the formation of such a heterodimer can be attributed to the mismatch in crystal structure and lattice constant between gold and iron. When iron atoms nucleate on the surface of a gold seed, a single nucleation site would be more favorable than multiple nucleation sites in minimizing the energy penalty caused by the mismatch.

Anisotropy and surface

Wolfgang Pauli once said that "God made solids, but surfaces were the work of the devil". Nevertheless, surface plays a pivotal role in determining the function, functional anisotropy, and applications of nanoparticles. For example, by simply decorating the surface of a spherical particle with a patch made of a different material or functional group, one would obtain a Janus particle with behaviors or properties completely different from the isotropic, spherical counterpart. On the other hand, the interface between two components has been utilized to grow highly anisotropic nanostructures such as nanorod and nanowires, as demonstrated by the vapor–liquid–solid (VLS) methods. A typical VLS process starts with the dissolution of gaseous reactants into nanometer-sized droplets of a catalyst metal, followed by the nucleation and growth of nanorods and then nanowires. The anisotropic growth is dictated by the droplets, whose sizes remain essentially the same during the entire growth process. Both physical (*e.g.*, laser ablation, thermal evaporation, and arc discharge) and chemical (*e.g.*, chemical vapor deposition) methods have been employed to generate the gaseous species. As a major requirement, there should exist a good solvent capable of forming alloy with the target material, ideally they should form eutectic compounds.

As mentioned earlier, it is nontrivial to characterize patchy nanoparticles. Even for spherical nanoparticles "uniformly" modified with molecular adsorbates, it is not always straightforward to characterize their surface due to the possible inhomogeneity hidden to our characterization tools. It is even more challenging if the adsorbates (*e.g.*, proteins) tend to alter their conformation upon adsorption onto a surface. At the meeting, Yarovsky (DOI:10.1039/c6fd00050a) discussed how to computationally model the adsorption of proteins onto a silica surface covered by organic ligands with varying degrees of chemical heterogeneity. It was demonstrated that the ligand arrangement that corresponded to the highest hydrophilic chain mobility and thereby the lack of significant hydrophobic patches had the highest anti-fouling efficacy toward hydrophobin. However, the presence of a flexible loop with amphipathic character on the protein surface would facilitate its adsorption onto all surfaces by enabling the protein to match the surface pattern. Mattoussi (DOI:10.1039/c6fd00017g) discussed the design and development of a multi-coordinating polymer ligand well-suited for functionalizing gold nanoparticles and nanorods by promoting steric stabilization. Gold nanoparticles and nanorods functionalized with this polymer exhibited excellent long-term colloidal stability over a broad range of conditions, including solutions with variable pH and growth media. Significantly, this polymer coating could prevent the formation of a protein corona, which has been a major issue in applying nanoparticles to biomedicine. Pasquato (DOI:10.1039/c6fd00016a) discussed the formation of self-assembled monolayers of hydrogenated and fluorinated alkanethiolates on the surface of gold nanoparticles. Their data suggest that the initial step in the formation of such mixed monolayers involved the nucleation and growth of fluorinated domains. The onset for the formation of domains was dependent on the structure of the fluorinated and hydrogenated species. The solubility of the nanoparticles covered by mixed monolayers displayed a marked discontinuity as a function of the monolayer composition. When the fluorinated component content was low, the nanoparticles were fully soluble in chloroform, at intermediate content the nanoparticles became soluble in hexane, and eventually they became soluble in fluorinated solvents only.

Concluding remarks

The last decade has witnessed tremendous progress in the design and synthesis of nanoparticles with morphological and/or functional anisotropy. As a testimony, Müller presented a perspective lecture on the rational design of ABC triblock terpolymers for the facile synthesis of anisotropic nanoparticles with controllable patchy morphology. Some of the structures discussed in his lecture are so complex in terms of architecture that it is not even an easy job to find the right words to describe them. Overall, it is much easier to obtain anisotropic nanoparticles comprised of organic or polymeric materials relative to the inorganic system.

Thanks to the progress in synthesis, now it is an exciting time to investigate the assembly of nanoparticles with morphological and/or functional anisotropy. To this end, people tend to make analogy between the particulate building blocks and atoms. In a sense, atoms can be considered as patchy particles, with the valence electrons corresponding to the patches for selective, anisotropic interactions. At the meeting, Duguet (DOI:10.1039/c6fd00022c) discussed a template-based method for the synthesis of silica beads with four spatially separated patches arranged in tetrahedral symmetry to mimic the carbon atom. They first used emulsion polymerization of styrene in the presence of 52 nm silica beads whose surface had been modified with methacryloxymethyltriethoxysilane (MMS) to obtain silica–polystyrene tetrapods and then fabricated silica particles with four dimples on the surface in a tetrahedral configuration. Owing to the presence of MMS grafts, polystyrene could be covalently attached to the silica surface within the dimples and further thiolated to induce the adsorption of 12 nm gold particles. The size of the gold nanoparticles could be enlarged through seed-mediated growth by introducing chloroauric acid, ascorbic acid, and sodium citrate. While some intriguing images of the tetrahedrally dimpsed particles were presented, the overall yield of this synthesis still needs to be substantially improved. Due to the involvement of multiple steps, it is not clear if the uniformity and overall yield can eventually reach an acceptable level. In contrast to the bottom-up method, the top-down approach might enable the mass production of mesoscale building blocks with desired precision and tunability in terms of surface patterning. At the meeting, Gracias (DOI:10.1039/c6fd00021e) discussed an interesting method for the fabrication of Self-folding Nanostructures with Imprint Patterned Surfaces (SNIPS). Using a variety of examples, it was demonstrated that SNIPS are mass producible, either individually or in ordered arrays, with the desired precision, tunability, and material heterogeneity.

It is worth pointing out that atoms are much more versatile and in a sense, smarter than patchy particles in that the valence electrons can be hybridized or reconfigured in different ways to alter the bonding architecture. Taking carbon as an example, its valence electrons can be hybridized in sp^3 and sp^2 configurations, respectively, to generate two completely different types of materials (as exemplified by diamond and graphite/graphene) depending on the thermodynamic conditions. To this end, patchy particles will probably never be able to match the capability and feasibility of atoms no matter how much the synthesis or fabrication methods will be advanced. Nevertheless, patchy particles (or more generally, particles with morphological or functional anisotropy) will provide endless opportunities for us to extend the complexity and functionality of the molecular world to the nanoscale and larger scales.

Notes and references

1 Y. Xia, X. Xia and H.-C. Peng, *J. Am. Chem. Soc.*, 2015, **137**, 7947–7966.

2 C. J. Murphy, A. M. Gole, J. W. Stone, P. N. Sisco, A. M. Alkilany, E. C. Goldsmith and S. C. Baxter, *Acc. Chem. Res.*, 2008, **41**, 1721–1730.

3 M. Mayer, L. Scarabelli, K. March, T. Altantzis, M. Tebbe, M. Kociak, S. Bals, F. J. García de Abajo, A. Fery and L. M. Liz-Marzań, *Nano Lett.*, 2015, **15**, 5427–5437.

4 S. Ye, A. R. Rathmell, Z. Chen, I. E. Stewart and B. J. Wiley, *Adv. Mater.*, 2014, **8**, 9673–9679.

5 Q. Chen, J. K. Whitmer, S. Jiang, S. C. Bae, E. Luijten and S. Granick, *Science*, 2011, **331**, 199–202.

6 X. Xia, M. Yang, J. Park, L. Zhang and Y. Xia, *ACS Nano*, 2012, **6**, 512–522.

7 Y. Xia, P. Yang, Y. Sun, Y. Wu, B. Mayers, B. Gates, Y. Yin, F. Kim and H. Yan, *Adv. Mater.*, 2003, **15**, 353–389.

8 H.-C. Peng, J. Park, L. Zhang and Y. Xia, *J. Am. Chem. Soc.*, 2015, **137**, 6643–6652.

Poster titles

A simple technique for cross-correlation of gold nanorods micrographs taken by laser scanning confocal and scanning electron microscopes, **F. Havel, F. Novotný, J. Proška, M. Žarská, M. Šrámek, K. Musílek, K. Kuča and J. Bártek**, *CTU in Prague, Czech Republic*

CdSe nanoplatelets living polymers, **S. Jana, P. Davidson and B. Abécassis**, CNRS, France

Homologous series of surfactant-free Janus nanoparticles – synthesis and application, **D. Wu and A. Honciuc**, *Zurich University of Applied Sciences, Switzerland*

In-situ generation of hybrid thermoresponsive nanogels with anisotropic gold nanoparticles as cross-linkers, **E. Glitscher, J. Bergueiro and M. Calderón**, *Freie Universität Berlin, Germany*

Shape effects of metal nanoparticles regarding their nonlinear optical properties, **S. Dengler, C. Hege and B. Eberle**, *Fraunhofer IOSB, Germany*

Silver nanowires on an optical fibre containing tilted fiber Bragg grating surface for heating and extraction of light, **D. O'Neill, W. Zhou, A. Laronche, V. Marquez, H. Jean-Ruel, J. Albert and A. Ianoul**, *Carleton University, Canada*

Tuning particle interactions of branched-shaped nanocrystals toward control over 2D self-assembled superstructures, **A. Castelli, R. Brescia, M. Prato, L. Manna and M. Arciniegas**, *Istituto Italiano di Tecnologia(IIT), Italy*

Ag@AgCl & urchin-like Ni hybrid, a magnetically separable visible-light-driven photocatalyst, **S. Zhong and W. Jiang**, *Nanjing University of Science and Technology, China*

Anisotropic core-shell plasmonic nanoparticles, **E. Jorgenson, J. Coyle, A. Bottomley, S. Barry and A. Ianoul**, *Carleton University, Canada*

Assembly of hairy nanoparticles in solvent and solvent-free systems: insight from dissipative particle dynamics, **M. Řeřicha and Z. Posel**, *Institute of Chemical Process Fundamentals of the CAS, v. v. i., Czech Republic*

Biological and chemical synthesis of highly anisotropic gold nanostructures, **M. Klekotko, M. Gołąb, K. Matczyszyn and M. Samoć**, *Wroclaw University of Technology, Poland*

Cell-penetrating-peptide/nanoparticle complex shows anisotropic charge regulation during membrane translocation, **J. Klug and M. Del Pópolo**, *Queen's University Belfast, UK*

Bimetallic nanocrystals: a unique probe for monitoring catalytic reactions by surface enhanced Raman spectroscopy, **Y. Wu, J. Li, J. Zhang, S. Winget and D. Qin**, *Georgia Institute of Technology, USA*

Colloidal chain stoppers for nanopolymers, **A. Klinkova, H. Thérien-Aubin, R. Choueiri, M. Rubinstein and E. Kumacheva**, *University of Toronto, Canada*

Combining functionalised nanoparticles and SERS for the detection of miRNA biomarkers related to type 2 diabetes(T2D), **L. Al Maqbali, K. Faulds and D. Graham**, *University of Strathclyde, UK*

Continuous production and spectral tuning of highly monodisperse and shape anisotropic nanostructures by using microfluidic synthesis approaches, **M. Thiele, C. Singh, A. Knauer, R. Müller, A. Csáki, T. Henkel, M. Köhler and W. Fritzsche**, *Leibniz Institute of Photonic Technology, Germany*

Cytotoxicity, antibacterial effects, and crystallographic properties of silver nanoparticles with different shapes, **K. Pappert, J. Helmlinger, C. Sengstock, C. Groß-Heitfeld, C. Mayer, T. A. Schildhauer, M. Köller, O. Prymak, K. Loza, M. Gocyla, M. Heggen and M. Epple**, *University of Duisburg-Essen, Germany*

Development of TiO_2/Au/TiO_2 plasmonic photocatalysts with enhanced activity and stability, **E. Kowalska, L. Rosa, S. Juodkazis and B. Ohtani**, *Hokkaido University, Japan*

Octahedral-shaped anatase titania photocatalyst particles, **B. Ohtani, Z. Wei and E. Kowalska**, *Hokkaido University, Japan*

Fabrication of a nano-sized $Ag_4V_2O_7$/attapulgite photocatalyst with enhanced visible-light photocatalytic activity and stability, **Y. Luo, T. Huang and X. Liu**, *Nanjing University of Science and Technology, China*

Femtosecond laser-controlled tip-to-tip assembly and welding of gold nanorods, **G. González, J. González-Izquierdo, L. Bañares, G. Tardajos, A. Rivera, T. Altantzis, S. Bals, O. Peña-Rodríguez, A. Guerrero-Martínez and L. Liz-Marzán**, *CICbioMA-GUNE, Spain*

"Green" gold nanotriangles: synthesis, purification by polyelectrolyte/micelle depletion flocculation and SERS performance, **F. Liebig, R. M. Sarhan, C. Prietzel, A. Reinecke and J. Koetz**, *University of Potsdam, Germany*

Large (3-aminopropyl)triethoxysilane monolith prepared by doping graphene quantum dots and its reinforcement in photoluminescence and strength, **R. Tian and W. Jiang**, *Nanjing University of Science and Technology, China*

Lipid coated gold nanoparticles for ultra-sensitive label free detection of protein adsorption kinetics, **D. J. Hamilton, J. D. Knight and S.M. Reed**, *University of Colorado Denver, USA*

Anisotropic properties of assembled monocrystalline nanorods, **J. Kim, L. Martinelli, K. Lahlil, E. Grelet, J.-P. Boilot, J. Peretti and T. Gacoin**, *CNRS/Ecole Polytechnique, France*

Nanocomposite films as a gas sensor for organic compounds, **S. Ali, B. Horrocks and A. Houlton**, *Newcastle University, UK*

Nd^{3+}-sensitized upconversion nanoparticles with epitaxial growth and optical properties, **S. Guo and W.-T. Wong**, *Hong Kong Polytechnic University, Hong Kong*

Precise synthesis of poly(methyl methacrylate) brush with well-controlled stereoregularity using surface initiated living anionic polymerization method, **T. Hirai, M. Sato, T. Kato and A. Takahara**, *Kyushu University, Japan*

Preparation and characterization of cubic Au nanoparticle and researches on its electrocatalytic activity, **H. Gao and W. Jiang**, *Nanjing University of Science and Technology, China*

Preparation and chemical characterization of Janus silica particles with grafted polysaccharides, **D. Barsi, M. Corricelli, M. L. Curri, A. Farah and M. Bertoldo**, *IPCF-CNR, Italy*

Preparing highly axial aligned endohedral fullerene derivatives for a controllable ensemble of potential electron spin qubits, **S. Zhou, K. Porfyrakis and A. Briggs**, *Oxford University, UK*

Programmable assembly of hybrid colloidal molecules, **S. Ni, J. Leemann, I. Buttinoni, L. Isa and H. Wolf**, *ETH Zurich, Switzerland*

Programmed assembly of anisotropic nanoparticles supporting surface chemical patches with spherical particles by fine tuning colloidal interactions, **A. Deák, S. Pothorszky, D. Zámbó and Z. Hajnal**, *MTA EK, Hungary*

Self-assembly of 1-dimensional zinc oxide nanoparticles: role of aspect ratio and polydispersity, **M. Gerigk, J. Bahner and S. Polarz**, *University Konstanz, Germany*

Preparation and self-assembly of patchy nanocubes, **E. Galati, R. M. Choueiri, A. Klinkova, H. Thérien-Aubin, E. Kumacheva, A. Querejeta-Fernández and M. Tebbe**, *University of Toronto, Canada*

Surface functionalization with Bovine Serum Albumin (BSA): surface plasmon resonance and molecular dynamics analyses, **K. Tokarczyk, K. Kubiak-Ossowska, P. A. Mulheran, I. Kralka and B. Jachimska**, *Polish Academy of Sciences, Poland*

Interaction of common body proteins with PAMAM dendrimers, **K. Tokarczyk and B. Jachimska**, *Polish Academy of Sciences, Poland*

Coupling between particle and emulsion phase behaviours in Pickering systems, **D. French, J. Fowler, P. Taylor and P. Clegg**, *University of Edinburgh, UK*

Synthesis and assembly of gold particles on an emulsion droplet; nanoparticles, nanosheets and core-shell particles, **S. Sachdev, S. Christie and M. Platt**, *Loughborough University, UK*

Synthesis and characterization of $Ag/Zn_XCd_{1-X}S$ with enhanced visible-light photocatalytic activity, **T. Huang and X. Liu**, *Nanjing University of Science and Technology, China*

Synthesis and characterization of ultrasmall gold nanorods with low toxicity, **F. Novotný, F. Havel, J. Proška, K. Musílek and K. Kuča**, *CTU in Prague, Czech Republic*

Towards a more efficient phototransducer: Precise control of branch growing on gold nanocages, **J. Bergueiro, L. M. Bouchet, E. A. Glitscher and M. Calderón**, *Freie Universität Berlin, Germany*

Versatile molten salt-mediated syntheses of metal boride nanocrystals, **D. Portehault, G. Gouget, P. Beaunier and C. Sanchez**, *Lab. Chimie de la Matière Condensée de Paris, France*

Graphene-decorated 3D $BiVO_4$ superstructure: highly reactive (040) facets formation and enhanced visible-light-induced photocatalytic oxidation of NO, **M. Ou and Q. Zhong**, *Nanjing University of Science and Technology, China*

Self-assembly of ultrathin gold nanowires: ligand–solvent interplay, **B. Reiser, D. Gerstner, L. Gonzalez-Garcia, J. Maurer, I. Kanelidis and T. Kraus**, *INM - Leibniz Institute for New Materials, Germany*

Central role of electricity in the molecule-like behavior of anisotropic nanoparticles, **A. Knauer and M. Koehler**, *Technical University Ilmenau, Germany*

Nanoparticle architectures: charging and controlled polarization as tools for the hierarchical assembling of nanoparticles in colloidal solutions, **A. Knauer, M. Koehler, N. Visaveliya and C. Serra**, *Technical University Ilmenau, Germany*

Multiresolution study of protein transport in realistically confined geometries by combination of MD and lattice boltzmann, **R. Faller and V. Ustach**, UC Davis, USA

Dynamic chiral nanoparticle assemblies and specific chiroplasmonic analysis of cancer cells, **Q. Song, Y. Zhao and C. Ran**, *Jiangnan University, China*

CLEAR! computational and experimental designing rules in tailoring the monolayer organization of fluorocarbon/hydrocarbon thiolates mixtures on the surface of gold nanoparticles, **P. Posocco, D. Marson, M. Sologan, S. Boccardo, P. Pengo, S. Pricl and L. Pasquato**, *University of Trieste, Italy*

QCM study of nanoparticle immobilisation and biomolecule recognition, **Y. Yang, H. Zheng, S. Guldin, F. Stellacci and N. Nianias,** *University College London, UK*

The applications of multifunctional gold nanorods in photothermal therapy and fluorescence imaging for cancer theranostics, **Y. Zhou, D. S. Elson, A. E. G. Cass, M. Singh, E. Nabaui, H. Zhao, D. Ma and G. Hanna**, *Imperial College London, UK*

Fabrication of anisotropic nanoparticle assemblies for metamaterials and SERS sensors, **C. Hanske, M. Tebbe, C. Kuttner, M. Chanana, T. König, A. Fery and L. Liz-Marzán,** *CIC biomaGUNE, Spain*

Structural transitions in nanoparticle assemblies governed by competing nanoscale forces, **R. Choueiri, A. Klinkova, H. Thérien-Aubin, M. Rubinstein and E. Kumacheva**, *University of Toronto, Canada*

Electro-orientation of non-spherical particles: the case of platelets and rods, **P. Arenas-Guerrero, Á. V. Delgado and M. L. Jiménez**, *University of Granada, Spain*

Preparation of polyaniline/Europium phthalocyanine Janus nanocomposite electrochemical probe for reagentless detection of glucose, **H. Al-Sagur and S. Komathi**, *Sheffield Hallam University, UK*

Beaker epitaxy: solution synthesis of 3-dimensionally (3D) organized M@MO_x (M = Ni, Pd & MO_x = γ-Fe_2O_3) superparticles with enhanced peroxidase mimetics, **M. N. Tahir, M. Klünker, R. Ragg, K. Korschelt and W. Tremel,** *Johannes Gutenberg University of Mainz, Germany*

Branched gold nanoprisms for biotechnological applications, **O. Rojas, P. Conejo, J. Bergueiro and M. Calderón**, *Nartional University, Costa Rica*

Observation of compositional domains within chalcopyrite $CuInS_2$ quantum dots by electron energy loss spectroscopy, **A. Harvie, M. Booth, R. Chantry, N. Hondow, D. Kepaptsoglou, Q. Ramasse, S. Evans and K. Critchley**, *University of Leeds, UK*

Magnetic cylindrical colloids at liquid interfaces exhibit non-volatile switching of their orientation in an external field, **D. M. Buzza and B. Newton,** *University of Hull, UK*

Colloidal molecules assembled from binary spheres under an AC electric field, **K. Heatley, N. Wu and F. Ma**, *Colorado School of Mines, USA*

Developing new routes for functionalizing gold nanorods with DNA to perform single nanoparticle imaging bioaffinity sensing, **M. Azab, A. Wark, M. M. Abou El-Alamin and M. A. Sultan**, *University of Strathclyde, UK*

Nanowire gas sensors for a range of analytes of industrial interest, **M. Y. Ibrahim, B. R. Horrocks and A. Houlton**, *Chemical Nanoscience Laboratory, UK*

Hybridization of isotropic and anisotropic metal nanoparticles into core/satellite nanoclusters by protein-assisted self-assembly, **C. Kuttner, R. P. M. Höller, M. Dulle, S. Thomä, M. Mayer, A. M. Steiner, T. A. F. König, S. Förster, A. Fery and M. Chanana**, *Leibniz-Institut für Polymerforschung Dresden e.V, Germany*

Dynamic covalent control of nanoparticle properties and self-assembly, **E. Kay, N. Marro and S. Borsley**, *University of St Andrews, UK*

Synthesis and characterisation of DNA-templated binary metal sulphide ($CdZnS_2$) for application in gas sensing, **A. B. Oshido, B. R. Hollocks and A. Houlton**, *Newcastle University, UK*

Macroscopic magnetic metasurfaces by colloidal gold nanorod self-assembly, **M. Mayer, T. König, M. Tebbe, M. Schnepf, C. Kuttner and A. Fery**, *Leibniz-Institut für Polymerforschung Dresden e. V, Germany*

Solubility and self-assembly of nanoparticles in a thermotropic liquid crystal, **N. Mac Fhionnlaoich, S. Guldin, S. Schrettl and F. Stellacci**, *University College London, UK*

Preparation of Pickering emulsions with silver nanoparticles, **O. Armstrong and J. Thomas,** *Bangor University, UK*

Development of a synergistic drug combination for a novel anti-melanoma nanotherapy, **V. Sherwood, C. Webster, K. Al-Yousef, P. S. Moreno, L. Lanfrancone and F. Baldelli Bombelli**, *University of Dundee, UK*

Control of symmetry breaking and its role in gold nanorod growth, **W. Tong, M. Walsh, J. Etheridge and A. Funston**, *Monash University, Australia*

Anisotropic nanoparticles as dopants for DNA liquid crystals, **K. Brach, K. Matczyszyn, J. Olesiak-Bańska and M. Samoć**, *Wroclaw University of Science and Technology, Poland*

The Skinner Prize for the best poster was jointly awarded to Elizabeth Galati of the University of Toronto, Canada, for her poster on preparation and self-assembly of patchy nanocubes and Songbo Ni of ETH Zurich, Switzerland, for his poster on programmable assembly of hybrid colloidal molecules.

The *Nanoscale Horizons* poster prize was awarded to Andrew Harvie of the University of Leeds, UK for his poster on observation of compositional domains within chalcopyrite $CuInS_2$ quantum dots by electron energy loss spectroscopy.

List of participants

Dr Benjamin Abécassis, *CNRS*, *France*
Mrs Laila Al Maqbali, *University of Strathclyde*, *United Kingdom*
Mr Hadi Al-Sagur, *Sheffield Hallam University*, *United Kingdom*
Mr Karamallah Al-Yousuf, *School of Medicine, University of Dundee*, *United Kingdom*
Miss Shams Ali, *School of Chemistry, Newcastle University*, *United Kingdom*
Dr Milena Arciniegas, *Istituto Italiano di Tecnologia (IIT)*, *Italy*
Miss Paloma Arenas-Guerrero, *Department of Applied Physics, University of Granada*, *Spain*
Mr Oliver Armstrong, *Bangor University*, *United Kingdom*
Miss Ana Maria Bago Rodriguez, *University of Hull*, *United Kingdom*
Mr Jochen Bahner, *University Konstanz*, *Germany*
Mr David Barsi, *IPCF-CNR*, *Italy*
Dr Julian Bergueiro, *Freie Universität Berlin*, *Germany*
Professor Bernard Binks, *University of Hull*, *United Kingdom*
Mrs Katarzyna Brach, *Wroclaw University of Science and Technology*, *Poland*
Professor Mathias Brust, *University of Liverpool*, *United Kingdom*
Dr David Buceta, *University of Santiago de Compostela*, *Spain*
Dr Martin Buzza, *University of Hull*, *United Kingdom*
Professor Eleanor Campbell, *University of Edinburgh*, *United Kingdom*
Mr Andrea Castelli, *Istituto Italiano di Tecnologia (IIT)*, *Italy*
Ms Rachelle Choueiri, *University of Toronto*, *Canada*
Dr Jason Corbett, *Malvern Instruments Ltd*, *United Kingdom*
Dr Kevin Critchley, *University of Leeds*, *United Kingdom*
Ms Stefanie Dengler, *Fraunhofer IOSB*, *Germany*
Dr András Deák, *MTA EK, Hungarian Academy of Sciences*, *Hungary*
Ms Petra Dral, *University of Twente*, *Netherlands*
Professor Calum Drummond, *RMIT University*, *Australia*
Professor Etienne Duguet, *CNRS/University of Bordeaux*, *France*
Professor Roland Faller, *UC Davis*, *USA*
Dr David French, *University of Edinburgh*, *United Kingdom*
Miss Elizabeth Galati, *University of Toronto*, *Canada*
Mr Han Gao, *Nanjing University of Science and Technology*, *China*
Mrs Melanie Gerigk, *University Konstanz*, *Germany*
Mr Emanuel Glitscher, *Freie Universität Berlin*, *Germany*
Dr Estefania Gonzalez Solveyra, *Northwestern University*, *USA*
Mr Guillermo González, *CIC biomaGUNE*, *Spain*
Professor David Gracias, *Johns Hopkins University*, *USA*
Professor Steve Granick, *Institute for Basic Science*, *South Korea*
Dr Marcin Grzelczak, *University of Liverpool*, *United Kingdom*
Miss Shuai Guo, *HongKong Polytechnic University*, *Hong Kong*
Miss Abigail Hallowes, *Royal Society of Chemistry*, *United Kingdom*
Mr Desmond Hamilton, *University of Colorado Denver*, *USA*
Dr Christoph Hanske, *CIC biomaGUNE*, *Spain*

Mr Andrew Harvie, *University of Leeds, United Kingdom*
Mr Filip Havel, *Faculty of Nuclear Sciences and Physical Engineering, CTU in Prague, Czech Republic*
Ms Kelley Heatley, *Colorado School of Mines, USA*
Mr Alex Hill, *University of Liverpool, United Kingdom*
Dr Tomoyasu Hirai, *Kyushu University/Institute for Materials Chemistry and Engineering, Japan*
Professor Andrei Honciuc, *Zurich University of Applied Sciences, Institute of Chemistry and Biotechnology, Switzerland*
Miss Ting Huang, *Nanjing University of Science and Technology, China*
Dr Castillo-Leon Jaime, *Sol Voltaics AB, Sweden*
Professor Richard Jones, *University of Texas, USA*
Ms Emma Jorgenson, *Carleton University, Canada*
Dr Euan Kay, *University of St Andrews, United Kingdom*
Mr Jongwook Kim, *CNRS/Ecole Polytechnique, France*
Mrs Magdalena Klekotko, *Wroclaw University of Technology, Poland*
Dr Anna Klinkova, *University of Toronto, Canada*
Dr Joaquin Klug, *Atomistic Simulation Centre, Queen's University Belfast, United Kingdom*
Dr Andrea Knauer, *Technische Universität Ilmenau, Germany*
Professor Nicholas Kotov, *University of Michigan, USA*
Professor Ewa Kowalska, *Hokkaido University, Japan*
Professor Eugenia Kumacheva, *University of Toronto, Canada*
Dr Christian Kuttner, *Leibniz-Institut für Polymerforschung Dresden e.V., Germany*
Dr Tobias König, *Leibniz-Institut für Polymerforschung Dresden e. V., Germany*
Professor Joerg Lahann, *Biointerfaces Institute, USA*
Mr Ferenc Liebig, *University of Potsdam, Germany*
Professor Luis Liz-Marzán, *CIC biomaGUNE, Spain*
Miss Kirsty Lunnon, *Royal Society of Chemistry, United Kingdom*
Miss Yuting Luo, *Nanjing University of Science and Technology, China*
Professor M. Arturo López-Quintela, *University of Santiago de Compostela, Spain*
Ms Niamh Mac Fhionnlaoich, *University College London, United Kingdom*
Dr Sandra Macaskill, *Royal Society of Chemistry, United Kingdom*
Mr Cai Mann, *Anton Paar, United Kingdom*
Professor Hedi Mattoussi, *Florida State University, USA*
Mr Martin Mayer, *Leibniz-Institut für Polymerforschung Dresden e.V., Germany*
Miss Anna Meehan, *Royal Society of Chemistry, United Kingdom*
Professor Axel Mueller, *Johannes Gutenberg University Mainz, Germany*
Professor Catherine Murphy, *University of Illinois at Urbana-Champaign, USA*
Mr Songbo Ni, *IBM Research - Zurich, Switzerland*
Dr Filip Novotný, *Faculty of Nuclear Sciences and Physical Engineering, CTU in Prague, Czech Republic*
Mr Devin O'Neill, *Carleton University, Canada*
Mr Atsinafe Oshido, *Newcastle University, United Kingdom*
Dr Man Ou, *Nanjing University of Science and Technology, China*
Mr Kevin Pappert, *University of Duisburg-Essen, Germany*
Dr Michael Parkes, *Shell Global Solutions, United Kingdom*
Professor Lucia Pasquato, *University of Trieste, Italy*
Dr Matthias Pauly, *Institut Charles Sadron (CNRS), France*
Dr Davide Peddis, *ISM-CNR, Italy*
Dr Matthew Penna, *RMIT, Australia*
Dr David Portehault, *Laboratoire Chimie de la Matière Condensée de Paris, France*

Dr Zbysek Posel, *Institute of Chemical Process Fundamentals of the CAS, v. v. i., Czech Republic*
Dr Paola Posocco, *University of Trieste, Italy*
Professor Dong Qin, *Georgia Institute of Technology, USA*
Mrs Guoxia Ran, *School of Chemical and Material Engineering, Jiangnan University, China*
Professor Scott Reed, *University of Colorado Denver, USA*
Dr Javier Reguera, *CIC biomaGUNE, Spain*
Mrs Beate Reiser, *INM- Leibniz Institute for New Materials, Germany*
Mr Michal Rericha, *University of J.E. Purkinje in Usti nad Labem, Czech Republic*
Dr Anna Roig, *Institut de Ciència de Materials de Barcelona (ICMAB-CSIC), Spain*
Dr Oscar Rojas-Carrillo, *Nartional University, Costa Rica, Costa Rica*
Mr Francesco Rossi, *UCL Department of Physics and Astronomy, United Kingdom*
Miss Suchanuch Sachdev, *Loughborough University, United Kingdom*
Dr Leonardo Scarabelli, *CIC biomaGUNE, Spain*
Dr Tomoko Sekine, *Shiseido Global Innovation Center, Japan*
Dr Francois Sicard, *University College London, United Kingdom*
Professor Qijun Song, *School of Chemical and Material Engineering, Jiangnan University, China*
Dr Michael Spencelayh, *Royal Society of Chemistry, United Kingdom*
Professor Molly Stevens, *Imperial College London, United Kingdom*
Mr Antonio Stocco, *Laboratoire Charles Coulomb CNRS, France*
Professor Alberto Striolo, *University College London, United Kingdom*
Dr Alla Synytska, *Leibniz Institute of Polymer Research Dresden, Germany*
Dr Lola Sánchez, *Faurecia Interiors Systems SALC España S.L., Spain*
Mr Luciano Tadiello, *Pirelli Tyre, Italy*
Dr Muhammad Nawaz Tahir, *Johannes Gutenberg University of Mainz, Germany*
Mr Matthias Thiele, *Leibniz Institute of Photonic Technology, Germany*
Dr Antoine Thill, *CEA, France*
Mr Renbing Tian, *Nanjing University of Science and Technology, China*
Mrs Karolina Tokarczyk, *Jerzy Haber Institute of Catalysis and Surface Chemistry Polish Academy of Sciences, Poland*
Mr Wenming Tong, *Monash University, Australia*
Professor Jan Vermant, *ETH Zürich, Switzerland*
Miss Paulina Westerlind, *Sol Voltaics AB, Sweden*
Dr Heiko Wolf, *IBM Research - Zurich, Switzerland*
Dr Dalin Wu, *Zurich University of Applied Sciences, Institute of Chemistry and Biotechnology, Switzerland*
Professor Younan Xia, *Georgia Institute of Technology, USA*
Mr Mansur Yahaya, *Newcastle University, United Kingdom*
Mr Ye Yang, *University College London, United Kingdom*
Professor Irene Yarovsky, *RMIT University, Australia*
Mr Hanrui Zheng, *University College London, United Kingdom*
Miss Suting Zhong, *Nanjing University of Science and Technology, China*
Mr Shen Zhou, *Oxford University, United Kingdom*
Mr Yu Zhou, *Imperial College London, United Kingdom*